Lecture Notes in Physics

Volume 1017

The series Lecture Notes in Physics (LNP), founded in 1969, reports new developments in physics research and teaching - quickly and informally, but with a high quality and the explicit aim to summarize and communicate current knowledge in an accessible way. Books published in this series are conceived as bridging material between advanced graduate textbooks and the forefront of research and to serve three purposes:

- to be a compact and modern up-to-date source of reference on a well-defined topic;
- to serve as an accessible introduction to the field to postgraduate students and non-specialist researchers from related areas;
- to be a source of advanced teaching material for specialized seminars, courses and schools.

Both monographs and multi-author volumes will be considered for publication. Edited volumes should however consist of a very limited number of contributions only. Proceedings will not be considered for LNP.
Volumes published in LNP are disseminated both in print and in electronic formats, the electronic archive being available at springerlink.com. The series content is indexed, abstracted and referenced by many abstracting and information services, bibliographic networks, subscription agencies, library networks, and consortia.
Proposals should be sent to a member of the Editorial Board, or directly to the responsible editor at Springer:

Dr Lisa Scalone
lisa.scalone@springernature.com

Christian Pfeifer • Claus Lämmerzahl
Editors

Modified and Quantum Gravity

From Theory to Experimental Searches on All Scales

Editors
Christian Pfeifer
ZARM
University of Bremen
Bremen, Germany

Claus Lämmerzahl
ZARM
University of Bremen
Bremen, Germany

ISSN 0075-8450 ISSN 1616-6361 (electronic)
Lecture Notes in Physics
ISBN 978-3-031-31519-0 ISBN 978-3-031-31520-6 (eBook)
https://doi.org/10.1007/978-3-031-31520-6

This Springer imprint is published by the registered company Springer Nature Switzerland AG
The registered company address is: Gewerbestrasse 11, 6330 Cham, Switzerland

Paper in this product is recyclable.

Preface

From an observational as well as theoretical perspective, it is evident that general relativity and the standard model of particle physics cannot be the final answer to our understanding of the gravitational interaction. The incompleteness is easily demonstrated by puzzling observations and theoretical obstacles such as: dark matter and dark energy, the rotation curves of galaxies, the accelerated expansion of the universe as well as the Hubble and density fluctuation tension in cosmology; the prediction of the existence of singularities and infinite tidal gravitational acceleration; still open consistency question between quantum theory and gravity, such as the information paradox or the vacuum energy, and the still elusive theory of quantum gravity.

The search for a better understanding of the gravitational interaction is an ongoing worldwide effort, and the theoretical ideas, as well as the experimental searches, are numerous.

The 740. Wilhelm and Else Heraeus Seminar "Experimental Tests and Signatures of Modified and Quantum Gravity", which took place virtually in February 2021, was a meeting where the problem of the gaps in our understanding of the gravitational interaction was discussed from a large variety of different viewpoints. The topics ranged from fundamental theoretical approaches to experimental searches for modified and quantum gravity effects on scales from earth-based laboratories to the whole universe.

One insight from the seminar was, that, to step forward towards a self-consistent complete picture of gravity, it is important to extend the communication between the different communities working on modified and quantum gravity on different scales. Theoreticians need to be aware of which kind of theoretical models are constrained by experiments, by how much and at what scales. On the other hand, experimentalists need to know more precisely where to look in their data for quantum gravity effects predicted by theory. In addition, the connection between experiments searching for deviations from general relativity on different scales can be improved.

The goal of this book is to give a comprehensible overview on the different angles and viewpoints on modified and quantum gravity for the different communities working on this topic. In particular, the aim is to give master and PhD students as well as postdocs and researchers, working on one specific subject in the field, a readable access and broad overview over different other approaches to the topic. To cover all possible aspects of deviations from general relativity in one volume is not possible; thus, we decided to demonstrate the variety of possible effects on selected topics. One selection criterion was to cover aspects from all scales and all regimes, i.e. from the local laboratory to the universe, as well as from classical to quantum.

We envision that this book serves as door opener between the different approaches to understand the fundamental nature of the gravitational interaction.

Bremen, Germany
March 2023

Christian Pfeifer
Claus Lämmerzahl

Editor Acknowledgments

We are very grateful to the Wilhelm and Else Heraeus Foundation for giving us the opportunity and the support to organize the 740. Wilhelm and Else Heraeus Seminar "Experimental Tests and Signatures of Modified and Quantum Gravity".

We acknowledge support and funding by the Deutsche Forschungsgemeinschaft (DFG, German Research Foundation) via the Excellence Cluster QuantumFrontiers funded under Germany's Excellence Strategy—EXC-2123 QuantumFrontiers, the Collaborative Research Center "Relativistic and quantum based geodesy (Terra Q)", the research training group GRK 1620 "Models of Gravity", as well as the Project "Momentum dependent spacetime geometries: Traces of quantum gravity and fields in media", Project Number 420243324, and all colleagues at the Center for Applied Space Technology and Microgravity ZARM, and at the above-mentioned research consortia.

Also, we would like to acknowledge networking support by the COST Action CA18108.

Contents

Contributors

Michele Arzano Dipartimento di Fisica "Ettore Pancini", Università di Napoli Federico II, Naples, Italy
INFN, Naples, Italy

Domenico Giulini Institute for Theoretical Physics, University of Hannover, Hannover, Germany
Center of Applied Space Technology and Microgravity, University of Bremen, Bremen, Germany

Jean-François Glicenstein IRFU, CEA Paris-Saclay, Paris, France

André Großardt Institute for Theoretical Physics, Friedrich Schiller University Jena, Jena, Germany

Giulia Gubitosi Dipartimento di Fisica "Ettore Pancini", Università di Napoli Federico II, Naples, Italy
INFN Naples, Italy

Carlos A. R. Herdeiro Departamento de Matemática da Universidade de Aveiro, Aveiro, Portugal

Sven Herrmann ZARM, University Bremen, Bremen, Germany

Manuel Hohmann Laboratory of Theoretical Physics, Institute of Physics, University of Tartu, Tartu, Estonia

Galina L. Klimchitskaya Central Astronomical Observatory at Pulkovo of the Russian Academy of Sciences, Saint Petersburg, Russia
Peter the Great Saint Petersburg Polytechnic University, Saint Petersburg, Russia

Jutta Kunz Institute of Physics, University of Oldenburg, Oldenburg, Germany

Claus Lämmerzahl Center of Applied Space Technology and Microgravity (ZARM), University of Bremen, Bremen, Germany

Elisa Maggio Max Planck Institute for Gravitational Physics, Albert Einstein Institute, Potsdam, Germany

Nick E. Mavromatos National Technical University of Athens, School of Applied Mathematical and Physical Sciences, Department of Physics, Athens, Greece
King's College London, Department of Physics, London, UK

Vladimir M. Mostepanenko Central Astronomical Observatory at Pulkovo of the Russian Academy of Sciences, Saint Petersburg, Russia
Peter the Great Saint Petersburg Polytechnic University, Saint Petersburg, Russia
Kazan Federal University, Kazan, Russia

Yuri N. Obukhov Nuclear Safety Institute, Russian Academy of Sciences, Moscow, Russia

Carlos Pérez de los Heros Department of Physics and Astronomy, Uppsala University, Uppsala, Sweden

Volker Perlick ZARM, University of Bremen, Bremen, Germany

Christian Pfeifer Center of Applied Space Technology and Microgravity (ZARM), University of Bremen, Bremen Germany

Dennis Rätzel ZARM, University Bremen, Bremen, Germany

José Javier Relancio Departamento de Física, Burgos, Spain

Philip K. Schwartz Institute for Theoretical Physics, University of Hannover, Hannover, Germany

Lijing Shao Kavli Institute for Astronomy and Astrophysics, Peking University, Beijing, China
National Astronomical Observatories, Chinese Academy of Sciences, Beijing, China

Yakov Shnir BLTP, JINR, Dubna, Russia

Tomislav Terzić University of Rijeka, Faculty of Physics, Rijeka, Croatia

J.W. van Holten Nikhef, Amsterdam, Netherlands
Lorentz Insitute, Leiden University, Leiden, Netherlands

Aneta Wojnar Institute of Physics, University of Tartu, Tartu, Estonia

Part I
Theoretical Models Beyond Special and General Relativity

Theoretical models for modified and quantum gravity are numerous. Some aim to be a fundamental theory of nature, like the standard model of particle physics, others aim for an approximate phenomenological description of physics, which is only valid on certain energy or length scales. This first part of the book is devoted to different theoretical approaches that aim to explain shortcomings of general relativity.

Chapter 1 discusses a path from one of the most famous and prominent fundamental approach to quantum gravity, namely String Theory, towards predictions of observables, by deducing a Chern-Simons term modified theory of gravity with torsion, in a limit of fundamental String Theory. The effective theory is then connected to Lorentz invariance violating terms appearing in the Standard Model Extension.

A different angle on deviations from local Lorentz invariance, caused by the quantum nature of gravity, is discussed in Chap. 2. The concept of a fundamental observer invariant energy or length scale (the Planck energy or Planck length for example), leads to quantum deformations of local Lorentz (or Poincaré) symmetry, which can be described by Hopf algebras or momentum dependent spacetime geometry. The new symmetry algebra discussed is known as κ-Poincré symmetry and the framework is called doubly (or deformed) special relativity.

Chapter 3 discusses gravity as a gauge theory of the Poincaé group, to set gravity on similar footings as how we describe the other three fundamental forces in physics. In contrast to pure general relativity, the geometry of spacetime can contain not only curvature but also torsion; the former being the field strength of the gauged Lorentz group, while the later is the field strength of the gauged group of translations.

The following Chapter, Chap. 4 picks these ideas up and considers general flat spacetime geometries with torsion or non-metricity as carrier of the information about the gravitational interaction. Most interestingly, one can understand already classical general relativity without the need to introduce curvature, but instead as theory of torsion or of non-metricity alone. These reformulations nicely serve as starting point for modified theories of gravity based on these more general geometric concepts.

In the final Chap. 5 of Part I, consequences from deformations or violations of local Lorentz invariance on gravitational lensing images are discussed. The phenomenological consequences are derived from the Hamiltonian, as well as the Lagrangian description of light propagation. For the former the starting point are modified dispersion relations, while the later considers Finsler spacetimes geometry. For certain (but certainly not all) models both approaches can be seen as dual to each other, one leading to a momentum dependent spacetime geometry, the other to a velocity dependent one.

Part II of this book, will deal with astrophysical systems and possible signatures of modified and quantum gravity in cosmic observations.

1 Lorentz Symmetry Violation in String-Inspired Effective Modified Gravity Theories

Nick E. Mavromatos

Abstract

We discuss situations under which Lorentz symmetry is violated in effective gravitational field theories that arise in the low-energy limit of string theory. In particular, we discuss spontaneous violation of the symmetry by the ground state of the system. In the flat space-time limit, the effective theory of the broken Lorentz Symmetry acquires a form that belongs to the general framework of the so-called Standard Model Extension (SME) formalism. A brief review of this formalism is given before we proceed to describe a concrete example, where we discuss a Lorentz-symmetry-Violating (LV) string-inspired cosmological model. The model is a gravitational field theory coupled to matter, which contains torsion, arising from the fundamental degrees of freedom of the underlying string theory. The latter, under certain conditions which we shall specify, can acquire a LV condensate, and lead, via the appropriate equations of motion, to solutions that violate Lorentz and CPT (Charge-Parity-Time-Reversal) symmetry. The model is described by a specific form of an SME effective theory, with specific LV and CPT symmetry Violating coefficients, which depend on the microscopic parameters of the underlying string theory, and thus can be bounded by current-era phenomenology.

N. E. Mavromatos (✉)
School of Applied Mathematical and Physical Sciences, National Technical University of Athens, Athens, Greece

Department of Physics, King's College London, London, UK
e-mail: mavroman@mail.ntua.gr

C. Pfeifer, C. Lämmerzahl (eds.), *Modified and Quantum Gravity*, Lecture Notes in Physics 1017, https://doi.org/10.1007/978-3-031-31520-6_1

1.1 Lorentz- and CPT Symmetries in Particle Physics and Cosmology and Their Potential Violation

Ignoring gravity, particle-physics theory and the respective phenomenology, as we understand them today, are based exclusively on Lorentz Symmetric formalisms. The Standard Model (SM) of Particle Physics, which is a mathematically consistent gauge field theory of the electromagnetic, weak and strong interactions, in flat Minkowski space-time background, is a relativistic (*i.e* Lorentz invariant), unitary quantum field theory, with local, renormalizable interactions. As such, it satisfies the important CPT theorem [1, 2], proved independently by Schwinger [3], Lüders [4], Pauli [5], Bell [6] and Jost [7], which states that such field theories are described by Lagrangian densities that are invariant under the successive action (in any order) of the generators of the discrete symmetries of Charge Conjugation (C), Parity (or spatial reflexion symmetry) (P) and Time Reversal (T). Although it is often stated that Lorentz invariance violation is somewhat fundamental in inducing CPT violation [8, 9], nonetheless there have been objections to this statement, through explicit examples given in [10, 11], which support the thesis that the aforementioned conditions for the validity of the CPT theorem, that is, locality, unitarity and Lorentz invariance, are truly independent, since, for instance, non-local but otherwise Lorentz invariant models could be explicitly constructed which violate CPT. Indeed, the proof of the theorem of [8, 9] necessitates well-defined time-ordered products and transfer (thus scattering) matrices, which exclude non-local or non-unitary models, for which scattering matrices are not well defined.

This CPT symmetry has important implications for particle physics in that it implies equality of masses m, lifetimes (or equivalently decay widths Γ), magnitude (with opposite sign) of electric charges $q^+ = -q^-$, and magnetic dipole moments g_m, between particles (matter) and antiparticles (antimatter). The most stringent experimental bound between particle-antiparticle mass differences to date refers to the neutral-Kaon system, $K^0, \overline{K}^0$ [12]:

$$\frac{m^{K^0} - m^{\overline{K}^0}}{m^{K^0} + m^{\overline{K}^0}} < 10^{-18}, \quad \text{with} \quad \frac{\Gamma^{K^0} - \Gamma^{\overline{K}^0}}{\frac{1}{2}(\Gamma^{K^0} + \Gamma^{\overline{K}^0})} < 10^{-17}, \tag{1.1}$$

where m^{K^0} is its (rest) mass, Γ^{K^0} its decay width, and the overline above a symbol denotes a quantity referring to the corresponding antiparticle. For completeness, we mention that the most stringent current upper bounds in the differences between proton (p)-antiproton($\overline{p}$) electric charges and electron (e^-)-positron(e^+) magnetic dipole moments are [12]

$$q(p) - q(\overline{p}) < 10^{-21}\, e, \quad \frac{g_m(e^+) - g_m(e^-)}{\frac{1}{2}(g_m(e^+) + g_m(e^-))} < 2 \times 10^{-12} \tag{1.2}$$

where e is the electron charge (at zero energy scale).

For atoms, CPT invariance means that the anti-matter atoms will have identical spectra with the corresponding matter atoms. We mention at this stage that, since antihydrogen has been produced in the Laboratory [13–15], it provides, together with other man-made antimatter atoms, such as antirprotonic helium [16–18], a playground for additional tests of CPT invariance, at an atomic spectra level [19,20].

The implications of CPT invariance for the evolution and state of the Universe are also of immense importance. If CPT symmetry characterises a (yet elusive though) quantum theory of gravity, which is believed to describe the birth and dynamics of our Universe immediately after the Big Bang (i.e. at times after the Big Bang of order of the Planck time $t_{\rm Pl} \sim 5.4 \times 10^{-44}$ s), then matter and antimatter would have been generated in equal amounts in the early Universe. The dominance of matter over antimatter in the Cosmos, however, is overwhelming. Indeed, a plethora of observations, including cosmic microwave background (CMB) ones [21], as well measurements on the abundance of elements in the Universe (Big-Bang-Nucleosynthesis (BBN) data) [22], yield the following matter-antimatter asymmetry (or baryon-asymmetry in the universe (BAU), as it is alternatively called, due to the dominance of baryonic matter among the observable matter) :

$$\Delta n = \frac{n_B - n_{\overline{B}}}{n_B + n_{\overline{B}}} \sim \frac{n_B - n_{\overline{B}}}{s} = (8.4 - 8.9) \times 10^{-11}, \tag{1.3}$$

at the early stages of the cosmic expansion, *i.e* at times $t \sim 10^{-6}$ s and temperatures $T \gtrsim 1$ GeV. In the above expression, s denotes the entropy density of the Universe, and $n_{B(\overline{B})}$ the baryon (antibaryon) number densities. The above number essentially implies the existence of one antiproton in 10^9 protons in the Universe.

In the framework of CPT-symmetric quantum field theories, in the absence of quantum gravity, which is a valid one at the regime of temperatures and times for which (1.3) applies, one could generate such an asymmetry, provided the following conditions, postulated by A.D. Sakharov [23], are met in the early Universe:

1. Baryon-number (B)-violating interactions that allow the generation of states with $B \neq 0$ starting from an initial state with $B = 0$;
2. Interactions capable of distinguishing between matter and antimatter. Assuming CPT symmetry, this would require violation of both C and CP;
3. Since matter-antimatter asymmetry is impossible in chemical equilibrium, one also requires some breakdown of chemical equilibrium during an epoch in the early Universe, otherwise any generated matter-antimatter asymmetry would be washed out by the reverse interaction.

In the Standard Model of particle physics, which is a Lorentz and CPT invariant, unitary, local quantum field theory, the above conditions are met but *only qualitatively*. Indeed, Baryon number violation occurs due to quantum chiral anomalies [24–28], as a consequence of non-perturbative (instanton) effects of the electroweak gauge group SU(2), which lead to non-conservation of the chiral Baryon-number current $J^{B\mu}$:

$$\partial_\mu J^{B\,\mu} \propto g^2 n_f \,\mathrm{Tr}(\mathbf{F}_{\mu\nu} \cdot \mathbf{F}_{\mu\nu}) + \text{Abelian weak hypercharge } \mathrm{U_Y}(1)\text{terms}\,, \tag{1.4}$$

where the Tr is over SU(2) gauge group indices, g is the SU(2) coupling, n_f is the flavour (generation) number, and $\mathbf{F}_{\mu\nu}$ is the field strength of the SU(2) gauge field. Due to the instanton effects, the system of the early Universe can tunnel through to a sector with non-zero baryon number from a state with a zero baryon number, and as a result there is induced B-number violation.[1]

Moreover, CP Violation (CPV) is known to charascterise the hadron sector of the Standard Model (it has been observed for the first time in the neutral Kaon system [29]). However, the order of the observed CP violation in the quark sector of the Standard Model is several orders of magnitude smaller than the one required to produce the BAU (1.3). Moreover, CPV has still not been observed in the lepton sector. For these reasons, physicists attempt to extend the Standard Model in order to discover new sources of CPV that could explain the BAU according to Sakharov's conditions (*e.g.* supersymmetric models, extra dimensions, including strings, models with right-handed neutrinos *etc.*).

Minimal, not necessarily supersymmetric, Lorentz and CPT invariant field-theoretic extensions of the Standard Model, in (3 + 1)-dimensional space time, that could provide extra sources of CPV, are the ones augmented with massive right-handed neutrinos (RHN) in their spectra. Such models, with three species of heavy sterile Majorana RHN, might be used as providers—via the seesaw mechanism [30–35]—of light masses for (at least two of) the active neutrinos of the Standard Model, as required by the observed flavour oscillations [36]. In such models, in the early Universe, there is lepton asymmetry generation (*Leptogenesis*), through appropriate one-loop corrected decays of the RHN into Standard Model particles and antiparticles [37, 38]. In such processes, which are CPT-conserving, the existence of a non-trivial CPV requires *more than one species* of Majorana neutrinos [38] and at least one-loop corrections in the appropriate decay processes. These features lead to a difference in the respective CPV decay widths of the Majorana neutrino into standard-model particles and antiparticles, thus producing a Lepton-number (L) violation at an appropriate cosmological freeze-out point. We stress that tree-level decays and cases with only one species of Majorana neutrino lead to zero lepton asymmetry in CPT invariant models.

Such lepton number asymmetry generation is then communicated to the baryon sector via equilibrated sphaleron processes [24, 26], which violate both B and L numbers, but preserve their difference B-L (*Baryogenesis*). This Baryogenesis

[1] We remark that the chiral anomalies also induce the same amount of lepton number (L) violation, since $\partial_\mu J^{B\,\mu} = \partial_\mu J^{L\,\mu}$.

through Leptogenesis mechanism is currently a very popular one for the generation of matter-antimatter asymmetry in the Universe [38].[2]

Although there is a well established theoretical understanding of the above processes in the context of more or less conventional (i.e Lorentz and CPT invariant) particle physics models, nonetheless the lack of experimental evidence for the existence of additional sources (beyond the Standard Model) of CP violation, and right-handed neutrinos, may be a hint that some other, less conventional mechanism is in operation to explain the big question as to why *we exist*, that is, why there is this overwhelming dominance of matter over antimatter in our observable Universe. In this respect, a question arises as to whether the above processes of generating matter-antimatter asymmetry in the Universe could have a *geometric origin*, possibly due to quantum fluctuations of space time (quantum gravity), which are strong in the early Universe, and such that they violate Lorentz and CPT symmetry, leading to unconventional origins and processes for Lepto/Baryogenesis.

The theory of quantum gravity is still elusive, despite several theoretical attempts in the past and current centuries. One of the biggest questions associated with a consistent quantum theory of space time concerns the dynamical emergence of spacetime itself, and therefore the background independence of the theory. In some background-independent modern approaches to quantum gravity, e.g. the so-called *spin foam models* [42, 43], one starts from a rather abstract discrete set of states, which eventually condense to form dynamically the space-time continuum. Lorentz invariance in such models of quantum gravity, at least in the way we are familiar with from particle physics, may thus not be sacrosanct. We also mention at this stage, that in more conventional models, where a background space time is assumed, Wheeler has conjectured, many years ago, that microscopic black-hole and other topologically non trivial fluctuations of space time, may themselves give space time a "foamy structure" at Planck length scales [44], which may not respect Lorentz symmetry. Such structures may also hinder information from a low-energy observer, who conducts scattering experiments, which may lead to an effective decoherence of quantum matter in such space times. In such systems, the quantum operator corresponding to the generator of CPT symmetry may not be well-defined in the effective low-energy theory [45], leading to intrinsic CPT violation, which may have distinguishing features [46, 47] as compared to conventional violation of CPT symmetry, the latter occurring, for instance, as a result of violation of Lorentz invariance in effective local field theories [8], in which the generator of CPT

[2] In most leptogenesis scenarios, the RHN are superheavy, of masses close to the Grand Unification scale, $m_N \gtrsim 10^{14}$ GeV, as required by microscopic seesaw models (including supersymmetric ones). Nonetheless, there are also non-supersymmetric models [39, 40], termed the νMinimal Standard Model (νMSM), according to which the sterile Majorana neutrinos have masses spanning the range from a few GeV to $\mathcal{O}(10)$ keV, with the lightest having very weak couplings to the Standard Model sector, so that it has a life time longer than the lifetime of the Universe, and as such it can provide a candidate for (warm) dark matter. Baryogenesis mechanisms in this latter framework have been discussed in [41].

symmetry is well defined, but does not commute with the Hamiltonian operator of the system.

In general, the idea that LV and/or CPT Violation (CPTV) might characterise some approaches to quantum gravity, and their effective low-energy field theories, which may lead to interesting phenomenology, has gained attention in recent years, as a result of the increased sensitivity of experiments, especially cosmic multi-messenger ones, to such violations. Although at present there is no experimental evidence for such violations, nonetheless the sensitivity of some experiments to some model parameters may reach Planck scale sensitivity, or even surpass it under some circumstances [48], thus approaching the regime of quantum gravity.

String theory [49], which is one of the most successful to date attempts to unify gravity with the rest of the fundamental interactions in nature, but so far has been developed as a space-time background-dependent approach, is based perturbatively on well-defined scattering matrices, and as such, most of its effective low-energy field theories so far are characterised by Lorentz and CPT invariance. There has also been an attempt to claim that non perturbative strings would also be characterised by some form of CPT invariance [50]. Nonetheless, there is no rigorous proof that non-perturbative string theory is not characterised by ground states which do violate Lorentz and/or CPT symmetries, leading to effective low-energy theories which are plagued by such violations. To the contrary, there are claims, supported by plausibility arguments, that such ground states do exist [51–54] in the landscape of (open) string vacua, thereby leading to the possibility of *spontaneous* Lorentz and CPT Violation in string theory (although it must be said that the non-perturbative stability of such vacua has not been rigorously established, as yet).

> Important

It is the purpose of this book chapter to discuss another scenario for the spontaneous violation of Lorentz and CPT symmetry in the closed string sector, which in fact will also involve gravitational anomalies. As we shall see, the condensation of the corresponding anomaly currents, in the presence of primordial gravitational waves, will result in the *spontaneous breaking* of Lorentz and CPT symmetry, with far reaching consequences for unconventional Baryogenesis through Leptogenesis in the respective string-inspired cosmologies, but also for inflation [55–58].

Before doing so, it is instructive to mention that a formalism for testing phenomenologically the predictions of local effective field theories with Lorentz and CPT Violation is the so-called SME [59–61], whose LV and CPTV parameters and their current bounds have been tabulated in [62]. An SME in the presence of gravitational backgrounds has also been formulated [63]. In the next section we review briefly the SME formalism in flat space-time backgrounds, which will be of relevance to us here.

1.2 The Standard-Model-Extension Effective Field Theory Formalism

The SME formalism [59–61] assumes that the spontaneous breakdown of Lorentz and/or CPT symmetries arises in effective local interacting field theories, which are initially Lorentz and CPT invariant, respecting unitarity and locality, and as such can be expressed in terms of (an infinite in principle) set of local quantum field theory operators, involving general-coordinate invariant (and thus also locally, in space-time, Lorentz invariant, on account of the equivalence principle) products of tensorial field operators $O^{\rm SM}_{\mu_1\mu_2\dots}$ (with $\mu_i = 0, \dots 3$ being (3 + 1)-dimensional space-time indices, $i = 1, 2, \dots,$), depending on the fields of the Standard Model, with field operators $C^{\mu_1\mu_2\dots}$ involving fields beyond the Standard Model. The spontaneous breaking of Lorentz and/or CPT symmetries arises from condensation of the latter operators, which in this way obtain non-trivial constant vacuum expectation values $\langle C^{\mu_1\mu_2\dots}\rangle = \text{constant} \neq 0$ ("background tensors"):

$$O^{\rm SM}_{\mu_1\mu_2\dots}\, C^{\mu_1\mu_2\dots} \quad \text{condensation}\Rightarrow \quad O^{\rm SM}_{\mu_1\mu_2\dots}\, \langle C^{\mu_1\mu_2\dots}\rangle \quad \mu_i = 0, \dots 3, \quad i = 1, 2, \dots . \tag{1.5}$$

The background tensors of (mass) dimension five and higher, are suppressed by appropriate inverse powers of the scale Λ of new physics, beyond the Standard Model, up to which the effective SME is valid. A complete classification of dimension five LV and CPTV operators in the fermion, scalar (Higgs) and gauge sectors of the Standard Model extension in flat space-time (Minkowski) backgrounds, including interactions among these sectors, as well as modifications of the respective kinetic terms, has been provided in [64].

Important

The following criteria for acceptable SME operators have been adopted (which also characterise operators of any dimension in the SME formalism):

1. The operators must be gauge invariant.
2. The operators must be Lorentz invariant, after contraction with a background tensor.
3. The operators must not be reduced to a total derivative (as this would imply that the respective operators would not contribute to the dynamics of the system).
4. The operators must not reduce to lower-dimension operators by the use of the Euler-Lagrange equations of motion.
5. The operators must couple to an irreducible background tensor.

The SME formalism should be viewed as an effective field theory formalism, providing a framework to perform calculations that are associated with (spontaneous) violation of Lorentz (and CPT) symmetry which can be used in the respective tests. It is not meant to delve into the microscopic way by means of which the symmetry violating background condensate tensors $\langle C^{\mu_1\mu_2\cdots}\rangle$ arise, if at all, as this is a feature of the underlying ultraviolet (UV) complete theory of quantum gravity. As already mentioned, the various background tensors constitute the parameters of the SME effective theory, whose experimental bounds from a plethora of diverse, terrestrial and extraterrestrial experiments/observations, including cosmological measurements, are tabulated and continuously updated in [62].

For our purposes below, we shall restrict ourselves to the free fermion sector, and in particular to the lowest order (and simplest) SME effective Lagrangian [59–61]:

$$\mathcal{L}_{\rm eff}^{\rm SME\ fermion} = \overline{\psi}(x)\Big(\frac{i}{2}\gamma^\mu \overset{\leftrightarrow}{\partial_\mu} - \mathcal{M}\Big)\psi(x), \quad \mathcal{M} = m\mathbf{1} + a_\nu\,\gamma^\nu + b_\mu\,\gamma_5\,\gamma^\nu, \tag{1.6}$$

where $\psi(x)$ denote a generic fermion, that could be a chiral spinor or even a Majorana one (in the case of right-handed neutrinos), m is its mass, the quantity $\mathbf{1}$ denotes the identity in spinor space, and $\gamma_5 = i\gamma^0\gamma^1\gamma^2\gamma^3$ is the chirality matrix. The coefficients a_μ and b_μ in the generalised mass term in (1.6) are both LV and CPTV background vectors.

? Exercise

1.1. Consider the Lagrangian of a Dirac fermion $\psi(x)$ of mass m, coupled to electromagnetic, $A_\mu(x)$, and axion (pseudoscalar) fields, $b(x)$:

$$\mathcal{L}_{A,b,\psi} = -\frac{1}{4}F_{\mu\nu}\,F^{\mu\nu} + \overline{\psi}(x)\Big(\frac{i}{2}\gamma^\mu \overset{\leftrightarrow}{\partial_\mu} - q\,A_\mu(x)\gamma^\mu - m\mathbf{1} - i\,g_{\rm ch}\,\partial_\mu b(x)\,\gamma^\mu\,\gamma_5\Big)\psi(x) \tag{1.7}$$

where e, $g_{\rm ch} \in \mathbb{R}$ denote the corresponding real couplings, and $F_{\mu\nu}$ is the Maxwell tensor.

(i) First, show that under Lorentz transformations, including improper ones, i.e. parity P and time reversal T, the quantities $\overline{\psi}\,\gamma^\mu\psi$ transform as $\Lambda^\mu{}_\nu\,\overline{\psi}\,\gamma^\nu\psi$, while $\overline{\psi}\,\gamma^\mu\,\gamma_5\psi$ transform as $\det(\Lambda)\,\Lambda^\mu{}_\nu\,\overline{\psi}\,\gamma^\nu\,\gamma_5\psi$, where $\det(\Lambda)$ denotes the determinant of the Lorentz transformations $\Lambda^\mu{}_\nu$ (including improper ones).

(ii) Then, by taking into account the way the vector $A_\mu(x)$ and pseudoscalar field $b(x)$ transform under such transformations, prove the Lorentz and CPT invariance of the Lagrangian (1.7).

(iii) Finally, by using the explicit transformations of the spinor *fields* under Parity (P), Charge conjugation (C) and Time reversal (T), and the corresponding transformations of the $A_\mu(x)$ and $b(x)$ fields, that you can find in standard quantum field theory books [1, 2], prove the CPT invariance of (1.7), under any order of the combined application of C, T, P. Pay special attention to argue that under the antiunitary T operation, the imaginary unit i that appears in the Lagrangian transforms as $T\,i\,T^{-1} = -i$.

(iv) Consider now the case of constant background vector fields $\langle A_\mu\rangle \equiv a_\mu =$ constant, and $\langle \partial_\mu b\rangle \equiv b_\mu =$ constant whose values remain constant under Lorentz (proper or improper) transformations. Show that these terms violate both Lorentz and CPT invariance.

Hint: To prove **(i)**, assume without proof that under a Lorentz transformation (schematically $x \to \Lambda x$) a Dirac spinor $\psi(x)$ transforms as $\psi'(x) = S\,\psi(\Lambda^{-1}x)$, with $S = e^{\frac{i}{2}\,\omega_{\alpha\beta}\,\Sigma^{\alpha\beta}}$, $\Sigma^{\alpha\beta} = \frac{i}{4}[\gamma^\alpha, \gamma^\beta]$, $\alpha, \beta = 0, \ldots 3$, $\omega_{\alpha\beta} = -\omega_{\beta\alpha}$ the six independent parameters of a (3+1)-dimensional Lorentz transformation, and γ^α the Dirac matrices. Then, by considering infinitesimal transformations, $\overline{\psi}'(x) = \overline{\psi}(\Lambda^{-1}x)\,S^{-1}$, and using standard properties of Dirac matrices [2], as well as the fact that $\Lambda^\alpha{}_\beta = \left(e^{-\frac{1}{2}\,\omega_{\mu\nu}\,M^{\mu\nu}}\right)^\alpha{}_\beta$, with $(M^{\mu\nu})^\alpha{}_\beta = \eta^{\mu\alpha}\,\delta^\nu{}_\beta - \eta^{\nu\alpha}\,\delta^\mu{}_\beta$, $\alpha, \beta, \mu, \nu = 0, \ldots 3$, prove that $S\,\gamma^\mu\,S^{-1} = \Lambda^\mu{}_\nu\,\gamma^\nu$.

In the context of our string-inspired model [55–58], we shall describe a mechanism for the dynamical generation of the LV and CPTV background b_μ in the effective field theory (1.6), through an appropriate condensate of gravitational waves in a string-inspired gravitational effective field theory with torsion and anomalies. In fact, as we shall see, the coefficient b_μ in this case will be associated with a condensation of the dual of a totally antisymmetric component of a torsion tensor in the $(3+1)$-dimensional spacetime arising from string compactitication. In our model, there is no generation of an a_μ background, so from now on we set this coefficient to zero.

We mention at this stage, that, phenomenologically, there are stringent bounds of the coefficient b_μ today, which amount to [62]:

$$|b_0| < 0.2\,\mathrm{eV}, \quad |b_i| < 10^{-31}\,\mathrm{GeV}. \tag{1.8}$$

We shall show that such bounds are quite naturally respected in our cosmological model, as a consequence of the cosmic (temperature) evolution of the LV and CPTV coefficients b_μ, which are generated during the inflationary period, and remain undiluted in the radiation era [55–58].

As we shall discuss, these background vectors b_μ play an important rôle in inducing phenomenologically-relevant Leptogenesis in string-inspired effective particle-physics models which involve RHN in their spectra [57, 65–68]. It is worth stressing already at this point that, unlike the conventional CPT and Lorentz invariant approaches [38], which require at least two species of (Majorana) RHN, and one-loop treatment for the respective decays, in order for the necessary CPV to be effective in producing the lepton asymmetry, this type of LV and CPTV Leptogenesis occurs at tree level, and one species of RHN suffices. As we shall see, it is the CPTV properties of the background vector b_μ, in the presence of which the RHN decays into standard-model particles take place, that guarantee this.

We stress that the association of b_μ to torsion provides a geometric origin to the cosmic matter-antimatter asymmetry generated in this way. We also remark that, given the universal coupling of the torsion to all fermion species, including lepton and quarks of the Standard Model, such a LV and CPTV mechanism through torsion condensation, may also lead directly to matter-antimatter asymmetries (LV and CPTV direct baryogenesis [69, 70]) in this Universe, without necessitating the presence of RHN, but we shall not explore these latter scenarios here.

The geometric concept of torsion in gravitational theories, exists also independent from its connection to string theory, as will be discussed in Chaps. 3 and 4 and Poincaré gauge theory and Teleparallel gravity.

1.3 A String-Inspired Gravitational Theory with Torsion and Anomalies

We are now well motivated to start employing string theory considerations that will lead us to the effective gravitational theory with the aforementioned LV and CPTV properties. To this end, we first remark that in closed string theory [49], the bosonic massless gravitational mutliplet consists of a spin-zero (scalar) field, the dilaton $\Phi(x)$, a spin-two symmetric tensor field, the graviton, $g_{\mu\nu}(x) = g_{\nu\mu}(x)$, where μ, ν are spacetime indices, and a spin-one antisymmetric tensor (or Kalb-Ramond (KR)) field $B_{\mu\nu}(x) = -B_{\nu\mu}(x)$. In the phenomenologically-relevant case of superstrings, this multiplet belongs to the ground state of string theory, which is augmented by the (local) supersymmetry partners of these fields. In our approach we shall not discuss those partners, and concentrate only on the aforementioned bosonic fields. In the scenario of [57], which we follow as a prototype model for our discussion in this chapter, we assume that supersymmetry is dynamically broken during a pre-inflationary epoch of the string-inspired Universe, and, as such, the supersymmetric partner fields acquire heavy masses, even close to the Planck scale. Therefore they decouple from the low-energy spectrum, which is of relevance to our subsequent discussion.

We next remark that, in the framework of perturbative strings, the closed-string σ-model deformation describing the propagation of the string in a KR field background $B_{\mu\nu}$, is given by the world-sheet expression [49]:

$$\Delta S^{\sigma}_{B} \equiv \int_{\Sigma^{(2)}} d^2\sigma \, B_{\mu\nu}(X) \, \varepsilon^{AB} \partial_A X^\mu \, \partial_B X^\nu \ , \ \ \mu, \nu = 0, \dots 3, \quad A, B = 1, 2 \,, \tag{1.9}$$

where the integral is over the surface $\Sigma^{(2)}$, which corresponds to the string-tree-level world-sheet with the topology of a two-dimensional sphere $S^{(2)}$. For our purposes, such lowest genus world-sheet topologies suffice, given that string loop corrections, which would be associated with higher-genus world-sheet surfaces, are subdominant for weak string couplings we assume throughout; the indices $A, B = 1, 2$ are world-sheet indices, $\varepsilon_{AB} = -\varepsilon_{BA}$ is the world-sheet covariant Levi-Civita antisymmetric tensor, and X^μ, $\mu = 0, \dots 3$, are world-sheet fields, whose zero modes play the rôle of target-space coordinates. We have assumed that consistent string compactification [49] to (3 + 1) spacetime dimensions has taken place, whose details will not be discussed here.

It can be seen straightforwardly (using Stokes theorem, and taking into account that the spherical-like surface $\Sigma^{(2)}$ has no boundary) that the integrand in (1.9) is invariant under the following U(1) gauge transformation in target space (which is not related to electromagnetism):

$$B_{\mu\nu} \rightarrow B_{\mu\nu} + \partial_\mu \theta_\nu(X) - \partial_\nu \theta_\mu(X), \quad \mu, \nu = 0, \ldots 3, \tag{1.10}$$

where $\theta_\mu(X)$, $\mu = 0, \ldots 3$, are gauge parameters.

? Exercise

1.2. Starting from the expression for the world-sheet deformation (1.9), prove its invariance under the gauge transformation (1.10).

This implies that the target-space effective action, which describes the low-energy limit of the string theory at hand, will be invariant under the U(1) gauge symmetry (1.10), and, as such, it will depend only on the field strength of $B_{\mu\nu}$:

$$H_{\mu\nu\rho} = \partial_{[\mu} B_{\nu\rho]} \,, \tag{1.11}$$

where the symbol [...] indicates total antisymmetrisation of the respective indices.

However, in string theory [49], cancellation between gauge and gravitational anomalies in the extra dimensional space requires the introduction of Green-Schwarz counterterms [71], which results in the modification of the field strength $H_{\mu\nu\rho}$ by the respective Chern-Simons (gravitational ("Lorentz", L) and gauge (Y)) anomalous terms :

$$\mathcal{H} = \mathbf{dB} + \frac{\alpha'}{8\kappa}\Big(\Omega_{3\mathrm{L}} - \Omega_{3\mathrm{Y}}\Big),$$
$$\Omega_{3\mathrm{L}} = \omega^a_{\ c} \wedge \mathbf{d}\omega^c_{\ a} + \frac{2}{3}\omega^a_{\ c} \wedge \omega^c_{\ d} \wedge \omega^d_{\ a}, \quad \Omega_{3\mathrm{Y}} = \mathbf{A} \wedge \mathbf{dA} + \mathbf{A} \wedge \mathbf{A} \wedge \mathbf{A}, \tag{1.12}$$

where we used differential form language, for notational convenience. In the above expression, $\mathcal{H}$ is a three-form, the symbol $\wedge$ denotes the exterior product among differential (k, ℓ) forms $(\mathbf{f}^{(k)} \wedge \mathbf{g}^{(\ell)} = (-1)^{k\,\ell}\,\mathbf{g}^{(\ell)} \wedge \mathbf{f}^{(k)})$, $\mathbf{A} \equiv \mathbf{A}_\mu\, dx^\mu$ denotes the Yang-Mills gauge field one form, and $\omega^a_{\ b} \equiv \omega^a_{\ \mu\, b}\, dx^\mu$ is the spin connection one form, with the Latin indices a, b, c, d being tangent space (SO(1,3)) indices. The quantity α' is the Regge slope $\alpha' = M_s^{-2}$, where M_s is the string mass scale, which is in general different from the reduced Planck scale in four space-time dimensions that enters the definition of the four-dimensional gravitational constant $\kappa = \sqrt{8\pi\, \mathrm{G}} = M_{\rm Pl}^{-1}$, with $M_{\rm Pl} = 2.43 \times 10^{18}$ GeV (we work in units of $\hbar = c = 1$ throughout this work).

To lowest (zeroth) order in a perturbative expansion in powers of the Regge slope α', *i.e.* to quadratic order in a derivative expansion, the low-energy effective four-dimensional action corresponding to the bosonic massless string multiplet, reads [49]:[3]

$$S_B = \int d^4x \sqrt{-g}\Big(\frac{1}{2\kappa^2}[-R + 2\,\partial_\mu \Phi\, \partial^\mu \Phi] - \frac{1}{6}\, e^{-4\Phi}\, \mathcal{H}_{\lambda\mu\nu}\mathcal{H}^{\lambda\mu\nu} + \dots\Big), \tag{1.13}$$

with the ellipses ... denoting higher-derivative terms, and possible dilaton potentials (arising from string loops or other mechanisms in effective string-inspired models, such as dilaton and non-critical-string cosmologies [72–74], and pre-Big-Bang scenarios [75].). The action (1.13) can be found by either matching the corresponding (lowest order in derivatives) string scattering amplitudes with those obtained from the action (1.13), or by considering the world-sheet conformal invariance conditions (*i.e.* the vanishing of the corresponding Weyl-anomaly coefficients [49]) of the corresponding two-dimensional σ model, which describes the propagation of strings in the backgrounds of Φ, $g_{\mu\nu}$ and $B_{\mu\nu}$, and identify them with the corresponding equations of motion stemming from the effective action (1.13).[4]

In our approach we shall consider the dilaton field as fixed to an appropriate constant value, corresponding to minimisation of its potential, so that the string coupling $g_s = \exp(\Phi)$ is fixed to phenomenologically acceptable values [49]. Without loss of generality, then, we may set from now on $\Phi = 0$. This is a self consistent procedure, as explained in [56] (see also Exercise 1.16 in Sect. 1.4.3) , which yields the $\Phi = 0$ configuration as a solution for the dilaton equation that acts as a constraint in this case.

The torsion [82] interpretation of $\mathcal{H}_{\mu\nu\rho}$ arises by noticing that one can combine the quadratic in $\mathcal{H}_{\mu\nu\rho}$ terms of (1.13) with the Einstein-Hilbert curvature scalar term R in a generalised curvature scalar $\overline{R}(\overline{\Gamma})$ with respect to a generalised connection, so that the action (1.13), with $\Phi = 0$, is equivalent to the action:

$$S_B = \int d^4x \sqrt{-g}\, \frac{1}{2\kappa^2}\Big(- R(\overline{\Gamma}) + \dots\Big), \tag{1.14}$$

[3] In this work we follow the convention for the signature of the metric $(+, -, -, -)$, and the definitions of the Riemann Curvature tensor $R^\lambda_{\ \ \mu\nu\sigma} = \partial_\nu\, \Gamma^\lambda_{\ \ \mu\sigma} + \Gamma^\rho_{\ \ \mu\sigma}\, \Gamma^\lambda_{\ \ \rho\nu} - (\nu \leftrightarrow \sigma)$, the Ricci tensor $R_{\mu\nu} = R^\lambda_{\ \ \mu\lambda\nu}$, and the Ricci scalar $R = R_{\mu\nu} g^{\mu\nu}$.

[4] There are, of course, well-known ambiguities in such processes [76–79], associated with local field redefinitions which leave the perturbative string scattering matrix invariant, according to the equivalence theorem of local quantum field theories [80, 81]. Such ambiguities, allow for instance, the string effective actions at quartic order in derivatives ($O(\alpha')$) to be cast in the dilaton-Gauss-Bonnet combination [76], which is free from gravitational ghosts.

where

$$\overline{\Gamma}^{\rho}_{\mu\nu} = \Gamma^{\rho}_{\mu\nu} + \frac{\kappa}{\sqrt{3}}\mathcal{H}^{\rho}_{\mu\nu} \neq \overline{\Gamma}^{\rho}_{\nu\mu} \tag{1.15}$$

where $\Gamma^{\rho}_{\mu\nu} = \Gamma^{\rho}_{\nu\mu}$ is the torsion-free Christoffel symbol. Since the KR field strength satisfies

$$\mathcal{H}^{\mu}_{\nu\rho} = -\mathcal{H}^{\mu}_{\rho\nu}\,, \tag{1.16}$$

it plays the rôle of contorsion [82]. This contorted geometry contains only a totally antisymmetric component of torsion [82].[5]

? Exercise

1.3. Prove the equivalence, up to total derivative terms, of the actions (1.13) and (1.14), taking into account (1.15) and (1.16).

The modification (1.12) leads to the Bianchi identity (in differential form language) [49]

$$\mathbf{d}\mathcal{H} = \frac{\alpha'}{8\,\kappa}\mathrm{Tr}\Big(\mathbf{R}\wedge\mathbf{R} - \mathbf{F}\wedge\mathbf{F}\Big) \tag{1.17}$$

where $\mathbf{R}^{a}{}_{b} = \mathbf{d}\omega^{a}{}_{b} + \omega^{a}{}_{c}\wedge\omega^{c}{}_{b}$ is the curvature two form, $\mathbf{F} = \mathbf{dA} + \mathbf{A}\wedge\mathbf{A}$ is the Yang-Mills field-strength two form, and the trace (Tr) is over Lorentz- and gauge-group indices, respectively. The non zero quantity on the right hand side of (1.17) is the "mixed (gauge and gravitational) quantum anomaly" [84]. In the (more familiar) component form, the identity (1.17), becomes:

$$\varepsilon_{abc}{}^{\mu}\,\mathcal{H}^{abc}{}_{;\mu} = \frac{\alpha'}{32\,\kappa}\sqrt{-g}\,\Big(R_{\mu\nu\rho\sigma}\,\widetilde{R}^{\mu\nu\rho\sigma} - F_{\mu\nu}\,\widetilde{F}^{\mu\nu}\Big) \equiv \sqrt{-g}\,\mathcal{G}(\omega,\mathbf{A}), \tag{1.18}$$

where the semicolon denotes gravitational covariant derivative with respect to the standard Christoffel connection, and

$$\varepsilon_{\mu\nu\rho\sigma} = \sqrt{-g}\,\epsilon_{\mu\nu\rho\sigma}\,, \qquad \varepsilon^{\mu\nu\rho\sigma} = \frac{\mathrm{sgn}(g)}{\sqrt{-g}}\,\epsilon^{\mu\nu\rho\sigma}\,, \tag{1.19}$$

[5] Using local field redefinition ambiguities [49,76–78,83] one can extend the torsion interpretation of $\mathcal{H}$ to $O(\alpha')$ effective actions, which include fourth-order derivative terms.

denote the gravitationally covariant Levi-Civita tensor densities, totally antisymmetric in their indices, with $\epsilon_{\mu\nu\rho\sigma}$ ($\epsilon_{0123} = +1$, etc.) the Minkowski-space-time Levi-Civita totally antisymmetric symbol. The symbol $\widetilde{(\dots)}$ over the curvature or gauge field strength tensors denotes the corresponding duals, defined as

$$\widetilde{R}^{\mu\nu\rho\sigma} \equiv \frac{1}{2}\varepsilon^{\mu\nu\lambda\pi} R_{\lambda\pi}{}^{\rho\sigma}\,, \qquad \widetilde{F}^{\mu\nu} \equiv \frac{1}{2}\varepsilon^{\mu\nu\rho\sigma} F_{\rho\sigma}\,, \tag{1.20}$$

respectively. The mixed-anomaly term is a total derivative term

$$\sqrt{-g}\,\mathcal{G}(\omega, \mathbf{A}) = \sqrt{-g}\,\mathcal{K}^{\mu}(\omega, \mathbf{A})_{;\mu} = \partial_{\mu}\Big(\sqrt{-g}\,\mathcal{K}^{\mu}(\omega, \mathbf{A})\Big) \tag{1.21}$$

as can be seen from

$$\begin{aligned} &\sqrt{-g}\,\Big(R_{\mu\nu\rho\sigma}\,\widetilde{R}^{\mu\nu\rho\sigma} - F_{\mu\nu}\,\widetilde{F}^{\mu\nu}\Big) \\ &= 2\,\partial_{\mu}\Big[\epsilon^{\mu\nu\alpha\beta}\,\omega_{\nu}^{ab}\,\Big(\partial_{\alpha}\,\omega_{\beta ab} + \frac{2}{3}\,\omega_{\alpha a}{}^{c}\,\omega_{\beta cb}\Big) \\ &\quad - 2\epsilon^{\mu\nu\alpha\beta}\,\Big(A^{i}_{\nu}\,\partial_{\alpha}A^{i}_{\beta} + \frac{2}{3}\,f^{ijk}\,A^{i}_{\nu}\,A^{j}_{\alpha}\,A^{k}_{\beta}\Big)\Big], \end{aligned} \tag{1.22}$$

where i, j, k denote gauge group indices, with f^{ijk} the gauge group structure constants.

? Exercise

1.4. Using the definitions of the curvature and gauge field strength differential forms (in a shorthand notation, for brevity), $\mathbf{R} = d\omega + \omega \wedge \omega$ and $\mathbf{F} = d\mathbf{A} + \mathbf{A} \wedge \mathbf{A}$, in terms of the spin connection ω and the gauge field connection $\mathbf{A}$, respectively, prove Eq. (1.17), by taking the exterior derivative of the three form $\mathcal{H}$ in (1.12).

> Important

In our four-dimensional cosmology [55–58] we shall *not* cancel the anomalies. In fact, we shall assume that only fields of the bosonic degrees of freedom of the massless gravitational string multiplet appear as external fields in the effective action describing the dynamics of the early Universe. Chiral fermionic and gauge matter are generated at the end of the inflationary period as we shall discuss later on. With the above assumptions, one may implement the Bianchi identity (1.17) as a *constraint* in a path-integral, via a pseudoscalar (axion-like) Lagrange multiplier field $b(x)$. After the $\mathcal{H}_{\mu\nu\rho}$ path-integration, then, one arrives at an effective action for the dynamics of the early epoch of the string-inspired Universe, which, upon the assumption of constant dilatons, contains only gravitons and the now dynamical field $b(x)$, canonically normalised, without potential, which corresponds to the massless string-model-independent gravitational (or KR) axion field [83, 85]:

$$\begin{aligned} S_B^{\rm eff} &= \int d^4x\, \sqrt{-g}\Big[-\frac{1}{2\kappa^2}\, R + \frac{1}{2}\, \partial_\mu b\, \partial^\mu b + \sqrt{\frac{2}{3}}\, \frac{\alpha'}{96\, \kappa}\, b(x)\, R_{\mu\nu\rho\sigma}\, \widetilde{R}^{\mu\nu\rho\sigma} + \dots \Big] \\ &= \int d^4x\, \sqrt{-g}\Big[-\frac{1}{2\kappa^2}\, R + \frac{1}{2}\, \partial_\mu b\, \partial^\mu b \Big] \\ &\quad - \int d^4x \sqrt{\frac{2}{3}}\, \frac{\alpha'}{96\, \kappa}\, b(x)\, R_{\mu\nu\rho\sigma}\, {}^{*}R^{\mu\nu\rho\sigma} + \dots \\ &= \int d^4x\, \sqrt{-g}\Big[-\frac{1}{2\kappa^2}\, R + \frac{1}{2}\, \partial_\mu b\, \partial^\mu b - \sqrt{\frac{2}{3}}\, \frac{\alpha'}{96\, \kappa}\, \mathcal{K}^\mu(\omega)\, \partial_\mu b(x) + \dots \Big]. \end{aligned} \tag{1.23}$$

In passing from the first to the second line of (1.23) we have used the definitions (1.19) and that sgn(g)=-1. The symbol ${}^{*}R^{\mu\nu\rho\sigma}$ denotes the dual with respect to the flat-space-time Levi-Civita totally antisymmetric symbol $\epsilon^{\mu\nu\rho\sigma}$, with $\epsilon^{0123} = +1$, *etc.*:

$${}^{*}R^{\mu\nu\rho\sigma} \equiv \frac{1}{2}\epsilon^{\mu\nu\lambda\pi}\, R_{\lambda\pi}^{\ \ \ \ \rho\sigma}\,, \tag{1.24}$$

In the last line of (1.23) we have used (1.22), setting $\mathbf{A} = 0$ ($\mathcal{K}^\mu(\omega) = \mathcal{K}^\mu(\omega, 0)$), and performed appropriately the integration by parts, taking into account that fields and their first derivatives vanish at space-time infinity. The action (1.23) is nothing other than the action describing the Chern-Simons modification of general relativity in the presence of axion fields [86, 87]. In fact, from this latter point of view, one may view this action as a generic Chern-Simons-modified-gravity action, beyond the specific context of string theory, in which case the coefficient of the Chern-Simons term should be replaced by a generic real parameter:

$$\sqrt{\frac{2}{3}}\frac{\alpha'}{96\, \kappa} \quad \Rightarrow \quad \mathcal{A}_{\rm CS} \in \mathbb{R}\,, \tag{1.25}$$

to be determined "phenomenologically" in various contexts (*e.g.*, rotating black holes and wormholes, beyond string theory, as in [88–90]).

? Exercise

1.5. Consider the path integral of the action (1.13) with respect to the field $\mathcal{H}_{\mu\nu\rho}$, setting the dilaton $\Phi = 0$:

$$\mathcal{Z}_{\mathcal{H}} = \int \mathcal{D}\mathcal{H} \exp(i S_{\mathrm{B}}) , \tag{1.26}$$

where $\mathcal{D}\mathcal{H}$ denotes the appropriate path-integration measure. Insert the Bianchi-identity (1.18), in the absence of gauge fields ($\mathbf{A} = 0$), as a δ-functional constraint,

$$\delta\Big(\varepsilon_{abc}{}^{\mu}\, \mathcal{H}^{abc}{}_{;\mu} - \frac{\alpha'}{32\,\kappa}\sqrt{-g}\,\Big(R_{\mu\nu\rho\sigma}\, \widetilde{R}^{\mu\nu\rho\sigma}\Big)\Big),$$

in the integrand of (1.26). By representing the $\delta(x)$ functional as an integral over a pseudoscalar Lagrange multiplier field, perform the $\mathcal{H}$-path integration, and normalise appropriately the Lagrange multiplier to link it to the field $b(x)$ appearing in the action (1.23), with canonical kinetic term, thus mapping (1.26) to a path integral over $b(x)$ corresponding to the action (1.23). Why the Lagrange multiplier field in this case has to be a pseudoscalar?

We note that classically, in (3+1) dimensional space-times, the duality between $\mathcal{H}_{\mu\nu\rho}$ and $b(x)$ is provided by the relation (corresponding to saddle points of the $\mathcal{H}$ path-integral (1.26) after the b-representation of the Bianchi-constraint-(1.18) δ-functional) [72, 83]

$$-\,3\sqrt{2}\,\partial_\sigma b = \sqrt{-g}\,\epsilon_{\mu\nu\rho\sigma}\,\mathcal{H}^{\mu\nu\rho} . \tag{1.27}$$

The ellipses ... in (1.23) denote subdominant, for our purposes, higher derivative terms (in fact an infinity of them), but also other axions, arising from compactification in string theory [85], which have been discussed in [58], but will not be the focus of our present study. The reader should notice the presence of anomalous CP-violating couplings of the KR axion to gravitational anomalies in the action (1.23). These will play an important role in inducing inflation in our string-inspired cosmology.

We note at this stage that, had we kept gauge fields in our early-universe cosmology as external fields, the KR axion field would also exhibit Lagrangian couplings of the form $\propto b(x)\mathrm{Tr}\Big(\mathbf{F}_{\mu\nu}\,\widetilde{\mathbf{F}}^{\mu\nu}\Big)$. Such terms would not contribute to the stress tensor, being topological.

? Exercise

1.6. Prove that the contributions of the term

$$b(x)\mathrm{Tr}\Big(\mathbf{F}_{\mu\nu}\,\widetilde{\mathbf{F}}^{\mu\nu}\Big)$$

where $\mathbf{F}_{\mu\nu}$ is the (non-Abelian, in general) gauge field strength, and $\widetilde{\mathbf{F}}^{\mu\nu}$ its dual (defined below Eq. (1.18)), to the stress-energy tensor of the theory (the variation of the action w.r.t. the metric) vanish identically.

This needs to be contrasted with the gravitational anomaly terms in (1.23), whose variation with respect to the metric field $g_{\mu\nu}$ yields non-trivial results [86, 87]:

$$\begin{aligned}\delta\Big[\int d^4x\sqrt{-g}\,b\,R_{\mu\nu\rho\sigma}\,\widetilde{R}^{\mu\nu\rho\sigma}\Big] &= 4\int d^4x\sqrt{-g}\,C^{\mu\nu}\,\delta g_{\mu\nu}\\ &= -4\int d^4x\sqrt{-g}\,C_{\mu\nu}\,\delta g^{\mu\nu}\,,\end{aligned} \tag{1.28}$$

where

$$C^{\mu\nu} \equiv -\frac{1}{2}\Big[u_\sigma\left(\varepsilon^{\sigma\mu\alpha\beta}\,R^{\nu}_{\ \beta;\alpha} + \varepsilon^{\sigma\nu\alpha\beta}\,R^{\mu}_{\ \beta;\alpha}\right) + u_{\sigma\tau}\left(\widetilde{R}^{\tau\mu\sigma\nu} + \widetilde{R}^{\tau\nu\sigma\mu}\right)\Big], \tag{1.29}$$

is the (tracelss) Cotton tensor [86],

$$g^{\mu\nu}\,C_{\mu\nu} = 0\,, \tag{1.30}$$

with $u_\sigma \equiv \partial_\sigma b = b_{;\sigma}$, $u_{\sigma\tau} \equiv u_{\tau;\sigma} = b_{;\tau;\sigma}$. Taking into account conservation properties of the Cotton tensor [86],

$$C^{\mu\nu}_{\ \ \ ;\mu} = \frac{1}{8}\,u^{\nu}\,R^{\alpha\beta\gamma\delta}\,\widetilde{R}_{\alpha\beta\gamma\delta}\,, \tag{1.31}$$

we observe that the gravitational field (or Einstein-Chern-Simons) equations stemming from (1.23) (and (1.25)) read:

$$R^{\mu\nu} - \frac{1}{2}\,g^{\mu\nu}\,R - \mathcal{A}_{\rm CS}\,C^{\mu\nu} = \kappa^2\,T^{\mu\nu}_{\rm matter}, \tag{1.32}$$

where $T^{\mu\nu}_{\rm matter}$ denotes a matter stress tensor, which in our early-Universe cosmology includes only the KR axion-like field [55–58]

$$T^b_{\mu\nu} = \partial_\mu b\, \partial_\nu b - \frac{1}{2}\Big(\partial_\alpha b\, \partial^\alpha b\Big)\,. \tag{1.33}$$

In more general situations, $T^{\mu\nu}_{\rm matter}$ contains all matter and radiation fields, but does *not* contain couplings to the curvature or derivatives of the metric tensor.

? Exercise

1.7. Prove the variational Eq. (1.28) and the properties (1.30) and (1.31) of the Cotton tensor, using its definition (1.29).

> Important

From the properties of the Einstein and Cotton tensors, stated above, we observe that the matter stress tensor is not conserved but it satisfies the conservation of an improved stress tensor in the form:

$$T^{\mu\nu}_{\rm improved\,;\mu} \equiv T^{\mu\nu}_{\rm matter\,;\mu} + \mathcal{A}_{\rm CS}\, C^{\mu\nu}{}_{;\mu} = 0\,. \tag{1.34}$$

The presence of the Cotton tensor in this conservation equation indicates exchange of energy between the KR axion field and the gravitational anomaly term, in a way consistent with diffeomorphism invariance and general covariance [55]. For Friedman-Lemaître-Robertson-Walker (FLRW) geometries, the gravitational anomaly terms vanish, but this is not the case for (chiral) fluctuations about the FLRW background which violate CP invariance, for instance chiral gravitational-wave (GW) perturbations, as we discuss below.

? Exercise

1.8. Prove (1.34).

1.4 Chiral Gravitational-Wave (Quantum) Fluctuations, Anomaly Condensates and Running-Vacuum-Model Inflation

In the string-inspired cosmological model of [55, 57], which is assumed to describe the dynamics of the early Universe, only fields from the massless gravitational string multiplet are assumed to appear in the effective gravitational action. The generation of chiral fermionic and gauge matter occurs at the end of the inflationary era, as we shall discuss later on. For constant dilatons, we assumed so far, this implies that the effective Chern-Simons-modified gravity action (1.23) is the relevant one for a discussion of inflation in such a Universe. In the presence of (chiral) gravitational-wave (GW) quantum fluctuations of spacetime the anomaly terms are non trivial. In the literature there have been essentially two ways of computing the effects of GW on the gravitational anomaly terms: one is through Green's functions [91], and the other through canonically quantised linearised gravity formalism [92], which we adopt below as it seems closer to our spirit that the anomaly condensates are induced by quantum gravitational fluctuations.

1.4.1 Chiral-Gravitational-Wave Quantized Perturbations

To this end, let one consider *quantised* tensor perturbations $h_{ij}(\eta, \mathbf{x})$ in a flat FLRW expanding Universe:

$$ds^2 = a^2(\eta)\Big(d\eta^2 - (\delta_{ij} + 2h_{ij}(\eta, \mathbf{x})\Big)dx^i dx^j, \quad i, j = 1, 2, 3, \tag{1.35}$$

where $a(\eta)$ is the scale factor of the FLRW Universe, and η is the conformal time [93], which is related in our approach to the Robertson-Walker time t via

$$a(\eta)d\eta = +dt, \tag{1.36}$$

(we consider the flow of both times in the same direction). The perturbations can be written in terms of their three-dimensional-space Fourier components, $h_p(\mathbf{k}, \eta)$, with $p =$ Left (L) or Right (R), as [92]:

$$h_{ij}(\mathbf{x}, \eta) = \frac{\sqrt{2}}{M_{\rm Pl}} \int \frac{d^3k}{(2\pi)^{3/2}} e^{i\mathbf{k}\cdot\mathbf{x}} \sum_{p={\rm L,\,R}} \epsilon^p_{ij}(\mathbf{k})\, h_p(\mathbf{k}, \eta), \tag{1.37}$$

with $\epsilon^{p}_{ij}(\mathbf{k})$ the polarisation tensors, satisfying:

$$k_i \epsilon^{p}_{ij}(\mathbf{k}) = 0, \quad \epsilon^{p\star}_{ij}(\mathbf{k})\,\epsilon^{p'}_{ij}(\mathbf{k}) = 2\,\delta_{pp'}\,,$$
$$\epsilon^{ilm}\,\epsilon^{L\star}_{ij}(\mathbf{k})\,\epsilon^{R}_{jl}(\mathbf{k}) = \epsilon^{ilm}\,\epsilon^{R\star}_{ij}(\mathbf{k})\,\epsilon^{L}_{jl}(\mathbf{k}) = 0\,,$$
$$\epsilon^{ilm}\,\epsilon^{L\star}_{ij}(\mathbf{k})\,\epsilon^{L}_{jl}(\mathbf{k}) = -\epsilon^{ilm}\,\epsilon^{R\star}_{ij}(\mathbf{k})\,\epsilon^{R}_{jl}(\mathbf{k}) = -2i\,\frac{k_m}{|\mathbf{k}|}\,, \tag{1.38}$$

where the $\star$ denotes complex conjugation.

We quantise the tensor perturbations, assumed weak, by writing the tensor perturbation as an operator $\widehat{h}_{ij}(\mathbf{k})$ in the Heisenberg picture, which implies that it satisfies the corresponding Einstein equation [92]:

$$\widehat{h}_{ij}(\mathbf{x}, \eta) = \frac{\sqrt{2}}{M_{\rm Pl}} \int \frac{d^3k}{(2\pi)^{3/2}} \sum_{p=L,\,R} \Big(e^{i\mathbf{k}\cdot\mathbf{x}}\,\epsilon^{p}_{ij}(\mathbf{k})\,\widehat{h}_p(\mathbf{k}, \eta) \Big)\,,$$
$$\widehat{h}_p(\mathbf{k}, \eta) = h_p(\mathbf{k}, \eta)\,\widehat{a}_p(\mathbf{k}) + h^{\star}_p(-\mathbf{k}, \eta)\,\widehat{a}^{\dagger}_p(-\mathbf{k})\,, \tag{1.39}$$

where the $\widehat{(\dots)}$ denotes a quantum operator, and for the creation, $\widehat{a}^{\dagger}_p(\mathbf{k})$, and annihilation operators, $\widehat{a}_p(\mathbf{k})$, we have the canonical commutation relations

$$\Big[\widehat{a}_p(\mathbf{k})\,,\,\widehat{a}^{\dagger}_{p'}(\mathbf{k}')\Big] = \delta^{(3)}(\mathbf{k} - \mathbf{k}') \tag{1.40}$$

and the hermitian conjugate relation, with all others zero. The time-independent vacuum state $|0\rangle$ is defined by its annihilation by $\widehat{a}_p(\mathbf{k})$, *i.e.*

$$\widehat{a}_p(\mathbf{k})|0\rangle = 0\,. \tag{1.41}$$

Assuming weak tensor perturbations h_{ij}, (1.35), we may show that, up to second order in such perturbations, the gravitational Chern Simons term assumes the form

$$R_{\mu\nu\rho\sigma}\,{}^*R^{\mu\nu\rho\sigma} = -\,\frac{8}{a(\eta)^4}\epsilon^{ijk}\Big(\frac{\partial^2}{\partial x^l\,\partial\eta}\,h_{jm}\,\frac{\partial^2}{\partial x^m\,\partial x^i}\,h_{kl}$$
$$-\,\frac{\partial^2}{\partial x^l\,\partial\eta}\,h_{jm}\,\frac{\partial^2}{\partial x^l\,\partial x^i}\,h_{km} + \frac{\partial^2}{\partial\eta^2}\,h_{jl}\,\frac{\partial^2}{\partial x^i\,\partial\eta}\,h_{lk}\Big), \tag{1.42}$$

where ϵ^{ijk}, $i,\,j,\,k = 1, 2, 3$ spatial indices, is the totally antisymmetric symbol in Euclidean three-dimensional space. We remind the reader that the dual tensor ${}^*R^{\mu\nu\rho\sigma}$ is defined (*cf.* (1.24)) with respect to the flat space-time Levi-Civita symbol $\epsilon^{\mu\nu\rho\sigma}$, with $\epsilon^{0ijk} = +1 \equiv \epsilon^{ijk}$, $i,\,j,\,k = 1, 2, 3$.

? Exercise

1.9. Starting from the metric (1.35), prove (1.42), up to second order in weak tensor perturbations h_{ij}.

Important

The classical quantity (1.42) becomes a Chern-Simons operator $R_{\mu\nu\rho\sigma}\,\widehat{\widetilde{R}^{\mu\nu\rho\sigma}}$ upon replacing $h_{ij}(\eta, \mathbf{x})$ by the corresponding quantum operators $\widehat{h}_{ij}(\eta, \mathbf{x})$ (1.39). Using (1.40) and (1.41), one can show [92] that the vacuum expectation value of the Chern-Simons operator

$$\langle 0|\widehat{R_{\mu\nu\rho\sigma}\,{}^*R^{\mu\nu\rho\sigma}}|0\rangle = \frac{16}{a(\eta)^4\,M_{\rm Pl}^2}\int\frac{d^3k}{(2\pi)^3}\Big[k^2 h_L^\star(k,\eta)\,h'_L(k,\eta) - k^2\,h_R^\star(k,\eta)\,h'_R(k,\eta) - h_L^{\star\prime}(k,\eta)\,h''_L(k,\eta) + h_R^{\star\prime}(k,\eta)\,h''_R(k,\eta)\Big], \tag{1.43}$$

where $k \equiv |\mathbf{k}|$, and the prime denotes derivative with respect to the conformal time η. Notice that the vacuum expectation value (1.43) vanishes in a Left-Right symmetric situation, therefore the result is only non zero when there is gravitational birefringence, *i.e.* chirality (in the sense of differences between left-right GW perturbations).

? Exercise

1.10. Inserting (1.39) into (1.42), and using canonical quantization properties, (1.40) and (1.41), prove the result (1.43) for the vacuum expectation value of the gravitational Chern Simons operator.

The physical momenta $k/(a(\eta))$ of the graviton modes should be cut-off at an UltraViolet (UV) scale μ, which means that terms in the ultraviolet regime dominate the integrals. It is at this point that to have a full understanding of the condensate one needs the UV complete theory of quantum gravity, such as the full string theory in this case.

In [91], it was assumed that in the evolution equation of the graviton modes one can keep only up to second order derivative terms. Although, as correctly remarked in [92], this is far from a satisfactory treatment within a quantum gravity regime, as required by the fact that the dominant part of the Fourier integration is near the UV cut-off, where quantum gravity is fully operational, nonetheless, for our purposes of discussing qualitatively the effects of gravitational anomaly condensates on inducing a running vacuum inflation, this will suffice [55–58], in view of the slow-roll of the weak axion KR field that characterises our cosmology model, as we shall discuss below.

Making this assumption, it can be easily seen that, in an *inflationary (de Sitter-like) space-time background* of interest to us here, with an approximately constant Hubble parameter $H \simeq$ constant, the normalised solutions of the Einstein-Chern-Simons gravitational field Eq. (1.32), in the presence of a weak anomaly term, assume the form:

$$h_{p=L,R}(k,\eta) \sim \exp(-i\,k\eta)\exp(\pm k\,\Theta(\eta-\eta_0))\,,$$
$$\Theta \equiv \frac{4}{a^2(\eta)}\Big(f''(b) + a(\eta)\,H\,f'(b)\Big). \tag{1.44}$$

where

$$f(b) \equiv \sqrt{\frac{2}{3}}\,\frac{1}{24}\,\frac{b(\eta)}{M_s^2\,M_{\rm Pl}} = \frac{3.4\times 10^{-2}}{M_s^2\,M_{\rm Pl}}\,b(\eta)\,, \tag{1.45}$$

with η_0 signalling the beginning of the inflationary phase in conformal-time-η coordinates. The $\pm$ in the exponent of the second factor on the right-hand side of (1.44) refer to left (L), right (R) movers respectively. In arriving at (1.44) we ignored the "matter" contributions to the Einstein equations, as these are associated with the slowly rolling stiff-matter KR axions in our context (see below), and thus yield subleading terms, *quadratic* in $\dot{b}$ [55, 57] (the dot denotes derivative with respect to the cosmic Robertson-Walker time t, which is related to the conformal time η via (1.36)). Thus, the quantity Θ is assumed weak in our approach, $|\Theta| \ll 1$, due to the slow-roll assumption for the KR axion field, which Θ depends upon (*cf.* (1.44),(1.45)). In our approximations below, therefore, we keep only terms linear in Θ.

On substituting (1.44) and (1.45) in (1.43), then, and performing the Fourier integrations, up to an UV cutoff μ [91, 92], such that $k \le a(\eta)\,\mu$, we obtain to first order in Θ: [92]:

$$\langle 0|\widehat{R_{\mu\nu\rho\sigma}\,{}^*R^{\mu\nu\rho\sigma}}|0\rangle = \frac{4\,H^2}{\pi^2\,M_{\rm Pl}^2}\,\Theta\,\mu^4\,. \tag{1.46}$$

Passing from a conformal to Robertson-Walker cosmic time, we write $\Theta = 4\ddot{f} + 8H\dot{f}$. In our case $\ddot{f} \ll H\dot{f}$ as a consequence of the slow roll nature of the KR axion, which arises dynamically in a self consistent way, upon formation of gravitational anomaly condensates [55–58], to be reviewed below. Thus, for our purposes, we may ignore the $\ddot{f}$ terms, and approximate:

$$\Theta \simeq 0.27 \frac{H}{M_s^2 M_{\rm Pl}} \dot{b}(t), \tag{1.47}$$

Hence, (1.46) yields

$$\langle 0| \widehat{R_{\mu\nu\rho\sigma}\, {}^\star R^{\mu\nu\rho\sigma}} |0\rangle \simeq \frac{1.1}{\pi^2} \left(\frac{H}{M_{\rm Pl}}\right)^3 \mu^4 \frac{\dot{b}(t)}{M_s^2}. \tag{1.48}$$

In the specific context of string theory, it is reasonable to define the effective theory below the string mass scale, M_s, which therefore should act as an UV cutoff, hence we can identify $\mu \sim M_s$. On the other hand, in generic Chern-Simons modified gravity theories, the string scale enters only in the coefficient of the Chern-Simons anomaly term $\mathcal{A}_{\rm CS}$ (1.25), which is viewed as a phenomenological parameter. In such a context, it is natural to assume that the UV cut-off scale of the graviton modes is the Planck scale $\mu \sim M_{\rm Pl}$, while the string scale (and thus the magnitude of the coefficient $\mathcal{A}_{\rm CS}$) can be determined phenomenologically by discussing conditions for the formation of the anomaly condensate [56], which we do in the next subsection. The reader should recall that our main aim in this work is to demonstrate that the formation of anomaly condensates will connect this Chern-Simons gravitational theory with the (Lorentz and CPT Violating) Standard Model Extension framework [59, 60], described in Sect. 1.2.

1.4.2 Gravitational-Anomaly Condensates and Spontaneous Violation of Lorentz Symmetry

In [55, 57] we have discussed the possibility of forming a condensate of the gravitational anomaly, in case there is a *macroscopic number* of sources of GW, with constructive interference. Below we shall clarify in some detail how we envisage the appearance of such a condensate, and what is its connection with the vacuum expectation value of the Chern Simons term operator (1.43) during inflation. It is understood that our approach will be phenomenological, providing only plausibility arguments for the condensate formation. The microscopic treatment requires a complete (non-perturbative) understanding of the underlying string theory or, more general, the UV complete theory of quantum gravity that characterises the effective theory, should one view the action (1.23) as a generic Chern-Simons modified gravity model, with a phenomenological Chern-Simons coefficient $\mathcal{A}_{\rm CS}$ (1.25).

To this end, we assume that the condensate is created by the collective effects of a (time-dependent) number $\mathcal{N}(t)$ of sources of GW per unit volume in the expanding Universe (in this notation, the total number of sources is given by $N = \int d^4x \mathcal{N}(t) = \int d^4x \sqrt{-g}\, \frac{\mathcal{N}(t)}{\sqrt{-g}}$, with $n_\star \equiv \frac{\mathcal{N}(t)}{\sqrt{-g}}$ the proper number density of sources). In this case, the induced tensor perturbations in (1.35) are replaced by the sum

$$h_{ij}(n, \mathbf{x}) \quad \Rightarrow \quad \sum_{I=1}^{\mathcal{N}(t)} h^I_{ij}(\eta, \mathbf{x}), \tag{1.49}$$

expressing the collective effect of sources, where the index I labels the source that produces a specific GW perturbation. In this effective, "phenomenological" approach, each individual $h^{(I)}_{ij}$ satisfies an Einstein-Chern-Simons field equation, but the collective metric induced under (1.49) does not, as a result of the dynamical (time dependent) nature of $\mathcal{N}(t)$, given that the dynamics of the formation of sources can only be dealt with in a full theory of UV quantum gravity, such as string theory *etc.*[6] Nonetheless, we may proceed in a rather agnostic, phenomenological approach, and quantize each individual h^I_{ij} by means of replacing it simply with operators $\widehat{h}^I_{ij}(\eta, \mathbf{x})$ (1.39), with the creation and annihilation operators, $\widehat{a}^{I\,\dagger}(\mathbf{k})$, $\widehat{a}^I(\mathbf{k})$, respectively, now carrying a "source" index I. For the vacuum we demand

$$a^I(\mathbf{k})|0\rangle = 0, \quad \text{and} \quad a^I(\mathbf{k})\, a^{J\,\dagger}(\mathbf{k}')\,|0\rangle = \delta_{IJ}\, \delta^{(3)}(\mathbf{k} - \mathbf{k}')|0\rangle,$$
$$I,\ J = 1, \ldots \mathcal{N}(t)\,, \tag{1.50}$$

where δ_{IJ} denotes a Kronecker delta. It is then immediately seen, that, upon assuming that the dominant GW perturbations $h^I_p(\mathbf{k}, \eta)$ coming from these set of sources have all the same magnitude, so that the index I can be omitted from the corresponding expression, we arrive at the analogue of (1.43) in this multi-source case:

$$\langle 0|\widehat{R_{\mu\nu\rho\sigma}\,{}^*R^{\mu\nu\rho\sigma}}|0\rangle_N = \mathcal{N}(t)\frac{16}{a^4\, M_{\rm Pl}^2} \int \frac{d^3k}{(2\pi)^3}\Big[k^2 h^\star_L(k,\eta)\, h'_L(k,\eta)$$
$$- k^2\, h^\star_R(k,\eta)\, h'_R(k,\eta) - h'^{\,\star}_L(k,\eta)\, h''_L(k,\eta) + h'^{\,\star}_R(k,\eta)\, h''_R(k,\eta)\Big], \tag{1.51}$$

that is, the collective effect is represented by a simple multiplication of the right-hand side of (1.43) by the number of sources. We stress again this is a plausible, but effective description, valid for weak GW perturbations from the

[6] Indeed, one may envisage that the primordial sources of GW appear dynamically as excitations of the ground state of the full quantum gravity system, and span the whole range, from non-spherically collapsing domain walls in a pre-inflationary epoch, to merging primordial black holes [57], which are themselves created from (quantum) gravitational vacuum perturbations.

various sources. This result, however, allows us now to represent the gravitational-anomaly condensate as

$$\langle R_{\mu\nu\rho\sigma}\, \widetilde{R}^{\mu\nu\rho\sigma} \rangle_{\text{condensate}\,\mathcal{N}} = \frac{1}{\sqrt{-g}}\, \langle 0 | \widehat{R_{\mu\nu\rho\sigma}\, {}^{*}R^{\mu\nu\rho\sigma}} | 0 \rangle_{\mathcal{N}}\,, \tag{1.52}$$

to quadratic order in the weak GW perturbations, where $\sqrt{-g}$ is the de Sitter *unperturbed* background metric, which was estimated above, *cf.* (1.48). In arriving at (1.52) we used (1.19) and (1.20). Using the estimate (1.48), we can thus estimate the magnitude of the gravitational anomaly condensate, induced by a macroscopic number of sources $\mathcal{N}(t)$, as:

$$\begin{aligned} \langle R_{\mu\nu\rho\sigma}\, \widetilde{R}^{\mu\nu\rho\sigma} \rangle_{\text{condensate}\,\mathcal{N}} &= \frac{\mathcal{N}(t)}{\sqrt{-g}}\, \frac{1.1}{\pi^2} \Big(\frac{H}{M_{\rm Pl}}\Big)^3 \mu^4\, \frac{\dot{b}(t)}{M_s^2} \\ &\equiv n_\star\, \frac{1.1}{\pi^2} \Big(\frac{H}{M_{\rm Pl}}\Big)^3 \mu^4\, \frac{\dot{b}(t)}{M_s^2}\,. \end{aligned} \tag{1.53}$$

The reader should recall that in the above expression $n_\star \equiv \frac{\mathcal{N}(t)}{\sqrt{-g}}$ denotes the number density (over the proper de Sitter volume) of the sources. Without loss of generality, we may take this density to be (approximately) time independent during inflation.

From the anomaly Eq. (1.22), which expresses the gravitational Chern-Simons term (1.53) as a divergence of an anomaly current, and assuming isotropy and homogeneity of the background space time (which we can justify microscopically in our framework through pre-inflationary epochs [57,58]) we may write, to leading order in GW pertubations during inflation [55–57]:

$$\langle \mathcal{K}^{\mu}{}_{;\,\mu} \rangle_{\text{condensate}\mathcal{N}} \simeq \frac{d}{dt} < \mathcal{K}^0 > +3\, H\, < \mathcal{K}^0 > \simeq n_\star\, \frac{1.1}{\pi^2} \Big(\frac{H}{M_{\rm Pl}}\Big)^3 \mu^4\, \frac{\dot{b}(t)}{M_s^2}\,, \tag{1.54}$$

where $< \mathcal{K}^0 >$ denotes the (dominant) average temporal component of the anomaly current in the de-Sitter background.

We next observe that from the Euler-Lagrange equations of the KR axion stemming from (1.23), one obtains:

$$\frac{1}{\sqrt{-g}}\, \partial_\mu \Big(\sqrt{-g} [\partial^\mu b - \mathcal{A}_{\rm CS}\, \mathcal{K}^\mu] \Big) = 0\,, \tag{1.55}$$

which for isotropic and homogeneous cosmological space times, leads to a solution (under the assumption of the formation of a condensate)

$$\dot{b} = \mathcal{A}_{\rm CS}\ < \mathcal{K}^0 >\,, \tag{1.56}$$

where in the case of strings, the Chern-Simons coefficient $\mathcal{A}_{\rm CS}$ is defined in (1.25) (see also (1.23)). A condensate should be a (approximately) time-independent solution during inflation, hence upon substituting (1.56) onto the evolution Eq. (1.54), we obtain that a (approximately) constant solution $\frac{d}{dt} < \mathcal{K}^0 > \simeq 0$ necessitates a constant H (inflation) and also the condition [55–57]:

$$0 \simeq 1 - n_\star \, \frac{1.1}{3\,\pi^2} \left(\frac{H}{M_{\rm Pl}}\right)^2 \frac{\mu^4\, \mathcal{A}_{\rm CS}}{M_s^2\, M_{\rm Pl}}\,, \tag{1.57}$$

during inflation. In the context of string theory (1.23), for which the coefficient $\mathcal{A}_{\rm CS}$ is given by $\frac{1}{96}\sqrt{\frac{2}{3}\frac{M_{\rm Pl}}{M_s^2}}$ (*cf.* (1.25), (1.23)), this condition translates to:

$$1 \simeq 3 \times 10^{-4}\, n_\star \left(\frac{H}{M_{\rm Pl}}\right)^2 \left(\frac{\mu}{M_s}\right)^4 \quad \Rightarrow \quad n_\star^{1/4}\, \frac{\mu}{M_s} \sim 7.6 \times \left(\frac{M_{\rm Pl}}{H}\right)^{1/2}. \tag{1.58}$$

? Exercise

1.11. Verify (1.57) and (1.58), starting from (1.54) and (1.56).

In [56], we have assumed $n_\star \simeq \mathcal{O}(1)$. In that case, upon taking the Planck data results for the upper bound of the inflationary Hubble scale H_I [21]

$$\frac{H_I}{M_{\rm Pl}} \lesssim 10^{-5}\,, \tag{1.59}$$

we obtain from Eq. (1.58), $\mu \gtrsim 2.4 \times 10^3\, M_s$. In [56–58] we assumed that the graviton modes are allowed to have momenta up to Planck scale, thus taking $\mu \sim M_{\rm Pl}$.[7] This determined the string scale at a high value and is consistent with the transplanckian conjecture, that no momenta of the effective field theory exceeds the Planck scale. However, from the point of view of a specific microscopic string theory, it may seem more appropriate to consider $\mu \sim M_s$, as mentioned above, and adopted in [92], but considering the proper number density of sources $n_\star$ as the adjustable parameter that will guarantee the formation of time-independent gravitational-anomaly condensates. In this scenario, one obtains from

[7] In [56,57] we followed the Green's function method of [91], instead of the Fourier method of [92] adopted here, in order to evaluate the condensate. The two methods cannot be directly compared, especially in view of the various approximations involved. Nonetheless, as we see by comparing (1.58) with the corresponding one in [56], in our model, the two methods yield qualitatively similar results, that agree in order of magnitude, as expected for consistency.

(1.58) and (1.59),

$$n_\star \gtrsim 3.3 \times 10^{13}\,, \tag{1.60}$$

which defines the macroscopic number of sources (per proper volume) needed to produce a gravitational anomaly condensate in the context of an effective Chern-Simons gravitational theory, inspired from strings, with the string scale M_s playing the rôle of the UV cutoff in the theory. In this second approach, the string scale is arbitrary and the conditions for the formation of the condensate translate into bounds on the number density of the sources of GW that lead to the condensate.

> Important

Thus, upon formation of the condensate during inflation ($H = H_I \simeq$ constant), we obtain a constant cosmic rate for the KR axion field, which we parametrise as [55]:

$$\dot{b} = \mathcal{A}_{\rm CS}\, \mathcal{K}^0 \simeq \text{constant} = \sqrt{2\epsilon}\; H\, M_{\rm Pl}. \tag{1.61}$$

The parameter $\epsilon \ll 1$ needs to be compatible with the slow-roll cosmological data [21] (see discussion below, Eq. (1.79)).

This background solution *violates spontaneously* Lorentz symmetry. This can be readily seen from the duality relation (1.27), which connects $\dot{b}$ with the dual of the Kalb-Ramond torsion field,

$$\text{constant} = \dot{b} \propto \epsilon^{ijk}\mathcal{H}_{ijk},\; i, j, k = 1, 2, 3. \tag{1.62}$$

This implies the dynamical selection of a preferred Lorentz frame by the ground state of the theory, in which the spatial components of the totally antisymmetric torsion of the system is constant. As we shall discuss later on, this property connects this gravitational theory with a SME effective field theory at the end of the inflationary period, when *chiral fermionic matter*, along with gauge fields, is assumed generated, according to the approach of [55–58].

We now remark that the creation of the anomaly condensate (1.53) produces in principle a linear potential for the KR axion (*cf.* (1.23))

$$V(b) = b(x)\, \mathcal{A}_{\rm CS}\, \langle\, R_{\mu\nu\rho\sigma}\, \widetilde{R}^{\mu\nu\rho\sigma} \rangle_{\text{condensate}\ \mathcal{N}}\,, \tag{1.63}$$

with

$$\begin{aligned}\mathcal{A}_{\rm CS}\, \langle\, R_{\mu\nu\rho\sigma}\, \widetilde{R}^{\mu\nu\rho\sigma} \rangle_{\text{condensate}\ \mathcal{N}} &\overset{\text{Eq. (1.61)}}{=} n_\star\, \frac{1.1}{\pi^2} \Big(\frac{H}{M_{\rm Pl}}\Big)^4 \mu^4\, \frac{\sqrt{2\epsilon}\, M_{\rm Pl}^2}{M_s^2} \sqrt{\frac{2}{3}}\, \frac{M_{\rm Pl}}{96\, M_s^2} \\ &\overset{\text{Eq. (1.59)}}{\lesssim} 4.3 \times 10^{-10}\, \sqrt{\epsilon}\, M_{\rm Pl}^3\,,\end{aligned} \tag{1.64}$$

where in the last inequality we saturated the bound (1.60), for concreteness.

The potential (1.63) is reminiscent of the linear axion-monodromy inflation potentials from appropriate brane compactifications (in, say, type IIB strings [94]). Such linear potentials have been argued to lead to slow-roll hill-top inflation. However, in our case the situation is very different. As we shall argue next, inflation in our scenario arises due to the non-linearities of the Running-Vacuum-Model (RVM)-type vacuum energy, in particular the condensate induced H^4 term, without the need for external fields. The linear axion potential then serves merely as a consistency check of the slow-roll KR axion (1.56) which characterises our case and leads to the parametrization (1.61). In [55–58] we have taken $\epsilon \sim \mathcal{O}(10^{-2})$, as a generic slow-roll parameter of the cosmological data [21], but this is not restrictive, given that b is not the inflaton, in the sense that it is not the linear potential of the KR axion that drives inflation in our case but the RVM non linearities. We shall discuss in the next Sect. 1.4.3 the RVM properties of our inflation, and come back to this issue of estimating theoretically the order of magnitude of the phenomenological parameter ϵ (see Eq. (1.79)).

For the moment, we proceed to estimate the condensate contribution to the vacuum energy density. Indeed, under the formation of a condensate (1.53), one may expand the effective action (1.23) about this condensate, by writing for the gravitational Chern Simons term:

$$b(x)\, R_{\mu\nu\rho\sigma}\, \widetilde{R}^{\mu\nu\rho\sigma} = \langle b(x)\, R_{\mu\nu\rho\sigma}\, \widetilde{R}^{\mu\nu\rho\sigma} \rangle_{\text{condensate } \mathcal{N}} + \, : b(x)\, R_{\mu\nu\rho\sigma}\, \widetilde{R}^{\mu\nu\rho\sigma} :, \tag{1.65}$$

where $: \cdots :$ denotes normal ordering (*i.e.* the creation operators appearing in the pertinent quantum-field correlation functions are placed on the of the left of the annihilation operators), which ensures that the vacuum expectation value of the second term vanishes, upon quantization. The condensate term behaves as a de-Sitter type cosmological constant in the following sense [55, 57]: the integrated solution of (1.61), implies $b(t) = \overline{b}(t_0) + \sqrt{2\epsilon}\, H\, (t - t_0)\, M_{\rm Pl}$, where t_0 denotes the beginning of inflation. The duration of inflation Δt is given by $H\Delta t = N_e$, where $N_e = \mathcal{O}(60 - 70)$ is the number of e-foldings [21], thus in order for $b(t)$ not to change order of magnitude during the entire inflationary period, we may require

$$|\overline{b}(t_0)| \gtrsim N_e\, \sqrt{2\epsilon}\, M_{\rm Pl} = \mathcal{O}(10^2)\, \sqrt{\epsilon}\, M_{\rm Pl}\,, \tag{1.66}$$

in which case the condensate term behaves approximately as a de Sitter (positive) cosmological constant term, provided (in our conventions) $\overline{b}(t_0) < 0$ [55].

The total vacuum energy, with contributions from the KR axion (b) terms, the gravitational Chern-Simons (non condensate) terms, proportional to the Cotton tensor (1.29), and the condensate itself (cond), can be obtained by the total stress energy tensor appearing in the appropriate Einstein-Chern-Simons equations (1.32), upon inclusion of a de Sitter term. It can be easily shown (using (1.61)) that the dominant term in the early Universe vacuum energy density is the one due to the condensate, which acquires the form

$$\rho_{\rm cond} = 1.3 \times 10^{-3}\,\sqrt{\epsilon}\,\frac{|\overline{b}(0)|}{M_{\rm Pl}}\left(\frac{\mu}{M_s}\right)^4 n_\star\, H^4\,, \tag{1.67}$$

where the various quantities appearing in (1.67) have been defined previously. The reader should recall that consistency of our approach requires the condition (1.58), which, on saturating the bounds (1.59), (1.60), for definiteness, yields (*cf.* (1.64)):

$$\rho_{\rm cond} \equiv \frac{\Lambda}{\kappa^2} \sim 4.3 \times 10^{10}\,\sqrt{\epsilon}\,\frac{|\overline{b}(0)|}{M_{\rm Pl}}\,H^4\,. \tag{1.68}$$

Given (1.66), this implies that the condensate term dominates over any other contributions to the vacuum energy coming from the KR axion field b or the Cotton tensor due to the gravitational Chern-Simons term [55–58]. We leave the verification of this as an exercise to the reader.

? Exercise

1.12. Start from the property (1.31) of the Cotton tensor, for the temporal component $\nu = 0$, and replace the Chern-Simons anomaly term $R_{\alpha\beta\gamma\delta}\,\widetilde{R}^{\alpha\beta\gamma\delta}$ on the right-hand-side by its condensate. Consider the left-hand side of (1.31) on a homogeneous and isotropic de Sitter background and thus argue, using also (1.30), that a constant C_{00} arises as a consistent solution of this equation. Show that the constant C_{00} is a negative quantity, but leads to subleading in magnitude contributions to the total vacuum energy compared to the condensate contributions. Also show that the energy density of the KR field, stemming from the stress tensor (1.33), is subleading to the energy-density contribution of the condensate term (you should make use of the parametrisation (1.61), assuming simply $\epsilon \ll 1$).

We also leave it as an exercise to the reader to prove that the equation of state of this cosmological fluid satisfies [58] is of de-Sitter type during the condensate phase, in the sense that the total pressure ($p^{\rm total}$) and energy ($\rho^{\rm total}$) density, including KR axion (b), gravitational Chern-Simons (Cotton tensor, $C_{\mu\nu}$, gravitational anomaly (gGS)) contributions, and condensate (1.68) contributions, satisfy:

$$p^b + p^{\rm gCS} = -(\rho^b + \rho^{\rm gCS}) > 0, \quad \text{and} \quad p^{\rm total} = -\rho^{\rm total} < 0\,, \tag{1.69}$$

where the superscript "total" denotes the (algebraic) sum of contributions from the b-axion, gCS and the (dominant) condensate Λ terms.

? Exercise

1.13. Consider the total (modified) stress-energy tensor in our string-inspired cosmology,

$$T^{\rm total}_{\mu\nu} = T^b_{\mu\nu} + \mathcal{A}_{\rm CS}\, C_{\mu\nu} + \Lambda g_{\mu\nu}. \tag{1.70}$$

Using the results of exercise 1.12, on the estimate of the approximately constant C_{00}, as well as Eq. (1.61), the trace property (1.30) and the conservation Eq. (1.34), for an inflationary background spacetime, prove the following:

$$\begin{aligned} p^b &= \rho^b \quad \text{(stiff massless axion matter)}, \quad p^\Lambda_{\rm cond} = -\rho^\Lambda_{\rm cond}\,, \\ p^{\rm gCS} &= \frac{1}{3}\rho^{\rm gCS}\,, \quad \rho^b = -\frac{2}{3}\rho^{\rm gCS}\,, \end{aligned} \tag{1.71}$$

where the pressure density terms $p^{\rm gCS}$ are associated with the spatial diagonal components of the Cotton tensor (1.29), C_{ii}, no sum over $i = 1, 2, 3$, whilst the energy-density $\rho^{\rm gCS}$ is linked to the temporal components C_{00} [55, 58]. From (1.71), then, prove

$$\begin{aligned} \rho^b + \rho^{\rm gCS} &= \frac{1}{3}\rho^{\rm gCS} = -\frac{1}{2}\rho^b \; < 0\,, \\ p^b + p^{\rm gCS} &= -(\rho^b + \rho^{\rm gCS}) > 0 \end{aligned} \tag{1.72}$$

thus proving the de-Sitter (RVM-type) equation of state (1.69).

A remark is in order regarding the contributions of the non-condensate anomaly terms in (1.69) (and (1.72)), which are negative and such that the total energy density of the KR axion plus the Cotton-tensor-dependent anomaly terms is negative, satisfying though a de-Sitter-like equation of state. Were it not for the condensate-Λ-(1.68) dominance, whose energy density is positive, the system would behave as an exotic one with "phantom matter" [95, 96]. The condensate dominance ensures that the vacuum of this string-inspired cosmology is characterised by a dominant *positive* vacuum energy of the form (1.68), with a de-Sitter-like equation of state (1.69). The reader should also observe from the result of the first line of (1.72), in combination with the estimate (1.61), that the total energy density of our cosmological fluid, including the condensate (1.68) reads [55–57]:

$$\rho^{\rm total} = \rho^b + \rho^{\rm gCS} + \rho^\Lambda_{\rm condensate} = -\frac{1}{2}\,\epsilon\, M_{\rm Pl}^2\, H^2 + 4.3 \times 10^{10}\,\sqrt{\epsilon}\,\frac{|\overline{b}(0)|}{M_{\rm Pl}}\, H^4\,. \tag{1.73}$$

This form of the energy density is that of the running vacuum model (RVM) of Cosmology [97–100], whose main features we review briefly in the next Sect. 1.4.3 for completeness. This will also clarify the type of inflation induced by the condensate, since so far we have simply assumed a constant Hubble parameter to estimate the anomaly condensate, without specifying the microscopic origin of inflation.

As we shall show below, the inflation in our case is due to the non-linearities of the condensate term (1.68), which depends on the fourth power of the Hubble parameter, and dominates in the early universe. No external inflaton fields are required. The KR axion field will provide though a slowly-moving pseudoscalar field during this RVM inflation [55–58], whose rate of change can be constrained by the cosmological data [21].

1.4.3 Condensates and Running-Vacuum-Model Inflation

The RVM cosmology [97–100] is an effective cosmological framework, with a cosmic-time-varying dark energy $\Lambda(t)$, which, nonetheless, is still characterised by an equation of state of de Sitter type:

$$p_{\rm RVM}(t) = -\rho_{\rm RVM}(t) \tag{1.74}$$

where p (ρ) denotes the vacuum pressure (energy) density. The energy density is a function of even powers of the Hubble parameter $H(t)$ as a result of general covariance [97–100]:[8]

$$\rho_{\rm RVM}(t) \equiv \frac{1}{\kappa^2}\Lambda(t) = \frac{3}{\kappa^2}\Big(c_0 + \nu\, H(t)^2 + \frac{\alpha}{H_I^2}\, H(t)^4 + \frac{\zeta}{H_I^4}\, H(t)^6 + \dots\Big), \tag{1.76}$$

in a standard parametrisation within the RVM framework, where H_I is a fixed inflationary scale (obtained from the data, (1.59)), and the ... denote higher powers of $H^2(t)$. The (dimensionless) coefficients $\nu, \alpha, \zeta, \dots$ can be determined either phenomenologically, by fitting the model with the data, especially at late eras, or can be computed within specific quantum field theory models [100–102]. The RVM framework provides a smooth cosmic evolution of the Universe [103, 104],

[8] The expression (1.76) is the integrated form of the initially proposed 'renormalization-group(RG)-like' evolution of the energy density, with $H(t)$ playing the rôle of the RG scale [97–99],

$$\frac{d}{d\ln H}\rho_{\rm RVM} = \sum_{i=1}^{\infty} c_i\, H^{2i}, \tag{1.75}$$

with c_i constant dimensionful in general coefficients (except the c_4 coefficient which is dimensionless in (3+1)-dimensions). In general, the expansion also includes terms $\dot{H}$, which however can be expressed in terms of H^2 and the deceleration parameter q. In most of the realistic applications, the various epochs of the Universe are characterised roughly by constant q's and as such the expansion in even powers of H^2 suffices.

explaining its thermodynamical and entropy production aspects, as a result of the decay of the running vacuum [105–107], and a viable alternative to the ΛCDM at late epoch, with in-principle observable deviations, compatible with the current phenomenology [108–110] (see also [111, 112] for fits of general Λ-varying cosmologies). The RVM framework also provides potential resolutions [113] to the recently observed, persisting tensions in the current-epoch cosmological data [114–116], provided the latter do not admit mundane astrophysical and/or statistical explanations [117].

Phenomenologically, truncation of the expansion of the right-hand side of (1.76) to terms of fourth power in $H(t)$ suffices to describe the entire Universe evolution from inflation at early epochs to the current era, where the rôle of the cosmological constant is played by the constant c_0, which appears as an integration constant when passing from the differential (1.75) to the integrated form (1.76) of the vacuum energy density, assuming, as is standard in RVM, that the entire evolution of the Universe is explained by (1.76), with constant coefficients c_0, ν, α. However, in the case of microscopic systems, such as the string-inspired one we discuss here, there may be phase transitions separating the various eras, and as a result the coefficients of the RVM evolution might change from era to era. Moreover, within local quantum field theory studies [100–102], at least in non-minimally coupled scalar fields to gravity examined in those works, there is no coefficient H^4 arising, but only H^2 and H^6 (and higher). As we discussed in [57] and review here, a term H^4 arises as a result exclusively of the condensation of gravitational anomalies in this string-inspired Chern-Simons modified theory. As we shall discuss below, the higher than H^2 non-linear terms provide inflation within the RVM framework, without the need for external inflaton fields.

Indeed, let us restrict our attention to the case (1.76) (relevant for our purposes here). Let us denote collectively quantities referring to matter and radiation with the suffix "m". The pertinent equation of state reads $p_m = \omega_m \rho_m$, which can be added to the RVM framework in such a way that the total energy and pressure densities, including the vacuum (RVM) contributions, are given by $p_{\rm total} = p_{\rm RVM} + p_m$, $\rho_{\rm total} = \rho_{\rm RVM} + \rho_m$. From the conservation of the total stress tensor of vacuum matter and radiation one obtains the following evolution equation for the Hubble parameter $H(t)$ [103, 104]:

$$\dot{H} + \frac{3}{2}\,(1+\omega_m)\,H^2\left(1 - \nu - \frac{c_0}{H^2} - \alpha\,\frac{H^2}{H_I^2}\right) = 0\,. \tag{1.77}$$

Ignoring c_0 (which, as we shall see, is a consistent assumption in our case), leads to a solution for $H(a)$ as a function of the scale factor a (in units of the present-era scale factor) and the equation of state ω_m of matter/radiation:

$$H(a) = \left(\frac{1-\nu}{\alpha}\right)^{1/2} \frac{H_I}{\sqrt{D\,a^{3(1-\nu)(1+\omega_m)} + 1}}\,, \tag{1.78}$$

where $D > 0$ is an integration constant. For the early Universe, $a \ll 1$, and thus one may assume without loss of generality that $D\, a^{3(1-\nu)(1+\omega_m)} \ll 1$. On account of (1.78), then, this leads to an (unstable) dynamical early de Sitter phase, characterised by an approximately constant Hubble parameter, $H_{\rm de\ Sitter} \simeq \left(\frac{1-\nu}{\alpha}\right)^{1/2} H_I$.

? Exercise

1.14. Starting from (1.77), and assuming $c_0 = 0$, prove that a solution for $H(a(t))$ is given by (1.78).

It can be seen that at the current epoch, where $a(t) \gg 1$, and one has matter dominance ($\omega_m \simeq 0$), there are in principle observable deviations from the ΛCDM model, still compatible though with the current phenomenology, due to the non-trivial νH^2 term in (1.76) which dominates today. Phenomenologically, by fitting the CMB, weak- and strong- lensing, and baryon-acoustic-oscillation data [108–112], one obtains $0 < \nu = \mathcal{O}(10^{-3})$ today, which incidentally is the order of magnitude of this parameter required by consistency of the RVM with BBN data [118].

In our string-inspired model, as discussed in the previous subsection, we observe that the dominant condensate term (1.68) is of the RVM form (1.76), with the constant $\alpha \sim \sqrt{\epsilon}\,\frac{|\overline{b}(0)|}{M_{\rm Pl}} \sim \mathcal{O}(10^2)\epsilon$, if one saturates the bound (1.66). In general, the total energy density of the vacuum (1.73) is of RVM form, with $c_0 = 0$. The coefficient of the H^2 term, though, is negative, in contrast to the conventional RVM. This is due to the effects of the Chern-Simons (quadratic in curvature) terms in the effective action (1.23). In the microscopic model of of [55], during the post inflationary period, cosmic electromagnetic background fields can switch the sign of this term to a positive one, thus recovering the conventional RVM form at late epochs. This RVM form during inflation is consistent with the condensate itself inducing inflation at early epochs of the Universe evolution, according to the arguments leading to (1.78). So the estimate of the condensate (1.68), in a constant inflationary background with H constant, is self consistent [55–58].

Let us now comment briefly on estimating the order of magnitude of ϵ [58]. To this end, we may assume that the presence of the condensate is compatible with the Freedman equation for this Universe, implying that, during inflation, one has:

$$
\begin{aligned}
\frac{3}{\kappa^2}\,H^2 &= \rho^{\rm total} \simeq \rho^{\Lambda}_{\rm condensate} = 4.3 \times 10^{10}\sqrt{\epsilon}\,\frac{|\overline{b}(0)|}{M_{\rm Pl}}\,H^4 \\
&\overset{\text{Eq. (1.68)}}{\Rightarrow} \quad \epsilon \sim 7 \times 10^{-3} = \mathcal{O}(10^{-2})\,,
\end{aligned}
\tag{1.79}
$$

where we saturated the bounds (1.66) and (1.59), for definiteness. The order of magnitude of ϵ is thus the same as the one assumed in [55, 58]. This should be considered as an allowed upper bound. On account of (1.66), for $N_e = O(60-70)$, this value imply transplanckian values for the magnitude of $\overline{b}(0)$, $|\overline{b}(0)| \gtrsim 8.4 M_{\rm Pl}$. This does not affect the transplanckian conjecture, since the effective action depends only on $\dot{b}$ which assumes sub-planckian values. It may be in conflict though with the so-called distance conjecture of swampland [119], which, however seems to affect also almost all single-field inflation models. This issue can only be resolved within the full UV complete string theory framework, and is beyond the effective field theory we are considering here, and beyond our purposes. Finally, we conclude this section by remarking that, with an $\epsilon = O(10^{-2})$, the RVM coefficient α (*cf.* (1.76)) in our model turns out to be of $O(1)$, and positive, while the ν coefficient assumes the value $\nu = -\frac{1}{6}\epsilon \sim -1.7 \times 10^{-3}$. Both coefficients are of the same order of magnitude as the corresponding ones in [58].[9]

We leave as a series of exercises for the reader to discuss the potentially drastic role of non-trivial, cosmic-time dependent, dilatons for the fate of the condensate, during RVM inflation, in a toy model.

? Exercise

1.15. Consider the string effective action in the presence of non-trivial dilatons Φ, under the constraint (1.18) in the absence of gauge fields, in its dual form, that is, the effective action written in terms of the (canonically normalised) Lagrange multiplier KR axion fields [120]:

$$S_B^{\rm eff} \simeq \int d^4x\,\sqrt{-g}\Big[\frac{1}{2\kappa^2}\,R - \frac{1}{2\,\kappa^2}\partial_\mu\Phi\,\partial^\mu\Phi - \frac{1}{2}\,e^{-2\Phi}\,\partial_\mu b\,\partial^\mu + \mathcal{A}_{\rm CS}\,\partial_\mu b(x)\,\mathcal{K}^\mu\; + \dots\Big], \tag{1.80}$$

where $\mathcal{K}^\mu$ is the gravitational anomaly current (1.22), and $\mathcal{A}_{\rm CS} = \frac{1}{96}\sqrt{\frac{2}{3}}\frac{M_{\rm Pl}}{M_s^2}$ (*cf.* (1.25)), with M_s the string scale.

(i) You should notice that there is no dilaton coupling in the Chern-Simons anomaly term. Explain briefly this feature.

(ii) Consider a homogeneous and isotropic cosmological model based on the action (1.80), and assume that a *slowly-varying with the cosmic time* condensate for the Chern-Simons gravitational anomaly has been formed, so that (1.53) is in operation, but in an RVM form, *i.e.* one should replace the constant H_I by a slowly

[9] The alert reader might have noticed different numerical factors in front of the H^4 terms in the vacuum energy density (1.73), as compared to those in [55–57]. This is due to the fact that in estimating the condensate (1.68) we followed here the method and normalizations of [92] instead of [91]. However, as we have just seen, and already remarked in Footnote 7, there are no qualitative or quantitative changes in the main phenomenological conclusions between these two frameworks.

varying $H(t)$:

$$\langle R_{\mu\nu\rho\sigma}\,\widetilde{R}^{\mu\nu\rho\sigma}\rangle_{\rm condensate} \simeq n_\star\,\frac{1.1}{\pi^2}\left(\frac{H(t)}{M_{\rm Pl}}\right)^3 \mu^4\,\frac{\dot{b}(t)}{M_s^2}$$

$$\simeq 3.5\times 10^{12}\left(\frac{H(t)}{M_{\rm Pl}}\right)^3 M_s^2\,\dot{b}(t)\;, \qquad (1.81)$$

where in the second (approximate) equality (which you should verify explicitly) we used, for definiteness, an $n_\star$ that satisfies (1.58) for a constant inflationary scale H_I saturating the upper bound of (1.59), as inferred from the data [21]. H_I here should not be identified with $H(t)$. This would ensure that in the absence of non trivial dilatons, one would recover the situation discussed previously, with $H \simeq H_I$.

(iii) Write down the dilaton and KR axion equations of motion, derived from the homogeneous and isotropic cosmological version of the action (1.80).

(iv) Show that

$$e^{-2\Phi(t)}\dot{b} = \mathcal{A}_{\rm CS}\,\mathcal{K}^0\;, \qquad (1.82)$$

is a solution of the KR axion equation of motion.

(v) Make the assumption that $\dot{b}$ in (1.81) can be replaced by the one in the solution (1.82) in terms of the dilaton $\Phi(t)$. Then, by approximating the anomaly-current equation in a Robertson-Walker background corresponding to $H(t)$ as

$$\langle\nabla_\mu\mathcal{K}^\mu\rangle \simeq \frac{d}{dt}\langle\mathcal{K}^0\rangle + 3\,H(t)\,\langle\mathcal{K}^0\rangle \simeq \langle R_{\mu\nu\rho\sigma}\,\widetilde{R}^{\mu\nu\rho\sigma}\rangle_{\rm condensate}, \qquad (1.83)$$

derive the condition for the validity of (1.83):

$$H(t)\,e^{\Phi(t)} \simeq 10^{-5}M_{\rm Pl}\,, \qquad (1.84)$$

for an approximately time-t-independent condensate $\langle\mathcal{K}^0\rangle$.

(vi) Using (1.84), show that the dilaton equation of motion stemming from the action (1.80) leads to:

$$3H^2\,\dot{H} - (\dot{H})^2 + H\,\ddot{H} \simeq 7.2\times 10^{-15}\,\frac{M_{\rm Pl}^2}{M_s^4}\,\langle\mathcal{K}^0\rangle^2\,. \qquad (1.85)$$

(vii) In this toy cosmological model, assume the validity of (1.85) from an initial cosmic time t_0 in which $H(t_0)\to H_I$, where H_I saturates the observational bound (1.59). Then, assuming slow roll for H, in which $H\,\ddot{H}$ and $(\dot{H})^2$ terms in (1.85) are subleading, show that one obtains an inflationary scenario with a higher-than-simple-exponential expansion for the scale factor $a(t)$ of this dilaton-dominated Universe, that is, show that

$$a(t)\sim \exp\Big(\frac{3H_I}{4c_1}\Big[\Big(1+c_1(t-t_0)\Big)^{4/3}-1\Big]\Big)\,, \qquad (1.86)$$

in units of $a(t_0)$, and determine the constant $c_1>0$. Interpret the boundary condition $H(t_0)=H_I$ in the context of the model of [57] discussed previously in this work.

(viii) Check and discuss the self consistency of the slow-roll assumption for $H(t)$ in part **(vii)**.

(ix) Finally, by assuming the validity of the Friedmann equation, provide an estimate of $\langle\mathcal{K}^0\rangle$ for this RVM universe, in which the condensate (1.81) dominates the total energy density. To answer this part of the question, first discuss the conditions under which the background axion field $b(t)$, satisfying (1.82), does not change order of magnitude under the entire duration of inflation.

The above exercise on a potential role of the dilaton on the induced RVM inflation does not constitute a complete treatment within string theory. In the case of bosonic (or Heterotic) strings, in addition to the anomaly four-derivative (order α') Chern-Simons term in the effective action, there are also four-derivative (quadratic in curvature) Gauss-Bonnet (GB) terms [76–79], which are non-trivial when the dilaton is non-trivial. The exception is the type IIB string, for which the GB terms are absent.

? Exercise

1.16. Consider the string-inspired effective action (1.80), but in the presence of a quintessence-type dilaton-Φ potential, arising, for instance, in non-critical string cosmologies [72, 73]: $V(\Phi) = C\exp(c_1\Phi)$, where $C, c_1 \in \mathbb{R}$ are appropriate real constants. Determine c_1 and C such that a dilaton $\Phi = 0$ is a consistent solution of the equations of motion, corresponding to the case studied in [55–58], in which an anomaly condensate is formed, with $\dot{b} = \mathcal{A}_{\rm CS}\langle\mathcal{K}^0\rangle = $ constant. Comparing your results with the studies in [72, 73], and using their definitions for super (sub) critical string, depending on the sign of C, determine which type of non-critical string this situation corresponds to.

1.5 Links with the Lorentz- and CPT- Violating Standard Model Extension, and Leptogenesis

We now come to the final, but also crucial, topic of our discussion, namely how the above results are linked to the Standard Model Extension [59, 60] with Lorentz and CPT Violation. This becomes possible if we consider the generation of fermionic (chiral) matter, which in our model occurs towards the end of the RVM inflationary period, as a consequence of the decay of the running vacuum.

As discussed in [55–58], in the context of the precursor string-theory model, (chiral) fermionic matter, represented by a generic fermion ψ for our purposes, for brevity, will couple to the (totally antisymmetric) torsion $\mathcal{H}_{\mu\nu\rho}$ via the gravitational covariant derivative. Adding the fermion action to the string action (1.13), with $\Phi = 0$, and performing the path-integration over the torsion field $\mathcal{H}_{\mu\nu\rho}$ in a curved background, with the δ-functional constraint (1.18), implemented as in Exercise 1.5, and being represented in terms of the pseudoscalar Lagrange multiplier field $b(x)$

(KR axion), one obtains the effective action:

$$
\begin{aligned}
S^{\rm eff} &= \int d^4x\sqrt{-g}\Big[-\frac{1}{2\kappa^2}\,R + \frac{1}{2}\,\partial_\mu b\,\partial^\mu b - \sqrt{\frac{2}{3}}\,\frac{\alpha'}{96\,\kappa}\,\partial_\mu b(x)\,\mathcal{K}^\mu\Big] \\
&+ S^{Free}_{\rm Dirac\ or\ Majorana} + \int d^4x\sqrt{-g} \\
&\times \Big[\Big(\mathcal{F}_\mu + \frac{\alpha'}{2\,\kappa}\sqrt{\frac{3}{2}}\,\partial_\mu b\Big)\,J^{5\mu} - \frac{3\alpha'^2}{16\,\kappa^2}\,J^5_\mu J^{5\mu}\Big] + \dots, \qquad (1.87)
\end{aligned}
$$

where $J^{5\mu} \equiv \sum \overline{\psi}\gamma^5\,\gamma^\mu\,\psi$ denotes the axial fermion current , $\mathcal{F}^d = \varepsilon^{abcd}\,e_{b\lambda}\,\partial_a\,e^\lambda_{\ c}$, with e^μ_c the vielbeins (with Latin indices pertaining to the tangent space of the space-time manifold at a given point, in a standard notation), $S^{Free}_{\rm Dirac\ or\ Majorana}$ denotes the free-fermion kinetic terms, and the ... in (1.87) indicate gauge field kinetic terms, as well as terms of higher order in derivatives. The action (1.87) is valid for both Dirac or Majorana fermions.[10] The reader is invited to take note of the presence in (1.87) of the CP-violating interactions of the derivative of the field b with the axial fermion current $J^{5\mu}$, as well as of the repulsive axial-fermion-current-current term, $-\dfrac{3\alpha'^2}{16\,\kappa^2}\,J^5_\mu J^{5\mu}$, which is characteristic of theories with Einstein-Cartan torsion [82, 121], as is our string-inspired model [83]. The proof of (1.87) is left as a set of exercises for the reader.

? Exercise

1.17. Consider for definiteness a Dirac fermion ψ in a curved space-time with a string-inspired totally antisymmetric torsion $\mathcal{H}_{\mu\nu\rho}$, as in (1.15). First, on using the definition of the gravitational covariant derivative acting on the fermions in terms of vielbeins $e^a_{\ \mu}$ and the torsionful spin connection $\overline{\omega}^a_{\ \mu\ b}$ corresponding to (1.15) (where Latin indices are tangent-space indices), write down the kinetic term of the corresponding Dirac Lagrangian density. Then, by means of properties of the product of three Dirac γ^μ matrices, prove the existence in the Lagrangian density of a linear coupling of the fermion axial current $\overline{\psi}\,\gamma^5\,\gamma^\mu\,\psi$ to $\varepsilon_{\mu\nu\rho\sigma}\,\mathcal{H}^{\nu\rho\sigma}$, where the covariant Levi-Civita tensor density $\varepsilon_{\mu\nu\rho\sigma}$ has been defined in (1.19), and determine its coefficient. Then, by adding this fermion action to (1.13), consider the constrained path-integration over $\mathcal{H}_{\mu\nu\rho}$, using a δ-functional constraint for (1.18), as in Exercise 1.5. On representing the δ-functional by means of the canonically normalised Lagrange multiplier field $b(x)$, then, prove (1.87). Show also that, for Robertson-Walker backgrounds, in the absence of perturbations, the quantity $\mathcal{F}^d = \varepsilon^{abcd}\,e_{b\lambda}\,\partial_a\,e^\lambda_{\ c}$ vanishes.

[10] In case of multifermion theories, as required in phenomenologically realistic models, one simply has to sum the appropriate effective action terms over all the fermion species, with the axial current reading as $J^{5\mu} = \sum_{i={\rm fermion\ species}} \overline{\psi}_i\gamma^5\gamma^\mu\psi_i$.

We next observe, that, in case of the spontaneous LV background (1.61), due to the anomaly condensate in our cosmology, the fermion-axial-current-KR-axion interaction in (1.13) leads to a LV and CPTV interaction with the background, which is of a SME type (1.6), with $a_\mu = 0$ and

$$b_\mu = M_{\rm Pl}^{-1}\,\dot{\overline{b}}\,\delta_{\mu\,0}\,, \quad \mu = 0, \ldots 3\,, \quad \dot{\overline{b}} = \text{constant}\,, \tag{1.88}$$

having only a temporal component, with $\overline{b}$ the solution to the KR equation of motion stemming from (1.23).

Important

In the presence of massive right-handed neutrinos, with standard portals, coupling the RHN sector to SM lepton and Higgs sectors, then, one can consider the following fermion action in the background (1.88):

$$\begin{aligned}\mathcal{L} = \mathcal{L}_{\rm SM} + i\overline{N}\,\gamma^\mu\,\partial_\mu\,N - \frac{m_N}{2}(\overline{N^c}N + \overline{N}N^c)\\ - \overline{N}\gamma^\mu\,b_\mu\,\gamma^5 N - \sum_f\, y_f \overline{L}_f \tilde{\phi}^d N + \text{h.c.}\end{aligned} \tag{1.89}$$

where h.c. denotes hermitian conjugate, $\mathcal{L}_{\rm SM}$ denotes the SM Lagrangian, N is the RHN field (with N^c its charge conjugate field), of (Majorana) mass m_N, $\tilde{\phi}$ is the SU(2) adjoint of the Higgs field ϕ ($\tilde{\phi}^d_i \equiv \varepsilon_{ij}\phi_j$, $i, j = 1, 2$, SU(2) indices), and L_f is a lepton (doublet) field of the SM sector, with f a generation index, $f = e, \mu, \tau$, in a standard notation for the three SM generations; y_f is a Yukawa coupling, which is non-zero and provides a non-trivial ("Higgs portal") interaction between the RHN and the SM sector, used in the seesaw mechanism for generation of SM neutrino masses. As discussed in [65–68], and we shall describe briefly below, such backgrounds can produce phenomenologically correct leptogenesis. In particular, we consider lepton-number asymmetry originating from *tree-level* decays of heavy sterile RHN into SM leptons.

Indeed, in the context of the model (1.89), a lepton asymmetry is generated due to the CPV and CPTV tree-level decays of the RHN N into SM leptons, in the presence of the background (1.88), through Channel I : $N \to l^- h^+$, $\nu\, h^0$, and Channel II : $N \to l^+ h^-$, $\overline{\nu}\, h^0$, where $\ell^\pm$ are charged leptons, ν ($\overline{\nu}$) are light, "active", neutrinos (antineutrinos) in the SM sector, h^0 is the neutral Higgs field, and $h^\pm$ are the charged Higgs fields, which, at high temperatures, above the spontaneous electroweak symmetry breaking, of interest in this scenario [57,65–68], do not decouple from the physical spectrum. As a result of the non-trivial $b_0 \neq 0$ background (1.88), the decay rates of the Majorana RHN between the channels I and II are different, resulting in a Lepton-number asymmetry.

? Exercise

1.18. Consider the tree-level decays of a massive Majorana neutrino N, of mass m_N, in the theory (1.89) into charged lepton and Higgs particles and antiparticles only, in the background (1.88). By following standard particle physics methods, prove that the tree-level decay rates Γ, $N \to \ell^- h^+$ and $N \to \ell^+ h^-$, are given, respectively, by:

$$\Gamma_{N\to\ell^-h^+} = \sum_{f=e,\mu,\tau} \frac{|y_f|^2}{32\pi^2} \frac{m_N^2}{\Omega} \frac{\Omega + b_0}{\Omega - b_0}, \qquad \Gamma_{N\to\ell^+h^-} = \sum_{f=e,\mu,\tau} \frac{|y_f|^2}{32\pi^2} \frac{m_N^2}{\Omega} \frac{\Omega - b_0}{\Omega + b_0},$$

$$\text{with} \quad \Omega = \sqrt{m_N^2 + b_0^2}\,. \tag{1.90}$$

The reader should observe that the difference between these two rates vanishes for vanishing background $b_0 \to 0$. To linear order in b_0, with $|b_0| \ll m_N$, argue that these decay rates may be interpreted as implying the presence of "effective" RHN masses $m_{N\,\text{eff}}^{\pm} = m_N \pm 2b_0$ in the $\ell^{\mp} h^{\pm}$ decay channels, respectively.

Such asymmetries in the decay rates produce lepton asymmetry, which can then be communicated [68] to the baryon sector by means of appropriate baryon(B) and lepton(L)-number violating but B-L conserving processes, e.g. sphalerons in the SM sector of the model [24–28], according to standard leptogenesis scenarios [37, 38].

Before closing, we remark that in the actual model of [55–58] the situation is a bit more complicated than the simplified scenario with a constant background (1.88) surviving in the radiation phase. The generation of chiral fermions at the end phase of the RVM inflation in the model leads to a cancellation of the primordial gravitational anomalies by the ones generated by the chiral fermions themselves, leaving only possible chiral anomalies (in the gauge sector) surviving in the post-inflationary period. This leads, as a consequence, to a temperature dependence for the background $b_0 \propto T^3$, during the post inflationary period, as explained in detail in [55]. For the short period of leptogenesis, such temperature dependent backgrounds are almost constant, and the resulting lepton asymmetry can be calculated analytically [65, 66, 68], leading to similar, qualitatively and quantitatively, conclusions as the simple constant-background (1.88) case, reviewed above.

The advantage of the T^3 temperature dependence of the axion background B_0, which survives the inflationary period, is that one can trace it to the current era (up to complications including chiral anomalies, which can change the T^3 behaviour, *e.g.* to T^2, as discussed in some detail in [55]). The current KR axion background is well below the current bounds (1.8) of this background [62]. Specifically one finds [55] $b_0|_{\text{today}} \sim 10^{-44}$ eV, if chiral anomalies are ignored (i.e., T^3 scaling), and $b_0|_{\text{today}} \sim 10^{-34}$ eV, if chiral anomalies take over at late epochs. Even if one takes into account the relative motion of our Earthly laboratory frames with respect to the CMB frame (with velocity v_i, $|\mathbf{v}| = O(390 \pm 60)$ km/sec [21, 122]), which leads to spatial components of the background $b_i = \gamma \frac{v_i}{c} b_0$, with $\gamma \sim 1$ the Lorentz factor, the resulting spatial components b_i of the LV and CPTV background lie comfortably within the existing bounds (1.8) [62].

The following exercise provides the reader with a simple way to understand the T^3 scaling of $\dot{b}$ in the absence of chiral anomalies, during the post-inflationary era of the cosmological model of [55–58].

? Exercise

1.19. Consider the KR axion $b(x)$ equation of motion stemming from the effective action (1.87) for the case of a conserved axial-fermion current $J^{5\mu}(x)$ in the absence of gravitational anomalies (*i.e.* set $\mathcal{K}^\mu = 0$).

(i) Show, that in a homogeneous and isotropic Robertson-Walker background, in the radiation-dominated era, there is a T^3 scaling of the cosmic rate of the $b(x)$ field, $\dot{b}$, where T is the cosmic temperature (use standard cosmology arguments [93] to relate the scale factor of the Universe to the cosmic temperature T).

(ii) By assuming [55] that the radiation era succeeds the inflationary one, during which the inflationary scale H_I is related to the (de-Sitter observer dependent) Gibbons-Hawking temperature [123], $T = H_I/2\pi$, determine the T-scaling proportionality constant in the expression for $\dot{b}$ of part **(i)**, using the parametrization (1.61) (with an $\epsilon = \mathcal{O}(10^{-2})$, and a H_I saturating the bound (1.59)) at the exit phase of inflation/beginning of the radiation era in the framework of the model of [55–58]. Thus show that this $\dot{b}$ satisfies the current LV and CPTV bounds (1.8).

1.6 Summary and Outlook

With the above remarks we conclude our discussion on how one can obtain Lorentz and/or CPT Violating terms, that appear phenomenologically in the SME effective Lagrangian, starting from a microscopic quantum gravity theory. Within our specific string-inspired cosmological field theory example, we have seen how condensates of primordial gravitational waves, of quantum-gravitational origin, can lead to spontaneous violation of Lorentz and CPT symmetries in low-energy effective theories, which contain terms of a form appearing in SME Lagrangians.

We have pointed out the crucial role of the UV complete theory of quantum gravity in leading to these condensates. The lack, however, of a complete understanding of energy regimes above the Planck scale, even in the context of UV complete theories, such as strings, has some consequences for the accuracy of the relative estimates. Nonetheless, we hope we made it clear to the reader that LV and CPTV processes might play a crucial role on the existence of our Universe, and thus ourselves, given, the potential link of the spontaneous violation of Lorentz and CPT symmetries by the condensates to the matter-antimatter asymmetry in our Universe, as we discussed above. An important aspect of our considerations is that the matter-antimatter asymmetry in the Cosmos might have a geometric origin, as a consequence of the close connection of LV to condensates of torsion (axion-like) fields, which characterise the massless gravitational multiplet of string theory (which is also the ground state of the phenomenologically relevant superstrings).

From a phenomenological point of view it would be interesting to explore further the profile of the primordial GW generated during our RVM inflation, as well as the densities of primordial black holes during that era, especially in models with non-trivial dilatons, such as (1.80) (in case of type-IIB-string inspired models), or extensions thereof, including Gauss-Bonnet-dilaton coupled combinations (in case of heterotic and bosonic strings). There is the possibility of enhanced gravitational perturbations and densities of primordial black holes in such models, which could affect the aforementioned GW profiles at post inflationary (radiation) eras, thus leading to observable in principle effects in interferometers. In addition, effects of the LV and CPTV SME-type background coefficients b_μ (1.88), which are linked to forbidden atomic transitions and other modifications of atomic spectra [20, 54, 59, 61], might affect BBN physics, given the increasing nature of this coefficient with the cosmic temperature, which might lead to further constraints. These are issues to be examined in the future, by extending, for instance, the LV analysis of [124] to incorporate appropriately the CPTV effects arising in our framework.

Acknowledgments The author is grateful to Prof. C. Lämmerzahl and Dr. C. Pfeifer for the invitation to contribute a chapter to this book on *Modified and Quantum Gravity - From theory to experimental searches on all scales—WEH 740*, based on an invited talk in the 740 Wilhelm-und-Else-Heraeus-Seminar, 1-5 February 2021. This work was supported partly by STFC (UK) Grant ST/T000759/1. NEM also acknowledges participation in the COST Association Action CA18108 "*Quantum Gravity Phenomenology in the Multimessenger Approach (QG-MM)*".

References

1. R. Streater, A.S. Wightman, *PCT, Spin & Statistics, and All That* (Benjamin, New York, 1964)
2. B.J. Bjorken, S.D. Drell, *Relativistic Quantum Fields* (McGraw-Hill, New York, 1965) ISBN-13: 978-0070054943
3. J. Schwinger, The theory of quantized fields. I. Phys. Rev. **82**, 914 (1951)
4. G. Lüders, On the equivalence of invariance under time reversal and under particle-antiparticle conjugation for relativistic field theories. Kongelige Danske Videnskabernes Selskab, Matematisk-Fysiske Meddelelser. **28**(5), 1–17 (1954)
5. W. Pauli, in *Niels Bohr and the Development of Physics*, ed. by W. Pauli, L. Rosenfelf, V. Weisskopf (London Pergamon Press/McGraw-Hill, London/New York, 1955). LCCN: 56040984, https://lccn.loc.gov/56040984
6. A. Whitaker, *John Stuart Bell and Twentieth-Century Physics* (Oxford University Press, Oxford, 2016). ISBN 978-0198742999
7. R. Jost, Helv. Phys. Acta **30**, 409 (1957)
8. O.W. Greenberg, CPT violation implies violation of Lorentz invariance. Phys. Rev. Lett. **89**, 231602 (2002). arXiv:hep-ph/0201258 [hep-ph]
9. O.W. Greenberg, Why is CPT fundamental? Found. Phys. **36**, 1535–1553 (2006). arXiv:hep-ph/0309309 [hep-ph]

10. M. Chaichian, A.D. Dolgov, V.A. Novikov, A. Tureanu, CPT violation does not lead to violation of lorentz invariance and vice versa. Phys. Lett. B **699**, 177–180 (2011). arXiv:1103.0168 [hep-th]
11. M. Chaichian, K. Fujikawa, A. Tureanu, Lorentz invariant CPT violation. Eur. Phys. J. C **73**(3), 2349 (2013). arXiv:1205.0152 [hep-th]
12. P.A. Zyla, et al., Particle Data Group, Review of particle physics. Progress Theor. Exp. Phys. **2020**(8), 083C01 (2020)
13. M. Amoretti, et al., ATHENA, The ATHENA antihydrogen apparatus. Nucl. Instrum. Meth. A **518**, 679–711 (2004)
14. G.B. Andresen, et al., ALPHA, Trapped antihydrogen. Nature **468**, 673–676 (2010)
15. G. Gabrielse,et al., ATRAP, Trapped antihydrogen in its ground state. Phys. Rev. Lett. **108**(11), 113002 (2012). arXiv:1201.2717 [physics.atom-ph]
16. E. Widmann, ASACUSA, Testing CPT with antiprotonic helium and antihydrogen: The ASACUSA experiment at CERN-AD. Nucl. Phys. A **752**, 87–96 (2005)
17. R.S. Hayano, M. Hori, D. Horvath, E. Widmann, Antiprotonic helium and CPT invariance. Rep. Prog. Phys. **70**, 1995–2065 (2007)
18. M. Hori, et al., Direct measurement of transition frequencies in isolated \$\bar{p}$-He+ atoms, and new CPT-violation limits on the antiproton charge and mass. Phys. Rev. Lett. **91**, 123401 (2003)
19. R. Bluhm, V.A. Kostelecky, N. Russell, CPT and Lorentz tests in hydrogen and anti-hydrogen. Phys. Rev. Lett. **82**, 2254–2257 (1999). arXiv:hep-ph/9810269 [hep-ph]
20. V.A. Kostelecký, A.J. Vargas, Lorentz and CPT tests with hydrogen, antihydrogen, and related systems. Phys. Rev. D **92**(5), 056002 (2015). arXiv:1506.01706 [hep-ph]
21. N. Aghanim, et al., Planck, Planck 2018 results. VI. Cosmological parameters. Astron. Astrophys. **641**, A6 (2020). erratum: Astron. Astrophys. **652**, C4 (2021). arXiv:1807.06209 [astro-ph.CO]
22. R.H. Cyburt, B.D. Fields, K.A. Olive, An update on the big bang nucleosynthesis prediction for Li-7: The problem worsens. J. Cosmol. Astro. Phys. **11**, 012 (2008). arXiv:0808.2818 [astro-ph]
23. A.D. Sakharov, Violation of CP invariance, C asymmetry, and baryon asymmetry of the universe. Pisma Zh. Eksp. Teor. Fiz. **5**, 32–35 (1967)
24. V.A. Kuzmin, V.A. Rubakov, M.E. Shaposhnikov, On the anomalous Electroweak baryon number nonconservation in the early universe. Phys. Lett. B **155**, 36 (1985)
25. M.E. Shaposhnikov, Baryon asymmetry of the universe in standard electroweak theory. Nucl. Phys. B **287**, 757–775 (1987)
26. V.A. Rubakov, M.E. Shaposhnikov, Electroweak baryon number nonconservation in the early universe and in high-energy collisions. Usp. Fiz. Nauk **166**, 493–537 (1996). arXiv:hep-ph/9603208 [hep-ph]
27. M.B. Gavela, P. Hernandez, J. Orloff, O. Pene, Mod. Phys. Lett. A **9**, 795–810 (1994). arXiv:hep-ph/9312215 [hep-ph]
28. M.B. Gavela, P. Hernandez, J. Orloff, O. Pene, C. Quimbay, Nucl. Phys. B **430**, 382–426 (1994). arXiv:hep-ph/9406289 [hep-ph]
29. J.H. Christenson, J.W. Cronin, V.L. Fitch, R. Turlay, Evidence for the 2π Decay of the K_2^0 Meson. Phys. Rev. Lett. **13**, 138–140 (1964)
30. P. Minkowski, Phys. Lett. B **67**, 421 (1977)
31. M. Gell-Mann, P. Ramond, R. Slansky, in *Supergravity*, eds. D.Z. Freedman, P. van Nieuwenhuizen (North-Holland, Amsterdam, 1979)
32. T. Yanagida, in *Proceedings of the Workshop on the Unified Theory and the Baryon Number in the Universe*, ed. by O. Sawada, A. Sugamoto (Tsukuba, 1979)
33. R.N. Mohapatra, G. Senjanovic, Neutrino mass and spontaneous parity violation. Phys. Rev. Lett. **44**, 912 (1980).
34. J. Schechter, J.W.F. Valle, Neutrino masses in SU(2) x U(1) theories. Phys. Rev. D **22**, 2227 (1980)
35. G. Lazarides, Q. Shafi, C. Wetterich, Proton lifetime and fermion masses in an SO(10) model. Nucl. Phys. B **181**, 287–300 (1981)

36. for a review see: T. Schwetz, M.A. Tortola, J.W.F. Valle, Three-flavour neutrino oscillation update. New J. Phys. **10**, 113011 (2008). arXiv:0808.2016 [hep-ph]
37. M. Fukugita, T. Yanagida, Baryogenesis without grand unification. Phys. Lett. B **174**, 45–47 (1986)
38. S. Davidson, E. Nardi, Y. Nir, Leptogenesis. Phys. Rept. **466**, 105–177 (2008). arXiv:0802.2962 [hep-ph] and references therein
39. T. Asaka, M. Shaposhnikov, The νMSM, dark matter and baryon asymmetry of the universe. Phys. Lett. B **620**, 17–26 (2005). arXiv:hep-ph/0505013 [hep-ph]
40. T. Asaka, S. Blanchet, M. Shaposhnikov, The nuMSM, dark matter and neutrino masses. Phys. Lett. B **631**, 151–156 (2005). arXiv:hep-ph/0503065 [hep-ph]
41. L. Canetti, M. Shaposhnikov, Baryon asymmetry of the universe in the NuMSM. J. Cosmol. Astro. Phys. **09**, 001 (2010). arXiv:1006.0133 [hep-ph]
42. A. Perez, Spin foam models for quantum gravity. Class. Quant. Grav. **20**, R43 (2003). arXiv:gr-qc/0301113 [gr-qc]
43. J.C. Baez, An introduction to spin foam models of BF theory and quantum gravity. Lect. Notes Phys. **543**, 25–93 (2000). arXiv:gr-qc/9905087 [gr-qc]
44. J.A. Wheeler, K. Ford, Geons, black holes, and quantum foam: a life in physics (W.W. Norton & Company, New York, 1998)
45. R.M. Wald, Quantum gravity and time reversibility. Phys. Rev. D **21**, 2742–2755 (1980)
46. J. Bernabeu, N.E. Mavromatos, J. Papavassiliou, Novel type of CPT violation for correlated EPR states. Phys. Rev. Lett. **92**, 131601 (2004). arXiv:hep-ph/0310180 [hep-ph]
47. N.E. Mavromatos, CPT violation and decoherence in quantum gravity. Lect. Notes Phys. **669**, 245–320 (2005). arXiv:gr-qc/0407005 [gr-qc] and references therein
48. A. Addazi, J. Alvarez-Muniz, R.A. Batista, G. Amelino-Camelia, V. Antonelli, M. Arzano, M. Asorey, J.L. Atteia, S. Bahamonde, F. Bajardi, et al., Quantum gravity phenomenology at the dawn of the multi-messenger era – A review. Prog. Part. Nucl. Phys. 103948 (2022). arXiv:2111.05659 [hep-ph] and referenfes therein
49. M.B. Green, J.H. Schwarz, E. Witten, *Superstring Theory*. Cambridge Monographs in Mathematical Physics, vols. 1 and 2, 25th anniversary edn. (Cambridge University Press, Cambridge, 2012)
50. M. Dine, M. Graesser, CPT and other symmetries in string/M theory. J. High Energy Phys. **01**, 038 (2005). arXiv:hep-th/0409209 [hep-th]
51. V.A. Kostelecky, S. Samuel, Spontaneous breaking of lorentz symmetry in string theory. Phys. Rev. D **39**, 683 (1989)
52. V.A. Kostelecky, R. Potting, CPT and strings. Nucl. Phys. B **359**, 545–570 (1991)
53. V.A. Kostelecky, R. Potting, Expectation values, Lorentz invariance, and CPT in the open bosonic string. Phys. Lett. B **381**, 89–96 (1996). arXiv:hep-th/9605088 [hep-th]
54. V.A. Kostelecky, R. Lehnert, Stability, causality, and Lorentz and CPT violation. Phys. Rev. D **63**, 065008 (2001). https://doi.org/10.1103/PhysRevD.63.065008. arXiv:hep-th/0012060 [hep-th]
55. S. Basilakos, N.E. Mavromatos, J. Solà Peracaula, Gravitational and chiral anomalies in the running vacuum universe and matter-antimatter asymmetry. Phys. Rev. D **101**(4), 045001 (2020). arXiv:1907.04890 [hep-ph]
56. S. Basilakos, N.E. Mavromatos, J. Solà Peracaula, Quantum anomalies in string-inspired running vacuum universe: inflation and axion dark matter. Phys. Lett. B **803**, 135342 (2020). arXiv:2001.03465 [gr-qc]
57. N.E. Mavromatos, J. Solà Peracaula, Stringy-running-vacuum-model inflation: from primordial gravitational waves and stiff axion matter to dynamical dark energy. Eur. Phys. J. ST **230**(9), 2077–2110 (2021). arXiv:2012.07971 [hep-ph]
58. N.E. Mavromatos, J. Solà Peracaula, Inflationary physics and trans-Planckian conjecture in the stringy running vacuum model: from the phantom vacuum to the true vacuum. Eur. Phys. J. Plus **136**(11), 1152 (2021). arXiv:2105.02659 [hep-th]
59. D. Colladay, V.A. Kostelecky, Lorentz violating extension of the standard model. Phys. Rev. D **58**, 116002 (1998). arXiv:hep-ph/9809521 [hep-ph]

60. D. Colladay, V.A. Kostelecky, CPT violation and the standard model. Phys. Rev. D **55**, 6760–6774 (1997). arXiv:hep-ph/9703464 [hep-ph]
61. D. Colladay, V.A. Kostelecky, Cross-sections and Lorentz violation. Phys. Lett. B **511**, 209–217 (2001). https://doi.org/10.1016/S0370-2693(01)00649-9. arXiv:hep-ph/0104300 [hep-ph]
62. V.A. Kostelecky, N. Russell, Data tables for Lorentz and CPT violation. Rev. Mod. Phys. **83**, 11–31 (2011). arXiv:0801.0287 [hep-ph]
63. V.A. Kostelecky, Gravity, lorentz violation, and the standard model. Phys. Rev. D **69**, 105009 (2004). arXiv:hep-th/0312310 [hep-th]
64. P.A. Bolokhov, M. Pospelov, Classification of dimension 5 Lorentz violating interactions in the standard model. Phys. Rev. D **77**, 025022 (2008). arXiv:hep-ph/0703291 [hep-ph]
65. T. Bossingham, N.E. Mavromatos, S. Sarkar, The role of temperature dependent string-inspired CPT violating backgrounds in leptogenesis and the chiral magnetic effect. Eur. Phys. J. C **79**(1), 50 (2019). arXiv:1810.13384 [hep-ph]
66. T. Bossingham, N.E. Mavromatos, S. Sarkar, Leptogenesis from heavy right-handed neutrinos in CPT violating backgrounds. Eur. Phys. J. C **78**(2), 113 (2018). arXiv:1712.03312 [hep-ph]
67. N.E. Mavromatos, S. Sarkar, Spontaneous CPT violation and quantum anomalies in a model for matter–antimatter asymmetry in the cosmos. Universe **5**(1), 5 (2018). arXiv:1812.00504 [hep-ph]
68. N.E. Mavromatos, S. Sarkar, Curvature and thermal corrections in tree-level CPT-Violating Leptogenesis. Eur. Phys. J. C **80**(6), 558 (2020). arXiv:2004.10628 [hep-ph]
69. O. Bertolami, D. Colladay, V.A. Kostelecky, R. Potting, CPT violation and baryogenesis. Phys. Lett. B **395**, 178–183 (1997). https://doi.org/10.1016/S0370-2693(97)00062-2. arXiv:hep-ph/9612437 [hep-ph]
70. N.J. Poplawski, Matter-antimatter asymmetry and dark matter from torsion. Phys. Rev. D **83**, 084033 (2011). https://doi.org/10.1103/PhysRevD.83.084033. arXiv:1101.4012 [gr-qc]
71. M.B. Green, J.H. Schwarz, Anomaly cancellation in supersymmetric D = 10 Gauge theory and superstring theory. Phys. Lett. B **149**, 117–122 (1984)
72. I. Antoniadis, C. Bachas, J.R. Ellis, D.V. Nanopoulos, An expanding universe in string theory. Nucl. Phys. B **328**, 117–139 (1989)
73. J.R. Ellis, N.E. Mavromatos, D.V. Nanopoulos, A microscopic Liouville arrow of time. Chaos Solitons Fractals **10**, 345–363 (1999). arXiv:hep-th/9805120 [hep-th]
74. A.B. Lahanas, N.E. Mavromatos, D.V. Nanopoulos, Smoothly evolving supercritical-string dark energy relaxes supersymmetric-dark-matter constraints. Phys. Lett. B **649**, 83–90 (2007). arXiv:hep-ph/0612152 [hep-ph]
75. M. Gasperini, G. Veneziano, The pre-big bang scenario in string cosmology. Phys. Rept. **373**, 1–212 (2003). arXiv:hep-th/0207130 [hep-th]
76. B. Zwiebach, Curvature squared terms and string theories. Phys. Lett. B **156**, 315–317 (1985)
77. D.J. Gross, J.H. Sloan, The quartic effective action for the heterotic string. Nucl. Phys. B **291**, 41–89 (1987)
78. R.R. Metsaev, A.A. Tseytlin, Order alpha-prime (two loop) equivalence of the string equations of motion and the sigma model Weyl invariance conditions: dependence on the Dilaton and the antisymmetric tensor. Nucl. Phys. B **293**, 385–419 (1987)
79. M.C. Bento, N.E. Mavromatos, Ambiguities in the low-energy effective actions of string theories with the inclusion of antisymmetric tensor and Dilaton fields. Phys. Lett. B **190**, 105–109 (1987)
80. R.E. Kallosh, I.V. Tyutin, The equivalence theorem and gauge invariance in renormalizable theories. Yad. Fiz. **17**, 190–209 (1973)
81. M.C. Bergere, Y.M.P. Lam, Equivalence theorem and Faddeev-Popov ghosts. Phys. Rev. D **13**, 3247–3255 (1976)
82. F.W Hehl, P. Von Der Heyde, G.D. Kerlick, J.M. Nester, General relativity with spin and torsion: foundations and prospects, Rev. Mod. Phys. **48**, 393 (1976)
83. M.J. Duncan, N. Kaloper, K.A. Olive, Axion hair and dynamical torsion from anomalies. Nucl. Phys. B **387**, 215–235 (1992)
84. L. Alvarez-Gaume, E. Witten, Gravitational anomalies. Nucl. Phys. B **234**, 269 (1984)

85. P. Svrcek, E. Witten, Axions in string theory. J. High Energy Phys. **06**, 051 (2006). arXiv:hep-th/0605206 [hep-th]
86. R. Jackiw, S.Y. Pi, Chern-Simons modification of general relativity. Phys. Rev. D **68**, 104012 (2003). arXiv:gr-qc/0308071 [gr-qc]
87. S. Alexander, N. Yunes, Chern-Simons modified general relativity. Phys. Rept. **480**, 1–55 (2009). arXiv:0907.2562 [hep-th]
88. N. Yunes, F. Pretorius, Dynamical Chern-Simons modified gravity. I. Spinning black holes in the slow-rotation approximation. Phys. Rev. D **79**, 084043 (2009). arXiv:0902.4669 [gr-qc]
89. K. Yagi, N. Yunes, T. Tanaka, Slowly rotating black holes in dynamical Chern-Simons gravity: deformation quadratic in the spin. Phys. Rev. D **86**, 044037 (2012). Erratum: Phys. Rev. D **89**, 049902 (2014). arXiv:1206.6130 [gr-qc]
90. N. Chatzifotis, P. Dorlis, N. E. Mavromatos and E. Papantonopoulos, Scalarization of Chern-Simons-Kerr black hole solutions and wormholes. Phys. Rev. D **105**(8), 084051 (2022). [arXiv:2202.03496 [gr-qc]].
91. S.H.S. Alexander, M.E. Peskin, M.M. Sheikh-Jabbari, Leptogenesis from gravity waves in models of inflation. Phys. Rev. Lett. **96**, 081301 (2006). arXiv:hep-th/0403069 [hep-th]
92. D.H. Lyth, C. Quimbay, Y. Rodriguez, Leptogenesis and tensor polarisation from a gravitational Chern-Simons term. J. High Energy Phys. **03**, 016 (2005). arXiv:hep-th/0501153 [hep-th]
93. E.W. Kolb, M.S. Turner, The early universe. Front. Phys. **69**, 1–547 (1990)
94. L. McAllister, E. Silverstein, A. Westphal, Gravity waves and linear inflation from axion monodromy. Phys. Rev. D **82**, 046003 (2010). arXiv:0808.0706 [hep-th]
95. J. Grande, J. Solà, H. Stefancic, LXCDM: a Cosmon model solution to the cosmological coincidence problem? J. Cosmol. Astro. Phys. **08**, 011 (2006). arXiv:gr-qc/0604057 [gr-qc]
96. J. Grande, A. Pelinson, J. Solà, Dark energy perturbations and cosmic coincidence. Phys. Rev. D **79**, 043006 (2009). arXiv:0809.3462 [astro-ph]
97. I.L. Shapiro, J. Solà, On the scaling behavior of the cosmological constant and the possible existence of new forces and new light degrees of freedom. Phys. Lett. B **475**, 236–246 (2000). arXiv:hep-ph/9910462 [hep-ph]
98. I.L. Shapiro, J. Solà, Scaling behavior of the cosmological constant: Interface between quantum field theory and cosmology. J. High Energy Phys. **02**, 006 (2002). arXiv:hep-th/0012227 [hep-th]
99. J. Solà, Dark energy: a quantum fossil from the inflationary universe? J. Phys. A **41**, 164066 (2008). arXiv:0710.4151 [hep-th]
100. J. Solà Peracaula, The cosmological constant problem and running vacuum in the expanding universe. arXiv:2203.13757 [gr-qc]
101. C. Moreno-Pulido, J. Solà Peracaula, Renormalizing the vacuum energy in cosmological spacetime: implications for the cosmological constant problem. arXiv:2201.05827 [gr-qc]
102. C. Moreno-Pulido, J. Solà Peracaula, Renormalized $\rho_{\rm vac}$ without m^4 terms. arXiv:2110.08070 [gr-qc]
103. J.A.S. Lima, S. Basilakos, J. Solà, Expansion history with decaying vacuum: a complete cosmological scenario. Mon. Not. R. Astron. Soc. **431**, 923–929 (2013). arXiv:1209.2802 [gr-qc]
104. E.L.D. Perico, J.A.S. Lima, S. Basilakos, J. Solà, Complete cosmic history with a dynamical $\Lambda = \Lambda(H)$ term. Phys. Rev. D **88**, 063531 (2013). arXiv:1306.0591 [astro-ph.CO]
105. J.A.S. Lima, S. Basilakos, J. Solà, Thermodynamical aspects of running vacuum models. Eur. Phys. J. C **76**(4), 228 (2016). arXiv:1509.00163 [gr-qc]
106. J.A.S. Lima, S. Basilakos, J. Solà, Nonsingular decaying vacuum cosmology and entropy production. Gen. Rel. Grav. **47**, 40 (2015). arXiv:1412.5196 [gr-qc]
107. J. Solà Peracaula, H. Yu, Particle and entropy production in the running vacuum universe. Gen. Rel. Grav. **52**(2), 17 (2020). arXiv:1910.01638 [gr-qc]
108. J. Solà Peracaula, J. de Cruz Pérez, A. Gómez-Valent, Dynamical dark energy vs. Λ = const in light of observations. Europhys. Lett. **121**(3), 39001 (2018). arXiv:1606.00450 [gr-qc]

109. J. Solà Peracaula, J. de Cruz Pérez, A. Gomez-Valent, Mon. Not. R. Astron. Soc. **478**(4), 4357–4373 (2018). arXiv:1703.08218 [astro-ph.CO]
110. J. Solà Peracaula, A. Gomez-Valent, J. de Cruz Pérez, Phys. Dark Univ. **25**, 100311 (2019). arXiv:1811.03505 [astro-ph.CO]
111. G. Papagiannopoulos, P. Tsiapi, S. Basilakos, A. Paliathanasis, Eur. Phys. J. C **80**(1), 55 (2020). arXiv:1911.12431 [gr-qc]
112. P. Tsiapi, S. Basilakos, Testing dynamical vacuum models with CMB power spectrum from Planck. Mon. Not. R. Astron. Soc. **485**(2), 2505–2510 (2019). arXiv:1810.12902 [astro-ph.CO]
113. J. Solà Peracaula, A. Gómez-Valent, J. de Cruz Perez, C. Moreno-Pulido, Running vacuum against the H_0 and σ_8 tensions. Europhys. Lett. **134**(1), 19001 (2021). arXiv:2102.12758 [astro-ph.CO]
114. L. Verde, T. Treu and A. G. Riess, Tensions between the Early and the Late Universe. Nature Astron. **3**, 891 (2019). [arXiv:1907.10625 [astro-ph.CO]].
115. E. Di Valentino, O. Mena, S. Pan, L. Visinelli, W. Yang, A. Melchiorri, D.F. Mota, A.G. Riess, J. Silk, In the realm of the Hubble tension—a review of solutions. Class. Quant. Grav. **38**(15), 153001 (2021). arXiv:2103.01183 [astro-ph.CO]
116. L. Perivolaropoulos, F. Skara, Challenges for ΛCDM: an update. arXiv:2105.05208 [astro-ph.CO]
117. W.L. Freedman, Cosmology at a crossroads. Nature Astron. **1**, 0121 (2017). arXiv:1706.02739 [astro-ph.CO], and references therein
118. P. Asimakis, S. Basilakos, N.E. Mavromatos, E.N. Saridakis, Big bang nucleosynthesis constraints on higher-order modified gravities. Phys. Rev. D **105**(8), 8 (2022). arXiv:2112.10863 [gr-qc]
119. H. Ooguri, E. Palti, G. Shiu, C. Vafa, Distance and de sitter conjectures on the swampland. Phys. Lett. B **788**, 180–184 (2019). arXiv:1810.05506 [hep-th]
120. B.A. Campbell, M.J. Duncan, N. Kaloper, K.A. Olive, Gravitational dynamics with Lorentz Chern-Simons terms. Nucl. Phys. B **351**, 778–792 (1991)
121. G. de Berredo-Peixoto, L. Freidel, I.L. Shapiro, C.A. de Souza, Dirac fields, torsion and Barbero-Immirzi parameter in cosmology. J. Cosmol. Astro. Phys. **06**, 017 (2012). arXiv:1201.5423 [gr-qc]
122. G.F. Smoot, M.V. Gorenstein, R.A. Muller, Detection of anisotropy in the cosmic black body radiation. Phys. Rev. Lett. **39**, 898 (1977). And references therein
123. G.W. Gibbons, S.W. Hawking, Cosmological event horizons, thermodynamics, and particle creation. Phys. Rev. D **15**, 2738–2751 (1977). https://doi.org/10.1103/PhysRevD.15.2738
124. G. Lambiase, Lorentz invariance breakdown and constraints from big-bang nucleosynthesis. Phys. Rev. D **72**, 087702 (2005). https://doi.org/10.1103/PhysRevD.72.087702. arXiv:astro-ph/0510386 [astro-ph]

Deformed Relativistic Symmetry Principles

2

Michele Arzano, Giulia Gubitosi, and José Javier Relancio

Abstract

We review the main features of models where relativistic symmetries are deformed at the Planck scale. We cover the motivations, links to other quantum gravity approaches, describe in some detail the most studied theoretical frameworks, including Hopf algebras, relative locality, and other scenarios with deformed momentum space geometry, discuss possible phenomenological consequences, and point out current open questions.

2.1 Introduction

The proposal that local space-time symmetries might be deformed at the Planck scale was put forward in the early 2000s [1, 2]. The motivation was given by phenomenological studies on the possibility that the energy-momentum dispersion

M. Arzano · G. Gubitosi (✉)
Dipartimento di Fisica "Ettore Pancini", Università di Napoli Federico II, Naples, Italy

INFN, Sezione di Napoli, Naples, Italy
e-mail: michele.arzano@na.infn.it; giulia.gubitosi@unina.it

J. J. Relancio
Departamento de Matemáticas y Computación, Universidad de Burgos, Burgos, Spain
e-mail: relancio@unizar.es

C. Pfeifer, C. Lämmerzahl (eds.), *Modified and Quantum Gravity*, Lecture Notes in Physics 1017, https://doi.org/10.1007/978-3-031-31520-6_2

relation of particles is deformed at the Planck scale $E_P \sim 10^{19}$ GeV, such that[1] [3–9]

$$m^2c^4 = E^2 - |\mathbf{p}|^2c^2 + \eta \frac{E^n}{E_P^n}|\mathbf{p}|^2c^2 \,. \tag{2.1}$$

Because this law is not covariant under the standard Lorentz transformations linking inertial observers in special relativity, the first studies proposing such a modified dispersion relation assumed that invariance under Lorentz symmetries was broken at ultra-high energies, where the modification appearing in (2.1) becomes relevant. As a consequence, a preferred reference frame would emerge, where the dispersion law would take the form (2.1), and this frame was typically identified with the rest frame with respect to the cosmic microwave background. This scenario is usually called Lorentz Invariance Violation (LIV), similar to what was discussed in Chap. 1 emerging from String Theory.

In the alternative framework proposed in [1, 2, 10, 11], a modified dispersion relation of the kind of (2.1) can take the same form in all inertial frames of reference, if the laws of transformation between these frames are in turn modified. In particular, they must admit two relativistic invariant quantities, the speed of light and the Planck energy E_P. For this reason, such framework was called Doubly Special Relativity (DSR).

The relation between the special relativistic and the DSR scenario can be understood in analogy to the transition from the Galilean relativity of Newtonian mechanics to special relativity [12]. In Galilean relativity the (kinetic) energy of a particle is related to its momentum $\mathbf{p} \equiv m\mathbf{v}$ by

$$E = \frac{|\mathbf{p}|^2}{2m} \,. \tag{2.2}$$

All observers moving with relative constant velocity see the same law, and are linked by Galilean boosts, whose generator is:

$$B_j^G = iE\frac{\partial}{\partial p_j} \,. \tag{2.3}$$

Notice that according to these laws spatial velocities add up linearly and there is no maximum speed. Special relativity can be seen as a deformation of Galilean relativity that emerges when considering large velocities. In special relativity the energy-momentum dispersion relation reads

$$E^2 = |\mathbf{p}|^2c^2 + m^2c^4 \,, \tag{2.4}$$

[1] This formula is to be understood as indicating the lowest-order correction to the standard special-relativistic expression in powers of the particle's energy over the Planck energy, where the order is given by the positive integer n and η is a dimensionless parameter indicating the strength of the effect at the Planck scale. In general, formulas considering all-order corrections go beyond this simple power-law expression, see e.g. Sect. 2.3.3.

where c is a velocity scale. This law is not covariant under Galilean boosts, so that in order for it to take the same form for all observers moving at constant relative speed the laws of transformation linking these observers need to be deformed. These transformations, the Lorentz boosts generated by,

$$B_j = i\frac{p_j}{c^2}\frac{\partial}{\partial E} + iE\frac{\partial}{\partial p_j}, \tag{2.5}$$

are a deformation of the Galilean boosts, such that (2.4) is covariant. They are such that the speed scale c is a relativistic invariant, identified with the speed of light, and that velocities no longer add linearly. Galilean boosts (and Galilean relativity in general) are recovered in the small velocity limit, $\frac{|\mathbf{v}|}{c} \to 0$ (see e.g. [13, 14]).

A further modification of the laws linking observers moving at constant relative speed, which generalizes the Lorentzian boosts, allows us to retain covariance when extending the dispersion relation from the special relativistic form (2.4) to the modified form (2.1). Just as the extension of boosts from the Galilean form to the Lorentzian form (necessary to describe a high-velocity regime) introduces an new invariant scale c, the extension of boosts to their DSR form (supposedly necessary to describe a very-high energy regime) introduces another relativistic invariant scale, the Planck energy E_P.[2] Explicit examples of these modified boost transformations are provided in Sects. 2.2 and 2.3.3.

While the basic ideas behind the DSR proposal are quite simple, the mathematical formalization and the study of phenomenological implications have already taken the efforts of many researchers over the past two decades. On the theoretical level, we have now several mathematical frameworks that can accommodate the DSR principles, some more developed than others. These include most notably quantum groups and Hopf algebras (see Sect. 2.3), curved momentum space models and modified phase space models (see Sects. 2.4 and 2.5). Each framework is more suited to study specific phenomenological implications, so it is worth pursuing all of them in parallel. On the phenomenological side, one of the most relevant advancements concerns the uncovering of the deep links between modified boost transformations and the loss of absolute locality. Just like the modification of boost transformations induced by the transition from Galilean to special relativity requires us to give up the absoluteness of simultaneity, in the transition from special to doubly special relativity we are required to give up the absoluteness of locality. This was understood at the beginning of the past decade, leading to the development of the relative locality proposal, whose implications are the center of a very active research programme (see Sect. 2.4.2).

[2] While in special relativity c is the maximum allowed speed, in DSR it is to be understood as the speed of low-energy massless particles. And the Planck energy is a relativistic invariant, but is not necessarily the maximum allowed energy. It might be the case in some specific models, but it is not true in general.

In this review, we aim at providing the reader with the current state of the art of this research field, highlighting the progress that has been made in the theoretical modelling and the phenomenological developments, and pointing out current open questions.

2.1.1 Link to More Fundamental QG Frameworks

As we have discussed above, DSR was motivated by phenomenological considerations relevant in searches for effects induced by Planck-scale physics. However, subsequent studies showed that deformations of relativistic symmetries can emerge in specific limits of more fundamental quantum gravity theories.

For example, it is now well established that departures from special relativity could arise in a "semiclassical" regime of quantum gravity, where the gravitational degrees of freedom are integrated out and leave an effective field theory for the matter fields. Of course, this cannot be done explicitly for the full quantum theory of gravity, but this was shown to be the case in 2+1 dimensions, where gravity can be quantized as a topological field theory and can be coupled to point particles, represented by topological defects [15–22].

In the loop quantum gravity approach, one can adopt a perspective suggesting deformations of relativistic symmetries in the regime where the large-scale (coarse-grained) space-time metric is flat [23–26]. This is done by studying the modifications to the hypersurface deformation algebra, which is the algebra of generators of invariance with respect to local diffeomorphisms. Such modifications provide a picture [25] that is consistent with deformations of the relativistic symmetries of the kind that are encountered in the κ-Minkowski non-commutative spacetime, described in Sect. 2.3. Besides these studies, more heuristic arguments supporting the emergence of deformed relativistic symmetries have also been put forward in the context of $3 + 1$-dimensional loop quantum gravity [18, 27] and of polymer quantization [28].

Finally, is has been established that deformed relativistic symmetries emerge in the context of non-commutative space-time geometry [29–40]. This point will be discussed in greater detail in Sect. 2.3, for the specific case of the κ-Minkowski spacetime.

2.2 Doubly Special Relativity—Phenomenological Models

The fundamental ingredients to define the phenomenolgy associated to a DSR kinematical model are

- the energy-momentum dispersion relation, which can be schematically denoted as[3]

$$C(E, \mathbf{p}) = \mu^2 , \tag{2.6}$$

where μ is (a function of) the mass of the particle;
- the conservation laws of energy and momenta in interactions, which for n incoming particles with momenta $p^{(i)}$ and m outgoing particles with momenta $p^{(o)}$ can be written as

$$\left(p_1^{(i)} \oplus p_2^{(i)} \oplus ...p_n^{(i)}\right)_\mu = \left(p_1^{(o)} \oplus p_2^{(o)} \oplus ...p_m^{(o)}\right)_\mu , \tag{2.7}$$

were $\oplus$ encodes a deformed addition law of energy and momenta;
- the laws of transformation between inertial observers, encoded via the action of boost transformations on energy and momenta, such that[4]

$$p_\mu \rightarrow (B^\xi \triangleright p)_\mu , \tag{2.8}$$

where ξ is the rapidity characterizing the magnitude of the boost.[5]

These ingredients need to combine into quite a rigid structure, constrained by the requirement that relativistic invariance is not spoiled [12, 41–43]. The observer independence of the dispersion relation can be stated as the requirement that it is invariant under boosts

$$C(B^\xi \triangleright E, B^\xi \triangleright \mathbf{p}) = C(E, \mathbf{p}) . \tag{2.9}$$

Covariance of the conservation laws is achieved if the sum of momenta of all incoming (outgoing) particles transforms as a momentum under boosts:

$$q = p_1 \oplus p_2 ... \oplus p_n \leftrightarrow B^\xi \triangleright q = B^\xi \triangleright (p_1 \oplus p_2 ... \oplus p_n) . \tag{2.10}$$

In special relativity (where momenta add linearly, $p_1 \oplus p_2 ... \oplus p_n = p_1 + p_2 ... + p_n$) the above condition is achieved by asking that

$$B^\xi \triangleright (p_1 \oplus p_2 ... \oplus p_n) = \left(B^\xi \triangleright p_1\right) \oplus \left(B^\xi \triangleright p_2\right) ... \oplus \left(B^\xi \triangleright p_n\right) , \tag{2.11}$$

[3] From now on we set $c = 1$.

[4] One might also consider deformations of the other relativistic symmetries, but here we will only focus on boosts for simplicity.

[5] In the following we will sometimes use a simplified notation omitting the explicit indication of the four-vector index μ.

and this is also the case for DSR models with a commutative law of addition of momenta $\oplus$. However, one may have DSR models where the deformed addition rule $\oplus$ is noncommutative (a notable example is provided by models based on the κ-Poincaré Hopf algebra, discussed in the following Section). In this case, covariance of the conservation law (2.7) can only be achieved if the boost acts on systems of interacting particles in a non-trivial way. Namely, the rapidity parameter with which different particles participating in the interaction transform depends on the momenta of the other particles [41,42,44,45]. Considering the addition of n momenta p_i, the action of boosts is given by

$$B^{\xi} \triangleright (p_1 \oplus p_2 ... \oplus p_n) \equiv \left(B^{\xi_1} \triangleright p_1\right) \oplus \left(B^{\xi_2} \triangleright p_2\right) ... \oplus \left(B^{\xi_n} \triangleright p_n\right) , \tag{2.12}$$

where $\xi_1 = \xi_1(p_2, ...p_n)$, $\xi_2 = \xi_2(p_1, p_3...p_n)$ and so on, such that $\xi_i = \xi$ when $p_1, ..., p_{i-1}, p_{i+1}, ..., p_n$ vanish. This notion of covariance of the conservation laws is a generalization of the one we are familiar with, based on the intuition we built working with special relativity. Even though this might seem counter-intuitive, it can be shown that it does not lead to the emergence of preferred observers (see a detailed discussion in [42], arXiv version). Very recently, issues related to what is called "history problem" have emerged in association to boost actions of the sort of (2.12). Since this is quite a subtle point which is currently under further study, we are not going to discuss the details here. The interested reader can refer to [45].

When considering only first order corrections to special relativity (e.g., the $n = 1$ case of (2.1)), the constraints we just discussed can be translated into constraints on the coefficients of the possible correction terms that can be added to the dispersion relation, the law of addition of momenta and the boost [12, 41, 43]. In particular, it can be shown that these constraints imply some "golden rules" on the allowed physical processes for theories without preferred frames: for example, photons cannot decay into electron-positron pairs and it must be possible for a photon of any arbitrarily low energy to produce electron-positron pairs when it interacts with a sufficiently high-energy photon [12]. These conditions guarantee that there is no threshold associated to a photon, a fundamental requirement for any DSR theory. In fact, because the energy of a photon could be tuned above or below the threshold with an appropriate boost, the existence of such threshold would identify a preferred frame.

Having exposed the general requirements that a DSR model needs to meet, we are going to provide one simple example, in order to see an application of the general concepts we have just exposed. Because the subtleties emerging when the addition law $\oplus$ is noncommutative will be discussed in detail in the following Section, here we are going to consider a $1 + 1$ dimensional DSR model with a commutative addition law. Specifically, let us consider a DSR model where all deviations from special relativity are only relevant at the first order in the ratio $\frac{E}{E_P}$ between the particles' energy and the Planck energy. This amounts to take $n = 1$ in the dispersion relation (2.1):

$$m^2 = E^2 - p^2 + \eta \frac{E}{E_P} p^2 \,. \tag{2.13}$$

In order for this equation to be covariant under Lorentz boosts, their action needs to be modified in such a way that energy and momentum transform non-linearly [12,41,46]:

$$\begin{aligned} B^{\xi} \triangleright E &= E + \xi p \,, \\ B^{\xi} \triangleright p &= p + \xi E + \xi \tfrac{\eta}{E_P} \left(E^2 + \tfrac{p^2}{2} \right) . \end{aligned} \tag{2.14}$$

? Exercise

2.1. Show that the dispersion relation (2.13) is not covariant under the action of standard special-relativistic boosts and verify its covariance with respect to the deformed boosts (2.14).

In turn, the modified boost transformations (2.14) are not compatible with standard conservation laws in interactions. Considering a process with two incoming particles a, b, and two outgoing particles c, d, a conservation law that is covariant under the transformation (2.14) is

$$\begin{aligned} E_a + E_b + \tfrac{\eta}{E_P} p_a p_b &= E_c + E_d + \tfrac{\eta}{E_P} p_c p_d \\ p_a + p_b + \tfrac{\eta}{E_P} \left(E_a p_b + E_b p_a \right) &= p_c + p_d + \tfrac{\eta}{E_P} \left(E_c p_d + E_d p_c \right) . \end{aligned} \tag{2.15}$$

This encodes a modified law of addition of energy and momenta:

$$\begin{aligned} E_a \oplus E_b &= E_a + E_b + \tfrac{\eta}{E_P} p_a p_b \,, \\ p_a \oplus p_b &= p_a + p_b + \tfrac{\eta}{E_P} \left(E_a p_b + E_b p_a \right) , \end{aligned} \tag{2.16}$$

which is covariant assuming the following action of boosts on the interacting particles:

$$\begin{aligned} B^{\xi} \triangleright (E_a \oplus E_b) &= \left(B^{\xi} \triangleright E_a \right) \oplus \left(B^{\xi} \triangleright E_b \right) , \\ B^{\xi} \triangleright (p_a \oplus p_b) &= \left(B^{\xi} \triangleright p_a \right) \oplus \left(B^{\xi} \triangleright p_b \right) . \end{aligned} \tag{2.17}$$

? Exercise

2.2. Verify that the addition law (2.16) is covariant with respect to the deformed boosts (2.14).

Notice, that since we are working at the first order in $\frac{E}{E_P}$, the modified boost action and addition law are not the unique possible choices to have a relativistic picture starting from the dispersion relation (2.13). Another possibility is discussed in Sect. 2.3.3. More thorough studies of the possibilities available at the first order can be found in [12, 41].

In closing this section, we want to remark that, while working in momentum space suffices to study the kinematics of interactions, other possible predictions of DSR models affect the propagation of particles (e.g., one might have an induced energy dependence in the travel time of massless particles, see Sect. 2.6). However, in order to study this kind of effects we are required to find a suitable way to describe the spacetime in which such particles propagate, or, by the least, if we want to compute the energy-dependent shift in the time of arrival of particles with different energies, we need to define a time coordinate. As we mentioned in the introduction, and will be discussed in greater detail in Sect. 2.4.2, in DSR models we expect departures from the observer-independent notion of locality that applies in special relativity, and one might have that the space-time picture depends on the energy of the particle used to probe it. So defining spacetime is a highly nontrivial task, and the various mathematical frameworks that are discussed in the following can be seen as different ways to implement a spacetime picture within the DSR scenario.

2.3 Hopf Algebras and the Example of κ-Poincaré

By far the most studied example of deformation of space-time symmetries is the κ-Poincaré algebra. Such model was introduced in the early 1990s [47–49] and it was historically the first attempt at modifying the algebraic structure of relativistic symmetries in order to introduce a fundamental energy scale using the theory of Hopf algebras. The study of κ-deformed relativistic kinematics as a candidate for an effective description of Planck-scale physics [31, 50, 51] paved the way to the formulation of DSR models [1, 2].

2.3.1 Emergence of Hopf Algebra Structures in Quantum Theory

In order to understand how Hopf algebra structures allow for the introduction of an additional invariant scale in the description of space-time symmetries, it will be necessary to first briefly review how Hopf algebraic structures emerge in the description of symmetries in physical systems (for a more extensive and pedagogical treatment we refer the reader to [52]).

In relativistic quantum theory invariance under the isometries of Minkowski spacetime requires that the states describing elementary particles carry a unitary irreducible representation of the Poincaré group. For a real scalar field, for example, we have a "one-particle" Hilbert space $\mathcal{H}$, whose elements can be given in terms of complex functions on the positive mass-shell in four-momentum space and whose elements we denote as kets labelled by the spatial momentum carried by the particle $|\mathbf{k}\rangle \in \mathcal{H}$. Multi-particle states will be described by (symmetrized) tensor products of such irreducible representations belonging to the Fock space [53]. Let us now look at how observables act on such states. We focus on the specific example of observables P_i, the generators of space translations of the Poincaré algebra. One-particle states labelled by the linear momentum above are diagonal under the action of these operators

$$P_i|\mathbf{k}\rangle = k_i|\mathbf{k}\rangle\,, \tag{2.18}$$

where k_i is the i-th component of the vector $\mathbf{k}$. The action of the observable P_i on generic Fock space elements is given by its *second quantized* version

$$\begin{aligned} d\Gamma(P_i) \equiv\; & 1 + P_i + (P_i\otimes 1 + 1\otimes P_i) \\ & +(P_i\otimes 1\otimes 1 + 1\otimes P_i\otimes 1 + 1\otimes 1\otimes P_i) + \ldots\,, \end{aligned} \tag{2.19}$$

where 1 is the identity operator. The additional information required to extend the representation of the Poincaré algebra from the one-particle Hilbert space to the Fock space, as encoded in (2.19), can be formalized in terms of an operation called the *coproduct*

$$\Delta P_i = P_i\otimes 1 + 1\otimes P_i\,, \tag{2.20}$$

in terms of which (2.19) can be written as

$$d\Gamma(P_i) \equiv 1 + P_i + \Delta P_i + \Delta_2 P_i + \ldots + \Delta_n P_i + \ldots\,, \tag{2.21}$$

where

$$\Delta_n \equiv (\Delta\otimes 1)\circ\Delta_{n-1}\,, \qquad n\geq 2\,, \tag{2.22}$$

with $\Delta_1 \equiv \Delta$. The coproduct (2.20) gives us the action of the observable P_i linear on a two-particle state

$$|\mathbf{k}\,\mathbf{l}\rangle \equiv \frac{1}{\sqrt{2}}\left(|\mathbf{k}\rangle\otimes|\mathbf{l}\rangle + |\mathbf{k}\rangle\otimes|\mathbf{l}\rangle\right). \tag{2.23}$$

In particular

$$\Delta P_i |\mathbf{k}\,\mathbf{l}\rangle = (k_i + l_i)|\mathbf{k}\,\mathbf{l}\rangle\,, \tag{2.24}$$

and thus, the eigenvalue of ΔP_i on a two-particle state is simply its total linear momentum. For our purposes it is important to notice that the coproduct encodes the property of *additivity of quantum numbers*. As we will see, the possibility to render *non-abelian* such property for quantum numbers associated to space-time symmetries is at the core of the concept of deformation we consider.

Before proceeding we need to introduce one more ingredient concerning the action of observables on the states $\langle\mathbf{k}|$, i.e., on elements of the dual[6] to the one-particle Hilbert space $\mathcal{H}^*$. Since $\mathcal{H}$ carries a representation of the Poincaré algebra on the space $\mathcal{H}^*$ can be defined a *dual* representation. Starting from the action (2.18), one defines the action of P_i on a vector $\langle\mathbf{k}'| \in \mathcal{H}^*$, so that the following equality holds

$$(P_i \langle\mathbf{k}'|)|\mathbf{k}\rangle = -\langle\mathbf{k}'|(P_i|\mathbf{k}\rangle)\,. \tag{2.25}$$

We thus see that the dual representation defines an action *from the left* of the translations generators on bras given by

$$P_i \langle\mathbf{k}| = -k_i \langle\mathbf{k}|\,. \tag{2.26}$$

Notice that such action is different from the action from the right, which is simply obtained by taking the hermitian adjoint of (2.18)

$$\langle\mathbf{k}|P_i \equiv (P_i|\mathbf{k}\rangle)^\dagger = \langle\mathbf{k}|k_i\,. \tag{2.27}$$

We thus see that the dual representation can be defined in terms of a map, known as the *antipode*

$$S(P_i) = -P_i\,, \tag{2.28}$$

connecting the left and right action of the generators $S(P_i)$ on dual states

$$P_i \langle\mathbf{k}| = -k_i \langle\mathbf{k}| = \langle\mathbf{k}|(-k_i) \equiv \langle\mathbf{k}|S(P_i)\,. \tag{2.29}$$

To understand the physical role of the antipode map let us recall that given the one-particle Hilbert space $\mathcal{H}$, the space describing anti-particles is given by the complex conjugate Hilbert space $\bar{\mathcal{H}}$. Such Hilbert space is isomorphic to the dual Hilbert space $\mathcal{H}^*$ [52], and thus, for example, for a complex scalar field the bras $\langle k|$ can

[6] By definition, elements of the dual of a Hilbert space $\mathcal{H}$ are continuous linear maps from $\mathcal{H}$ to $\mathbb{C}$. Given the inner product $\langle\mathbf{k}'|\mathbf{k}\rangle$ on $\mathcal{H}$, it is evident that bra $\langle\mathbf{k}|$ is an element of the dual space.

be identified with *antiparticle states*. This shows that the antipode map introduced above describes the way observables act on antiparticle states.

At the algebraic level, the coproduct and the antipode maps are additional ingredients which (together with certain consistency conditions, the interested reader can consult [52] for details) equip the algebra of generators of space-time symmetries (and as a matter of fact of any quantum observable) with the structure of a Hopf algebra. We thus see that Hopf algebra structures are not just an abstract mathematical construct but are used in our everyday quantum field theory when we look at the action of observables on antiparticles or on systems with more than one particle.

2.3.2 The κ-Poincaré Hopf Algebra and Spacetime Relativistic Symmetries

After our brief detour concerning the algebraic structures underlying the action of symmetry generators on the states of a relativistic quantum system, we are now ready to introduce the κ-Poincaré Hopf algebra. The best way to understand the structure of such deformation of the Poincaré algebra is to start from a four-momentum space which is no longer a vector space, as in ordinary relativistic systems, but its geometry is that of a *non-abelian Lie group* which admits an action of the Lorentz group. In the κ-deformed context such group is denoted by $AN(3)$ and it is defined by the *Iwasawa decomposition* of the five-dimensional Lorentz group $SO(4, 1)$. Such decomposition is better understood by starting from the Lie algebra $\mathfrak{so}(1, 4)$ written as a direct sum of subalgebras

$$\mathfrak{so}(1,4) = \mathfrak{so}(1,3) \oplus \mathfrak{n} \oplus \mathfrak{a}\,, \tag{2.30}$$

where the algebra $\mathfrak{a}$ is generated by the element

$$H = \begin{bmatrix} 0&0&0&0&1 \\ 0&0&0&0&0 \\ 0&0&0&0&0 \\ 0&0&0&0&0 \\ 1&0&0&0&0 \end{bmatrix}, \tag{2.31}$$

and the algebra $\mathfrak{n}$ by the elements

$$\mathfrak{n}_i = \begin{bmatrix} 0 & (\epsilon_i)^T & 0 \\ \epsilon_i & 0 & \epsilon_i \\ 0 & -(\epsilon_i)^T & 0 \end{bmatrix}, \tag{2.32}$$

where ϵ_i are unit vectors in i-th direction ($\epsilon_1 = (1, 0, 0)$, etc), and T denotes transposition.

We now introduce a constant parameter, which carries dimensions of energy, denoted by κ. We can use such constant to define non-commuting objects with dimension of length[7]

$$X^0 = -\frac{i}{\kappa} H, \qquad X^i = \frac{i}{\kappa} \mathfrak{n}_i . \tag{2.33}$$

These *non-commuting coordinates* obey the commutator

$$[X^0, X^i] = \frac{1}{\kappa} X^i \qquad [X^i, X^j] = 0, \tag{2.34}$$

known in the literature as the κ-Minkowski non-commutative spacetime [54].

From (2.30) it follows that every element λ of the group $SO(1,4)$ can be decomposed as follows

$$\lambda = (Kna) \quad \text{or} \quad \lambda = (K\,\vartheta\, na), \tag{2.35}$$

where $K \in SO(1,3)$, the element a belongs to group the A generated by H

$$A = \exp\left(ik_0 X^0\right) = \exp\left(\frac{k_0}{\kappa} H\right) = \begin{bmatrix} \cosh\frac{k_0}{\kappa} & 0\;0\;0 & \sinh\frac{k_0}{\kappa} \\ 0 & 1\;0\;0 & 0 \\ 0 & 0\;1\;0 & 0 \\ 0 & 0\;0\;1 & 0 \\ \sinh\frac{k_0}{\kappa} & 0\;0\;0 & \cosh\frac{k_0}{\kappa} \end{bmatrix}, \tag{2.36}$$

and the element n belongs to the group N generated by the matrices $\mathfrak{n}_i$

$$N = \exp\left(ik_i X^i\right) = \exp\left(-\frac{1}{\kappa} k_i \mathfrak{n}_i\right) = \begin{bmatrix} 1 + \frac{1}{2\kappa^2}\mathbf{k}^2 & \frac{k_1}{\kappa} & \frac{k_2}{\kappa} & \frac{k_3}{\kappa} & \frac{1}{2\kappa^2}\mathbf{k}^2 \\ \frac{k_1}{\kappa} & 0 & 0 & 0 & \frac{k_1}{\kappa} \\ \frac{k_2}{\kappa} & 0 & 0 & 0 & \frac{k_2}{\kappa} \\ \frac{k_3}{\kappa} & 0 & 0 & 0 & \frac{k_3}{\kappa} \\ -\frac{1}{2\kappa^2}\mathbf{k}^2 & -\frac{k_1}{\kappa} & -\frac{k_2}{\kappa} & -\frac{k_3}{\kappa} & 1 - \frac{1}{2\kappa^2}\mathbf{k}^2 \end{bmatrix}, \tag{2.37}$$

and ϑ=diag$(-1,1,1,1,-1)$.

[7] As we will explain in the following section, in some approaches to DSR based purely on the geometry of momentum space one assumes to be in a "semiclassical" regime of quantum gravity, such that the Planck constant $\hbar$ and the Newton constant G vanish, but their ratio is fixed and finite. In this regime, one can build an energy scale E_P but not a length scale $L_P \to 0$. In the context of Hopf algebra and non-commutative geometry, this is not the regime that is considered, since one needs a constant with dimensions of length to govern space-time noncommutativity as in (2.34).

The group $AN(3)$ can be identified with the product group NA. Elements of $AN(3)$ can be expressed as *plane waves* on the non-commutative κ-Minkowski spacetime with the time component appearing to the right [31]

$$\hat{e}_k = \exp(ik_i X_i)\exp\left(ik_0 X^0\right). \tag{2.38}$$

The four-momenta k_μ are coordinate functions on the group $AN(3)$ known as *bicrossporduct coordinates*. The quotient group structure of $AN(3) \simeq SO(1,4)/SO(1,3)$ allows us to obtain another system of coordinates. Indeed, it is well known that the quotient of Lie groups $SO(1,4)/SO(1,3)$ is the de Sitter space. Acting with the subgroup $AN(3)$ of $SO(1,4)$ on the vector O

$$O = \begin{bmatrix} 0 \\ 0 \\ 0 \\ 0 \\ \kappa \end{bmatrix}, \tag{2.39}$$

we obtain the coordinates

$$p_0 = \kappa \sinh\frac{k_0}{\kappa} + \frac{\mathbf{k}^2}{2\kappa} e^{\frac{k_0}{\kappa}},$$

$$p_i = k_i e^{\frac{k_0}{\kappa}},$$

$$p_4 = \kappa \cosh\frac{k_0}{\kappa} - \frac{\mathbf{k}^2}{2\kappa} e^{\frac{k_0}{\kappa}}, \tag{2.40}$$

on de Sitter space defined as the submanifold of the five-dimensional Minkowski space by the equation

$$-p_0^2 + p_i^2 + p_4^2 = \kappa^2.$$

? Exercise

2.3. Using the relations (2.40) show that the coordinates $k_0, \mathbf{k}$ correspond to comoving coordinates on the de Sitter manifold, according to which the line element is $ds^2 = dk_0^2 - e^{2k_0/\kappa} d\mathbf{k}^2$. Use [44] as guidance.

Notice how such *embedding* coordinates only cover "half" of de Sitter manifold determined by the inequality

$$p_0 + p_4 > 0. \tag{2.41}$$

Considering two sets of coordinates

$$p_0 = \pm(\kappa \sinh\frac{k_0}{\kappa} + \frac{\mathbf{k}^2}{2\kappa}e^{\frac{k_0}{\kappa}}),$$

$$p_i = \pm k_i e^{\frac{k_0}{\kappa}},$$

$$p_4 = \mp(\kappa \cosh\frac{k_0}{\kappa} - \frac{\mathbf{k}^2}{2\kappa}e^{\frac{k_0}{\kappa}}), \tag{2.42}$$

we can cover the entire de Sitter manifold, and this is reflected in the fact that the group SO(1,4) can be written in the form

$$SO(1,4) = KNA \cup K\vartheta NA. \tag{2.43}$$

In fact, non-trivial geometrical properties of momentum space are a generic feature of DSR models. The relative locality proposal, see Sect. 2.4, takes this observation as a starting point.

We are now ready to discuss the algebra structure of κ-deformed symmetries. We start with translation generators that can be defined as acting like ordinary derivatives on non-commutative plane waves, once we have chosen an ordering of the non-commuting factors [35]. Let us focus on the time-to-the-right ordered plane waves (2.38) and define translation generators associated to bicrossproduct coordinates as

$$\tilde{P}_\mu\, \hat{e}_k \equiv k_\mu \hat{e}_k. \tag{2.44}$$

The eigenvalues k_μ are coordinates on the $AN(3)$ group manifold, and thus, just real numbers, from which we can deduce that the generators of translations commute

$$[\tilde{P}_\mu, \tilde{P}_\nu] = 0. \tag{2.45}$$

In order to determine the other commutators of the κ-Poincaré algebra we need to define the action of the Lorentz group on the $AN(3)$ momentum space.

Let us introduce the following notation for the Iwasawa decomposition for an element of $\lambda \in SO(4,1)$

$$\lambda = K_g\, g, \tag{2.46}$$

with $K_g \in SO(3,1)$ and $g \in AN(3)$. Uniqueness of the Iwasawa decomposition guarantees that given the Lorentz group element K_g and g, there are unique elements $K'_{g'}$, g', satisfying

$$K_g\, g = g'\, K'_{g'}, \tag{2.47}$$

so that one *defines* the Lorentz transformed group valued momentum as

$$g' = K_g\, g (K'_{g'})^{-1} . \tag{2.48}$$

In order to derive the action of such Lorentz transformation on momenta, and thus the commutators with the generators of translations, one can write the following expression for infinitesimal transformations

$$K_g \approx 1 + i\xi^a \mathfrak{k}_a \,, \quad K'_{g'} \approx 1 + i\xi^a\, h_a^b(g)\mathfrak{k}_b \,, \tag{2.49}$$

where $\mathfrak{k}_a$ are the generators of the Lorentz algebra $\mathfrak{so}(1,3)$, and $h_a^b(g)$ is a matrix function of the momentum. We can write the momentum group element as a matrix in terms of embedding coordinates (2.39)

$$g = \begin{pmatrix} \tilde{p}_4 & \mathbf{p}/p_+ & p_0 \\ \mathbf{p} & 1 & \mathbf{p} \\ \tilde{p}_0 & -\mathbf{p}/p_+ & p_4 \end{pmatrix} , \tag{2.50}$$

where 1 is the unit 3×3 matrix and $p_+ = p_0 + p_4$. Plugging such a matrix expression for g and g', and (2.49) in (2.48), one remarkably obtains [55] that the action of the Lorentz generators (and thus also of the Lorentz group) on the four-momenta p_μ is the *ordinary one*. Thus, defining the set of translation generators associated to embedding coordinates through the following action on plane waves

$$P_\mu\, \hat{e}_k \equiv p_\mu \hat{e}_k \,, \tag{2.51}$$

(notice how this differs from (2.44) by the different choice of momentum space coordinates p and k) we have that they obey the ordinary commutators with generators of rotations M_i and N_i:

$$\begin{aligned} [M_i, P_j] &= i\, \epsilon_{ijk} p_k, \quad [M_i, P_0] = 0 \\ \left[N_i, P_j\right] &= i\, \delta_{ij} P_0, \;\; [N_i, P_0] = i\, P_i\, . \end{aligned} \tag{2.52}$$

Now we come to one of the key points which distinguishes the κ-deformed algebra from the standard Poincaré algebra. As the reader might have noticed, we have already introduced two different sets of translation generators: the ones associated to bicrossproduct coordinates, which we denoted by $\tilde{P}_\mu$, and those associated to embedding coordinates P_μ. In the literature, these different sets of generators are known as different *bases* of the κ-Poincaré algebra. The generators P_μ (together with the generators of rotations and boosts) are known as the "classical basis" of the κ-Poincaré algebra [56], since at the Lie algebra level they just reproduce the standard Poincaré algebra. The generators $\tilde{P}_\mu$ determine the so-called "bicrossproduct basis" [54]. Using the coordinate transformation (2.40), one can easily see that the commutators between the bicrossproduct generators of translations and the Lorentz ones are *deformed*

$$
\begin{aligned}
&[M_i, \tilde{P}_j] = i\,\epsilon_{ijk}\tilde{P}_k, \quad [M_i, \tilde{P}_0] = 0\,, \\
&\left[N_i, \tilde{P}_j\right] = i\,\delta_{ij}\left(\frac{\kappa}{2}\left(1 - e^{-2\tilde{P}_0/\kappa}\right) + \frac{\tilde{\mathbf{P}}^2}{2\kappa}\right) - i\,\frac{1}{\kappa}\,\tilde{P}_i\tilde{P}_j, \ \left[N_i, \tilde{P}_0\right] = i\,\tilde{P}_i\,.
\end{aligned}
\tag{2.53}
$$

? Exercise

2.4. Compute the commutators (2.53) starting from (2.52) and using the change of basis (2.40).

Notice how in the classical basis since the algebra is undeformed so is the mass Casimir (an element of the algebra commuting with all other elements)

$$
C_\kappa(P) = P_0^2 - \mathbf{P}^2\,, \tag{2.54}
$$

reflecting the invariance under $SO(3,1)$ transformations of the subspaces $p_4 = const.$ of de Sitter space. Of course, the same invariant object written in bicrossproduct coordinates will have a very complicated non-linear form. Thus, we see that the curved manifold structure of momentum space makes it possible to have non-linear energy-momentum dispersion relations which at leading order are formally analogous to the modifications of the energy-momentum dispersion relations characterizing models in which Lorentz invariance is broken. Here, however, such modifications are fully compatible with the action of Lorentz transformations which are now deformed according, e.g., to the modified commutators (2.53). This feature is at the basis of the general ideas of DSR models, in which the deformation parameter κ is seen as a fundamental, observer independent, Planckian energy scale, see Sects. 2.1, 2.2 and 2.3.3. The price to pay for the introduction of such a scale in a way in which Poincaré symmetries are preserved is to renounce to the Abelian additivity of quantum numbers associated to such symmetries, as we show below. It is important to notice that this last feature is also present in the classical basis. So this basis is not equivalent to special relativity, despite having a trivial algebra and Casimir.

Before proceeding, let us summarize the results obtained so far: in the classical basis, the κ-Poincaré algebra is nothing but the ordinary Poincaré algebra. In the bicrossproduct basis, the commutators are given by [52]

$$[\tilde{P}_\mu, \tilde{P}_\nu] = 0\,, \tag{2.55}$$

$$[M_i, \tilde{P}_j] = i\,\epsilon_{ijk}\tilde{P}_k, \quad [M_i, \tilde{P}_0] = 0\,,$$

$$\left[N_i, \tilde{P}_j\right] = i\,\delta_{ij}\left(\frac{\kappa}{2}\left(1 - e^{-2\tilde{P}_0/\kappa}\right) + \frac{\tilde{\mathbf{P}}^2}{2\kappa}\right) - i\,\frac{1}{\kappa}\,\tilde{P}_i\tilde{P}_j, \; \left[N_i, \tilde{P}_0\right] = i\,\tilde{P}_i \tag{2.56}$$

$$[M_i, M_j] = i\,\epsilon_{ijk}M_k, \;\; [M_i, N_j] = i\,\epsilon_{ijk}N_k, \;\; [N_i, N_j] = -i\,\epsilon_{ijk}M_k\,, \tag{2.57}$$

while the Casimir invariant is given by

$$C_\kappa(\tilde{P}) = \left(2\kappa\,\sinh\frac{\tilde{P}_0}{2\kappa}\right)^2 - \tilde{\mathbf{P}}^2 e^{\frac{\tilde{P}_0}{\kappa}}\,. \tag{2.58}$$

Notice that the relation between the classical basis Casimir, $C_\kappa(P)$, and the bicrossproduct one, $C_\kappa(\tilde{P})$, is [57] $C_\kappa(P) = C_\kappa(\tilde{P})\left(1 + \frac{1}{4\kappa^2}C_\kappa(\tilde{P})\right)$. As a matter of fact, the presence of the invariant energy scale κ in the model renders any function of $C_\kappa(P)$ a good candidate for the invariant mass Casimir. Historically, (a variant of) the bicrossproduct basis Casimir $C_\kappa(\tilde{P})$ was first derived in the literature using contraction techniques on a q-deformed anti-de Sitter algebra [48]. For this reason, the vast majority of works focusing on the applications of the κ-Poincaré algebra adopted such Casimir to define a κ-deformed energy-momentum dispersion relation.

We are now ready to discuss the so-called *co-algebra* sector of the κ-Poincaré algebra, namely, the generalization to the κ-deformed setting of the coproduct and antipode maps discussed at the beginning of this section. As we have seen, working in the classical basis one can establish a Lie algebra isomorphism between the κ-Poincaré algebra and the ordinary Poincaré algebra. Thus, at the one-particle level, irreducible representations of the κ-Poincaré algebra can be identified with those of the ordinary Poincaré algebra [58]. As in ordinary quantum field theory, such irreducible representations for a scalar field can be constructed starting from plane waves. In the κ-deformed case, we deal with non-commutative plane waves (2.38) which, as in the standard case, can be put in correspondence with kets labelled by the eigenvalues associated to space-time translation generators. States characterized by on-shell momenta provide irreducible representations of the Lie algebra. We focus on bicrossproduct generators,

$$\tilde{P}_\mu|k\rangle = k_\mu|k\rangle\,. \tag{2.59}$$

Such kets can be put in correspondence with *ordinary* plane waves

$$\langle x|k\rangle \sim e_k \tag{2.60}$$

equipped with a non-commutative $\star$-product, such that

$$e_k \star e_l \equiv \hat{e}_k \hat{e}_l \,, \tag{2.61}$$

where $\hat{e}_k$ and $\hat{e}_l$ are the ordered plane waves (2.38). To any choice of ordering it will correspond a different choice of $\star$-product for ordinary plane waves [35]. The non-commutative nature of the $\star$-product is simply a reflection of the non-Abelian structure of momentum space. Indeed, the product of non-commutative plane waves $\hat{e}_k$ and $\hat{e}_l$, results in an ordered plane wave

$$\hat{e}_k \hat{e}_l = \hat{e}_{k\oplus l} \,, \tag{2.62}$$

where $k \oplus l$ is a non-Abelian addition law for the four-momenta k and l. Its form can be derived explicitly (see e.g. [59]) by re-ordering the factors in the product $\hat{e}_k \hat{e}_l$ in such a way to restore the normal ordering with all the factors containing X^0 to the right using the Baker-Campbell-Hausdorff formula for the Lie algebra (2.34). The resulting addition law is

$$k \oplus l = (k_0 + l_0, \mathbf{k} + e^{-k_0/\kappa}\, \mathbf{l}) \,. \tag{2.63}$$

Another way to derive such non-Abelian addition law would be to read-off the bicrossproduct four-momentum of the product matrix $\hat{e}_{k\oplus l} = \hat{e}_k \hat{e}_l$ obtained using the matrix expressions (2.36) and (2.37). Looking at the plane wave $\hat{e}_{k\oplus l}$, we can write

$$\hat{e}_{k\oplus l} \sim \langle x|k \oplus l\rangle = \star(\langle x|k\rangle \otimes \langle x|l\rangle) \,, \tag{2.64}$$

where the $\star$-product is seen as a map defined on the tensor product of two copies of the space of functions on Minkowski space-time. Now, since $\tilde{P}_\mu \hat{e}_{k\oplus l} = (k \oplus l)\hat{e}_{k\oplus l}$, we can derive the co-product for the bicrossproduct translation generators through the identity [60]

$$\tilde{P}_\mu \langle x|k \oplus l\rangle = \star(\langle x|(\Delta \tilde{P}_\mu (|k\rangle \otimes l\rangle)) \,, \tag{2.65}$$

obtaining

$$\Delta \tilde{P}_0 = \tilde{P}_0 \otimes 1 + 1 \otimes \tilde{P}_0 \,, \qquad \Delta \tilde{P}_i = \tilde{P}_i \otimes 1 + e^{-\tilde{P}_0/\kappa} \otimes \tilde{P}_i \,. \tag{2.66}$$

This shows that, in the κ-deformed context, the non-Abelian composition law of four-momenta is translated into a non-tivial coproduct for translation generators.

This in turn can be seen as a non-Abelian generalization of the Leibniz rule for the action of such generators on tensor product states. The non-trivial structure of the κ-deformed coproduct is intimately related to the non-Abelian product of two momentum group elements. The operation of taking the *inverse* of a momentum group element translates instead into a deformation of the antipode map. In terms of non-commutative plane waves, we can define a new plane wave labelled by a new momentum $\ominus k$ such that

$$\hat{e}_k \hat{e}_{\ominus k} = 1\,, \tag{2.67}$$

in other words, $\hat{e}_{\ominus k} \equiv (\hat{e}_k)^{-1}$. It is easy to see that from the definitions above we have $k \oplus (\ominus k) = 0$. One can easily show (e.g. by inverting the matrix expression for $\hat{e}_k$ and reading off the coordinates) that

$$\ominus\, k = (-k_0, -e^{k_0/\kappa}\, \mathbf{k})\,. \tag{2.68}$$

Recalling our definition of antipode map (2.29), this immediately reflects on the non-trivial antipode map on translation generators

$$S(\tilde{P}_\mu) = (-\tilde{P}_0, -e^{\tilde{P}_0/\kappa}\, \tilde{\mathbf{P}})\,. \tag{2.69}$$

The last ingredients needed to complete our "derivation" of the κ-Poincaré Hopf algebra are the co-products and antipodes for the Lorentz generators. In analogy with the generators of translations, the coproducts for the Lorentz generators can be obtained from the action of the $SO(3,1)$ group on the product of two momentum $AN(3)$ group elements (an alternative derivation in terms of the so-called Weyl maps can be found in [35]). Such action is given by a generalization of (2.48)

$$K_g\, g\, h\, K'^{-1}_{(gh)'} = (gh)'. \tag{2.70}$$

We immediately see that $(gh)' \neq g' h'$, which shows that the Lorentz group action on momentum space is not Leibnizian. For the antipodes, one looks at the action of Lorentz transformations on inverse group elements, namely

$$(g^{-1})' = K'_{g'}\, g^{-1}\, K_g^{-1}\,. \tag{2.71}$$

Using the infinitesimal form of the transformations (2.49), together with matrix representation of the $AN(3)$ group elements and of the Lorentz generators, one can show (see [55] for details) that the coproduct and antipode for rotation generators remain trivial

$$\Delta M_i = M_i \otimes 1 + 1 \otimes M_i\,, \qquad S(M_i) = -M_i\,, \tag{2.72}$$

while for boost generators one finds

$$\Delta(N_i) = N_i \otimes 1 + P_+^{-1} \otimes N_i + \epsilon_{ijk} \frac{1}{\kappa} P_j P_+^{-1} \otimes M_k \,, \tag{2.73}$$

$$S(N_i) = -N_i P_+ + \epsilon_{ijk} \frac{1}{\kappa} P_j M_k \,, \tag{2.74}$$

where $P_+ = P_0 + P_4$, with $P_4 = \sqrt{\kappa^2 + P_0^2 - \mathbf{P}^2}$. Written in terms of the bicrossproduct generators such coproduct and antipode read

$$\Delta(N_i) = N_i \otimes 1 + e^{-\tilde{P}_0/\kappa} \otimes N_i + \frac{1}{\kappa} \epsilon_{ijk} \tilde{P}_j \otimes M_k \,, \tag{2.75}$$

$$S(N_i) = -e^{\frac{\tilde{P}_0}{\kappa}} (N_i - \frac{1}{\kappa} \epsilon_{ijk} \tilde{P}_j M_k) \,. \tag{2.76}$$

These complete our description of the co-algebra structure of the κ-Poincaré Hopf algebra.

? Exercise

2.5. Using the change of basis (2.40), compute the co-algebra structure in terms of the generators P_μ of the classical basis.

2.3.3 Link to DSR Phenomenological Models

The Hopf algebra structure just described can be used to define a phenomenological model for DSR of the kind discussed in Sect. 2.2. In fact, because in the bicrossproduct basis the translation generators form a Hopf-subalgebra, they can be represented as an algebra of functions over momentum space [44, 61, 62], such that the translation generators correspond to the coordinate functions p_μ,

$$\tilde{P}_\mu(p) = p_\mu \,. \tag{2.77}$$

Then, the Casimir (2.58) can be used to read a deformed dispersion relation. Given that the Casimir is by definition an invariant of the symmetry generators, it can be equated to (a function of) the squared mass of the particle μ^2, so that the energy-momentum dispersion relation reads

$$\mu^2 = \left(2\kappa \sinh \frac{p_0}{2\kappa}\right)^2 - \mathbf{p}^2 e^{\frac{p_0}{\kappa}} \simeq p_0^2 - \mathbf{p}^2 - \frac{1}{\kappa} p_0 \mathbf{p}^2 \,, \tag{2.78}$$

where we have also indicated the first-order expansion in $\frac{1}{\kappa}$. In this scenario, the parameter κ gives the relativistically invariant energy scale. Invariance of this dispersion relation can be verified explicitly by performing a boost transformation according to:[8]

$$B_j^\xi \triangleright p_0 \equiv p_0 + \xi\{N_j, p_0\} = p_0 + \xi p_j \,,$$
$$B_j^\xi \triangleright p_i \equiv p_i + \xi\{N_j, p_i\} = p_i + \xi\delta_{ij}\left[\frac{\kappa}{2}\left(1 - e^{-2p_0/\kappa}\right) - \frac{1}{2\kappa}\mathbf{p}^2\right] - \xi\frac{1}{\kappa} p_i p_j \,,$$

in which ξ is the rapidity parameter.

When comparing the first-order expansion with (2.13), we see that the two expressions are equivalent upon setting $\frac{1}{\kappa} = \frac{\eta}{E_P}$. However, we are now going to show that the other ingredients of the DSR model inspired from the κ-Poincaré Hopf algebra, namely, the composition law and the deformed boosts, when considered at the first order in the deformation parameter, are not the same as the ones of the model discussed in Sect. 2.2, despite the two models sharing the same first-order dispersion relation. This is possible because when working to the first order in the deformation parameter the relativistic constraints leave open some degrees of freedom in the definition of the model [12,41], so that starting from the same deformed dispersion relation one can construct different relativistic models. The model we are discussing in this section is in principle valid to all orders in $\frac{1}{\kappa}$, even though it might still be the case that it only describes physics in a limited energy range (one could imagine that Nature is such that DSR models only describe the relativistic symmetries in a limited energy range, above which one might have a full breakdown of symmetries, or a restoration of special relativistic symmetries).

Other structures of the Hopf sub-algebra of translations are linked to the properties of the momentum space. Specifically, a deformed composition law of momenta is read off from the Hopf algebra coproduct:

$$\Delta(P_\mu)(p, q) = (p \oplus q)_\mu \,, \tag{2.79}$$

[8] We adopt a semiclassical approximation, so that symmetry generators act on the momentum space coordinates via Poisson brackets. The properties of the generators of the Hopf algebra are inherited by the Poisson brackets with the convention that, if $[G, f(P_\mu)] = ih(P_\mu)$, then $\{G, f(p_\mu)\} = h(p_\mu)$, for any generator G of the Hopf algebra. The functions f, h, take as argument the translation generators P_μ in the first case, and the momentum space coordinates p_μ in the second one. This approximation is justified in the "semiclassical" limit we mentioned in the previous footnote and further described in Sect. 2.4.

so that the deformed addition law $\oplus$ reads

$$\begin{aligned} E_a \oplus E_b &= E_a + E_b \,, \\ \mathbf{p}_a \oplus \mathbf{p}_b &= \mathbf{p}_a + e^{E_a/\kappa}\mathbf{p}_b \,. \end{aligned} \tag{2.80}$$

Notice that this is also compatible with the interpretation of the momentum space as a group manifold, so that the momentum composition is defined by the group multiplication law (2.62)–(2.63).

In contrast to the example considered in Sect. 2.2, here we have a composition law for spatial momenta that is noncommutative (because of the noncocommutativity of the coproduct) and associative (because of the coassociativity of the coproduct).

As we already mentioned briefly in Sect. 2.2, when the addition rule $\oplus$ is noncommutative then the momenta of each particle are boosted with rapidities that depend on the other particles' momenta, and this guarantees covariance of the addition law. The model we are describing in this section is an example of such behaviour. In fact, it is now well understood [44,45,63] that the addition law (2.80) is not covariant if each momentum is boosted with the same rapidity ξ:

$$p_a \to B^\xi \triangleright p_a \,, \quad p_b \to B^\xi \triangleright p_b \,. \tag{2.81}$$

In fact, calling $q \equiv p_a \oplus p_b$, it can be shown that:

$$B^\xi \triangleright q \neq (B^\xi \triangleright p_a) \oplus (B^\xi \triangleright p_b) \,. \tag{2.82}$$

? Exercise

2.6. Show explicitly that

$$B^\xi \triangleright q \neq (B^\xi \triangleright p_a) \oplus (B^\xi \triangleright p_b) \,. \tag{2.83}$$

What does work to achieve covariance of the addition law is to account for a "backreaction" of the individual momenta onto the rapidity ξ [44,45].[9] This is such

[9] A similar feature as the one we are discussing here for boosts exists for rotation transformations, see [44].

that the rapidity with which the second momentum transforms is affected by the first momentum:[10]

$$B^{\xi} \triangleright q = (B^{\xi} \triangleright p_a) \oplus (B^{\xi \triangleleft p_a} \triangleright p_b)\,, \tag{2.84}$$

where $\xi \triangleleft p_a \equiv e^{-E_a/\kappa}\xi$. As discussed in detail in [42], such backreaction does not identify a preferred frame of reference and is fully compatible with relativistic invariance.

? Exercise

2.7. Verify the covariance of the conservation law (2.80) under the transformation (2.84). Use [45] as guidance.

Since the model we are discussing in this section is linked to a Hopf algebra, we can understand such action of boosts on composed momenta in terms of the properties of the coproduct of the boost generator (2.75). In fact, one can interpret the backreaction (2.84) in terms of a law of "addition" of boost generators that dictates how composed momenta transform. The "total boost" generator

$$N_{[p_a \oplus p_b]} = N_{[p_a]} + e^{-E_a/\kappa} N_{[p_b]}, \tag{2.85}$$

is defined by the coproduct of the boost generator in the underlying Hopf algebra, Eq. (2.75). Here, the notation $N_{[p_a]}$ indicates that the transformation only acts on p_a and not on p_b. The "total boost" of rapidity ξ then has the following action on the momenta of each of the two interacting particles:

$$\begin{aligned} B^{\xi} \triangleright p_a &= p_a + \xi\{N_{[p_a \oplus p_b]}, p_a\} = p_a + \xi\{N_{[p_a]}, p_a\}\,, \\ B^{\xi} \triangleright p_b &= p_b + \xi\{N_{[p_a \oplus p_b]}, p_b\} = p_b + \xi e^{-E_a/\kappa}\{N_{[p_b]}, p_b\} \\ &= p_b + (\xi \triangleleft p_a)\{N_{[p_b]}, p_b\}\,. \end{aligned}$$

When considering the composition of several momenta, the rapidity acting on each of them receives a backreaction from all the other momenta that come before it in the composition law. Specifically, considering the addition of n momenta

$$p^{(1)} \oplus \ldots \oplus p^{(n)}\,, \tag{2.86}$$

[10] A more general expression applies when considering finite transformations [44]; however, here we only discuss the first order in ξ.

the rapidity with which the particle with momentum p^k is boosted reads:

$$\xi^{[p^{(k)}]} = \xi \triangleleft p^{(1)} \triangleleft \ldots \triangleleft p^{(k-1)} \equiv \xi \triangleleft (p^{(1)} \oplus \ldots \oplus p^{(k-1)})\,. \tag{2.87}$$

Again, we can interpret the backreaction in terms of the action of a total boost. Boosting each momentum with its own boost generator, $N_{[p^{(i)}]}$, and incorporating the backreaction on the rapidity, as explained above, is completely equivalent to boosting each momentum with the total boost generator [45]

$$N_{[\bigoplus_{i=1}^{n} p^{(i)}]} = N_{[p^{(1)}]} + e^{-E^{(1)}/\kappa} N_{[p^{(2)}]} + \cdots + e^{-\left(\sum_{i=1}^{n-1} E_0^{(i)}\right)/\kappa} N_{[p^{(n)}]}\,. \tag{2.88}$$

2.3.4 DSR on Curved Spacetime—The κ-(Anti) de Sitter Example

Most of the currently available DSR models, including the κ-Poincaré algebra described in this section, describe deformations of the relativistic transformations in flat spacetime. However, as we will discuss in Sect. 2.6, the most interesting phenomenological applications refer to the propagation of particles over cosmological distances, where the flat spacetime approximation is no longer valid.

Phenomenological studies aimed at extending DSR models to de Sitter or even Friedmann-Robertson-Walker spacetimes have been undertaken in relatively recent times [64–67], while some first exploratory studies were already performed more than a decade ago [18, 68]. An interesting line of investigation concerns the the generalization of the κ-Poincaré Hopf algebra to allow for a non-vanishing cosmological constant Λ. This leads to a quantum-deformed (Anti)-de Sitter Hopf algebra, known as κ-(A)dS [69, 70] and its associated non-commutative spacetime [71, 72].

These investigations agree on the fact that in general one should expect a nontrivial interplay between effects due to the quantum deformation and those due to spacetime curvature. For example, once the quantum deformation is taken into account the effects that are classically associated to space-time curvature acquire a new energy-dependence [64, 65, 68, 73, 74]. Moreover, when space-time curvature is present, the description of the geometrical properties of momentum space is non-trivial [75–78] and it turns out that one needs to account for an enlarged momentum space, which includes additional coordinates associated to "hyperbolic angular momentum". In $3+1$ dimensions, the geometry of these momentum spaces is half of the $(6+1)$- dimensional de Sitter space in the case of κ-dS, and half of a space with $SO(4, 4)$ invariance for κ-AdS [76].

We are not going to revisit the details of the derivation of the results we mentioned, since this would go beyond the scope of these notes. A thorough review and additional references can be found in [75].

Here we recall only some of the more interesting features concerning the interplay between the quantum deformation and curvature parameters, in order to illustrate the previous remarks. The rotations sector is deformed into a quantum $so(3)$ algebra with deformation parameter given by $\eta/\kappa = \sqrt{-\Lambda}/\kappa$:

$$\Delta(J_3) = J_3 \otimes 1 + 1 \otimes J_3,$$
$$\Delta(J_1) = J_1 \otimes e^{\frac{\eta}{\kappa}J_3} + 1 \otimes J_1, \qquad \Delta(J_2) = J_2 \otimes e^{\frac{\eta}{\kappa}J_3} + 1 \otimes J_2, \tag{2.89}$$

and whose deformed brackets read[11]

$$\{J_1, J_2\} = \frac{e^{2\frac{\eta}{\kappa}J_3} - 1}{2\eta/\kappa} - \frac{\eta}{2\kappa}\left(J_1^2 + J_2^2\right), \qquad \{J_1, J_3\} = -J_2, \qquad \{J_2, J_3\} = J_1\,. \tag{2.90}$$

The translations sector, that provides the deformed composition law for momenta in the corresponding DSR model, as seen for the κ-Poincaré case in the previous section, reads

$$\begin{aligned}
\Delta(P_0) &= P_0 \otimes 1 + 1 \otimes P_0,\\
\Delta(P_1) &= P_1 \otimes \cosh(\eta J_3/\kappa) + e^{-P_0/\kappa} \otimes P_1 - \eta K_2 \otimes \sinh(\eta J_3/\kappa)\\
&\quad - \frac{\eta}{\kappa} P_3 \otimes J_1 + \frac{\eta^2}{\kappa} K_3 \otimes J_2 + \frac{\eta^2}{\kappa^2}(\eta K_1 - P_2) \otimes J_1 J_2 e^{-\frac{\eta}{\kappa}J_3}\\
&\quad - \frac{\eta^2}{\kappa^2}(\eta K_2 + P_1) \otimes \left(J_1^2 - J_2^2\right) e^{-\frac{\eta}{\kappa}J_3},\\
\Delta(P_2) &= P_2 \otimes \cosh(\eta J_3/\kappa) + e^{-P_0/\kappa} \otimes P_2 + \eta K_1 \otimes \sinh(\eta J_3/\kappa)\\
&\quad - \frac{\eta}{\kappa} P_3 \otimes J_2 - \frac{\eta^2}{\kappa} K_3 \otimes J_1 - \frac{\eta^2}{\kappa^2}(\eta K_2 + P_1) \otimes J_1 J_2 e^{-\frac{\eta}{\kappa}J_3}\\
&\quad - \frac{1}{2}\frac{\eta^2}{\kappa^2}(\eta K_1 - P_2) \otimes \left(J_1^2 - J_2^2\right) e^{-\frac{\eta}{\kappa}J_3},\\
\Delta(P_3) &= P_3 \otimes 1 + e^{-P_0/\kappa} \otimes P_3 + \frac{1}{\kappa}\left(\eta^2 K_2 + \eta P_1\right) \otimes J_1 e^{-\frac{\eta}{\kappa}J_3}\\
&\quad - \frac{1}{\kappa}\left(\eta^2 K_1 - \eta P_2\right) \otimes J_2 e^{-\frac{\eta}{\kappa}J_3}\,.
\end{aligned} \tag{2.91}$$

Notice that the deformed composition law for momenta involves the full Lorentz sector, which indicates that the construction of the associated momentum needs to include the Lorentz sector, as we discussed at the beginning of this subsection.

[11] Here we are using Poisson brackets instead of commutators because we are taking the semiclassical limit which turns a Hops algebra into a Poisson-Lie algebra.

Moreover, the deformed brackets describe both non-commutativity due to spacetime curvature ($\eta \neq 0$) and quantum deformation:

$$
\begin{aligned}
\{P_1, P_2\} &= -\eta^2 \frac{\sinh(2\frac{\eta}{\kappa}J_3)}{2\eta/\kappa} - \frac{\eta}{2\kappa}\left(2P_3^2 + \eta^2(J_1^2 + J_2^2)\right) \\
&\quad - \frac{\eta^5}{4\kappa^3} e^{-2\frac{\eta}{\kappa}J_3}\left(J_1^2 + J_2^2\right)^2 , \\
\{P_1, P_3\} &= \frac{1}{2}\eta^2 J_2 \left(1 + e^{-2\frac{\eta}{\kappa}J_3}\left[1 + \frac{\eta^2}{\kappa^2}\left(J_1^2 + J_2^2\right)\right]\right) + \frac{\eta}{\kappa} P_2 P_3 , \\
\{P_2, P_3\} &= -\frac{1}{2}\eta^2 J_1 \left(1 + e^{-2\frac{\eta}{\kappa}J_3}\left[1 + \frac{\eta^2}{\kappa^2}\left(J_1^2 + J_2^2\right)\right]\right) - \frac{\eta}{\kappa} P_1 P_3 .
\end{aligned}
\tag{2.92}
$$

Finally, the Casimir of the algebra reads

$$
\begin{aligned}
C_{\kappa,\eta} &= 2\kappa^2\left[\cosh(P_0/\kappa)\cosh\left(\frac{\eta}{\kappa}J_3\right) - 1\right] + \eta^2\cosh(P_0/\kappa)(J_1^2 + J_2^2)e^{-\frac{\eta}{\kappa}J_3} \\
&\quad - e^{P_0/\kappa}\left(\mathbf{P}^2 + \eta^2\mathbf{K}^2\right)\left[\cosh\left(\frac{\eta}{\kappa}J_3\right) + \frac{\eta^2}{2\kappa^2}(J_1^2 + J_2^2)e^{-\frac{\eta}{\kappa}J_3}\right] \\
&\quad + 2\eta^2 e^{P_0/\kappa}\left[\frac{\sinh(\frac{\eta}{\kappa}J_3)}{\eta}\mathcal{R}_3 + \frac{1}{\kappa}\left(J_1\mathcal{R}_1 + J_2\mathcal{R}_2 + \frac{\eta}{2\kappa}(J_1^2 + J_2^2)\mathcal{R}_3\right)e^{-\frac{\eta}{\kappa}J_3}\right],
\end{aligned}
\tag{2.93}
$$

where $\mathcal{R}_a = \epsilon_{abc}K_b P_c$. As expected, in the $\kappa \to \infty$ limit we obtain the de Sitter Casimir, and in the $\eta \to 0$ limit, we obtain the κ-Poincaré Casimir in the bicrossproduct basis (2.58).

A derivation of the DSR model that would correspond to the κ-(A)dS algebra in $3+1$ dimensions is still missing, due to the difficulty of defining the phase space of particles when coordinates and momenta are intertwined in such a way. Preliminary studies on the propagation of free particles in $1+1$ dimensions were recently performed [73], see also Sect. 2.6.

2.4 Relative Locality

We have seen in the previous sections that DSR models are most naturally described in momentum space rather than in spacetime. Indeed, a commonly accepted view, also supported by the results concerning the emergence of DSR within more fundamental quantum gravity theories discussed in Sect. 2.1.1, is that DSR may characterize a semi-classical and "non-gravitational" regime of quantum gravity. That is, heuristically, the limit in which Newton and Planck constants are negligible, $G_N \to 0$ and $\hbar \to 0$, so that both quantum and gravitational effects are small, but their ratio $\hbar/G_N$ remains constant [79, 80]. In this regime, which is labeled the "relative-locality regime", for reasons that will be made clear in the following, modifications of standard physics governed by the Planck energy $E_P \sim \sqrt{\hbar/G_N} \neq 0$ can be present. Given the presence of this energy scale, it is natural to take as fundamental notion that of momentum space, assumed to be a pseudo-Riemannian manifold with an origin, a metric $g_{\mu\nu}(p)$, and a connection $\Gamma_\mu^{\nu\sigma}$, which can have torsion and non-metricity. The energy scale is then linked to the curvature scale of such manifold.

On the other hand, in this limit the Planck length $L_P \sim \sqrt{\hbar G_N}$ goes to zero, which smooths out possible small-scale properties of spacetime, such as noncommutativity. Nevertheless, this does not mean that we can describe spacetime as we usually do in general relativity or in quantum physics. In particular, the notion of locality becomes observer-dependent: the fact that two events take place at the same space-time point can only be established by observers close to the events themselves. The introduction of an observer-independent energy scale in DSR implies relativity of locality just as the introduction of an observer-independent speed scale in special relativity implies relativity of simultaneity.

The relative locality proposal [79, 80] resulted from a deepening in the understanding of the fate of the locality principle in DSR models [81–85]. Besides clarifying this important issue, the proposal also provided a more physical framework to understand the interpretation of the group manifold underlying the κ-Poincaré Hopf algebra as a curved momentum space, which was discussed in the previous section [44,45,61]. In fact, the relative-locality framework provides an interpretation based on the geometry of momentum space for deformations of on-shellness and of conservation laws of energy-momentum: the metric on momentum space is linked to the on-shell relation while the affine connection on momentum space is related to the law of composition of momenta, which enters into the laws of conservation of energy-momentum.

2.4.1 Geometry of Momentum Space

One-particle system measurements allow the observer to determine the metric of momentum space through the dispersion relation, linked to the square of the geodesic distance from the origin to a point p in momentum space corresponding to the momentum of the particle:

$$m^2 = D^2(0, p) = \int_0^1 ds \sqrt{g_{\mu\nu} \dot{p}^\mu \dot{p}^\nu}, \tag{2.94}$$

where p^μ is the solution of the geodesic equation for the metric $g_{\mu\nu}$ [61].

In order to allow us to construct a relativistic model for particle kinematics, the metric $g_{\mu\nu}$ must be maximally symmetric, leading to only three options: Minkowski, de Sitter, or anti-de Sitter metrics. In most cases studied so far, the metric is that of de Sitter [42, 44, 61]. Anti-de Sitter momentum spaces have been explored [86, 87], but it is not clear whether they lead to viable models. However, as we mentioned above, the momentum space manifold can have a non-metric connection, so that its geometry is not completely determined by the metric.

Determination of the addition rules of energy-momentum in particles interactions can be used to define the connection of the momentum space. This connection is in general non-metrical, in the sense that it is not the Levi-Civita connection given by the metric defined by the deformed dispersion relation.

Considering the momenta of two particles, the original proposal of [79,80] relies on the parallel transport of the momentum of one particle to the point in momentum space corresponding to the momentum of the other particle [42], and defines the connection via

$$\Gamma^{\tau\lambda}_{\nu}(k) = -\left.\frac{\partial^2 (p \oplus_k q)_\nu}{\partial p_\tau \partial q_\lambda}\right|_{p=q=k}, \tag{2.95}$$

where

$$(p \oplus_k q) \doteq k \oplus ((\ominus k \oplus p) \oplus (\ominus k \oplus q)), \tag{2.96}$$

and $\ominus$ is the so-called antipode operation of $\oplus$, which we already encountered in the previous section, such that $(\ominus p) \oplus p = (\oplus p) \ominus p = 0$. The antisymmetric part of this connection is the torsion, which is linked to the noncommutativity of the composition law

$$T^{\tau\lambda}_{\nu}(k) = -\left.\frac{\partial^2 \left((p \oplus_k q) - (q \oplus_k p)\right)_\nu}{\partial p_\tau \partial q_\lambda}\right|_{p=q=k}, \tag{2.97}$$

while non-associativity of the composition law determines the connection curvature

$$R^{\mu\nu\rho}_{\sigma}(k) = 2\left.\frac{\partial^3 \left((p \oplus_k q) \oplus_k r - p \oplus_k (q \oplus_k r)\right)_\sigma}{\partial p_{[\mu} \partial q_{\nu]} \partial r_\rho}\right|_{p=q=r=k}, \tag{2.98}$$

where the bracket denotes the anti-symmetrization. In [79, 88], the nonmetricity, defined from the metric and the connection as

$$N^{\mu\nu\rho} = \nabla^\rho g^{\mu\nu}(k), \tag{2.99}$$

is claimed to be responsible for the leading order time-delay effect in the arrival of photons from distant sources, see also Sect. 2.6.

Notice that because of the derivatives of the addition law appearing in the definition of the torsion, one may have non-commutative composition laws that still produce a symmetric (i.e. torsionless) connection. This can be traced back to the fact that, when going beyond the first order in the deformation, the definition (2.95) does not allow for a unique identification of the composition law based on a given connection (and in particular, symmetric connections can be associated to non-commutative composition laws). Motivated by this, an alternative proposal for the connection was provided in [42]. In this alternative proposal, there is a one-to-one correspondence between the composition law and the connection (at least up to second order in the deformation) and symmetric connections correspond to commutative composition laws. The drawback is that not all possible composition laws can be mapped to a connection, but only those satisfying a cyclic property [42]. The two proposals for the connection are equivalent when applied to the κ-Poincaré kinematics described in Sect. 2.3.3.

? Exercise

2.8. Compute the connection (2.95) associated to the composition law of the κ-Poincaré model, Eq. (2.80). Use [42, 44] as guidance.

The reason why a geometrical description of the relativistic kinematics is useful is that it may allow to characterize the property of kinematics on momentum space of being (DSR-)relativistic in terms of constraints on the geometry. We have already discussed one such example in the case of the dispersion relation and the momentum space metric. Concerning the connection, a thorough analysis is provided in [42].

2.4.2 Spacetime and Relativity of Locality

Some of the most relevant phenomenological applications of the deformed kinematics encoded in the relative locality proposal require that some notion of spacetime is provided. This is particularly important for studies of the time of flight, where one looks for a difference in the arrival time of particles with different energies emitted simultaneously by some astrophysical source (see Sect. 2.6).

The relative locality framework provides a proposal for a description of spacetime suited for this purpose, that is compatible with the deformed relativistic symmetries of the momentum space and allows for a description of the phase space of a single free particle, as well as in the more complex case of interacting particles. Notice that this last point is especially nontrivial. In fact, when the momentum space is curved, one can take the momentum space as the base manifold and construct spacetime as the cotangent space to the momentum space at a given point p.[12] Such notion is well defined in the case of one free particle, because the particle lives on one point of the curved momentum manifold, p. Then one can define the free particle dynamics in a canonical way, with the role of spacetime and momentum

[12] This is a completely analogous construction to the one of general relativity where momentum space is the cotangent space of the space-time manifold at a point in spacetime.

space exchanged with respect to the usual construction: spacetime coordinates x^μ are canonically conjugated to momenta via Poisson brackets,[13]

$$\{x^u, p_\nu\} = \delta^\mu_\nu \,, \tag{2.100}$$

and the dynamics of a free particle is described by the action

$$S^{\text{free}} = \int d\lambda \left(-x^\mu \dot{p}_\mu + \mathcal{N}\left(D(p)^2 - m^2 \right) \right) . \tag{2.101}$$

The over-dot indicates the derivative with respect to the affine parameter λ and the parameter $\mathcal{N}$ is a Lagrange multiplier enforcing on-shellness $D(p)^2 - m^2 = 0$ (see Eq. (2.94)). Variation of (2.101) with respect to x^μ and p_μ yields, respectively, conservation of momentum

$$\dot{p}_\mu = 0 \,, \tag{2.102}$$

and the evolution equation for the space-time coordinates:

$$\dot{x}^\mu = -\mathcal{N} \frac{\partial C}{\partial p_\mu} \,, \tag{2.103}$$

where $C(p) \equiv D(p)^2 - m^2$.

? Exercise

2.9. Show that, in $1+1$ dimensions, the on-shell relation and the constraint equations in the case of the κ-Poincaré model are (use [44, 45] for guidance):

$$m = \kappa \, \text{arccosh} \left(\cosh \frac{p_0}{\kappa} - e^{\frac{p_0}{\kappa}} \frac{(p_1)^2}{2\kappa^2} \right) , \tag{2.104}$$

$$\dot{p}_\mu = 0 \,, \tag{2.105}$$

$$\frac{\partial x^1}{\partial x^0} \equiv \frac{\dot{x}^1}{\dot{x}^0} = \frac{2\kappa p_1}{\kappa^2 \left(e^{-2p_0/\kappa} - 1 \right) + (p_1)^2} \,. \tag{2.106}$$

In the case of the κ-Poincaré model, after integrating the coordinates evolution (2.106) and using the on-shell relation (2.104), one finds the worldline [44]:

$$x^1(x^0) = x^1(0) + v(p) x^0 \,, \qquad v(p) = \frac{e^{p_0/\kappa} \sqrt{e^{2p_0/\kappa} + 1 - 2\, e^{p_0/\kappa} \cosh(m/\kappa)}}{1 - e^{p_0/\kappa} \cosh(m/\kappa)} \,. \tag{2.107}$$

[13] See [89] for alternative, but physically equivalent, prescriptions.

Because the momentum p_μ is a constant of motion, the spatial velocity $v(p)$ is constant as well. The linearity of the worldlines with respect to space-time coordinates indicates that spacetime is flat. The deformed expression for $v(p)$ can be attributed to the non-trivial geometry of momentum space. Notice that in the $\kappa \to \infty$ limit $v(p) = \frac{\sqrt{p_0^2 - m^2}}{p_0}$ as expected.

When several interacting particles are considered, it is not obvious what momentum to use in order to build the spacetime at the interaction event. For each particle I with momentum p^I, the construction outlined above requires a different set of space-time coordinates x_I^μ, each living on the cotangent space of the momentum manifold at a different point p^I and canonically conjugate to the associated momentum. If the particles are not interacting, the total action is given by the sum of the free actions of each particle:

$$S^{tot} = \sum_I S_I^{\text{free}} ,$$

$$S_I^{\text{free}} = \int_{-\infty}^{\infty} d\lambda \left(-x_I^\mu \dot{p}_\mu^I + N_I \left(D(p^I)^2 - m_I^2 \right) \right) . \tag{2.108}$$

If the particles are interacting, it does not make sense to ask that the coordinates x_I^μ take the same value for all I's at the interaction event, because the space-time coordinates of each particle live in different cotangent spaces. The solution provided within the relative locality framework is to introduce a boundary interaction term in the action, with a constraint that enforces momentum conservation at the interaction [90]. In the case of a single vertex (interaction among n incoming and m outgoing particles) the total action is:

$$S^{tot} = \sum_{I=1}^{n+m} S_I^{\text{free}} + S^{int} ,$$

$$S_I^{\text{free}} = \pm \int_{\lambda_0^I}^{\pm\infty} d\lambda \left(-x_I^\mu \dot{p}_\mu^I + N_I \left(D(p^I)^2 - m_I^2 \right) \right) ,$$

$$S^{\text{int}} = z^\mu \mathcal{K}_\mu (p_1(\lambda_0^1), \ldots, p_n(\lambda_0^n), p_{n+1}(\lambda_0^{n+1}), \ldots, p_m(\lambda_0^m)) , \tag{2.109}$$

where the $\pm$ sign is chosen according to whether the I-th particle is outgoing or incoming, λ_0^I is the value of the affine parameter at the endpoint of the worldline of each particle where the interaction occurs, and z^μ is a Lagrange multiplier enforcing the conservation law $\mathcal{K}_\mu(p_1(\lambda_0^1), \ldots, p_n(\lambda_0^n), p_{n+1}(\lambda_0^{n+1}), \ldots, p_m(\lambda_0^m)) = 0$. K_μ accounts for the deformed composition of momenta

$$p_1 \oplus \cdots \oplus p_n = p_{n+1} \oplus \cdots \oplus p_m , \tag{2.110}$$

and can take different forms compatible with this relation (see [45, 89]).

From varying the action one gets similar constraints for each interacting particle as those found for the free particle, Eqs. (2.104)–(2.106). Additionally, the interaction term yields an additional constraint on the endpoints of the worldlines at the interaction,

$$x_I^\mu(\lambda_0^I) = \mp z^\nu \left.\frac{\partial \mathcal{K}_\nu}{\partial p_\mu^I}\right|_{\lambda=\lambda_0}, \tag{2.111}$$

where the upper (lower) sign is for outgoing (incoming) particles.[14] In the case of special relativity, $\mathcal{K}_\mu = p_1 + \cdots + p_n - (p_{n+1} + \cdots + p_m)$, and all the worldlines simply end up at the interaction point $x_I^\mu = z^\mu$, so that the interaction is local. If the nonlinearity of momentum space induces nonlinear corrections to the composition law of momenta (as e.g. in (2.80)), then the worldlines will have in general different endpoints, since $\frac{\partial \mathcal{K}_\nu}{\partial p_\mu^I} \neq \frac{\partial \mathcal{K}_\nu}{\partial p_\mu^J}$. In this case, only if the interaction happens at $z^\mu = 0$ then all worldlines end at $x_I^\mu = z^\mu = 0$. If instead $z^\mu \neq 0$, then each worldline ends at a different value of x_I^μ. This is a manifestation of relative locality: only a local observer, $z^\mu = 0$, sees the interaction as local, while other observers, $z^\mu \neq 0$, see each worldline ending at a slightly different point. A more in-depth discussion can be found in [61], Section V.

It can be shown that this space-time picture is compatible with the deformed relativistic symmetries, see [45, 61]. In particular, the worldlines transform covariantly under translations and boosts, if the corresponding generators are taken to be the "total generators" in the sense already discussed in Sect. 2.3.3, see Eq. (2.88). Moreover, the "interaction coordinates" z^μ transform as the space-time coordinates of a single particle with momentum given by the total momentum of the vertex.

Notice that the action (2.109) can be further expanded to include several interaction vertices, sharing some of the particles involved. In doing so, however, one runs into what is known as "history problem" (sometimes also called "spectator problem"). Because the action is invariant under the action of "total generators", one needs to know the whole sequence of causally connected vertices in order to correctly define such operators and transform any individual vertex [45]. Understanding how to solve this issue is still an open problem currently under study.

2.5 Deformed Kinematics on Curved Momentum Space

In the previous section we have discussed a possible way to describe a curved momentum space which takes into account a relativistic deformed kinematics. In that approach, the starting ingredients are the free particle energy-momentum dispersion relation and the addition law of momenta in interactions. From these,

[14] This result was recently rederived using a line element in phase space for a multi-particle system in [91].

one can derive the geometrical properties of the momentum space (metric and connection) following the prescriptions discussed in Sect. 2.4.1. In this section we develop a different perspective [92]: we start from the geometry of a maximally symmetric curved momentum space and derive all the ingredients of a deformed kinematics, preserving a relativity principle [12,41–43].

2.5.1 Definition of the Deformed Kinematics

A relativistic deformed kinematics (we will see in Sect. 2.5.2 that this construction preserves a relativity principle) can be obtained by identifying the isometries of the momentum space metric with the composition law and the Lorentz transformations in the one-particle system, fixing the dispersion relation.

It is well known that in a four dimensional maximally symmetric space, there are 10 isometries [93]. An isometry is a transformation $k \to k'$ such that, when acting on a momentum metric $g_{\mu\nu}(k)$, does not changes the form of the metric, i.e.,

$$g_{\mu\nu}(k') = \frac{\partial k'_\mu}{\partial k_\rho} \frac{\partial k'_\nu}{\partial k_\sigma} g_{\rho\sigma}(k) . \tag{2.112}$$

By choosing a system of coordinates such that $g_{\mu\nu}(0) = \eta_{\mu\nu}$, we can write the isometries as

$$k'_\mu = [T_a(k)]_\mu = T_\mu(a, k) , \qquad k'_\mu = [J_\omega(k)]_\mu = J_\mu(\omega, k) , \tag{2.113}$$

where a is a set of four parameters and ω of six, and

$$T_\mu(a, 0) = a_\mu , \qquad J_\mu(\omega, 0) = 0 . \tag{2.114}$$

Here $J_\mu(\omega, k)$ are the 6 Lorentz isometries (three rotations and three boosts) which form a subgroup (Lorentz algebra), leaving the origin in momentum space invariant. On the other hand, $T_\mu(a, k)$ are the other 4 isometries associated to translations which transform the origin. This idea was also considered in [87] but, as we will see, there is some arbitrariness that needs to be fixed in order to obtain the desired kinematics.

Therefore, the isometries $k'_\mu = J_\mu(\omega, k)$ are the Lorentz transformations of the one-particle system, being ω the six parameters of a Lorentz transformation. In order to define the dispersion relation $C_\kappa(k)$, we can use any arbitrary function of the distance from the origin to a point k, in such a way that special relativity is recovered when taking the limit in which the high-energy scale tends to infinity[15]. Since the distance is invariant under a Lorentz transformation, the equality $C_\kappa(k) = C_\kappa(k')$

[15] Note that in the previous section the squared distance was identified with the squared of the distance in momentum space, but any function of the Casimir will be also a Casimir.

holds, which allows us to determine the Casimir directly from $J_\mu(\omega, k)$ without computing the explicit form of the distance:

$$\frac{\partial C_\kappa(k)}{\partial k_\mu} \mathcal{J}^{\alpha\beta}_\mu(k) = 0. \tag{2.115}$$

The other 4 isometries $k'_\mu = T_\mu(a, k)$ are related with translations and define the composition law $p \oplus q$ of two momenta p, q through

$$(p \oplus q)_\mu \doteq T_\mu(p, q). \tag{2.116}$$

Indeed one can see that this composition is related with the composition of translations

$$p \oplus q = T_p(q) = T_p(T_q(0)) = (T_p \circ T_q)(0). \tag{2.117}$$

In the following, we will discuss the possible different definitions of translations for a given metric.

Then, the deformed kinematics can be obtained from a momentum metric by

$$\begin{aligned} g_{\mu\nu}(T_a(k)) &= \frac{\partial T_\mu(a,k)}{\partial k_\rho} \frac{\partial T_\nu(a,k)}{\partial k_\sigma} g_{\rho\sigma}(k), \\ g_{\mu\nu}(J_\omega(k)) &= \frac{\partial J_\mu(\omega,k)}{\partial k_\rho} \frac{\partial J_\nu(\omega,k)}{\partial k_\sigma} g_{\rho\sigma}(k). \end{aligned} \tag{2.118}$$

These equations must be satisfied for any a, ω. One can see, from the limit $k \to 0$ in (2.118)

$$\begin{aligned} g_{\mu\nu}(a) &= \left[\lim_{k\to 0} \frac{\partial T_\mu(a,k)}{\partial k_\rho}\right] \left[\lim_{k\to 0} \frac{\partial T_\nu(a,k)}{\partial k_\sigma}\right] \eta_{\rho\sigma}, \\ \eta_{\mu\nu} &= \left[\lim_{k\to 0} \frac{\partial J_\mu(\omega,k)}{\partial k_\rho}\right] \left[\lim_{k\to 0} \frac{\partial J_\nu(\omega,k)}{\partial k_\sigma}\right] \eta_{\rho\sigma}, \end{aligned} \tag{2.119}$$

that

$$\lim_{k\to 0} \frac{\partial T_\mu(a,k)}{\partial k_\rho} = \delta^\rho_\alpha e^\alpha_\mu(a), \qquad \lim_{k\to 0} \frac{\partial J_\mu(\omega,k)}{\partial k_\rho} = L^\rho_\mu(\omega), \tag{2.120}$$

where $e^{\alpha}_{\mu}(k)$ is the inverse of the tetrad of the momentum space metric,[16] and $L^{\rho}_{\mu}(\omega)$ is the standard Lorentz transformation matrix with parameters ω. From Eq. (2.116) and Eq. (2.120), one obtains

$$\lim_{k\to 0} \frac{\partial (a\oplus k)_\mu}{\partial k_\rho} = \delta^{\rho}_{\alpha} e^{\alpha}_{\mu}(a) \,, \tag{2.121}$$

which establishes a relationship between the composition law and the tetrad.

We can write, for infinitesimal transformations

$$T_\mu(\epsilon, k) = k_\mu + \epsilon_\alpha \mathcal{T}^{\alpha}_{\mu}(k) \,, \qquad J_\mu(\epsilon, k) = k_\mu + \epsilon_{\beta\gamma} \mathcal{J}^{\beta\gamma}_{\mu}(k) \,, \tag{2.122}$$

then Eq. (2.118) becomes

$$\frac{\partial g_{\mu\nu}(k)}{\partial k_\rho} \mathcal{T}^{\alpha}_{\rho}(k) = \frac{\partial \mathcal{T}^{\alpha}_{\mu}(k)}{\partial k_\rho} g_{\rho\nu}(k) + \frac{\partial \mathcal{T}^{\alpha}_{\nu}(k)}{\partial k_\rho} g_{\mu\rho}(k) \,, \tag{2.123}$$

$$\frac{\partial g_{\mu\nu}(k)}{\partial k_\rho} \mathcal{J}^{\beta\gamma}_{\rho}(k) = \frac{\partial \mathcal{J}^{\beta\gamma}_{\mu}(k)}{\partial k_\rho} g_{\rho\nu}(k) + \frac{\partial \mathcal{J}^{\beta\gamma}_{\nu}(k)}{\partial k_\rho} g_{\mu\rho}(k) \,, \tag{2.124}$$

which define the Killing vectors $\mathcal{J}^{\beta\gamma}$, but do not completely determine $\mathcal{T}^{\alpha}$. This can be understood from the fact that if $\mathcal{T}^{\alpha}$, $\mathcal{J}^{\beta\gamma}$ are a solution of the Killing equations (2.123)–(2.124), then $\mathcal{T}'^{\alpha} = \mathcal{T}^{\alpha} + c^{\alpha}_{\beta\gamma}\mathcal{J}^{\beta\gamma}$ is also a solution of Eq. (2.123) for any arbitrary constants $c^{\alpha}_{\beta\gamma}$, and then, $T'_\mu(\epsilon, 0) = T_\mu(\epsilon, 0) = \epsilon_\mu$, where $T'_\mu(\epsilon, k) = k_\mu + \epsilon_\alpha \mathcal{T}'^{\alpha}_{\mu}(k)$. We can eliminate this ambiguity by taking into account that the isometry generators close an algebra [94]. Therefore we can ask the isometry generators, written as

$$T^{\alpha} = x^{\mu} \mathcal{T}^{\alpha}_{\mu}(k), \qquad J^{\alpha\beta} = x^{\mu} \mathcal{J}^{\alpha\beta}_{\mu}(k) \,, \tag{2.125}$$

which lead to the Poisson brackets

$$\{T^{\alpha}, T^{\beta}\} = x^{\rho} \left(\frac{\partial \mathcal{T}^{\alpha}_{\rho}(k)}{\partial k_\sigma} \mathcal{T}^{\beta}_{\sigma}(k) - \frac{\partial \mathcal{T}^{\beta}_{\rho}(k)}{\partial k_\sigma} \mathcal{T}^{\alpha}_{\sigma}(k) \right) , \tag{2.126}$$

$$\{T^{\alpha}, J^{\beta\gamma}\} = x^{\rho} \left(\frac{\partial \mathcal{T}^{\alpha}_{\rho}(k)}{\partial k_\sigma} \mathcal{J}^{\beta\gamma}_{\sigma}(k) - \frac{\partial \mathcal{J}^{\beta\gamma}_{\rho}(k)}{\partial k_\sigma} \mathcal{T}^{\alpha}_{\sigma}(k) \right) , \tag{2.127}$$

to close a particular algebra. Note that x^{μ} are canonically conjugated variables of k_ν, satisfying the Poisson brackets of (2.100). This ambiguity in defining the translations is just the ambiguity in the choice of the isometry algebra, leading

[16] Note that the metric $g_{\mu\nu}$ is the inverse of $g^{\mu\nu}$.

each choice to a different composition law, and therefore to different relativistic deformed kinematics. Note that the dispersion relation is univocally defined once the metric is given, while the composition law can take different forms depending on the choice of the generators of translations T^α, leading to different kind of deformed kinematics (see [41, 43] for a systematic way of constructing deformed kinematics order by order in the high-energy scale), as we will see in the following.

2.5.2 Relativistic Deformed Kinematics

In this part we demonstrate that the previously defined kinematics are in fact relativistic. This can be understood from the following diagram:

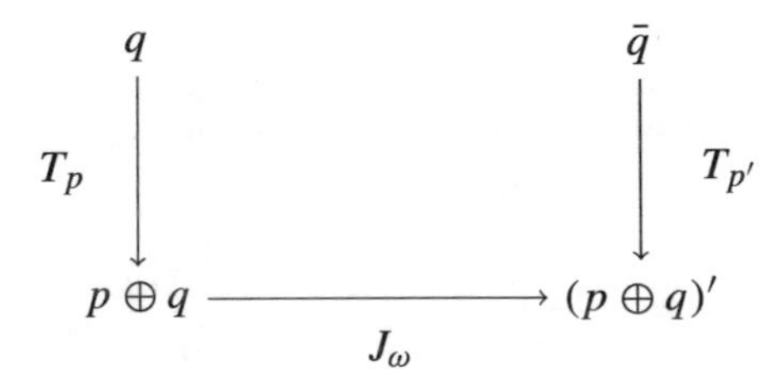

where the momentum with prime denotes the transformation through $\mathcal{J}_\omega$, and T_p, $T_{p'}$ are the translations with parameters p and p', respectively. We can define $\bar{q}$ as a point in momentum spaces satisfying

$$(p \oplus q)' = (p' \oplus \bar{q}) \,. \tag{2.128}$$

From this definition it is easy to note that for $q = 0$, also $\bar{q} = 0$, and for any other value $q \neq 0$, the point $\bar{q}$ is obtained from q by an isometry, being this a composition of the translation T_p, a Lorentz transformation J_ω, and the inverse of the translation $T_{p'}$. This transformation is obviously an isometry, due to the fact that the isometries are a group of transformations, and therefore, any composition of isometries is also an isometry. Since we have proved that there is an isometry from $q \to \bar{q}$, leaving the origin invariant, then the distance of both points to the origin are the same, which is tantamount to say

$$C_\kappa(q) = C_\kappa(\bar{q}) \,. \tag{2.129}$$

From Eqs. (2.128)–(2.129) one can see that the deformed kinematics with ingredients C and $\oplus$ is a relativistic deformed kinematics when identifying the momenta $(p', \bar{q})$ as the two-particle Lorentz transformation of (p, q). Indeed, Eq. (2.128) implies that the composition law is invariant under the previously defined Lorentz transformation and Eq. (2.129), together with $C_\kappa(p) = C_\kappa(p')$, that the deformed dispersion relation of both momenta is also Lorentz invariant. From this definition of the two-particle Lorentz transformations, one of the points (p) transforms as a single momentum, but for the other one (q) the transformation generally depends of both momenta (which indeed is the case of several examples obtained within Hopf algebras [44,45,55]).

In the literature, it was also considered a deformed dispersion relation, and some Lorentz transformations (in the one-particle system) compatible with it, in the geometrical context. In particular, one way to consider such modification of the special relativistic kinematics is through a velocity or momentum dependent spacetime, known as Finsler [95] and Hamilton [96] geometries, respectively. The case of Finsler geometries was considered in [97–99], while [100–102] were devoted to Hamilton spaces. In this geometrical constructions a clear connection with a deformed composition law and two-particle Lorentz transformations is still missing.

2.5.3 κ-Poincaré Relativistic Kinematics

Here we show how the kinematics of κ-Poincaré defined in Sect. 2.3.3 can be obtained from the previous prescription, and explain how other models can also be defined in this context.

Let us consider an isotropic kinematics, for which the general form of the algebra of the generators of isometries must be

$$\{T^0, T^i\} = \frac{c_1}{\kappa} T^i + \frac{c_2}{\kappa^2} J^{0i}, \qquad \{T^i, T^j\} = \frac{c_2}{\kappa^2} J^{ij}, \tag{2.130}$$

where we impose the generators $J^{\alpha\beta}$ to satisfy the standard Lorentz algebra, and because of the fact that isometries are a group, the Poisson brackets of T^α and $J^{\beta\gamma}$ are fixed by Jacobi identities. Different algebras of the generators of translations, i.e., different choices of the coefficients (c_1/κ) and (c_2/κ^2) will lead to different composition laws.

For the simple case where $c_2 = 0$ in Eq. (2.130), the generators of translations close a subalgebra[17]

$$\{T^0, T^i\} = \pm\frac{1}{\kappa} T^i . \tag{2.131}$$

A well known result of differential geometry (see Ch.6 of Ref. [103]) is that, when the generators of left-translations T^α transforming $k \to T_a(k) = (a \oplus k)$ form a Lie algebra, also the generators of right-translations $\tilde{T}^\alpha$ transforming $k \to (k \oplus a)$, close the same algebra but with a different sign

$$\{\tilde{T}^0, \tilde{T}^i\} = \mp\frac{1}{\kappa} \tilde{T}^i . \tag{2.132}$$

[17] We have reabsorbed the coefficient c_1 in the scale κ.

We have found the explicit relation between the infinitesimal right-translations and the tetrad of the momentum metric in Eq. (2.121), which gives

$$(k \oplus \epsilon)_\mu = k_\mu + \epsilon_\alpha e^\alpha_\mu \equiv \tilde{T}_\mu(k, \epsilon)\,. \tag{2.133}$$

Comparing with Eqs. (2.122) and (2.125), we see that right-translation generators are given by

$$\tilde{T}^\alpha = x^\mu e^\alpha_\mu(k)\,. \tag{2.134}$$

Both algebras (2.131)–(2.132) satisfy κ-Minkowski noncommutativity (2.34), so the problem of finding a tetrad $e^\alpha_\mu(k)$ fulfilling the algebra of Eq. (2.132) is tantamount to obtaining a representation of this noncommutativity written in terms of canonical coordinates of the phase space. As a particular solution, one can see that the following choice of the tetrad

$$e^0_0(k) = 1\,, \qquad e^0_i(k) = e^i_0(k) = 0\,, \qquad e^i_j(k) = \delta^i_j e^{\mp k_0/\kappa}\,, \tag{2.135}$$

leads to a representation of κ-Minkowski noncommutativity.

For obtaining the finite translations $T_\mu(a, k)$, which form a subgroup inside the isometry group, Eq. (2.120) can be generalized to defining a transformation that does not change the form of the tetrad:

$$e^\alpha_\mu(T(a, k)) = \frac{\partial T_\mu(a, k)}{\partial k_\nu} e^\alpha_\nu(k)\,. \tag{2.136}$$

If $T_\mu(a, k)$ is a solution to the previous equation, leaving invariant the form of the tetrad, the metric will also be invariant, so it is then an isometry. These previously defined translations form a group, since the composition of two transformations also leaves the tetrad invariant. Indeed, we can solve Eq. (2.136) for the tetrad in Eq. (2.135), obtaining

$$T_0(a, k) = a_0 + k_0, \qquad T_i(a, k) = a_i + k_i e^{\mp a_0/\kappa}\,, \tag{2.137}$$

so the deformed addition (or composition) law (DCL) is

$$(p \oplus q)_0 = T_0(p, q) = p_0 + q_0\,, \qquad (p \oplus q)_i = T_i(p, q) = p_i + q_i e^{\mp p_0/\kappa}\,, \tag{2.138}$$

which is the one obtained in the bicrossproduct basis of κ-Poincaré kinematics (2.66) (up to a sign depending on the choice of the initial sign of κ in Eq. (2.135)).

? Exercise

2.10. Derive Eqs. (2.137), (2.138) from (2.136) and (2.135) explicitly.

From (2.115) one can obtain the dispersion relation, where $\mathcal{J}^{\alpha\beta}$ are the infinitesimal Lorentz transformations satisfying Eq. (2.124) with the metric $g_{\mu\nu}(k) = e^{\alpha}_{\mu}(k)\eta_{\alpha\beta}e^{\beta}_{\nu}(k)$ defined by the tetrad (2.135):

$$0 = \frac{\mathcal{J}^{\alpha\beta}_0(k)}{\partial k_0}, \quad 0 = -\frac{\mathcal{J}^{\alpha\beta}_0(k)}{\partial k_i}e^{\mp 2k_0/\kappa} + \frac{\mathcal{J}^{\alpha\beta}_i(k)}{\partial k_0}, \\ \pm\frac{2}{\kappa}\mathcal{J}^{\alpha\beta}_0(k)\delta_{ij} = -\frac{\partial\mathcal{J}^{\alpha\beta}_i(k)}{\partial k_j} - \frac{\partial\mathcal{J}^{\alpha\beta}_j(k)}{\partial k_i}. \tag{2.139}$$

One finally gets

$$\mathcal{J}^{0i}_0(k) = -k_i, \qquad \mathcal{J}^{0i}_j(k) = \pm\delta^i_j\frac{\kappa}{2}\left[e^{\mp 2k_0/\kappa} - 1 - \frac{\mathbf{k}^2}{\kappa^2}\right] \pm \frac{k_i k_j}{\kappa}, \tag{2.140}$$

which is equivalent to (2.53),and so

$$C_\kappa(k) = \kappa^2\left(e^{k_0/\kappa} + e^{-k_0/\kappa} - 2\right) - e^{\pm k_0/\kappa}\mathbf{k}^2, \tag{2.141}$$

which is the same function of the momentum which defines the dispersion relation of κ-Poincaré kinematics in the bicrossproduct basis (2.58) (up to the sign in κ).

Finally, using the diagram of Sect. 2.5.2, we can find $\bar{q}$ satisfying

$$(p \oplus q)' = p' \oplus \bar{q}. \tag{2.142}$$

Equating both expressions and taking only the linear terms in $\epsilon_{\alpha\beta}$ (parameters of the infinitesimal Lorentz transformation) one arrives to the equation

$$\epsilon_{\alpha\beta}\mathcal{J}^{\alpha\beta}_\mu(p \oplus q) = \epsilon_{\alpha\beta}\frac{\partial(p \oplus q)_\mu}{\partial p_\nu}\mathcal{J}^{\alpha\beta}_\nu(p) + \frac{\partial(p \oplus q)_\mu}{\partial q_\nu}(\bar{q}_\nu - q_\nu). \tag{2.143}$$

From the composition law of (2.138) with the minus sign, we find

$$\frac{\partial(p\oplus q)_0}{\partial p_0}=1\,,\quad \frac{\partial(p\oplus q)_0}{\partial p_i}=0\,,$$
$$\frac{\partial(p\oplus q)_i}{\partial p_0}=-\frac{q_i}{\kappa}e^{-p_0/\kappa}\,,\quad \frac{\partial(p\oplus q)_i}{\partial p_j}=\delta_i^j\,, \tag{2.144}$$

$$\frac{\partial(p\oplus q)_0}{\partial q_0}=1\,,\quad \frac{\partial(p\oplus q)_0}{\partial q_i}=0\,,\quad \frac{\partial(p\oplus q)_i}{\partial q_0}=0\,,\quad \frac{\partial(p\oplus q)_i}{\partial q_j}=\delta_i^j\,e^{-p_0/\kappa}\,. \tag{2.145}$$

Therefore, we obtain

$$\begin{aligned}\bar{q}_0 &= q_0+\epsilon_{\alpha\beta}\left[\mathcal{J}_0^{\alpha\beta}(p\oplus q)-\mathcal{J}_0^{\alpha\beta}(p)\right],\\ \bar{q}_i &= q_i+\epsilon_{\alpha\beta}\,e^{p_0/\kappa}\left[\mathcal{J}_i^{\alpha\beta}(p\oplus q)-\mathcal{J}_i^{\alpha\beta}(p)+\frac{q_i}{\kappa}e^{-p_0/\kappa}\mathcal{J}_0^{\alpha\beta}(p)\right],\end{aligned} \tag{2.146}$$

and one can check that this is the Lorentz transformation of the two-particle system of κ-Poincaré in the bicrossproduct basis (2.75).

For the choice of the tetrad in Eq. (2.135), the metric in momentum space reads[18]

$$g_{00}(k)=1\,,\qquad g_{0i}(k)=g_{i0}(k)=0\,,\qquad g_{ij}(k)=-\delta_{ij}e^{\mp 2k_0/\kappa}\,. \tag{2.147}$$

? Exercise

2.11. Show that (2.147) is a de Sitter momentum space metric with curvature $(12/\kappa^2)$.

This shows that the κ-Poincaré kinematics in the bicrossproduct basis [104] can be completely obtained from the geometric ingredients of a de Sitter momentum space with the choice of the tetrad of Eq. (2.135). By using different choices of tetrad, such that the generators of Eq. (2.134) close the algebra Eq. (2.132), one can find the κ-Poincaré kinematics in different bases. Therefore, the different bases of κ-Poincaré can be geometrically interpreted as different choices of coordinates in de Sitter space.

Different relativistic kinematics, outside the Hopf algebra scheme, can be obtained in the aforementioned framework.

[18] This is the de Sitter metric written in the comoving coordinate system used in Refs. [44, 92].

? Exercise

2.12. Snyder kinematics is a very particular kinematics from the point of view of Lorentz symmetry. Indeed, it is compatible with linear Lorentz invariance in both one- and two-particle systems. The deformed addition law is given by [105]

$$(p\oplus q)_\mu = p_\mu\left(\sqrt{1+\frac{q^2}{\Lambda^2}}+\frac{p_\rho\eta^{\rho\nu}q_\nu}{\Lambda^2\left(1+\sqrt{1+p^2/\Lambda^2}\right)}\right)+q_\mu\,. \tag{2.148}$$

Show that Snyder kinematics can be derived imposing $c_1 = 0$ in Eq. (2.130).

Moreover, the kinematics known as hybrid models [106] can be obtained when c_1, c_2 are non-zero. As a final note, it is important to notice that, with the construction discussed here, different kinematics (with different composition laws) are related to the same metric, and therefore, also to the same dispersion relation.

2.5.3.1 Comparison with Previous Works

In this section, we will compare the prescription followed in this section with the one proposed in Ref. [90]. This comparison can only be carried out for the κ-Poincaré kinematics, since as we will see, the associativity property of the composition law plays a crucial role.

In order to make the comparison, we consider the derivative of Eq. (2.136) with respect to p_τ , and display it in terms of the deformed addition law

$$\frac{\partial e^\alpha_\nu(p\oplus q)}{\partial p_\tau}=\frac{\partial e^\alpha_\nu(p\oplus q)}{\partial(p\oplus q)_\sigma}\frac{\partial(p\oplus q)_\sigma}{\partial p_\tau}=\frac{\partial^2(p\oplus q)_\nu}{\partial p_\tau\partial q_\rho}e^\alpha_\rho(q)\,. \tag{2.149}$$

One can find the second derivative of the deformed addition law

$$\frac{\partial^2(p\oplus q)_\nu}{\partial p_\tau\partial q_\rho}=e^\rho_\alpha(q)\frac{\partial e^\alpha_\nu(p\oplus q)}{\partial(p\oplus q)_\sigma}\frac{\partial(p\oplus q)_\sigma}{\partial p_\tau}\,, \tag{2.150}$$

where e^ν_α is the inverse of e^α_ν, $e^\alpha_\nu e^\mu_\alpha=\delta^\mu_\nu$. But also using Eq. (2.136), one has

$$e^\rho_\alpha(q)=\frac{\partial(p\oplus q)_\mu}{\partial q_\rho}e^\mu_\alpha(p\oplus q)\,, \tag{2.151}$$

and then

$$\frac{\partial^2(p\oplus q)_\nu}{\partial p_\tau\partial q_\rho}+\Gamma^{\sigma\mu}_\nu(p\oplus q)\frac{\partial(p\oplus q)_\sigma}{\partial p_\tau}\frac{\partial(p\oplus q)_\mu}{\partial q_\rho}=0\,, \tag{2.152}$$

where

$$\Gamma^{\sigma\mu}_{\nu}(k) \doteq -e^{\mu}_{\alpha}(k)\,\frac{\partial e^{\alpha}_{\nu}(k)}{\partial k_\sigma}\,. \tag{2.153}$$

? Exercise

2.13. Check that the combination of tetrads and derivatives appearing in Eq. (2.153) in fact transforms like a connection [93] under a change of momentum coordinates.

In Ref. [42], it is proposed another way to define a connection and a DCL in momentum space through parallel transport, establishing a link between these two ingredients. It is easy to check that the DCL obtained in this way satisfies Eq. (2.152). This equation only determines the DCL for a given connection if one imposes the associativity property of the composition. Comparing with the previous reference, one then concludes that the DCL obtained from translations that leaves the form of the tetrad invariant is the associative composition law one finds by parallel transport, with the connection constructed from a tetrad and its derivatives as in Eq. (2.153).

Finally, if the DCL is associative, then Eq. (2.96) reduces to

$$(p \oplus_k q) = p \oplus \hat{k} \oplus q\,. \tag{2.154}$$

Replacing q by $(\hat{k} \oplus q)$ in Eq. (2.152), which is valid for any momenta (p, q), one obtains

$$\frac{\partial^2 (p \oplus \hat{k} \oplus q)_\nu}{\partial p_\tau \partial(\hat{k} \oplus q)_\rho} + \Gamma^{\sigma\mu}_{\nu}(p \oplus \hat{k} \oplus q)\frac{\partial(p \oplus \hat{k} \oplus q)_\sigma}{\partial p_\tau}\frac{\partial(p \oplus \hat{k} \oplus q)_\mu}{\partial(\hat{k} \oplus q)_\rho} = 0\,. \tag{2.155}$$

Multiplying by $\partial(\hat{k} \oplus q)_\rho/\partial q_\lambda$, one finds

$$\frac{\partial^2 (p \oplus \hat{k} \oplus q)_\nu}{\partial p_\tau \partial q_\lambda} + \Gamma^{\sigma\mu}_{\nu}(p \oplus \hat{k} \oplus q)\frac{\partial(p \oplus \hat{k} \oplus q)_\sigma}{\partial p_\tau}\frac{\partial(p \oplus \hat{k} \oplus q)_\mu}{\partial q_\lambda} = 0\,. \tag{2.156}$$

Taking $p = q = k$ in Eq. (2.156), one finally gets

$$\Gamma^{\tau\lambda}_{\nu}(k) = -\left.\frac{\partial^2 (p \oplus_k q)_\nu}{\partial p_\tau \partial q_\lambda}\right|_{p,q\to k}, \tag{2.157}$$

which is the same expression of Eq. (2.95) proposed in Ref. [90]. This concludes that the connection of Eq. (2.153) constructed from the tetrad is the same connection given by the prescription developed in Ref. [90] when the DCL is associative.

2.6 Phenomenological Consequences

Given the current stage of development of DSR, which we have described in the previous sections, phenomenological studies can rely on a framework to describe kinematics, while a full theory capable of describing the dynamical features of DSR is still missing. In this context, the two main avenues of investigation concern particle propagation effect and effects due to the modified kinematics in interactions.

Particle propagation effects have played a prominent role in the birth and the development of phenomenological studies in quantum gravity in general [3, 107, 108], and are very relevant for the phenomenology of DSR models in particular [109, 110]. As we are going to discuss more in detail in the following, the deformed kinematics described by DSR may lead to in-vacuo dispersion in particle propagation, producing shifts in the time of arrival of photons with different energies emitted simultaneously by the same source. Even though these effects will be in general suppressed by the ratio between the particle's energy and the Planck energy, they could be amplified significantly over large propagation distances, such as those characterizing astrophysical sources. In fact, sensitivity of astrophysical observations allows for meaningful constraints on Planck-scale suppressed time shifts, see Chap. 6. Another class of propagation effects which may be present in DSR models is known as "dual lensing" [111–113], and is such that the apparent direction from which astrophysical particles are emitted depends on their energy. This kind of effect has until now received less attention than the time shift, because current experiments have a limited source localisation capabilities. However, future multi-satellite telescopes will significantly improve space-time localization of sources with respect to traditional telescopes, possibly opening a window on dual lensing investigation.

Since the time shift effect is at the moment the one that is receiving the largest attention in theoretical and phenomenological studies, we will focus on that in the following before we briefly discuss modifications of interaction effects.

2.6.1 Propagation Effects: Time Shift

As we mentioned in the introduction, DSR models where originally conceived to provide a relativistic framework to encode Planck-scale modified energy-momentum dispersion relation which can induce potentially observable corrections to particles' speed of propagation.

In the subsequent theoretical developments it was progressively understood that the emergence of the sort of effects that provided the original motivation is not a necessary consequence of DSR, and even when such time shift effects do emerge they can take different quantitative dependence on the relevant quantities at play, including space-time curvature, which was ignored in the first studies.

So while the original idea concerned quite a specific kind of propagation effect, these more recent findings provide us with a range of possibilities for effects we can search for in astrophysical data. On the one hand, it is important to be aware of all the possibilities, in order to make sure not to miss a discovery opportunity. On the other hand, when some of these effects are excluded by experimental analyses, we get an essential guidance on the construction of a consistent DSR model that is compatible with observations. While one might worry that the variety of possibilities that at the moment seem to be compatible with the DSR framework might imply a lack of predictivity, one should consider that longitudinal propagation effects are not the only ones that can emerge in this framework (we already mentioned transverse propagation effects and effects on interactions), and that relativistic compatibility imposes compatibility conditions between the different effects.

The first DSR phenomenological studies were based on models such as the ones described in Sect. 2.2, and in that context the velocity of particles was simply deduced from the modified energy-momentum dispersion via $v = \partial E/\partial p$. The first studies of the phenomenological consequence of Hopf-algebra deformations of relativistic symmetries [31, 36] observed that the group velocity of plane waves defined in this framework would imply a momentum-dependent velocity for massless particles. This conclusion, which relied on the assumption that in the Hopf algebra framework one could still define the group velocity as $v = \frac{\partial E}{\partial p}$, was challenged by studies claiming that alternative definitions of the velocity should be used [114–117], leading to standard propagation velocity. The validity of $v = \partial E/\partial p$ ultimately relies on the assumption that a Hamiltonian description is still available, such that $v \equiv dx/dt = \{x, H(p)\}$, where $\{...\}$ are Poisson brackets, and that phase space coordinates satisfy the usual relation $\{x, p\} = 1$, so that $x = \partial/\partial p$.

For this reason, subsequent studies [83] relied on a covariant Hamiltonian formalism to derive the particle's worldline from

$$\dot{x}^0 = \{x^0, C\}, \tag{2.158}$$

$$\dot{x}^i = \{x^i, C\}, \tag{2.159}$$

where the role of the Hamiltonian is played by the (possibly deformed) Casimir of the DSR algebra (both in the Hopf algebra setting and in more general phenomenological models), e.g. given by (2.58). The use of this formalism allows us to account for a possibly deformed symplectic structure of the phase space, namely a deformed bracket $\{x, p\}$ between coordinates and momenta. And in fact it turned out that, depending on the choice of this bracket (different choices being linked by momentum-dependent redefinitions of the coordinates), one could alternatively get standard [116] or momentum dependent velocities for massless particles within the same momentum-space DSR description.

A solution to this puzzle came [83, 85, 118] when it was pointed out that looking at the expression of the coordinate velocity is not enough to state whether a time shift effect is to be expected. In fact, in presence of relativity of locality the equations of motion (leading to the worldlines) written by an observer are affected by coordinate artifacts when they are used to infer the behaviour of particles far away from the observer. One needs to compare the observations made by observers local to the emission and to the detection of the particle whose propagation time is being computed. By doing this, one accounts for the possibly nontrivial action of the translation generators on the spacetime coordinates. In particular, for the model inspired by the κ-Poincaré algebra in the bicrossproduct basis (see Sect. 2.3.3) it turned out that such nontrivial action compensates the effect of momentum-dependent redefinition of coordinates, so that the time shift effect cannot be reabsorbed in such a way [61, 85].[19] This observation set the ground for the investigations that led to the relative locality proposal, discussed in Sect. 2.4.

By performing the Hamiltonian analysis, and properly transforming from the reference frame of the emitter to that of the observer, one can show that for the κ-Poincaré model of Sect. 2.3.3 the expected time shift between a low-energy massless particle (not affected by DSR effects) and a high-energy massless particle (for which DSR effects are relevant) seen by the observer local to the detection is [61, 73, 85, 124]:

$$\Delta t = -L\left(1 - e^{-\frac{E}{\kappa}}\right) \simeq -L\frac{E}{\kappa}, \tag{2.160}$$

where $L = T$ is the distance/time of flight between the source and the detector according to low energy particles[20] and we also gave the first-order expression in powers of the particle energy over the energy deformation scale κ.

? Exercise

2.14. Derive Eq. (2.160), following e.g. [124]. Notice that the first order expression in (2.160) coincides with the one that would be derived from the modified dispersion relation (2.13) by using $v = \frac{\partial E}{\partial p}$ and $\frac{\eta}{E_P} = \frac{1}{\kappa}$.

In recent studies on the relative locality framework, the importance of specifying the emission/detection mechanism has emerged [61, 125]. This is due to the fact

[19] Of course one might reach different conclusions concerning time shift when using different bases of the κ-Poincaré algebra. For example, using the the classical basis of κ-Poincaré, there could be an absence of time shifts for massless particles with different energies [119]. Within the relative locality framework this can be understood in terms of the non-invariance of physical predictions under momentum space diffeomorphisms [120]. Moreover, depending on the effective scheme used for studying this effect, different time delay formulas are obtained, and may not lead to a time delay [119, 121–123].

[20] Remember that we set the low-energy particle velocity $c = 1$.

that in order for the the action describing the interaction vertex (2.109) to be covariant, one needs to consider the translation/boost generators associated to the total momentum of the vertex [45, 61], and this depends on the particles entering the interaction. Because of this, one may or may not predict a time shift effect, depending on the interactions at play [61]. Whether this is a specific feature of the relative locality framework, due to the insistence on the use of a Lagrangian formalism, or whether a similar behaviour is to be expected in any DSR model where a Hamiltonian approach is used is still matter of investigation (see [45] for a discussion on the conceptual drawbacks of using a "total momentum" or "total boost" generator).

A somewhat parallel line of investigation has been focusing on how to include the effects of space-time curvature on the DSR-induced time shifts. Expressions such as (2.160) are valid for situations where space-time curvature can be neglected. However, given the cosmological distance of sources used to test this kind of effect, curvature should be taken into account, and it is generically expected to induce a dependence of the time shift on the redshift z of the source. Until very recently, the great majority of studies assumed that one can simply replace the momenta appearing in the modified velocity of propagation with the physical momenta $p \to p/a$, where a is the scale factor in a Friedmann-Robertson-Walker (FRW) metric. Then the time shift between a low-energy and a high-energy particle emitted simultaneously can be found by asking that they travelled the same comoving distance, and reads, for the $n = 1$ case of (2.1):

$$\Delta t = \eta \frac{E_0}{E_P} D(z) \,, \tag{2.161}$$

$$D(z) = \frac{1}{H_0} \int_0^z d\zeta \frac{(1+\zeta)}{\sqrt{\Omega_m (1+\zeta)^3 + \Omega_\Lambda}} \,, \tag{2.162}$$

with H_0, Ω_m and Ω_Λ denoting, respectively, the Hubble parameter, the matter fraction and the cosmological constant in a FRW universe. E_0 is the energy of the high-energy particle measured today.

? Exercise

2.15. Compute the time shift (2.161) using the criteria described in the above paragraph. Use [126] as guidance.

In [64, 127] it was pointed out that one could in principle consider a more general dependence of the physical momenta on the scale factor. Moreover, in

[64] it was observed that the formula (2.161) is valid in the DSR scenario only when translational invariance is not deformed, otherwise one gets a more general dependence on the redshift of the source.[21]. In [73] the time shift expected in a de Sitter spacetime with Hopf-algebra deformation of the relativistic symmetries was computed.

2.6.2 Modified Interaction Effects

As we mentioned, the DSR framework is not currently embedded into a theory providing the dynamics of particles interactions. For this reason, we can only make arguments on the allowed interactions based on kinematical constraints.

Both modifications of the dispersion relation and of the energy-momentum conservation law have a relevant role in this kind of analysis. As we discussed in Sect. 2.2, relativistic consistency generally provides quite a rigid structure on the allowed combinations of dispersion relation and conservation laws, and this affects crucially the kinematical analysis of interactions, making any possible modification very mild.

In order to make this point clearer, let us start from an example where the relativistic consistency is violated, because there is a modified dispersion relation but everything else is the same as in special relativity. Clearly in this case the modified dispersion relation identifies a preferred frame, where it takes the specific form considered. This kind of scenario belongs to the framework of Lorentz Invariance Violation (LIV).

We are not going to review the LIV framework in detail. Here we simply want to characterize the phenomenological differences between this and the DSR framework as far as interactions are concerned. Notice that since in both models one can envisage the emergence of modified dispersion relations, one expects in both cases propagation effects for particles traveling from astrophysical sources.[22]

In the LIV framework, modified dispersion relations can have significant implications for certain decay processes [128, 129]. For example, massless particles would be allowed to decay, so that a process like photon decay into an electron-positron pair ($\gamma \to e^+ + e^-$) would become possible. Using the dispersion relation (2.1) with $n = 1$, and assuming that the law of energy-momentum conservation is unmodified with respect to the one of special relativity, as usually done in LIV studies, one finds

[21] The result of [64] was obtaining starting from a deformation of the relativistic transformations in de Sitter spacetime, which is maximally symmetric, and then deducing the time shift in FRW via a slicing procedure, first devised in [68].

[22] As we mentioned in the previous subsection, even propagation effects might allow to distinguish between the two frameworks, since there could be a different dependence on the redshift of the source.

a relation between the opening angle θ between the outgoing electron-positron pair and their energies E_+, E_-

$$\cos\theta = \frac{2E_+E_- + 2m_e^2 - 2\frac{\eta}{E_P}(E_+E_-^2 + E_-E_+^2)}{2E_+E_- - m_e^2\left(\frac{E_+}{E_-} + \frac{E_-}{E_+}\right)} \tag{2.163}$$

? Exercise

2.16. Derive Eq. (2.163), see [109] for guidance.

For $\eta < 0$, the process is always forbidden ($\cos\theta > 1$), as in special relativity, but, for positive η and $E_\gamma >> (m_e^2 E_P/|\eta|)^{1/3}$, one finds that one may have $\cos\theta < 1$. Notice that the energy scale $(m_e^2 E_P)^{1/3} \sim 10^{13}$ eV is within reach of astrophysics observations, and, in fact, strong constraints on η have been set using the fact that we see photons of higher energies coming from astrophysical sources [128].

Conversely, any DSR model must have stable massless particles. By establishing the existence of a threshold for photon decay one could therefore falsify the DSR idea. In fact, an energy threshold for the decay of massless particles cannot be introduced as an observer-independent law, and is therefore incompatible with the DSR principles. Massless particles which are below a threshold energy value for one observer will be above that threshold for other boosted observers, and this would allow to identify a preferred frame. Let us see how this works in practice when considering the photon decay process discussed above when the kinematics is given by the DSR model considered in Sect. 2.2, straightforwardly generalized to the 3+1 dimensions.

From the conservation of spatial momenta encoded in the addition law (2.16) one finds that[23]

$$\begin{aligned} p_\gamma^2 &= p_+^2 + p_-^2 + 2p_+p_-\cos\theta + 2\frac{\eta}{E_P}(E_+E_-^2 + E_-E_+^2) \\ &\quad + 2\frac{\eta}{E_P}(E_+^2E_- + E_-^2E_+)\cos\theta\,, \end{aligned} \tag{2.164}$$

[23] We write all formulas up to the first order in $\frac{\eta}{E_P}$. For ultra-relativistic electrons and positrons one can consider $m_e\frac{\eta}{E_P} \simeq 0$.

where p_+, p_- and p_γ are, respectively, the spatial momentum of the positron, the electron and the photon. Using the dispersion relation (2.13) this becomes

$$\begin{aligned} E_\gamma^2 + \frac{\eta}{E_P} E_\gamma^3 =& E_+^2 + E_-^2 - 2m_e^2 + 2\left(E_+E_- - m_e^2\left(\frac{E_+}{E_-} + \frac{E_-}{E_+}\right)\right) \\ & \cos\theta + \frac{\eta}{E_P}(E_+^3 + E_-^3) \\ & + 2\frac{\eta}{E_P}(E_+E_-^2 + E_-E_+^2) + 3\frac{\eta}{E_P}\left(E_+ + E_-\right)E_+E_-\cos\theta\,. \end{aligned} \tag{2.165}$$

Using the conservation of energy encoded in (2.16) one obtains:

$$\begin{aligned} & E_+^2 + E_-^2 + 2E_+E_- + \frac{\eta}{E_P}(E_+ + E_-)^3 + 2\frac{\eta}{E_P}(E_+ + E_-)E_+E_-\cos\theta = \\ & E_+^2 + E_-^2 - 2m_e^2 + 2\left(E_+E_- - m_e^2\left(\frac{E_+}{E_-} + \frac{E_-}{E_+}\right)\right)\cos\theta + \frac{\eta}{E_P}(E_+^3 + E_-^3) \\ & + 2\frac{\eta}{E_P}(E_+E_-^2 + E_-E_+^2) + 3\frac{\eta}{E_P}\left(E_+ + E_-\right)E_+E_-\cos\theta\,. \end{aligned} \tag{2.166}$$

From this, one finds that the correction terms cancel out and one recovers the standard special-relativistic result

$$\cos\theta = \frac{2E_+E_- + 2m_e^2}{2E_+E_- - m_e^2\left(\frac{E_+}{E_-} + \frac{E_-}{E_+}\right)}\,, \tag{2.167}$$

that signals the impossibility of such decay (one always has $\cos\theta > 1$).

While massless particles decays are forbidden in DSR, one might still have modifications of the threshold for the decay of massive particles. However, it turns out that in this case the modification of the thresholds is only appreciable for energies of the order of the Planck scale E_P, which render them unobservable in practice. Nevertheless, since in some DSR models there could be an absence of time delays [119, 121–123], one might assume that in these scenarios the effective Planck energy is much lower, without being incompatible with current observations in particle accelerators [130, 131] and in astroparticle physics [132, 133]

For similar reasons as the ones discussed above, in DSR it is also forbidden to have a threshold below which a photon cannot produce electron-positron pairs in interactions with another sufficiently high-energy photon. Such process is indeed always allowed in special relativity, regardless of the energy of the low-energy photon, and this must be the case also in DSR scenarios. This is a necessary consequence of the fact that two relatively boosted observers attribute different energy to a given photon, and if there was a threshold then it would be possible to distinguish frames where the interaction can take place from those where the interaction cannot take place.

? Exercise

2.17. Using again the 3+1 version of the model of Sect. 2.2 as done above, perform the kinematical analysis of the process $\gamma\gamma \rightarrow e^+e^-$ focusing on a collinear process. Given some low energy ϵ for the low-energy photon, find the minimum energy E_{min} of the high-energy photon for the process to take place. Show that within this model one recovers the same result as in special relativity, $E_{min} = \frac{m_e^2}{\epsilon}$. Use [12] as guidance.

References

1. G. Amelino-Camelia, Int. J. Mod. Phys. D **11**, 35 (2002). https://doi.org/10.1142/S0218271802001330
2. G. Amelino-Camelia, Phys. Lett. B **510**, 255 (2001). https://doi.org/10.1016/S0370-2693(01)00506-8
3. G. Amelino-Camelia, J.R. Ellis, N.E. Mavromatos, D.V. Nanopoulos, S. Sarkar, Nature **393**, 763 (1998). https://doi.org/10.1038/31647
4. R. Gambini, J. Pullin, Phys. Rev. D **59**, 124021 (1999). https://doi.org/10.1103/PhysRevD.59.124021
5. J. Alfaro, H.A. Morales-Tecotl, L.F. Urrutia, Phys. Rev. Lett. **84**, 2318 (2000). https://doi.org/10.1103/PhysRevLett.84.2318
6. B.E. Schaefer, Phys. Rev. Lett. **82**, 4964 (1999). https://doi.org/10.1103/PhysRevLett.82.4964
7. S.D. Biller et al., Phys. Rev. Lett. **83**, 2108 (1999). https://doi.org/10.1103/PhysRevLett.83.2108
8. R. Aloisio, P. Blasi, P.L. Ghia, A.F. Grillo, Phys. Rev. **D62**, 053010 (2000). https://doi.org/10.1103/PhysRevD.62.053010
9. G. Amelino-Camelia, T. Piran, Phys. Rev. D **64**, 036005 (2001). https://doi.org/10.1103/PhysRevD.64.036005
10. G. Amelino-Camelia, Int. J. Mod. Phys. D **11**, 1643 (2002). https://doi.org/10.1142/S021827180200302X
11. G. Amelino-Camelia, Nature **418**, 34 (2002). https://doi.org/10.1038/418034a
12. G. Amelino-Camelia, Phys. Rev. D **85**, 084034 (2012). https://doi.org/10.1103/PhysRevD.85.084034
13. A. Ballesteros, G. Gubitosi, F.J. Herranz, Class. Quant. Grav. **37**(19), 195021 (2020). https://doi.org/10.1088/1361-6382/aba668
14. A. Ballesteros, G. Gubitosi, I. Gutierrez-Sagredo, F.J. Herranz, Phys. Lett. B **805**, 135461 (2020). https://doi.org/10.1016/j.physletb.2020.135461
15. H.J. Matschull, M. Welling, Class. Quant. Grav. **15**, 2981 (1998). https://doi.org/10.1088/0264-9381/15/10/008
16. F.A. Bais, N.M. Muller, B.J. Schroers, Nucl. Phys. B **640**, 3 (2002). https://doi.org/10.1016/S0550-3213(02)00572-2
17. C. Meusburger, B.J. Schroers, Class. Quant. Grav. **20**, 2193 (2003). https://doi.org/10.1088/0264-9381/20/11/318
18. G. Amelino-Camelia, L. Smolin, A. Starodubtsev, Class. Quant. Grav. **21**, 3095 (2004). https://doi.org/10.1088/0264-9381/21/13/002

19. L. Freidel, J. Kowalski-Glikman, L. Smolin, Phys. Rev. D **69**, 044001 (2004). https://doi.org/10.1103/PhysRevD.69.044001
20. L. Freidel, E.R. Livine, Phys. Rev. Lett. **96**, 221301 (2006). https://doi.org/10.1103/PhysRevLett.96.221301
21. G. Amelino-Camelia, M. Arzano, S. Bianco, R.J. Buonocore, Class. Quant. Grav. **30**, 065012 (2013). https://doi.org/10.1088/0264-9381/30/6/065012
22. G. Rosati, Phys. Rev. D **96**(6), 066027 (2017). https://doi.org/10.1103/PhysRevD.96.066027
23. M. Bojowald, G.M. Paily, Phys. Rev. D **87**(4), 044044 (2013). https://doi.org/10.1103/PhysRevD.87.044044
24. J. Mielczarek, EPL **108**(4), 40003 (2014). https://doi.org/10.1209/0295-5075/108/40003
25. G. Amelino-Camelia, M.M. da Silva, M. Ronco, L. Cesarini, O.M. Lecian, Phys. Rev. D **95**(2), 024028 (2017). https://doi.org/10.1103/PhysRevD.95.024028
26. F. Cianfrani, J. Kowalski-Glikman, D. Pranzetti, G. Rosati, Phys. Rev. D **94**(8), 084044 (2016). https://doi.org/10.1103/PhysRevD.94.084044
27. L. Smolin, Nucl. Phys. B **742**, 142 (2006). https://doi.org/10.1016/j.nuclphysb.2006.02.017
28. G. Amelino-Camelia, M. Arzano, M.M. Da Silva, D.H. Orozco-Borunda, Phys. Lett. B **775**, 168 (2017). https://doi.org/10.1016/j.physletb.2017.10.071
29. S. Majid, Lect. Notes Phys. **541**, 227 (2000)
30. G. Amelino-Camelia, S. Majid, Int. J. Mod. Phys. A **15**, 4301 (2000). https://doi.org/10.1142/S0217751X00002777
31. G. Amelino-Camelia, S. Majid, Int. J. Mod. Phys. A **15**, 4301 (2000). https://doi.org/10.1142/S0217751X00002777
32. N.R. Bruno, G. Amelino-Camelia, J. Kowalski-Glikman, Phys. Lett. B **522**, 133 (2001). https://doi.org/10.1016/S0370-2693(01)01264-3
33. J. Kowalski-Glikman, S. Nowak, Int. J. Mod. Phys. D **12**, 299 (2003). https://doi.org/10.1142/S0218271803003050
34. J. Kowalski-Glikman, S. Nowak, Int. J. Mod. Phys. **D12**, 299 (2003). https://doi.org/10.1142/S0218271803003050
35. A. Agostini, G. Amelino-Camelia, F. D'Andrea, Int. J. Mod. Phys. A **19**, 5187 (2004). https://doi.org/10.1142/S0217751X04020919
36. G. Amelino-Camelia, F. D'Andrea, G. Mandanici, J. Cosmol. Astropart. Phys. **09**, 006 (2003). https://doi.org/10.1088/1475-7516/2003/09/006
37. A. Agostini, G. Amelino-Camelia, M. Arzano, A. Marciano, R.A. Tacchi, Mod. Phys. Lett. A **22**, 1779 (2007). https://doi.org/10.1142/S0217732307024280
38. G. Amelino-Camelia, G. Gubitosi, A. Marciano, P. Martinetti, F. Mercati, Phys. Lett. B **671**, 298 (2009). https://doi.org/10.1016/j.physletb.2008.12.032
39. G. Amelino-Camelia, F. Briscese, G. Gubitosi, A. Marciano, P. Martinetti, F. Mercati, Phys. Rev. D **78**, 025005 (2008). https://doi.org/10.1103/PhysRevD.78.025005
40. G. Amelino-Camelia, G. Gubitosi, A. Marciano, P. Martinetti, F. Mercati, D. Pranzetti, R.A. Tacchi, Prog. Theor. Phys. Suppl. **171**, 65 (2007). https://doi.org/10.1143/PTPS.171.65
41. J.M. Carmona, J.L. Cortes, F. Mercati, Phys. Rev. D **86**, 084032 (2012). https://doi.org/10.1103/PhysRevD.86.084032
42. G. Amelino-Camelia, G. Gubitosi, G. Palmisano, Int. J. Mod. Phys. **D25**(02), 1650027 (2016). https://doi.org/10.1142/S0218271816500279
43. J.M. Carmona, J.L. Cortes, J.J. Relancio, Phys. Rev. D **94**(8), 084008 (2016). https://doi.org/10.1103/PhysRevD.94.084008
44. G. Gubitosi, F. Mercati, Class. Quant. Grav. **30**, 145002 (2013). https://doi.org/10.1088/0264-9381/30/14/145002
45. G. Gubitosi, S. Heefer, Phys. Rev. **D99**(8), 086019 (2019). https://doi.org/10.1103/PhysRevD.99.086019
46. G. Amelino-Camelia, Symmetry **2**, 230 (2010). https://doi.org/10.3390/sym2010230
47. J. Lukierski, H. Ruegg, A. Nowicki, V.N. Tolstoi, Phys. Lett. **B264**, 331 (1991). https://doi.org/10.1016/0370-2693(91)90358-W

48. J. Lukierski, A. Nowicki, H. Ruegg, Phys. Lett. **B293**, 344 (1992). https://doi.org/10.1016/0370-2693(92)90894-A
49. S. Majid, H. Ruegg, Phys. Lett. B **334**, 348 (1994). https://doi.org/10.1016/0370-2693(94)90699-8
50. J. Lukierski, H. Ruegg, W.J. Zakrzewski, Ann. Phys. **243**, 90 (1995). https://doi.org/10.1006/aphy.1995.1092
51. G. Amelino-Camelia, J. Lukierski, A. Nowicki, Phys. Atom. Nucl. **61**, 1811 (1998)
52. M. Arzano, J. Kowalski-Glikman, *Deformations of Spacetime Symmetries: Gravity, Group-Valued Momenta, and Non-commutative Fields*. Lecture Notes in Physics, vol. 986 (2021). https://doi.org/10.1007/978-3-662-63097-6
53. R. Geroch, *Mathematical Physics*. Chicago Lectures in Physics (The University of Chicago Press, 1985)
54. S. Majid, H. Ruegg, Phys. Lett. **B334**, 348 (1994). https://doi.org/10.1016/0370-2693(94)90699-8
55. M. Arzano, J. Kowalski-Glikman, A group theoretic description of the κ-Poincaré Hopf algebra, Phys. Lett. B **835**, 137535 (2022). https://doi.org/10.1016/j.physletb.2022.137535
56. A. Borowiec, A. Pachol, J. Phys. **A43**, 045203 (2010). https://doi.org/10.1088/1751-8113/43/4/045203
57. M. Arzano, T. Trzesniewski, Phys. Rev. D **89**(12), 124024 (2014). https://doi.org/10.1103/PhysRevD.89.124024
58. H. Ruegg, V.N. Tolstoi, Lett. Math. Phys. **32**, 85 (1994). https://doi.org/10.1007/BF00739419
59. P. Kosinski, J. Lukierski, P. Maslanka, Czech. J. Phys. **50**, 1283 (2000). https://doi.org/10.1023/A:1022821310096
60. C. Guedes, D. Oriti, M. Raasakka, J. Math. Phys. **54**, 083508 (2013). https://doi.org/10.1063/1.4818638
61. G. Amelino-Camelia, M. Arzano, J. Kowalski-Glikman, G. Rosati, G. Trevisan, Class. Quant. Grav. **29**, 075007 (2012). https://doi.org/10.1088/0264-9381/29/7/075007
62. J. Kowalski-Glikman, S. Nowak, Class. Quant. Grav. **20**, 4799 (2003). https://doi.org/10.1088/0264-9381/20/22/006
63. G. Amelino-Camelia, Phys. Rev. **D85**, 084034 (2012). https://doi.org/10.1103/PhysRevD.85.084034
64. G. Rosati, G. Amelino-Camelia, A. Marciano, M. Matassa, Phys. Rev. **D92**(12), 124042 (2015). https://doi.org/10.1103/PhysRevD.92.124042
65. G. Amelino-Camelia, A. Marciano, M. Matassa, G. Rosati, Phys. Rev. D **86**, 124035 (2012). https://doi.org/10.1103/PhysRevD.86.124035
66. J.J. Relancio, S. Liberati, Phys. Rev. D **101**(6), 064062 (2020). https://doi.org/10.1103/PhysRevD.101.064062
67. J.J. Relancio, S. Liberati, Class. Quant. Grav. **38**(13), 135028 (2021). https://doi.org/10.1088/1361-6382/ac05d7
68. A. Marciano, G. Amelino-Camelia, N.R. Bruno, G. Gubitosi, G. Mandanici, A. Melchiorri, J. Cosmol. Astropart. Phys. **06**, 030 (2010). https://doi.org/10.1088/1475-7516/2010/06/030
69. A. Ballesteros, F.J. Herranz, M.A.D. Olmo, M. Santander, J. Phys. A: Math. Gen. **27**(4), 1283 (1994). https://doi.org/10.1088/0305-4470/27/4/021. https://doi.org/10.1088/0305-4470/27/4/021
70. A. Ballesteros, F.J. Herranz, F. Musso, P. Naranjo, Phys. Lett. B **766**, 205 (2017). https://doi.org/10.1016/j.physletb.2017.01.020
71. A. Ballesteros, I. Gutierrez-Sagredo, F.J. Herranz, Phys. Lett. B **796**, 93 (2019). https://doi.org/10.1016/j.physletb.2019.07.038
72. C. Pfeifer, J.J. Relancio, Eur. Phys. J. C **82**(2), 150 (2022). https://doi.org/10.1140/epjc/s10052-022-10066-w
73. L. Barcaroli, G. Gubitosi, Phys. Rev. D **93**(12), 124063 (2016). https://doi.org/10.1103/PhysRevD.93.124063
74. P. Aschieri, A. Borowiec, A. Pacho, J. Cosmol. Astropart. Phys. **04**, 025 (2021). https://doi.org/10.1088/1475-7516/2021/04/025

75. A. Ballesteros, G. Gubitosi, F. Mercati, Symmetry **13**(11), 2099 (2021). https://doi.org/10.3390/sym13112099
76. A. Ballesteros, G. Gubitosi, I. Gutiérrez-Sagredo, F.J. Herranz, Phys. Rev. **D97**(10), 106024 (2018). https://doi.org/10.1103/PhysRevD.97.106024
77. A. Ballesteros, G. Gubitosi, I. Gutiérrez-Sagredo, F.J. Herranz, Phys. Lett. **B773**, 47 (2017). https://doi.org/10.1016/j.physletb.2017.08.008
78. I. Gutierrez-Sagredo, A. Ballesteros, G. Gubitosi, F.J. Herranz *Quantum groups, non-commutative Lorentzian spacetimes and curved momentum spaces*, In Spacetime Physics 1907–2017. C. Duston and M. Holman (Eds). Minkowski Institute Press, Montreal (2019), pp. 261–290
79. G. Amelino-Camelia, L. Freidel, J. Kowalski-Glikman, L. Smolin, Phys. Rev. D **84**, 084010 (2011). https://doi.org/10.1103/PhysRevD.84.084010
80. G. Amelino-Camelia, L. Freidel, J. Kowalski-Glikman, L. Smolin, Gen. Rel. Grav. **43**, 2547 (2011). https://doi.org/10.1142/S0218271811020743
81. S. Hossenfelder, Phys. Rev. Lett. **104**, 140402 (2010). https://doi.org/10.1103/PhysRevLett.104.140402
82. L. Smolin, Gen. Rel. Grav. **43**, 3671 (2011). https://doi.org/10.1007/s10714-011-1235-1
83. G. Amelino-Camelia, M. Matassa, F. Mercati, G. Rosati, Phys. Rev. Lett. **106**, 071301 (2011). https://doi.org/10.1103/PhysRevLett.106.071301
84. U. Jacob, F. Mercati, G. Amelino-Camelia, T. Piran, Phys. Rev. D **82**, 084021 (2010). https://doi.org/10.1103/PhysRevD.82.084021
85. G. Amelino-Camelia, N. Loret, G. Rosati, Phys. Lett. B **700**, 150 (2011). https://doi.org/10.1016/j.physletb.2011.04.054
86. M. Arzano, G. Gubitosi, J.a. Magueijo, G. Amelino-Camelia, Phys. Rev. D **92**(2), 024028 (2015). https://doi.org/10.1103/PhysRevD.92.024028
87. I.P. Lobo, G. Palmisano, Int. J. Mod. Phys. Conf. Ser. **41**, 1660126 (2016). https://doi.org/10.1142/S2010194516601265
88. L. Freidel, L. Smolin, Gamma ray burst delay times probe the geometry of momentum space, (2011) arxiv:hep-th/arXiv:1103.5626
89. G. Amelino-Camelia, M. Arzano, J. Kowalski-Glikman, G. Rosati, G. Trevisan, Class. Quant. Grav. **29**, 075007 (2012). https://doi.org/10.1088/0264-9381/29/7/075007
90. G. Amelino-Camelia, L. Freidel, J. Kowalski-Glikman, L. Smolin, Phys. Rev. **D84**, 084010 (2011). https://doi.org/10.1103/PhysRevD.84.084010
91. J.J. Relancio, Phys. Rev. D **104**(2), 024017 (2021). https://doi.org/10.1103/PhysRevD.104.024017
92. J.M. Carmona, J.L. Cortés, J.J. Relancio, Phys. Rev. **D100**(10), 104031 (2019). https://doi.org/10.1103/PhysRevD.100.104031
93. S. Weinberg, *Gravitation and Cosmology* (Wiley , New York, 1972). http://www-spires.fnal.gov/spires/find/books/www?cl=QC6.W431
94. S.L. Dubovsky, S.M. Sibiryakov, Phys. Lett. **B638**, 509 (2006). https://doi.org/10.1016/j.physletb.2006.05.074
95. P. Finsler. Über Kurven und Flächen in allgemeinen Räumen. Göttingen, Zürich: O. Füssli, 120 S. 8° (1918)
96. R. Miron, D. Hrimiuc, H. Shimada, S. Sabau, *The Geometry of Hamilton and Lagrange Spaces*. Fundamental Theories of Physics (Springer, 2001). https://books.google.es/books?id=l3JNMzL14SAC
97. F. Girelli, S. Liberati, L. Sindoni, Phys. Rev. **D75**, 064015 (2007). https://doi.org/10.1103/PhysRevD.75.064015
98. G. Amelino-Camelia, L. Barcaroli, G. Gubitosi, S. Liberati, N. Loret, Phys. Rev. D **90**(12), 125030 (2014). https://doi.org/10.1103/PhysRevD.90.125030
99. M. Letizia, S. Liberati, Phys. Rev. **D95**(4), 046007 (2017). https://doi.org/10.1103/PhysRevD.95.046007
100. L. Barcaroli, L.K. Brunkhorst, G. Gubitosi, N. Loret, C. Pfeifer, Phys. Rev. D **95**(2), 024036 (2017). https://doi.org/10.1103/PhysRevD.95.024036

101. L. Barcaroli, L.K. Brunkhorst, G. Gubitosi, N. Loret, C. Pfeifer, Phys. Rev. D **92**(8), 084053 (2015). https://doi.org/10.1103/PhysRevD.92.084053
102. L. Barcaroli, L.K. Brunkhorst, G. Gubitosi, N. Loret, C. Pfeifer, Phys. Rev. D **96**(8), 084010 (2017). https://doi.org/10.1103/PhysRevD.96.084010
103. S.S. Chern, W.H. Chen, K.S. Lam, *Lectures on Differential Geometry* (World Scientific, Singapore, 1999). See Eqs. (1.30) and (1.31) of Chapter 6
104. J. Kowalski-Glikman, S. Nowak, Phys. Lett. **B539**, 126 (2002). https://doi.org/10.1016/S0370-2693(02)02063-4
105. M.V. Battisti, S. Meljanac, Phys. Rev. D **82**, 024028 (2010). https://doi.org/10.1103/PhysRevD.82.024028
106. S. Meljanac, D. Meljanac, A. Samsarov, M. Stojic (2009) [arXiv:0909.1706];
107. J.R. Ellis, N.E. Mavromatos, D.V. Nanopoulos, A.S. Sakharov, E.K.G. Sarkisyan, Astropart. Phys. **25**, 402 (2006). https://doi.org/10.1016/j.astropartphys.2007.12.003 [Erratum: Astropart. Phys. 29, 158–159 (2008)]
108. D. Mattingly, Living Rev. Rel. **8**, 5 (2005)
109. G. Amelino-Camelia, Living Rev. Rel. **16**, 5 (2013). https://doi.org/10.12942/lrr-2013-5
110. A. Addazi et al., Prog. Part. Nucl. Phys. **125**, 103948 (2022). https://doi.org/10.1016/j.ppnp.2022.103948
111. G. Amelino-Camelia, L. Barcaroli, N. Loret, Int. J. Theor. Phys. **51**, 3359 (2012). https://doi.org/10.1007/s10773-012-1216-5
112. G. Amelino-Camelia, L. Barcaroli, S. Bianco, L. Pensato, Adv. High Energy Phys. **2017**, 6075920 (2017). https://doi.org/10.1155/2017/6075920
113. G. Amelino-Camelia, L. Barcaroli, G. D'Amico, N. Loret, G. Rosati, Int. J. Mod. Phys. D **26**(08), 1750076 (2017). https://doi.org/10.1142/S0218271817500766
114. P. Kosinski, P. Maslanka, Phys. Rev. D **68**, 067702 (2003). https://doi.org/10.1103/PhysRevD.68.067702
115. S. Mignemi, Phys. Lett. A **316**, 173 (2003). https://doi.org/10.1016/S0375-9601(03)01176-9
116. M. Daszkiewicz, K. Imilkowska, J. Kowalski-Glikman, Phys. Lett. A **323**, 345 (2004). https://doi.org/10.1016/j.physleta.2004.02.046
117. J. Kowalski-Glikman, in *10th Marcel Grossmann Meeting on Recent Developments in Theoretical and Experimental General Relativity, Gravitation and Relativistic Field Theories (MG X MMIII)* (2003), pp. 1169–1182
118. G. Rosati, N. Loret, G. Amelino-Camelia, J. Phys. Conf. Ser. **343**, 012105 (2012). https://doi.org/10.1088/1742-6596/343/1/012105
119. J.M. Carmona, J.L. Cortes, J.J. Relancio, Class. Quant. Grav. **35**(2), 025014 (2018). https://doi.org/10.1088/1361-6382/aa9ef8
120. G. Amelino-Camelia, S. Bianco, G. Rosati, Phys. Rev. D **101**(2), 026018 (2020). https://doi.org/10.1103/PhysRevD.101.026018
121. J.M. Carmona, J.L. Cortés, J.J. Relancio, Symmetry **10**(7), 231 (2018). https://doi.org/10.3390/sym10070231
122. J.M. Carmona, J.L. Cortes, J.J. Relancio, Symmetry **11**, 1401 (2019). https://doi.org/10.3390/sym11111401
123. J. Relancio, S. Liberati, Phys. Rev. D **102**(10), 104025 (2020). https://doi.org/10.1103/PhysRevD.102.104025
124. G. Amelino-Camelia, L. Barcaroli, G. Gubitosi, N. Loret, Class. Quant. Grav. **30**, 235002 (2013). https://doi.org/10.1088/0264-9381/30/23/235002
125. G. Amelino-Camelia, S. Bianco, F. Brighenti, R.J. Buonocore, Phys. Rev. **D91**(8), 084045 (2015). https://doi.org/10.1103/PhysRevD.91.084045
126. U. Jacob, T. Piran, J. Cosmol. Astropart. Phys. **01**, 031 (2008). https://doi.org/10.1088/1475-7516/2008/01/031
127. G. Amelino-Camelia, G. Rosati, S. Bedić, Phys. Lett. B **820**, 136595 (2021). https://doi.org/10.1016/j.physletb.2021.136595
128. T. Jacobson, S. Liberati, D. Mattingly, Phys. Rev. D **66**, 081302 (2002). https://doi.org/10.1103/PhysRevD.66.081302

129. G. Amelino-Camelia, Phys. Lett. B **528**, 181 (2002). https://doi.org/10.1016/S0370-2693(02)01223-6
130. G. Albalate, J.M. Carmona, J.L. Cortés, J.J. Relancio, Symmetry **10**, 432 (2018). https://doi.org/10.3390/sym10100432
131. J.M. Carmona, J.L. Cortés, J.J. Relancio, Symmetry **13**, 1266 (2021). https://doi.org/10.3390/sym13071266
132. J.M. Carmona, J.L. Cortés, L. Pereira, J.J. Relancio, Symmetry **12**(8), 1298 (2020). https://doi.org/10.3390/sym12081298
133. J.M. Carmona, J.L. Cortés, J.J. Relancio, M.A. Reyes, A. Vincueria, Eur. Phys. J. Plus **137**, 768 (2022). https://10.1140/epjp/s13360-022-02920-3

Poincaré Gauge Gravity Primer

3

Yuri N. Obukhov

Abstract

We give an introductory overview of the classical Poincaré gauge theory of gravity formulated on the spacetime manifold that carries the Riemann-Cartan geometry with nontrivial curvature and torsion. After discussing the basic mathematical structures at an elementary level in the framework of the standard tensor analysis, we formulate the general dynamical scheme of Poincaré gauge gravity for the class of Yang-Mills type models, and consider a selected number of physically interesting consequences of this theory.

3.1 Introduction

The gauge-theoretic approach in classical field theory has a long history (going back to the early works of Weyl, Cartan, Fock, for an overview see [1–3]) and it underlies the modern understanding of the nature of the physical interactions [4–6]. The original Yang-Mills [7] treatment of the internal symmetry groups was subsequently extended to the spacetime symmetries by Utiyama [8], Sciama [9] and Kibble [10]. The detailed review of the development of the gauge approach in gravity theory and the corresponding mathematical structures can be found in [11–22]. It is worthwhile to mention that the book [22] provides an essentially complete bibliography on this subject. Here we do not intend to present an exhaustive and comprehensive review of the Poincaré gauge gravity theory, and give a rather concise and elementary introduction into the subject. One may view this paper as a continuation of the earlier work [23–25].

Y. N. Obukhov (✉)
Nuclear Safety Institute, Russian Academy of Sciences, Moscow, Russia
e-mail: obukhov@ibrae.ac.ru

C. Pfeifer, C. Lämmerzahl (eds.), *Modified and Quantum Gravity*, Lecture Notes in Physics 1017, https://doi.org/10.1007/978-3-031-31520-6_3

The Poincaré gauge (PG) gravity is a natural extension of Einstein's general relativity (GR) theory. Being based on gauge-theoretic principles, it takes into account the spin (commonly viewed as a microstructural property of matter) as an additional physical source of the gravitational field on an equal footing with the energy and momentum (naturally viewed as macroscopic properties of matter). The corresponding spacetime structure is then adequately described by the Riemann-Cartan geometry with curvature and torsion. From the mathematical point of view, the PG formalism arises as a special case of the metric-affine gravity (MAG) theory [16] that provides a unified framework for the study of alternative theories based on post-Riemannian geometries [26, 27]. Other special cases of MAG include the geometries of Riemann of GR [28], Weyl [29], Weitzenböck [30] (which will also be discussed further in Chap. 4), etc.

The Poincaré gauge gravity occupies a prominent place in the colorful landscape of modified gravitational theories that generalize or extend the physical and mathematical structure of Einstein's GR. Among such theories it is worthwhile to highlight the large classes of $f(R)$ and $f(T)$ models, and of theories with nonminimal coupling to matter, developed mainly in the context of relativistic cosmology, see [31–37]. The so-called Palatini approach represents another class of widely discussed theories in which metric and connection are treated as independent variables in the action principle [38–40].

Our basic notation and conventions are consistent with [16, 41]. In particular, Greek indices $\alpha, \beta, \dots = 0, \dots, 3$, denote the anholonomic components (for example, of an orthonormal frame e_α), while the Latin indices $i, j, \dots = 0, \dots, 3$, label the holonomic components (e.g., the world coordinate basis e_i). Spatial components are numbered by Latin indices from the beginning of the alphabet $a, b, \dots = 1, 2, 3$. To distinguish separate holonomic components from the anholonomic ones, we put hats over the latter indices: e.g., $e_\alpha = \{e_{\hat{0}}, e_{\hat{1}}, e_{\hat{2}}, e_{\hat{3}}\}$ vs. $e_i = \{e_0, e_1, e_2, e_3\}$. The totally antisymmetric Levi-Civita tensor is denoted η_{ijkl}. The Minkowski metric is $g_{\alpha\beta} = \mathrm{diag}(c^2, -1, -1, -1)$, while the coordinate components of a spacetime metric are denoted by g_{ij}. All the objects related to the parity-odd sector (coupling constants, irreducible pieces of the curvature, etc.) are marked by an overline, to distinguish them from the corresponding parity-even objects. Partial derivatives are denoted $\partial^n_{i_1 \dots . i_n} = \frac{\partial^n}{\partial x^{i_1} \dots . \partial x^{i_n}}$.

3.2 Riemann-Cartan Geometry

We model spacetime as a four-dimensional smooth manifold M, and leaving aside the global (topological) aspects, we focus only on local issues. The local coordinates x^i, $i = 0, 1, 2, 3$, are introduced in the neighborhood of an arbitrary point of the spacetime manifold. The geometrical (gravitational) and physical (material) variables are then fields of different nature (both tensors and nontensors) over the spacetime. They are characterized by their components and transformation properties under local diffeomorphisms $x^i \to x'^i(x^k)$. An infinitesimal diffeomorphism

$x^i \to x^i + \delta x^i,$

$$\delta x^i = \xi^i(x), \tag{3.1}$$

is thus parametrized by the four arbitrary functions $\xi^i(x)$.

3.2.1 Geometrical Structures

In the framework of what can be quite generally called an Einsteinian approach (with the principles of equivalence and general coordinate covariance as the cornerstones), the gravitational phenomena are described by the two fundamental geometrical structures on a spacetime manifold: the metric g_{ij} and connection $\Gamma_{ki}{}^j$. As Einstein himself formulated [28], the crucial achievement of his theory was the elimination of the notion of inertial systems as preferred ones among all possible coordinate systems.

From the geometrical point of view, the metric introduces lengths and angles of vectors, and thereby determines the distances (intervals) between points on the spacetime manifold. The connection introduces the notion of parallel transport and defines the covariant differentiation ∇_k of tensor fields. In the metric-affine theory of gravity, the connection is not necessarily symmetric and compatible with the metric. Under infinitesimal diffeomorphisms (3.1), these geometrical variables transform as

$$\delta g_{ij} = -\,(\partial_i \xi^k)\, g_{kj} - (\partial_j \xi^k)\, g_{ik}, \tag{3.2}$$

$$\delta \Gamma_{ki}{}^j = -\,(\partial_k \xi^l)\, \Gamma_{li}{}^j - (\partial_i \xi^l)\, \Gamma_{kl}{}^j + (\partial_l \xi^j)\, \Gamma_{ki}{}^l - \partial^2_{ki} \xi^j. \tag{3.3}$$

The *Riemann-Cartan geometry* of a spacetime manifold is characterized by two tensors: the curvature and the torsion which are defined [27] as

$$R_{kli}{}^j := \partial_k \Gamma_{li}{}^j - \partial_l \Gamma_{ki}{}^j + \Gamma_{kn}{}^j \Gamma_{li}{}^n - \Gamma_{ln}{}^j \Gamma_{ki}{}^n, \tag{3.4}$$

$$T_{kl}{}^i := \Gamma_{kl}{}^i - \Gamma_{lk}{}^i, \tag{3.5}$$

whereas the nonmetricity vanishes:

$$Q_{kij} := -\,\nabla_k g_{ij} = -\,\partial_k g_{ij} + \Gamma_{ki}{}^l g_{lj} + \Gamma_{kj}{}^l g_{il} = 0. \tag{3.6}$$

The curvature and the torsion tensors determine the commutator of the covariant derivatives. For a tensor $A^{i_1\dots i_p}{}_{j_1\dots j_q}$ of arbitrary rank and index structure:

$$(\nabla_k \nabla_l - \nabla_l \nabla_k) A^{i_1\dots i_p}{}_{j_1\dots j_q} = -\,T_{kl}{}^n \nabla_n A^{i_1\dots i_p}{}_{j_1\dots j_q}$$
$$+ \sum_{r=1}^{p} R_{kln}{}^{i_r} A^{i_1\dots n\dots i_p}{}_{j_1\dots j_q} - \sum_{r=1}^{q} R_{klj_r}{}^n A^{i_1\dots i_p}{}_{j_1\dots n\dots j_q}. \tag{3.7}$$

By applying the covariant derivative ∇_l to the metricity condition (3.6), and evaluating the commutator of covariant derivatives, we find

$$R_{lk(ij)} = 0, \tag{3.8}$$

i.e., the curvature tensor is skew-symmetric in both pairs of its indices.

The Riemannian (Levi-Civita) connection $\overset{\circ}{\Gamma}{}_{kj}{}^{i}$ is uniquely determined by the conditions of vanishing torsion and nonmetricity which yield explicitly

$$\overset{\circ}{\Gamma}{}_{kj}{}^{i} = \frac{1}{2} g^{il} (\partial_j g_{kl} + \partial_k g_{lj} - \partial_l g_{kj}). \tag{3.9}$$

Here and in the following, a circle over a symbol denotes a Riemannian object (such as the curvature tensor) or a Riemannian operator (such as the covariant derivative) constructed from the Christoffel symbols (3.9). The deviation of the Riemann-Cartan geometry from the Riemannian one is then conveniently described by the *contortion* tensor

$$K_{kj}{}^{i} := \overset{\circ}{\Gamma}{}_{kj}{}^{i} - \Gamma_{kj}{}^{i}. \tag{3.10}$$

The system (3.5) and (3.6) allows to find the contortion tensor in terms of the torsion:

$$K_{kj}{}^{i} = -\frac{1}{2}(T_{kj}{}^{i} + T^{i}{}_{kj} + T^{i}{}_{jk}). \tag{3.11}$$

From this we can check the skew symmetry in the two last indices, $K_{k(ij)} = 0$, which is also seen directly when we use (3.10) in (3.6). Furthermore, combining (3.10) with (3.5) one can express the torsion tensor in terms of the contortion,

$$T_{kj}{}^{i} = -2K_{[kj]}{}^{i}. \tag{3.12}$$

Substituting (3.10) into (3.4), we find the relation between the non-Riemannian and the Riemannian curvature tensors

$$R_{kli}{}^{j} = \overset{\circ}{R}{}_{kli}{}^{j} - \overset{\circ}{\nabla}_k K_{li}{}^{j} + \overset{\circ}{\nabla}_l K_{ki}{}^{j} + K_{kn}{}^{j} K_{li}{}^{n} - K_{ln}{}^{j} K_{ki}{}^{n}. \tag{3.13}$$

Applying the covariant derivative to (3.4)–(3.6) and antisymmetrizing, we derive the Bianchi identities [27]:

$$\nabla_{[n} R_{kl]i}{}^{j} = T_{[kl}{}^{m} R_{n]mi}{}^{j}, \tag{3.14}$$

$$\nabla_{[n} T_{kl]}{}^{i} = R_{[kln]}{}^{i} + T_{[kl}{}^{m} T_{n]m}{}^{i}. \tag{3.15}$$

3.2.2 Special Cases

When the torsion vanishes,

$$T_{ij}{}^{k} = 0, \tag{3.16}$$

the Riemann-Cartan spacetime reduces to the Riemannian geometry of Einstein's GR, which is characterized by the curvature $\overset{\circ}{R}_{kli}{}^{j}$ constructed from the Christoffel symbols (3.9). In this case, the connection is no longer an independent dynamical variable.

Quite remarkably, the vanishing curvature condition

$$R_{kli}{}^{j} = 0, \tag{3.17}$$

also produces a meaningful spacetime structure. This is known as the *Weitzenböck geometry* [30] which is characterized by the property of a distant parallelism [42, 43]: the result of a parallel transport of a vector from a point x to a point y does not depend on a path along which it is transported. The Weitzenböck geometry underlies another interesting gauge gravity theory which is based on the group of spacetime translations [44–58], also known as teleparallel gravity, which will be the topic of Sect. 3.5 and Chap. 4.

When both conditions (3.16) and (3.17) are satisfied, the spacetime reduces to a flat Minkowski geometry. Figure 3.1 summarizes the landscape of special cases of the Riemann-Cartan geometry.

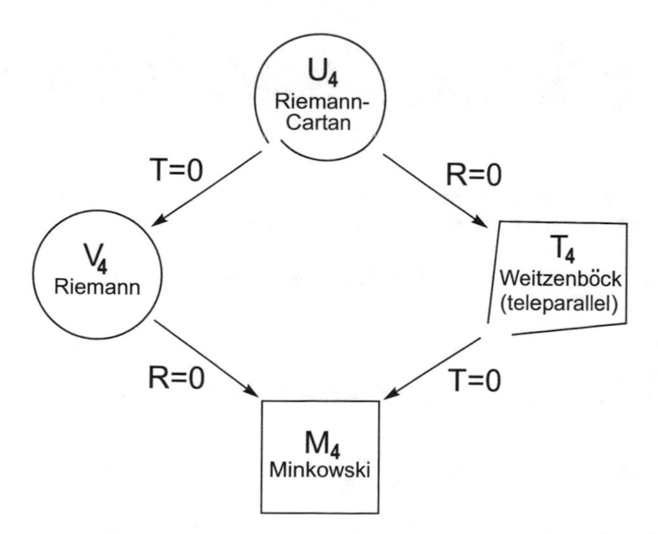

Fig. 3.1 A Riemann-Cartan space U_4 with torsion T and curvature R and its different limits (nonmetricity vanishes: $Q_{kij} := -\nabla_k g_{ij} = 0$), see [19], p. 174. ©Imperial College Press 2013, reproduced with permission. All rights reserved

3.2.3 Local Lorentz Structures (Frame Formalism)

The components of geometrical objects above are defined with respect to a coordinate basis e_i of the tangent space, which is composed of four vectors tangential to the coordinate lines x^i. Under the change of the local coordinates $x^i \to x'^i(x^k)$ the coordinate basis transforms as

$$e'_k = \frac{\partial x^i}{\partial x'^k}\, e_i, \tag{3.18}$$

and this yields the corresponding transformation of the metric, connection, and other geometrical objects. The vectors of the coordinate basis e_i are neither unit, nor orthogonal; their lengths and mutual angles are encoded in the components of the metric via the definition of the scalar product: $g_{ij} = (e_i, e_j)$.

It is then reasonable to introduce, at every point of the spacetime manifold, a local orthonormal frame e_α for which the scalar product $(e_\alpha, e_\beta) = g_{\alpha\beta}$ is equal to the Minkowski metric. To distinguish the local Lorentz frame from the coordinate basis, we label their legs by Greek letters instead of the Latin ones. Decomposing e_α with the respect to e_i,

$$e_\alpha = e^i_\alpha\, e_i\,, \tag{3.19}$$

we find the frame (tetrad, or vierbein) components $e^i_\alpha(x)$. Repeating the same for the coframe (basis of the cotangent space), we eventually obtain the inverse orthonormal coframe with components $e^\alpha_i(x)$. By construction we have

$$g_{ij} = e^\alpha_i\, e^\beta_j\, g_{\alpha\beta}, \tag{3.20}$$

? Exercise

3.1. Find an orthonormal frame for the Kerr metric of a black hole with the total mass M and the angular momentum J, which is described by the line element

$$ds^2 = g_{ij}dx^i dx^j = g_{tt}dt^2 + 2g_{t\phi}dtd\phi + g_{\theta\theta}d\theta^2 + g_{\phi\phi}d\phi^2\,, \tag{3.21}$$

where

$$\begin{aligned} g_{tt} &= c^2\left(1 - \frac{2mr}{\Sigma}\right), \qquad g_{t\varphi} = j_0 c \sin^2\theta\,\frac{2mr}{\Sigma}, \qquad g_{rr} = -\frac{\Sigma}{\Delta}, \\ g_{\theta\theta} &= -\Sigma\,, \qquad g_{\varphi\varphi} = -\frac{\sin^2\theta}{\Sigma}\left((r^2 + j_0^2)^2 - j_0^2\Delta\sin^2\theta\right), \end{aligned} \tag{3.22}$$

with $\Sigma = r^2 + j_0^2\cos^2\theta$, $\Delta = r^2 + j_0^2 - 2mr$, $m = GM/c^2$ and $j_0 = J/Mc$. The Schwarzschild black hole is recovered in the non-rotating limit $j_0 \to 0$.

which explains why the coframe is sometimes called a “square root” of the metric. Normally, a local Lorentz frame is anholonomic since it cannot be constructed from partial derivatives $e_i^\alpha \neq \partial_i f^\alpha$ of some functions $f^\alpha(x)$, and this is measured by the anholonomity object

$$C_{ij}{}^\alpha = \partial_i e_j^\alpha - \partial_j e_i^\alpha. \tag{3.23}$$

In order to describe the parallel transport of vectors with respect to local Lorentz frames, one needs to transform the connection components accordingly:

$$\Gamma_{k\alpha}{}^\beta = e_\alpha^i e_j^\beta \Gamma_{ki}{}^j + e_i^\beta \partial_k e_\alpha^i. \tag{3.24}$$

It is worthwhile to notice that we can rewrite this as

$$\partial_k e_i^\alpha + \Gamma_{k\beta}{}^\alpha e_i^\beta - \Gamma_{ki}{}^j e_j^\alpha = 0, \tag{3.25}$$

which in the literature is sometimes called a “postulate” of the vanishing of the “total” covariant derivative of the coframe. This is an unfortunate misunderstanding (sadly, a widely spread one). Both relations are not postulated, they merely describe a transformation law of the connection. The inverse transformation is straightforwardly derived from (3.24):

$$\Gamma_{ki}{}^j = e_\beta^j e_i^\alpha \Gamma_{k\alpha}{}^\beta + e_\beta^j \partial_k e_i^\beta. \tag{3.26}$$

? Exercise

3.2. Verify (3.26).

By definition (3.18), the tetrad relates the world coordinate and the local Lorentz components of geometrical objects, e.g., $V^\alpha = e_i^\alpha V^i$. By making use of (3.24)–(3.26), we then also relate the covariant derivatives: $e_j^\alpha \nabla_i V^j = D_i V^\alpha = \partial_i V^\alpha + \Gamma_{i\beta}{}^\alpha V^\beta$. It is convenient to distinguish notationally the covariant derivative in the world coordinates ∇_i from the covariant derivative D_i in the local Lorentz frames.

In particular, recasting the metricity condition (3.6) into the local Lorentz disguise we can demonstrate the skew symmetry of the local Lorentz connection:

$$-e_\alpha^i e_\beta^j \nabla_k g_{ij} = -D_k g_{\alpha\beta} = \Gamma_{k\alpha}{}^\gamma g_{\gamma\beta} + \Gamma_{k\beta}{}^\gamma g_{\alpha\gamma} = \Gamma_{k\alpha\beta} + \Gamma_{k\beta\alpha} = 0. \tag{3.27}$$

3.2.4 Symmetries in Riemann-Cartan Space: Generalized Killing Vectors

As is well known, symmetries of a Riemannian spacetime are generated by Killing vector fields. Each such field defines a so-called *motion* of the spacetime manifold, that is a diffeomorphism which preserves the metric g_{ij}. The Lie derivative $\mathcal{L}_\zeta$ is defined along any vector field $\zeta = \zeta^i e_i$ and it maps tensors into tensors of the same rank. Let us recall the explicit form of the Lie derivative of the metric and connection [59,60], which can be derived directly from the transformation laws (3.2) and (3.3):

$$\mathcal{L}_\zeta g_{ij} = \zeta^k \partial_k g_{ij} + (\partial_i \zeta^k) g_{kj} + (\partial_j \zeta^k) g_{ik}, \tag{3.28}$$

$$\mathcal{L}_\zeta \Gamma_{kj}{}^i = \zeta^n \partial_n \Gamma_{kj}{}^i + (\partial_k \zeta^n) \Gamma_{nj}{}^i + (\partial_j \zeta^n) \Gamma_{kn}{}^i - (\partial_n \zeta^i) \Gamma_{kj}{}^n + \partial_k (\partial_j \zeta^i). \tag{3.29}$$

The latter quantity measures the noncommutativity of the Lie derivative with the covariant derivative

$$(\mathcal{L}_\zeta \nabla_k - \nabla_k \mathcal{L}_\zeta) A^{i_1 \dots i_p}{}_{j_1 \dots j_q}$$
$$= \sum_{r=1}^{p} (\mathcal{L}_\zeta \Gamma_{kl}{}^{i_r}) A^{i_1 \dots l \dots i_p}{}_{j_1 \dots j_q} - \sum_{r=1}^{q} (\mathcal{L}_\zeta \Gamma_{kj_r}{}^{l}) A^{i_1 \dots j_p}{}_{j_1 \dots l \dots j_q}. \tag{3.30}$$

A vector field ζ is called a *Killing vector field* if $\mathcal{L}_\zeta g_{ij} = 0$. This condition can be recast, using (3.28), into an equivalent form

$$\overset{\circ}{\nabla}_i \zeta_j + \overset{\circ}{\nabla}_j \zeta_i = 0, \tag{3.31}$$

which is called a Killing equation. By covariant differentiation with respect to the Riemannian connection, after some algebra we derive from this

$$\mathcal{L}_\zeta \overset{\circ}{\Gamma}_{ki}{}^j = 0, \qquad \mathcal{L}_\zeta \overset{\circ}{R}_{kli}{}^j = 0. \tag{3.32}$$

That is, the Lie derivatives along the Killing vector field ζ vanish for all Riemannian geometrical objects. Moreover, one can show that the same is true for all higher covariant derivatives of the Riemannian curvature tensor [60]

$$\mathcal{L}_\zeta \left(\overset{\circ}{\nabla}_{n_1} \dots \overset{\circ}{\nabla}_{n_N} \overset{\circ}{R}_{kli}{}^j \right) = 0. \tag{3.33}$$

Let us generalize the notion of a symmetry to the Riemann-Cartan spacetime. We begin by noticing that for an arbitrary $\lambda_\alpha{}^\beta$ (Greek indices mean that this object is defined with respect to a local Lorentz frame e_i^α), we can recast (3.28) and (3.29) into

$$\begin{aligned}\mathcal{L}_\zeta g_{ij} =& \left(\mathcal{L}_\zeta e_i^\alpha - \lambda_\gamma{}^\alpha e_i^\gamma\right) e_j^\beta g_{\alpha\beta} + \left(\mathcal{L}_\zeta e_j^\beta - \lambda_\gamma{}^\beta e_j^\gamma\right) e_i^\alpha g_{\alpha\beta} \\ &+ e_i^\alpha e_j^\beta \left(\mathcal{L}_\zeta g_{\alpha\beta} + \lambda_{\alpha\beta} + \lambda_{\beta\alpha}\right), \end{aligned} \tag{3.34}$$

$$\begin{aligned}\mathcal{L}_\zeta \Gamma_{ki}{}^j =& \left(\mathcal{L}_\zeta e_\alpha^j + \lambda_\alpha{}^\gamma e_\gamma^j\right) D_k e_i^\alpha + e_\alpha^j D_k \left(\mathcal{L}_\zeta e_i^\alpha - \lambda_\gamma{}^\alpha e_i^\gamma\right) \\ &+ e_\alpha^j e_i^\beta \left(\mathcal{L}_\zeta \Gamma_{k\beta}{}^\alpha + D_k \lambda_\beta{}^\alpha\right). \end{aligned} \tag{3.35}$$

This is straightforwardly derived from (3.20) and (3.26) by making use of the standard definitions of the Lie derivatives of the coframe and the local Lorentz connection (which are, geometrically, both covectors), $\mathcal{L}_\zeta e_k^\alpha = \zeta^i \partial_i e_k^\alpha + (\partial_k \zeta^i) e_i^\alpha$ and $\mathcal{L}_\zeta \Gamma_{k\beta}{}^\alpha = \zeta^i \partial_i \Gamma_{k\beta}{}^\alpha + (\partial_k \zeta^i) \Gamma_{i\beta}{}^\alpha$, whereas for the world scalar $\mathcal{L}_\zeta g_{\alpha\beta} = \zeta^i \partial_i g_{\alpha\beta}$.

Accordingly, a natural definition of the symmetry of the Riemann-Cartan manifold can be formulated as the set of conditions

$$\mathcal{L}_\zeta g_{\alpha\beta} = -\lambda_{\alpha\beta} - \lambda_{\beta\alpha}, \tag{3.36}$$

$$\mathcal{L}_\zeta e_i^\alpha = \lambda_\beta{}^\alpha e_i^\beta, \tag{3.37}$$

$$\mathcal{L}_\zeta \Gamma_{k\beta}{}^\alpha = -D_k \lambda_\beta{}^\alpha, \tag{3.38}$$

where the possible form of $\lambda_\beta{}^\alpha(x)$ is eventually determined by the vector field ζ, which is naturally called a *generalized Killing vector* of the Riemann-Cartan spacetime. We will solve these equations explicitly for the spherical symmetry in Sect. 3.4.2). Since the Minkowski metric $g_{\alpha\beta}$ has constant components, (3.36) yields the skew symmetry $\lambda_{\alpha\beta} = -\lambda_{\beta\alpha}$. As a result, from (3.34) and (3.35) we find

$$\mathcal{L}_\zeta g_{ij} = 0, \tag{3.39}$$

$$\mathcal{L}_\zeta \Gamma_{ki}{}^j = 0. \tag{3.40}$$

Thereby, the generalized Killing vector ζ generates a diffeomorphism of the spacetime manifold that is simultaneously an isometry (3.39) and an isoparallelism (3.40).

Combining (3.40), (3.32) and (3.10), we derive $\mathcal{L}_\zeta K_{ki}{}^j = 0$ for the contortion, and we accordingly conclude that the generalized Killing vector leaves the torsion and the Riemann-Cartan curvature tensors invariant

$$\mathcal{L}_\zeta T_{ij}{}^k = 0, \qquad \mathcal{L}_\zeta R_{klj}{}^i = 0. \tag{3.41}$$

? Exercise

3.3. Verify (3.41).

It is also straightforward to demonstrate that

$$\mathcal{L}_\zeta\left(\nabla_{n_1}\dots\nabla_{n_N}T_{ij}{}^k\right) = 0, \qquad \mathcal{L}_\zeta\left(\nabla_{n_1}\dots\nabla_{n_N}R_{klj}{}^i\right) = 0\,, \tag{3.42}$$

for any number of covariant derivatives of the torsion and the curvature.

3.2.5 Matter Variables

Without specializing the discussion of matter to any particular physical field, we can describe matter by a generalized field ψ^A. The range of the indices $A, B, \dots$ is not important in our study. However, we do need to know the behavior of the matter field under spacetime diffeomorphisms (3.1):

$$\delta\psi^A = -\,(\partial_i\xi^j)\,(\sigma_j{}^i)^A{}_B\,\psi^B. \tag{3.43}$$

Here $(\sigma_j{}^i)^A{}_B$ are the generators of general coordinate transformations that satisfy the commutation relations

$$(\sigma_j{}^i)^A{}_C(\sigma_l{}^k)^C{}_B - (\sigma_l{}^k)^A{}_C(\sigma_j{}^i)^C{}_B = (\sigma_l{}^i)^A{}_B\,\delta^k_j - (\sigma_j{}^k)^A{}_B\,\delta^i_l. \tag{3.44}$$

We immediately recognize in (3.44) the Lie algebra of the general linear group $GL(4, R)$. This fact is closely related to the standard gauge-theoretic interpretation [16] of metric-affine gravity as the gauge theory of the general affine group $GA(4, R)$, which is a semidirect product of spacetime translation group times $GL(4, R)$.

The transformation properties (3.43) determine the form of the covariant and the Lie derivative of a matter field:

$$\nabla_k\psi^A := \partial_k\psi^A - \Gamma_{ki}{}^j\,(\sigma_j{}^i)^A{}_B\,\psi^B, \tag{3.45}$$

$$\mathcal{L}_\zeta\psi^A := \zeta^k\partial_k\psi^A + (\partial_i\zeta^j)(\sigma_j{}^i)^A{}_B\,\psi^B. \tag{3.46}$$

The commutators of these differential operators read

$$(\nabla_k\nabla_l - \nabla_l\nabla_k)\psi^A = -\,R_{klj}{}^i\,(\sigma_i{}^j)^A{}_B\psi^B - T_{kl}{}^i\,\nabla_i\psi^A, \tag{3.47}$$

$$(\mathcal{L}_\zeta\nabla_k - \nabla_k\mathcal{L}_\zeta)\psi^A = -\,(\mathcal{L}_\zeta\Gamma_{kj}{}^i)(\sigma_i{}^j)^A{}_B\psi^B. \tag{3.48}$$

3.2.6 Irreducible Decomposition of Curvature and Torsion

In order to establish the dynamical scheme of the Poincaré gauge gravity in a most transparent way, and also to understand more clearly the coupling of the gravitational field to the physical sources of different physical nature, it is convenient to decompose the Poincaré gauge field strengths, the curvature and the torsion, into irreducible parts.

With the help of the metric g_{ij} and the totally antisymmetric Levi-Civita tensor η_{ijkl}, one can construct a number of contractions of the curvature. In particular, we introduce the *Ricci tensor* and the *co-Ricci tensor* as

$$R_{ij} := R_{kij}{}^{k}, \qquad \overline{R}^{ij} := \frac{1}{2}\, R_{klm}{}^{i}\, \eta^{klmj}, \tag{3.49}$$

respectively. By definition, the former is a parity-even object, whereas the latter is a parity-odd one. We can split (3.49) into the skew-symmetric and symmetric pieces

$$R_{ij} = R_{[ij]} + R_{(ij)}, \qquad \overline{R}^{ij} = \overline{R}^{[ij]} + \overline{R}^{(ij)}, \tag{3.50}$$

and, furthermore, extract the traceless parts from the latter

$$\not{R}_{ij} := R_{(ij)} - \frac{1}{4} R g_{ij}, \qquad \not{\overline{R}}{}^{ij} := \overline{R}^{(ij)} - \frac{1}{4}\overline{R} g^{ij}. \tag{3.51}$$

Here the curvature scalar and pseudoscalar arise naturally as the traces

$$R = g^{ij}\, R_{ij} = R_{ij}{}^{ji}, \qquad \overline{R} = g_{ij}\,\overline{R}^{ij} = \frac{1}{2}\, R_{ijkl}\, \eta^{ijkl}. \tag{3.52}$$

With the help of a straightforward algebra we can verify that the antisymmetric Ricci parts are related via

$$\overline{R}^{[ij]} = \frac{1}{2} \eta^{ijkl}\, R_{[kl]}, \tag{3.53}$$

and it will be convenient to denote the skew-symmetric tensor as

$$\check{R}_{ij} := R_{[kl]}. \tag{3.54}$$

Then irreducible parts of the curvature tensor are as follows:

$${}^{(2)}R_{kl}{}^{ij} = \not{\overline{R}}_{m}{}^{[i}\,\eta^{j]m}{}_{kl}, \tag{3.55}$$

$${}^{(3)}R_{kl}{}^{ij} = -\frac{1}{12}\,\overline{R}\,\eta_{kl}{}^{ij}, \tag{3.56}$$

$${}^{(4)}R_{kl}{}^{ij} = -2\not{R}_{[k}{}^{[i}\,\delta^{j]}_{l]}, \tag{3.57}$$

$$^{(5)}R_{kl}{}^{ij} = -2\check{R}_{[k}{}^{[i}\,\delta_{l]}^{j]}, \tag{3.58}$$

$$^{(6)}R_{kl}{}^{ij} = -\frac{1}{6}\,R\,\delta_{[k}{}^{i}\,\delta_{l]}^{j}. \tag{3.59}$$

In the literature, the five objects (3.55)–(3.59) are known [16] as the "paircom", "pscalar", "ricsymf", "ricanti", and "scalar" parts, respectively. Finally, the Weyl part is defined as

$$^{(1)}R_{kl}{}^{ij} = R_{kl}{}^{ij} - \sum_{I=2}^{6}{}^{(I)}R_{kl}{}^{ij} \tag{3.60}$$

Introducing the trace vector and the axial trace vector, respectively,

$$T_j := T_{ij}{}^{i}, \qquad \overline{T}^j = \frac{1}{2}T_{kli}\eta^{klij}, \tag{3.61}$$

the torsion tensor decomposition reads $T_{kl}{}^{i} = {}^{(1)}T_{kl}{}^{i} + {}^{(2)}T_{kl}{}^{i} + {}^{(3)}T_{kl}{}^{i}$, with

$$^{(2)}T_{kl}{}^{i} = \frac{2}{3}\,\delta_{[k}^{i}T_{l]}, \tag{3.62}$$

$$^{(3)}T_{kl}{}^{i} = -\frac{1}{3}\eta_{kl}{}^{ij}\overline{T}_j, \tag{3.63}$$

$$^{(1)}T_{kl}{}^{i} = T_{kl}{}^{i} - {}^{(2)}T_{kl}{}^{i} - {}^{(3)}T_{kl}{}^{i}. \tag{3.64}$$

3.3 General Structure of Poincaré Gauge Gravity

The gauging of the Poincaré symmetry group is well understood within the framework of a general gauge-theoretic approach which is formulated as a heuristic scheme in the Lagrange formalism in the Minkowski space of special relativity for the purpose of deriving a new interaction from a conserved Noether current associated with rigid symmetry group [4–6]. Such a new gauge interaction arises from the requirement that the rigid (global) symmetry should be extended to a local symmetry.

In contrast to the standard theories of electroweak and strong interactions, which are based on the gauging of internal symmetries, the gauge theory of the gravitational interaction is underlied by external symmetry groups of the spacetime. In the absence of gravity, the fundamental spacetime symmetry of the flat Minkowski space is its group of motions, namely, the Poincaré group $T_4 \rtimes SO(1,3)$, the semi-direct product of the translations group T_4 (four parameters ε^α) and the Lorentz group $SO(1,3)$ (six parameters $\varepsilon^{\alpha\beta} = -\varepsilon^{\beta\alpha}$). The corresponding Lagrange-Noether treatment of the invariance of Minkowski space under rigid (global) Poincaré transformations gives rise to the conservation laws of the canonical energy-momentum $\Sigma_\alpha{}^i$ and spin angular momentum $\tau_{\alpha\beta}{}^i = -\tau_{\beta\alpha}{}^i$ currents. In relation with this, it is worthwhile to recall Wigner's classification [61] of quantum mechanical systems in a Minkowski space according to *mass and spin.*

An appropriate gauge-theoretic formalism that extends the approach of Yang and Mills [7] from the case of internal symmetries to the spacetime symmetry groups was developed by Utiyama, Sciama and Kibble [8–10]. Up-to-date reviews of the Poincaré gauge theory of gravity can be found in [3, 24, 25], and for more historic and technical details readers may refer to [12, 16, 19, 21, 22]. Here we briefly outline the most essential notions and constructions.

3.3.1 Poincaré Gauge Gravity Kinematics

Following the general Yang-Mills-Utiyama-Sciama-Kibble gauge-theoretic scheme, the 10-parameter Poincaré group $T_4 \rtimes SO(1,3)$ gives rise to the 10-plet of the gauge potentials which are consistently identified with the orthonormal coframe e_i^α (4 potentials corresponding to the translation subgroup T_4) and the local connection $\Gamma_i{}^{\alpha\beta} = -\Gamma_i{}^{\beta\alpha}$ (6 potentials for the Lorentz subgroup $SO(1,3)$). The corresponding field strengths of translations and Lorentz rotations arise as covariant "curls"

$$T_{ij}{}^\alpha = \partial_i e_j{}^\alpha - \partial_j e_i{}^\alpha + \Gamma_{i\beta}{}^\alpha e_j{}^\beta - \Gamma_{j\beta}{}^\alpha e_i{}^\beta, \tag{3.65}$$

$$R_{ij\alpha}{}^\beta = \partial_i \Gamma_{j\alpha}{}^\beta - \partial_j \Gamma_{i\alpha}{}^\beta + \Gamma_{i\gamma}{}^\beta \Gamma_{j\alpha}{}^\gamma - \Gamma_{j\gamma}{}^\beta \Gamma_{i\alpha}{}^\gamma. \tag{3.66}$$

Comparing these with (3.4) and (3.5), respectively, with an account of (3.26), we immediately identify the Poincaré gauge field strengths (3.65) and (3.66) with the torsion $T_{ij}{}^\alpha = e_k^\alpha T_{ij}{}^k$ and the curvature $R_{ij\alpha}{}^\beta = e_\alpha^k e_k^\beta R_{ijk}{}^l$ of the Riemann-Cartan geometry on the spacetime manifold.

In accordance with the heuristic gauging scheme, the gravitational spin-connection interaction is derived from the rigid Lorentz symmetry of a matter field ψ^A which belongs to a representation of the Lorentz group with the generators $(\rho_{\alpha\beta})^A{}_B$:

$$\delta\psi^A = -\frac{1}{2}\varepsilon^{\alpha\beta}(\rho_{\alpha\beta})^A{}_B\,\psi^B. \tag{3.67}$$

When the transformation is extended to the local one with infinitesimal parameters $\varepsilon^{\alpha\beta} = \varepsilon^{\alpha\beta}(x)$, the covariant derivative is introduced by

$$D_i\psi^A = \partial_i\psi^A - \frac{1}{2}\Gamma_i{}^{\alpha\beta}(\rho_{\alpha\beta})^A{}_B\,\psi^B. \tag{3.68}$$

The Poincaré gauge field strengths satisfy the Bianchi identities, cf. (3.14) and (3.15):

$$D_{[k}T_{ij]}{}^{\alpha} = e_{[k}{}^{\beta}R_{ij]\beta}{}^{\alpha}, \tag{3.69}$$

$$D_{[k}R_{ij]}{}^{\alpha\beta} = 0. \tag{3.70}$$

Note that the rigid Lie algebra of the Poincaré group is extended to a so-called deformed, soft, or local "Lie algebra" (D_α and $\rho_{\alpha\beta} = -\rho_{\beta\alpha}$ generate translations and Lorentz transformations, respectively):

$$\left.\begin{aligned} [D_\alpha, D_\beta] &= -T_{\alpha\beta}{}^{\gamma}D_\gamma + R_{\alpha\beta}{}^{\gamma\delta}\rho_{\delta\gamma} \\ [\rho_{\alpha\beta}, D_\gamma] &= -g_{\gamma\alpha}D_\beta + g_{\gamma\beta}D_\alpha \\ [\rho_{\alpha\beta}, \rho_{\mu\nu}] &= -g_{\alpha\mu}\rho_{\beta\nu} + g_{\alpha\nu}\rho_{\beta\mu} + g_{\beta\mu}\rho_{\alpha\nu} - g_{\beta\nu}\rho_{\alpha\mu} \end{aligned}\right\}. \tag{3.71}$$

The rigid Lie algebra of Minkowski space is recovered for $T_{\alpha\beta}{}^{\gamma} = 0$ and $R_{\alpha\beta}{}^{\gamma\delta} = 0$, when $D_\alpha \to \partial_a$ in Cartesian coordinates, for details see [12].

3.3.2 Poincaré Gauge Gravity Dynamics: Yang-Mills Type Models

Assuming the standard minimal coupling, the total Lagrangian of interacting gravitational and matter fields reads

$$L = V(g_{ij}, R_{ijk}{}^{l}, T_{ki}{}^{j}) + L_{\rm mat}(g_{ij}, \psi^A, \nabla_i\psi^A). \tag{3.72}$$

In general, the gravitational Lagrangian V is constructed as a diffeomorphism invariant function of the curvature and torsion. The matter Lagrangian $L_{\rm mat}$ depends on the matter fields ψ^A and their covariant derivatives $\nabla_i\psi^A$.

Let us now specialize to the general quadratic model with the Lagrangian that contains all possible quadratic invariants of the torsion and the curvature:

$$\begin{aligned} V = -\frac{1}{2\kappa c}\Bigg\{ & a_0 R + \overline{a}_0\overline{R} + 2\lambda_0 + \frac{1}{2}\sum_{I=1}^{3}\Big[a_I\,{}^{(I)}T^{kl}{}_{i}\,T_{kl}{}^{i} - \frac{\overline{a}_I}{2}\,{}^{(I)}T_{mni}\,T_{kl}{}^{i}\,\eta^{mnkl}\Big] \\ & + \frac{\ell_\rho^2}{2}\sum_{I=1}^{6}\Big[b_I\,{}^{(I)}R^{kl}{}_{ij}\,R_{kl}{}^{ij} - \frac{\overline{b}_I}{2}\,{}^{(I)}R_{mnij}\,R_{kl}{}^{ij}\,\eta^{mnkl}\Big]\Bigg\}. \end{aligned} \tag{3.73}$$

Here $\kappa = \frac{8\pi G}{c^4}$ is Einstein's gravitational constant, so the dimension of $[\kappa c] = \mathrm{s\,kg^{-1}}$. $G = 6.67 \times 10^{-11}\ \mathrm{m^3\,kg^{-1}\,s^{-2}}$ is Newton's gravitational constant. The speed of light $c = 2.9 \times 10^8$ m/s.

Besides the linear "Hilbert type" part characterized by a_0 and $\overline{a}_0$, the Lagrangian (3.73) contains several additional coupling constants which fix the "Yang-Mills type" part: $a_1, a_2, a_3, \overline{a}_1, \overline{a}_2, \overline{a}_3, b_1, \cdots, b_6, \overline{b}_1, \cdots, \overline{b}_6$, and ℓ_ρ^2. The latter has the dimension $[\ell_\rho^2] = $[area] so that $[\ell_\rho^2/\kappa c] = [\hbar]$, whereas a_I, $\overline{a}_I$, b_I and $\overline{b}_I$ are dimensionless. Moreover, not all of these constants are independent: in the parity-odd sector we take $\overline{b}_2 = \overline{b}_4$ and $\overline{b}_3 = \overline{b}_6$ because the two pairs of terms in (3.73) are the same:

$$ {}^{(2)}\!R_{mnij}\, R_{kl}{}^{ij}\, \eta^{mnkl} = {}^{(4)}\!R_{mnij}\, R_{kl}{}^{ij}\, \eta^{mnkl} = {}^{(2)}\!R_{mnij}\, {}^{(4)}\!R_{kl}{}^{ij}\, \eta^{mnkl}, \qquad (3.74) $$

$$ {}^{(3)}\!R_{mnij}\, R_{kl}{}^{ij}\, \eta^{mnkl} = {}^{(6)}\!R_{mnij}\, R_{kl}{}^{ij}\, \eta^{mnkl} = {}^{(3)}\!R_{mnij}\, {}^{(6)}\!R_{kl}{}^{ij}\, \eta^{mnkl}, \qquad (3.75) $$

whereas ${}^{(1)}\!R_{mnij}\, R_{kl}{}^{ij}\, \eta^{mnkl} = {}^{(1)}\!R_{mnij}\, {}^{(1)}\!R_{kl}{}^{ij}\, \eta^{mnkl}$ and ${}^{(5)}\!R_{mnij}\, R_{kl}{}^{ij}\, \eta^{mnkl} = {}^{(5)}\!R_{mnij}\, {}^{(5)}\!R_{kl}{}^{ij}\, \eta^{mnkl}$. Similarly, for the torsion one finds

$$ {}^{(2)}T_{mni}\, T_{kl}{}^{i}\, \eta^{mnkl} = {}^{(3)}T_{mni}\, T_{kl}{}^{i}\, \eta^{mnkl} = {}^{(2)}T_{mni}\, {}^{(3)}T_{kl}{}^{i}\, \eta^{mnkl}, \qquad (3.76) $$

which leads to a constraint $\overline{a}_2 = \overline{a}_3$.

? Exercise

3.4. Use the definitions (3.55)–(3.60) and (3.62)–(3.64), to prove (3.74), (3.75) and (3.76).

For completeness, we included the cosmological constant λ_0 with the dimension of the inverse area, $[\lambda_0] = [\ell^{-2}]$. In the special case $a_0 = \overline{a}_0 = 0$ the purely quadratic model is obtained without the Hilbert-Einstein linear term in the Lagrangian.

In the literature, the quadratic Poincaré gravity theories are often formulated in terms of the standard tensor objects which are not decomposed into irreducible parts. In order to be able to compare (3.73) to the models studied in the literature, let us rewrite the Lagrangian V explicitly:

$$ \begin{aligned} V = -\frac{1}{2\kappa c}\Big\{ & a_0 R + \overline{a}_0 \overline{R} + 2\lambda_0 \\ & + \alpha_1\, T_{kl}{}^{i}\, T^{kl}{}_{i} + \alpha_2\, T_i\, T^i + \alpha_3\, T_{kl}{}^{i}\, T_i{}^{kl} \\ & + \overline{\alpha}_1\, \eta^{klmn}\, T_{kli}\, T_{mn}{}^{i} + \overline{\alpha}_2\, \eta^{klmn}\, T_{klm}\, T_n \\ + \ell_\rho^2 \Big(& \beta_1\, R_{ijkl} R^{ijkl} + \beta_2\, R_{ijkl} R^{ikjl} + \beta_3\, R_{ijkl} R^{klij} \\ & + \beta_4\, R_{ij} R^{ij} + \beta_5\, R_{ij} R^{ji} + \beta_6\, R^2 \\ & + \overline{\beta}_1\, \eta^{klmn}\, R_{klij} R_{mn}{}^{ij} + \overline{\beta}_2\, \eta^{klmn}\, R_{kl} R_{mn} \\ & + \overline{\beta}_3\, \eta^{klmn}\, R_{klm}{}^{i}\, R_{ni} + \overline{\beta}_4\, \eta^{klmn}\, R_{klmn}\, R \Big) \Big\}. \end{aligned} \qquad (3.77) $$

Using the definitions of the irreducible torsion (3.62)–(3.64) and curvature (7.43)–(3.60) parts, we find the relation between the coupling constants:

$$a_1 = 2\alpha_1 - \alpha_3, \quad a_2 = 2\alpha_1 + 3\alpha_2 - \alpha_3, \quad a_3 = 2\alpha_1 + 2\alpha_3, \tag{3.78}$$

$$\overline{a}_1 = -4\overline{\alpha}_1, \quad \overline{a}_2 = \overline{a}_3 = -4\overline{\alpha}_1 - 3\overline{\alpha}_2, \tag{3.79}$$

$$b_1 = 2\beta_1 + \beta_2 + 2\beta_3, \tag{3.80}$$

$$b_2 = 2\beta_1 - 2\beta_3, \tag{3.81}$$

$$b_3 = 2\beta_1 - 2\beta_2 + 2\beta_3, \tag{3.82}$$

$$b_4 = 2\beta_1 + \beta_2 + 2\beta_3 + \beta_4 + \beta_5, \tag{3.83}$$

$$b_5 = 2\beta_1 - 2\beta_3 + \beta_4 - \beta_5, \tag{3.84}$$

$$b_6 = 2\beta_1 + \beta_2 + 2\beta_3 + 3\beta_4 + 3\beta_5 + 12\beta_6, \tag{3.85}$$

$$\overline{b}_1 = -4\overline{\beta}_1, \quad \overline{b}_2 = \overline{b}_4 = -4\overline{\beta}_1 + \overline{\beta}_3, \tag{3.86}$$

$$\overline{b}_5 = -4\overline{\beta}_1 - 2\overline{\beta}_2 + 2\overline{\beta}_3, \quad \overline{b}_3 = \overline{b}_6 = -4\overline{\beta}_1 + 3\overline{\beta}_3 + 12\overline{\beta}_4. \tag{3.87}$$

The inverse of (3.78) reads

$$\alpha_1 = \frac{2a_1 + a_3}{6}, \quad \alpha_2 = \frac{a_2 - a_1}{3}, \quad \alpha_3 = \frac{a_3 - a_1}{3}. \tag{3.88}$$

Not all terms in the Lagrangian (3.77) are independent, since the expressions

$$V_{\rm GB} = -\frac{1}{4}\,\eta^{klmn}\eta_{ijpq}R_{kl}{}^{ij}R_{mn}{}^{pq} = R_{klij}R^{ijkl} - 4R_{ij}R^{ji} + R^2, \tag{3.89}$$

$$V_{\rm PC} = \frac{1}{2}\,\eta^{klmn}R_{klij}R_{mn}{}^{ij}, \tag{3.90}$$

$$V_{\rm NY} = \overline{R} + \frac{1}{2}\,\eta^{klmn}T_{kli}T_{mn}{}^{i}, \tag{3.91}$$

are the total divergences. Integrating these scalar quantities (with appropriate normalization factors) over the spacetime manifold, one obtains the topological invariants [62–64] known as the Euler, Pontryagin (or Chern), and Nieh-Yan characteristics, respectively. Therefore, some of the constants $\beta_3, \beta_5, \beta_6, \overline{\beta}_1, \overline{a}_0$, and $\overline{\alpha}_1$, may be eliminated (same applies to the set of constants $b_I, \overline{b}_I, \overline{a}_J$). However, here this possibility is not used.

The Poincaré gauge gravity field equations arise from the variation of the total action with respect to the coframe and connection. They read explicitly

$$a_0\Big(R_i{}^j - \frac{1}{2}R\delta_i^j\Big) - \overline{a}_0\overline{R}_i{}^j - \lambda_0\delta_i^j$$
$$+ \overset{(T)}{q}{}_i{}^j + \ell_\rho^2\overset{(R)}{q}{}_i{}^j - \Big[(\nabla_l - T_l)h^{jl}{}_i + \frac{1}{2}T_{mn}{}^j h^{mn}{}_i\Big] = \kappa\Sigma_i{}^j, \tag{3.92}$$

$$a_0\left(T_{ij}{}^k + 2T_{[i}\delta^k_{j]}\right) - \frac{\overline{a}_0}{2}\,\eta_{ij}{}^{mn}\left(T_{mn}{}^k + 2T_{[m}\delta^k_{n]}\right) - 2h^k{}_{[ij]}$$
$$-2\ell_\rho^2\Big[(\nabla_l - T_l)h^{kl}{}_{ij} + \frac{1}{2}T_{mn}{}^k h^{mn}{}_{ij}\Big] = \kappa c\tau_{ij}{}^k. \tag{3.93}$$

Here the gravitational momenta are described by

$$h^{ij}{}_k = \sum_{I=1}^{3} a_I\,{}^{(I)}T^{ij}{}_k - \frac{1}{2}\eta^{ij}{}_{mn}\sum_{I=1}^{3}\overline{a}_I\,{}^{(I)}T^{mn}{}_k, \tag{3.94}$$

$$h^{ij}{}_{kl} = \sum_{I=1}^{6} b_I\,{}^{(I)}R^{ij}{}_{kl} - \frac{1}{2}\eta^{ij}{}_{mn}\sum_{I=1}^{6}\overline{b}_I\,{}^{(I)}R^{mn}{}_{kl}, \tag{3.95}$$

from which we construct the two objects quadratic in the torsion and the curvature

$$\overset{(T)}{q}{}_i{}^j = T_{in}{}^k h^{jn}{}_k - \frac{1}{4}\delta^j_i\, T_{mn}{}^k h^{mn}{}_k, \tag{3.96}$$

$$\overset{(R)}{q}{}_i{}^j = R_{in}{}^{kl} h^{jn}{}_{kl} - \frac{1}{4}\delta^j_i\, R_{mn}{}^{kl} h^{mn}{}_{kl}. \tag{3.97}$$

3.3.2.1 Example: Matter with Spin

In order to give an explicit example of a typical physical matter source of the gravitational field in PG theory, we recall the classical model of spinning fluid [65]. This model was first worked out by Weyssenhoff and Raabe [66] as a direct development of the ideas of Cosserats [67] who proposed to describe the microstructure properties of a medium by attaching a rigid material frame to every element of a continuum. Using the variational principle for the spinning fluid [68], one derives the canonical energy-momentum and spin tensors:

$$\Sigma_j{}^i = u^i\mathcal{P}_j - p\Big(\delta^i_j - \frac{1}{c^2}u_j u^i\Big), \tag{3.98}$$

$$\tau_{ij}{}^k = u^k\mathcal{S}_{ij}, \tag{3.99}$$

where u^i is the 4-velocity of the fluid and p is the pressure. Fluid elements are characterized by their microstructural properties: the energy density ε, the intrinsic spin density $\mathcal{S}_{ij} = -\mathcal{S}_{ji}$ (subject to the Frenkel supplementary condition $\mathcal{S}_{ij}u^j = 0$), and the momentum density

$$\mathcal{P}_i = \frac{1}{c^2}\left[\varepsilon u_i + u^j(\nabla_k - T_k)(\mathcal{S}_{ij}u^k)\right]. \tag{3.100}$$

3.3.3 Parity-Even Model: Particle Spectrum

To streamline the subsequent discussion, we now specialize to the *purely parity-even model* and hence assume $\overline{a}_0 = 0$, $\overline{a}_I = 0$, and $\overline{b}_I = 0$ in the rest of the paper.

The number of graviton modes in PG theory (that mediate the gravitational interaction) is much larger than in Einstein's theory. The analysis of the particle spectrum for the quadratic model (3.73) reveals [69–71] that the dynamics of graviton modes in different J^P (spin^{parity}) sectors is determined by the following combinations of the coupling constants:

$$\left.\begin{aligned} 2^+ : \Lambda_1 &= b_1 + b_4, \\ 2^- : \Lambda_2 &= b_1 + b_2, \\ 0^+ : \Lambda_3 &= b_4 + b_6, \\ 0^- : \Lambda_4 &= b_2 + b_3, \\ 1^+ : \Lambda_5 &= b_2 + b_5, \\ 1^- : \Lambda_6 &= b_4 + b_5, \end{aligned}\right\} \tag{3.101}$$

$$\left.\begin{aligned} \mu_1 &= -a_0 + 2a_3, \\ \mu_2 &= -2a_0 + a_2, \\ \mu_3 &= -a_0 - a_1, \end{aligned}\right\} \tag{3.102}$$

which specify the corresponding kinetic and mass terms for these modes. The mapping (3.101) between the two sets of coupling constants b_I and Λ_I is not one-to-one. Namely, whereas $b_I = 0$ yields $\Lambda_I = 0$, inverse is not true, and from $\Lambda_I = 0$ we find $b_1 = -b_2 = b_3 = -b_4 = b_5 = b_6$, which brings the curvature square part of the gravitational Lagrangian (3.77) to $b_1 V_{\rm GB}$, cf. (3.89).

The general analysis of the particle spectrum with both parity-even and parity-odd sectors included can be found in the recent papers [72, 73].

3.4 Physical Consequences of Poincaré Gauge Gravity

In this introductory review, we do not aim to deal with all the aspects of PG theory. Instead, we will focus on the issues of correspondence of PG and GR for the whole class of models (3.73) and then will discuss in some more detail several specific models that were studied in the literature.

3.4.1 Correspondence with GR: Torsionless Solutions

Let us discuss the correspondence of the vacuum field equations (3.92)–(3.93) for the Yang-Mills type quadratic Lagrangian (3.73) and Einstein's general relativity under the assumption of *vanishing torsion*, $T_{kl}{}^{i} = 0$. Then we find $h^{ij}{}_{k} = 0$, hence $\overset{(T)}{q}{}_{i}{}^{j} = 0$, whereas $\overline{R}_{i}{}^{j} = 0$ and the curvature has only three nontrivial parts (3.57), (3.59), (3.60). A direct computation yields

$$\overset{(R)}{q}{}_{i}{}^{j} = \Lambda_1 {}^{(1)}R_{ikl}{}^{j}\, \not{R}^{kl} + \frac{\Lambda_3}{6} R\, \not{R}_{i}{}^{j}, \tag{3.103}$$

and in vacuum (when $\Sigma_{i}{}^{j} = 0$ and $\tau_{ij}{}^{k} = 0$) the field equations (3.92)–(3.93) reduce to

$$a_0\Big(R_{i}{}^{j} - \frac{1}{2}R\delta_{i}^{j}\Big) - \lambda_0\delta_{i}^{j} + \ell_\rho^2\Lambda_1 {}^{(1)}R_{ikl}{}^{j}\, \not{R}^{kl} + \frac{\ell_\rho^2\Lambda_3}{6} R\, \not{R}_{i}{}^{j} = 0, \tag{3.104}$$

$$\Lambda_1\nabla_l {}^{(1)}R^{kl}{}_{ij} + \frac{\Lambda_3}{6}\,\nabla_{[i} R\,\delta_{j]}^{k} = 0. \tag{3.105}$$

As we see, only two coupling constants Λ_1 and Λ_3 enter the field equations. This is consistent with the fact that they essentially determine the structure of the effective Lagrangian obtained from (3.77) for the vanishing torsion, $T_{kl}{}^{i} = 0$:

$$\begin{aligned} V = -\frac{1}{2\kappa c}\Big(a_0 R + 2\lambda_0 + \frac{\ell_\rho^2}{12}\Big\{(4\Lambda_1 - \Lambda_3)\, R_{ijkl}R^{ijkl} + 4(\Lambda_3 - \Lambda_1)\, R_{ij}R^{ij}\Big\} \\ + \ell_\rho^2\beta_6\Big\{R_{ijkl}R^{klij} - 4R_{ij}R^{ji} + R^2\Big\}\Big). \end{aligned} \tag{3.106}$$

The last line does not contribute to the field equations, being a total divergence of the Euler (Gauss-Bonnet) topological term (3.89). When $\Lambda_1 = 0$ and $\Lambda_3 = 0$, the field equations reduce to Einstein's equation with the cosmological term.

Contracting (3.104), we find

$$a_0 R = -4\lambda_0, \tag{3.107}$$

and, provided $a_0 \neq 0$, the system (3.104)–(3.105) is recast into

$$a_0^{\text{eff}}\, \not{R}_{i}{}^{j} + \ell_\rho^2\Lambda_1 {}^{(1)}R_{ikl}{}^{j}\, \not{R}^{kl} = 0, \tag{3.108}$$

$$\Lambda_1\Big(\nabla_i \not{R}_{j}{}^{k} - \nabla_j \not{R}_{i}{}^{k}\Big) = 0. \tag{3.109}$$

where the effective constant is introduced by

$$a_0^{\rm eff} = a_0 + \frac{\ell_\rho^2 \Lambda_3 R}{6} = a_0 - \frac{2\ell_\rho^2 \Lambda_3 \lambda_0}{3a_0}. \tag{3.110}$$

The last equation (3.109) follows from the Bianchi identity (3.14).

It is obvious that the vacuum Einstein spaces [74] with a cosmological term (3.107)

$$\not{R}_{ij} = 0, \qquad \text{i.e.,} \qquad R_{ij} = \frac{1}{4} R\, g_{ij}\,, \tag{3.111}$$

are solutions of the system (3.108)–(3.109). The questions is: are there other solutions, or (3.111) represents the unique solution? When $a_0 \neq 0$, there are several situations depending on the values of Λ_1 and Λ_3.

When $\Lambda_1 = 0$, the system (3.104)–(3.105) reduces to

$$a_0^{\rm eff}\, \not{R}_{ij} = 0. \tag{3.112}$$

Then we have one of the two possibilities. If $a_0^{\rm eff} \neq 0$ the system (3.112) coincides with Einstein's field equations (3.111). In the special case $a_0^{\rm eff} = 0$, equations (3.108)–(3.109) are fulfilled identically, and the solutions are arbitrary spaces which satisfy $a_0 R = -4\lambda_0$.

If $\Lambda_1 \neq 0$, we introduce

$$\xi := \frac{a_0^{\rm eff}}{\ell_\rho^2 \Lambda_1} \tag{3.113}$$

and the system (3.108)–(3.109) is then recast into

$${}^{(1)}R_{iklj}\, \not{R}^{kl} = -\,\xi\, \not{R}^{kl}\,, \tag{3.114}$$

$$\nabla_i\, \not{R}_{jk} - \nabla_j\, \not{R}_{ik} = 0. \tag{3.115}$$

One can prove (for technical details and the references, see [71]) that the only solutions of the system (3.108)–(3.109) are Einstein spaces (3.111), provided

$$\xi \neq \left\{0,\ -\frac{2\lambda_0}{3a_0},\ \frac{4\lambda_0}{3a_0}\right\}. \tag{3.116}$$

When ξ takes one of the exceptional values listed in (3.116), the system (3.108)–(3.109) admits solutions with $\not{R}_{ij} \neq 0$ which are not Einstein spaces [75].

For completeness, let us mention that similar conclusions can be derived for the purely quadratic model with $a_0 = 0$. In this case the cosmological constant should vanish $\lambda_0 = 0$, and we find $\xi = \Lambda_3 R/6\Lambda_1$.

3.4.2 Correspondence with GR: Birkhoff Theorem

Let us now return to the general case, and discuss the correspondence of PG and GR without assuming vanishing torsion. In view of the fact that the fundamental gravitational experiments in our Solar system are perfectly consistent with the Schwarzschild geometry, a natural question arises: to which extent the solutions of the field equations in PG theory may deviate from the Schwarzschild spacetime?

Quite generally, spherically symmetric solutions are of particular interest in all field-theoretic models. In Einstein's general relativity theory, the Schwarzschild metric is a unique solution of the gravitational field equations under the assumption of a spherical symmetry of the spacetime geometry and matter source distribution. This remarkable result is known as the Birkhoff theorem.

In order to discuss the validity of the *generalized Birkhoff theorem* in Poincaré gauge gravity, we need to clarify how the spherical symmetry is described for the gravitational gauge fields [76–78]. Following Sect. 3.2.4, in the local coordinate system $x^i = (t, r, \theta, \varphi)$, the most general spherically symmetric spacetime interval

$$ds^2 = A^2 dt^2 - B^2 dr^2 - C^2(d\theta^2 + \sin^2\theta d\varphi^2) \tag{3.117}$$

depends on the three arbitrary functions $A = A(t, r)$, $B = B(t, r)$, $C = C(t, r)$. An $SO(3)$ rotation motion of the manifold M is generated by three vector fields

$$\zeta^i_{\{x\}} = \begin{pmatrix} 0 \\ 0 \\ \sin\varphi \\ \cos\varphi\cot\theta \end{pmatrix}, \quad \zeta^i_{\{y\}} = \begin{pmatrix} 0 \\ 0 \\ -\cos\varphi \\ \sin\varphi\cot\theta \end{pmatrix}, \quad \zeta^i_{\{z\}} = \begin{pmatrix} 0 \\ 0 \\ 0 \\ -1 \end{pmatrix}, \tag{3.118}$$

and the spherical invariance is manifest in the vanishing Lie derivative of the metric $\mathcal{L}_\zeta g_{ij} = 0$ under the action $\zeta : SO(3) \times M \to M$.

In the framework of Poincaré gauge gravity, the general spherically symmetric configuration for the gravitational gauge field potentials $(e_i^\alpha, \Gamma_{i\beta}{}^\alpha)$ reads

$$e_i^\alpha = \begin{pmatrix} A & 0 & 0 & 0 \\ 0 & B & 0 & 0 \\ 0 & 0 & C & 0 \\ 0 & 0 & 0 & C\sin\theta \end{pmatrix}, \tag{3.119}$$

for the coframe, and for the connection (using an obvious matrix notation):

$$\Gamma_{0\beta}{}^\alpha = \begin{pmatrix} 0 & f & 0 & 0 \\ f & 0 & 0 & 0 \\ 0 & 0 & 0 & \overline{f} \\ 0 & 0 & -\overline{f} & 0 \end{pmatrix}, \quad \Gamma_{1\beta}{}^\alpha = \begin{pmatrix} 0 & g & 0 & 0 \\ g & 0 & 0 & 0 \\ 0 & 0 & 0 & \overline{g} \\ 0 & 0 & -\overline{g} & 0 \end{pmatrix}, \tag{3.120}$$

$$\Gamma_{2\beta}{}^{\alpha} = \begin{pmatrix} 0 & 0 & p & \overline{q} \\ 0 & 0 & q & -\overline{p} \\ p & -q & 0 & 0 \\ \overline{q} & \overline{p} & 0 & 0 \end{pmatrix}, \quad \Gamma_{3\beta}{}^{\alpha} = \sin\theta \begin{pmatrix} 0 & 0 & -\overline{q} & p \\ 0 & 0 & \overline{p} & q \\ -\overline{q} & -\overline{p} & 0 & -\cot\theta \\ p & -q & \cot\theta & 0 \end{pmatrix}. \tag{3.121}$$

This increases the number of arbitrary functions by eight more variables: $f = f(t,r)$, $g = g(t,r)$, $p = p(t,r)$, $q = q(t,r)$, and $\overline{f} = \overline{f}(t,r)$, $\overline{g} = \overline{g}(t,r)$, $\overline{p} = \overline{p}(t,r), \overline{q} = \overline{q}(t,r)$.

The gauge potentials (3.119)–(3.121) are constructed in full agreement with the analysis in Sect. 3.2.4, and they satisfy the generalized invariance conditions (3.37)–(3.38):

$$\mathcal{L}_\zeta\, e_i^\alpha = \overset{\zeta}{\lambda}_\beta{}^\alpha\, e_i^\beta, \qquad \mathcal{L}_\zeta\, \Gamma_{i\beta}{}^\alpha = -\, D_i \overset{\zeta}{\lambda}_\beta{}^\alpha, \tag{3.122}$$

where the Lorentz algebra-valued $\overset{\zeta}{\lambda}{}^{\alpha\beta} = -\overset{\zeta}{\lambda}{}^{\beta\alpha}$ parameter is determined by vector fields which generate symmetries. For the rotation symmetry generators (3.118) we have explicitly

$$\overset{\zeta_{\{x\}}}{\lambda}{}_\beta{}^\alpha = \frac{\cos\varphi}{\sin\theta}\begin{pmatrix} 0 & 0 & 0 & 0 \\ 0 & 0 & 0 & 0 \\ 0 & 0 & 0 & 1 \\ 0 & 0 & -1 & 0 \end{pmatrix}, \quad \overset{\zeta_{\{y\}}}{\lambda}{}_\beta{}^\alpha = \frac{\sin\varphi}{\sin\theta}\begin{pmatrix} 0 & 0 & 0 & 0 \\ 0 & 0 & 0 & 0 \\ 0 & 0 & 0 & 1 \\ 0 & 0 & -1 & 0 \end{pmatrix}, \quad \overset{\zeta_{\{z\}}}{\lambda}{}_\beta{}^\alpha = 0.$$

The general spherically symmetric configuration is only invariant under the group of proper rotations $SO(3)$, however, it is not invariant under spatial reflections when the parity-odd functions $\overline{f}$, $\overline{g}$, $\overline{p}$, $\overline{q}$ are nonzero. By demanding also the invariance under reflections, one extends the symmetry group from $SO(3)$ to the full rotation group $O(3)$, and such an extension imposes an additional condition on field configurations, which forbids parity-odd variables: $\overline{f} = \overline{g} = \overline{p} = \overline{q} = 0$.

The generalized Birkhoff theorem in the Poincaré gauge gravity is much more nontrivial [79–85] than its Riemannian analogue in GR, since besides the metric (coframe) variables A, B, C there are additional connection variables f, g, p, q and $\overline{f}, \overline{g}, \overline{p}, \overline{q}$, and the torsion is not assumed to be zero. To prove the generalized Birkhoff theorem, one needs to plug the spherically symmetric ansatz (3.119)–(3.121) into the field equations (3.92)–(3.93) and then establish the conditions under which these field equations admit only solutions with the vanishing torsion and the Schwarzschild metric. There are different types of conditions: some of them may restrict the coupling constants (hence, refine the structure of the Lagrangian V), other conditions may impose constraints on the geometric properties of spacetime. Among the latter are: an asymptotic flatness condition which requires that in the limit of $r \to \infty$ the metric approaches the Minkowski line element, i.e. $A \to 1$, $B \to 1$, $C \to r$, or an assumption of the vanishing scalar curvature.

There are two versions of the generalized Birkhoff theorem in the Poincaré gauge gravity [71,78]: the strong (*SB*) and the weak (*WB*) ones. The *strong* $SO(3)$ *theorem* reads: under the assumption of the spherical symmetry in the sense of invariance under the proper rotation $SO(3)$ group, the Schwarzschild (Kottler, in general, when the cosmological constant is nontrivial) spacetime with zero torsion is a unique solution of the vacuum field equations. This result holds for the four families in the class of quadratic models (3.73):

(SB1) No curvature square terms, $b_I = 0$ (thus $\Lambda_I = 0$), provided $\mu_1\mu_2\mu_3 \neq 0$.
(SB2) $\Lambda_1 = \Lambda_2 = \Lambda_5 = \Lambda_6 = 0$, $\Lambda_3 \neq 0$, $\Lambda_4 \neq 0$, provided the scalar curvature R is constant, and $\mu_1^{\rm eff}\mu_2^{\rm eff}\mu_3^{\rm eff} \neq 0$, where $\mu_I^{\rm eff}$ are obtained from (3.102) by replacing a_0 with an effective coupling constant (3.110).
(SB3) $\Lambda_I = 0$, provided $\mu_1\mu_2\mu_3 \neq 0$. This case is close to (*SB1*) but not quite equivalent, see the remark below (3.102).
(SB4) No torsion square terms, $a_1 = a_2 = a_3 = 0$, but $a_0 \neq 0$ and $\Lambda_1 = \Lambda_2 = \Lambda_4 = \Lambda_5 = \Lambda_6 = 0$, $\Lambda_3 \neq 0$, provided $a_0^{\rm eff} \neq 0$.

It is important to notice that in the latter case the gravitational field equations yield for the curvature scalar $a_0 R = -4\lambda_0$, and thereby the condition $a_0^{\rm eff} \neq 0$ actually means that the constant $\xi \neq 0$, cf. (3.116).

The *weak* $O(3)$ *version of the Birkhoff theorem* reads: under the assumption of the spherical symmetry in the sense of invariance under the full rotation $O(3)$ group, when spatial reflections are included along with proper rotations, the Schwarzschild (Kottler, in general) spacetime without torsion is a unique solution of the vacuum field equations. In addition to the above cases, the weak theorem holds for the following families in the class of quadratic models (3.73):

(WB1) No curvature square terms, $b_I = 0$ (thus $\Lambda_I = 0$), provided $\mu_2\mu_3 \neq 0$.
(WB2) $\Lambda_1 = \Lambda_6 = 0$, $\Lambda_2 \neq 0$, $\Lambda_3 \neq 0$, $\Lambda_4 \neq 0$, $\Lambda_5 \neq 0$, provided the scalar curvature $R(\Gamma)$ is constant, and $\mu_2^{\rm eff}\mu_3^{\rm eff} \neq 0$.
(WB3) $\Lambda_1 = \Lambda_3 = \Lambda_6 = 0$, $\Lambda_2 \neq 0$, $\Lambda_4 \neq 0$, $\Lambda_5 \neq 0$, provided $\mu_2\mu_3 \neq 0$.
(WB4) $a_1 = a_2 = 0$, but $a_0 \neq 0$ and $\Lambda_1 = \Lambda_6 = 0$, $\Lambda_2 \neq 0$, $\Lambda_3 \neq 0$, $\Lambda_4 \neq 0$, $\Lambda_5 \neq 0$, provided $a_0^{\rm eff} \neq 0$.
(WB5) $a_1 = a_2 = 0$, but $a_0 \neq 0$ and $\Lambda_6 = 0$, $\Lambda_1 \neq 0$, $\Lambda_2 \neq 0$, $\Lambda_3 \neq 0$, $\Lambda_4 \neq 0$, $\Lambda_5 \neq 0$, provided the ξ constant (3.113) satisfies the condition (3.116).
(WB6) $a_1 = a_2 = 0$, but $a_0 \neq 0$ and $\Lambda_1 = \Lambda_6 \neq 0$, $\Lambda_2 \neq 0$, $\Lambda_3 \neq 0$, $\Lambda_4 \neq 0$, $\Lambda_5 \neq 0$, provided the ξ constant (3.113) satisfies the condition (3.116).
(WB7) $a_1 = a_2 = 0$, but $a_0 \neq 0$ and arbitrary Λ_I, under the condition of asymptotic flatness.
(WB8) $\Lambda_1 = \Lambda_2 = \Lambda_4 = \Lambda_5 = \Lambda_6 = 0$, $\Lambda_3 \neq 0$, provided $\mu_1\mu_2\mu_3 \neq 0$ and the curvature scalar vanishing condition.

3.4.3 Correspondence with GR: Dynamical Torsion Beyond Einstein

The class of Yang-Mills type quadratic Lagrangians (3.73) encompasses many interesting and physically viable models. Since Einstein's general relativity theory

(GR) is convincingly supported by experiments in terrestrial laboratories and astrophysical observations both in the Solar system and on the extra-galactic scales, it is important to investigate the relation between the general Poincaré gravity and GR.

3.4.3.1 Einstein-Cartan Theory

The most well known is the so-called Einstein-Cartan theory which is the closest extension of GR. The corresponding Lagrangian is obtained from (3.73) by dropping all quadratic terms, for the coupling constants $a_0 = 1$, $a_1 = a_2 = a_3 = 0$, and $b_I = 0$:

$$V_{\rm EC} = -\frac{1}{2\kappa c}\left(R + 2\lambda_0\right). \tag{3.123}$$

The gravitational field equations (3.92)–(3.93) then reduce to

$$R_i{}^j - \frac{1}{2}R\delta_i^j - \lambda_0\delta_i^j = \kappa\Sigma_i{}^j, \tag{3.124}$$

$$T_{ij}{}^k + 2T_{[i}\delta^k_{j]} = \kappa c\tau_{ij}{}^k. \tag{3.125}$$

The Einstein-Cartan theory represents a certain degenerate case of the Poincaré gauge gravity in the sense that the second field equation (3.125) describes an algebraic coupling between the spin of matter and the torsion. This means that the torsion is a non-dynamical field which vanishes outside the matter sources, and thereby the first equation (3.124) reduces to Einstein's field equation of GR.

Resolving (3.125), one can express the torsion in terms of the matter spin, and plugging it in (3.124), it is possible to recast the latter into an effective Einstein field equation

$$\overset{\circ}{R}_{ij} - \frac{1}{2}\overset{\circ}{R}g_{ij} - \lambda_0 g_{ij} = \kappa\overset{\rm eff}{\Sigma}_{ij}, \tag{3.126}$$

where the original canonical energy-momentum tensor is replaced by the effective energy-momentum tensor that includes additional contributions of the spin. For the particular case of the spinning fluid (3.98)–(3.99) one finds

$$\overset{\rm eff}{\Sigma}_{ij} = -p^{\rm eff}\Big(g_{ij} - \frac{1}{c^2}u_iu_j\Big) + \frac{\varepsilon^{\rm eff}}{c^2}u_iu_j + \Big(g^{kl} + \frac{1}{c^2}u^ku^l\Big)\overset{\circ}{\nabla}_k\left(u_{(i}\mathcal{S}_{j)l}\right), \tag{3.127}$$

where the effective pressure and energy density depend on spin:

$$p^{\rm eff} = p - \frac{\zeta\kappa c^2}{8}\mathcal{S}_{ij}\mathcal{S}^{ij}, \tag{3.128}$$

$$\varepsilon^{\rm eff} = \varepsilon - \frac{\zeta\kappa c^2}{8}\mathcal{S}_{ij}\mathcal{S}^{ij}. \tag{3.129}$$

Here the numeric constant $\zeta = 1$.

A qualitatively equivalent model (which can be called a generalized Einstein-Cartan theory "EC+") is obtained as a natural extension of the Lagrangian (3.123) when we include all possible torsion quadratic terms:

$$V_{\mathrm{EC+}} = -\frac{1}{2\kappa c}\Big(R + 2\lambda_0 + \frac{1}{2}\sum_{I=1}^{3} a_I\, {}^{(I)}T^{kl}{}_i\, T_{kl}{}^i\Big). \tag{3.130}$$

The resulting field equations then read

$$\begin{aligned} R_i{}^j - \frac{1}{2}R\delta_i^j - \lambda_0\delta_i^j + T_{in}{}^k h^{jn}{}_k - \frac{1}{4}\delta_i^j\, T_{mn}{}^k h^{mn}{}_k \\ - (\nabla_l - T_l)h^{jl}{}_i - \frac{1}{2}T_{mn}{}^j h^{mn}{}_i = \kappa\Sigma_i{}^j, \end{aligned} \tag{3.131}$$

$$T_{ij}{}^k + 2T_{[i}\delta_{j]}^k - 2h^k{}_{[ij]} = \kappa c\tau_{ij}{}^k, \tag{3.132}$$

where, recalling (3.94),

$$h^{ij}{}_k = a_1{}^{(1)}T^{ij}{}_k + a_2{}^{(2)}T^{ij}{}_k + a_3{}^{(3)}T^{ij}{}_k\,. \tag{3.133}$$

Since the second field equation (3.132) still describes an algebraic coupling of the matter spin and the torsion, one can resolve the latter and recast (3.131) into an effective Einstein field equation (3.126). In this model, however, the effective energy-momentum then picks up a dependence on the coupling constants a_I. For the case of the spinning fluid (3.98)–(3.99) one again finds (3.127)–(3.129), but with a more nontrivial constant

$$\zeta = \frac{4}{3(1+a_1)} - \frac{1}{3(1-2a_3)}. \tag{3.134}$$

It is worthwhile to note that a_2 does not contribute in view of the Frenkel condition $S_{ij}u^j = 0$ imposed on the spin density. When the torsion-square terms are absent, $a_1 = a_2 = a_3 = 0$, we recover the value $\zeta = 1$ of the Einstein-Cartan theory. It is interesting to note that there exists a large class of models with the torsion quadratic Lagrangians (3.130) which yield $\zeta = 0$.

Qualitatively, the EC and EC+ models are very much alike, because they both can be recast into the form of the effective Einstein theory (3.126) with the energy-momentum tensor modified by the spin contributions. The magnitude of the terms quadratic in spin in $\overset{\mathrm{eff}}{\Sigma}_{ij}$ becomes comparable with the original canonical energy-momentum tensor Σ_{ij} at densities $\rho \geq \rho_{\mathrm{cr}} = \frac{m^2c^4}{\hbar^2 G}$ of the spinning matter built of particles with the mass m [11]. For the mass of a nucleon, the critical density $\rho_{\mathrm{cr}} \approx 10^{57}$ kg/m^3 is still much smaller than the Planck density $\rho_{\mathrm{Pl}} \sim 10^{97}$ kg/m^3 at which the quantum-gravitational effects are expected to start dominating. Consequently, the torsion can be essential already at the level of the classical theory of the gravitational interactions. In particular, this may avert the singularity in the early universe, predicting a finite minimum for the cosmological scale factor reached at the critical matter density [86].

A thorough analysis of the observational cosmology in the Einstein-Cartan theory with the Weyssenhoff spinning fluid can be found in [87–91]. Other physical consequences of the Einstein-Cartan theory are discussed in great detail in [11, 17, 22].

3.4.3.2 Einstein's GR as a Special Case of Poincaré Gravity Theory

In the Einstein-Cartan theory above, we assumed the minimal coupling of the matter to the Poincaré gauge fields, which is a natural assumption in the framework of the gauge-theoretic approach.

Quite remarkably, however, one can also view Einstein's general relativity theory as a special case in the framework of the Poincaré gauge gravity theory under the assumption of a suitable nonminimal coupling of matter to the Riemann-Cartan geometry of spacetime.

In order to demonstrate this, we start with the extended Einstein-Cartan Lagrangian (3.130) and fix the torsion coupling constants as

$$a_1 = -1, \qquad a_2 = 2, \qquad a_3 = \frac{1}{2}. \tag{3.135}$$

As a result, (3.133) reduces to

$$h^{ij}{}_k = -{}^{(1)}T^{ij}{}_k + 2{}^{(2)}T^{ij}{}_k + \frac{1}{2}{}^{(3)}T^{ij}{}_k = -K_k{}^{ij} - 2T^{[i}\delta_k^{j]}, \tag{3.136}$$

where we used (3.62)–(3.64) and (3.10). A direct computation then yields a remarkable simplification of the left-hand sides of the field equations (3.131) and (3.132):

$$\overset{\circ}{R}_i{}^j - \frac{1}{2}\overset{\circ}{R}\delta_i^j = \kappa\Sigma_i{}^j, \tag{3.137}$$

$$0 = \kappa c \tau_{ij}{}^k. \tag{3.138}$$

The last equation would be an obviously contradictory relation for the case of the minimal coupling, allowing only for the spinless matter. Nevertheless, this equation becomes meaningful under the assumption of a special type of nonminimal coupling, when the Lagrangian $L_{\rm mat} = L_{\rm mat}(\psi^A, D_i\psi^A, e_i^\alpha, T_{ij}{}^\alpha)$ of the matter fields ψ^A depends on the torsion tensor $T_{ij}{}^\alpha$ that, however, may enter the Lagrangian only in a combination

$$D_i\psi^A - \frac{1}{2}K_i{}^{\alpha\beta}(\rho_{\alpha\beta})^A{}_B\,\psi^B, \tag{3.139}$$

Technically, this means that the spinning matter couples only to the Levi-Civita connection. Then one can demonstrate [92] that the source on the right-hand side of (3.137) has the form

$$\Sigma_i{}^j = \overset{\rm m}{\Sigma}{}_i{}^j + \frac{c}{2}\overset{\circ}{\nabla}_k\left(\overset{\rm m}{\tau}{}^{jk}{}_i + \overset{\rm m}{\tau}{}^j{}_i{}^k + \overset{\rm m}{\tau}{}_i{}^{kj}\right), \tag{3.140}$$

where $\overset{\rm m}{\Sigma}{}_i{}^j$ is the canonical energy-momentum tensor, and

$$c\overset{\rm m}{\tau}{}_{\alpha\beta}{}^k = (\rho_{\alpha\beta})^A{}_B\,\psi^B\,\frac{\partial L_{\rm mat}}{\partial D_k\psi^A}, \qquad \overset{\rm m}{\tau}{}_{ij}{}^k = \overset{\rm m}{\tau}{}_{\alpha\beta}{}^k e_i^\alpha e_j^\beta, \tag{3.141}$$

is the canonical spin tensor. We immediately recognize in (3.140) the well known metrical energy-momentum tensor symmetrized by means of the Belinfante-Rosenfeld procedure.

3.4.3.3 Von der Heyde Model

As we saw above, the Einstein-Cartan theory is the closest generalization of Einstein's GR which takes into account the spin of matter as a source of the gravitational field. This changes the geometrical structure of the spacetime manifold, but the torsion remains a non-dynamical field which disappears in the absence of spin. This motivates the study of more general Poincaré gravity models in which the torsion becomes a dynamical field. Here we briefly consider two such models.

The von der Heyde (VdH) model [13, 93] attracted considerable attention in the literature. It is described by the Lagrangian that does not contain a linear in the curvature Hilbert-Einstein term, and is purely quadratic in the Poincaré gauge field strengths:

$$V_{\rm VdH} = -\frac{1}{2\kappa c}\left(-\frac{1}{2}T_{ij}{}^kT^{ij}{}_k + T_iT^i + \frac{\ell_\rho^2}{2}R_{ij}{}^{kl}R^{ij}{}_{kl}\right). \tag{3.142}$$

One thus recovers a special case of the general Lagrangian (3.73) with $a_0 = 0$, $b_I = 1$, $I = 1, \dots, 6$, and

$$a_1 = -1, \qquad a_2 = 2, \qquad a_3 = -1. \tag{3.143}$$

A peculiar feature of the VdH model is that it demonstrates a remarkable compatibility with GR despite the absence of the Hilbert-Einstein term in the Lagrangian. Technically, this is explained by an almost the same set of the torsion coupling constants, cf. (3.143) and (3.135).

Unlike the Einstein-Cartan theory, the VdH model predicts nontrivial dynamical torsion effects. To demonstrate this, let us specialize to the spherically symmetric ansatz (3.117), (3.119)–(3.121). Inspecting the Poincaré gauge field equations (3.92)–(3.93) for the Lagrangian (3.142), we then find an exact solution for the metric variables

$$A^2 = 1 - \frac{2m}{r} + \frac{r^2}{4\ell_\rho^2}, \qquad B = \frac{1}{A}, \qquad C = r, \tag{3.144}$$

whereas the anholonomic torsion components $T_{\mu\nu}{}^{\alpha} = e^i_\mu e^j_\nu e^\alpha_k T_{ij}{}^k$ read

$$T_{\hat{1}\hat{0}}{}^{\hat{0}} = T_{\hat{1}\hat{0}}{}^{\hat{1}} = T_{\hat{0}\hat{2}}{}^{\hat{2}} = T_{\hat{2}\hat{1}}{}^{\hat{2}} = T_{\hat{0}\hat{3}}{}^{\hat{3}} = T_{\hat{3}\hat{1}}{}^{\hat{3}} = \frac{m}{A\,r^2}. \tag{3.145}$$

Here $m = GM/c^2$ is an integration constant, with M interpreted as a total mass of the field configuration. From the point of view of the Riemannian geometry, the line element (3.117), (3.144) describes the Schwarzschild-de Sitter (or Kottler) GR solution, where the dynamical torsion induces a "fake" cosmological term.

One can extend this result to the axially symmetric case, and demonstrate that the Kerr-de Sitter metric of a massive and rotating field configuration with dynamical torsion is an exact solution in the VdH model [13, 94]. Moreover, a systematic analysis [95] reveals the existence of such solutions in a more general class of Poincaré gauge models with the Yang-Mills type Lagrangian (3.73).

3.4.3.4 Cembranos-Valcarcel Model

While in VdH model a "fake" cosmological term arises from the dynamical torsion, the latter can manifest even more nontrivial effects in the Cembranos-Valcarcel model [96, 97]. The corresponding Lagrangian is a special case of (3.73), where $a_0 = 1$, and the torsion coupling constants are fixed by (3.135), whereas the curvature coupling sector reads

$$b_3 = -\,b_2, \quad b_5 = -\frac{b_2}{3}, \qquad b_1 = b_4 = b_6 = 0. \tag{3.146}$$

Specializing again to the spherically symmetric ansatz (3.117), (3.119)–(3.121), one then finds an exact solution for the metric variables

$$A^2 = 1 - \frac{2m}{r} - \frac{\lambda_0 r^2}{3} + \frac{Q^2}{r^2}, \qquad B = \frac{1}{A}, \qquad C = r, \tag{3.147}$$

where an arbitrary integration constant σ_0 enters

$$Q^2 = \frac{2b_2\ell_\rho^2\sigma_0^2}{3}, \tag{3.148}$$

and determines the structure of the dynamical torsion. The latter (as before, in anholonomic components $T_{\mu\nu}{}^{\alpha} = e^i_\mu e^j_\nu e^\alpha_k T_{ij}{}^k$) reads explicitly:

$$T_{\hat{1}\hat{0}}{}^{\hat{0}} = T_{\hat{1}\hat{0}}{}^{\hat{1}} = \frac{dA}{dr}, \tag{3.149}$$

$$T_{\hat{2}\hat{0}}{}^{\hat{2}} = T_{\hat{1}\hat{2}}{}^{\hat{2}} = T_{\hat{3}\hat{0}}{}^{\hat{3}} = T_{\hat{1}\hat{3}}{}^{\hat{3}} = \frac{A}{2r}, \tag{3.150}$$

$$T_{\hat{0}\hat{3}}{}^{\hat{2}} = T_{\hat{3}\hat{1}}{}^{\hat{2}} = T_{\hat{2}\hat{0}}{}^{\hat{3}} = T_{\hat{1}\hat{2}}{}^{\hat{3}} = \frac{\sigma_0}{A\,r}. \tag{3.151}$$

As we now see, from the point of view of the Riemannian geometry, the line element (3.117), (3.147) describes the Reissner-Nordström-de Sitter GR solution, where the dynamical torsion induces a "fake" electric charge (3.148), and the torsion plays a role of a fictitious electromagnetic field.

An extension of the Cembranos-Valcarcel results for a more general models with both the parity-even and parity-odd sectors included was discussed in [78]. It is worthwhile to notice that the generalized Birkhoff theorem is not valid for the von der Heyde and the Cembranos-Valcarcel models, and precisely this fact underlies the existence of the spherically symmetric solutions with nontrivial torsion.

3.4.4 Gravitational Waves

In conclusion, it is instructive to discuss a possible physical manifestation of the rich graviton spectrum (3.101) and (3.102) of Poincaré gauge gravity theory in the form of the gravitational waves.

The study of gravitational waves is of fundamental importance in physics, that became an even more significant issue after the purely theoretical research in this area was finally supported by the first experimental evidence [98, 99]. The plane-fronted gravitational waves represent an important class of exact solutions [100–102] which generalize the basic properties of electromagnetic waves in flat spacetime to the case of curved spacetime geometry.

Let us discuss the gravitational wave solutions in the PG model with the general quadratic Lagrangian (3.73) for the case without the cosmological constant $\lambda_0 = 0$. We start with the flat Minkowski geometry described by the coframe and connection $\widehat{e}^\alpha_i = \delta^\alpha_i$, $\widehat{\Gamma}_{i\beta}{}^\alpha = 0$, where $x^i = (x^0, x^1, x^2, x^3)$ are Cartesian coordinates. Differentiating the phase variable $\sigma = x^0 - x^1$, we introduce the wave covector $k_i = \partial_i \sigma = (1, -1, 0, 0)$. With $k_\alpha = \widehat{e}^i_\alpha k_i$, the gravitational wave ansatz is then introduced as a Kerr-Schild deformation of the flat background:

$$e^\alpha_i = \widehat{e}^\alpha_i + \frac{1}{2} U\, k^\alpha k_i, \tag{3.152}$$

$$\Gamma_{i\beta}{}^\alpha = \widehat{\Gamma}_{i\beta}{}^\alpha + (k_\beta W^\alpha - k^\alpha W_\beta)\, k_i. \tag{3.153}$$

The resulting line element (with $\rho = x^0 + x^1$)

$$ds^2 = g_{\alpha\beta} e_i^\alpha e_j^\beta dx^i dx^j = d\sigma d\rho + U d\sigma^2 - \delta_{AB} dx^A dx^B, \tag{3.154}$$

represents the plane-fronted wave in the form of Brinkmann [103, 104]. By construction, $k_\alpha = (1, -1, 0, 0)$, so that $k^\alpha = (1, 1, 0, 0)$. Therefore, this is a null vector field, $k_\alpha k^\alpha = 0$.

The gravitational wave configuration (3.152) and (3.153) is described by the two unknown variables U and W^a which determine wave's profile, they are functions $U = U(\sigma, x^A)$ and $W^\alpha = W^\alpha(\sigma, x^A)$ of the phase σ and the transversal coordinates $x^A = (x^2, x^3)$. From now on, the indices from the beginning of the Latin alphabet $a, b, c, \dots = 0, 1$, whereas the capital Latin indices run $A, B, C \dots = 2, 3$. In addition, we assume the orthogonality $k_\alpha W^\alpha = 0$, which is guaranteed if we choose

$$W^\alpha = \begin{cases} W^a = 0, & a = 0, 1, \\ W^A = W^A(\sigma, x^B), & A = 2, 3. \end{cases} \tag{3.155}$$

Obviously, $\partial_i k^\alpha = 0$, and $D_i k^\alpha = 0$ for the wave covector with constant components, and we straightforwardly find the torsion and the curvature:

$$T_{kl}{}^i = -2k^i k_{[k}\Theta_{l]}, \qquad R_{kl}{}^{ij} = -4k_{[k}k^{[i}\Omega_{l]}{}^{j]}. \tag{3.156}$$

Here we constructed the two objects from the derivatives of $U = U(\sigma, x^A)$ and $W^\alpha = W^\alpha(\sigma, x^A)$ with respect to the transversal coordinates $x^A = (x^2, x^3)$:

$$\Theta_i = \left\{\Theta_a = 0, \quad \Theta_A = \frac{1}{2}\partial_A U - \delta_{AB} W^B\right\}, \tag{3.157}$$

$$\Omega_j{}^i = \left\{\Omega_b{}^a = 0,\ \Omega_B{}^a = 0,\ \Omega_b{}^A = 0,\ \Omega_B{}^A = \partial_B W^A\right\}. \tag{3.158}$$

The translational and rotational Poincaré gauge field strengths (3.156) have qualitatively the same structure as the electromagnetic field strength F_{ij} of a plane wave, that has the properties $k^j F_{ij} = 0$, $k_{[i}F_{jk]} = 0$, $F_{ij}F^{ij} = 0$. In complete analogy, the Poincaré gauge field strengths of a gravitational wave satisfy

$$k^j T_{ij}{}^k = 0, \qquad k_{[i}T_{jk]}{}^l = 0, \qquad T_{ij}{}^k T^{ij}{}_l = 0, \tag{3.159}$$

$$k^j R_{ij}{}^{kl} = 0, \qquad k_{[i}R_{jk]}{}^{mn} = 0, \qquad R_{ij}{}^{kl} R^{ij}{}_{mn} = 0. \tag{3.160}$$

In addition, however, for the gravitational Poincaré gauge field strengths we find

$$k_l T_{ij}{}^l = 0, \qquad k_l R_{ij}{}^{kl} = 0. \tag{3.161}$$

The torsion (3.156) vanishes when $\Theta_i = 0$, which means that $W^A = \frac{1}{2}\delta^{AB}\partial_B U$, Then U remains the only nontrivial variable and the solution reduces to the usual plane gravitational wave of the Riemannian GR. By noticing this, it is convenient to express the wave profile vector variable in terms of potentials

$$W^A = \frac{1}{2}\delta^{AB}\partial_B(U + V) + \frac{1}{2}\eta^{AB}\partial_B\overline{V}, \tag{3.162}$$

where $\eta_{AB} = -\eta_{BA}$ is the totally antisymmetric Levi-Civita tensor on the two-dimensional space of the wave front. This brings us to the physically transparent representation of the plane wave in the Poincaré gauge gravity in terms of the three scalar variables $U = U(\sigma, x^A)$, and $V = V(\sigma, x^A)$, $\overline{V} = \overline{V}(\sigma, x^A)$, where the first one is a Riemannian mode, and the two last ones account for the torsion wave modes.

The explicit gravitational wave solution is constructed as follows [105, 106]. Substituting the wave ansatz (3.152), (3.153) and (3.162) into the gravitational field equations (3.92) and (3.93), the latter reduce to the system of three linear differential equations

$$a_0\,\Delta\,U - \mu_3\,\Delta\,V = 0, \tag{3.163}$$

$$\ell_\rho^2\,\Lambda_1(a_0 + \mu_3)\,\Delta\,V - a_0\mu_3\,V = 0, \tag{3.164}$$

$$\ell_\rho^2\,\Lambda_2\,\Delta\,\overline{V} - \mu_3\,\overline{V} = 0. \tag{3.165}$$

Here $\Delta = \delta^{AB}\partial_A\partial_B$ is the two-dimensional Laplacian on the (x^2, x^3) space. Using the solutions of (3.164) and (3.165) for the torsion waves in (3.163), we can find the Riemannian mode U from the resulting inhomogeneous equation.

Quite remarkably, all the three wave modes are massless when $\mu_3 = 0$.

3.5 Teleparallel Gravity

Presently, considerable research efforts are focused on the teleparallel theory of the gravitational field. From the gauge-theoretic point of view, the latter is based on the gauging of the group of spacetime translations, and it is worthwhile to mention that the fundamental relation of translation symmetry to gravity was clear already at the beginning of the 1960s to Sakurai, Glashow, Gell-Mann, and Feynman (see the historic account in [3]). The conserved energy-momentum current of matter is associated to translations via the Noether theorem, and it naturally arises as a physical source of the corresponding gauge gravitational field.

The structure of the teleparallel gravity (TG) as a translational gauge theory became essentially established since 1970s, see [44–58]. A revival of interest to the gauge-theoretic subtleties underlying TG has lead to a recent highly enlightening discussion [48, 107–110], in particular within the fruitful framework of Tartu

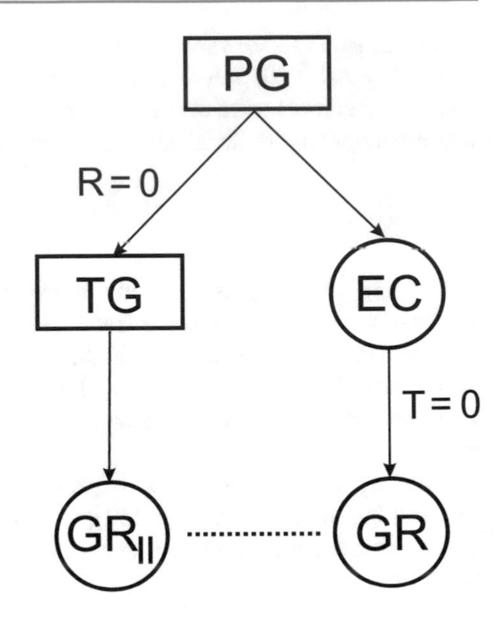

Fig. 3.2 Classification of Poincaré gauge theories of gravity (see the frontispiece of [19]. ©Imperial College Press 2013, reproduced with permission. All rights reserved): **PG** = Poincaré gauge gravity, **EC** = Einstein-Cartan theory, **GR** = Einstein's general relativity, **TG** = translation gauge theory (teleparallel theory), $\mathbf{GR}_{||}$ = a specific TG known as teleparallel equivalent of GR. The symbols denote here: rectangle—general class of theories; circle—viable models

conferences. Since the teleparallel gravity theory is considered in full depth in the comprehensive review of Manuel Hohmann in Chap. 4 of this volume, here we merely highlight the main features of TG as a special case of PG, see Fig. 3.2.

The gauging of the group of translations yields the condition (3.17) that introduces the distant parallelism geometry of Weitzenböck on the spacetime manifold.

As a result, the general Yang-Mills type Lagrangian (3.73) reduces to (we confine attention to the parity-even case, and assume zero cosmological constant, for now) [57, 111, 112]

$$V_{\rm TG} = -\frac{1}{4\kappa c}\left(a_1\,{}^{(1)}T^{kl}{}_i\,T_{kl}{}^i + a_2\,{}^{(2)}T^{kl}{}_i\,T_{kl}{}^i + a_3\,{}^{(3)}T^{kl}{}_i\,T_{kl}{}^i\right). \tag{3.166}$$

The gravitational field equations can be derived from the action principle either by implementing the teleparallel condition (3.17) by means of the Lagrange multiplier, or by making use of the gauge $\Gamma_{i\alpha}{}^{\beta} = 0$ which means that the connection (3.26) takes the Weitzenböck form

$$\Gamma_{ki}{}^{j} = e^{j}_{\alpha}\partial_k e^{\alpha}_{i}. \tag{3.167}$$

Then the torsion reduces to the anholonomity object (3.23).

The dynamical contents of a general TG model (3.166) strongly depends on the values of the coupling constants a_1, a_2, a_3. In particular, in generic case, black hole solutions are absent in this theory [47, 48], and there is no consistency with GR.

However, for a very special case when the coupling constants a_1, a_2, a_3 take the values (3.135), the dynamics of the gravitational field is fully consistent with Einstein's theory. Then the Lagrangian is simplified to

$$V_{\mathrm{GR}_{||}} = -\frac{1}{2\kappa c}\Big(-\frac{1}{4}T_{kl}{}^{i}\,T^{kl}{}_{i} + T_i\,T^i + \frac{1}{2}T_{kl}{}^{i}\,T_i{}^{kl}\Big), \tag{3.168}$$

and this model is called a *teleparallel equivalent of GR*.

3.6 Conclusion and Outlook

In this review we presented, at an elementary level using the standard tensor language, the formulation of the theory of gravitational interaction as a gauge theory of the Poincaré symmetry group. This approach is developed along the lines of a heuristic scheme in which a new physical interaction is derived in the Lagrange-Noether formalism from a conserved current corresponding to the rigid symmetry group by extending the latter to a local symmetry. Leaving aside the derivation of the relevant conservation laws, which was thoroughly discussed earlier in [3,24,25], we have formulated here the general dynamical scheme of Poincaré gauge gravity for the class of the Yang-Mills type models (3.73) and considered a selected number of particular physically interesting models.

A more mathematically elaborated formulation of the gauge gravity approach in terms of the modern differential geometry language of the affine frame bundle can be found in [16, 21–23]. We did not intend to give a detailed review of the physical contents of the Poincaré gauge gravity theory. This subject was intensively studied in the past and the relevant results are available in the classic reviews [11, 13, 17]. At present we are again observing a considerable growth of interest to the gauge gravitational issues. The search and analysis of exact solutions of the gravitational field equations is at the center of the current research, which is essential for improvement of understanding of the nature of the gravitational interaction [71,94–97,113].

It is worthwhile to mention that the recent advances in the modern cosmological science have seriously warmed up the interest in the thorough revision of universe's evolution in the broad framework of the modified gravity theories and, in particular, in Poincaré gauge gravity. The early predictions [86, 114, 115] of a possible avertion of singularity in the early universe, and more recent proposals of possible modifications of the late stage of cosmological evolution [116–121], are currently revisited and extended with an aim to better understand the role of the torsion in the early universe and to resolve the problem of the dark energy [122–125], furthermore, the inclusion of parity-odd sector was critically evaluated in [126–131].

The last but not least remarks are in order about the direct experimental tests and estimates of the torsion effects to probe possible deviations of the spacetime structure beyond the Riemannian geometry, in accordance with Einstein's [132] statement that "... the question whether this continuum has a Euclidean,

Riemannian, or any other structure is a question of physics proper which must be answered by experience, and not a question of a convention to be chosen on grounds of mere expediency." The consistent analysis [93, 133–135] of the propagation equations, derived from the conservation laws of PG theory in the framework of the multipole expansion approach, demonstrates that the torsion couples only to the intrinsic spin and never to the orbital angular momentum of test particles. The predicted spin-torsion effects are expected to be quite small, and no spacetime torsion effects were directly observed so far. From the analysis of the data available from precision experiments with spinning particles in high energy physics and astrophysical observations, performed in the numerous theoretical studies [136–149], one typically finds a rather strong bound $|T| \lesssim 10^{-15}\,\mathrm{m}^{-1}$ for the magnitude of the spacetime torsion.

Acknowledgments I am grateful to Friedrich Hehl for the careful reading of the manuscript and helpful comments.

References

1. L. O'Raifeartaigh, N. Straumann, Gauge theory: historical origins and some modern developments. Rev. Mod. Phys. **72**, 1–23 (2000)
2. N. Straumann, Hermann Weyl's space-time geometry and the origin of gauge theory 100 years ago, in *One Hundred Years of Gauge Theory*, ed. by S. De Bianchi, C. Kiefer. Fundamental Theories of Physics, vol. 199 (Springer, Cham, 2020), pp. 3–23
3. F.W. Hehl, Yu.N. Obukhov, Conservation of energy-momentum of matter as the basis for the gauge theory of gravitation, in *One Hundred Years of Gauge Theory*, ed. by S. De Bianchi, C. Kiefer. Fundamental Theories of Physics, vol. 199 (Springer, Cham, 2020), pp. 217–252
4. L. O'Raifeartaigh, *Group Structure of Gauge Theories* (Cambridge University Press, Cambridge, 1986)
5. G. Mack, Physical principles, geometrical aspects, and locality properties of gauge field theories. Fortsch. Phys. **29**, 135–185 (1981)
6. M. Chaichian, N.F. Nelipa, *Introduction to Gauge Field Theories* (Springer, Berlin, 1984)
7. C.N. Yang, R.L. Mills, Conservation of isotopic spin and isotopic gauge invariance. Phys. Rev. **96**, 191–195 (1954)
8. R. Utiyama, Invariant theoretical interpretation of interaction. Phys. Rev. **101**, 1597–1607 (1956)
9. D.W. Sciama, The analogy between charge and spin in general relativity, in *Recent Developments in General Relativity, Festschrift for L. Infeld* (Pergamon Press, Oxford; PWN, Warsaw, 1962), pp. 415–439
10. T.W.B. Kibble, Lorentz invariance and the gravitational field. J. Math. Phys. **2**, 212–221 (1961)
11. F.W. Hehl, P. von der Heyde, G.D. Kerlick, J.M. Nester, General relativity with spin and torsion: foundations and prospects. Rev. Mod. Phys. **48**, 393–416 (1976)
12. F.W. Hehl, Four lectures on Poincaré gauge field theory, in *Proc. of the 6th Course of the School of Cosmology and Gravitation on Spin, Torsion, Rotation, and Supergravity*, Erice, May 1979, ed. by P.G. Bergmann, V. de Sabbata (Plenum, New York, 1980), pp. 5–61
13. F.W. Hehl, J. Nitsch, P. von der Heyde, Gravitation and Poincaré gauge field theory with quadratic Lagrangian, in *General Relativity and Gravitation—One Hundred Years After the Birth of Albert Einstein*, ed. by A. Held, vol. 1 (Plenum Press, New York, 1980), pp. 329–355

14. A. Trautman, Yang-Mills theory and gravitation: a comparison, in *Geometric Techniques in Gauge Theories*, ed. by R. Martini, E.M. de Jager, Lect. Notes Math., vol. 926 (Springer, Berlin, 1982), pp. 179–189
15. J.M. Nester, Gravity, torsion and gauge theory, in *An Introduction to Kaluza-Klein Theories*, ed. by H.C. Lee (World Scientific, Singapore, 1984), pp. 83–115
16. F.W. Hehl, J.D. McCrea, E.W. Mielke, Y. Ne'eman, Metric affine gauge theory of gravity: field equations, Noether identities, world spinors, and breaking of dilation invariance. Phys. Rep. **258**, 1–171 (1995)
17. I.L. Shapiro, Physical aspects of the space-time torsion. Phys. Rep. **357**, 113–213 (2002)
18. M. Blagojević, *Gravitation and Gauge Symmetries* (Institute of Physics, Bristol, 2002)
19. M. Blagojević, F.W. Hehl (eds.), *Gauge Theories of Gravitation: A Reader with Commentaries* (Imperial College Press, London, 2013)
20. J.M. Nester, C.-M. Chen, Gravity: a gauge theory perspective. Int. J. Mod. Phys. D **25**, 1645002 (2016)
21. E.W. Mielke, *Geometrodynamics of Gauge Fields: On the Geometry of Yang-Mills and Gravitational Gauge Theories*, 2nd edn. (Springer, Cham, 2017)
22. V.N. Ponomarev, A.O. Barvinsky, Yu.N. Obukhov, *Gauge Approach and Quantization Methods in Gravity Theory* (Nauka, Moscow, 2017)
23. F.W. Hehl, Yu.N. Obukhov, Élie Cartan's torsion in geometry and in field theory, an essay. Ann. Fond. Louis de Broglie **32**, 157–194 (2007)
24. Yu.N. Obukhov, Poincaré gauge gravity: selected topics. Int. J. Geom. Meth. Mod. Phys. **3**, 95–137 (2006)
25. Yu.N. Obukhov, Poincaré gauge gravity: an overview. Int. J. Geom. Meth. Mod. Phys. **15**(Suppl. 1), 1840005 (2018)
26. E. Schrödinger, *Space-Time Structure* (Cambridge University Press, London, 1960)
27. J.A. Schouten, *Ricci-Calculus. An Introduction to Tensor Analysis and Its Geometric Applications*, 2nd edn. (Springer, Berlin, 1954)
28. A. Einstein, *The Meaning of Relativity*, 5th revised edn. (Princeton University Press, Princeton, 1956)
29. H. Weyl, *Raum-Zeit-Materie* (Springer, Berlin, 1923)
30. R. Weitzenböck, *Invariantentheorie* (Noordhoff, Groningen, 1923)
31. T. Harko, Thermodynamic interpretation of the generalized gravity models with geometry-matter coupling. Phys. Rev. D **90**, 044067 (2014)
32. H.J. Schmidt, Fourth order gravity: equations, history, and applications to cosmology. Int. J. Geom. Meth. Mod. Phys. **4**, 209–248 (2007)
33. O. Bertolami, C.G. Böhmer, T. Harko, F.S.N. Lobo, Extra force in $f(R)$ modified theories of gravity. Phys. Rev. D. **75**, 104016 (2007)
34. N. Straumann, Problems with modified theories of gravity, as alternatives to dark energy, in *Beyond Einstein*, ed. by D.E. Rowe, T. Sauer, S.A. Walter. Einstein Studies, vol. 14 (Springer Nature, New York, 2018), pp. 243–259
35. S. Nojiri, S.D. Odintsov, Unified cosmic history in modified gravity: from $F(R)$ theory to Lorentz non-invariant models. Phys. Rep. **505**, 59–144 (2011)
36. V. Faraoni, S. Capozziello (eds.), *Beyond Einstein Gravity: A Survey of Gravitational Theories for Cosmology and Astrophysics* (Springer, Dordrecht, 2011)
37. S. Bahamonde, K.F. Dialektopoulos, M. Hohmann, J.L. Levi Said, Teleparallel gravity: foundations and cosmology, in *Modified Gravity and Cosmology*, ed. by E.N. Saridakis, R. Lazkoz, V. Salzano, P.V. Moniz, S. Capozziello, J.B. Jiménez, M. De Laurentis, G.J. Olmo (Springer, Cham, 2021), pp. 191–242
38. F.W. Hehl, G.D. Kerlick, Metric-affine variational principles in general relativity. I. Riemannian spacetime. Gen. Relat. Grav. **9**, 691–710 (1978)
39. F.W. Hehl, G.D. Kerlick, Metric-affine variational principles in general relativity. II. Relaxation of the Riemannian constraint. Gen. Relat. Grav. **13**, 1037–1056 (1981)
40. T.P. Sotiriou, V. Faraoni, $f(R)$ theories of gravity. Rev. Mod. Phys. **82**, 451–497 (2010)

41. F.W. Hehl, Yu.N. Obukhov, *Foundations of Classical Electrodynamics: Charge, Flux, and Metric* (Birkhäuser, Boston, 2003)
42. R. Weitzenböck, Differentialinvarianten in der Einsteinschen Theorie des Fernparallelismus. Sitzungsber. Preuss. Akad. Wiss. Berlin, Phys.-math. Klasse (1928), pp. 466–474
43. A. Einstein, Riemann-Geometrie mit Aufrechterhaltung des Begriffes des Fernparallelismus. Sitzungsber. Preuss. Akad. Wiss. Berlin, Phys.-math. Klasse (1928), pp. 217–221
44. K. Hayashi, T. Nakano, Extended translation invariance and associated gauge fields. Prog. Theor. Phys. **38**, 491–507 (1967)
45. Y.M. Cho, Einstein Lagrangian as the translational Yang-Mills Lagrangian. Phys. Rev. D **14**, 2521–2525 (1976)
46. J. Nitsch, F.W. Hehl, Translational gauge theory of gravity: postNewtonian approximation and spin precession. Phys. Lett. B **90**, 98–102 (1980)
47. K. Hayashi, T. Shirafuji, New general relativity. Phys. Rev. D **19**, 3524–3553 (1979)
48. Yu.N. Obukhov, J.G. Pereira, Metric affine approach to teleparallel gravity. Phys. Rev. D **67**, 044016 (2003)
49. J.G. Pereira, Yu.N. Obukhov, Gauge structure of teleparallel gravity. Universe **5**(6), 139 (2019)
50. T. Koivisto, M. Hohmann, T. Złośnik, The general linear Cartan khronon. Universe **5**(6), 168 (2019)
51. R. Aldrovandi, J.G. Pereira, *Teleparallel Gravity: An Introduction* (Springer, Dordrecht, 2013)
52. J.W. Maluf, F.F. Faria, Teleparallel gauge theory of gravity. Ann. Phys. (Berlin) **524**, 366–370 (2012)
53. J.W. Maluf, The teleparallel equivalent of general relativity. Ann. Phys. (Berlin) **525**, 339–357 (2013)
54. Y. Itin, Energy momentum current for coframe gravity. Class. Quantum Grav. **19**, 173 (2002)
55. F.W. Hehl, Y. Itin, Yu.N. Obukhov, On Kottler's path: origin and evolution of the premetric program in gravity and in electrodynamics. Int. J. Mod. Phys. D **25**, 1640016 (2016)
56. Y. Itin, Yu.N. Obukhov, J. Boos, F.W. Hehl, Premetric teleparallel theory of gravity and its local and linear constitutive law. Eur. Phys. J. C **78**, 907 (2018)
57. Y. Itin, F.W. Hehl, Yu.N. Obukhov, Premetric equivalent of general relativity: teleparallelism. Phys. Rev. D **95**, 084020 (2017)
58. M. Krššák, R.J. van den Hoogen, J.G. Pereira, C.G. Boehmer, A.A. Coley, Teleparallel theories of gravity: illuminating a fully invariant approach. Class. Quantum Grav. **36**, 183001 (2019)
59. A. Lichnerowicz, *Geometry of Groups of Transformations* (Noordhoff International Publishing, Leyden, 1977)
60. K. Yano, *The Theory of Lie Derivatives and Its Applications* (North-Holland, Amsterdam, 1955)
61. E.P. Wigner, On unitary representations of the inhomogeneous Lorentz group. Ann. Math. **40**, 149–204 (1939)
62. T. Eguchi, P.B. Gilkey, A.J. Hanson, Gravitation, gauge theories and differential geometry. Phys. Rep. **66**, 213–393 (1980)
63. F.W. Hehl, W. Kopczyński, J.D. McCrea, E.W. Mielke, Chern-Simons terms in metric-affine spacetime: Bianchi identities as Euler-Lagrange equations. J. Math. Phys. **32**, 2169–2180 (1991)
64. O. Chandia, J. Zanelli, Torsional topological invariants (and their relevance for real life). AIP Conf. Proc. **419**, 251–264 (1998)
65. F.W. Hehl, On the energy tensor of spinning massive matter in classical field theory and general relativity. Rep. Math. Phys. **9**, 55–82 (1976)
66. J. Weyssenhoff, A. Raabe, Relativistic dynamics of spin-fluids and spin-particle. Acta Phys. Pol. **9**, 7–18 (1947)
67. E. Cosserat, F. Cosserat, *Theorie des corps deformables* (Hermann, Paris, 1909)
68. Yu.N. Obukhov, V.A. Korotky, The Weyssenhoff fluid in Einstein-Cartan theory. Class. Quantum Grav. **4**, 1633–1657 (1987)

69. D.E. Neville, Spin-2 propagating torsion. Phys. Rev. D **23**, 1244–1249 (1981)
70. E. Sezgin. P. van Nieuwenhuizen, New ghost-free gravity Lagrangians with propagating torsion. Phys. Rev. D **21**, 3269–3280 (1980)
71. Yu.N. Obukhov, V.N. Ponomariev, V.V. Zhytnikov, Quadratic Poincaré gauge theory of gravity: a comparison with the general relativity theory. Gen. Relat. Grav. **21**, 1107–1142 (1989)
72. G.K. Karananas, The particle spectrum of parity-violating Poincaré gravitational theory. Class. Quantum Grav. **32**, 055012 (2015). Corrigendum: Class. Quantum Grav. **32**, 089501 (2015)
73. M. Blagojević, B. Cvetković, General Poincaré gauge theory: Hamiltonian structure and particle spectrum. Phys. Rev. D **98**, 104018 (2018)
74. A.Z. Petrov, *Einstein Spaces* (Pergamon, Oxford, 1969)
75. W.-T. Ni, Yang's gravitational field equations. Phys. Rev. Lett. **35**, 319–320 (1975)
76. M. Hohmann, Spacetime and observer space symmetries in the language of Cartan geometry. J. Math. Phys. **57**, 082502 (2016)
77. M. Hohmann, Metric-affine geometries with spherical symmetry. Symmetry **12**, 453 (2020)
78. Yu.N. Obukhov, Generalized Birkhoff theorem in the Poincaré gauge gravity theory. Phys. Rev. D **102**, 104059 (2020)
79. S. Ramaswamy, P. Yasskin, Birkhoff theorem for an $R + R^2$ theory of gravity with torsion. Phys. Rev. D **19**, 2264–2267 (1979)
80. R.T. Rauch, S.J. Shaw, H.T. Nieh, Birkhoff's theorem for ghost-free tachyon-free $R+R^2+Q^2$ theories with torsion. Gen. Relat. Grav. **14**, 331–354 (1982)
81. R. Rauch, H.T. Nieh, Birkhoff's theorem for general Riemann-Cartan $R + R^2$ theories of gravity. Phys. Rev. D **24**, 2029–2048 (1981)
82. R.T. Rauch, Asymptotic flatness, reflection symmetry, and Birkhoff's theorem for $R + R^2$ actions containing quadratic torsion. Phys. Rev. D **25**, 577–580 (1982)
83. D.E. Neville, Gravity Lagrangian with ghost-free curvature-squared terms. Phys. Rev. D **18**, 3535–3543 (1978)
84. D.E. Neville, Birkhoff theorems for $R + R^2$ gravity theories with torsion. Phys. Rev. D **21**, 2770–2775 (1980)
85. A. de la Cruz-Dombriz, F.J.M. Torralba, Birkhoff's theorem for stable torsion theories. J. Cosmol. Astropart. Phys. **03**, 002 (2019)
86. A. Trautman, Spin and torsion may avert gravitational singularity. Nat. Phys. Sci. **242**, 7–8 (1973)
87. D. Palle, On certain relationships between cosmological observables in the Einstein-Cartan gravity. Nuovo Cim. B **111**, 671–675 (1996)
88. D. Palle, On primordial cosmological density fluctuations in the Einstein-Cartan gravity and COBE data. Nuovo Cim. B **114**, 853–860 (1999)
89. S.D. Brechet, M.P. Hobson, A.N. Lasenby, Weyssenhoff fluid dynamics in general relativity using a $1 + 3$ covariant approach. Class. Quantum Grav. **24**, 6329–6348 (2007)
90. S.D. Brechet, M.P. Hobson, A.N. Lasenby, Classical big-bounce cosmology: dynamical analysis of a homogeneous and irrotational Weyssenhoff fluid. Class. Quantum Grav. **25**, 245016 (2008)
91. D. Palle, On the Einstein-Cartan cosmology vs. Planck data. J. Exp. Theor. Phys. **118**, 587–592 (2014)
92. Yu.N. Obukhov, F.W. Hehl, General relativity as a special case of Poincaré gauge gravity. Phys. Rev. D **102**, 044058 (2020)
93. P. von der Heyde, Is gravitation mediated by the torsion of spacetime? Z. Naturf. **31a**, 1725–1726 (1976)
94. C. Heinicke, F.W. Hehl, Schwarzschild and Kerr Solutions of Einstein's Field Equation – an introduction. Int. J. Mod. Phys. D **24**, 1530006 (2014)
95. Yu.N. Obukhov, Exact solutions in Poincaré gauge gravity theory. Universe **5**(5), 127 (2019)
96. J.A.R. Cembranos, J.G. Valcarcel, New torsion black hole solutions in Poincaré gauge theory. J. Cosmol. Astropart. Phys. **01**, 014 (2017)

97. J.A.R. Cembranos, J.G. Valcarcel, Extended Reissner-Nordström solutions sourced by dynamical torsion. Phys. Lett. B **779**, 143–150 (2018)
98. B.P. Abbott et al., (LIGO Scientific Collaboration and Virgo Collaboration), Observation of gravitational waves from a binary black hole merger. Phys. Rev. Lett. **116**, 061102 (2016)
99. C.-M. Chen, J.M. Nester, W.-T. Ni, A brief history of gravitational wave research. Chin. J. Phys. **55**, 142–169 (2017)
100. V.D. Zakharov, *Gravitational Waves in Einstein's Theory* (Halsted Press, New York, 1973)
101. J.B. Griffiths, *Colliding Plane Waves in General Relativity* (Clarendon Press, Oxford, 1991)
102. H. Stephani, D. Kramer, M. MacCallum, C. Hoenselaers, E. Herlt, *Exact Solutions of Einstein's Field Equations*, 2nd edn., Secs. 24, 31 (Cambridge University Press, Cambridge, 2003)
103. H.W. Brinkmann, On Riemann spaces conformal to Euclidean space. Proc. Natl. Acad. Sci. **9**, 1–3 (1923)
104. H.W. Brinkmann, Einstein spaces which are mapped conformally on each other. Math. Ann. **94**, 119–145 (1925)
105. Yu.N. Obukhov, Gravitational waves in Poincaré gauge gravity theory. Phys. Rev. D **95**, 084028 (2017)
106. M. Blagojević, B. Cvetković, Yu.N. Obukhov, Generalized plane waves in Poincaré gauge theory of gravity. Phys. Rev. D **96**, 064031 (2017)
107. M. Fontanini, E. Huguet, M. Le Delliou, Teleparallel gravity equivalent of general relativity as a gauge theory: translation or Cartan connection? Phys. Rev. D **99**, 064006 (2019)
108. M. Le Delliou, E. Huguet, M. Fontanini, Teleparallel theory as a gauge theory of translations: remarks and issues. Phys. Rev. D **101**, 024059 (2020)
109. E. Huguet, M. Le Delliou, M. Fontanini, Cartan approach to teleparallel equivalent to general relativity: a review. Int. J. Geom. Meth. Mod. Phys. **18**(Supp. 01), 2140004 (2021)
110. E. Huguet, M. Le Delliou, M. Fontanini, Z.-C. Lin, Teleparallel gravity as a gauge theory: coupling to matter using the Cartan connection. Phys. Rev. D **103**, 044061 (2021)
111. C. Pellegrini, J. Plebanski, Tetrad fields and gravitational fields. Mat. Fys. Skr. Dan. Vid. Selsk. **2**(4), 1–39 (1963)
112. F.A. Kaempffer, Vierbein field theory of gravitation. Phys. Rev. **165**, 1420–1423 (1968)
113. P. Spindel, Dynamical torsion gravity backgrounds. Phys. Rev. D **103**, 124054 (2021)
114. A.V. Minkevich, Generalized cosmological Friedmann equations without gravitational singularity. Phys. Lett. A **80**, 232–234 (1980)
115. A.V. Minkevich, Towards the theory of regular accelerating Universe in Riemann-Cartan space-time. Int. J. Mod. Phys. A **31**, 1641011 (2016)
116. J. Magueijo, T.G. Złośnik, T.W.B. Kibble, Cosmology with a spin. Phys. Rev. D **87**, 063504 (2013)
117. J. Magueijo, T.G. Złośnik, Parity violating Friedmann universes. Phys. Rev. D **100**, 084036 (2019)
118. N. Popławski, Big bounce from spin and torsion. Gen. Relat. Gravit. **44**, 1007–1014 (2012)
119. N. Popławski, Cosmological consequences of gravity with spin and torsion. Astron. Rev. **8**, 108–115 (2013)
120. N. Popławski, The simplest origin of the big bounce and inflation. Int. J. Mod. Phys. D **27**, 1847020 (2018)
121. D. Puetzfeld, Status of non-Riemannian cosmology. New Astron. Rev. **49**, 59–64 (2005)
122. H. Zhang, L. Xu, Late-time acceleration and inflation in a Poincaré gauge cosmological model. J. Cosmol. Astropart. Phys. **09**, 050 (2019)
123. D. Kranas, C.G. Tsagas, J.D. Barrow, D. Iosifidis, Friedmann-like universes with torsion. Eur. Phys. J. C **79**, 341 (2019)
124. J.D. Barrow, C.G. Tsagas, G. Fanaras, Friedmann-like universes with torsion: a dynamical system approach. Eur. Phys. J. C **79**, 764 (2019)
125. A.N. Ivanov, M. Wellenzohn, Einstein-Cartan gravity with torsion field serving as an origin for the cosmological constant or dark energy density. Astrophys. J. **829**, 47 (2016)

126. P. Baekler, F.W. Hehl, Beyond Einstein-Cartan gravity: quadratic torsion and curvature invariants with even and odd parity including all boundary terms. Class. Quantum Grav. **28**, 215017 (2011)
127. P. Baekler, F.W. Hehl, J.M. Nester, Poincaré gauge theory of gravity: Friedman cosmology with even and odd parity modes: analytic part. Phys. Rev. D **83**, 024001 (2011)
128. H. Chen, F.-H. Ho, J.M. Nester, C.-H. Wang, H.-J. Yo, Cosmological dynamics with propagating Lorentz connection modes of spin zero. J. Cosmol. Astropart. Phys. **10**, 027 (2009)
129. F.H. Ho, J.M. Nester, Poincaré gauge theory with coupled even and odd parity spin-0 modes: cosmological normal modes. Ann. d. Physik (Berlin) **524**, 97–106 (2012)
130. F.H. Ho, J.M. Nester, Poincaré gauge theory with coupled even and odd parity dynamic spin-0 modes: dynamical equations for isotropic Bianchi cosmologies. Int. J. Mod. Phys. D **20**, 2125–2138 (2011)
131. F.H. Ho, H. Chen, J.M. Nester, H.J. Yo, General Poincaré gauge theory cosmology. Chin. J. Phys. **53**, 110109 (2015)
132. A. Einstein, Geometrie und Erfahrung. Sitzungsber. Preuss. Akad. Wiss. Phys.-math. Klasse **1**, 123–130 (1921)
133. P.B. Yasskin, W.R. Stoeger, Propagating equations for test bodies with spin and rotation in theories of gravity with torsion. Phys. Rev. D **21**, 2081–2094 (1980)
134. F.W. Hehl, Yu.N. Obukhov, D. Puetzfeld, On Poincaré gauge theory of gravity, its equations of motion, and Gravity Probe B. Phys. Lett. A **377**, 1775–1781 (2013)
135. Yu.N. Obukhov, D. Puetzfeld, Multipolar test body equations of motion in generalized gravity theories, in *Fundamental Theories of Physics*, vol. 179 (Springer, Cham, 2015), pp. 67–119
136. F.W. Hehl, How does one measure torsion of space-time? Phys. Lett. A **36**, 225–226 (1971)
137. W. Adamowicz, A. Trautman, The principle of equivalence for spin. Bull. Acad. Pol. Sci., Sér. Sci. Math. Astron. Phys. **23**, 339–342 (1975)
138. J. Audretsch, C. Lämmerzahl, Neutron interference: general theory of the influence of gravity, inertia and space-time torsion. J. Phys. A: Math. Gen. **16**, 2457–2477 (1983)
139. H. Rumpf, Quasiclassical limit of the Dirac equation and the equivalence principle in the Riemann-Cartan geometry, in *Cosmology and Gravitation: Spin, Torsion, Rotation and Supergravity*, ed. by P.G. Bergmann, V. de Sabbata (Plenum, New York, 1980), pp. 93–104
140. J. Audretsch, Dirac electron in space-times with torsion. Spinor propagation, spin precession, and nongeodesic orbit. Phys. Rev. D **24**, 1470–1477 (1981). Erratum: Phys. Rev. D **25**, 605 (1982)
141. C. Lämmerzahl, Constraints on space-time torsion from Hughes-Drever experiments. Phys. Lett. A **228**, 223–231 (1997)
142. W.-T. Ni, Searches for the role of spin and polarization in gravity. Rep. Prog. Phys. **73**, 056901 (2010)
143. V.A. Kostelecký, R. Russell, J.D. Tasson, Constraints on torsion from bounds on Lorentz violation. Phys. Rev. Lett. **100**, 111102 (2008)
144. V.A. Kostelecký, Z. Li, Searches for beyond-Riemann gravity. Phys. Rev. D **104**, 044054 (2021)
145. R. Lehnert, W.M. Snow, H. Yan, A first experimental limit on in-matter torsion from neutron spin rotation in liquid ^{4}He. Phys. Lett. B **730**, 353–356 (2014). Corrigendum: Phys. Lett. B **744**, 415 (2015)
146. A.N. Ivanov, W.M. Snow, Parity-even and time-reversal-odd neutron optical potential in spinning matter induced by gravitational torsion. Phys. Lett. B **764**, 186–189 (2017)
147. A.N. Ivanov, M. Wellenzohn, H. Abele, Quantum gravitational states of ultracold neutrons as a tool for probing of beyond-Riemann gravity. Phys. Lett. B **822**, 136640 (2021)
148. Yu.N. Obukhov, A.J. Silenko, O.V. Teryaev, Spin-torsion coupling and gravitational moments of Dirac fermions: theory and experimental bounds. Phys. Rev. D **90**, 124068 (2014)
149. M.I. Trukhanova, Yu.N. Obukhov, Quantum hydrodynamics of spinning particles in electromagnetic and torsion fields. Universe **7**(12), 498 (2021)

Teleparallel Gravity

4

Manuel Hohmann

Abstract

In general relativity, the only dynamical field describing the gravitational interaction of matter, is the metric. It induces the causal structure of spacetime, governs the motion of physical bodies through its Levi-Civita connection, and mediates gravity via the curvature of this connection. While numerous modified theories of gravity retain these principles, it is also possible to introduce another affine connection as a fundamental field, and consider its properties—curvature, torsion, nonmetricity—as the mediators of gravity. In the most general case, this gives rise to the class of metric-affine gravity theories, while restricting to metric-compatible connections, for which nonmetricity vanishes, comprises the class of Poincaré gauge theories. Alternatively, one may also consider connections with vanishing curvature. This assumption yields the class of *teleparallel* gravity theories. This chapter gives a simplified introduction to teleparallel gravity, with a focus on performing practical calculations, as well as an overview of the most commonly studied classes of teleparallel gravity theories.

4.1 Introduction

In his original work, Einstein formulated the general theory of relativity in terms of the metric tensor as the fundamental field variable of the gravitational field, which describes gravity by the curvature of its Levi-Civita connection. Numerous modified gravity theories depart from this formulation, either keeping the metric as the only fundamental field variable and modifying its dynamics through a modified

---

M. Hohmann (✉)
Laboratory of Theoretical Physics, Institute of Physics, University of Tartu, Tartu, Estonia
e-mail: manuel.hohmann@ut.ee

C. Pfeifer, C. Lämmerzahl (eds.), *Modified and Quantum Gravity*, Lecture Notes in Physics 1017, https://doi.org/10.1007/978-3-031-31520-6_4

action, or by adding further fundamental field which couple non-minimally to the curvature [1]. However, there exist also other classes of gravity theories, in which the curvature of the Levi-Civita connection plays a less prominent role, and another, independent connection is introduced as a fundamental field variable next to the metric. Unlike the Levi-Civita connection, this connection is assumed to have vanishing curvature, but instead one allows for non-vanishing torsion or nonmetricity, or both. Gravity theories of this type are known as *teleparallel* gravity theories.

In fact, already Einstein studied the possibility to describe gravity in terms of the torsion of a flat, metric-compatible connection instead of curvature [2], in an attempt to unify gravity and electromagnetism. While this attempt was not successful, it gave rise to a new class of gravity theories, now known as metric teleparallel gravity theories [3, 4], in which gravity is mediated by torsion instead of curvature. Only much later another class of gravity theories was introduced, which attributes gravity to the nonmetricity of a flat, torsion-free (i.e., symmetric) connection, and is hence known as symmetric teleparallel gravity [5]. Finally, allowing for both torsion and nonmetricity leads to the realm of general teleparallel gravity [6]. It is worth mentioning that these theories are embedded in the much wider and well-studied framework of metric-affine gravity theories [7, 8], for the metric-compatible case also in the framework of Poincaré gauge theories [9, 10], as it was discussed in the previous chapter, Chap. 3. However, a full account of this relationship and the historic development and studies of teleparallel gravity theories would by far exceed the scope of this chapter.

Despite the long-standing history of teleparallel gravity theories and the studies of their fundamental properties and underlying structure for several decades, a renewed and growing interest in teleparallel modifications and extension of general relativity and their phenomenology has arisen only recently with the growing number of unexplained observations and tensions in cosmology. Numerous theories have been constructed as possible candidates to explain the early and late accelerating phases of the universe, known as inflation and dark energy eras, to resolve the question of singularities and the information paradox of black holes, and to provide alternative pathways towards a quantization of gravity and a unification with other fundamental forces. The phenomenology of these theories greatly differs and depends on their choice of dynamical fields and action, so that a full account would, again, exceed the scope of this chapter, and we must limit ourselves to a more general discussion of the class of teleparallel gravity theories, and leave specific theories and their phenomenological properties for further reading [4].

The aim of this chapter is to provide a practical introduction to teleparallel gravity. In Sect. 4.2 we give a simplified summary of the general structure and underlying mathematical foundations of teleparallel gravity theories in their three flavors—general, symmetric and metric. In particular, we discuss the fundamental fields in these theories, the general form of the action and the field equations. This practical introduction continues in Sect. 4.3, where we explain how to formulate physical principles and perform common calculations necessary to solve the gravitational field equations of teleparallel gravity theories. We discuss how the invariance of the action under diffeomorphisms leads to the conservation of the matter currents, and show how to construct teleparallel geometries with spacetime symmetries and their perturbations, which can be used to solve the field equations of a given theory of gravity and thus study its phenomenology. Finally, Sect. 4.4 gives an overview of the most commonly studied classes of teleparallel gravity theories and their field equations, and briefly summarizes their common properties.

There are many interesting aspects of teleparallel gravity which cannot be covered in this chapter, as they would by far exceed its scope and its aim towards performing practical calculations. In particular, we do not discuss the role of gauge symmetries in teleparallel gravity, which allow its interpretation as a gauge theory of the translation group. In relation to this, we do not discuss its formulation in terms of a tetrad. Throughout the chapter, we use only the tensor notation, which is more widespread in relating gravity to observations, and avoid the use of differential form language, which is often more concise and thus preferred by theorists, but less common in practical calculations of phenomenology. Further, we cannot cover fundamental questions such as the number of degrees of freedom of these theories, which is studied in their Hamiltonian formulation, and hints towards theoretical issues known under the term strong coupling. The interested reader is encouraged to follow the references provided in this chapter for a more detailed account of these mathematical foundations, their applications and possible issues.

We use the convention that spacetime coordinate indices are labeled with lowercase Greek letters (observe the difference to the convention in the previous chapter, Chap. 3) and take the values $(0, 1, 2, 3)$, as well as the metric signature $(-1, +1, +1, +1)$.

4.2 Dynamical Fields, Action and Field Equations

In this introductory section we give an overview of the dynamical fields and their properties, the general structure of the action, and the variational methods used to obtain their field equations. Here we focus on three different flavors of teleparallel theories: general teleparallel theories, in which both torsion and nonmetricity are allowed to be non-vanishing, are discussed in Sect. 4.2.1; we then restrict the theories to symmetric teleparallel gravity by imposing vanishing torsion in Sect. 4.2.2, and to metric teleparallel gravity by imposing vanishing nonmetricity in Sect. 4.2.3.

4.2.1 General Teleparallel Gravity

We start our discussion of teleparallel gravity theories from the viewpoint of metric-affine gravity, in which next to the metric $g_{\mu\nu}$ a connection with coefficients $\Gamma^{\mu}{}_{\nu\rho}$ is introduced as a fundamental field on the spacetime manifold M, which is independent of the Levi-Civita connection. To distinguish these two connections, we write the latter, and all derived quantities such as the covariant derivative and the curvature tensor, with a circle on top, i.e.,

$$\overset{\circ}{\Gamma}^{\mu}{}_{\nu\rho} = \frac{1}{2} g^{\mu\sigma} \left(\partial_\nu g_{\sigma\rho} + \partial_\rho g_{\nu\sigma} - \partial_\sigma g_{\nu\rho} \right) . \tag{4.1}$$

Given another, independent connection, their difference can always be written in the form

$$\Gamma^{\mu}{}_{\nu\rho} - \overset{\circ}{\Gamma}^{\mu}{}_{\nu\rho} = M^{\mu}{}_{\nu\rho} = K^{\mu}{}_{\nu\rho} + L^{\mu}{}_{\nu\rho} . \tag{4.2}$$

Here, $M^{\mu}{}_{\nu\rho}$ is called the *distortion*: it is the difference between two connection coefficients, and hence a tensor field. If one of these two connections is the Levi-Civita connection of a metric, the distortion decomposes further into the *contortion* $K^{\mu}{}_{\nu\rho}$ and the *disformation* $L^{\mu}{}_{\nu\rho}$, which can be obtained as follows. First, define the *torsion*

$$T^{\mu}{}_{\nu\rho} = \Gamma^{\mu}{}_{\rho\nu} - \Gamma^{\mu}{}_{\nu\rho} , \tag{4.3}$$

as well as the *nonmetricity*

$$Q_{\mu\nu\rho} = \nabla_\mu g_{\nu\rho} = \partial_\mu g_{\nu\rho} - \Gamma^{\sigma}{}_{\nu\mu} g_{\sigma\rho} - \Gamma^{\sigma}{}_{\rho\mu} g_{\nu\sigma} . \tag{4.4}$$

These are, again, tensor fields. Using the metric to raise and lower indices, one then obtains the contortion

$$K^{\mu}{}_{\nu\rho} = \frac{1}{2} \left(T_{\nu}{}^{\mu}{}_{\rho} + T_{\rho}{}^{\mu}{}_{\nu} - T^{\mu}{}_{\nu\rho} \right) , \tag{4.5}$$

as well as the disformation

$$L^{\mu}{}_{\nu\rho} = \frac{1}{2} \left(Q^{\mu}{}_{\nu\rho} - Q_{\nu}{}^{\mu}{}_{\rho} - Q_{\rho}{}^{\mu}{}_{\nu} \right) . \tag{4.6}$$

Hence, in the presence of a metric, an independent connection can always uniquely be specified in terms of its torsion and nonmetricity, which determine its deviation from the Levi-Civita connection.

The dynamical fields then enter the action of the theory, which is of the general form

$$S[g, \Gamma, \psi] = S_{\mathrm{g}}[g, \Gamma] + S_{\mathrm{m}}[g, \Gamma, \psi] , \tag{4.7}$$

where the gravitational part S_{g} of the action depends only on the metric and the connection, while the matter part S_{m} also depends on some set of matter fields ψ^I, whose components we do not specify further and simply label them with an index I. By variation with respect to these matter fields, the matter action determines the matter field equations, which govern the dynamics of the matter fields in a given gravitational field background. In general, this background depends both on the metric $g_{\mu\nu}$ and the connection $\Gamma^{\mu}{}_{\nu\rho}$. Further, varying the matter action with respect to the metric and the connection gives rise to the *energy-momentum* $\Theta^{\mu\nu}$ and *hypermomentum* $H_{\mu}{}^{\nu\rho}$ defined by the variation [7]

$$\delta S_{\mathrm{m}} = \int_M \left(\frac{1}{2}\Theta^{\mu\nu}\delta g_{\mu\nu} + H_{\mu}{}^{\nu\rho}\delta\Gamma^{\mu}{}_{\nu\rho} + \Psi_I \delta\psi^I\right)\sqrt{-g}\mathrm{d}^4x\,, \tag{4.8}$$

where $\Psi_I = 0$ are the matter field equations. The specific form of $\Theta^{\mu\nu}$ and $H_{\mu}{}^{\nu\rho}$ depends on the type of matter under consideration and its coupling to the background geometry. These terms will act as the source of the gravitational field equations. To obtain the latter, one writes the variation of the gravitational part of the action in the similar form

$$\delta S_{\mathrm{g}} = -\int_M \left(\frac{1}{2}W^{\mu\nu}\delta g_{\mu\nu} + Y_{\mu}{}^{\nu\rho}\delta\Gamma^{\mu}{}_{\nu\rho}\right)\sqrt{-g}\mathrm{d}^4x\,, \tag{4.9}$$

where any necessary integration by parts has been carried out in order to eliminate derivatives acting on the variations. This variation defines two further tensor fields, which we denote $W^{\mu\nu}$ and $Y_{\mu}{}^{\nu\rho}$, and which will enter as the dynamical part of the gravitational field equations.

The action and variation given above constitute the general form for a metric-affine theory of gravity. In teleparallel gravity, however, the connection is further restricted to have vanishing curvature,

$$R^{\mu}{}_{\nu\rho\sigma} = \partial_\rho\Gamma^{\mu}{}_{\nu\sigma} - \partial_\sigma\Gamma^{\mu}{}_{\nu\rho} + \Gamma^{\mu}{}_{\tau\rho}\Gamma^{\tau}{}_{\nu\sigma} - \Gamma^{\mu}{}_{\tau\sigma}\Gamma^{\tau}{}_{\nu\rho} \equiv 0\,. \tag{4.10}$$

Note that this condition involves both the connection coefficients and their derivatives. In the context of Lagrange theory, such type of condition constitutes a non-holonomic constraint. Different possibilities exist to implement this constraint [11]. One possibility is to add another term of the form

$$S_{\mathrm{r}} = \int_M \tilde{r}_{\mu}{}^{\nu\rho\sigma} R^{\mu}{}_{\nu\rho\sigma}\mathrm{d}^4x\,, \tag{4.11}$$

where the tensor density $\tilde{r}_{\mu}{}^{\nu\rho\sigma}$ acts as a Lagrange multiplier, and can be taken to be antisymmetric in its last two indices, $\tilde{r}_{\mu}{}^{\nu\rho\sigma} = \tilde{r}_{\mu}{}^{\nu[\rho\sigma]}$, since the contraction of its symmetric part with the antisymmetric indices of the curvature tensor vanishes and thus does not contribute to the action. Variation with respect to $\tilde{r}_{\mu}{}^{\nu\rho\sigma}$ then yields the constraint equation $R^{\mu}{}_{\nu\rho\sigma} = 0$. In order to derive the variation with respect to the connection coefficients, note that the variation of the curvature can be expressed as

$$\delta R^{\mu}{}_{\nu\rho\sigma} = \nabla_\rho\delta\Gamma^{\mu}{}_{\nu\sigma} - \nabla_\sigma\delta\Gamma^{\mu}{}_{\nu\rho} + T^{\tau}{}_{\rho\sigma}\delta\Gamma^{\mu}{}_{\nu\tau}\,. \tag{4.12}$$

With the help of this expression, as well as performing integration by parts, one obtains the variation of the Lagrange multiplier term S_{r} in the action with respect to the connection as

$$\begin{aligned}\delta_\Gamma S_{\mathrm{r}} &= \int_M \tilde{r}_\mu{}^{\nu\rho\sigma}\left(\nabla_\rho \delta\Gamma^\mu{}_{\nu\sigma} - \nabla_\sigma \delta\Gamma^\mu{}_{\nu\rho} + T^\tau{}_{\rho\sigma}\delta\Gamma^\mu{}_{\nu\tau}\right) \mathrm{d}^4x \\ &= \int_M \left(T^\sigma{}_{\sigma\rho}\tilde{r}_\mu{}^{\nu\rho\tau} - T^\rho{}_{\rho\sigma}\tilde{r}_\mu{}^{\nu\tau\sigma} + T^\tau{}_{\rho\sigma}\tilde{r}_\mu{}^{\nu\rho\sigma}\right. \\ &\left. -\nabla_\rho \tilde{r}_\mu{}^{\nu\rho\tau} + \nabla_\sigma \tilde{r}_\mu{}^{\nu\tau\sigma}\right)\delta\Gamma^\mu{}_{\nu\tau}\mathrm{d}^4x\,.\end{aligned} \tag{4.13}$$

Combining all terms, one finds that the gravitational field equations are given by the metric field equation

$$W_{\mu\nu} = \Theta_{\mu\nu}\,, \tag{4.14}$$

as well as the connection field equation

$$\tilde{Y}_\mu{}^{\nu\tau} = \tilde{H}_\mu{}^{\nu\tau} + T^\sigma{}_{\sigma\rho}\tilde{r}_\mu{}^{\nu\rho\tau} - T^\rho{}_{\rho\sigma}\tilde{r}_\mu{}^{\nu\tau\sigma} + T^\tau{}_{\rho\sigma}\tilde{r}_\mu{}^{\nu\rho\sigma} - \nabla_\rho \tilde{r}_\mu{}^{\nu\rho\tau} + \nabla_\sigma \tilde{r}_\mu{}^{\nu\tau\sigma}\,, \tag{4.15}$$

where it is convenient to define the tensor densities

$$\tilde{Y}_\mu{}^{\nu\tau} = Y_\mu{}^{\nu\tau}\sqrt{-g}\,, \quad \tilde{H}_\mu{}^{\nu\tau} = H_\mu{}^{\nu\tau}\sqrt{-g}\,. \tag{4.16}$$

Note that the connection equation still contains the undetermined Lagrange multiplier $\tilde{r}_\mu{}^{\nu\rho\sigma}$. However, the latter can be eliminated using the following procedure. First, we calculate the divergence

$$\begin{aligned}\nabla_\tau \tilde{Y}_\mu{}^{\nu\tau} = \nabla_\tau \tilde{H}_\mu{}^{\nu\tau} &+ \nabla_\tau\left(T^\sigma{}_{\sigma\rho}\tilde{r}_\mu{}^{\nu\rho\tau} - T^\rho{}_{\rho\sigma}\tilde{r}_\mu{}^{\nu\tau\sigma} + T^\tau{}_{\rho\sigma}\tilde{r}_\mu{}^{\nu\rho\sigma}\right) \\ &- \nabla_\tau\nabla_\rho \tilde{r}_\mu{}^{\nu\rho\tau} + \nabla_\tau\nabla_\sigma \tilde{r}_\mu{}^{\nu\tau\sigma}\,.\end{aligned} \tag{4.17}$$

The last two terms can be simplified by realizing that the Lagrange multiplier $\tilde{r}_\mu{}^{\nu\rho\sigma}$ is antisymmetric in its last two indices, so that one can apply the commutator of covariant derivatives given by

$$\begin{aligned}2\nabla_{[\rho}\nabla_{\sigma]}\tilde{r}_\mu{}^{\nu\rho\sigma} &= -T^\tau{}_{\rho\sigma}\nabla_\tau \tilde{r}_\mu{}^{\nu\rho\sigma} \\ &- R^\tau{}_{\mu\rho\sigma}\tilde{r}_\tau{}^{\nu\rho\sigma} + R^\nu{}_{\tau\rho\sigma}\tilde{r}_\mu{}^{\tau\rho\sigma} + R^\rho{}_{\tau\rho\sigma}\tilde{r}_\mu{}^{\nu\tau\sigma} + R^\sigma{}_{\tau\rho\sigma}\tilde{r}_\mu{}^{\nu\rho\tau} - R^\tau{}_{\tau\rho\sigma}\tilde{r}_\mu{}^{\nu\rho\sigma}\,.\end{aligned} \tag{4.18}$$

Also using the vanishing curvature (4.10), the only remaining term is given by

$$2\nabla_{[\rho}\nabla_{\sigma]}\tilde{r}_\mu{}^{\nu\rho\sigma} = -T^\tau{}_{\rho\sigma}\nabla_\tau \tilde{r}_\mu{}^{\nu\rho\sigma}\,. \tag{4.19}$$

Further, one can use the antisymmetry of the Lagrange multiplier to write

$$\nabla_\tau \left(T^\sigma{}_{\sigma\rho}\tilde{r}_\mu{}^{\nu\rho\tau} - T^\rho{}_{\rho\sigma}\tilde{r}_\mu{}^{\nu\tau\sigma} + T^\tau{}_{\rho\sigma}\tilde{r}_\mu{}^{\nu\rho\sigma}\right) = 3\nabla_{[\tau}\left(T^\tau{}_{\rho\sigma]}\tilde{r}_\mu{}^{\nu\rho\sigma}\right). \tag{4.20}$$

The derivative of the torsion tensor can be rewritten by making use of the curvature-free Bianchi identity

$$\nabla_{[\nu}T^\mu{}_{\rho\sigma]} + T^\mu{}_{\tau[\nu}T^\tau{}_{\rho\sigma]} = R^\mu{}_{[\nu\rho\sigma]} = 0, \tag{4.21}$$

from which after contraction follows

$$3\nabla_{[\tau}T^\tau{}_{\rho\sigma]} = -3T^\tau{}_{\omega[\tau}T^\omega{}_{\rho\sigma]} = T^\tau{}_{\tau\omega}T^\omega{}_{\rho\sigma}. \tag{4.22}$$

By combining all terms, one finds that the divergence (4.17) of the connection field Eq. (4.15) reads

$$\begin{aligned}\nabla_\tau\tilde{Y}_\mu{}^{\nu\tau} &= \nabla_\tau\tilde{H}_\mu{}^{\nu\tau} + T^\tau{}_{\tau\omega}T^\omega{}_{\rho\sigma}\tilde{r}_\mu{}^{\nu\rho\sigma} + 3T^\tau{}_{[\rho\sigma}\nabla_{\tau]}\tilde{r}_\mu{}^{\nu\rho\sigma} - T^\tau{}_{\rho\sigma}\nabla_\tau\tilde{r}_\mu{}^{\nu\rho\sigma}\\ &= \nabla_\tau\tilde{H}_\mu{}^{\nu\tau} + T^\tau{}_{\tau\omega}T^\omega{}_{\rho\sigma}\tilde{r}_\mu{}^{\nu\rho\sigma} + 2T^\tau{}_{\tau[\rho}\nabla_{\sigma]}\tilde{r}_\mu{}^{\nu\rho\sigma}.\end{aligned} \tag{4.23}$$

Similarly, contracting the field Eq. (4.15) with the trace of the torsion tensor, one obtains

$$T^\omega{}_{\omega\tau}\tilde{Y}_\mu{}^{\nu\tau} = T^\omega{}_{\omega\tau}\tilde{H}_\mu{}^{\nu\tau} + T^\omega{}_{\omega\tau}T^\tau{}_{\rho\sigma}\tilde{r}_\mu{}^{\nu\rho\sigma} + 2T^\omega{}_{\omega[\rho}\nabla_{\sigma]}\tilde{r}_\mu{}^{\nu\rho\sigma}. \tag{4.24}$$

Subtracting these two equations, the Lagrange multiplier terms cancel, and one obtains the connection field equations

$$\nabla_\tau\tilde{Y}_\mu{}^{\nu\tau} - T^\omega{}_{\omega\tau}\tilde{Y}_\mu{}^{\nu\tau} = \nabla_\tau\tilde{H}_\mu{}^{\nu\tau} - T^\omega{}_{\omega\tau}\tilde{H}_\mu{}^{\nu\tau}. \tag{4.25}$$

This equation can also be rewritten by eliminating the density factors using

$$\nabla_\mu\sqrt{-g} = \frac{1}{2}g^{\nu\rho}\nabla_\mu g_{\nu\rho}\sqrt{-g} = \frac{1}{2}Q_{\mu\nu}{}^\nu\sqrt{-g} = M^\nu{}_{\nu\mu}\sqrt{-g}, \tag{4.26}$$

where the last expression follows from rewriting the covariant derivative of the metric in terms of its (vanishing) covariant derivative with respect to the Levi-Civita connection using the decomposition (4.2). The latter can also be used to write the torsion as

$$T^\mu{}_{\nu\rho} = M^\mu{}_{\rho\nu} - M^\mu{}_{\nu\rho}. \tag{4.27}$$

Using these relations and the definition (4.16) of the densities $\tilde{Y}_\mu{}^{\nu\rho}$ and $\tilde{H}_\mu{}^{\nu\rho}$, the field equations become

$$\nabla_\tau Y_\mu{}^{\nu\tau} - M^\omega{}_{\tau\omega} Y_\mu{}^{\nu\tau} = \nabla_\tau H_\mu{}^{\nu\tau} - M^\omega{}_{\tau\omega} H_\mu{}^{\nu\tau} \,. \tag{4.28}$$

Equations (4.14) and (4.28) constitute the field equations for the dynamical fields in teleparallel gravity. In Sect. 4.4 we will derive these field equations explicitly for selected gravity theories.

? Exercise

4.1. Check the integration by parts performed in Eq. (4.13). Recall that the covariant derivative of a tensor density receives an extra term, which is not present for pure tensors.

Besides the method of Lagrange multipliers, the teleparallel field equations can also be obtained by using the method of restricted variation. Using this method, no Lagrange multiplier is introduced, the constraint equation (4.10) of vanishing curvature is imposed to restrict the connection $\Gamma^\mu{}_{\nu\rho}$, and the variation $\delta\Gamma^\mu{}_{\nu\rho}$ is restricted in order to preserve this constraint. Using the expression (4.12), one finds that the variation of the connection must be of the form

$$\delta\Gamma^\mu{}_{\nu\rho} = \nabla_\rho \xi^\mu{}_\nu \tag{4.29}$$

for a tensor field $\xi^\mu{}_\nu$. Indeed, for the curvature perturbation one then finds

$$\delta R^\mu{}_{\nu\rho\sigma} = \nabla_\rho \nabla_\sigma \xi^\mu{}_\nu - \nabla_\sigma \nabla_\rho \xi^\mu{}_\nu + T^\tau{}_{\rho\sigma} \nabla_\tau \xi^\mu{}_\nu = 0 \,, \tag{4.30}$$

using the formula for the commutator of covariant derivatives in the absence of curvature. It follows that the variation of the action takes the form

$$\begin{aligned} \delta_\Gamma S &= \int_M \left(\tilde{H}_\mu{}^{\nu\rho} - \tilde{Y}_\mu{}^{\nu\rho}\right) \nabla_\rho \xi^\mu{}_\nu \mathrm{d}^4x \\ &= \int_M \left(T^\sigma{}_{\sigma\rho} \tilde{H}_\mu{}^{\nu\rho} - \nabla_\rho \tilde{H}_\mu{}^{\nu\rho} - T^\sigma{}_{\sigma\rho} \tilde{Y}_\mu{}^{\nu\rho} + \nabla_\rho \tilde{Y}_\mu{}^{\nu\rho}\right) \xi^\mu{}_\nu \mathrm{d}^4x \,, \end{aligned} \tag{4.31}$$

where the second line follows from integration by parts. Hence, one finds the same connection field equation (4.25).

4.2.2 Symmetric Teleparallel Gravity

The class of teleparallel gravity theories discussed in the previous section, in which the affine connection $\Gamma^{\mu}{}_{\nu\rho}$ is restricted only by the flatness condition (4.10), is also known as *general* teleparallel gravity, and is the youngest among the different classes of teleparallel gravity theories. Two other classes of teleparallel gravity theories can be obtained by demanding that either the torsion (4.3) or the nonmetricity (4.4) vanishes. We will start with the former condition, which yields the class of *symmetric* teleparallel gravity theories, which refers to the fact that the coefficients of a torsion-free connection are symmetric in their lower two indices. In order to implement the condition of vanishing torsion, one may proceed in full analogy to the flatness condition in the previous section, by adding another Lagrange multiplier term

$$S_{\mathrm{t}} = \int_M \tilde{t}_{\mu}{}^{\nu\rho} T^{\mu}{}_{\nu\rho} \mathrm{d}^4x\,, \tag{4.32}$$

where variation with respect to the tensor density $\tilde{t}_{\mu}{}^{\nu\rho}$ leads to the constraint equation $T^{\mu}{}_{\nu\rho} = 0$. In order to derive the field equations, one then proceeds as in the previous section, by varying the full action and eliminating the Lagrange multipliers from the resulting field equations. This calculation is rather lengthy, but straightforward, and so we will not show it here. Instead, we will follow the alternative procedure of restricted variation of the action, by considering only variations $\delta\Gamma^{\mu}{}_{\nu\rho}$ which maintain the vanishing curvature and torsion of the connection. We can use the fact that the flatness is maintained by the variation (4.29), and further restrict the form of $\xi^{\mu}{}_{\nu}$. It turns out that this is achieved by setting $\xi^{\mu}{}_{\nu} = \nabla_{\nu}\zeta^{\mu}$ for some vector field ζ^{μ}, and thus

$$\delta\Gamma^{\mu}{}_{\nu\rho} = \nabla_{\rho}\nabla_{\nu}\zeta^{\mu}\,. \tag{4.33}$$

Using the fact that covariant derivatives commute in the absence of curvature and torsion, one now immediately sees

$$\delta T^{\mu}{}_{\nu\rho} = \delta\Gamma^{\mu}{}_{\rho\nu} - \delta\Gamma^{\mu}{}_{\nu\rho} = \nabla_{\nu}\nabla_{\rho}\zeta^{\mu} - \nabla_{\rho}\nabla_{\nu}\zeta^{\mu} = 0\,. \tag{4.34}$$

The variation of the action with respect to the connection is then simply given by

$$\begin{aligned} \delta_{\Gamma} S &= \int_M \left(\tilde{H}_{\mu}{}^{\nu\rho} - \tilde{Y}_{\mu}{}^{\nu\rho}\right) \nabla_{\rho}\nabla_{\nu}\zeta^{\mu}\mathrm{d}^4x \\ &= -\int_M \nabla_{\rho}\left(\tilde{H}_{\mu}{}^{\nu\rho} - \tilde{Y}_{\mu}{}^{\nu\rho}\right) \nabla_{\nu}\zeta^{\mu}\mathrm{d}^4x \\ &= \int_M \nabla_{\nu}\nabla_{\rho}\left(\tilde{H}_{\mu}{}^{\nu\rho} - \tilde{Y}_{\mu}{}^{\nu\rho}\right) \zeta^{\mu}\mathrm{d}^4x\,, \end{aligned} \tag{4.35}$$

where integration by parts simplifies due to the vanishing torsion. The connection field equation thus becomes

$$\nabla_\nu \nabla_\rho \tilde{Y}_\mu{}^{\nu\rho} = \nabla_\nu \nabla_\rho \tilde{H}_\mu{}^{\nu\rho} \,. \tag{4.36}$$

Together with the metric field equation (4.14), it constitutes the field equations of symmetric teleparallel gravity.

4.2.3 Metric Teleparallel Gravity

We finally come to the remaining class of theories, which are defined by imposing the condition of vanishing nonmetricity, so that the connection becomes metric-compatible. This class of theories is therefore known as *metric* teleparallel gravity, or simply as teleparallel gravity, since it was conceived first among the three different classes we discuss here. To derive its field equations, one can also in this case either introduce a Lagrange multiplier

$$S_{\mathrm{q}} = \int_M \tilde{q}^{\mu\nu\rho} Q_{\mu\nu\rho} \mathrm{d}^4x \,, \tag{4.37}$$

and vary with respect to the tensor density $\tilde{q}^{\mu\nu\rho}$ to obtain $Q_{\mu\nu\rho} = 0$, or find a suitable restriction on the connection variation. Here we will follow once again the latter approach. From the definition (4.4) of the nonmetricity, one obtains its variation

$$\delta Q_{\mu\nu\rho} = \nabla_\mu \delta g_{\nu\rho} - g_{\sigma\rho}\delta\Gamma^\sigma{}_{\nu\mu} - g_{\nu\sigma}\delta\Gamma^\sigma{}_{\rho\mu} = \nabla_\mu(\delta g_{\nu\rho} - 2\xi_{(\nu\rho)}) \,, \tag{4.38}$$

provided that the variation of the connection is chosen to implement the flatness condition (4.29). Here we also used the metric compatibility of the connection to commute lowering an index with the covariant derivative. It turns out that the condition of vanishing nonmetricity imposes a relation

$$\delta g_{\mu\nu} = 2\xi_{(\mu\nu)} \tag{4.39}$$

between the variations of the metric and the connection. Since both are now expressed in terms of the tensor field $\xi_{\mu\nu}$, the field equations follow from the total variation

$$\begin{aligned}
\delta S &= \int_M \left(\Theta^{\mu\nu}\xi_{(\mu\nu)} + H^{\mu\nu\rho}\nabla_\rho\xi_{\mu\nu} - W^{\mu\nu}\xi_{(\mu\nu)} - Y^{\mu\nu\rho}\nabla_\rho\xi_{\mu\nu}\right)\sqrt{-g}\mathrm{d}^4x \\
&= \int_M \Big(\Theta^{(\mu\nu)} - \nabla_\rho H^{\mu\nu\rho} + H^{\mu\nu\rho}T^\tau{}_{\tau\rho} - W^{(\mu\nu)} \\
&\quad + \nabla_\rho Y^{\mu\nu\rho} - Y^{\mu\nu\rho}T^\tau{}_{\tau\rho}\Big)\xi_{\mu\nu}\sqrt{-g}\mathrm{d}^4x \,,
\end{aligned} \tag{4.40}$$

after performing integration by parts, and using the metric compatibility of the connection to obtain $\nabla_\mu\sqrt{-g} = 0$. Keeping in mind that $W^{\mu\nu}$ and $\Theta^{\mu\nu}$ are defined by the variation of the action with respect to the metric, and thus symmetric by definition, one obtains the field equation

$$W^{\mu\nu} - \nabla_\rho Y^{\mu\nu\rho} + Y^{\mu\nu\rho}T^\tau{}_{\tau\rho} = \Theta^{\mu\nu} - \nabla_\rho H^{\mu\nu\rho} + H^{\mu\nu\rho}T^\tau{}_{\tau\rho} \,. \tag{4.41}$$

This single field equation therefore conveys the dynamics in metric teleparallel gravity.

4.3 Physical Aspects and Formalisms in Teleparallel Geometry

To be able to make contact with phenomenology and observations, it is necessary to discuss a few general physical principles in the framework of teleparallel gravity. The first principle, which we discuss in Sect. 4.3.1, is the conservation of the matter currents, which are energy-momentum and hypermomentum, which follows from the invariance of the action under diffeomorphisms. We then continue with spacetime symmetries in Sect. 4.3.2, which can be used to obtain solutions of teleparallel gravity theories, such as black holes, whose phenomenology can subsequently be studied. In particular, we focus on the case of homogeneous and isotropic teleparallel spacetimes, and derive the dynamical variables which appear in teleparallel cosmology. Finally, we discuss the theory of perturbations of teleparallel geometries in Sect. 4.3.3. These form the basis of testing teleparallel gravity theories using gravitational waves and high-precision post-Newtonian observations.

4.3.1 Energy-Momentum-Hypermomentum Conservation

In order to be independent of the choice of coordinates, the different components S_g and S_m of the action discussed in the previous sections are demanded to be independently invariant under diffeomorphisms. Note that an infinitesimal diffeomorphism generated by a vector field $X = X^\mu\partial_\mu$ changes the metric by

$$\delta_X g_{\mu\nu} = (\mathcal{L}_X g)_{\mu\nu} = X^\rho\partial_\rho g_{\mu\nu} + \partial_\mu X^\rho g_{\rho\nu} + \partial_\nu X^\rho g_{\mu\rho} = 2\overset{\circ}{\nabla}_{(\mu}X_{\nu)} \,, \tag{4.42}$$

while the connection is changed by

$$\begin{aligned}\delta_X\Gamma^\mu{}_{\nu\rho} &= (\mathcal{L}_X\Gamma)^\mu{}_{\nu\rho}\\ &= X^\sigma\partial_\sigma\Gamma^\mu{}_{\nu\rho} - \partial_\sigma X^\mu\Gamma^\sigma{}_{\nu\rho} + \partial_\nu X^\sigma\Gamma^\mu{}_{\sigma\rho} + \partial_\rho X^\sigma\Gamma^\mu{}_{\nu\sigma} + \partial_\nu\partial_\rho X^\mu\\ &= \nabla_\rho\nabla_\nu X^\mu - X^\sigma R^\mu{}_{\nu\rho\sigma} - \nabla_\rho(X^\sigma T^\mu{}_{\nu\sigma}) \,,\end{aligned} \tag{4.43}$$

compare with the discussion in the context of Riemann-Cartan spaces in Sect. 3.2.4 of the previous chapter. Note that both expressions are tensor fields, despite the fact that the connection coefficients are not tensor fields. Their variation, however, being an infinitesimal difference between connection coefficients, is a tensor field. In the teleparallel case, the curvature tensor vanishes. Using these formulas, it is now easy to calculate the change of the gravitational part S_{g} of the action, which reads

$$\begin{aligned}\delta_X S_{\mathrm{g}} &= -\int_M \left(\frac{1}{2}\sqrt{-g}W^{\mu\nu}\delta_X g_{\mu\nu} + \tilde{Y}_\mu{}^{\nu\rho}\delta_X\Gamma^\mu{}_{\nu\rho}\right)\mathrm{d}^4x \\ &= -\int_M \left\{\sqrt{-g}W^{\mu\nu}\mathring{\nabla}_\mu X_\nu + \tilde{Y}_\mu{}^{\nu\rho}\left[\nabla_\rho\nabla_\nu X^\mu - \nabla_\rho(X^\sigma T^\mu{}_{\nu\sigma})\right]\right\}\mathrm{d}^4x \\ &= \int_M \Big[\sqrt{-g}\mathring{\nabla}_\nu W_\mu{}^\nu + T^\sigma{}_{\mu\nu}(\nabla_\rho\tilde{Y}_\sigma{}^{\nu\rho} - T^\tau{}_{\tau\rho}\tilde{Y}_\sigma{}^{\nu\rho}) \\ &\quad - \nabla_\nu(\nabla_\rho\tilde{Y}_\mu{}^{\nu\rho} - T^\tau{}_{\tau\rho}\tilde{Y}_\mu{}^{\nu\rho}) + T^\omega{}_{\omega\nu}(\nabla_\rho\tilde{Y}_\mu{}^{\nu\rho} - T^\tau{}_{\tau\rho}\tilde{Y}_\mu{}^{\nu\rho})\Big]X^\mu\mathrm{d}^4x\,.\end{aligned} \tag{4.44}$$

Assuming that the gravitational part S_{g} of the action is invariant under diffeomorphisms, this variation must vanish identically for arbitrary vector fields X^μ. Hence, it follows that the terms $W^{\mu\nu}$ and $\tilde{Y}_\mu{}^{\nu\rho}$ obtained from the variation of the action satisfy

$$\begin{aligned}&\sqrt{-g}\mathring{\nabla}_\nu W_\mu{}^\nu + T^\sigma{}_{\mu\nu}(\nabla_\rho\tilde{Y}_\sigma{}^{\nu\rho} - T^\tau{}_{\tau\rho}\tilde{Y}_\sigma{}^{\nu\rho}) \\ &\quad - \nabla_\nu(\nabla_\rho\tilde{Y}_\mu{}^{\nu\rho} - T^\tau{}_{\tau\rho}\tilde{Y}_\mu{}^{\nu\rho}) + T^\sigma{}_{\sigma\nu}(\nabla_\rho\tilde{Y}_\mu{}^{\nu\rho} - T^\tau{}_{\tau\rho}\tilde{Y}_\mu{}^{\nu\rho}) = 0\,.\end{aligned} \tag{4.45}$$

Alternatively, one can also write this relation without density factors, and finds

$$\begin{aligned}&\mathring{\nabla}_\nu W_\mu{}^\nu + T^\sigma{}_{\mu\nu}(\nabla_\rho Y_\sigma{}^{\nu\rho} - M^\tau{}_{\rho\tau}Y_\sigma{}^{\nu\rho}) \\ &\quad - \nabla_\nu(\nabla_\rho Y_\mu{}^{\nu\rho} - M^\tau{}_{\rho\tau}Y_\mu{}^{\nu\rho}) + M^\sigma{}_{\nu\sigma}(\nabla_\rho Y_\mu{}^{\nu\rho} - M^\tau{}_{\rho\tau}Y_\mu{}^{\nu\rho}) = 0\,.\end{aligned} \tag{4.46}$$

This equation is derived from a purely geometric property of the gravitational part of the action, and so it is a geometric identity, i.e., it holds for any field configuration of the metric $g_{\mu\nu}$ and the connection $\Gamma^\mu{}_{\nu\rho}$, independently of whether these satisfy the gravitational field equations or not[1]. Such a relation is therefore also said to hold *off-shell*. This is to be contrasted with the variation of the matter action S_{m}, which

[1] This equation takes the same role as $\mathring{\nabla}_\nu G_\mu{}^\nu = 0$ for the Einstein tensor, which is satisfied identically as a consequence of the Bianchi identities.

reads

$$\begin{aligned}\delta_X S_g &= \int_M \left(\frac{1}{2}\sqrt{-g}\Theta^{\mu\nu}\delta_X g_{\mu\nu} + \tilde{H}_\mu{}^{\nu\rho}\delta_X\Gamma^\mu{}_{\nu\rho} + \tilde{\Psi}_I\delta_X\psi^I\right)\mathrm{d}^4x \\ &= \int_M \Big\{\sqrt{-g}\Theta^{\mu\nu}\mathring{\nabla}_\mu X_\nu + \tilde{H}_\mu{}^{\nu\rho}\left[\nabla_\rho\nabla_\nu X^\mu - \nabla_\rho(X^\sigma T^\mu{}_{\nu\sigma})\right] \\ &\quad + \tilde{\Psi}_I\mathcal{L}_X\psi^I\Big\}\,\mathrm{d}^4x\,. \end{aligned} \tag{4.47}$$

Here, $\tilde{\Psi}_I = 0$ (or equivalently $\Psi_I = 0$, without using densities) are the matter field equations. If these are satisfied, and only then, demanding that the matter action is invariant under diffeomorphisms generated by an arbitrary vector field X^μ leads to the energy-momentum-hypermomentum conservation law

$$\begin{aligned}&\sqrt{-g}\mathring{\nabla}_\nu\Theta_\mu{}^\nu + T^\sigma{}_{\mu\nu}(\nabla_\rho\tilde{H}_\sigma{}^{\nu\rho} - T^\tau{}_{\tau\rho}\tilde{H}_\sigma{}^{\nu\rho}) \\ &- \nabla_\nu(\nabla_\rho\tilde{H}_\mu{}^{\nu\rho} - T^\tau{}_{\tau\rho}\tilde{H}_\mu{}^{\nu\rho}) + T^\sigma{}_{\sigma\nu}(\nabla_\rho\tilde{H}_\mu{}^{\nu\rho} - T^\tau{}_{\tau\rho}\tilde{H}_\mu{}^{\nu\rho}) = 0\,,\end{aligned} \tag{4.48}$$

or, again in the version without densities,

$$\begin{aligned}&\mathring{\nabla}_\nu\Theta_\mu{}^\nu + T^\sigma{}_{\mu\nu}(\nabla_\rho H_\sigma{}^{\nu\rho} - M^\tau{}_{\rho\tau}H_\sigma{}^{\nu\rho}) \\ &- \nabla_\nu(\nabla_\rho H_\mu{}^{\nu\rho} - M^\tau{}_{\rho\tau}H_\mu{}^{\nu\rho}) + M^\sigma{}_{\nu\sigma}(\nabla_\rho H_\mu{}^{\nu\rho} - M^\tau{}_{\rho\tau}H_\mu{}^{\nu\rho}) = 0\,.\end{aligned} \tag{4.49}$$

Since this relation does not hold for arbitrary field configurations of the gravitational and matter field, but only for those which satisfy the matter field equations $\Psi_I = 0$, it is said to hold *on-shell*. Note that we have not made any assumptions on the properties of the connection except for vanishing curvature. In particular, we have not imposed vanishing torsion or nonmetricity. It follows that the geometric identity and energy-momentum-hypermomentum law given above hold for all three classes of teleparallel gravity theories (but their expressions will simplify in the symmetric and metric cases, as we will see below). Finally, we remark that in the case of vanishing hypermomentum, i.e., for matter which couples only to the metric and not to the connection, which is most commonly considered in the context of teleparallel gravity, the conservation law reduces to

$$\mathring{\nabla}_\nu\Theta_\mu{}^\nu = 0\,, \tag{4.50}$$

which is the well-known energy-momentum conservation.

? Exercise

4.2. Show that the conservation law (4.49) can also be derived from the geometric identity (4.46), by imposing the gravitational field equations.

This is most straightforward for the general teleparallel gravity class, whose gravitational field equations are (4.14) and (4.28). One easily sees that the terms appearing in the identity (4.46) are exactly the left-hand sides of the gravitational field equations. Replacing them with the respective right-hand sides, one obtains the energy-momentum-hypermomentum conservation law (4.49). Of course, the same holds true also if one uses the tensor density version of these equations.

A similar derivation can also be used in the case of symmetric teleparallel gravity, where one assumes vanishing torsion, $T^{\mu}{}_{\nu\rho} = 0$. In this case, it is most convenient to start from the density version (4.45), which simplifies to become

$$\sqrt{-g}\overset{\circ}{\nabla}_{\nu} W_{\mu}{}^{\nu} - \nabla_{\nu}\nabla_{\rho}\tilde{Y}_{\mu}{}^{\nu\rho} = 0\,. \tag{4.51}$$

Using the metric field equation (4.14) and the connection field equation (4.36), one thus immediately obtains the conservation law

$$\sqrt{-g}\overset{\circ}{\nabla}_{\nu}\Theta_{\mu}{}^{\nu} - \nabla_{\nu}\nabla_{\rho}\tilde{H}_{\mu}{}^{\nu\rho} = 0\,, \tag{4.52}$$

which agrees with the general form (4.48) in the absence of torsion. What is most remarkable in the case of symmetric teleparallel gravity is the fact that one can also proceed in a different order: by imposing the matter field equations $\Psi_I = 0$, from which follows the conservation law (4.52), further imposing the metric field equation (4.14), and using the identity (4.51), one obtains the connection field equation (4.36). In other words, any field configuration of the matter and gravitational fields, which satisfies the matter and metric field equations, automatically satisfies also the connection field equation. For this reason, one often omits the latter when it comes to solving the field equations.

Finally, we study the energy-momentum-hypermomentum conservation also in the metric teleparallel setting. In this case, one can omit the density factors in the geometric identity (4.45), since the connection is metric-compatible, so that it becomes

$$\begin{aligned}&\overset{\circ}{\nabla}_{\nu} W_{\mu}{}^{\nu} + T^{\sigma}{}_{\mu\nu}(\nabla_{\rho}Y_{\sigma}{}^{\nu\rho} - T^{\tau}{}_{\tau\rho}Y_{\sigma}{}^{\nu\rho})\\ &\quad - \nabla_{\nu}(\nabla_{\rho}Y_{\mu}{}^{\nu\rho} - T^{\tau}{}_{\tau\rho}Y_{\mu}{}^{\nu\rho}) + T^{\sigma}{}_{\sigma\nu}(\nabla_{\rho}Y_{\mu}{}^{\nu\rho} - T^{\tau}{}_{\tau\rho}Y_{\mu}{}^{\nu\rho}) = 0\,.\end{aligned} \tag{4.53}$$

Further, we impose the metric teleparallel gravity field equation (4.41), which we will write in the form

$$W^{\mu\nu} - \Theta^{\mu\nu} = A^{\mu\nu}\,, \tag{4.54}$$

where we have defined the abbreviation

$$A^{\mu\nu} = \nabla_\rho Y^{\mu\nu\rho} - Y^{\mu\nu\rho} T^\tau{}_{\tau\rho} - \nabla_\rho H^{\mu\nu\rho} + H^{\mu\nu\rho} T^\tau{}_{\tau\rho}\,. \tag{4.55}$$

Note that the left hand side of the field equation (4.54) is symmetric by definition. Hence, when the equation holds, also the right hand side must be symmetric, and thus $A^{[\mu\nu]} = 0$. We then take the Levi-Civita covariant derivative of this equation, which reads

$$\overset{\circ}{\nabla}_\nu W^{\mu\nu} - \overset{\circ}{\nabla}_\nu \Theta^{\mu\nu} = \overset{\circ}{\nabla}_\nu A^{\mu\nu}\,. \tag{4.56}$$

On the right-hand side, we can use the relation

$$\begin{aligned}
\overset{\circ}{\nabla}_\nu A^{\mu\nu} &= \nabla_\nu A^{\mu\nu} - K^\mu{}_{\rho\nu} A^{\rho\nu} - K^\nu{}_{\rho\nu} A^{\mu\rho} \\
&= \nabla_\nu A^{\mu\nu} - \frac{1}{2}\big[(T_\rho{}^\mu{}_\nu + T_\nu{}^\mu{}_\rho - T^\mu{}_{\rho\nu}) A^{\rho\nu} \\
&\quad -(T_\rho{}^\nu{}_\nu + T_\nu{}^\nu{}_\rho - T^\nu{}_{\rho\nu}) A^{\mu\rho}\big] \\
&= \nabla_\nu A^{\mu\nu} - T_\rho{}^\mu{}_\nu A^{\rho\nu} - T^\nu{}_{\nu\rho} A^{\mu\rho}\,,
\end{aligned} \tag{4.57}$$

where we have used the symmetry $A^{[\mu\nu]} = 0$ to obtain the last line. Now combining the geometric identity (4.53), the divergence (4.56) of the gravitational field equation and the result (4.57), one finally arrives at

$$\begin{aligned}
&\overset{\circ}{\nabla}_\nu \Theta_\mu{}^\nu + T^\sigma{}_{\mu\nu}(\nabla_\rho H_\sigma{}^{\nu\rho} - T^\tau{}_{\tau\rho} H_\sigma{}^{\nu\rho}) \\
&\quad - \nabla_\nu(\nabla_\rho H_\mu{}^{\nu\rho} - T^\tau{}_{\tau\rho} H_\mu{}^{\nu\rho}) + T^\sigma{}_{\sigma\nu}(\nabla_\rho H_\mu{}^{\nu\rho} - T^\tau{}_{\tau\rho} H_\mu{}^{\nu\rho}) = 0\,,
\end{aligned} \tag{4.58}$$

which agrees with (4.49) in the case of vanishing nonmetricity.

4.3.2 Spacetime Symmetries and Cosmology

In the previous section we have made use of the transformation laws (4.42) of the metric and (4.43) of the connection under infinitesimal diffeomorphisms generated by a vector field X. The same transformation laws also find application in the discussion of symmetric spacetimes, i.e., teleparallel geometries, which are invariant under the action of particular vector fields, $\delta_X g_{\mu\nu} = 0$ and $\delta_X \Gamma^\mu{}_{\nu\rho} = 0$ [12, 13]. The choice of these vector fields depends on the physical situation under consideration. A few common examples can be expressed most conveniently in spherical coordinates $(t, r, \varphi, \vartheta)$: a *stationary* spacetime is invariant under the (timelike) vector field ∂_t; *spherical* symmetry is conveyed by the three rotation generators

$$\sin\varphi\partial_\vartheta + \frac{\cos\varphi}{\tan\vartheta}\partial_\varphi\,, \quad -\cos\varphi\partial_\vartheta + \frac{\sin\varphi}{\tan\vartheta}\partial_\varphi\,, \quad -\partial_\varphi\,; \tag{4.59}$$

finally, *cosmological* symmetry comprises of invariance under both rotations as given above and translations, defined by the vector fields

$$\chi\sin\vartheta\cos\varphi\partial_r + \frac{\chi}{r}\cos\vartheta\cos\varphi\partial_\vartheta - \frac{\chi\sin\varphi}{r\sin\vartheta}\partial_\varphi\,, \tag{4.60a}$$

$$\chi\sin\vartheta\sin\varphi\partial_r + \frac{\chi}{r}\cos\vartheta\sin\varphi\partial_\vartheta + \frac{\chi\cos\varphi}{r\sin\vartheta}\partial_\varphi\,, \tag{4.60b}$$

$$\chi\cos\vartheta\,\partial_r - \frac{\chi}{r}\sin\vartheta\,\partial_\vartheta\,, \tag{4.60c}$$

where we used the abbreviation $\chi = \sqrt{1-(ur)^2}$, and u can be any real or imaginary number, so that the sign of $u^2 \in \mathbb{R}$ determines the curvature of the spatial hypersurfaces of constant time t. For $u^2 > 0$, their spatial curvature is positive, while $u^2 < 0$ corresponds to negative spatial curvature. Finally, $u^2 = 0$ is the spatially flat case.

Symmetric spacetimes are often considered as potential solutions to the field equations of a given theory, since they are completely characterized by fewer functions than there are components of the dynamical fields, and these functions depend on a smaller number of coordinates, hence leading to a simple ansatz for solving the field equations. As a simple and physically well motivated example, we show this for the case of cosmological symmetry in the teleparallel geometry. It is well known that the most general metric which is homogeneous and isotropic is the Friedmann-Lemaître-Robertson-Walker metric

$$g_{\mu\nu} = -n_\mu n_\nu + h_{\mu\nu}\,, \tag{4.61}$$

where the hypersurface conormal

$$n_\mu \mathrm{d}x^\mu = -N\mathrm{d}t \tag{4.62}$$

and spatial metric

$$h_{\mu\nu}\mathrm{d}x^\mu \otimes \mathrm{d}x^\nu = A^2\left[\frac{\mathrm{d}r\otimes\mathrm{d}r}{\chi^2} + r^2(\mathrm{d}\vartheta\otimes\mathrm{d}\vartheta + \sin^2\vartheta\mathrm{d}\varphi\otimes\mathrm{d}\varphi)\right] \tag{4.63}$$

are fully determined by two functions of time, known as the lapse function $N = N(t)$ and scale factor $A = A(t)$. Using this metric, we can apply the decomposition (4.2) of the affine connection, and we find that the most general homogeneous and isotropic connection is characterized through its torsion and nonmetricity

$$T^{\mu}{}_{\nu\rho} = \frac{2}{A}(\mathcal{T}_1 h^{\mu}_{[\nu} n_{\rho]} + \mathcal{T}_2 n_{\sigma} \varepsilon^{\sigma\mu}{}_{\nu\rho}), \tag{4.64a}$$

$$Q_{\rho\mu\nu} = \frac{2}{A}(\mathcal{Q}_1 n_{\rho} n_{\mu} n_{\nu} + 2\mathcal{Q}_2 n_{\rho} h_{\mu\nu} + 2\mathcal{Q}_3 h_{\rho(\mu} n_{\nu)}), \tag{4.64b}$$

by five further functions $\mathcal{T}_1, \mathcal{T}_2, \mathcal{Q}_1, \mathcal{Q}_2, \mathcal{Q}_3$ of time, and $\varepsilon_{\mu\nu\rho\sigma}$ is the totally antisymmetric tensor normalized such that

$$\varepsilon_{0123} = \sqrt{-g} = \frac{NA^3 r^2 \sin\vartheta}{\chi}. \tag{4.65}$$

Note that in general the curvature of this connection does not vanish, and so one must impose additional constraints on the aforementioned functions. Before discussing these constraints, it is most convenient to introduce the conformal time derivative

$$F' = \frac{A}{N}\frac{\mathrm{d}F}{\mathrm{d}t} \tag{4.66}$$

acting on any time-dependent scalar function $F = F(t)$, as well as the conformal Hubble parameter

$$\mathcal{H} = \frac{A'}{A} = \frac{1}{N}\frac{\mathrm{d}A}{\mathrm{d}t}. \tag{4.67}$$

With the help of these definitions, the conditions on the parameter functions under which the curvature tensor vanishes become

$$\mathcal{T}_2(\mathcal{H} - \mathcal{T}_1 + \mathcal{Q}_2) = 0, \tag{4.68a}$$

$$\mathcal{T}_2(\mathcal{H} - \mathcal{T}_1 + \mathcal{Q}_2 - \mathcal{Q}_3) = 0, \tag{4.68b}$$

$$(\mathcal{H} - \mathcal{T}_1 + \mathcal{Q}_2)(\mathcal{H} - \mathcal{T}_1 + \mathcal{Q}_2 - \mathcal{Q}_3) - \mathcal{T}_2^2 + u^2 = 0, \tag{4.68c}$$

$$(\mathcal{Q}_1 + \mathcal{Q}_2)(\mathcal{H} - \mathcal{T}_1 + \mathcal{Q}_2) + (\mathcal{H} - \mathcal{T}_1 + \mathcal{Q}_2)' = 0, \tag{4.68d}$$

$$(\mathcal{Q}_1 + \mathcal{Q}_2)(\mathcal{H} - \mathcal{T}_1 + \mathcal{Q}_2 - \mathcal{Q}_3) - (\mathcal{H} - \mathcal{T}_1 + \mathcal{Q}_2 - \mathcal{Q}_3)' = 0, \tag{4.68e}$$

$$\mathcal{T}_2' = 0. \tag{4.68f}$$

Note that u appears in only one of these equations; nevertheless, it plays an important role for the solutions of this system, as we will show now. For this purpose, consider first the case $u^2 \neq 0$. In this case the condition (4.68c) implies

$$(\mathcal{H} - \mathcal{T}_1 + \mathcal{Q}_2)(\mathcal{H} - \mathcal{T}_1 + \mathcal{Q}_2 - \mathcal{Q}_3) \neq \mathcal{T}_2^2 \,. \tag{4.69}$$

From the two conditions (4.68a) and (4.68b) then further follows that either the right hand side vanishes, or both factors on the left hand side vanish. We first consider this latter case. The condition (4.68c) then requires $\mathcal{T}_2 = \pm u$, and we find that all remaining equations are solved by

$$\mathcal{T}_2 = \pm u \,, \quad \mathcal{T}_1 - \mathcal{Q}_2 = \mathcal{H} \,, \quad \mathcal{Q}_3 = 0 \,. \tag{4.70}$$

Also we see that we must demand u to be real in order to obtain a real value of the connection coefficients; hence, this solution is valid only for positive spatial curvature $u^2 > 0$. Alternatively, the first two equations (4.68a) and (4.68b) can also be solved by setting $\mathcal{T}_2 = 0$. From the remaining equations then follows the solution

$$\begin{aligned} \mathcal{T}_2 = 0 \,, \quad (\mathcal{H} - \mathcal{T}_1 + \mathcal{Q}_2)(\mathcal{H} - \mathcal{T}_1 + \mathcal{Q}_2 - \mathcal{Q}_3) = -u^2 \,, \\ \mathcal{Q}_1 + \mathcal{Q}_2 = -\frac{\mathcal{H}' - \mathcal{T}_1' + \mathcal{Q}_2'}{\mathcal{H} - \mathcal{T}_1 + \mathcal{Q}_2} \,. \end{aligned} \tag{4.71}$$

Now we see that both signs of u^2 are allowed. These are the only two possibilities to solve the first three equations, and so we may turn our attention to the case $u = 0$. In this case the third Eq. (4.68c) mandates

$$(\mathcal{H} - \mathcal{T}_1 + \mathcal{Q}_2)(\mathcal{H} - \mathcal{T}_1 + \mathcal{Q}_2 - \mathcal{Q}_3) = \mathcal{T}_2^2 \,, \tag{4.72}$$

and we see that both sides of this equation must vanish in order to satisfy the conditions (4.68a) and (4.68b), so that all solutions will have $\mathcal{T}_2 = 0$. For the left hand side, we are free to choose at most one of the two factors to be non-vanishing. This leads to the three possible solutions

$$\mathcal{T}_2 = 0 \,, \quad \mathcal{T}_1 - \mathcal{Q}_2 = \mathcal{H} \,, \quad \mathcal{Q}_3 = 0 \tag{4.73}$$

if both factors vanish,

$$\mathcal{T}_2 = 0 \,, \quad \mathcal{T}_1 - \mathcal{Q}_2 + \mathcal{Q}_3 = \mathcal{H} \,, \quad \mathcal{Q}_1 + \mathcal{Q}_2 = -\frac{\mathcal{Q}_3'}{\mathcal{Q}_3} \tag{4.74}$$

if only the second factor vanishes, as well as

$$\mathcal{T}_2 = 0 \,, \quad \mathcal{T}_1 - \mathcal{Q}_2 = \mathcal{H} \,, \quad \mathcal{Q}_1 + \mathcal{Q}_2 = \frac{\mathcal{Q}_3'}{\mathcal{Q}_3} \tag{4.75}$$

if only the first factor vanishes. These are the only possible homogeneous and isotropic teleparallel geometries. Note that for each solution one has three conditions on the five parameter functions, so that two of them can be freely chosen to parametrize the solution, and must be determined alongside the scale factor A and lapse N by solving the field equations of a given teleparallel gravity theory.

In the discussion above we have assumed a general teleparallel geometry, for which both torsion and nonmetricity are allowed to be non-vanishing. From the solutions we have found one can now easily deduce the symmetric and metric teleparallel geometries. We start with the former, by imposing the additional condition $\mathcal{T}_1 = \mathcal{T}_2 = 0$. One immediately sees that this condition is not compatible with the first solution (4.70), which explicitly demands $\mathcal{T}_2 \neq 0$, and so this solution cannot be restricted to symmetric teleparallel gravity. This is different for the remaining solutions. From the solution (4.71), one obtains the spatially curved case

$$(\mathcal{H}+\mathcal{Q}_2)(\mathcal{H}+\mathcal{Q}_2-\mathcal{Q}_3) = -u^2\,, \quad \mathcal{Q}_1+\mathcal{Q}_2 = -\frac{\mathcal{H}'+\mathcal{Q}_2'}{\mathcal{H}+\mathcal{Q}_2}\,, \tag{4.76}$$

while the three spatially flat solutions become

$$\mathcal{Q}_2 = -\mathcal{H}\,, \qquad \mathcal{Q}_3 = 0\,, \tag{4.77a}$$

$$\mathcal{Q}_2 - \mathcal{Q}_3 = -\mathcal{H}\,, \qquad \mathcal{Q}_1+\mathcal{Q}_2 = -\frac{\mathcal{Q}_3'}{\mathcal{Q}_3}\,, \tag{4.77b}$$

$$\mathcal{Q}_2 = -\mathcal{H}\,, \qquad \mathcal{Q}_1+\mathcal{Q}_2 = \frac{\mathcal{Q}_3'}{\mathcal{Q}_3}\,. \tag{4.77c}$$

For each of these solutions one has two conditions on the three scalar functions $\mathcal{Q}_{1,2,3}$ which parametrize the nonmetricity, so that one of them remains undetermined by the symmetry condition, and is left to be determined by the gravitational field equations [14, 15].

In a similar fashion, one can also restrict the general teleparallel cosmologies to the metric teleparallel geometry, by imposing the conditions $\mathcal{Q}_1 = \mathcal{Q}_2 = \mathcal{Q}_3 = 0$ of vanishing nonmetricity. For the first solution (4.70), this leads to

$$\mathcal{T}_2 = \pm u\,, \quad \mathcal{T}_1 = \mathcal{H}\,, \tag{4.78}$$

while the second solution (4.71) becomes

$$\mathcal{T}_2 = 0\,, \quad \mathcal{T}_1 = \mathcal{H} \pm iu\,. \tag{4.79}$$

For the latter, we have explicitly solved the appearing quadratic equation. In this case we see that u must be imaginary in order to obtain a real torsion, and so we are restricted to the case $u^2 < 0$ of negative spatial curvature. Finally, the three solutions for $u = 0$ reduce to the common case

$$\mathcal{T}_2 = 0, \quad \mathcal{T}_1 = \mathcal{H}. \tag{4.80}$$

In all three cases, the two free functions in the torsion scalar are fixed by the conditions of cosmological symmetry and vanishing curvature, so that the field equations are fully expressed in terms of the scale factor A and the lapse N [16].

? Exercise

4.3. Determine the homogeneous and isotropic symmetric teleparallel geometries (4.76) and (4.77), as well as the homogeneous and isotropic metric teleparallel geometries (4.78),(4.79) and (4.80).

4.3.3 Perturbation Theory

Besides the use of symmetric spacetimes as shown in the previous section, another common approach to simplify the (in general non-linear) field equations of a given teleparallel gravity theory is to start from a known, usually highly symmetric solution of the field equations, given by a metric $\bar{g}_{\mu\nu}$ and a flat affine connection with coefficients $\bar{\Gamma}^{\mu}{}_{\nu\rho}$, and perform a perturbative expansion of the dynamical fields and their governing field equations around this solution. For this purpose, one conventionally introduces a perturbation parameter ϵ on which the solution will depend, and which can be related, for example, to the gravitational constant for a weak-field approximation, or the inverse speed of light for a low-velocity approximation. The full solution $g_{\mu\nu}(\epsilon)$ and $\Gamma^{\mu}{}_{\nu\rho}(\epsilon)$, is then expanded in a Taylor series

$$g_{\mu\nu} = \sum_{k=0}^{\infty} \frac{\epsilon^k}{k!} \frac{\mathrm{d}^k}{\mathrm{d}\epsilon^k} g_{\mu\nu}\bigg|_{\epsilon=0}, \quad \Gamma^{\mu}{}_{\nu\rho} = \sum_{k=0}^{\infty} \frac{\epsilon^k}{k!} \frac{\mathrm{d}^k}{\mathrm{d}\epsilon^k} \Gamma^{\mu}{}_{\nu\rho}\bigg|_{\epsilon=0} \tag{4.81}$$

around the background solution $\bar{g}_{\mu\nu} = g_{\mu\nu}(0)$ and $\bar{\Gamma}^{\mu}{}_{\nu\rho} = \Gamma^{\mu}{}_{\nu\rho}(0)$. Different conventions are abundant for the terms in this Taylor expansion, either for the coefficients

$$\delta^k g_{\mu\nu} = \frac{\mathrm{d}^k}{\mathrm{d}\epsilon^k} g_{\mu\nu}\bigg|_{\epsilon=0}, \quad \delta^k \Gamma^{\mu}{}_{\nu\rho} = \frac{\mathrm{d}^k}{\mathrm{d}\epsilon^k} \Gamma^{\mu}{}_{\nu\rho}\bigg|_{\epsilon=0}, \tag{4.82}$$

or for the full terms

$$\overset{k}{g}_{\mu\nu} = \frac{\epsilon^k}{k!} \frac{\mathrm{d}^k}{\mathrm{d}\epsilon^k} g_{\mu\nu}\bigg|_{\epsilon=0}, \quad \overset{k}{\Gamma}{}^{\mu}{}_{\nu\rho} = \frac{\epsilon^k}{k!} \frac{\mathrm{d}^k}{\mathrm{d}\epsilon^k} \Gamma^{\mu}{}_{\nu\rho}\bigg|_{\epsilon=0}. \tag{4.83}$$

In the following, we will make use of the latter, as it turns out to be shorter for the examples we consider. It follows from the fact that the metric $g_{\mu\nu}$ is a symmetric tensor field that the same property holds also for all terms $\overset{k}{g}_{\mu\nu}$ in its perturbative expansion. For the connection coefficients, one similarly concludes from the fact that both $\Gamma^{\mu}{}_{\nu\rho}$ and $\bar{\Gamma}^{\mu}{}_{\nu\rho}$ are connection coefficients that the remaining terms $\overset{k}{\Gamma}{}^{\mu}{}_{\nu\rho}$ with $k > 0$ are tensor fields. In order to determine these terms, one performs a similar Taylor expansion of the gravitational field equations in the perturbation parameter ϵ. It is the main virtue of this expansion that at each perturbation order k, the corresponding terms of the field equations comprise a linear equation for the field terms $\overset{k}{g}_{\mu\nu}$ and $\overset{k}{\Gamma}{}^{\mu}{}_{\nu\rho}$ at the same order, which contain the lower order terms as a source; hence, they can be solved subsequently for increasing orders, where the previously found solutions for lower orders are used in each further order to be solved.

In the case of teleparallel gravity, it is important to keep in mind that next to the gravitational field equations also the constraint (4.10) of vanishing curvature must be satisfied at any perturbation order. In the symmetric and metric teleparallel classes of theories, also either the torsion (4.3) or nonmetricity (4.4) must vanish at each order. While it is possible to simply consider these constraints as additional equations which must be solved next to the field equations at each order, one may also pose the question whether it is possible to find a general perturbative solution to these constraints, which is independent of the gravity theory under consideration, and which can then be inserted into the perturbed field equations of any specific gravity theory. To obtain this solution, one needs to perform a perturbative expansion of the corresponding constraint equations. We start by showing this procedure for the flatness constraint (4.10). At the zeroth order, this simply becomes the vanishing of the curvature

$$0 = \bar{R}^{\mu}{}_{\nu\rho\sigma} = \partial_\rho \bar{\Gamma}^{\mu}{}_{\nu\sigma} - \partial_\sigma \bar{\Gamma}^{\mu}{}_{\nu\rho} + \bar{\Gamma}^{\mu}{}_{\tau\rho} \bar{\Gamma}^{\tau}{}_{\nu\sigma} - \bar{\Gamma}^{\mu}{}_{\tau\sigma} \bar{\Gamma}^{\tau}{}_{\nu\rho} \tag{4.84}$$

for the background connection $\bar{\Gamma}^{\mu}{}_{\nu\rho}$, which we assume to be satisfied from now on. For the first-order perturbation of the curvature, one finds the condition

$$0 = \overset{1}{R}{}^{\mu}{}_{\nu\rho\sigma} = \bar{\nabla}_\rho \overset{1}{\Gamma}{}^{\mu}{}_{\nu\sigma} - \bar{\nabla}_\sigma \overset{1}{\Gamma}{}^{\mu}{}_{\nu\rho} + \bar{T}^{\tau}{}_{\rho\sigma} \overset{1}{\Gamma}{}^{\mu}{}_{\nu\tau}, \tag{4.85}$$

where all quantities which are calculated with respect to the background connection are denoted with a bar. Now it is helpful to recall that the commutator of covariant derivatives is given by

$$\bar{\nabla}_\rho \bar{\nabla}_\sigma \lambda^{\mu}{}_{\nu} - \bar{\nabla}_\sigma \bar{\nabla}_\rho \lambda^{\mu}{}_{\nu} = \bar{R}^{\mu}{}_{\tau\rho\nu} \lambda^{\tau}{}_{\nu} - \bar{R}^{\tau}{}_{\nu\rho\nu} \lambda^{\mu}{}_{\tau} - \bar{T}^{\tau}{}_{\rho\sigma} \bar{\nabla}_\tau \lambda^{\mu}{}_{\nu} \tag{4.86}$$

for a tensor field $\lambda^{\mu}{}_{\nu}$, where the two curvature terms on the right hand side vanish. Hence, we can solve the flatness condition at the first perturbation order by setting

$$\overset{1}{\Gamma}{}^{\mu}{}_{\nu\rho} = \bar{\nabla}_{\rho}\overset{1}{\lambda}{}^{\mu}{}_{\nu} \tag{4.87}$$

with an arbitrary first-order tensor field $\overset{1}{\lambda}{}^{\mu}{}_{\nu}$. To illustrate the further procedure, we calculate the second order curvature perturbation

$$0 = \overset{2}{R}{}^{\mu}{}_{\nu\rho\sigma} = \bar{\nabla}_{\rho}\overset{2}{\Gamma}{}^{\mu}{}_{\nu\sigma} - \bar{\nabla}_{\sigma}\overset{2}{\Gamma}{}^{\mu}{}_{\nu\rho} + \bar{T}^{\tau}{}_{\rho\sigma}\overset{2}{\Gamma}{}^{\mu}{}_{\nu\tau} + 2\overset{1}{\Gamma}{}^{\mu}{}_{\tau[\rho}\overset{1}{\Gamma}{}^{\tau}{}_{|\nu|\sigma]}\,, \tag{4.88}$$

where we now also need to take into account the first-order connection perturbation. A naive ansatz $\bar{\nabla}_{\rho}\overset{2}{\lambda}{}^{\mu}{}_{\nu}$ for $\overset{2}{\Gamma}{}^{\mu}{}_{\nu\rho}$ is therefore not sufficient. In order to cancel the term arising from the first-order connection perturbation, one also needs to include terms which are quadratic in $\overset{1}{\lambda}{}^{\mu}{}_{\nu}$, and must contain one derivative. One finds that a possible solution is given by

$$\overset{2}{\Gamma}{}^{\mu}{}_{\nu\rho} = \bar{\nabla}_{\rho}\overset{2}{\lambda}{}^{\mu}{}_{\nu} - \overset{1}{\lambda}{}^{\mu}{}_{\tau}\bar{\nabla}_{\rho}\overset{1}{\lambda}{}^{\tau}{}_{\nu}\,. \tag{4.89}$$

Note that this solution is not unique. Alternatively, one could choose, for example,

$$\overset{2}{\Gamma}{}^{\mu}{}_{\nu\rho} = \bar{\nabla}_{\rho}\overset{2}{\lambda}{}^{\mu}{}_{\nu} + \bar{\nabla}_{\rho}\overset{1}{\lambda}{}^{\mu}{}_{\tau}\overset{1}{\lambda}{}^{\tau}{}_{\nu}\,. \tag{4.90}$$

This becomes clear by realizing that these solutions differ only by a term $\bar{\nabla}_{\rho}(\overset{1}{\lambda}{}^{\mu}{}_{\tau}\overset{1}{\lambda}{}^{\tau}{}_{\nu})$, which can be absorbed by a redefinition of $\overset{2}{\lambda}{}^{\mu}{}_{\nu}$. By a similar procedure, one can also subsequently solve the flatness condition at any higher perturbation order.

For the symmetric and metric teleparallel cases, one can make use of the already determined solution of the perturbative flatness condition, and further restrict the perturbation tensor fields $\overset{k}{\lambda}{}^{\mu}{}_{\nu}$ in order to achieve a connection with vanishing torsion or nonmetricity at any perturbation. We start with the former, which means that the background connection coefficients $\bar{\Gamma}^{\mu}{}_{\nu\rho}$ as well as the perturbations $\overset{k}{\Gamma}{}^{\mu}{}_{\nu\rho}$ must be symmetric in their lower two indices. To obtain this property, one can make use of the result that the flat connection perturbations can always be parametrized in the form

$$\overset{k}{\Gamma}{}^{\mu}{}_{\nu\rho} = \bar{\nabla}_{\rho}\overset{k}{\lambda}{}^{\mu}{}_{\nu} + \sum_{j=1}^{k-1}\overset{j,k}{\Lambda}{}^{\mu}{}_{\tau}\overset{j}{\Gamma}{}^{\tau}{}_{\nu\rho}\,, \tag{4.91}$$

where $\overset{j,k}{\Lambda}{}^{\mu}{}_{\tau}$ is determined by solving the flatness condition at the k'th perturbation order. Indeed, we have seen this form explicitly for the first order (4.87), as well as the second order (4.89), where for the latter $\overset{1,2}{\Lambda}{}^{\mu}{}_{\tau} = -\overset{1}{\lambda}{}^{\mu}{}_{\tau}$. Using this parametrization, it follows that once we have solved the condition of vanishing torsion up to the perturbation order $k - 1$, the condition for the k'th order simply becomes

$$\bar{\nabla}_{[\rho}\overset{k}{\lambda}{}^{\mu}{}_{\nu]} = 0\,. \tag{4.92}$$

Further using the fact that the covariant derivatives with respect to the background connection commute in the absence of curvature and torsion, we can thus write the solution as

$$\overset{k}{\lambda}{}^{\mu}{}_{\nu} = \bar{\nabla}_{\nu}\overset{k}{\zeta}{}^{\mu}\,, \tag{4.93}$$

where we introduced the perturbation parameters $\overset{k}{\zeta}{}^{\mu}$. Note, however, that also in this case numerous other parametrizations of the connection coefficients can be found. A parametrization which turns out particularly convenient for practical calculations arises from the fact that two flat, torsion-free connections, are locally related by a diffeomorphism. It follows that also the perturbed connection $\Gamma^{\mu}{}_{\nu\rho}$, which depends on the perturbation parameter ϵ, and the background $\bar{\Gamma}^{\mu}{}_{\nu\rho}$, are locally related by a family Φ_{ϵ} of diffeomorphisms parametrized by ϵ, such that

$$\Gamma^{\mu}{}_{\nu\rho}(\epsilon) = \Phi_{\epsilon}^{*}\bar{\Gamma}^{\mu}{}_{\nu\rho}\,. \tag{4.94}$$

Performing a Taylor expansion in order to obtain the perturbation terms $\overset{k}{\Gamma}{}^{\mu}{}_{\nu\rho}$, we see that on the right hand side $\bar{\Gamma}^{\mu}{}_{\nu\rho}$ remains fixed, and we need to expand the diffeomorphism Φ_{ϵ} into a corresponding Taylor series. It turns out that such an expansion gives rise to a series of vector fields $\overset{k}{\xi}{}^{\mu}$ for $k > 0$, in terms of which the Taylor expansion reads [17–19]

$$\overset{k}{\Gamma}{}^{\mu}{}_{\nu\rho} = \sum_{l_1+2l_2+\ldots=k} \frac{1}{l_1!l_2!\cdots}\left(\mathcal{L}^{l_1}_{\overset{1}{\xi}}\cdots\mathcal{L}^{l_j}_{\overset{j}{\xi}}\cdots\bar{\Gamma}\right)^{\mu}{}_{\nu\rho}\,. \tag{4.95}$$

It is instructive to calculate the lower order terms explicitly, using the formula (4.43) with vanishing curvature and torsion. For the background, the expansion trivially reduces to

$$\overset{0}{\Gamma}{}^{\mu}{}_{\nu\rho} = \bar{\Gamma}^{\mu}{}_{\nu\rho}\,. \tag{4.96}$$

For the first order, only one term appears on the right-hand side, which reads

$$\overset{1}{\Gamma}{}^{\mu}{}_{\nu\rho} = \left(\mathcal{L}_{\overset{1}{\xi}}\bar{\Gamma}\right)^{\mu}{}_{\nu\rho} = \bar{\nabla}_{\nu}\bar{\nabla}_{\rho}\overset{1}{\xi}{}^{\mu}\,, \tag{4.97}$$

while for the second order one finds the two terms

$$\begin{aligned}\overset{2}{\Gamma}^{\mu}{}_{\nu\rho} &= \left(\mathcal{L}_{\overset{2}{\xi}}\bar{\Gamma}\right)^{\mu}{}_{\nu\rho} + \frac{1}{2}\left(\mathcal{L}_{\overset{1}{\xi}}\mathcal{L}_{\overset{1}{\xi}}\bar{\Gamma}\right)^{\mu}{}_{\nu\rho} \\ &= \bar{\nabla}_{\nu}\bar{\nabla}_{\rho}\overset{2}{\xi}^{\mu} + \bar{\nabla}_{(\nu}\overset{1}{\xi}^{\sigma}\bar{\nabla}_{\rho)}\bar{\nabla}_{\sigma}\overset{1}{\xi}^{\mu} - \frac{1}{2}\bar{\nabla}_{\nu}\bar{\nabla}_{\rho}\overset{1}{\xi}^{\sigma}\bar{\nabla}_{\sigma}\overset{1}{\xi}^{\mu} + \frac{1}{2}\overset{1}{\xi}^{\sigma}\bar{\nabla}_{\nu}\bar{\nabla}_{\rho}\bar{\nabla}_{\sigma}\overset{1}{\xi}^{\mu} .\end{aligned} \tag{4.98}$$

It is also helpful to compare these formulas with the perturbations (4.87) and (4.89), together with the substitution (4.93). For the first perturbation order, the perturbation (4.87) becomes

$$\overset{1}{\Gamma}^{\mu}{}_{\nu\rho} = \bar{\nabla}_{\nu}\bar{\nabla}_{\rho}\overset{1}{\zeta}^{\mu} , \tag{4.99}$$

which agrees with (4.97) for $\overset{1}{\zeta}^{\mu} = \overset{1}{\xi}^{\mu}$. At the second order, the perturbation (4.89) becomes

$$\overset{2}{\Gamma}^{\mu}{}_{\nu\rho} = \bar{\nabla}_{\nu}\bar{\nabla}_{\rho}\overset{2}{\zeta}^{\mu} - \bar{\nabla}_{\sigma}\overset{1}{\zeta}^{\mu}\bar{\nabla}_{\nu}\bar{\nabla}_{\rho}\overset{1}{\zeta}^{\sigma} , \tag{4.100}$$

which agrees with the result (4.98) for

$$\overset{2}{\zeta}^{\mu} = \overset{2}{\xi}^{\mu} + \frac{1}{2}\overset{1}{\xi}^{\sigma}\bar{\nabla}_{\sigma}\overset{1}{\xi}^{\mu} . \tag{4.101}$$

Also one easily checks that the curvature and torsion vanish at any perturbation order. This type of perturbative expansion is used, for example, to determine the propagation of gravitational waves [20] and the post-Newtonian limit [21, 22].

Finally, we take a look at the form of the perturbations in the metric teleparallel case, which means that the nonmetricity must vanish at all perturbation orders. This holds in particular for the background,

$$0 = \bar{Q}_{\mu\nu\rho} = \bar{\nabla}_{\mu}\bar{g}_{\nu\rho} , \tag{4.102}$$

and so raising and lowering indices of the perturbations with the metric commutes with the covariant derivative, which greatly simplifies the calculations. We make use of this fact for calculating the conditions on the connection perturbation $\overset{k}{\lambda}^{\mu}{}_{\nu}$ which we need to satisfy in order to obtain vanishing nonmetricity. At the linear order, the perturbation of the nonmetricity is given by

$$\overset{1}{Q}_{\mu\nu\rho} = \bar{\nabla}_{\mu}\left(\overset{1}{g}_{\nu\rho} - 2\overset{1}{\lambda}_{(\nu\rho)}\right) , \tag{4.103}$$

and so it vanishes if we fix the symmetric part of the connection perturbation by the condition

$$2\overset{1}{\lambda}_{(\mu\nu)} = \overset{1}{g}_{\mu\nu} \,. \tag{4.104}$$

One could naively conclude that the same formula holds identically also for higher orders, replacing the first perturbation order with an arbitrary order k. However, this is not the case, as follows from a perturbative expansion of the nonmetricity (4.4), whose higher than linear orders contain products of the lower order perturbations of the metric and the connection. For example, for the second order we have

$$\begin{aligned}\overset{2}{Q}_{\mu\nu\rho} &= \bar{\nabla}_\mu \left(\overset{2}{g}_{\nu\rho} - 2\overset{2}{\lambda}_{(\nu\rho)}\right) - 2\nabla_\mu \overset{1}{\lambda}{}^\sigma{}_{(\nu} \left(\overset{1}{g}_{\rho)\sigma} - \overset{1}{\lambda}_{\rho)\sigma}\right) \\ &= \bar{\nabla}_\mu \left(\overset{2}{g}_{\nu\rho} - 2\overset{2}{\lambda}_{(\nu\rho)} - \bar{g}_{\sigma\omega}\overset{1}{\lambda}{}^\sigma{}_\nu \overset{1}{\lambda}{}^\omega{}_\rho\right) ,\end{aligned} \tag{4.105}$$

after substituting $\overset{1}{g}_{\nu\rho}$ from the first order result, so that we can easily read off the condition for the second order perturbation. Following the same procedure also for higher order perturbations, one arrives at the general formula

$$2\overset{k}{\lambda}_{(\mu\nu)} = \overset{k}{g}_{\mu\nu} - \bar{g}_{\rho\sigma} \sum_{j=1}^{k-1} \overset{j}{\lambda}{}^\rho{}_\mu \overset{k-j}{\lambda}{}^\sigma{}_\nu \,. \tag{4.106}$$

We see that the condition of vanishing nonmetricity links the symmetric part of the connection perturbation to the perturbation of the metric, and so the latter can always be expressed in terms of the former, leaving $\overset{k}{\lambda}_{\mu\nu}$ as the only independent perturbation variable. Like in the case of symmetric teleparallel gravity, this perturbative expansion is used for the calculation of gravitational waves [23] and the post-Newtonian limit [24], but also in the cosmological perturbation theory around flat [25] and general [26] cosmological backgrounds.

In order to get further acquainted with the perturbations theory formalism, that is presented in this section please solve the following exercise step by step.

? Exercise

4.4. Calculate the perturbations of the curvature, torsion and nonmetricity tensors up to the second order for an arbitrary perturbation around a flat connection and metric. What are the conditions arising on the perturbations if one demands that the perturbed connection remains flat at any order? Which further conditions arise if one also demands that either torsion or nonmetricity vanish?

4.4 Teleparallel Gravity Theories

In this final section, we discuss a few selected classes of teleparallel gravity theories, their actions and field equations. These theories constitute modifications of general relativity, which depart from a reformulation of the Einstein-Hilbert action in terms of teleparallel geometries, known as the teleparallel equivalent of general relativity, which we discuss in Sect. 4.4.1. A simple modification is then obtained by replacing the Lagrangian of these theories by a free function thereof, as we show in Sect. 4.4.2. Another modification arises by considering the most general action which is quadratic in torsion and nonmetricity, and which we show in Sect. 4.4.3. Moreover, modified theories can be obtained by considering a scalar field as another dynamical variable in addition to the metric and the flat connection; we discuss theories of this type in Sect. 4.4.4, and see how a particular subclass of them is connected to a previously discussed class of theories in Sect. 4.4.5.

4.4.1 The Teleparallel Equivalents of General Relativity

In the previous sections we have discussed the general form of the action and field equations for teleparallel gravity theories, but we have not yet considered any particular theories. As a starting point for the construction of modified teleparallel gravity theories, we now pose the question how the well-known general relativity action and field equations can be cast into the teleparallel framework. The crucial observation to answer this question is the fact that the decomposition (4.2) of the independent connection with respect to the Levi-Civita connection of the metric induces a related decomposition of the curvature given by

$$R^{\mu}{}_{\nu\rho\sigma} = \overset{\circ}{R}{}^{\mu}{}_{\nu\rho\sigma} + \overset{\circ}{\nabla}_{\rho} M^{\mu}{}_{\nu\sigma} - \overset{\circ}{\nabla}_{\sigma} M^{\mu}{}_{\nu\rho} + M^{\mu}{}_{\tau\rho} M^{\tau}{}_{\nu\sigma} - M^{\mu}{}_{\tau\sigma} M^{\tau}{}_{\nu\rho} \,. \tag{4.107}$$

Keeping in mind that the curvature (4.10) of the teleparallel connection is imposed to vanish, one can solve for the curvature tensor of the Levi-Civita connection, and finds

$$\overset{\circ}{R}{}^{\mu}{}_{\nu\rho\sigma} = -\overset{\circ}{\nabla}_{\rho} M^{\mu}{}_{\nu\sigma} + \overset{\circ}{\nabla}_{\sigma} M^{\mu}{}_{\nu\rho} - M^{\mu}{}_{\tau\rho} M^{\tau}{}_{\nu\sigma} + M^{\mu}{}_{\tau\sigma} M^{\tau}{}_{\nu\rho} \,. \tag{4.108}$$

This allows us to replace the Ricci scalar $\overset{\circ}{R}$ in the Einstein-Hilbert action

$$S_{\mathrm{g}} = \frac{1}{2\kappa^2} \int_M \overset{\circ}{R} \sqrt{-g} \mathrm{d}^4 x \tag{4.109}$$

by

$$\overset{\circ}{R} = -G + B \,, \tag{4.110}$$

where we defined the terms

$$G = 2M^{\mu}{}_{\tau[\mu}M^{\tau\nu}{}_{\nu]}\,, \quad B = 2\mathring{\nabla}_{\mu}M^{[\nu\mu]}{}_{\nu}\,. \tag{4.111}$$

One can see that B becomes a boundary term in the action, and therefore does not contribute to the field equations. Omitting this term from the action, one thus obtains [6]

$$S_{\mathrm{g}} = -\frac{1}{2\kappa^2}\int_M G\sqrt{-g}\mathrm{d}^4x\,. \tag{4.112}$$

This is the action of the *general teleparallel equivalent of general relativity* (GTEGR). To study the nature of this equivalence, we calculate the gravitational field equations. Note that the variation of the distortion tensor is given by

$$\delta M^{\mu}{}_{\nu\rho} = \delta\Gamma^{\mu}{}_{\nu\rho} - \delta\mathring{\Gamma}^{\mu}{}_{\nu\rho} = \delta\Gamma^{\mu}{}_{\nu\rho} - \frac{1}{2}g^{\mu\sigma}\left(\mathring{\nabla}_{\nu}\delta g_{\sigma\rho} + \mathring{\nabla}_{\rho}\delta g_{\nu\sigma} - \mathring{\nabla}_{\sigma}\delta g_{\nu\rho}\right)\,, \tag{4.113}$$

and so the variation of the gravity scalar becomes

$$\delta G = U^{\mu\nu}\delta g_{\mu\nu} + V^{\rho\mu\nu}\mathring{\nabla}_{\rho}\delta g_{\mu\nu} + Z_{\mu}{}^{\nu\rho}\delta\Gamma^{\mu}{}_{\nu\rho}\,, \tag{4.114}$$

where we have introduced the abbreviations

$$U^{\mu\nu} = M^{\rho\sigma(\mu}M_{\sigma}{}^{\nu)}{}_{\rho} - M^{\rho(\mu\nu)}M^{\sigma}{}_{\rho\sigma} \tag{4.115a}$$

$$V^{\rho\mu\nu} = M^{\rho(\mu\nu)} - M^{\sigma(\mu}{}_{\sigma}g^{\nu)\rho} - M^{[\rho\sigma]}{}_{\sigma}g^{\mu\nu} \tag{4.115b}$$

$$Z_{\mu}{}^{\nu\rho} = M^{\nu\sigma}{}_{\sigma}\delta^{\rho}_{\mu} + M^{\sigma}{}_{\mu\sigma}g^{\nu\rho} - M^{\nu\rho}{}_{\mu} - M^{\rho}{}_{\mu}{}^{\nu}\,. \tag{4.115c}$$

This allows us to calculate the variation of the action (4.112) and perform integration by parts in order to eliminate the derivatives acting on the metric perturbation $\delta g_{\mu\nu}$. The resulting variation then takes the form (4.9) with

$$\begin{aligned} W_{\mu\nu} &= \frac{1}{\kappa^2}\left(U_{\mu\nu} - \mathring{\nabla}_{\rho}V^{\rho}{}_{\mu\nu} + \frac{1}{2}Gg_{\mu\nu}\right) \\ &= \frac{1}{\kappa^2}\Big[\mathring{\nabla}_{(\mu}M^{\rho}{}_{\nu)\rho} - \mathring{\nabla}_{\rho}M^{\rho}{}_{(\mu\nu)} + M^{\rho}{}_{\sigma(\mu}M^{\sigma}{}_{\nu)\rho} - M^{\rho}{}_{\sigma\rho}M^{\sigma}{}_{(\mu\nu)} \\ &\quad - \frac{1}{2}\left(\mathring{\nabla}_{\rho}M^{\sigma\rho}{}_{\sigma} - \mathring{\nabla}_{\rho}M^{\rho\sigma}{}_{\sigma} + M^{\rho\sigma\omega}M_{\omega\rho\sigma} - M^{\rho}{}_{\omega\rho}M^{\omega\sigma}{}_{\sigma}\right)g_{\mu\nu}\Big] \end{aligned} \tag{4.116}$$

and

$$Y_\mu{}^{\nu\rho} = \frac{1}{2\kappa^2} Z_\mu{}^{\nu\rho} = \frac{1}{2\kappa^2} (M^{\nu\sigma}{}_\sigma \delta^\rho_\mu + M^\sigma{}_{\mu\sigma} g^{\nu\rho} - M^{\nu\rho}{}_\mu - M^\rho{}_\mu{}^\nu) . \quad (4.117)$$

By comparing with the relation (4.108), one finds that the first equation can be rewritten as

$$W_{\mu\nu} = \frac{1}{\kappa^2} \left(\overset{\circ}{R}_{\mu\nu} - \frac{1}{2} \overset{\circ}{R} g_{\mu\nu} \right) , \quad (4.118)$$

and so the metric equation resembles Einstein's equation

$$\overset{\circ}{R}_{\mu\nu} - \frac{1}{2} \overset{\circ}{R} g_{\mu\nu} = \kappa^2 \Theta_{\mu\nu} . \quad (4.119)$$

We also need to consider the connection field equation (4.28), which becomes

$$\frac{1}{\kappa^2} \left(\overset{\circ}{\nabla}_{[\mu} M^{\nu\rho}{}_{\rho]} + \overset{\circ}{\nabla}{}^{[\nu} M_{\rho\mu}{}^{\rho]} + M^\nu{}_{\rho[\mu} M^{\rho\sigma}{}_{\sigma]} + M_\rho{}^{\sigma[\nu} M_{\sigma\mu}{}^{\rho]} \right) = \nabla_\tau H_\mu{}^{\nu\tau} - M^\omega{}_{\tau\omega} H_\mu{}^{\nu\tau} . \quad (4.120)$$

Once again making use of the relation (4.108), the term in brackets becomes

$$\overset{\circ}{R}{}^{\nu\rho}{}_{\mu\rho} + \overset{\circ}{R}_{\rho\mu}{}^{\nu\rho} = \overset{\circ}{R}{}^\nu{}_\mu - \overset{\circ}{R}_\mu{}^\nu = 0 , \quad (4.121)$$

which vanishes, since the Ricci tensor of the Levi-Civita connection is symmetric. Hence, one is left with the equation

$$\nabla_\tau H_\mu{}^{\nu\tau} - M^\omega{}_{\tau\omega} H_\mu{}^{\nu\tau} = 0 \quad (4.122)$$

for the hypermomentum, which must be satisfied for any matter which is compatible with the gravitational action (4.112). Note that the connection does not appear anywhere on the gravitational side of the field equations, due to the fact that it enters into the action only through a total derivative term. For consistency, one conventionally assumes that it does not couple to the matter fields, so that the hypermomentum vanishes, and so the constraint (4.122) is satisfied identically, see also the discussion in the context of Poincaré gauge theory in Chap. 3 around Eq. (3.139). The only non-trivial field equation is then Einstein's equation (4.119), and so the field equations of GTEGR are equivalent to those of general relativity, hence justifying the name teleparallel equivalent.

Since the connection only has a spurious appearance in the action (4.112), one may expect that it will not enter the field equations also in the symmetric and metric classes of teleparallel gravity theories. This is not obvious from the Lagrange multiplier approach of deriving the field equations, since the Lagrange multiplier terms (4.32) and (4.37) are not total derivatives, and so the connection enters the field equations obtained by variation with respect to the Lagrange multipliers. Nevertheless, keeping in mind that this approach yields the same field equations as the approach of restricted variation, and that in the latter the variation of the connection appears only through a total derivative in the action, one may still expect to obtain an equivalent of general relativity. This is particularly easy to see in the case of symmetric teleparallel gravity, since its metric field equation (4.14) takes the same form as in the case of general teleparallel gravity, and hence once again resembles the Einstein equations (4.119), irrespective of the constraint $T^{\mu}{}_{\nu\rho} = 0$ imposed on the connection. For the remaining field equation (4.36), it is helpful to recall that for the variation (4.117) the left hand side of the field equation (4.28) vanishes identically, and hence does the left hand side of the equivalent field equation (4.25). In the absence of torsion, the torsion term vanishes, and one is left with

$$\nabla_{\rho}\tilde{Y}_{\mu}{}^{\nu\rho} = 0\,. \tag{4.123}$$

Hence, also the left hand side of the symmetric teleparallel field equation (4.36) vanishes identically, leaving only the hypermomentum constraint

$$\nabla_{\nu}\nabla_{\rho}\tilde{H}_{\mu}{}^{\nu\rho} = 0\,, \tag{4.124}$$

which can be satisfied by demanding vanishing hypermomentum.

A similar argument holds in the case of metric teleparallel gravity. Once again, one can make use of the fact that the left hand side of the field equation (4.25) vanishes identically for the variation (4.117). In the absence of nonmetricity, the covariant derivative (4.26) of the density factor $\sqrt{-g}$ vanishes, and so this factor can be canceled from the equations. One is then left with the equation

$$\nabla_{\tau}Y_{\mu}{}^{\nu\tau} - T^{\omega}{}_{\omega\tau}Y_{\mu}{}^{\nu\tau} = 0\,. \tag{4.125}$$

Using this result, the metric teleparallel field equation (4.41) reduces to

$$W^{\mu\nu} = \Theta^{\mu\nu} - \nabla_{\rho}H^{\mu\nu\rho} + H^{\mu\nu\rho}T^{\tau}{}_{\tau\rho}\,. \tag{4.126}$$

Demanding once again vanishing hypermomentum, one therefore obtains Einstein's equation (4.119) also in this case.

In order to gain more insight into the underlying structure of the different teleparallel equivalents of general relativity, it is helpful to decompose the gravity scalar G and the boundary term B into the individual contributions from the torsion and the nonmetricity. Using the connection decomposition (4.2), the contortion (4.5) and the disformation (4.6), one finds

$$
\begin{aligned}
G = &\frac{1}{4}Q^{\mu\nu\rho}Q_{\mu\nu\rho} - \frac{1}{2}Q^{\mu\nu\rho}Q_{\rho\mu\nu} - \frac{1}{4}Q^{\rho\mu}{}_{\mu}Q_{\rho\nu}{}^{\nu} + \frac{1}{2}Q^{\mu}{}_{\mu\rho}Q^{\rho\nu}{}_{\nu} \\
&+\frac{1}{4}T^{\mu\nu\rho}T_{\mu\nu\rho} + \frac{1}{2}T^{\mu\nu\rho}T_{\rho\nu\mu} - T^{\mu}{}_{\mu\rho}T_{\nu}{}^{\nu\rho} + T^{\mu\nu\rho}Q_{\nu\rho\mu} - T^{\mu}{}_{\rho\mu}Q^{\rho\nu}_{\nu} + T^{\mu}{}_{\rho\mu}Q^{\nu\rho}_{\nu}\,,
\end{aligned}
\tag{4.127}
$$

as well as

$$
B = \mathring{\nabla}_{\mu}(2T_{\nu}{}^{\nu\mu} + Q_{\nu}{}^{\nu\mu} - Q^{\mu\nu}{}_{\nu})\,. \tag{4.128}
$$

? Exercise

4.5. Make sure you understand the expansions of G and B into torsion and non-metricity tensors. Reproduce the above expressions.

If either torsion or nonmetricity vanish, these expressions simplify. In particular, the gravity scalar (4.127) reduces to the nonmetricity scalar

$$
\begin{aligned}
Q &= \frac{1}{2}Q_{\rho\mu\nu}P^{\rho\mu\nu} \\
&= \frac{1}{4}Q^{\mu\nu\rho}Q_{\mu\nu\rho} - \frac{1}{2}Q^{\mu\nu\rho}Q_{\rho\mu\nu} - \frac{1}{4}Q^{\rho\mu}{}_{\mu}Q_{\rho\nu}{}^{\nu} + \frac{1}{2}Q^{\mu}{}_{\mu\rho}Q^{\rho\nu}{}_{\nu}
\end{aligned}
\tag{4.129}
$$

or the torsion scalar

$$
\begin{aligned}
T &= \frac{1}{2}T^{\rho}{}_{\mu\nu}S_{\rho}{}^{\mu\nu} \\
&= \frac{1}{4}T^{\mu\nu\rho}T_{\mu\nu\rho} + \frac{1}{2}T^{\mu\nu\rho}T_{\rho\nu\mu} - T^{\mu}{}_{\mu\rho}T_{\nu}{}^{\nu\rho}\,,
\end{aligned}
\tag{4.130}
$$

respectively, where we have introduced the nonmetricity conjugate

$$
P^{\rho\mu\nu} = L^{\rho\mu\nu} - \frac{1}{2}g^{\mu\nu}(Q^{\rho\sigma}{}_{\sigma} - Q_{\sigma}{}^{\sigma\rho}) + \frac{1}{2}g^{\rho(\mu}Q^{\nu)\sigma}{}_{\sigma} \tag{4.131}
$$

and the superpotential

$$
S_{\rho}{}^{\mu\nu} = K^{\mu\nu}{}_{\rho} - \delta^{\mu}_{\rho}T_{\sigma}{}^{\sigma\nu} + \delta^{\nu}_{\rho}T_{\sigma}{}^{\sigma\mu}\,. \tag{4.132}
$$

In terms of these scalars, the action of the *symmetric teleparallel equivalent of general relativity* (STEGR) becomes [5]

$$S_g = -\frac{1}{2\kappa^2}\int_M Q\sqrt{-g}\mathrm{d}^4x\,, \tag{4.133}$$

while for the *metric teleparallel equivalent of general relativity* (MTEGR[2]) one has [28]

$$S_g = -\frac{1}{2\kappa^2}\int_M T\sqrt{-g}\mathrm{d}^4x\,. \tag{4.134}$$

Within their respective class of teleparallel gravity theories, these actions yield the same metric field equation as general relativity, and are thus common starting points for the construction of modified gravity theories, as we will see in the following sections. Note, however, that the construction of teleparallel equivalent theories is not confined to general relativity; in fact, the same procedure of replacing the Riemann tensor of the Levi-Civita connection using the relation (4.108), and possibly omitting boundary terms, can be applied to the action of any gravity theory whose action uses the metric as a fundamental variable [27].

4.4.2 The $f(G)$ Classes of Modified Theories

After discussing in the previous section a number of teleparallel gravity theories, whose metric field equation reproduces Einstein's field equation of general relativity for matter without hypermomentum, we now turn our focus towards modifications of these gravity theories. For the Einstein-Hilbert action (4.109), a well-known and thoroughly studied class of gravity theories is obtained by replacing the Ricci scalar $\overset{\circ}{R}$ by $f(\overset{\circ}{R})$, where f is an arbitrary real function of one variable, which is chosen such that the phenomenology of the resulting theory matches with observations, e.g., in cosmology. The same procedure can also be applied to the teleparallel equivalent theories [29]. Starting with the GTEGR action (4.112), one thus obtains the action

$$S_g = -\frac{1}{2\kappa^2}\int_M f(G)\sqrt{-g}\mathrm{d}^4x\,. \tag{4.135}$$

In order to derive the field equations, one proceeds as shown in the previous section, by variation of the action and integration by parts, so that the gravitational part S_g

[2] In the literature, the abbreviation TEGR is more common, since it was developed prior to the other equivalent theories. Another proposed nomenclature is "antisymmetric teleparallel equivalent of general relativity" (ATEGR) [27], since the distortion tensor becomes antisymmetric in its first two indices. However, the term "metric" or "metric-compatible" is more abundant in the contemporary literature on teleparallel gravity to denote the case of vanishing nonmetricity.

takes the form (4.9), with

$$W_{\mu\nu} = \frac{1}{\kappa^2}\left[f'U_{\mu\nu} - \overset{\circ}{\nabla}_\rho(f'V^\rho{}_{\mu\nu}) + \frac{1}{2}fg_{\mu\nu}\right] \tag{4.136}$$

and

$$Y_\mu{}^{\nu\rho} = \frac{1}{2\kappa^2}f'Z_\mu{}^{\nu\rho}\,. \tag{4.137}$$

where we wrote $f, f', \ldots$ as a shorthand for $f(G), f'(G), \ldots$, and used the abbreviations (4.115) we introduced for the variation of the gravity scalar G. Hence, it follows that the gravitational field equations are given by the metric equation

$$f'U_{\mu\nu} - \overset{\circ}{\nabla}_\rho(f'V^\rho{}_{\mu\nu}) + \frac{1}{2}fg_{\mu\nu} = \kappa^2\Theta_{\mu\nu} \tag{4.138}$$

and the connection equation

$$\nabla_\rho(f'Z_\mu{}^{\nu\rho}) - f'M^\omega{}_{\rho\omega}Z_\mu{}^{\nu\rho} = 2\kappa^2(\nabla_\rho H_\mu{}^{\nu\rho} - M^\omega{}_{\rho\omega}H_\mu{}^{\nu\rho})\,. \tag{4.139}$$

These equations can be written more explicitly as follows. First, recall that

$$U_{\mu\nu} - \overset{\circ}{\nabla}_\rho(V^\rho{}_{\mu\nu}) + \frac{1}{2}Gg_{\mu\nu} = \overset{\circ}{R}_{\mu\nu} - \frac{1}{2}\overset{\circ}{R}g_{\mu\nu} \tag{4.140}$$

is the left hand side of the GTEGR field equation. Using this fact, the metric field equation (4.138) becomes

$$f'\left(\overset{\circ}{R}_{\mu\nu} - \frac{1}{2}\overset{\circ}{R}g_{\mu\nu}\right) - V^\rho{}_{\mu\nu}\overset{\circ}{\nabla}_\rho f' + \frac{1}{2}(f - f'G)g_{\mu\nu} = \kappa^2\Theta_{\mu\nu}\,. \tag{4.141}$$

Finally, substituting $V^\rho{}_{\mu\nu}$ using the variation (4.115), one obtains

$$f'\left(\overset{\circ}{R}_{\mu\nu} - \frac{1}{2}\overset{\circ}{R}g_{\mu\nu}\right) - M^\rho{}_{(\mu\nu)}\overset{\circ}{\nabla}_\rho f' + \overset{\circ}{\nabla}_{(\mu}f'M^\sigma{}_{\nu)\sigma} + M^{[\rho\sigma]}{}_\sigma g_{\mu\nu}\overset{\circ}{\nabla}_\rho f'$$
$$+ \frac{1}{2}(f - f'G)g_{\mu\nu} = \kappa^2\Theta_{\mu\nu}\,. \tag{4.142}$$

Similarly, one can use the fact that the left hand side of the GTEGR connection equation vanishes, and hence

$$\nabla_\rho(Z_\mu{}^{\nu\rho}) - M^\omega{}_{\rho\omega}Z_\mu{}^{\nu\rho} = 0\,, \tag{4.143}$$

to write the connection field equation as

$$Z_\mu{}^{\nu\rho}\nabla_\rho f' = 2\kappa^2(\nabla_\rho H_\mu{}^{\nu\rho} - M^\omega{}_{\rho\omega}H_\mu{}^{\nu\rho})\,, \tag{4.144}$$

and substituting the variation (4.115),

$$M^{\nu\rho}{}_\rho\nabla_\mu f' + M^\sigma{}_{\mu\sigma}g^{\nu\rho}\nabla_\rho f' - M^{\nu\rho}{}_\mu\nabla_\rho f' - M^\rho{}_\mu{}^\nu\nabla_\rho f' \\ = 2\kappa^2(\nabla_\rho H_\mu{}^{\nu\rho} - M^\omega{}_{\rho\omega}H_\mu{}^{\nu\rho})\,. \tag{4.145}$$

The most important difference which distinguishes the $f(G)$ class of theories from GTEGR is the fact that for $f'' \neq 0$ the connection contribution to the action is no longer a total derivative, and so the connection remains as a dynamical field in the field equations, which now also contain a non-trivial connection field equation. It follows in particular that these field equations are not equivalent to those of $f(\mathring{R})$ gravity, since the latter has the metric as its only dynamical field, and its field equations are of fourth derivative order[3]. In contrast, the field equations of $f(G)$ gravity are of second derivative order.

In analogy to the GTEGR action (4.112), which is based on the general teleparallel geometry containing both torsion and nonmetricity, also the teleparallel equivalent theories based on more restricted geometries can be generalized by introducing a free function into their respective actions (4.133) and (4.134). Equivalently, one can take the action (4.135) and impose the vanishing torsion or nonmetricity either by introducing Lagrange multipliers or by imposing the constraint alongside a restricted variation. It turns out that the resulting field equations can be simplified, as we will see in the following. We start with the symmetric teleparallel gravity action [30]

$$S_g = -\frac{1}{2\kappa^2}\int_M f(Q)\sqrt{-g}\mathrm{d}^4x\,. \tag{4.146}$$

The variation of this action is still given by the expressions (4.136) and (4.137), but these simplify due to the vanishing torsion, and can be expressed in terms of the nonmetricity. Also the field equation simplify, which one can see as follows, starting with the connection equation. From the general form (4.36) follows that only the symmetric part $\tilde{Y}_\mu{}^{(\nu\rho)}$ contributes, since the covariant derivatives commute in the absence of curvature and torsion. Using (4.137), this means that only the symmetric part $Z_\mu{}^{(\nu\rho)}$ contributes, which is given by

$$Z_\mu{}^{(\nu\rho)} = Q_\mu{}^{\nu\rho} - \frac{1}{2}g^{\nu\rho}Q_{\mu\sigma}{}^\sigma + \frac{1}{2}\delta_\mu^{(\nu}Q^{\rho)\sigma}{}_\sigma - Q_\sigma{}^{\sigma(\nu}\delta_\mu^{\rho)} = -2P^{(\nu\rho)}{}_\mu\,. \tag{4.147}$$

[3] They can be reduced to second order by introducing an auxiliary scalar field.

The connection equation therefore becomes

$$-\nabla_\nu\nabla_\rho(f'\tilde{P}^{\nu\rho}{}_\mu) = \kappa^2\nabla_\nu\nabla_\rho\tilde{H}_\mu{}^{\nu\rho}\,, \tag{4.148}$$

where $\tilde{P}^{\nu\rho}{}_\mu = \sqrt{-g}P^{\nu\rho}{}_\mu$, and we omitted the symmetrization brackets around the indices, which are redundant due to the contraction with the commuting derivatives. Similarly, we can also simplify the metric field equation, which takes the same form (4.138), and can equivalently be written as

$$f'U^\mu{}_\nu - \frac{\overset{\circ}{\nabla}_\rho(\sqrt{-g}f'V^{\rho\mu}{}_\nu)}{\sqrt{-g}} + \frac{1}{2}f\delta^\mu_\nu = \kappa^2\Theta^\mu{}_\nu\,, \tag{4.149}$$

using the fact that the Levi-Civita connection is metric compatible, so that we can raise and lower indices and introduce the density factor $\sqrt{-g}$ inside the derivative. Changing this covariant derivative to the independent connection, one has

$$\begin{aligned}&\overset{\circ}{\nabla}_\rho(\sqrt{-g}f'V^{\rho\mu}{}_\nu) = \nabla_\rho(\sqrt{-g}f'V^{\rho\mu}{}_\nu)\\&+\sqrt{-g}f'(L^\sigma{}_{\sigma\rho}V^{\rho\mu}{}_\nu - L^\rho{}_{\sigma\rho}V^{\sigma\mu}{}_\nu - L^\mu{}_{\sigma\rho}V^{\rho\sigma}{}_\nu + L^\sigma{}_{\nu\rho}V^{\rho\mu}{}_\sigma)\,.\end{aligned} \tag{4.150}$$

Calculating the variation terms

$$U^{\mu\nu} = \frac{1}{4}Q^{\mu\rho\sigma}Q^\nu{}_{\rho\sigma} + \frac{1}{4}(Q^{\rho\mu\nu} - Q^{(\mu\nu)\rho})Q_{\rho\sigma}{}^\sigma - Q^{\rho\sigma\mu}Q_{[\rho\sigma]}{}^\nu \tag{4.151}$$

and

$$V^{\rho\mu\nu} = \frac{1}{2}Q^{\rho\mu\nu} - Q^{(\mu\nu)\rho} - \frac{1}{2}g^{\mu\nu}(Q^{\rho\sigma}{}_\sigma - Q_\sigma{}^{\sigma\rho}) + \frac{1}{2}g^{\rho(\mu}Q^{\nu)\sigma}{}_\sigma = P^{\rho\mu\nu}\,, \tag{4.152}$$

one finds that they combine into

$$\begin{aligned}&U^\mu{}_\nu - L^\sigma{}_{\sigma\rho}V^{\rho\mu}{}_\nu + L^\rho{}_{\sigma\rho}V^{\sigma\mu}{}_\nu + L^\mu{}_{\sigma\rho}V^{\rho\sigma}{}_\nu - L^\sigma{}_{\nu\rho}V^{\rho\mu}{}_\sigma\\&= \frac{1}{2}(Q^{\mu[\rho}{}_\rho Q_{\nu\sigma}{}^{\sigma]} - Q_{(\nu}{}^{\mu\rho}Q_{\rho)\sigma}{}^\rho + Q^{\rho\mu\sigma}Q_{\nu\rho\sigma}) = -\frac{1}{2}P^{\mu\rho\sigma}Q_{\nu\rho\sigma}\,.\end{aligned} \tag{4.153}$$

Combining these results, the metric field equation finally becomes

$$-\frac{\nabla_\rho(\sqrt{-g}f'P^{\rho\mu}{}_\nu)}{\sqrt{-g}} - \frac{f'}{2}P^{\mu\rho\sigma}Q_{\nu\rho\sigma} + \frac{1}{2}f\delta^\mu_\nu = \kappa^2\Theta^\mu{}_\nu\,. \tag{4.154}$$

Note, however, that this form changes if one raises or lowers indices, which appear also under the metric-incompatible covariant derivative ∇_ρ.

Finally, we also take a closer look at the metric teleparallel case, by imposing vanishing nonmetricity. Under this restriction, the general action (4.135) becomes [31–33]

$$S_{\mathrm{g}} = -\frac{1}{2\kappa^2}\int_M f(T)\sqrt{-g}\mathrm{d}^4x\,. \tag{4.155}$$

In this case we need to consider only the single field equation (4.41), whose left hand side now takes the form

$$\begin{aligned} &W^{\mu\nu} - \nabla_\rho Y^{\mu\nu\rho} + Y^{\mu\nu\rho}T^\tau{}_{\tau\rho} \\ &= \frac{1}{\kappa^2}\left[f'U^{\mu\nu} - \mathring{\nabla}_\rho(f'V^{\rho\mu\nu}) + \frac{1}{2}fg^{\mu\nu} - \frac{1}{2}\nabla_\rho(f'Z^{\mu\nu\rho}) + \frac{1}{2}f'Z^{\mu\nu\rho}T^\tau{}_{\tau\rho}\right]. \end{aligned} \tag{4.156}$$

using the variation expressions (4.136) and (4.137). In order to simplify this expression, we transform the covariant derivative with respect to the independent connection to that of the Levi-Civita connection, and find

$$\nabla_\rho(f'Z^{\mu\nu\rho}) - f'Z^{\mu\nu\rho}T^\tau{}_{\tau\rho} = \mathring{\nabla}_\rho(f'Z^{\mu\nu\rho}) + f'(K^\mu{}_{\sigma\rho}Z^{\sigma\nu\rho} + K^\nu{}_{\sigma\rho}Z^{\mu\sigma\rho})\,, \tag{4.157}$$

where the trace of the torsion tensor cancels with a trace of the contortion tensor. Now we can combine the two covariant derivatives, and evaluate

$$V^{\rho\mu\nu} + \frac{1}{2}Z^{\mu\nu\rho} = 2T_\sigma{}^{\sigma[\rho}g^{\nu]\mu} + T^{[\nu\rho]\mu} - \frac{1}{2}T^{\mu\nu\rho} = -S^{\mu\nu\rho}\,. \tag{4.158}$$

We are left with the terms

$$U^{\mu\nu} - \frac{1}{2}(K^\mu{}_{\sigma\rho}Z^{\sigma\nu\rho} + K^\nu{}_{\sigma\rho}Z^{\mu\sigma\rho}) = 2K^{\nu\rho[\sigma}K_{\rho\sigma}{}^{\mu]} = S^{\rho\sigma\nu}K_{\rho\sigma}{}^\mu\,. \tag{4.159}$$

Combining all terms and lowering indices, which commutes with all covariant derivatives since these are now metric-compatible, we can write the field equation as

$$\mathring{\nabla}_\rho(f'S_{\mu\nu}{}^\rho) + f'S^{\rho\sigma}{}_\nu K_{\rho\sigma\mu} + \frac{1}{2}fg_{\mu\nu} = \kappa^2(\Theta_{\mu\nu} - \nabla_\rho H_{\mu\nu}{}^\rho + H_{\mu\nu}{}^\rho T^\tau{}_{\tau\rho})\,. \tag{4.160}$$

Also this equation can be brought into various other forms by using the identities which hold for the contortion and the torsion.

4.4.3 The General Quadratic Lagrangians

The GTEGR action (4.112) has the appealing property that the gravity scalar (4.111), unlike the Ricci scalar, is quadratic in first order derivatives of the dynamical fields, and hence more reminiscent of the kinetic energy of a gauge field. This invites for another class of modified teleparallel gravity theories, by considering an action which is an arbitrary linear combination of all possible scalars which can be obtained by contracting the product of the distortion tensor $M^{\mu}{}_{\nu\rho}$ with itself. One easily checks that there are 11 possible terms: five terms arise from contracting $M^{\mu}{}_{\nu\rho}$ with a second copy carrying the same indices in an arbitrary permutation, and six terms arising from contracting two arbitrary traces of the distortion tensor with each other, where in both cases terms which are distinguished only by the order of the factors are counted only once, since they are identical. This gives rise to the generalized gravity scalar [6]

$$\begin{aligned}\mathcal{G} = {} & M^{\mu\nu\rho}(k_1 M_{\mu\nu\rho} + k_2 M_{\nu\rho\mu} + k_3 M_{\mu\rho\nu} + k_4 M_{\rho\nu\mu} + k_5 M_{\nu\mu\rho}) \\ & + k_6 M_{\rho\mu}{}^{\mu} M^{\rho\nu}{}_{\nu} + k_7 M_{\mu\rho}{}^{\mu} M^{\nu\rho}{}_{\nu} + k_8 M^{\mu}{}_{\mu\rho} M_{\nu}{}^{\nu\rho} \\ & + k_9 M_{\mu\rho}{}^{\mu} M_{\nu}{}^{\nu\rho} + k_{10} M^{\mu}{}_{\mu\rho} M^{\rho\nu}{}_{\nu} + k_{11} M_{\rho\mu}{}^{\mu} M^{\nu\rho}{}_{\nu}\end{aligned} \tag{4.161}$$

with arbitrary constants $k_1, \ldots, k_{11}$. Equivalently, one could also start from the expression (4.127), and consider the most general scalar which is quadratic in the torsion and nonmetricity tensors. Again one finds 11 possible terms, so that their most general linear combination is of the form

$$\begin{aligned}\mathcal{G} = {} & a_1 T^{\mu\nu\rho} T_{\mu\nu\rho} + a_2 T^{\mu\nu\rho} T_{\rho\nu\mu} + a_3 T^{\mu}{}_{\mu\rho} T_{\nu}{}^{\nu\rho} \\ & - b_1 Q^{\mu\nu\rho} T_{\rho\nu\mu} - b_2 Q^{\rho\mu}{}_{\mu} T^{\nu}{}_{\nu\rho} - b_3 Q_{\mu}{}^{\mu\rho} T^{\nu}{}_{\nu\rho} \\ & + c_1 Q^{\mu\nu\rho} Q_{\mu\nu\rho} + c_2 Q^{\mu\nu\rho} Q_{\rho\mu\nu} + c_3 Q^{\rho\mu}{}_{\mu} Q_{\rho\nu}{}^{\nu} \\ & + c_4 Q^{\mu}{}_{\mu\rho} Q_{\nu}{}^{\nu\rho} + c_5 Q^{\mu}{}_{\mu\rho} Q^{\rho\nu}{}_{\nu} ,\end{aligned} \tag{4.162}$$

where we introduced the arbitrary constants $a_1, \ldots, a_3, b_1, \ldots, b_3, c_1, \ldots, c_5$. Demanding that both expressions agree, one easily checks that these two sets of constants are related to each other by

$$\begin{gathered} k_1 = 2a_1 - b_1 + 2c_1 , \quad k_2 = -2a_2 + b_1 + 2c_2 , \quad k_9 = -2a_3 + 2b_2 - b_3 + 2c_5 , \\ k_4 = a_2 + c_2 , \quad k_5 = a_2 - b_1 + 2c_1 , \quad k_6 = c_4 , \quad k_7 = a_3 + b_3 + c_4 , \\ k_8 = a_3 - 2b_2 + 4c_3 , \quad k_3 = -2a_1 + b_1 + c_2 , \\ k_{10} = -b_3 + 2c_5 , \quad k_{11} = b_3 + 2c_4 . \end{gathered} \tag{4.163}$$

Further, choosing the values of these constants to be

$$k_{11} = -k_2 = 1\,, \quad k_1 = k_3 = k_4 = k_5 = k_6 = k_7 = k_8 = k_9 = k_{10} = 0\,, \tag{4.164}$$

one finds that the scalar $\mathcal{G}$ reduces to G. Hence, one may expect that the class of modified gravity theories defined by the action

$$S_g = -\frac{1}{2\kappa^2}\int_M \mathcal{G}\sqrt{-g}\mathrm{d}^4x \tag{4.165}$$

has a well-defined limit towards GTEGR, which is achieved if the constant parameters in the action take the aforementioned values. In order to derive the field equations, one can proceed in full analogy to the GTEGR field equations we discussed before. First, it is helpful to calculate the variation of the scalar (4.161), and write it in the form

$$\delta\mathcal{G} = \mathcal{U}^{\mu\nu}\delta g_{\mu\nu} + \mathcal{V}^{\rho\mu\nu}\mathring{\nabla}_\rho\delta g_{\mu\nu} + \mathcal{Z}_\mu{}^{\nu\rho}\delta\Gamma^\mu{}_{\nu\rho}\,. \tag{4.166}$$

Here we have made use of the abbreviations

$$\begin{aligned}\mathcal{U}^{\mu\nu} &= k_1(M^{\mu\rho\sigma}M^\nu{}_{\rho\sigma} - M^{\rho\mu}{}_\sigma M_\rho{}^{\nu\sigma} - M_{\rho\sigma}{}^\mu M^{\rho\sigma\mu}) - k_2 M_\rho{}^{\sigma(\mu}M_\sigma{}^{\nu)\rho}\\ &+ k_3(M^{\mu\rho\sigma}M^\nu{}_{\sigma\rho} - 2M^{\rho\sigma(\mu}M_\rho{}^{\nu)\sigma}) - k_4 M^{\rho\mu}{}_\sigma M^{\sigma\nu}{}_\rho - k_5 M^{\rho\sigma\mu}M_{\sigma\rho}{}^\nu\\ &+ k_6 M^{\mu\rho}{}_\rho M^{\nu\sigma}{}_\sigma - k_7 M^{\rho\mu}{}_\rho M^{\sigma\nu}{}_\sigma - k_8 M_\rho{}^{\rho\mu}M_\sigma{}^{\sigma\nu} - k_9 M_\rho{}^{\rho(\mu}M_\sigma{}^{\nu)\sigma}\\ &- (2k_6 M_{\rho\sigma}{}^\sigma + k_{11}M_{\sigma\rho}{}^\sigma + k_{10}M^\sigma{}_{\sigma\rho})M^{\rho(\mu\nu)}\,,\end{aligned} \tag{4.167a}$$

as well as

$$\begin{aligned}\mathcal{V}^{\rho\mu\nu} &= -2k_6 g^{\rho(\mu}M^{\nu)\sigma}{}_\sigma - k_{11}g^{\rho(\mu}M_\sigma{}^{\nu)\sigma} - k_{10}M_\sigma{}^{\sigma(\mu}g^{\nu)\rho}\\ &+\frac{1}{2}g^{\mu\nu}\big[(2k_6 - k_{10} - k_{11})M^{\rho\sigma}{}_\sigma + (k_{11} - 2k_7 - k_9)M^{\sigma\rho}{}_\sigma + (k_{10} - 2k_8 - k_9)M_\sigma{}^{\sigma\rho}\big]\\ &+ (k_4 - k_5 - k_1 - k_3)M^{(\mu\nu)\rho} + (k_5 - k_4 - k_1 - k_3)M^{(\mu|\rho|\nu)}\\ &+ (k_1 - k_2 + k_3 - k_4 - k_5)M^{\rho(\mu\nu)}\end{aligned} \tag{4.167b}$$

and

$$\begin{aligned}\mathcal{Z}_\mu{}^{\nu\rho} &= 2k_1 M_\mu{}^{\nu\rho} + k_2(M^{\nu\rho}{}_\mu + M^\rho{}_\mu{}^\nu) + 2k_3 M_\mu{}^{\rho\nu} + 2k_4 M^{\rho\nu}{}_\mu + 2k_5 M^\nu{}_\mu{}^\rho\\ &+ 2k_6 M_{\mu\sigma}{}^\sigma g^{\nu\rho} + 2k_7 M^{\sigma\nu}{}_\sigma\delta^\rho_\mu + 2k_8 M_\sigma{}^{\sigma\rho}\delta^\nu_\mu + k_9(M_\sigma{}^{\rho\sigma}\delta^\nu_\mu + M_\sigma{}^{\sigma\nu}\delta^\rho_\mu)\\ &+ k_{10}(M^{\rho\sigma}{}_\sigma\delta^\nu_\mu + M^\sigma{}_{\sigma\mu}g^{\nu\rho}) + k_{11}(M^{\nu\sigma}{}_\sigma\delta^\rho_\mu + M^\sigma{}_{\mu\sigma}g^{\nu\rho})\,.\end{aligned} \tag{4.167c}$$

By inserting the variation (4.166) into the variation of the action (4.165) and integration by parts, one obtains the form (4.9), with

$$W_{\mu\nu} = \frac{1}{\kappa^2}\left(\mathcal{U}_{\mu\nu} - \overset{\circ}{\nabla}_\rho \mathcal{V}^\rho{}_{\mu\nu} + \frac{1}{2}\mathcal{G}g_{\mu\nu}\right) \tag{4.168}$$

and

$$Y_\mu{}^{\nu\rho} = \frac{1}{2\kappa^2}\mathcal{Z}_\mu{}^{\nu\rho}\,. \tag{4.169}$$

Hence, by comparing with the corresponding GTEGR expressions (4.116) and (4.117), we see that these have the same form, and one simply replaces the terms derived by variation of G with those obtained from $\mathcal{G}$ in its place. One therefore finds the metric field equation

$$\mathcal{U}_{\mu\nu} - \overset{\circ}{\nabla}_\rho(\mathcal{V}^\rho{}_{\mu\nu}) + \frac{1}{2}\mathcal{G}g_{\mu\nu} = \kappa^2\Theta_{\mu\nu}\,, \tag{4.170}$$

as well as the connection field equation

$$\nabla_\tau \mathcal{Z}_\mu{}^{\nu\tau} - M^\omega{}_{\tau\omega}\mathcal{Z}_\mu{}^{\nu\tau} = 2\kappa^2(\nabla_\tau H_\mu{}^{\nu\tau} - M^\omega{}_{\tau\omega}H_\mu{}^{\nu\tau})\,, \tag{4.171}$$

with the abbreviations (4.167).

A more comprehensible set of field equations is obtained for the more restricted geometries, in which we impose either vanishing torsion or vanishing nonmetricity. This can most easily be seen from the expression (4.162), which shows that numerous terms vanish identically in either of these two cases. We first consider the symmetric teleparallel case of vanishing torsion. In this case, $\mathcal{G}$ reduces to the generalized nonmetricity scalar [30]

$$\begin{aligned}\mathcal{Q} &= \frac{1}{2}Q_{\rho\mu\nu}\mathcal{P}^{\rho\mu\nu} \\ &= c_1 Q^{\mu\nu\rho}Q_{\mu\nu\rho} + c_2 Q^{\mu\nu\rho}Q_{\rho\mu\nu} + c_3 Q^{\rho\mu}{}_\mu Q_{\rho\nu}{}^\nu \\ &+ c_4 Q^\mu{}_{\mu\rho}Q_\nu{}^{\nu\rho} + c_5 Q^\mu{}_{\mu\rho}Q^{\rho\nu}{}_\nu\,,\end{aligned} \tag{4.172}$$

and only the five constant parameters $c_1, \ldots, c_5$ remain present in the action. In place of the nonmetricity conjugate (4.131) we now have the generalized expression

$$\begin{aligned}\mathcal{P}^{\rho\mu\nu} = 2c_1 Q^{\rho\mu\nu} + 2c_2 Q^{(\mu\nu)\rho} + 2c_3 g^{\mu\nu}Q^{\rho\sigma}{}_\sigma \\ + 2c_4 Q_\sigma{}^{\sigma(\mu}g^{\nu)\rho} + c_5(g^{\mu\nu}Q_\sigma{}^{\sigma\rho} + g^{\rho(\mu}Q^{\nu)\sigma}{}_\sigma)\,.\end{aligned} \tag{4.173}$$

For the corresponding class of gravity theories depending on these parameters, whose action reads

$$S_g = -\frac{1}{2\kappa^2}\int_M Q\sqrt{-g}\mathrm{d}^4x\,, \tag{4.174}$$

the term "Newer General Relativity" has been coined. Its field equations can be obtained in great analogy to the other symmetric teleparallel gravity theories we have encountered before. First, we derive the connection field equation (4.36), and use the fact that only the symmetric part $Y_\mu{}^{(\nu\rho)}$ contributes. Using the variation (4.169), we thus calculate

$$\begin{aligned}\mathcal{Z}_\mu{}^{(\nu\rho)} &= -2c_2 Q_\mu{}^{\nu\rho} - 2(2c_1 + c_2)Q^{(\nu\rho)}{}_\mu - 2g^{\nu\rho}(2c_4 Q^\sigma{}_{\sigma\mu} + c_5 Q_{\mu\sigma}{}^\sigma)\\ &- 4(c_4 + c_5)Q_\sigma{}^{\sigma(\nu}\delta_\mu^{\rho)} - 2(4c_3 + c_5)\delta_\mu^{(\nu}Q^{\rho)\sigma}{}_\sigma = -2\mathcal{P}^{(\nu\rho)}{}_\mu\,,\end{aligned} \tag{4.175}$$

which generalizes the similar relation (4.147). Hence, we find that the connection field equation can be written in the simple form

$$-\nabla_\nu\nabla_\rho(\tilde{\mathcal{P}}^{\nu\rho}{}_\mu) = \kappa^2\nabla_\nu\nabla_\rho\tilde{H}_\mu{}^{\nu\rho}\,, \tag{4.176}$$

using the tensor density $\tilde{\mathcal{P}}^{\nu\rho}{}_\mu$ built from the generalized nonmetricity conjugate (4.173). We then proceed with the metric equation, which still takes the general form (4.170) also in the symmetric teleparallel case, but can be simplified as follows. Raising one index and introducing a density factor, it can equivalently be written as

$$\mathcal{U}^\mu{}_\nu - \frac{\overset{\circ}{\nabla}_\rho(\sqrt{-g}\mathcal{V}^{\rho\mu}{}_\nu)}{\sqrt{-g}} + \frac{1}{2}\mathcal{G}\delta_\nu^\mu = \kappa^2\Theta^\mu{}_\nu\,. \tag{4.177}$$

The covariant derivative with respect to the Levi-Civita connection can be transformed to the independent connection, by using the relation

$$\begin{aligned}\overset{\circ}{\nabla}_\rho(\sqrt{-g}\mathcal{V}^{\rho\mu}{}_\nu) &= \nabla_\rho(\sqrt{-g}\mathcal{V}^{\rho\mu}{}_\nu)\\ &+ \sqrt{-g}(L^\sigma{}_{\sigma\rho}\mathcal{V}^{\rho\mu}{}_\nu - L^\rho{}_{\sigma\rho}\mathcal{V}^{\sigma\mu}{}_\nu - L^\mu{}_{\sigma\rho}\mathcal{V}^{\rho\sigma}{}_\nu + L^\sigma{}_{\nu\rho}\mathcal{V}^{\rho\mu}{}_\sigma)\,.\end{aligned} \tag{4.178}$$

To proceed further, we need the terms

$$\begin{aligned}\mathcal{U}^{\mu\nu} &= (2c_1 Q^{\rho\sigma\mu} + c_2 Q^{\sigma\rho\mu})Q_{\rho\sigma}{}^\nu - (2c_1 + c_2)Q^{\rho\sigma(\mu}Q^{\nu)}{}_{\rho\sigma} - (c_1 + c_2)Q^{\mu\rho\sigma}Q^\nu{}_{\rho\sigma}\\ &- c_4 Q_\rho{}^{\rho(\mu}Q^{\nu)\sigma}{}_\sigma + 2c_4 Q_\rho{}^{\rho\mu}Q_\sigma{}^{\sigma\nu} - \left(c_3 + \frac{c_5}{2}\right)Q^{\mu\rho}{}_\rho Q^{\nu\sigma}{}_\sigma\\ &+ (2c_4 Q^\sigma{}_{\sigma\rho} + c_5 Q_{\rho\sigma}{}^\sigma)\left(\frac{1}{2}Q^{\rho\mu\nu} - Q^{(\mu\nu)\rho}\right)\end{aligned} \tag{4.179}$$

and

$$\begin{aligned}\mathcal{V}^{\rho\mu\nu} &= 2c_1 Q^{\rho\mu\nu} + 2c_2 Q^{(\mu\nu)\rho} + 2c_3 g^{\mu\nu} Q^{\rho\sigma}{}_{\sigma} \\ &\quad + 2c_4 Q_{\sigma}{}^{\sigma(\mu} g^{\nu)\rho} + c_5 (g^{\mu\nu} Q_{\sigma}{}^{\sigma\rho} + g^{\rho(\mu} Q^{\nu)\sigma}{}_{\sigma}) , \end{aligned} \tag{4.180}$$

which are obtained from the more general expressions (4.167) by imposing vanishing torsion. A tedious, but straightforward calculation shows that the resulting terms can be combined to yield

$$\begin{aligned}&\mathcal{U}^{\mu}{}_{\nu} - L^{\sigma}{}_{\sigma\rho}\mathcal{V}^{\rho\mu}{}_{\nu} + L^{\rho}{}_{\sigma\rho}\mathcal{V}^{\sigma\mu}{}_{\nu} + L^{\mu}{}_{\sigma\rho}\mathcal{V}^{\rho\sigma}{}_{\nu} - L^{\sigma}{}_{\nu\rho}\mathcal{V}^{\rho\mu}{}_{\sigma} \\ &= -(c_1 Q^{\mu\rho\sigma} + c_2 Q^{\rho\sigma\mu}) Q_{\nu\rho\sigma} - c_3 Q^{\mu\rho}{}_{\rho} Q_{\nu\sigma}{}^{\sigma} - c_4 Q^{\rho}{}_{\rho\sigma} Q_{\nu}{}^{\mu\sigma} - c_5 Q_{(\nu}{}^{\mu\rho} Q_{\rho)\sigma}{}^{\sigma} \\ &= -\frac{1}{2}\mathcal{P}^{\mu\rho\sigma} Q_{\nu\rho\sigma} . \end{aligned} \tag{4.181}$$

This finally yields the metric field equation

$$-\frac{\nabla_{\rho}(\sqrt{-g}\mathcal{P}^{\rho\mu}{}_{\nu})}{\sqrt{-g}} - \frac{1}{2}\mathcal{P}^{\mu\rho\sigma} Q_{\nu\rho\sigma} + \frac{1}{2}\mathcal{Q}\delta^{\mu}_{\nu} = \kappa^2 \Theta^{\mu}{}_{\nu} \tag{4.182}$$

for the Newer General Relativity class of gravity theories, where we now also used the relation $\mathcal{G} = \mathcal{Q}$ in the absence of torsion. Note that a special case is obtained when the parameters take the values (4.164), for which we have

$$c_1 = \frac{1}{4} , \quad c_2 = -\frac{1}{2} , \quad c_3 = -\frac{1}{4} , \quad c_4 = 0 , \quad c_5 = \frac{1}{2} . \tag{4.183}$$

In this case we find $\mathcal{Q} = Q$ and $\mathcal{P}^{\mu\nu\rho} = P^{\mu\nu\rho}$, so that the theory reduces to STEGR.

Finally, also in the metric teleparallel geometry we can find a general class of gravity theories, whose action is now quadratic in the torsion tensor. By imposing vanishing nonmetricity, the scalar (4.161) becomes the generalized torsion scalar [34] (see also Chap. 3 around Eq. (3.166)),

$$\begin{aligned}\mathcal{T} &= \frac{1}{2} T^{\rho}{}_{\mu\nu} \mathcal{S}_{\rho}{}^{\mu\nu} \\ &= a_1 T^{\mu\nu\rho} T_{\mu\nu\rho} + a_2 T^{\mu\nu\rho} T_{\rho\nu\mu} + a_3 T^{\mu}{}_{\mu\rho} T_{\nu}{}^{\nu\rho} , \end{aligned} \tag{4.184}$$

where the generalized superpotential is now given by

$$\mathcal{S}_{\rho}{}^{\mu\nu} = 2a_1 T_{\rho}{}^{\mu\nu} + 2a_2 T^{[\nu\mu]}{}_{\rho} + 2a_3 T_{\sigma}{}^{\sigma[\nu} \delta^{\mu]}_{\rho} . \tag{4.185}$$

The resulting class of gravity theories, which is now defined by the action

$$S_{\mathrm{g}} = -\frac{1}{2\kappa^2} \int_M \mathcal{T} \sqrt{-g} \mathrm{d}^4 x , \tag{4.186}$$

is known as "New General Relativity"[4]. In this case, the left hand side of the field equations (4.41) becomes

$$W^{\mu\nu} - \nabla_\rho Y^{\mu\nu\rho} + Y^{\mu\nu\rho}T^\tau{}_{\tau\rho}$$
$$= \frac{1}{\kappa^2}\left[\mathcal{U}^{\mu\nu} - \overset{\circ}{\nabla}_\rho \mathcal{V}^{\rho\mu\nu} + \frac{1}{2}\mathcal{G}g^{\mu\nu} - \frac{1}{2}\nabla_\rho \mathcal{Z}^{\mu\nu\rho} + \frac{1}{2}\mathcal{Z}^{\mu\nu\rho}T^\tau{}_{\tau\rho}\right]. \tag{4.187}$$

with the help of the formulas (4.168) and (4.169). In order to combine the two derivative terms, we convert the covariant derivative ∇_ρ with respect to the independent connection to a Levi-Civita covariant derivative $\overset{\circ}{\nabla}_\rho$, using the relation

$$\nabla_\rho \mathcal{Z}^{\mu\nu\rho} - \mathcal{Z}^{\mu\nu\rho}T^\tau{}_{\tau\rho} = \overset{\circ}{\nabla}_\rho \mathcal{Z}^{\mu\nu\rho} + K^\mu{}_{\sigma\rho}\mathcal{Z}^{\sigma\nu\rho} + K^\nu{}_{\sigma\rho}\mathcal{Z}^{\mu\sigma\rho}. \tag{4.188}$$

Now the two terms under the derivative combine into

$$\mathcal{V}^{\rho\mu\nu} + \frac{1}{2}\mathcal{Z}^{\mu\nu\rho} = -2a_1 T^{\mu\nu\rho} - 2a_2 T^{[\rho\nu]\mu} - 2a_3 T_\sigma{}^{\sigma[\rho}g^{\nu]\mu} = -\mathcal{S}^{\mu\nu\rho}. \tag{4.189}$$

The remaining terms take, once again, a very simple form, which is given by

$$\mathcal{U}^{\mu\nu} - \frac{1}{2}(K^\mu{}_{\sigma\rho}\mathcal{Z}^{\sigma\nu\rho} + K^\nu{}_{\sigma\rho}\mathcal{Z}^{\mu\sigma\rho})$$
$$= [(a_2 - 2a_1)K^{\rho\sigma\nu} + (3a_2 - 2a_1)K^{\nu\rho\sigma}]K_{\rho\sigma}{}^\mu + a_3 K_{\rho\sigma}{}^\sigma K^{\nu\rho\mu} = \mathcal{S}^{\rho\sigma\nu}K_{\rho\sigma}{}^\mu. \tag{4.190}$$

Hence, the full field equations of New General Relativity become

$$\overset{\circ}{\nabla}_\rho(\mathcal{S}_{\mu\nu}{}^\rho) + \mathcal{S}^{\rho\sigma}{}_\nu K_{\rho\sigma\mu} + \frac{1}{2}\mathcal{T}g_{\mu\nu} = \kappa^2(\Theta_{\mu\nu} - \nabla_\rho H_{\mu\nu}{}^\rho + H_{\mu\nu}{}^\rho T^\tau{}_{\tau\rho}). \tag{4.191}$$

Also for this class of theories a special case is obtained by choosing the parameter values (4.164), which now implies

$$a_1 = \frac{1}{4}, \quad a_2 = \frac{1}{2}, \quad a_3 = -1. \tag{4.192}$$

In this case, the theory reduces to MTEGR, with $\mathcal{T} = T$ and $\mathcal{S}_\rho{}^{\mu\nu} = S_\rho{}^{\mu\nu}$.

[4] This term is also, more commonly, used for a particular subclass of theories, in which $2a_1 + a_2 = 0$ and $a_3 = -1$, so that there is only one free parameter besides the gravitational constant κ [34].

4.4.4 Scalar-Teleparallel Theories

While the classes of modified teleparallel gravity theories we considered so far were constructed purely from the metric and the flat affine connection as fundamental fields, we now consider a class of theories in which in addition a scalar field is introduced as a fundamental field variable. Also, this class of theories can be motivated by analogy with a scalar-tensor modification of the Einstein-Hilbert action (4.109) of general relativity, which takes the general form

$$S_{\mathrm{g}} = \frac{1}{2\kappa^2}\int_M \left[\mathcal{A}(\phi)\overset{\circ}{R} - \mathcal{B}(\phi)g^{\mu\nu}\overset{\circ}{\nabla}_\mu\phi\overset{\circ}{\nabla}_\nu\phi - 2\kappa^2\mathcal{V}(\phi)\right]\sqrt{-g}\mathrm{d}^4x\,, \tag{4.193}$$

where $\mathcal{A}, \mathcal{B}, \mathcal{V}$ are free functions of the scalar field ϕ. Here we work in the so-called Jordan frame, which means that we assume no direct coupling between the scalar field and any matter fields. Recalling that the Ricci scalar $\overset{\circ}{R}$ can be written in the form (4.110), one may expect that replacing $\overset{\circ}{R}$ by $-G + B$ one obtains a teleparallel equivalent of the scalar-curvature theory, while using only $-G$ instead leads to an inequivalent scalar-teleparallel theory, since the omitted term is not a boundary term due to the non-minimal coupling term $\mathcal{A}(\phi)$. One can cover both cases by considering the action

$$S_{\mathrm{g}} = \frac{1}{2\kappa^2}\int_M \left[-\mathcal{A}(\phi)G - \mathcal{B}(\phi)g^{\mu\nu}\overset{\circ}{\nabla}_\mu\phi\overset{\circ}{\nabla}_\nu\phi - \hat{C}(\phi)B - 2\kappa^2\mathcal{V}(\phi)\right]\sqrt{-g}\mathrm{d}^4x\,, \tag{4.194}$$

where we introduced another free function $\hat{C}$ of the scalar field. Keeping in mind that B is a boundary term, i.e., a total divergence, we see that the field equations do not change if we add an arbitrary constant to $\hat{C}$. To resolve this ambiguity, we can use integration by parts,

$$\hat{C}\overset{\circ}{\nabla}_\mu M^{[\nu\mu]}{}_\nu = \overset{\circ}{\nabla}_\mu(\hat{C}M^{[\nu\mu]}{}_\nu) - \hat{C}'M^{[\nu\mu]}{}_\nu\overset{\circ}{\nabla}_\mu\phi\,, \tag{4.195}$$

and omit the boundary term. Defining a new parameter function $C = \hat{C}'$, we then have

$$S_{\mathrm{g}} = \frac{1}{2\kappa^2}\int_M \Big[-\mathcal{A}(\phi)G - \mathcal{B}(\phi)g^{\mu\nu}\overset{\circ}{\nabla}_\mu\phi\overset{\circ}{\nabla}_\nu\phi + 2C(\phi)M^{[\nu\mu]}{}_\nu\overset{\circ}{\nabla}_\mu\phi - 2\kappa^2\mathcal{V}(\phi)\Big]\sqrt{-g}\mathrm{d}^4x\,. \tag{4.196}$$

Note that for $\mathcal{A}' + C = 0$, the action becomes equivalent to the scalar-curvature action (4.193). To derive the field equations for this generalized class of theories, we proceed by varying the action as with the previous examples. Due to the presence of an additional fundamental field, also the variation (4.9) is enhanced by an additional term, and becomes

$$\delta S_g = -\int_M \left(\frac{1}{2}W^{\mu\nu}\delta g_{\mu\nu} + Y_\mu{}^{\nu\rho}\delta\Gamma^\mu{}_{\nu\rho} + \Phi\delta\phi\right)\sqrt{-g}\mathrm{d}^4x\,, \tag{4.197}$$

after eliminating derivatives of the variations using integration by parts. Varying the action (4.196), we find the terms

$$\begin{aligned} W_{\mu\nu} = \frac{1}{\kappa^2}\Big\{ & \mathcal{A}\mathring{R}_{\mu\nu} - \frac{\mathcal{A}}{2}\mathring{R}g_{\mu\nu} + \mathcal{C}\mathring{\nabla}_\mu\mathring{\nabla}_\nu\phi - (\mathcal{B}-\mathcal{C}')\mathring{\nabla}_\mu\phi\mathring{\nabla}_\nu\phi \\ & + (\mathcal{A}'+\mathcal{C})\left(\mathring{\nabla}_{(\mu}\phi M^\rho{}_{\nu)\rho} - M^\rho{}_{(\mu\nu)}\mathring{\nabla}_\rho\phi + M^{[\rho\sigma]}{}_\sigma\mathring{\nabla}_\rho\phi g_{\mu\nu}\right) \\ & + \left[\left(\frac{\mathcal{B}}{2}-\mathcal{C}'\right)\mathring{\nabla}_\rho\phi\mathring{\nabla}^\rho\phi - \mathcal{C}\mathring{\nabla}_\rho\mathring{\nabla}^\rho\phi + \kappa^2\mathcal{V}\right]g_{\mu\nu}\Big\}\,, \end{aligned} \tag{4.198a}$$

$$\begin{aligned} Y_\mu{}^{\nu\rho} = \frac{1}{2\kappa^2}\Big[& \mathcal{A}(g^{\nu\rho}M^\sigma{}_{\mu\sigma} + \delta^\rho_\mu M^{\nu\sigma}{}_\sigma - M^{\nu\rho}{}_\mu - M^\rho{}_\nu{}^\mu) \\ & + \mathcal{C}(g^{\nu\rho}\mathring{\nabla}_\mu\phi - \delta^\rho_\mu\mathring{\nabla}^\nu\phi)\Big]\,, \end{aligned} \tag{4.198b}$$

$$\Phi = \frac{1}{2\kappa^2}\left[-2\mathcal{B}\mathring{\nabla}_\mu\mathring{\nabla}^\mu\phi - \mathcal{B}'\mathring{\nabla}_\mu\phi\mathring{\nabla}^\mu\phi + \mathcal{C}B + \mathcal{A}'G\right] + \mathcal{V}\,, \tag{4.198c}$$

where we have made use of the relations (4.108) and (4.111), and from now on we omit the argument ϕ of the parameter functions for brevity. We can then read off the field equations and study their properties. We start with the metric field equation (4.14), which reads

$$\begin{aligned} & \mathcal{A}\mathring{R}_{\mu\nu} - \frac{\mathcal{A}}{2}\mathring{R}g_{\mu\nu} + \mathcal{C}\mathring{\nabla}_\mu\mathring{\nabla}_\nu\phi + (\mathcal{A}'+\mathcal{C})\Big(\mathring{\nabla}_{(\mu}\phi M^\rho{}_{\nu)\rho} - M^\rho{}_{(\mu\nu)}\mathring{\nabla}_\rho\phi \\ & \qquad + M^{[\rho\sigma]}{}_\sigma\mathring{\nabla}_\rho\phi g_{\mu\nu}\Big) \\ & -(\mathcal{B}-\mathcal{C}')\mathring{\nabla}_\mu\phi\mathring{\nabla}_\nu\phi + \left[\left(\frac{\mathcal{B}}{2}-\mathcal{C}'\right)\mathring{\nabla}_\rho\phi\mathring{\nabla}^\rho\phi - \mathcal{C}\mathring{\nabla}_\rho\mathring{\nabla}^\rho\phi + \kappa^2\mathcal{V}\right]g_{\mu\nu} = \kappa^2\Theta_{\mu\nu}\,. \end{aligned} \tag{4.199}$$

It is most remarkable that in the case $\mathcal{A}' + \mathcal{C} = 0$ the only term containing the flat, affine connection vanishes from these field equations, and one finds that they indeed resemble the field equations of scalar-curvature gravity in this case. To check whether this property holds also for the connection field equation (4.28), we calculate

$$\begin{aligned} & \nabla_\tau Y_\mu{}^{\nu\tau} - M^\omega{}_{\tau\omega}Y_\mu{}^{\nu\tau} \\ & = \frac{\mathcal{A}'+\mathcal{C}}{2\kappa^2}\left[M^{\nu\rho}{}_\rho\mathring{\nabla}_\mu\phi + M^\rho{}_{\mu\rho}\mathring{\nabla}^\nu\phi - (M^{\nu\rho}{}_\mu + M^\rho{}_\mu{}^\nu)\mathring{\nabla}_\rho\phi\right]\,, \end{aligned} \tag{4.200}$$

where any terms involving the covariant derivative of the distortion $M^{\mu}{}_{\nu\rho}$ cancel as a consequence of the flatness of the connection. We see that this expression becomes trivial for $\mathcal{A}' + \mathcal{C} = 0$. In that case, the connection field equation

$$(\mathcal{A}' + \mathcal{C})\left[M^{\nu\rho}{}_{\rho}\overset{\circ}{\nabla}_{\mu}\phi + M^{\rho}{}_{\mu\rho}\overset{\circ}{\nabla}^{\nu}\phi - (M^{\nu\rho}{}_{\mu} + M^{\rho}{}_{\mu}{}^{\nu})\overset{\circ}{\nabla}_{\rho}\phi\right] \\ = 2\kappa^2(\nabla_{\tau}H_{\mu}{}^{\nu\tau} - M^{\omega}{}_{\tau\omega}H_{\mu}{}^{\nu\tau}) \tag{4.201}$$

becomes a constraint for the hypermomentum. Finally, we study the scalar field equation

$$-2\mathcal{B}\overset{\circ}{\nabla}_{\mu}\overset{\circ}{\nabla}^{\mu}\phi - \mathcal{B}'\overset{\circ}{\nabla}_{\mu}\phi\overset{\circ}{\nabla}^{\mu}\phi + \mathcal{C}B + \mathcal{A}'G + 2\kappa^2\mathcal{V}' = 0\,. \tag{4.202}$$

Here the right hand side vanishes, since we do not consider any direct coupling between the scalar field and matter. Note that if $\mathcal{C} = -\mathcal{A}'$, i.e., in the case of the scalar-curvature equivalent, the two terms $\mathcal{C}B + \mathcal{A}'G$ combine to $-\mathcal{A}'\overset{\circ}{R}$, and the equation becomes independent of the teleparallel connection, as one would expect, and as we have seen for the remaining field equations. Further, one finds that the scalar field equation contains second order derivatives of both the scalar field and the metric, where the latter enter through the boundary term. In order to eliminate these metric derivatives from the equation, it is common to apply a "debraiding" procedure by adding a suitable multiple of the trace of the matter field equation. The latter reads

$$-\mathcal{A}\overset{\circ}{R} - 3\mathcal{C}\overset{\circ}{\nabla}_{\mu}\overset{\circ}{\nabla}^{\mu}\phi + 2(\mathcal{A}' + \mathcal{C})M^{[\mu\nu]}{}_{\nu}\overset{\circ}{\nabla}_{\mu}\phi \\ + (\mathcal{B} - 3\mathcal{C}')\overset{\circ}{\nabla}_{\mu}\phi\overset{\circ}{\nabla}^{\mu}\phi + 4\kappa^2\mathcal{V} = \kappa^2\Theta_{\mu}{}^{\mu}\,. \tag{4.203}$$

Hence, calculating the linear combination

$$\mathcal{C}W_{\mu}{}^{\mu} + 2\mathcal{A}\Phi = \frac{1}{\kappa^2}\Big[-(2\mathcal{A}\mathcal{B} + 3\mathcal{C}^2)\overset{\circ}{\nabla}_{\mu}\overset{\circ}{\nabla}^{\mu}\phi + (\mathcal{B}\mathcal{C} - 3\mathcal{C}\mathcal{C}' - \mathcal{A}\mathcal{B}')\overset{\circ}{\nabla}_{\mu}\phi\overset{\circ}{\nabla}^{\mu}\phi \\ + (\mathcal{A}' + \mathcal{C})\left(\mathcal{A}G + 2\mathcal{C}M^{[\mu\nu]}{}_{\nu}\overset{\circ}{\nabla}_{\mu}\phi\right)\Big] + 2\mathcal{A}\mathcal{V}' + 4\mathcal{C}\mathcal{V}\,, \tag{4.204}$$

we find that the debraided scalar field equation

$$-(2\mathcal{A}\mathcal{B} + 3\mathcal{C}^2)\overset{\circ}{\nabla}_{\mu}\overset{\circ}{\nabla}^{\mu}\phi + (\mathcal{B}\mathcal{C} - 3\mathcal{C}\mathcal{C}' - \mathcal{A}\mathcal{B}')\overset{\circ}{\nabla}_{\mu}\phi\overset{\circ}{\nabla}^{\mu}\phi \\ + (\mathcal{A}' + \mathcal{C})\left(\mathcal{A}G + 2\mathcal{C}M^{[\mu\nu]}{}_{\nu}\overset{\circ}{\nabla}_{\mu}\phi\right) + 2\kappa^2(\mathcal{A}\mathcal{V}' + 2\mathcal{C}\mathcal{V}) = \kappa^2\mathcal{C}\Theta_{\mu}{}^{\mu} \tag{4.205}$$

does not contain any derivatives of the independent connection, and has only first order derivatives of the metric tensor, which enter through the distortion and the Christoffel symbols contained in the covariant derivative. Also here we see that the teleparallel connection does not contribute to the field equation for $\mathcal{A}' + \mathcal{C} = 0$. Further, one finds that the trace of the energy-momentum tensor acts as the matter source for the scalar field.

? Exercise

4.6. Understand that for $\mathcal{A}' + \mathcal{C} = 0$, the field equations are independent of the teleparallel connection, i.e. one recovers scalar-tensor gravity theories.

It is now easy to study how the field equations change if we consider the symmetric or metric teleparallel geometries instead of the general teleparallel geometry we have used to construct the scalar-teleparallel gravity theory discussed above. We start with the former, which yields a class of scalar-nonmetricity theories of gravity, whose action is given by [35, 36]

$$S_{\mathrm{g}} = \frac{1}{2\kappa^2} \int_M \Big[-\mathcal{A}(\phi) Q - \mathcal{B}(\phi) g^{\mu\nu} \mathring{\nabla}_\mu \phi \mathring{\nabla}_\nu \phi + \mathcal{C}(\phi)(Q_\nu{}^{\nu\mu} - Q^{\mu\nu}{}_\nu) \mathring{\nabla}_\mu \phi - 2\kappa^2 \mathcal{V}(\phi) \Big] \sqrt{-g} \mathrm{d}^4 x \,. \tag{4.206}$$

For the metric field equations, which retain the general form (4.14), we see that the only change compared to the general teleparallel case arises from those terms which involve the teleparallel affine connection. These terms greatly simplify and become

$$\mathring{\nabla}_{(\mu} \phi M^\rho{}_{\nu)\rho} - M^\rho{}_{(\mu\nu)} \mathring{\nabla}_\rho \phi + M^{[\rho\sigma]}{}_\sigma \mathring{\nabla}_\rho \phi g_{\mu\nu} = -P^\rho{}_{\mu\nu} \mathring{\nabla}_\rho \phi \,, \tag{4.207}$$

using the nonmetricity conjugate (4.131). The metric field equations therefore read

$$\mathcal{A} \mathring{R}_{\mu\nu} - \frac{\mathcal{A}}{2} \mathring{R} g_{\mu\nu} + \mathcal{C} \mathring{\nabla}_\mu \mathring{\nabla}_\nu \phi - (\mathcal{B} - \mathcal{C}') \mathring{\nabla}_\mu \phi \mathring{\nabla}_\nu \phi - (\mathcal{A}' + \mathcal{C}) P^\rho{}_{\mu\nu} \mathring{\nabla}_\rho \phi + \left[\left(\frac{\mathcal{B}}{2} - \mathcal{C}' \right) \mathring{\nabla}_\rho \phi \mathring{\nabla}^\rho \phi - \mathcal{C} \mathring{\nabla}_\rho \mathring{\nabla}^\rho \phi + \kappa^2 \mathcal{V} \right] g_{\mu\nu} = \kappa^2 \Theta_{\mu\nu} \,. \tag{4.208}$$

We then continue with the connection equation, which now takes the form (4.36). Here we can make use of several simplifications we have employed before. First, using the variation (4.115) of the gravity scalar G, we write the variation (4.198) as

$$\begin{aligned} Y_\mu{}^{\nu\rho} &= \frac{1}{2\kappa^2}\left[\mathcal{A}Z_\mu{}^{\nu\rho} + C\left(g^{\nu\rho}\mathring{\nabla}_\mu\phi - \delta^\rho_\mu\mathring{\nabla}^\nu\phi\right)\right] \\ &= \frac{1}{2\kappa^2}\left[\mathcal{A}Z_\mu{}^{\nu\rho} + C\left(g^{\nu\rho}\nabla_\mu\phi - \delta^\rho_\mu g^{\nu\sigma}\nabla_\sigma\phi\right)\right], \end{aligned} \tag{4.209}$$

where we used the fact that any covariant derivative acts equally on the scalar field ϕ. Next, we introduce a density factor $\sqrt{-g}$ and take a covariant derivative, to calculate

$$\begin{aligned} \nabla_\rho\tilde{Y}_\mu{}^{\nu\rho} &= \frac{1}{2\kappa^2}\nabla_\rho\left[\mathcal{A}\tilde{Z}_\mu{}^{\nu\rho} + \sqrt{-g}C\left(g^{\nu\rho}\nabla_\mu\phi - \delta^\rho_\mu g^{\nu\sigma}\nabla_\sigma\phi\right)\right] \\ &= \frac{\sqrt{-g}}{2\kappa^2}\left[\mathcal{A}'Z_\mu{}^{\nu\rho}\nabla_\rho\phi + \left(\frac{1}{2}Q_{\rho\tau}{}^\tau C + C'\nabla_\rho\phi\right)\left(g^{\nu\rho}\nabla_\mu\phi - \delta^\rho_\mu g^{\nu\sigma}\nabla_\sigma\phi\right)\right. \\ &\quad \left. + C\left(g^{\nu\rho}\nabla_\rho\nabla_\mu\phi - g^{\nu\sigma}\nabla_\mu\nabla_\sigma\phi - Q_\rho{}^{\rho\nu}\nabla_\mu\phi + Q_\mu{}^{\nu\sigma}\nabla_\sigma\phi\right)\right] \\ &= \frac{1}{2\kappa^2}(\mathcal{A}' + C)\tilde{Z}_\mu{}^{\nu\rho}\nabla_\rho\phi \\ &= \frac{1}{2\kappa^2}\nabla_\rho[(\mathcal{A} + \hat{C})\tilde{Z}_\mu{}^{\nu\rho}], \end{aligned} \tag{4.210}$$

where we used the identity $\nabla_\rho\tilde{Z}_\mu{}^{\nu\rho} = 0$ we found in deriving the STEGR field equations, and the fact that numerous terms involving the scalar field cancel, while the remaining terms combine to a very compact form. Here $\hat{C}$ is defined by $\hat{C}' = C$ only up to an irrelevant constant. To obtain the connection field equations, we apply another covariant derivative, and use the relation (4.147) to finally obtain

$$\nabla_\nu\nabla_\rho\tilde{Y}_\mu{}^{\nu\rho} = \frac{1}{2\kappa^2}\nabla_\nu\nabla_\rho[(\mathcal{A} + \hat{C})\tilde{Z}_\mu{}^{\nu\rho}] = -\frac{1}{\kappa^2}\nabla_\nu\nabla_\rho[(\mathcal{A} + \hat{C})\tilde{P}^{\nu\rho}{}_\mu]. \tag{4.211}$$

Hence, we see that the left hand side of the connection field equations

$$-\nabla_\nu\nabla_\rho[(\mathcal{A} + \hat{C})\tilde{P}^{\nu\rho}{}_\mu] = \kappa^2\nabla_\nu\nabla_\rho\tilde{H}_\mu{}^{\nu\rho} \tag{4.212}$$

vanishes identically for $\mathcal{A}' + C = 0$. At last, we come to the scalar field equation, which we consider in its debraided form (4.205). Imposing vanishing torsion, the only affected term is given by

$$\mathcal{A}G + 2CM^{[\mu\nu]}{}_\nu\mathring{\nabla}_\mu\phi = \mathcal{A}Q + 2CQ^{[\mu\nu]}{}_\nu\mathring{\nabla}_\mu\phi, \tag{4.213}$$

and so the scalar field equation undergoes the trivial change to become

$$-(2\mathcal{A}\mathcal{B}+3\mathcal{C}^2)\mathring{\nabla}_\mu\mathring{\nabla}^\mu\phi+(\mathcal{B}\mathcal{C}-3\mathcal{C}\mathcal{C}'-\mathcal{A}\mathcal{B}')\mathring{\nabla}_\mu\phi\mathring{\nabla}^\mu\phi$$
$$+(\mathcal{A}'+\mathcal{C})\left(\mathcal{A}Q+2\mathcal{C}Q^{[\mu\nu]}{}_\nu\mathring{\nabla}_\mu\phi\right)+2\kappa^2(\mathcal{A}\mathcal{V}'+2\mathcal{C}\mathcal{V})=\kappa^2\mathcal{C}\Theta_\mu{}^\mu\,. \tag{4.214}$$

This completes the field equations for the scalar-nonmetricity class of gravity theories.

We finally also take a brief look at the metric teleparallel case, and study the field equations of a class of scalar-torsion theories defined by the action [37–39]

$$S_g=\frac{1}{2\kappa^2}\int_M\Big[-\mathcal{A}(\phi)T-\mathcal{B}(\phi)g^{\mu\nu}\mathring{\nabla}_\mu\phi\mathring{\nabla}_\nu\phi$$
$$+2\mathcal{C}(\phi)T_\nu{}^{\nu\mu}\mathring{\nabla}_\mu\phi-2\kappa^2\mathcal{V}(\phi)\Big]\sqrt{-g}\mathrm{d}^4x\,, \tag{4.215}$$

which directly follows from the action (4.196) by imposing vanishing nonmetricity. Recall that under this condition the single field equation obtained by simultaneous variation of the metric and connection is given by (4.41). Using the variation (4.198), these field equations become

$$\mathcal{A}\mathring{R}_{\mu\nu}-\frac{\mathcal{A}}{2}\mathring{R}g_{\mu\nu}+\mathcal{C}\mathring{\nabla}_\mu\mathring{\nabla}_\nu\phi-(\mathcal{B}-\mathcal{C}')\mathring{\nabla}_\mu\phi\mathring{\nabla}_\nu\phi+(\mathcal{A}'+\mathcal{C})S_{\mu\nu}{}^\rho\mathring{\nabla}_\rho\phi$$
$$+\left[\left(\frac{\mathcal{B}}{2}-\mathcal{C}'\right)\mathring{\nabla}_\rho\phi\mathring{\nabla}^\rho\phi-\mathcal{C}\mathring{\nabla}_\rho\mathring{\nabla}^\rho\phi+\kappa^2\mathcal{V}\right]g_{\mu\nu}$$
$$=\kappa^2(\Theta_{\mu\nu}-\nabla_\rho H_{\mu\nu\rho}+H_{\mu\nu\rho}T^\tau{}_{\tau\rho})\,. \tag{4.216}$$

These equations are supplemented by the scalar field equation, which follows from the general teleparallel equation (4.205) by using

$$\mathcal{A}G+2\mathcal{C}M^{[\mu\nu]}{}_\nu\mathring{\nabla}_\mu\phi=\mathcal{A}T-2\mathcal{C}T_\nu{}^{\nu\mu}\mathring{\nabla}_\mu\phi\,, \tag{4.217}$$

in the absence of nonmetricity. Hence, the (debraided) scalar field equation takes the form

$$-(2\mathcal{A}\mathcal{B}+3\mathcal{C}^2)\mathring{\nabla}_\mu\mathring{\nabla}^\mu\phi+(\mathcal{B}\mathcal{C}-3\mathcal{C}\mathcal{C}'-\mathcal{A}\mathcal{B}')\mathring{\nabla}_\mu\phi\mathring{\nabla}^\mu\phi$$
$$+(\mathcal{A}'+\mathcal{C})\left(\mathcal{A}T-2\mathcal{C}T_\nu{}^{\nu\mu}\mathring{\nabla}_\mu\phi\right)+2\kappa^2(\mathcal{A}\mathcal{V}'+2\mathcal{C}\mathcal{V})=\kappa^2\mathcal{C}\Theta_\mu{}^\mu \tag{4.218}$$

in the metric teleparallel gravity setting.

4.4.5 Scalar-Teleparallel Representation of $f(G)$ Theories

Among the general classes of scalar-teleparallel theories of gravity discussed in the previous section there is a particular subclass of theories, defined by a suitable choice of the parameter functions $\mathcal{A}, \mathcal{B}, \mathcal{C}, \mathcal{V}$, whose field equations turn out to be equivalent to those of the $f(G)$ class of theories. Note that for a given function f, the choice of the parameter functions in the scalar-teleparallel representation is not unique, and different choices are connected by redefinitions of the scalar field. For the general teleparallel geometry, a straightforward procedure is to start from the action (4.135), and to rewrite it, similarly to the $f(\overset{\circ}{R})$ class of theories [40], in the form

$$S_{\mathrm{g}} = -\frac{1}{2\kappa^2}\int_M [f(\phi) - \psi(\phi - G)]\sqrt{-g}\mathrm{d}^4x\,, \tag{4.219}$$

thereby introducing two scalar fields ψ and ϕ. Here ψ is a Lagrange multiplier, and imposes the constraint

$$\phi = G \tag{4.220}$$

for the scalar field ϕ. Variation with respect to the latter yields another constraint

$$\psi = f'(\phi)\,, \tag{4.221}$$

which can then be used to solve for the scalar field ψ. The remaining field equations are the metric field equation

$$\psi\left(\overset{\circ}{R}_{\mu\nu} - \frac{1}{2}\overset{\circ}{R}g_{\mu\nu}\right) + \left(\overset{\circ}{\nabla}_{(\mu}\psi M^{\rho}{}_{\nu)\rho} - M^{\rho}{}_{(\mu\nu)}\overset{\circ}{\nabla}_{\rho}\psi + M^{[\rho\sigma]}{}_{\sigma}\overset{\circ}{\nabla}_{\rho}\psi g_{\mu\nu}\right) + \frac{1}{2}[f(\phi) - \phi\psi]g_{\mu\nu} = \kappa^2\Theta_{\mu\nu}\,, \tag{4.222}$$

as well as the connection field equation

$$M^{\nu\rho}{}_{\rho}\overset{\circ}{\nabla}_{\mu}\psi + M^{\rho}{}_{\mu\rho}\overset{\circ}{\nabla}^{\nu}\psi - (M^{\nu\rho}{}_{\mu} + M^{\rho}{}_{\mu}{}^{\nu})\overset{\circ}{\nabla}_{\rho}\psi = 2\kappa^2(\nabla_{\tau}H_{\mu}{}^{\nu\tau} - M^{\omega}{}_{\tau\omega}H_{\mu}{}^{\nu\tau})\,. \tag{4.223}$$

Together with the constraints (4.220) and (4.221), one finds that these reproduce the $f(G)$ field equations (4.142) and (4.145).

Instead of keeping two scalar fields, one can take one further step and substitute the constraint (4.221) in the action (4.219), which then becomes

$$S_{\mathrm{g}} = -\frac{1}{2\kappa^2}\int_M [f(\phi) - f'(\phi)(\phi - G)]\sqrt{-g}\mathrm{d}^4x\,. \tag{4.224}$$

Note that this does not change the metric and connection field equations. Variation with respect to the scalar field now yields the field equation

$$(G - \phi)f'' = 0\,, \tag{4.225}$$

which resembles the constraint (4.220) for $f'' \neq 0$. By comparison with the general scalar-teleparallel action (4.196), one reads off the relations

$$\mathcal{A}(\phi) = f'(\phi)\,, \quad \mathcal{B}(\phi) = 0\,, \quad C(\phi) = 0\,, \quad \mathcal{V}(\phi) = \frac{f(\phi) - \phi f'(\phi)}{2\kappa^2}\,. \tag{4.226}$$

Alternatively, if the constraint (4.221) is invertible, one may also solve it for ϕ instead, which yields a different parametrization. The resulting action then takes the form

$$S_g = -\frac{1}{2\kappa^2}\int_M [\psi G - 2\kappa^2\mathcal{U}(\psi)]\sqrt{-g}\mathrm{d}^4x\,, \tag{4.227}$$

where $\mathcal{U}$ is implicitly defined by

$$\mathcal{U}(\psi) = \mathcal{V}(\phi)\,. \tag{4.228}$$

To obtain a more explicit relation, one may differentiate with respect to ϕ on both sides, which yields

$$f''(\phi)\mathcal{U}'(\psi) = -\frac{\phi f''(\phi)}{2\kappa^2}\,. \tag{4.229}$$

This shows that $f(\phi)$ and $\mathcal{U}(\psi)$ are related by a Legendre transformation. In this case the scalar field equation becomes

$$G = -2\kappa^2\mathcal{U}'(\psi)\,, \tag{4.230}$$

once again reproducing the constraint (4.220), up to a change of parametrization.

It is easy to check that for the values (4.226) of the parameter functions (and hence also for the equivalent parametrization via ψ) indeed yield a class of theories whose field equations reproduce those of the $f(G)$, $f(Q)$ [35] and $f(T)$ [41] classes of gravity theories, if suitable restrictions are imposed on the torsion or nonmetricity of the connection. Substituting the values (4.226) and the constraint (4.220) into the metric field equation (4.199) yields

$$f'\mathring{R}_{\mu\nu} - \frac{f'}{2}\mathring{R}g_{\mu\nu} + f''\left(\mathring{\nabla}_{(\mu}GM^{\rho}{}_{\nu)\rho} - M^{\rho}{}_{(\mu\nu)}\mathring{\nabla}_{\rho}G + M^{[\rho\sigma]}{}_{\sigma}\mathring{\nabla}_{\rho}Gg_{\mu\nu}\right) \\ + \frac{1}{2}(f - f'G)g_{\mu\nu} = \kappa^2\Theta_{\mu\nu}\,, \tag{4.231}$$

which, using $f''\mathring{\nabla}_{\mu}G = \mathring{\nabla}_{\mu}f'$, reproduces the field equation (4.142). The same relation is used to show that the connection field equation (4.201), which becomes

$$f''\left[M^{\nu\rho}{}_{\rho}\mathring{\nabla}_{\mu}G + M^{\rho}{}_{\mu\rho}\mathring{\nabla}^{\nu}G - (M^{\nu\rho}{}_{\mu} + M^{\rho}{}_{\mu}{}^{\nu})\mathring{\nabla}_{\rho}G\right] \\ = 2\kappa^2(\nabla_{\tau}H_{\mu}{}^{\nu\tau} - M^{\omega}{}_{\tau\omega}H_{\mu}{}^{\nu\tau})\,, \tag{4.232}$$

resembles the connection field equation (4.145). We then continue with the symmetric teleparallel case. Here, the connection equation (4.212) becomes

$$-\nabla_{\nu}\nabla_{\rho}(f'\tilde{P}^{\nu\rho}{}_{\mu}) = \kappa^2\nabla_{\nu}\nabla_{\rho}\tilde{H}_{\mu}{}^{\nu\rho}\,, \tag{4.233}$$

which is obviously identical to the corresponding equation (4.148). The metric field equation (4.208) takes the form

$$f'\mathring{R}_{\mu\nu} - \frac{f'}{2}\mathring{R}g_{\mu\nu} - f''P^{\rho}{}_{\mu\nu}\mathring{\nabla}_{\rho}Q + \frac{1}{2}(f - f'Q)g_{\mu\nu} = \kappa^2\Theta_{\mu\nu}\,. \tag{4.234}$$

To bring this to the familiar form, one raises one index, and uses the fact that the left hand side of the STEGR field equation satisfies

$$-\frac{\nabla_{\rho}(\sqrt{-g}P^{\rho\mu}{}_{\nu})}{\sqrt{-g}} - \frac{1}{2}P^{\mu\rho\sigma}Q_{\nu\rho\sigma} + \frac{1}{2}Q\delta^{\mu}_{\nu} = R^{\mu}{}_{\nu} - \frac{1}{2}R\delta^{\mu}_{\nu}\,. \tag{4.235}$$

This can be used to replace the Einstein tensor, so that the scalar-nonmetricity field equation becomes

$$-\frac{f'\nabla_{\rho}(\sqrt{-g}P^{\rho\mu}{}_{\nu})}{\sqrt{-g}} - \frac{f'}{2}P^{\mu\rho\sigma}Q_{\nu\rho\sigma} - P^{\rho\mu}{}_{\nu}\mathring{\nabla}_{\rho}f' + \frac{1}{2}f\delta^{\mu}_{\nu} = \kappa^2\Theta^{\mu}{}_{\nu}\,. \tag{4.236}$$

Observe that the two derivative terms can be combined into a single term, which yields the field equation (4.154). A similar procedure can be applied to the scalar-torsion case, whose field equations (4.216) now read

$$f'\overset{\circ}{R}_{\mu\nu} - \frac{f'}{2}\overset{\circ}{R}g_{\mu\nu} + f''S_{\mu\nu}{}^{\rho}\overset{\circ}{\nabla}_{\rho}T + \frac{1}{2}(f - f'T)g_{\mu\nu}$$
$$= \kappa^2(\Theta_{\mu\nu} - \nabla_{\rho}H_{\mu\nu\rho} + H_{\mu\nu\rho}T^{\tau}{}_{\tau\rho})\,. \tag{4.237}$$

Here one uses the left hand side of the MTEGR field equation, which can be written as

$$\overset{\circ}{\nabla}_{\rho}(S_{\mu\nu}{}^{\rho}) + S^{\rho\sigma}{}_{\nu}K_{\rho\sigma\mu} + \frac{1}{2}Tg_{\mu\nu} = \overset{\circ}{R}_{\mu\nu} - \frac{1}{2}\overset{\circ}{R}g_{\mu\nu}\,. \tag{4.238}$$

Using this relation to replace the Einstein tensor, one finds

$$f'\overset{\circ}{\nabla}_{\rho}(S_{\mu\nu}{}^{\rho}) + f'S^{\rho\sigma}{}_{\nu}K_{\rho\sigma\mu} + S_{\mu\nu}{}^{\rho}\overset{\circ}{\nabla}_{\rho}f' + \frac{1}{2}fg_{\mu\nu}$$
$$= \kappa^2(\Theta_{\mu\nu} - \nabla_{\rho}H_{\mu\nu\rho} + H_{\mu\nu\rho}T^{\tau}{}_{\tau\rho})\,. \tag{4.239}$$

Once again the two derivative terms can be combined, and one has the field equation (4.160).

4.5 Summary, Outlook and Open Questions

In this chapter we have given an introduction to teleparallel gravity theories, their underlying geometric structure which defines the fundamental fields of the theory, the general form of the field equations and the actions and field equations of a few selected classes of teleparallel gravity theories. We have seen that the main difference between teleparallel and curvature-based gravity theories such as general relativity is the existence of an independent, curvature-free connection, which appears as one of the fundamental fields mediating the gravitational interaction. From a mathematical point of view, this additional connection, together with the metric tensor, forms the foundation of teleparallel geometry. From a phenomenological point of view, the teleparallel connection is simply another field which enters the field equations and must be taken into account when these equations and their solutions are studied, e.g., for exact solutions exhibiting spherical or cosmological symmetry, or when performing perturbation theory. The coupling of this new field to other fields such as the metric and matter fields is determined by the particular teleparallel gravity action under consideration. Since these can be vastly different across the whole class of teleparallel gravity theories, they also lead to a plethora of potential new phenomenology.

Teleparallel gravity theories are an active field of research and many questions are yet unanswered at the time of writing of this chapter. One of the most prominent open questions is known as the "strong coupling problem" [42, 43]. It refers to the fact that both the Hamiltonian analysis and higher order perturbation theory predict the presence of additional degrees of freedom compared to general relativity in several classes of teleparallel gravity, which are not manifest as propagating modes in the linear perturbation theory. Such modes are called strongly coupled, and their presence hints towards possible instabilities and a lack of predictability, which potentially renders the perturbation theory around such background solutions invalid. Among the most common approaches to clarify the nature and severity of these issues is the Hamiltonian analysis and the study of constraints.

Besides fundamental questions, also the phenomenology of teleparallel gravity theories leaves numerous possibilities for further studies, which can potentially lead to new experimental tests. Active fields at the time of writing this chapter include the study of cosmology using the method of dynamical systems, cosmological perturbations, black holes and other exotic compact objects, as well as their shadows and their perturbations, which are closely related to the emission of gravitational waves. Hence, it is reasonable to expect numerous future developments in this field.

Acknowledgments The author thanks Claus Lämmerzahl and Christian Pfeifer for the kind invitation to contribute this book chapter. He acknowledges the full financial support of the Estonian Ministry for Education and Science through the Personal Research Funding Grant PRG356, as well as the European Regional Development Fund through the Center of Excellence TK133 "The Dark Side of the Universe".

References

1. E.N. Saridakis et al., *Modified Gravity and Cosmology: An Update by the CANTATA Network* (Springer, Cham, 2021). arxiv.org/abs/2105.12582, https://doi.org/10.1007/978-3-030-83715-0
2. A. Einstein, Riemann-Geometrie mit Aufrechterhaltung des Begriffes des Fernparallelismus. Sitzber. Preuss. Akad. Wiss., 217–221 (1928). http://echo.mpiwg-berlin.mpg.de/MPIWG:YP5DFQU1, https://doi.org/10.1002/3527608958.ch36
3. R. Aldrovandi, J. Geraldo Pereira, *Teleparallel Gravity*, vol. 173 (Springer, Dordrecht, 2013). https://doi.org/10.1007/978-94-007-5143-9
4. S. Bahamonde, K.F. Dialektopoulos, C. Escamilla-Rivera, G. Farrugia, V. Gakis, M. Hendry, M. Hohmann, J. Levi-Said, J. Mifsud, E. Di Valentino, Teleparallel gravity: from theory to cosmology. Rept. Prog. Phys. **86**(2), 026901 (2023). https://doi.org/10.1088/1361-6633/ac9cef
5. J.M. Nester, H.-J. Yo, Symmetric teleparallel general relativity. Chin. J. Phys. **37**, 113 (1999). http://www.ps-taiwan.org/cjp/download.php?type=paper&vol=37&num=2&page=113, arXiv:gr-qc/9809049
6. J. Beltrán Jiménez, L. Heisenberg, D. Iosifidis, A. Jiménez-Cano, T.S. Koivisto, General teleparallel quadratic gravity. Phys. Lett. B **805**, 135422 (2020). arXiv:1909.09045, https://doi.org/10.1016/j.physletb.2020.135422
7. F.W. Hehl, J.D. McCrea, E.W. Mielke, Y. Ne'eman, Metric affine gauge theory of gravity: field equations, Noether identities, world spinors, and breaking of dilation invariance. Phys. Rep. **258**, 1–171 (1995). arXiv:gr-qc/9402012, https://doi.org/10.1016/0370-1573(94)00111-F
8. M. Blagojevic, *Gravitation and Gauge Symmetries* (Institute of Physics, Bristol, 2002)
9. F.W. Hehl, P. Von Der Heyde, G.D. Kerlick, J.M. Nester, General relativity with spin and torsion: foundations and prospects. Rev. Mod. Phys. **48**, 393–416 (1976). https://doi.org/10.1103/RevModPhys.48.393

10. M. Blagojević, F.W. Hehl (eds.), *Gauge Theories of Gravitation: A Reader with Commentaries* (World Scientific, Singapore, 2013). https://doi.org/10.1142/p781
11. M. Hohmann, Variational principles in teleparallel gravity theories. Universe **7**(5), 114 (2021). arXiv:2104.00536, https://doi.org/10.3390/universe7050114
12. M. Hohmann, Spacetime and observer space symmetries in the language of Cartan geometry. J. Math. Phys. **57**(8), 082502 (2016). arXiv:1505.07809, https://doi.org/10.1063/1.4961152
13. M. Hohmann, L. Järv, M. Krššák, C. Pfeifer, Modified teleparallel theories of gravity in symmetric spacetimes. Phys. Rev. D **100**(8), 084002 (2019). arXiv:1901.05472, https://doi.org/10.1103/PhysRevD.100.084002
14. F. D'Ambrosio, L. Heisenberg, S. Kuhn, Revisiting cosmologies in teleparallelism. Class. Quant. Grav. **39**(2), 025013 (2022). arXiv:2109.04209, https://doi.org/10.1088/1361-6382/ac3f99
15. M. Hohmann, General covariant symmetric teleparallel cosmology. Phys. Rev. D **104**(12), 124077 (2021). arXiv:2109.01525, https://doi.org/10.1103/PhysRevD.104.124077
16. M. Hohmann, Complete classification of cosmological teleparallel geometries. Int. J. Geom. Meth. Mod. Phys. **18**(supp01), 2140005 (2021). arXiv:2008.12186, https://doi.org/10.1142/S0219887821400053
17. M. Bruni, S. Matarrese, S. Mollerach, S. Sonego, Perturbations of space-time: Gauge transformations and gauge invariance at second order and beyond. Class. Quant. Grav. **14**, 2585–2606 (1997). arXiv:gr-qc/9609040, https://doi.org/10.1088/0264-9381/14/9/014
18. S. Sonego, M. Bruni, Gauge dependence in the theory of nonlinear space-time perturbations. Commun. Math. Phys. **193**, 209–218 (1998). arXiv:gr-qc/9708068, https://doi.org/10.1007/s002200050325
19. M. Bruni, S. Sonego, Observables and gauge invariance in the theory of nonlinear space-time perturbations: Letter to the editor. Class. Quant. Grav. **16**, L29–L36 (1999). arXiv:gr-qc/9906017, https://doi.org/10.1088/0264-9381/16/7/101
20. M. Hohmann., C. Pfeifer, J. Levi Said, U. Ualikhanova, Propagation of gravitational waves in symmetric teleparallel gravity theories. Phys. Rev. D **99**(2), 024009 (2019). arXiv:1808.02894, https://doi.org/10.1103/PhysRevD.99.024009
21. K. Flathmann, M. Hohmann, Post-Newtonian limit of generalized symmetric teleparallel gravity. Phys. Rev. D **103**(4), 044030 (2021). arXiv:2012.12875, https://doi.org/10.1103/PhysRevD.103.044030
22. M. Hohmann, Gauge-invariant post-newtonian perturbations in symmetric teleparallel gravity. Astron. Rep. **65**(10), 952–956 (2021). arXiv:2111.06255, https://doi.org/10.1134/S1063772921100140
23. M. Hohmann, M. Krššák, C. Pfeifer, U. Ualikhanova, Propagation of gravitational waves in teleparallel gravity theories. Phys. Rev. D **98**(12), 124004 (2018). arXiv:1807.04580, https://doi.org/10.1103/PhysRevD.98.124004
24. U. Ualikhanova, M. Hohmann, Parametrized post-Newtonian limit of general teleparallel gravity theories. Phys. Rev. D **100**(10), 104011 (2019). arXiv:1907.08178, https://doi.org/10.1103/PhysRevD.100.104011
25. A. Golovnev, T. Koivisto, Cosmological perturbations in modified teleparallel gravity models. J. Cosmol. Astropart. Phys. **11**, 012 (2018). arXiv:1808.05565, https://doi.org/10.1088/1475-7516/2018/11/012
26. S. Bahamonde, K.F. Dialektopoulos, M. Hohmann, J. Levi Said, C. Pfeifer, E.N. Saridakis, Perturbations in non-flat cosmology for $f(T)$ gravity. Eur. Phys. J. C **83**(3), 193 (2023) https://doi.org/10.1140/epjc/s10052-023-11322-3
27. A. Baldazzi, O. Melichev, R. Percacci, Metric-affine gravity as an effective field theory. Ann. Phys. **438**, 168757 (2022). arXiv:2112.10193, https://doi.org/10.1016/j.aop.2022.168757
28. J.W. Maluf, The teleparallel equivalent of general relativity. Ann. Phys. **525**, 339–357 (2013). arXiv:1303.3897, https://doi.org/10.1002/andp.201200272
29. C.G. Boehmer, E. Jensko, Modified gravity: a unified approach. Phys. Rev. D **104**(2), 024010 (2021). arXiv:2103.15906, https://doi.org/10.1103/PhysRevD.104.024010

30. J. Beltrán Jiménez, L. Heisenberg, T. Koivisto, Coincident general relativity. Phys. Rev. D **98**(4), 044048 (2018). arXiv:1710.03116, https://doi.org/10.1103/PhysRevD.98.044048
31. G.R. Bengochea, R. Ferraro, Dark torsion as the cosmic speed-up. Phys. Rev. D **79**, 124019 (2009). arXiv:0812.1205, https://doi.org/10.1103/PhysRevD.79.124019
32. E.V. Linder, Einstein's other gravity and the acceleration of the Universe. Phys. Rev. D **81**, 127301 (2010). [Erratum: Phys. Rev. D 82, 109902 (2010)]. arXiv:1005.3039, https://doi.org/10.1103/PhysRevD.81.127301
33. M. Krššák, E.N. Saridakis, The covariant formulation of f(T) gravity. Class. Quant. Grav. **33**(11), 115009 (2016). arXiv:1510.08432, https://doi.org/10.1088/0264-9381/33/11/115009
34. K. Hayashi, T. Shirafuji, New general relativity. Phys. Rev. D **19**, 3524–3553 (1979). [Addendum: Phys. Rev. D 24, 3312–3314 (1982)]. https://doi.org/10.1103/PhysRevD.19.3524
35. L. Järv, M. Rünkla, M. Saal, O. Vilson, Nonmetricity formulation of general relativity and its scalar-tensor extension. Phys. Rev. D **97**(12), 124025 (2018). arXiv:1802.00492, https://doi.org/10.1103/PhysRevD.97.124025
36. M. Rünkla, O. Vilson, Family of scalar-nonmetricity theories of gravity. Phys. Rev. D **98**(8), 084034 (2018). arXiv:1805.12197, https://doi.org/10.1103/PhysRevD.98.084034
37. C.-Q. Geng, C.-C. Lee, E.N. Saridakis, Y.-P. Wu, "Teleparallel" dark energy. Phys. Lett. B **704**, 384–387 (2011). arXiv:1109.1092, https://doi.org/10.1016/j.physletb.2011.09.082
38. M. Hohmann., L. Järv, U. Ualikhanova, Covariant formulation of scalar-torsion gravity. Phys. Rev. D **97**(10), 104011 (2018). arXiv:1801.05786, https://doi.org/10.1103/PhysRevD.97.104011
39. M. Hohmann, Scalar-torsion theories of gravity III: analogue of scalar-tensor gravity and conformal invariants. Phys. Rev. D **98**(6), 064004 (2018). arXiv:1801.06531, https://doi.org/10.1103/PhysRevD.98.064004
40. T.P. Sotiriou, V. Faraoni, f(R) Theories of gravity. Rev. Mod. Phys. **82**, 451–497 (2010). arXiv: 0805.1726, https://doi.org/10.1103/RevModPhys.82.451
41. R.-J. Yang, Conformal transformation in $f(T)$ theories. Europhys. Lett. **93**(6), 60001 (2011). arXiv:1010.1376, https://doi.org/10.1209/0295-5075/93/60001
42. J. Beltrán Jiménez, K.F. Dialektopoulos, Non-linear obstructions for consistent new general relativity. J. Cosmol. Astropart. Phys. **01**, 018 (2020). arXiv:1907.10038, https://doi.org/10.1088/1475-7516/2020/01/018
43. A. Golovnev, M.-J. Guzmán, Foundational issues in f(T) gravity theory. Int. J. Geom. Methods Mod. Phys. **18**(supp01), 2140007 (2021). arXiv:2012.14408, https://doi.org/10.1142/S0219887821400077

Gravitational Lensing in Theories with Lorentz Invariance Violation

5

Jean-François Glicenstein and Volker Perlick

Abstract

Theories with Lorentz Invariance Violation are motivated e.g. by ideas from quantum gravity. In such theories the light rays are no longer given as the lightlike geodesics of a Lorentzian metric, therefore gravitational lensing can be used for confronting such theories with observations. In the first part of this chapter we discuss lensing in the DSR model, in the second part we summarize what is known about light rays in Finsler spacetimes.

5.1 Motivation

Several phenomenological approaches to quantum gravity, semi-classical (low-energy) limits of fundamental quantum gravity theories, modified gravity theories, and the standard model extension, predict deformations or violations of local Lorentz invariance. In this book we encountered already two examples of this kind, one emerging from string theory, see Chap. 1, and one from the doubly special relativity framework, see Chap. 2. In this chapter we will discuss how deformations and violations of Lorentz invariance would manifest themselves in gravitational lensing images. We approach this topic from two sides: In the first part we use the Hamiltonian approach, which is particularly convenient for determining lensing features from modified dispersion relations directly. In the second part,

J.-F. Glicenstein
IRFU, CEA Paris-Saclay, Paris, France
e-mail: glicens@cea.fr

V. Perlick (✉)
ZARM, University of Bremen, Bremen, Germany
e-mail: perlick@zarm.uni-bremen.de

C. Pfeifer, C. Lämmerzahl (eds.), *Modified and Quantum Gravity*, Lecture Notes in Physics 1017, https://doi.org/10.1007/978-3-031-31520-6_5

where we concentrate on Finsler modifications of General Relativity, we start from the Lagrangian approach which is dual to the Hamiltonian approach and sometimes advantageous, e.g. for describing light rays by a variational principle. Both approaches are equally appropriate for the formulation of gravitational theories with violation of local Lorentz invariance.

More explicitly, several quantum gravity theories predict the appearance of energy-momentum dependent terms in space-time metrics. The presence of these terms can be experimentally tested with photon propagation. In particular, this will lead to modifications of the Schwarzschild metric, used for relativity tests in the solar system and modifications of extensions of this metric which are commonly used to describe the gravitational lensing of light from distant galaxies by foreground galaxies. For this purpose it is convenient to use a Hamiltonian description of the motion of test particles (photons in our case) in the background cosmic object. Among other things, this is advantageous since Hamiltonians give an immediate access to dispersion relations, which are modified by quantum gravity effects. If the Hamiltonian is homogeneous with respect to the momenta, this induces a Finsler geometric structure [39].

Section 5.2 gives a brief introduction to gravitational lensing and some definitions. Section 5.3 discusses lensing in the DSR model introduced in [4]. A similar formalism can be applied to lensing in rainbow gravity models. We will show that some current constraints on that model can be improved by studying high-energy lensing by distant objects. Section 5.4 is devoted to light propagation in Finsler spacetimes. Whereas up to now a full-fledged theory of gravitational lensing in Finsler spacetimes does not exist, several mathematical results with relevance to lensing have been found. These results are summarized in this section.

5.2 Gravitational Lensing

Lensing is the scattering of photons from distant sources by a mass distribution. Light propagation in general relativistic spacetimes is discussed by Perlick [36]. A standard reference on gravitational lensing is the book by Schneider, Ehlers and Falco [44]. A generic property of gravitational lenses is the existence of multiple propagation paths (multiple images) as is the case for mirages. Gravitational lensing is widely used to constrain mass distributions in the universe. Notable examples are the distribution of compact dark matter such as MACHOs or primordial black holes, and the large scale dark matter distribution. Gravitation models such as MOND or modified gravity can be constrained by studying the lensing of distant objects.

Different lensing regimes are defined depending on the value of the lens mass M or the Schwarzschild radius which is defined as $r_S = 2\frac{GM}{c^2} = 2.95\frac{M}{M_\odot}\text{km}$. An order of magnitude of the deflection angle is $\alpha \sim \frac{r_S}{s}$ where s is the typical size of the lens (assuming it is not a black hole). The typical delay between images induced by lenses is $\Delta T \sim c r_S$. Notable lensing regimes are

1. *femtolensing,* which is the lensing of light (wavelength λ) by asteroid-like masses with $\lambda > r_S$. The various propagation paths interfere and produce spectral oscillations.
2. *microlensing,* which is the lensing of light by stellar type objects. Deflection angles and time delays are too small to be observable by current instruments for most lenses. The main observable in this regime is the light magnification as a function of time.
3. *macro-(or strong) lensing,* which occurs in the lensing of background galaxies by intervening galaxies. Galaxies have masses in the $10^{11} - 10^{12} M_\odot$ range and sizes in the 1–10 kpc range. The typical angular size of a macrolens is

$$\theta_L \simeq 6'' \left(\frac{M_L}{10^{12} M_\odot}\right)^{1/2} \left(\frac{1\text{Gpc}}{D_{OL}}\right).$$

   Time delays between images are of the order of 1 month. Images location and time delays are thus observable. In principle, the magnification of images is also observable, however it is affected by microlensing by stars in the lens galaxy.
4. *weak lensing,* which is the lensing by galaxy clusters or large scale structures. The shape of background galaxies is distorted, leading to measurements of shear and magnification.

In the rest of this chapter, we will be exclusively concerned with macrolensing.

5.2.1 Lens Equation

The lensing geometry is shown in Fig. 5.1.

In the "thin lens approximation", the photon moves on a straight line until it gets deflected towards the observer. The angular distance to the lens, located at redshift z_L is D_{OL}, the angular distance from lens to observer is D_{LS}. Coordinates are taken in the plane perpendicular to the line of sight (the "lens plane"). The projected source position on the lens plane is at η, and the impact parameter of the particle trajectory at ζ.

Lens models predict the dependence of the deflection angle $\delta\theta$ on the geometry and model parameters. The simplest example is the relation

$$\delta\theta = \frac{2r_S}{\eta}. \tag{5.1}$$

for Schwarzschild lenses of General Relativity. The relation between the distances and $\delta\theta$ (called the *lens equation*) is shown by a simple heuristic argument to be:

$$\theta_O + \theta_S = \frac{D_{LS} + D_{OL}}{D_{LS} D_{OL}} (\zeta - \eta) = \delta\theta \tag{5.2}$$

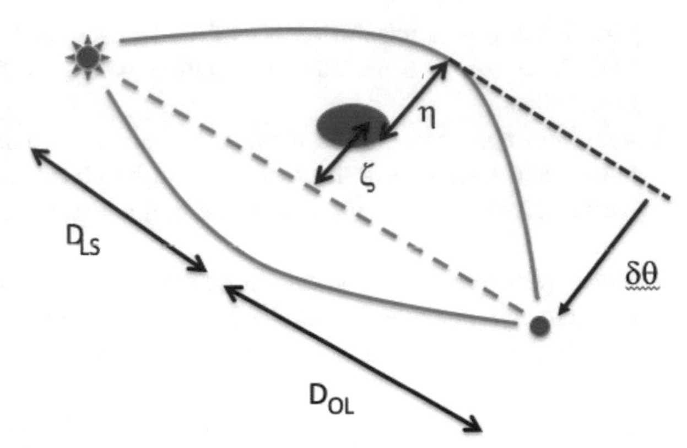

Fig. 5.1 Lensing geometry. Here D_{LS} is the angular distance between the lens and the source, D_{OL} is the angular distance between the lense and the observer, η is the position of the source projected to the lens plane, ζ is the impact parameter of the light ray, and $\delta\Theta$ is the deflection angle, that is predicted from the propagation of light rays through spacetime

5.3 Lensing in DSR

In this section we demonstrate explicitly how lensing images change, compared to the GR, when photons are subject to modified dispersion relations. Lensing is studied in the Hamiltonian formalism. The starting point is the most general Lorentz Invariance breaking, space rotation invariant Hamiltonian in DSR, given in Sect. 5.3.1. The simplest model, Schwarzschild lenses is studied in Sect. 5.3.2. Section 5.3.3 describes a more realistic galactic mass model: the Single Isothermal Lens (SIL) model. Lensing is best studied in isotropic coordinates instead of the more usual Schwarzschild coordinates. Sections 5.3.4, 5.3.5 and 5.3.6 discuss the transformation to isotropic coordinates. Galaxy lensing is then studied in Sect. 5.3.7. The main result of Sect. 5.3.7 is Eq. (5.85) which gives the lensing delay as a function of photon energy. Known high-energy lensing systems are presented in Sects. 5.3.8, 5.3.9, and 5.3.10. Finally, Eq. (5.85) is used in Sect. 5.3.10 to constrain the LIV energy scale.

5.3.1 Hamiltonian of a Single Particle in the Field of a Massive Object

The most general Lorentz Invariance breaking, space rotation invariant Hamiltonian in DSR is [4]: .

$$H = -\frac{4}{\epsilon l_P^2} \sinh\left(\frac{\epsilon l_P}{2}(cp_t + dp_r)\right)^2$$

$$+ e^{\epsilon l_P(cp_t+dp_r)}\left[\left(-h + c^2\right)p_t^2 + 2cdp_rp_t + \left(g + d^2\right)p_r^2 + \frac{1}{r^2}w^2\right]. \qquad (5.3)$$

where $w^2 = p_\theta^2 + \frac{p_\phi^2}{\sin^2\theta}$, p_t, p_r, p_θ, p_ϕ are the usual momenta conjugate to t, r, θ, ϕ and $\epsilon = \pm 1$ depending on whether the particle motion is super- or supraluminic. The Hamiltonian (5.3) depends on the unknown length scale of LIV, which is denoted by l_P, and assumed to be of the order of the inverse of the Planck mass. In principle, the functions c, d can be chosen arbitrarily, as long as they obey the normalization condition

$$\frac{c(t,r)^2}{g(t,r)} + \frac{d(t,r)^2}{h(t(r)} = -1 \tag{5.4}$$

where g and h are the tt- and rr-component of a spherically symmetric spacetime metric.

Gravitational lensing, our main application, assumes that sources and observers are located in "flat" space, at large distances from the lens, so that

$$\lim_{r\to\infty} g(r) = \lim_{r\to\infty} h(r) = 1 \tag{5.5}$$

When $r \to \infty$, the Hamiltonian (5.3) should tend to the κ-Poincaré Hamiltonian

$$H_\kappa(x,p) = -\frac{4}{{l_P}^2}\sinh\left(\frac{l_P}{2}p_t\right)^2 + e^{l_P p_t}\mathbf{p}^2\,, \tag{5.6}$$

which corresponds to the $c = 1, d = 0$ limit of Hamiltonian (5.3).

? Exercise

5.1. Investigate whether one can find a coordinate transformation such that the Hamiltonian (5.3) tends to (5.6) in the large distance limit.

Solution Using condition (5.5) in the large distance limit, one has

$$-(c_\infty)^2 + (d_\infty)^2 = -1$$

where

$$\lim_{r\to\infty} c = c_\infty$$
$$\lim_{r\to\infty} d = d_\infty$$

Using the point transformation

$$t' = c_\infty t - d_\infty r$$
$$r' = -d_\infty t + c_\infty r,$$

the canonical transformation from the old to the new coordinates is generated by:

$$F_2 = (c_\infty t - d_\infty r)(p_t)' + (-d_\infty t + c_\infty r)(p_r)' \tag{5.7}$$

$(p_t)'$, $((p_r)'$ are the new momenta. The relation between the old and new momenta is

$$(p_t)' = c_\infty p_t + d_\infty p_r \tag{5.8}$$

$$(p_r)' = c_\infty p_r + d_\infty p_t \tag{5.9}$$

In the large r limit, Hamiltonion (5.3) is thus transformed into

$$H_\kappa(x', p') = -\frac{4}{l_P{}^2} \sinh\left(\frac{l_P}{2}(p_t)'\right)^2 + e^{l_P (p_t)'}((p_r)'^2 + w^2/r^2), \tag{5.10}$$

which is identical to Hamiltonian (5.6) only if $r = r'$. This happens when $c \simeq c_\infty = \pm 1$ and $d \simeq d_\infty = 0$. From now on, $c = 1$ and $d = 0$ are assumed.

In the small l_P limit ($l_P p_t \ll 1$), the Hamiltonian (5.3) simplifies further to

$$H = -h(r)p_t^2 + (1 + \epsilon l_P p_t \sqrt{h(r)})\left[g(r)p_r^2 + \frac{1}{r^2}w^2\right]. \tag{5.11}$$

5.3.2 Deformed Schwarzschild Hamiltonian

Stellar objects, in particular the Sun, can be modelled as Schwarzschild lenses. The g and h functions of the Schwarzschild Hamiltonian are given by $f(r) = g(r) = 1 - \frac{r}{r_S}$ and $h(r) = 1/f(r)$. $r_S = \frac{2GM_L}{c^2}$ is the Schwarzschild radius and M_L is the mass of the lensing object. The deflection of light from distant stars by the Sun and the time delay of a signal sent from Earth and reflected off a planet are well known solar system tests on General Relativity.

5.3.2.1 Hamilton's Equations

The deflection angle and time-delay are usually directly derived from the Schwarzschild metric (see e.g [46]). Here we follow an alternative Hamiltonian approach. Only the main steps are sketched, the reader should refer to [19] for details.

? Exercise

5.2. Show from Hamilton's equations that p_t and p_ϕ are left invariant in the motion. If in addition $p_\phi = 0$, then p_θ is also invariant.

Introducing the affine parameter λ, the other Hamilton's equations are derived from Eq. (5.11):

$$\frac{dt}{d\lambda} = \frac{\partial H}{\partial p_t} = -2\frac{p_t}{f(r)} + \left(\frac{\epsilon l_P}{\sqrt{f(r)}}\right)\left[f(r)p_r^2 + \frac{1}{r^2}{p_\theta}^2\right]$$

$$\frac{dr}{d\lambda} = \frac{\partial H}{\partial p_r} = 2f(r)p_r\left(1 + \epsilon l_P\,\frac{p_t}{\sqrt{f(r)}}\right)$$

$$\frac{d\theta}{d\lambda} = \frac{\partial H}{\partial p_\theta} = \frac{2}{r^2}p_\theta\left(1 + \epsilon l_P\,\frac{p_t}{\sqrt{f(r)}}\right)$$

This is completed by the requirement that $H = 0$, since we are interested in photon trajectories. This allows eliminating p_r by

$$f(r)p_r = \sqrt{\frac{p_t^2}{\left(1 + \epsilon l_P\,\frac{p_t}{\sqrt{f(r)}}\right)} - \frac{f(r)}{r^2}{p_\theta}^2}. \tag{5.12}$$

Next, λ is then eliminated, followed by a change of variable from r to $u = 1/r$. Finally, one obtains the equations

$$\frac{dt}{du} \simeq \frac{p_t(1 - (\frac{3\epsilon l_P p_t}{2\sqrt{f(u)}}))}{u^2 f(u)^2 p_r} \tag{5.13}$$

$$\frac{d\theta}{du} = -\frac{p_\theta}{\sqrt{\frac{p_t^2}{(1+\epsilon l_P(\frac{p_t}{\sqrt{f(u)}}))} - f(u)u^2{p_\theta}^2}} \tag{5.14}$$

Equations (5.14) and (5.13) are the basis for the calculation of the deflection angle and the time-delay for DSR deformed Schwarzschild lenses. But before moving to the deflection angle calculation, we need to properly define the *impact parameter* which is the quantity labelled as η in Eq. (5.1).

5.3.2.2 Impact Parameter

Let $u_0 = \frac{1}{R_0}$ be a solution of the equation:

$$\frac{p_t^2}{(1 + \epsilon l_P(\frac{p_t}{\sqrt{f(u)}}))} - f(u)u^2{p_\theta}^2 = 0 \tag{5.15}$$

Following [46], chapter 6.3, the impact parameter β is defined as

$$\frac{1}{\beta^2} = \frac{p_t^2}{p_\theta^2(1 + \epsilon l_P p_t)} \tag{5.16}$$

β does not depend on r_S and is kept fixed in the calculation.

? Exercise

5.3. Show that the impact parameter β can be expressed as a function of u_0, namely that, to first order in the small parameter $u_0 r_S$, the relation between β and u_0 is

$$\frac{1}{\beta} = u_0 + \frac{r_S u_0^2 (1/2 \epsilon l_P p_t - 1)}{2} \tag{5.17}$$

Compare with equation 3.3.6 from [46].
Hint Introduce β in Eq. (5.15) to obtain

$$\frac{1}{\beta^2}(1 - 1/2 \epsilon l_P p_t r_S u_0) - u_0^2 + r_S u_0^3 = 0. \tag{5.18}$$

To the order zero in $r_S u_0$, the relation between u_0 and β is $\frac{1}{\beta} = u_0$. Write $\frac{1}{\beta} = u_0(1 + r_S u_0 x)$ with $x \ll 1$ and expand Eq. (5.18) to first order in x and $r_S u_0$.

5.3.2.3 Deflection Angle

Introducing u_0, and using Eq. (5.18), the square root term from Eqs. (5.12) and (5.14) can be expressed as

$$\sqrt{\frac{p_t^2}{(1 + \epsilon l_P (\frac{p_t}{\sqrt{f(u)}}))} - f(u) u^2 p_\theta{}^2}$$

$$= p_\theta \sqrt{u_0^2 - u^2 + r_S(u^3 - u_0^3) + \frac{\epsilon l_P p_t r_S u_0^2}{2}(u_0 - u)} \tag{5.19}$$

Expanding the right side of (5.14) and keeping only first order terms in $r_S u_0$ gives

$$\frac{d\theta}{du} = -\frac{1}{\sqrt{u_0^2 - u^2 + r_S(u^3 - u_0^3) + \frac{\epsilon l_P p_t r_S u_0^2}{2}(u_0 - u)}} \tag{5.20}$$

$$= -\left(\frac{1}{\sqrt{u_0^2 - u^2}} - \frac{r_S}{2}\left(\frac{(u^3 - u_0^3)}{(u_0^2 - u^2)^{3/2}} + \frac{\epsilon l_P p_t}{2} \frac{u_0^2(u_0 - u)}{(u_0^2 - u^2)^{3/2}}\right)\right) \tag{5.21}$$

? Exercise

5.4. Integrate u between 0 and $\pi/2$ and show that:

$$\int_0^{\pi/2} \frac{d\theta}{du} du = -\frac{\pi}{2} + r_S u_0 (1 - \frac{\epsilon l_P p_t}{4}) \tag{5.22}$$

Hint Change variable $u = u_0 \cos(\phi)$ to obtain elementary integrals. The value of the primitives at $\phi = 0$ can be obtained from l'Hospital rule.

In formula (5.22), u_0 can be replaced by $\frac{1}{\beta}$ since we are expanding θ to first order in the parameter $r_S u_0$.

The total deflection angle takes also into account the angular change between the lens and the observer. This amounts to multiply by a factor of 2 the deflection. Finally, the deflection angle is

$$\delta\theta = \frac{2r_S}{\beta} \left(1 - \frac{\epsilon l_P p_t}{4}\right) \tag{5.23}$$

The deflection angle depends linearly on the energy p_t of the photon and is larger than the Schwarzschild angle in the supraluminous case ($\epsilon = -1$)

5.3.2.4 Propagation Time

Equation (5.13) is the starting point to obtain the propagation time on individual photons paths. Introducing β from Eq. (5.16), $\frac{dt}{du}$ can be separated into 2 terms:

$$\frac{dt}{du} \simeq \left(\frac{(1 + \frac{\epsilon l_P p_t}{2})(1 - (\frac{3\epsilon l_P p_t}{2\sqrt{f(u)}}))}{\beta f(u)}\right) \left(\frac{p_\theta}{u^2 f(u) p_r}\right). \tag{5.24}$$

Recalling that $f(u) = 1 - r_S u$, the left parenthesis can be expressed as

$$\frac{1}{\beta} \frac{\left(1 - \frac{3\epsilon l_P p_t}{2\sqrt{f(u)}}\right)(1 + \epsilon l_P p_t 2)}{f} = \frac{1}{\beta}\left((1 - \epsilon l_P p_t) + r_S u (1 - \frac{7}{4} \epsilon l_P p_t)\right) \tag{5.25}$$

and using

$$f(u) p_r = p_\theta \sqrt{u_0^2 - u^2 + r_S(u^3 - u_0^3) + \frac{\epsilon l_P p_t r_S u_0^2}{2}(u_0 - u)} \tag{5.26}$$

Eq. (5.24) transforms into

$$\frac{dt}{du} \simeq \frac{(1-\epsilon l_P p_t)+r_S u(1-\frac{7}{4}\epsilon l_P p_t)}{\beta} \frac{1}{u^2\sqrt{u_0^2-u^2+r_S(u^3-u_0^3)+\frac{\epsilon l_P p_t r_S u_0^2}{2}(u_0-u)}}$$

$$\simeq \frac{(1-\epsilon l_P p_t)+r_S u(1-\frac{7}{4}\epsilon l_P p_t)}{\beta}\left(\frac{1}{u^2\sqrt{u_0^2-u^2}} - \frac{r_S(u^3-u_0^3)}{2u^2(u_0^2-u^2)^{3/2}} - \frac{r_S\epsilon l_P p_t u_0^2(u_0-u)}{4u^2(u_0^2-u^2)^{3/2}}\right)$$

? Exercise

5.5. Keeping the first order term in $r_S u_0$, show that

$$\frac{dt}{du} = J_1 + J_2 + J_3 \tag{5.27}$$

with

$$J_1 = \frac{(1-\epsilon l_P p_t)}{\beta}\left(\frac{1}{u^2\sqrt{u_0^2-u^2}}\right), \tag{5.28}$$

$$J_2 = \frac{r_S}{\beta}\left(\frac{1}{u\sqrt{u_0^2-u^2}} - \frac{1}{2}\frac{(u^3-u_0^3)}{u^2(u_0^2-u^2)^{3/2}}\right) \tag{5.29}$$

and

$$J_3 = -\frac{r_S\epsilon l_P p_t}{\beta}\left(\frac{7}{4u\sqrt{u_0^2-u^2}} + \frac{1}{2}\frac{(u_0^3-u^3)}{u^2(u_0^2-u^2)^{3/2}} + \frac{1}{4}\frac{{u_0}^2(u_0-u)}{u^2(u_0^2-u^2)^{3/2}}\right) \tag{5.30}$$

The impact parameter β has, from Eq. (5.17), a $r_S u_0$ dependence which needs to be taken into account in the propagation time calculation.

? Exercise

5.6. Using Eq. (5.17), transform Eq. (5.27) into

$$\frac{dt}{du} = J_1' + J_2' + J_3' \tag{5.31}$$

with

$$J_1' = (1 - \epsilon l_P p_t)\frac{u_0}{u^2\sqrt{u_0^2 - u^2}}, \tag{5.32}$$

$$J_2' = r_S u_0 \left(\frac{1}{u\sqrt{u_0^2 - u^2}} - \frac{1}{2}\frac{(u^3 - u_0^3)}{u^2(u_0^2 - u^2)^{3/2}} - \frac{1}{2}\frac{u_0}{u^2\sqrt{u_0^2 - u^2}}\right) \tag{5.33}$$

and

$$J_3' = \tag{5.34}$$

$$-r_S u_0 \epsilon l_P p_t \left(\frac{7}{4u\sqrt{u_0^2 - u^2}} + \frac{1}{2}\frac{(u_0^3 - u^3)}{u^2(u_0^2 - u^2)^{3/2}} + \frac{1}{4}\frac{{u_0}^2(u_0 - u)}{u^2(u_0^2 - u^2)^{3/2}} - \frac{3}{4}\frac{u_0}{u^2\sqrt{u_0^2 - u^2}}\right)$$

Equation (5.31) can now be integrated. Defining the angle ϕ by $u = u_0 \cos\phi$ as in Sect. (5.3.2.3), the integrals $\int J_1' du$, $\int J_2' du$, $\int J_3' du$, can be expressed as a function of ϕ.

? Exercise

5.7. Show that

$$\int J_3' du = -r_S \epsilon l_P p_t \left(\frac{3}{4}\frac{1 - \cos\phi}{\sin\phi} + \frac{3}{2}\ln\frac{1 + \sin\phi}{\cos\phi}\right) \tag{5.35}$$

Summing the three contributions, the propagation time is

$$\int \frac{dt}{du} du = (1 - \epsilon l_P p_t)\frac{1}{u_0}(\tan\phi)$$
$$+ r_S\left(1 - \frac{3}{2}\epsilon l_P p_t\right)\left(\ln\frac{1 + \sin\phi}{\cos\phi} + \frac{1}{2}\frac{1 - \cos\phi}{\sin\phi}\right) \tag{5.36}$$

To obtain the propagation time of the light emitted by a source at S to an observer at O (see Fig. 5.1), expression (5.36) is first applied to the LS part of the photon trajectory, then to the OL part. On the LS part $\phi = \arccos R_0/D_{LS}$, and Eq. (5.36) gives:

$$T_{LS} = (1 - \epsilon l_P p_t)\sqrt{D_{LS}^2 - R_0^2} \tag{5.37}$$

$$+ r_S(1 - \frac{3}{2}\epsilon l_P p_t)\left(\ln \frac{D_{LS} + \sqrt{D_{LS}^2 - R_0^2}}{R_0} + \frac{1}{2}\sqrt{\frac{D_{LS} - R_0}{D_{LS} + R_0}}\right)$$

A similar expression holds for the OL part of the photon trajectory. The total time delay measured with a satellite in the solar system is obtained by multiplying by a factor of 2 to account for the return trip of the photon.

$$\Delta T = 2(T_{LS} + T_{OL}) = 2\Big((1 - \epsilon l_P p_t)(\sqrt{D_{LS}^2 - R_0^2} + \sqrt{D_{OL}^2 - R_0^2})$$

$$+ r_S(1 - \frac{3}{2}\epsilon l_P p_t)\left(\ln \frac{D_{LS} + \sqrt{D_{LS}^2 - R_0^2}}{R_0} + \ln \frac{D_{OL} + \sqrt{D_{OL}^2 - R_0^2}}{R_0}\right.$$

$$\left. + \frac{1}{2}\left(\sqrt{\frac{D_{LS} - R_0}{D_{LS} + R_0}} + \sqrt{\frac{D_{OL} - R_0}{D_{OL} + R_0}}\right)\right)\Big). \tag{5.38}$$

5.3.3 More Realistic Lenses

Most cosmic lenses, especially macro-lenses, are not well described by the Schwarzschild model. A simple and popular model used by astronomers is the singular isothermal lens (SIL) model. In the SIL model, the intervening galaxy has a mass distribution with density:

$$\rho(r) = \frac{\rho_0 a^2}{r^2} \tag{5.39}$$

where a is a characteristic length. The Gauss theorem gives the gravitational field $g(r)$ as a function of density ρ_0 and scale a :

$$4\pi r^2 g(r) = -4\pi G M(r) = -4\pi G(4\pi \rho_0 a^2 r) \tag{5.40}$$

Hence:

$$g(r) = -4\pi G \rho_0 \frac{a^2}{r} \tag{5.41}$$

The velocity dispersion at radius a is $v_c^2 = 2\sigma_v^2 = \frac{GM(a)}{a} = 4\pi G\rho_0 a^2$ (v_c is the circular velocity, it has been assumed that the relation between the circular velocity and the velocity dispersion is $v_c^2 = 2\sigma_v^2$), so that finally

$$g(r) = -\frac{2\sigma_v^2}{r} \tag{5.42}$$

The gravitational potential of the SIL lens is thus

$$U_{\rm SIL}(r) = 2\sigma_v^2 \ln r \tag{5.43}$$

Except for the Schwarzschild lens, lensing models are generally given in isotropic coordinates (for more details and the motivation for this choice, see section 4.2 of [44]). Working with isotropic coordinates instead of a Schwarzschild-like metric greatly simplifies lensing calculations. In the next section, we will discuss how moving to isotropic coordinates affects the non Lorentz invariant part of the Hamiltonian (5.11).

5.3.4 Isotropic Coordinates

To move to isotropic coordinates and keep equations of motions invariant, one needs to make a canonical transformation.

The relevant coordinate transformation is

$$t' = t \tag{5.44}$$

$$r' = r'(r) = f_r(r) \tag{5.45}$$

$$\theta' = \theta \tag{5.46}$$

$$\phi' = \phi \tag{5.47}$$

and the associated canonical transformation is

$$F_2 = tp_t' + f_r(r)p_r' + \theta p_{\theta'} + \phi p_{\phi'} \tag{5.48}$$

The initial momenta p_t, p_r are related to the transformed momenta p_t', p_r' by

$$p_t = p_t' \tag{5.49}$$

$$p_r = \frac{\partial F_2}{\partial r} = \frac{\partial (r' p_r')}{\partial r} = p_r' \frac{\partial r'}{\partial r} \tag{5.50}$$

Before moving to Lorentz non-conserving Hamiltonians, it is worth discussing the transformation with a Lorentz-conserving Schwarzschild lens.

5.3.4.1 Lorentz Conserving Schwarzschild Lenses

Starting from

$$H = \frac{1}{2}\left(\frac{p_t^2}{f(r)} - f(r)p_r^2 - \frac{1}{r^2}(p_\theta^2 + \frac{1}{\sin\theta^2}p_\phi^2)\right) \tag{5.51}$$

we want to find the coordinate transformation (5.45) which transforms the spatial part of the momenta into an isotropic form. That is H is changed to

$$H' = \frac{1}{2}\left(\frac{p_t^2}{h(r')} - g(r')\left(p_{r'}^2 - \frac{1}{r'^2}(p_\theta^2 + \frac{1}{\sin\theta^2}p_\phi^2)\right)\right) \tag{5.52}$$

where g and h are functions to be determined.

Identifying the corresponding terms:

$$g(r')p_{r'}^2 = f(r)p_r^2 = f(r)p_{r'}^2\left(\frac{dr'}{dr}\right)^2 \tag{5.53}$$

$$\frac{g(r')}{r'^2} = \frac{1}{r^2} \tag{5.54}$$

and eliminating $g(r')$, one gets

$$\left(\frac{dr'}{r'}\right)^2 = \left(\frac{dr}{r}\right)^2\frac{1}{f(r)} \tag{5.55}$$

Using $f(r) = (1 - \frac{r_S}{r})$ for a Schwarzschild lens, one has

$$\frac{dr'}{r'} = \frac{dr}{r\sqrt{(1 - \frac{r_S}{r})}} = \frac{dy}{\sqrt{(y^2 - y)}} \tag{5.56}$$

with $y = r/r_S$.

Writing $y^2 - y = (y - 1/2)^2 - 1/4$ and setting $y - 1/2 = 1/2\cosh t$, one has $dy = 1/2\sinh t dt$ and $y^2 - y = 1/4\sinh^2 t$. Then

$$\ln(r'/r_0) = t = \cosh^{-1}(2y - 1) = \cosh^{-1}\left(\frac{2r}{r_S} - 1\right) \tag{5.57}$$

Using $\cosh^{-1}(x) = \ln(x + \sqrt{x^2 - 1})$, Eq. (5.57) becomes

$$\frac{r'}{r_0} = \frac{2r}{r_S} - 1 + \frac{2}{r_S}\sqrt{r^2 - rr_S} \tag{5.58}$$

The value of r_0 is found by imposing that $\lim_{r\to\infty} r' = r$. One finds $r_0 = r_S/4$ and

$$2r' = r - \frac{r_S}{2} + \sqrt{r^2 - rr_S}\,. \tag{5.59}$$

One has further the identity

$$(r - \frac{r_S}{2} + \sqrt{r^2 - rr_S})(r - \frac{r_S}{2} - \sqrt{r^2 - rr_S}) = \frac{r_S^2}{4} \tag{5.60}$$

Dividing Eq. (5.60) by Eq. (5.59), one obtains

$$r - \frac{r_S}{2} - \sqrt{r^2 - rr_S} = \frac{r_S^2}{8r'}\,, \tag{5.61}$$

so that

$$r = \frac{r_S}{2} + r' + \frac{r_S^2}{16r'} = r'\left(1 + \frac{r_S}{4r'}\right)^2\,, \tag{5.62}$$

(for a derivation directly from the Schwarzschild metric, see [42], chap. 14).

The h and g functions are

$$g(r') = \frac{1}{(1 + \frac{r_S}{4r'})^4} = 1 - \frac{r_S}{r'} + o(\frac{r_S}{r'}) \tag{5.63}$$

$$h(r') = 1 - \frac{r_S}{r'(1 + \frac{r_S}{4r'})^4} = 1 - \frac{r_S}{r'} + o(\frac{r_S}{r'}) \tag{5.64}$$

g and h turn out to be equal to f to first order in r_S. This is not true in general as we now see.

5.3.5 Other Lorentz-Invariant Lenses

The general case is

$$g(r') = (1 - V(r')) \tag{5.65}$$

where V is related to the gravitational potential $U(r)$ through $V = \frac{2U}{c^2}$. Then from Eq. (5.54),

$$r = \frac{r'}{\sqrt{1 - V(r')}} \tag{5.66}$$

Using now Eq. (5.54), $\frac{dr}{dr'} = \frac{1-V(r')+r'V'(r')/2}{(\sqrt{1-V(r')})^3}$ and

$$f(r) = g(r')(\frac{dr}{dr'})^2 = \frac{(1 - V(r') + r'V'(r')/2)^2}{(1 - V(r'))^2} \tag{5.67}$$

To gain some intuition on the solutions of these equation, let's take an example. If $V(r) = C/r^\beta$, then $1 - V(r) + r\frac{dV}{dr} = 1 - (1 + \beta/2)V(r)$ and

$$f(r) = (1 - \beta V(r')) = (1 - \beta V(r) + O(V^2)) \tag{5.68}$$

This example can be readily generalized. If $V(r) = \sum_i C_i/r^{\beta_i}$, then $V'(r) = -\sum_i C_i\beta_i/r^{\beta_i+1}$ and

$$f(r) = (1 - \sum_i C_i\beta_i/r^{\beta_i} + O(V^2)) \tag{5.69}$$

It is clear from this example that in the general case moving from Schwarzschild to isotropic coordinates changes the expression of the f (or V) function.

5.3.6 Lorentz Non-invariant Lenses

The derivation of the non-Lorentz invariant Hamiltonian in isotropic form is now left as exercise for the reader.

? Exercise

5.8. In the case of a Schwarzschild lens, show that the canonical transformation from Eq. (5.48) changes Hamiltonian (5.11) to an isotropic form.

Solution Using the canonical transformation to isotropic coordinates (Eq. (5.48)) changes Hamiltonian (5.11) to

$$H = -\frac{p_t^2}{f(r)} + \left(1 + \epsilon l_P \frac{p_t}{\sqrt{f(r)}}\right) f(r)\mathbf{P}^2. \tag{5.70}$$

where $\mathbf{P}^2 = p_r^2 + \frac{1}{r^2}p_\theta{}^2$.

As noted before, lensing metrics (or equivalently Hamiltonians) are generally given in isotropic coordinates (see [44]), with functions $g(r) = 1 - \frac{2U(r)}{c^2}$ and $h(r) = \frac{1}{1+\frac{2U(r)}{c^2}} \simeq g$. Based on the Schwarzschild lens example, we now assume that Lorentz non-conserving Hamiltonians in isotropic coordinates can be described by Hamitonian (5.70) with $f(r) = 1 - \frac{2U(r)}{c^2}$.

5.3.7 Galaxy Lensing

This section applies the formalism described in [19] on Hamiltonian (5.70). The aim is to describe the modification of galaxy lensing due to Lorentz Invariance violation in the photon propagation. Hence the function $U(r)$ in $f(r) = 1 - \frac{2U(r)}{c^2}$ is the gravitational potential of a galaxy. Hamilton's equation are obtained by derivating w.r.t the affine parameter λ giving

$$\begin{aligned}
\frac{dp_t}{d\lambda} &= -\frac{\partial H}{\partial t} = 0 \\
\frac{dt}{d\lambda} &= \frac{\partial H}{\partial p_t} = -2\frac{p_t}{f(r)} + (\frac{\epsilon l_P}{\sqrt{f(r)}}))\mathbf{P}^2 \\
\frac{dx^\alpha}{d\lambda} &= \frac{\partial H}{\partial p_\alpha} = 2f(r)p^\alpha(1 + \epsilon l_P(\frac{p_t}{\sqrt{f(r)}})) \\
\frac{dp_\alpha}{d\lambda} &= -\frac{\partial H}{\partial x^\alpha} = -\partial_\alpha f(r)\left(\frac{p_t^2}{f(r)^2} + \mathbf{P}^2(1 + \epsilon l_P(\frac{p_t}{2\sqrt{f(r)}}))\right)
\end{aligned}$$

The mass constraint relevant to photon motion is $H = 0$. It gives a relation between $\mid \mathbf{P} \mid$, $f(r)$ and p_t

$$\frac{p_t}{\mid \mathbf{P} \mid} = f\sqrt{1 + \frac{\epsilon l_P p_t}{\sqrt{f}}} \simeq f\left(1 + \frac{\epsilon l_P p_t}{2\sqrt{f}}\right) \tag{5.71}$$

Hamilton's equation can be transformed using the euclidian line element $dl^2 = \sum_1^3 dx_\alpha{}^2$ to give the evolution of t and p_α with l.

The relation between l and λ is

$$\frac{dl}{d\lambda} = -2f(r) \mid \mathbf{P} \mid \left(1 + \epsilon l_P \frac{p_t}{\sqrt{f(r)}}\right) \tag{5.72}$$

The next step is to eliminate λ.

$$\frac{dp_\alpha}{dl} = \partial_\alpha f(r) \mid \mathbf{P} \mid \frac{\left(2 + \epsilon l_P(\frac{3p_t}{2\sqrt{f(r)}})\right)}{2f(r)\left(1 + \epsilon l_P \frac{p_t}{\sqrt{f(r)}}\right)} \tag{5.73}$$

$$\frac{dt}{dl} = \frac{p_t}{f^2(r) \mid \mathbf{P} \mid}\left(1 - \frac{\epsilon l_P p_t}{\sqrt{f}}\right) - \frac{\epsilon l_P \mid \mathbf{P} \mid}{2\sqrt{f(r)}}\left(1 - \frac{\epsilon l_P p_t}{\sqrt{f}}\right) \tag{5.74}$$

In Eq. (5.73) $\partial_\alpha f(r) = 2\partial_\alpha U(r)/c^2$ is a first order quantity in U/c^2. To first order in U/c^2, the evolution of momenta is thus given by

$$\frac{dp_\alpha}{dl} = (2/c^2)\partial_\alpha U(r) \mid \mathbf{P} \mid \left(1 - \epsilon l_P \frac{p_t}{4}\right) \tag{5.75}$$

The right side of Eq. (5.74) is the sum of two terms:

$$\begin{aligned} I_1 &= \frac{p_t}{f^2(r) \mid \mathbf{P} \mid}(1 - \frac{\epsilon l_P p_t}{\sqrt{f}}) \\ &= (1 - \frac{\epsilon l_P p_t}{2}) - 2U/c^2(1 - \frac{3}{4}\epsilon l_P p_t) + O(U^2/c^4) \end{aligned}$$

and

$$\begin{aligned} I_2 &= -(\frac{\epsilon l_P \mid \mathbf{P} \mid}{2f(r)^{1/2}})(1 - \frac{\epsilon l_P p_t}{\sqrt{f}}) = -(\frac{\epsilon l_P p_t}{2f(r)^{3/2}}) + O({l_P}^2 {p_t}^2) \\ &= -(\frac{\epsilon l_P p_t}{2})(1 - 3U/c^2)\,. \end{aligned}$$

Summing the 2 contributions gives

$$\frac{dt}{dl} = (I_1 + I_2) = (1 - \epsilon l_P p_t) - 2U/c^2(1 - \frac{3}{2}\epsilon l_P p_t)\,. \tag{5.76}$$

The travel time T_{OS} is obtained by integrating $\frac{dt}{dl}$ over the photon trajectory from source to observer. This is done by approximating the photon trajectory (Fig. 5.1) as two straight lines, one from the source to the lens and the other from the lens to the observer, in accordance with the thin lens approximation. The result is

$$T_{OS} = \int dl \left((1 - \epsilon l_P p_t) - 2U/c^2(1 - \frac{3}{2}\epsilon l_P p_t)\right). \tag{5.77}$$

The structure of the T_{OS} integral ((5.77)) reminds of the *Fermat potential* encountered in usual photon lensing [44]. Using the potential $U(r)$ of the SIL model (Eq. (5.43)) in Eq. (5.77) gives:

$$\begin{aligned} T_{OS} = (z_L + 1)\Big(&\frac{1}{2}(1 - \epsilon l_P p_t)(\frac{1}{D_{OL}} + \frac{1}{D_{LS}}) \\ &(\zeta - \eta)^2 - 2\pi\sigma_v^2(1 - \frac{3}{2}\epsilon l_P p_t)|\zeta| + T_0\Big) \end{aligned} \tag{5.78}$$

The $(z_L + 1)$ factor has to be included in order to compare Eq. (5.78) to delays of cosmological lenses. See [44], section 4.6, for a derivation.

The absolute value of the deflection angle is obtained from Eq. (5.75) (see [18] for details).

$$\alpha = 2\pi\sigma_v^2 \left(1 - \epsilon l_P \frac{p_t}{4}\right) \mathrm{sgn}(\zeta). \tag{5.79}$$

The lens equation for the SIL model is:

$$\zeta - \eta = \mathrm{sgn}(\zeta) l_E, \tag{5.80}$$

with the Einstein length l_E defined by

$$l_E = \frac{4\pi\sigma_v^2 \left(1 - \epsilon l_P (\frac{p_t}{4})\right) D_{OL} D_{LS}}{(D_{OL} + D_{LS})} \tag{5.81}$$

Equation (5.80) has 2 solutions for $|\eta| \leq l_E$. These solutions are

$$\zeta_+ = l_E + \eta \tag{5.82}$$

$$\zeta_- = \eta - l_E \tag{5.83}$$

Since $(\zeta_+ - \eta)^2 = (\zeta_- - \eta)^2 = l_E^2$, only the second term in Eq. (5.78) contributes to the time delay between images. This term depends on

$$|\zeta_+| - |\zeta_-| = 2\eta. \tag{5.84}$$

Using Eq. (5.78), the time delay between the $\zeta_\pm$ images is

$$\Delta T = -4(z_L + 1)\pi\sigma_v^2 (1 - \frac{3}{2}\epsilon l_P p_t)\eta = \Delta T(p_t = 0)(1 - \frac{3}{2}\epsilon l_P p_t)\,. \tag{5.85}$$

A similar expression for the time delay between images was derived by a completely different method in reference [11]. In contrast, the expression for the Einstein length (Eq. (5.81)) differs from that obtained in [11].

Other gravity models predict a similar variation of the delay between lens images with energy. Since ΔT scales linearly with σ_v^2, hence with G, an equation similar to Eq. (5.85) would be obtained in gravity models with an energy-dependent G. In rainbow gravity models [29], the components of the Schwarzchild metric become energy-dependent:

$$d\tau^2 = \frac{1 - \frac{2GM}{r}}{f(E)^2} dt^2 - \frac{1}{g(E)^2} \left(\frac{dr^2}{1 - \frac{2GM}{r}} + r^2 d\Omega^2 \right).$$

The photon velocity is $v = \frac{g(E)}{f(E)}$, so that the time delay scales as the ratio $k(E) = \frac{f(E)}{g(E)}$. The time delay in rainbow gravity is thus

$$\Delta T = \Delta T(p_t = 0)\frac{G(p_t)}{G(p_t = 0)} = \Delta T(p_t = 0) \cdot k(p_t). \tag{5.86}$$

Following reference [10], the $k(E)$ function is parametrized as

$$k(E) = \frac{1}{\sqrt{1 - \alpha(\frac{E}{M_P})^s}} = 1 + \frac{\alpha}{2}\left(\frac{E}{M_P}\right)^s, \tag{5.87}$$

where $M_P = 1.22\ 10^{19}$ GeV is the Planck mass.

5.3.8 High Energy Lenses

As mentioned in Sect. 5.2, time-delays and image locations are routinely measured for strong lensing systems. Data are taken in passbands ranging from radio to high or very-high energies. To constrain quantum gravity models, the most straightforward idea is to exploit the potential change of image location with energy [15]. However present day high-energy instruments have angular resolutions of the order of arc-minutes, while the typical separation of lens images is one or two order of magnitude smaller, as explained in Sect. 5.2.

On the other hand, arrival delays between images of strong lensing systems are easily measurable, but with only a limited precision of the order of a few percent. Delay measurements at different energies are needed to exploit Eq. (5.85). While many strong lensing systems have delay measurements at several wavelength, only 2 strong lensing systems are known to have high-energy emission, namely blazars PKS1830-211 and JVAS B0218+357.

5.3.9 PKS 1830-211

PKS 1830-211 is a bright quasar at redshift $z_S = 2.5$, lensed by galaxy at $z_L = 0.89$. Two very clear compact images (A and B) separated by an Einstein ring are seen on radio images. The time-delay between the compact images has been measured in radio and microwaves. A third faint image was found recently using the ALMA array of radiotelescopes [34]. The lens structure has been elucidated and the delays between images have been predicted to be $T_{AB} = 26 - 29$ days and T_{AC} somewhat larger [34]. A time delay $T_{AB} = 26^{4}_{-5}$ days has been measured in the radio passband at 8.6 GHz [28] and a time delay $T_{AB} = 24^{5}_{-4}$ days in the millimeter wavelengths [47]. Evidence for $T_{AB} = 27.1 \pm 0.5$ day has been found in the high-energy passband

Table 5.1 Measurements of the time-delay of JVAS B0218. Data are reprinted with permission

Passband	Energy (eV)	Time-delay (days)	Data from reference
Radio (8.4 GHz)	$3.5\ 10^{-5}$	11.25 ± 0.55	[12] ©2018 the author(s).
Radio (15 GHz)	$6.2\ 10^{-5}$	11.3 ± 0.4	[12] ©2018 the author(s).
High energy	$10^8 - 3\ 10^{11}$	11.46 ± 0.16	[13] ©2014 AAS. All rights reserved.
Very high energy	$5\ e^{10} - 3e^{11}$	11.9 ± 0.4	Present work[a]

[a] See main text

[5], but was not confirmed by later studies [2, 6]. The situation concerning blazar PKS 1830-211 is thus confuse.

5.3.10 JVAS B0218+357

JVAS B0218+357 is a bright quasar located at redshift $z_S = 0.944$ and lensed by galaxy at $z_L = 0.685$. Radio images show 2 compact images and an Einstein ring. The time delay between the compact images is well measured in several radio pass-bands (8, 15, 5 GHz) and high energy gamma-rays (Fermi-LAT). The photon energies in these measurements span over 15 orders of magnitude. The measured values of time delays are listed in Table 5.1.

The MAGIC collaboration observed JVAS B0218+357 in response to a Fermi-LAT alert. They found a delayed flare with a slightly different shape. The lensing time-delay in the 100-GeV passband can be estimated by comparing the time of maximum of the MAGIC flare to the time of maximum of the Fermi-LAT flare. The value is shown in Table 5.1.

? Exercise

5.9. Constrain the linear and quadratic dependence of the lensing time-delay with the values of Table 5.1.

Solution The linear fit gives a slope value of $\alpha = 5\ 10^{-3} \pm 10^{-3}$ day/GeV. The slope is significant, but the measurements have large systematic errors. At the 90% confidence level (C.L.), one has

$$\alpha < 7\ 10^{-3}\text{day/GeV}. \tag{5.88}$$

The quadratic fit gives a slope value of $\alpha' = 5.2\ 10^{-5} \pm 1.210^{-5}$ day/GeV^2. At the 90%CL, one has $\alpha' < 7.510^{-5}$ day/GeV^2.

JVAS B0218 is well modelled by a SIL model [7]. Using Eq. (5.85), the limit on the linear slope (5.88) translates into a limit on l_P, or alternatively on the Planck scale. The limit obtained from Eq. (5.88) for a linear dependence and $\epsilon = -1$ is

$$\frac{1}{l_P} > 2.4\ \mathrm{TeV}(90\%\mathrm{C.L.}) \tag{5.89}$$

Since the value of α' is significantly positive, the data exclude completely a linear dependence for $\epsilon = 1$. However, it may be more appropriate to refer to the *expected limit* which is the limit based on the errors. The expected limit is

$$\frac{1}{l_P} > 17\ \mathrm{TeV}(90\%\mathrm{C.L.}) \tag{5.90}$$

for both the super- and supra-luminic case. The same limit can be translated into a limit on rainbow models parameters by using parametrization (5.87). Using $s = 1$, one obtains $\alpha < 5 \cdot 10^{16}$ at the 90% C.L. The quadratic case is also interesting for rainbow models. Using $s = 2$, one obtains $\alpha < 2 \cdot 10^{34}$ at the 90% C.L.

5.3.11 Comparison to Other Astrophysical Limits

The limits obtained with JVAS B0218+357 are smaller than the limits obtained by the direct observations of bursts from active galactic nuclei (see for instance [1]) or gamma-ray bursts by 11 to 16 orders of magnitude. In the studies of bursts from active galactic nuclei, the travel time of light is very large, typically 10^{17} light-seconds. Even when the uncertainty on the emission time of various wavelengths is large (of the order of 1 day for instance), the limit on $\frac{1}{l_p}$ is of the order of 10^{16} GeV for a TeV observation. In the study of lensing systems, the emission time is well defined and the theory can be well motivated, as in the case of DSR. However lensing time-delays are at most of the order of 10^8 light-seconds, Even with a very good measurement of the delay of the order of 0.1 day, the limit on $\frac{1}{l_p}$ can be at most of the order of 10^7 GeV.

Several authors (for instance [10, 14, 16]) have used solar system tests of General Relativity to put constraints on rainbow or similar models. The drawback of this approach is that only radio or visible light measurements can be used. The limit then relies on the accuracy of the deflection or time delay measurements. Taking for instance the accuracy of $2 \cdot 10^{-5}$ obtained on the delay measurement by the Cassini mission and a comparison of radio and visible light observations to separate the General Relativity and quantum gravity contribution, one would expect, from Eq. (5.38), limits on $\frac{1}{l_p}$ of the order of a few hundred keV at most.

Other authors (for instance [15, 20]) have advocated the using position measurements of lensed images to put constraints on quantum gravity models. However accurate image position measurements with photons of more than a few keV do not seem possible with the present technology. Reference [15] puts 95% C.L. limits of order 10 keV on $\frac{1}{l_p}$ with observations from the Chandra X-ray instrument. In contrast with accurate position measurements, precise time-delays between images have been measured up to the 100 GeV energy range. The $> 10^7$ fold increase in energy of the observed photons explains the factor of 10^8 improvement of limit (5.90) over previous results.

This concludes the discussion of the influence of non-Lorentz invariant dispersion relations in the Hamiltonian picture. It has been demonstrated how to determine features of lensing images from non-Lorentz invariant Hamilton functions explicitly.

Next, the influence of deviations from local Lorentz invariance on Lensing images is discussed, on the basis of Finsler geometry.

5.4 Towards a Theory of Gravitational Lensing in Finsler Spacetimes

In standard general relativity gravity is coded in a pseudo-Riemannian metric $g_{\mu\nu}(x)dx^\mu dx^\mu$ of Lorentzian signature. Correspondingly, the geodesics in this spacetime are the solutions of the Euler-Lagrange equations determined by a Lagrangian

$$L(x, v) = \frac{1}{2} g_{\mu\nu} v^\mu v^\nu \,. \tag{5.91}$$

This has the consequence that on each tangent space we have the symmetry of the ordinary Lorentz group of (special) relativity. Now Finsler geometry is based on the idea of replacing the Lagrangian (5.91) by a more general Lagrangian that is still homogenous with respect to the velocity coordinates $v = \left(v^0, v^1, v^2, v^3\right)$ but not necessarily a quadratic form. Obviously, in a Finsler spacetime there is no invariance with respect to the Lorentz group. As a matter of fact, the symmetry group of a Finsler spacetime can be very complicated, see Gallego and Piccione [17].

A full-fledged theory of gravitational lensing has not yet been developed for Finsler spacetimes. However, several mathematical results are known which are relevant in view of lensing. In the following we give an overview on these results.

5.4.1 Definition and Basic Properties of Finsler Spacetimes

The following is an appropriate working definition of a Finsler spacetime.

Definition 5.1. A Finsler spacetime is a 4-dimensional manifold M with a Lagrangian function L that satisfies the following properties:

(a) L is a real-valued function on the tangent bundle TM minus the zero section, i.e., $L(x, v)$ is defined for all $(x, v) \in TM$ with $v \neq 0$.
(b) L is positively homogeneous of degree two with respect to v, i.e.,

$$L(x, kv) = k^2 L(x, v) \quad \text{for all } k \in \mathbb{R} \text{ with } k > 0\,. \tag{5.92}$$

(c) The Finsler metric

$$g_{\mu\nu}(x, v) = \frac{1}{2} \frac{\partial^2 L(x, v)}{\partial v^\mu \partial v^\nu} \tag{5.93}$$

is well-defined and has Lorentzian signature $(-+++)$ for all (x, v) with $v \neq 0$.

Whenever working with a Finsler spacetime we use the summation convention for greek indices that take the values 0,1,2,3.

We call L the *Finsler Lagrangian* henceforth. A vector v at x is called *timelike* if $L(x, v) < 0$, *lightlike* if $L(x, v) = 0$ and *spacelike* if $L(x, v) > 0$.

It is a standard exercise to check that condition (b) of Definition 5.1 implies the identities

$$\frac{\partial L(x, v)}{\partial v^\mu} v^\mu = 2 L(x, v) \, , \tag{5.94}$$

$$\frac{\partial L(x, v)}{\partial v^\mu} = g_{\mu\nu}(x, v) \, v^\nu \, , \tag{5.95}$$

$$L(x, v) = \frac{1}{2} g_{\mu\nu}(x, v) \, v^\mu \, v^\nu \, . \tag{5.96}$$

The non-degeneracy of the Finsler metric implies that the Euler-Lagrange equations

$$\frac{d}{ds} \frac{\partial L(x, \dot{x})}{\partial \dot{x}^\mu} - \frac{\partial L(x, \dot{x})}{\partial x^\mu} = 0 \tag{5.97}$$

admit a unique solution $x(s)$ for each initial condition $(x(0), \dot{x}(0))$ with $\dot{x}(0) \neq 0$. These solution curves are called the (affinely parametrized) *geodesics* of the Finsler spacetime. Homogeneity of the Lagrangian implies that L is a constant of motion, so one can classify the geodesics as timelike if $L < 0$, lightlike if $L = 0$ and spacelike if $L > 0$. The timelike geodesics are to be interpreted as the worldlines of freely falling particles and the lightlike geodesics are to be interpreted as light rays. With this interpretation, which can be motivated by a modified version of the Ehlers-Pirani-Schild axiomatic approach to spacetime theory, see Bernal, Javaloyes and Sánchez [9], Finsler spacetimes provide a geometric framework for a theory of gravity that includes general relativity as a special case. Of course, for a complete theory one would also need a Finsler generalization of Einstein's field equation. Such generalizations have been suggested by various authors and are still a matter of debate, see in particular Rutz [43], Pfeifer and Wohlfarth [41] and Hohmann, Pfeifer and Voicu [23]. The latter paper clarifies the relation between the field equations suggested in the other two. In the following, we will restrict to kinematic considerations, i.e., we will not consider any field equation.

Finsler geometry was introduced originally by Paul Finsler in his 1918 doctoral dissertation. He considered Finsler metrics that are positive definite and for several decades the entire mathematical literature on Finsler geometry was restricted to this case. The definition of a Finsler structure with indefinite metric, in particular with Lorentzian signature, was first given by Beem [8] in 1970. If restricted to the physically interesting case of a Finsler structure with Lorentzian signature on a 4-dimensional manifold, our Definition 5.1 coincides with Beem's. However, over the last years it was found that for several applications it is actually desirable or even necessary to modify this definition by requiring that the Finsler metric is well-defined and of Lorentzian signature *almost everywhere*, rather than everywhere, on the tangent bundle minus the zero section. As outlined by Laemmerzahl et al. [27], this modification is necessary if one wants to include a certain class of static Finsler spacetimes where one encounters the situation that the Finsler metric fails to be well-defined on a set of measure zero. Another modification of Beem's definition was brought forward by Pfeifer and Wohlfarth [40] and used in several follow-up papers. They also allow for a violation of regularity on a set of measure zero and in addition they modify condition (b) by requiring the Lagrangian to be positively homogeneous of *some* degree which may be different from two.

We mention that there are alternative approaches to Finsler spacetimes. There is a number of papers where the Lagrangian is defined only on a conical subset of the tangent space at each point. This approach was pioneered by Asanov [3]. In the original work of Asanov, the Finsler Lagrangian is defined only for "admissible" vectors, i.e., on vectors inside a cone which is to be interpreted as the set of timelike vectors. This is clearly too weak for a theory of lensing because there is no satisfactory definition of lightlike vectors in this approach. We mention, however, that the Asanov definition has been modified by Javaloyes and Sanchez [24] to resolve this problem.

In a Finsler spacetime we can distinguish a particular parametrization for each timelike curve which generalizes the notion of proper time well known from general relativity. In the Finsler case, we say that a timelike curve $\gamma(\tau)$ is parametrized by proper time if

$$g_{\mu\nu}\big(\gamma(\tau), d\gamma(\tau)/d\tau\big)\frac{d\gamma^{\mu}(\tau)}{d\tau}\frac{d\gamma^{\nu}(\tau)}{d\tau} = -c^2 \tag{5.98}$$

where c is the vacuum speed of light. Clearly, if the Finsler metric is independent of v, this reduces to the standard definition of proper time in general relativity.

We have defined a Finsler spacetime here in terms of a Lagrangian. However, the non-degeneracy of the Finsler metric allows us to pass to a completely equivalent description in terms of a Hamiltonian, thereby meeting the formalism that was used in the first part of this paper. To that end we have to introduce the canonical momenta

$$p_{\mu} = g_{\mu\nu}(x, v)\, v^{\nu} \tag{5.99}$$

and to define the Hamiltonian

$$H(x,p) = v^\mu \frac{\partial L(x,v)}{\partial v^\mu} - L(x,v) = L(x,v)\,, \tag{5.100}$$

where (x, p) and (x, v) are related by (5.99). H is positively homogeneous of degree two, $H(x, kp) = k^2 H(x, p)$ for $k > 0$, and its Hessian $g^{\mu\nu}(x, p) = \partial^2 H(x, p)/(\partial p_\mu \partial p_\nu)$ is the inverse of $g_{\mu\nu}(x, v)$, thus non-degenerate with Lorentzian signature. The projections to M of the solutions of Hamilton's equations with $H = 0$ are precisely the affinely parametrized lightlike geodesics. So we may work in a Hamiltonian formalism on the cotangent bundle, rather than in a Lagrangian formalism on the tangent bundle, whenever we wish to do so.

The Hamiltonian formulation is particularly useful for treating symmetries. In general relativity, symmetries are described in terms of Killing vector fields. An appropriately generalized notion can be defined on Finsler spacetimes in the following way. We call a vector field $K^\mu(x)\partial_\mu$ a *Killing vector field* if $K^\mu(x)p_\mu$ has vanishing Poisson bracket with the Hamiltonian,

$$\{H(x,p), K^\mu(x)p_\mu\} = 0\,. \tag{5.101}$$

Here the Poisson bracket is defined in the familiar way for any two functions $f(x, p)$ and $h(x, p)$,

$$\{f(x,p), h(x,p)\} = \frac{\partial f(x,p)}{\partial x^\mu}\frac{\partial h(x,p)}{\partial p_\mu} - \frac{\partial h(x,p)}{\partial x^\mu}\frac{\partial f(x,p)}{\partial p_\mu}\,. \tag{5.102}$$

If written out, the Finslerian Killing Eq. (5.101) reads

$$K^\mu \frac{\partial g_{\rho\sigma}}{\partial x^\mu} + \frac{\partial K^\tau}{\partial x^\nu}\dot{x}^\nu \frac{\partial g_{\rho\sigma}}{\partial \dot{x}^\tau} + \frac{\partial K^\tau}{\partial x^\rho} g_{\tau\sigma} + \frac{\partial K^\tau}{\partial x^\sigma} g_{\rho\tau} = 0\,. \tag{5.103}$$

It was first given (for positive definite Finsler metrics) by Knebelman [25].

Clearly, (5.101) is equivalent to the condition that $K^\mu(x)p_\mu$ is constant along any geodesic. More generally, one calls $K^\mu(x)$ a *conformal* Killing vector field if

$$\{e^{\Omega(x,p)} H(x,p), K^\mu(x)p_\mu\} = 0 \tag{5.104}$$

with some function $\Omega(x, p)$. In this case $K^\mu(x)p_\mu$ is constant along every *lightlike* geodesic, although not in general along timelike or spacelike geodesics.

? Exercise

5.10. Prove the last statement. Given that (5.104) holds, show that $K^{\mu}(x)p_{\mu}$ is constant along lightlike geodesics of L.

5.4.2 Light Cone Structure and Fermat's Principle

In a Finsler spacetime, at each point x the set of all tangent vectors $v \neq 0$ with $L(x, v) < 0$ may have arbitrarily many connected components; correspondingly, there may be arbitrarily many "light cones". The light cones have the following properties.

Proposition 5.1. *Fix a point x in a Finsler spacetime (M, L). Let T_xM be the tangent space at x and $\overset{o}{T}_xM = \{(x, v) \in T_xM | v \neq 0\}$. Let Z_xM be a connected component of $\{(x, v) \in \overset{o}{T}_xM \big| L(x, v) < 0\}$ and let C_xM be the boundary of Z_xM in $\overset{o}{T}_xM$. Then the following is true.*

(a) *Z_xM is an open convex cone in T_xM.*
(b) *C_xM is a cone in T_xM and a closed C^{∞} submanifold of codimension one in $\overset{o}{T}_xM$.*
(c) *$g_{\mu\nu}(x, w)\, w^{\mu}\, u^{\nu} < 0$ for all $(x, w) \in C_xM$ and $(x, u) \in Z_xM$.*

For a proof we refer to Perlick [37]. Minguzzi [32] has analysed the light cone structure in Finsler spacetimes more deeply and he found that, under surprisingly mild conditions, there are exactly two light cones, one future and one past light-cone, at each point. Minguzzi's results also show that it is not a meaningful approach to model birefringence by a single Finsler structure. Birefringence is well known to occur in crystals and it is indeed possible to model the light rays in a uniaxial crystal as the lightlike geodesics of *two* Finsler structures, see Perlick [36], but not as the lightlike geodesics of a single Finsler structure with multiple light cones. Analogously, one would have to use two Finsler structures in a hypothetical Finsler gravity theory where birefringence occurs in vacuum.

For gravitational lensing it is often useful to characterize the light rays by a variational principle. For an arbitrary general-relativistic spacetime it is indeed possible to prove that light rays satisfy a version of Fermat's principle: If one fixes a point p (observation event) and a timelike curve γ (worldline of a light source), among all past-oriented lightlike curves from p to γ the lightlike geodesics make the arrival time extremal. Here the arrival time refers to the parametrization of γ

which may be chosen arbitrarily. For this general-relativistic Fermat princple we refer to Temple [45] for a version restricted to a local normal neighbourhood, to Kovner [26] for a formulation of the general principle and to Perlick [35] for a complete proof. The relevance of Fermat's principle in view of lensing is discussed e.g. in the book by Schneider, Ehlers and Falco [44]. Here we are interested in the question of whether a similar variational principle holds true for light rays in a Finsler spacetime.

To that end we fix, in a Finsler spacetime (M, L), a point p and a timelike curve γ. In applications to lensing the parametrization of γ, which may be arbitrary, is to be interpreted as past-oriented. At each point of γ, the tangent vector determines one connected component of the light cone. As the trial curves for our variational principle we choose all lightlike curves from p to γ whose tangent vector, on arrival at γ, belong to the connected component of the light cone that is selected by the tangent vector of γ. (Here part (c) of Proposition 5.1 is relevant.) On the set of trial curves we define the *arrival time functional* by assigning to each trial curve the parameter value of γ at the point of arrival. Then the following version of Fermat's principle is true:

Theorem 5.1. *A trial curve is a geodesic if and only if it is a stationary point of the arrival time functional.*

Loosely speaking, this means that among all ways to go from p to γ (backwards in time) at the speed of light, the actual light rays choose those paths that make the travel time extremal. For a precise mathematical formulation and a proof we refer to Perlick [37].

Obviously, the mathematical theorem is independent of whether the parametrization of γ is interpreted as past-ponting or as future-pointing. If one wants to formulate the variational principle in analogy to the classical Fermat principle, one would interpret the parametrization of γ as future-pointing. In applications to lensing, however, we are interested in light rays that arrive at the observation event p; that is why we prefer to think of the parametrization as past-pointing.

If the Finsler metric is independent of v, the theorem reduces to the one that was formulated by Kovner [26] and proven by Perlick [35]. As shown in the latter paper, in this case only local minima and saddles, but no local maxima, of the arrival time may occur. It is likely that the same is true in all Finsler spacetimes, but as far as we know this has not been proven so far.

As an important technical detail it should be mentioned that the results reviewed in this subsection rely on the above-given definition of Finsler spacetimes, i.e., it is assumed that the Finsler metric is well-defined and of Lorentzian signature *everywhere*, and not only *almost everywhere*, on the tangent bundle minus the zero section. If this regularity condition of the Finsler metric fails to hold on a set of measure zero that includes a lightlike vector w at x, the set of all lightlike vectors at x may fail to be a differentiable submanifold near w and the proof of Fermat's principle given in Perlick [37] has to be modified a bit. One would expect that the theorem is still true but this has not actually been worked out.

5.4.3 Measuring Angles in the Sky

For any theoretical description of gravitational lensing it is indispensable to assign an angle to a pair of light rays that arrive at the observer. In general relativity this notion is given in a very natural way: The set of all light rays that arrive at a point p is in one-to-one correspondence with the set of all one-dimensional lightlike subspaces of the tangent space at p. This set, which we will call S_p, is diffeomorphic to the 2-sphere S^2 and can be interpreted as the *sky* at p. If an observer with 4-velocity u has been chosen at p, we can orthogonally project each one-dimensional lightlike subspace into the orthocomplement of u which carries the usual 3-dimensional Euclidean geometry. In this way we can identify the sky S_p with the set of unit vectors orthogonal to u, see Fig. 5.2 where one of the three spatial dimensions is omitted, so the sphere S_p is represented by a circle. We can then assign an angle to any pair of points in S_p, just by using the ordinary notion of an angle between vectors in Euclidean geometry. Clearly, all observers at p receive the same light rays, so they see the same sky. However, they assign different angles to any two points of the sky. The transformation is provided by the standard aberration formula

$$\cos\vartheta' = \frac{\cos\vartheta - \frac{v}{c}}{1 - \frac{v}{c}\cos\vartheta} \tag{5.105}$$

which is derived in any text-book on special relativity. Here ϑ denotes the angle between a light ray and the direction of relative motion as measured by one observer and ϑ' denotes the analogous angle as measured by the other observer; v is the relative velocity.

If we try to carry this construction over into the Finsler setting, we have to face two problems. Firstly, the set of timelike vectors at a point may have more than two connected components, so there may be multiple light cones. If we define the sky S_p along the same lines as above, then the points of S_p will not be in a one-to-

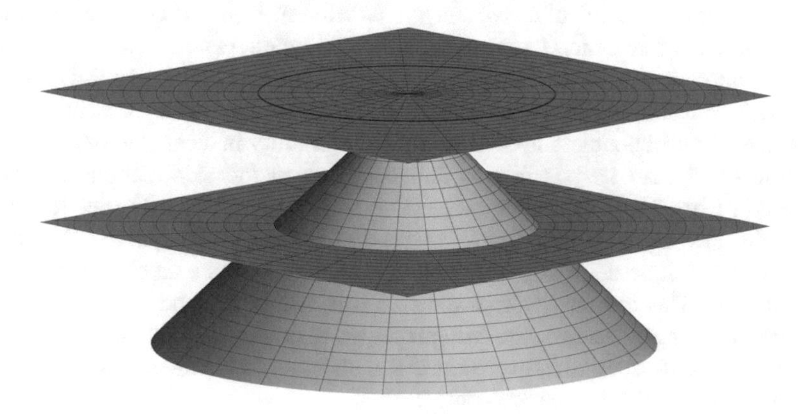

Fig. 5.2 The sky of an observer

one correspondence to the points of the two-sphere S^2. Probably there is only one way of resolving this problem: One has to restrict to those Finsler spacetimes where the set of timelike vectors at a point $p \in M$ does have exactly two components, one being interpreted as the set of past-pointing timelike vectors and the other one as the set of future-pointing timelike vectors. It was already mentioned that this restriction is comparatively mild. We can then define the sky S_p as the set of all one-dimensional lightlike subspaces at p that are tangent to the boundary of the past cone. This makes sure that the points of S_p are in one-to-one correspondence to the points of the two-sphere S^2. Secondly, even in this case it is not at all clear how the choice of an observer would allow to assign an angle to any two points of S_p: The method from general relativity does not carry over because in a Finsler spacetime the definition of an "orthocomplement" is ambiguous. This second problem is a major stumbling block on the way towards a viable theory of gravitational lensing in Finsler spacetimes.

At least from a mathematical point of view, it seems natural to define the orthocomplement of a vector u at x as the set of all v such that $g_{\mu\nu}(x, u)u^{\mu}v^{\nu} = 0$. For any non-zero u the orthocomplement of u is indeed a 3-dimensional vector space. However, it is not true that the orthocomplement of a timelike u carries a Euclidean metric. Moreover, this definition has the somewhat unwanted property that u is not necessarily in the orthocomplement of v if v is in the orthocomplement of u. Even more importantly, it is difficult to associate this definition with an operational procedure. In general relativity, the orthocomplement of a timelike vector gets its physical meaning from Einstein's radar method: If one fixes a timelike curve parametrized by proper time, one calls an event q off the worldline *simultaneous* with an event p on the worldline if the time a light ray needs from the worldline to q and back to the worldline is divided into two equal halves by p. Then the tangent space to the 3-dimensional manifold of all events that are simultaneous to a certain event on the worldline is the orthocomplent of the tangent vector to the worldline at this point. It was shown by Pfeifer [38] that in a Finsler spacetime this radar construction can also be carried through but that then the "radar orthocomplement" of a vector is quite different from the above given formal orthocomplement. In general, the radar orthocomplement is not even a vector space. (Pfeifer's paper is based on a definition of Finsler spacetimes that is slightly different from our's, but this is not relevant here.) So it seems fair to say that in a Finsler spacetime it is still not clear how to define orthogonality in a satisfactory way.

However, it may be possible to define angles in the sky without any direct reference to orthogonality. Minguzzi [33] has shown that in a Finsler spacetime where the sky S_p is diffemorphic to the two-dimensional sphere S^2 one can construct a (positive definite) Riemannian metric on S_p once an observer at p has been chosen. The distance with respect to this metric can be viewed as a way of measuring angles in the sky. Minguzzi has also shown that the metric in S_p undergoes a conformal transformation if the observer is changed. This implies, in particular, that circles are mapped onto circles.

From a mathematical point of view Minguzzi's construction is quite satisfactory. What is still missing is a link with an operational procedure, i.e., a prescription of how to measure in practice angles in the sky.

5.4.4 Light Deflection in Spherically Symmetric and Static Finsler Spacetimes

According to general relativity the vacuum spacetime around a spherically symmetric body is necessarily static and described by the Schwarzschild metric. Calculating the light deflection in this spacetime is one of the basic exercises in the theory of gravitational lensing. If one takes the Finsler modification of the spacetime metric seriously, one has to replace the Schwarzschild metric by a Finsler spacetime metric and to investigate how this affects the light deflection formula. In this subsection we review some results on this score that have been found by Laemmerzahl et al. [27].

The basic idea is to add a Finsler perturbation to the Schwarzschild metric that preserves spherical symmetry and staticity. More precisely, the Finsler Lagrangian is assumed to be given in coordinates $x = (t, r, \vartheta, \varphi)$ and to be of the form

$$2\,L(x,\dot{x})=\big(h_{tt}+c^2\psi_0\big)\dot{t}^2+\Big(\big(h_{ij}h_{kl}+\psi_{ijkl}\big)\dot{x}^i\dot{x}^j\dot{x}^k\dot{x}^l\Big)^{\frac{1}{2}} \tag{5.106}$$

where $h_{\mu\nu}dx^{\mu}dx^{\nu}$ is the Schwarzschild metric,

$$h_{\mu\nu}dx^{\mu}dx^{\nu} = -\Big(1-\frac{2GM}{c^2r}\Big)c^2dt^2+\Big(1-\frac{2GM}{c^2r}\Big)^{-1}dr^2+r^2\Big(\sin^2\vartheta\,d\varphi^2+d\vartheta^2\Big) \tag{5.107}$$

and the summation convention is used for latin indices that take values 1, 2, 3. Here c denotes the vacuum speed of light, G denotes Newton's gravitational constant and M is the mass of the gravitating body according to the unperturbed metric. The time perturbation ψ_0 is a function of r only and the spatial perturbation ψ_{ijkl} depends on r, φ and ϑ in the following form:

$$\psi_{ijkl}\dot{x}^i\dot{x}^j\dot{x}^k\dot{x}^l = \psi_1(r)\dot{r}^4+\psi_2(r)r^2\dot{r}^2\big(\sin^2\vartheta\,\dot{\varphi}^2+\dot{\vartheta}^2\big)+\psi_3(r)r^4\big(\sin^2\vartheta\,\dot{\varphi}^2+\dot{\vartheta}^2\big)^2\,. \tag{5.108}$$

This is the general form the ψ_{ijkl} must have in order to preserve spherical symmetry, i.e., in order to assure that the usual generators of spatial rotations form a Lie algebra of Killing vector fields that is isometric to so(3).

? Exercise

5.11. Show that

$$X_1 = \sin\varphi\partial_\vartheta + \cot\vartheta\cos\varphi\partial_\varphi,\ X_2 = -\cos\varphi\partial_\vartheta + \cot\vartheta\sin\varphi\partial_\varphi,\ X_3 = \partial_\varphi\,, \tag{5.109}$$

are Killing vector fields of L given in (5.106), with (5.107) and (5.108). For guidance see also [41, Sec. V.B]. For a more detailed analysis of spherically symmetric Finsler spacetimes we refer to McCarthy and Rutz [30, 31]

In addition to preserving spherical symmetry, the perturbations also preserve the staticity of the metric, i.e., ∂_t is a timelike Killing vector field and there are no spatio-temporal cross-terms.

Note that the fourth-order term $\psi_{ijkl}\dot{x}^i\dot{x}^j\dot{x}^k\dot{x}^l$ can be viewed as the leading-order term in a general Finsler power–law perturbation of the spatial part of the metric. One could of course also consider a third-order term but this seems not desirable because it would break the reflection symmetry $v^i \to -v^i$ on the tangent space. In this sense our ansatz gives the lowest-order non-trivial Finsler perturbation that preserves the symmetries of the spacetime.

As the Schwarzschild metric is experimentally well tested, the perturbation functions $\psi_1(r)$, $\psi_2(r)$ and $\psi_3(r)$ and their derivatives $\psi_1'(r)$, $\psi_2'(r)$ and $\psi_3'(r)$ are certainly small. Differentiability and smallness of the $\psi_A(r)$ guarantee that the Lagrangian $L(x, v)$ is real-valued, and the Finsler metric (5.93) is non-degenerate with Lorentzian signature for *almost* all (x, v) with $v \neq 0$. The only points where this condition is violated are the points where the *spatial* velocity components are all zero, $(v^r, v^\vartheta, v^\varphi) = (0, 0, 0)$, but $v^t \neq 0$. At these points, the Finsler metric gives undetermined expressions. However, it was shown by Laemmerzahl et al. [27] that even if we choose an initial velocity for which the metric is undetermined, the Euler-Lagrange equation (5.97) has a unique solution which is determined from neighbouring solutions by continuous extension.

Note that, by (5.107) and (5.108), the two-dimensional spacelike manifold $\{t = \text{const.}, r = \text{const.}\}$ is a "round sphere", i.e., it is isometric to the standard 2-sphere in Euclidean 3-space. In the unperturbed Schwarzschild spacetime, this sphere has area $4\pi r^2$ whereas in the perturbed spacetime it has area $4\pi r^2(1 + \psi_3)$.

However, as we are free to change the radius coordinate it is no restriction of generality if we require that also in the perturbed spacetime the sphere at radius r has area $4\pi r^2$. Then $\psi_3 = 0$ and only three perturbation functions ψ_0, ψ_1 and ψ_2 are left.

For calculating the light deflection in the perturbed spacetime we have to consider a lightlike geodesic that comes in from infinity and goes to infinity. Because of the symmetry it is no restriction of generality if we assume that the geodesics is in the equatorial plane, $\vartheta = \pi/2$. Moreover, as we assume that the perturbation functions are small, we may linearize the Lagrangian with respect to them. This results in

$$2L(x, \dot{x}) = (1 + \phi_0)h_{tt}\dot{t}^2 + (1 + \phi_1)h_{rr}\dot{r}^2 + r^2\dot{\varphi}^2 + \frac{\phi_2 h_{rr} r^2 \dot{r}^2 \dot{\varphi}^2}{h_{rr}\dot{r}^2 + r^2\dot{\varphi}^2} \tag{5.110}$$

where we have introduced modified perturbation functions

$$\phi_0 = \frac{c^2\psi_0}{h_{tt}}, \quad \phi_1 = \frac{\psi_1}{2h_{rr}^2}, \quad \phi_2 = \frac{\psi_2 h_{rr} - \psi_1}{2h_{rr}^2}. \tag{5.111}$$

Clearly, the Lagrangian (5.110) comes from a pseudo-Riemannian metric if and only if $\phi_2 = 0$. In other words, ϕ_2 is a measure for the "Finslerity" of the perturbed spacetime.

For a light ray that starts at a source at radius r_S, goes through a minimum radius r_m and reaches an observer at radius r_O, the deflection angle $\Delta\varphi_0$ in the unperturbed Schwarzschild spacetime is well known to be given by the integral formula

$$\pi + \Delta\varphi_0 = \left(\int_{r_m}^{r_S} + \int_{r_m}^{r_O} \right) \frac{dr}{\sqrt{A(r)}}, \tag{5.112}$$

where

$$A(r) = r^4 \left(p(r_m) - p(r) \right), \quad p(r) = r^{-2} \left(1 - \frac{2GM}{c^2 r} \right). \tag{5.113}$$

If the observer and the source are both far away from the centre, the angle $\Delta\varphi_0$ can be measured, and has been measured in the gravitational field of the Sun during the famous 1919 Solar eclipse expedition led by Arthur Eddington, in the following way: When comparing the position of light sources near the Sun with their positions half a year earlier, when the Sun was not in the neighbourhood, one notices a radial displacement by $\Delta\varphi_0$ in the positions. The same method carries over to the Finsler case provided that the spacetime is still asymptotically flat, i.e., provided that the perturbation functions fall off sufficiently quickly for $r \to 0$. When the observer is in a region where the spacetime can be assumed as being flat (i.e., Minkowski spacetime) the problem with measuring angles that was discussed in the preceding subsection does of course not exist.

As proved by Laemmerzahl et al. [27], the deflection angle $\Delta\varphi$ in the perturbed spacetime equals

$$\Delta\varphi = \Delta\varphi_0 + \left(\int_{r_m}^{r_S} + \int_{r_m}^{r_O} \right) \frac{\alpha(r)\, dr}{2\sqrt{A(r)}}. \tag{5.114}$$

where

$$\alpha(r) = \phi_1(r) - \phi_2(r)\frac{2p(r_m) - 3p(r)}{p(r_m)} - \big(\phi_0(r_m) - \phi_0(r)\big)\frac{p(r_m)}{p(r_m) - p(r)} . \tag{5.115}$$

? Exercise

5.12. Use the Finsler geodesic equation and the constants of motion energy and angular momentum in spherical symmetry for the Finsler Lagrangian (5.110) to derive (5.114).

We see that by measuring the deflection angle alone one cannot disentangle the Finslerity ϕ_2 from the other perturbation functions. By comparing with observations Laemmerzahl et al. [27] found that in the gravitational field of the Sun

$$|\alpha(r)| \lessapprox 2 \times 10^{-9} . \tag{5.116}$$

In the same paper an analogous integral formula was derived for the travel time of light. Also in this case, all three perturbation functions ϕ_0, ϕ_1 and ϕ_2 occur.

5.4.5 The Redshift of Light and Applications to Cosmology

In general relativity there is a universal formula for the redshift under which an observer sees a light source. This formula has a very natural generalization for Finsler spacetimes that was found by Hasse and Perlick [21]. For reviewing this result, we need the notion of proper time in a Finsler spacetime, recall (5.98).

Let us assume that we have two timelike curves, $\gamma(\tau)$ and $\tilde{\gamma}(\tilde{\tau})$, both parametrized by proper time, and an affinely parametrized lightlike geodesic $x(s)$ from an event $x(s_1) = \gamma(\tau_1)$ on γ to an event $x(s_2) = \tilde{\gamma}(\tilde{\tau}_2)$ on $\tilde{\gamma}$, see Fig. 5.3. Then the redshift z assigned to this light ray is

$$1 + z = \frac{g_{\mu\nu}\big(x(s_1), \dot{x}(s_1)\big)\dot{x}^\nu(s_1)\dfrac{d\gamma^\mu}{d\tau}(\tau_1)}{g_{\rho\sigma}\big(x(s_2), \dot{x}(s_2)\big)\dot{x}^\sigma(s_2)\dfrac{d\tilde{\gamma}^\rho}{d\tilde{\tau}}(\tilde{\tau}_2)} . \tag{5.117}$$

This formula for the redshift differs from the corresponding formula in general relativity only by the fact that now the metric has a second argument. It was derived in two different ways by Hasse and Perlick [21]. For the first method we have to recall that we have assigned a canonical momentum $p(s)$, see (5.99), to a light ray

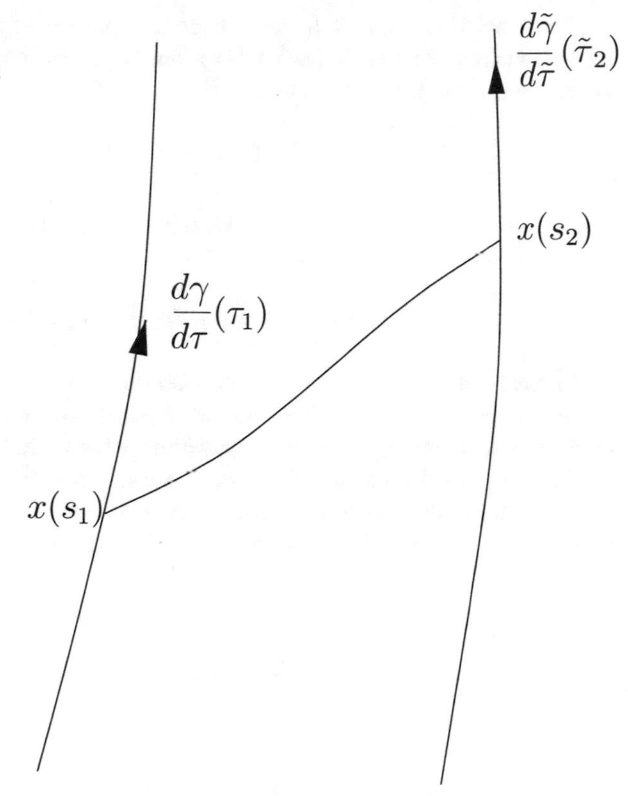

Fig. 5.3 Illustration of the redshift formula.The picture is taken from [21]. Copyright by the American Physical Society, all rights reserved

$x(s)$ when we considered the Hamiltonian formalism. Now assume that an observer, whose worldline $\gamma(\tau)$ is parametrized by proper time, crosses this light ray at an event $x(s_1) = \gamma(\tau_1)$. Then we call

$$\omega(s_1) = -p_\mu(s_1)\frac{d\gamma^\mu}{d\tau}(\tau_1) \tag{5.118}$$

the *frequency* of the light ray with respect to γ. Similarly, if another observer, whose worldline $\tilde{\gamma}(\tilde{\tau})$ is again parametrized by proper time, crosses the light ray at an event $x(s_2) = \tilde{\gamma}(\tilde{\tau}_2)$, we call

$$\omega(s_2) = -p_\mu(s_2)\frac{d\tilde{\gamma}^\mu}{d\tilde{\tau}}(\tilde{\tau}_2) \tag{5.119}$$

the frequency of the light ray with respect to $\tilde{\gamma}$. The right-hand side of (5.117) is precisely the ratio of the frequencies $\omega(s_2)/\omega(s_1)$. The second method is more geometrical. One considers two light rays emitted by γ at proper times τ and $\tau + \Delta\tau$. They will be received by $\tilde{\gamma}$ at proper times $\tilde{\tau}$ and $\tilde{\tau} + \Delta\tilde{\tau}$. The right-hand side of (5.117) equals the ratio $\Delta\tilde{\tau}/\Delta\tau$ in the limit that $\Delta\tau \to 0$.

The Finslerian redshift formula becomes particularly simple if γ and $\tilde{\gamma}$ are integral curves of a vector field $V^\mu(x)$ that is proportional to a conformal Killing vector field $K^\mu(x)$, recall (5.104),

$$V^\mu(x) = e^{f(x)} K^\mu(x) . \tag{5.120}$$

Then we can use the fact that $K^\mu\big(x(s)\big) p_\mu(s) = g_{\mu\nu}\big(x(s), \dot{x}(s)\big) K^\mu\big(x(s)\big) \dot{x}^\nu(s)$ is independent of s and the redshift formula can be rewritten as

$$\ln(1+z) = f\big(x(s_2)\big) - f\big(x(s_1)\big) . \tag{5.121}$$

In this case we say that f is a *redshift potential.*

Hasse and Perlick [21] illustrate the redshift formula with two examples. One of them is a cosmological Finsler spacetime where a redshift potential exists. The model is a perturbation of a Robertson-Walker spacetime, constructed in analogy to the model in the preceding section that was a perturbation of the Schwarzschild spacetime. The Finsler Lagrangian is given in coordinates $x = (t, r, \vartheta, \varphi)$ and reads

$$\begin{aligned} 2L(x, \dot{x}) = & -c^2 \dot{t}^2 \big(1 + \phi_0(t)\big) \\ & + S(t)^2 \Big(\dot{r}^2 + \Sigma(r)^2 (\dot{\vartheta}^2 + \sin^2\vartheta\, \dot{\varphi}^2)\Big) \big(1 + \phi_1(t)\big) \\ & + \frac{\phi_2(t) S^2 c^2 \dot{t}^2 \Big(\dot{r}^2 + \Sigma(r)^2 (\dot{\vartheta}^2 + \sin^2\vartheta\, \dot{\varphi}^2)\Big)}{S(t)^2 \Big(\dot{r}^2 + \Sigma(r)^2 (\dot{\vartheta}^2 + \sin^2\vartheta\, \dot{\varphi}^2)\Big) + c^2 \dot{t}^2} \end{aligned} \tag{5.122}$$

where Σ is one of the following three functions of r:

$$\Sigma(r)^2 = \begin{cases} k^{-1} \sin^2\big(\sqrt{k} r\big) & \text{for } k > 0 \\ r^2 & \text{for } k = 0 \\ |k|^{-1} \sinh^2\big(\sqrt{|k|}\, r\big) & \text{for } k < 0 \end{cases} \tag{5.123}$$

The perturbations $\phi_0(t)$, $\phi_1(t)$ and $\phi_2(t)$ preserve the symmetries of the underlying Robertson-Walker spacetime, i.e., there are six spacelike Killing vector fields which express the fact that the spacetime is spatially homogeneous and isotropic and ∂_t is a conformal Killing vector field. The latter implies the existence of a redshift potential which greatly simplifies the calculations.

Whereas the perturbation $\phi_0(t)$ changes the time measurement, the perturbation $\phi_1(t)$ changes the length measurement. By contrast, $\phi_2(t)$ is a genuine Finsler perturbation which may again be called the “Finslerity”. As we are free to change the time coordinate, we may require $\phi_0(t) = 0$ without loss of generality. Then t

gives proper time along the t-lines in the perturbed as well as in the unperturbed spacetime.

In complete analogy to the unperturbed situation, in the perturbed spacetime the notions of area distance D_A and luminosity distance D_L can be defined for an emitter and a receiver whose worldlines are integral curves of ∂_t. With the help of the redshift potential it can then be shown that they are related by the equation

$$D_L = (1+z)^2 \left(1 - \frac{\phi_2(t_2)}{4} + \frac{\phi_2(t_1)}{4}\right) D_A \tag{5.124}$$

where t_1 is the time when the light signal is emitted and t_2 is the time when it is received, see Hasse and Perlick [21] for a proof. Equation (5.124) generalizes the well known *reciprocity theorem*, also known as *Etherington's law*. Note that the deviation from the traditional version is determined by the Finslerity ϕ_2 alone.

Hasse and Perlick [21] also derive the generalization of the Lemaître-Hubble law, i.e., of the linearized relation beween (area or luminosity) distance and redshift. It reads

$$D_A = \frac{c\,S(t_2)}{S'(t_2)}\left(1 - \frac{\phi_2(t_2)}{4} - \frac{S'(t_2)}{S(t_2)}\left(\frac{\phi_1'(t_2)}{2} + \frac{\phi_2'(t_2)}{4}\right)z + O(z^2)\,. \tag{5.125}$$

For D_L the $O(z^2)$ terms are different, but the linearized law is the same. So in contrast to the Etherington law, the distance-redshift relation is modified not only by the Finslerity but also by $\phi_1(t)$.

For further results on the distance-redshift relation in Finsler spacetimes we refer to Hohmann and Pfeifer [22]. Whereas our approach was purely kinematical, considering perturbations of a Robertson-walker spacetime without assuming the validity of a particular field equation, the work by Hohmann and Pfeifer makes use of the field equation that was introduced by Pfeifer and Wohlfarth [41].

Summary and Outlook

In this chapter we have demonstrated how the Hamiltonian and the Lagrangian approach can be used for studying gravitational lensing in theories with violation of Lorentz invariance. In the first part we have learned how to calculate the image position and delay of lenses in the context of DSR models. We have obtained results for Schwarzschild lenses and for the more realistic SIL model, which describes galaxy lensing. The time-delay between the images has been shown to depend linearly on the energy (Eq. 5.85). Several lensing systems had their delays between images measured in different passbands. The delay between the compact images was measured in the GeV and TeV passbands in only one lensing system, JVAS B0218+357. In the future, other lensing system may be detected in the GeV and TeV band and could be monitored in upcoming facilities such as the

Cherenkov Telescope Array (CTA) [19]. Measurements of time-delays will thus provide constraints on DSR models.

In the second part we have concentrated on gravitational lensing features in Finsler space-times. We have seen that the Lagrangian—or Hamiltonian—approach is well suited for investigating these features. Some important partial results in this direction have already been achieved; this includes the characterization of light rays in terms of a variational principle (i.e., a Finsler version of Fermat's principle) and a general redshift formula. However, there are still open questions which have to be answered before a full-fledged theory of Finsler lensing can be formulated. At the center of the problems is the difficulty of operationally defining how angles in the sky are to be measured. This has to be left for future studies.

References

1. H. Abdalla et al., The 2014 TeV γ-ray flare of Mrk 501 seen with H.E.S.S.: temporal and spectral constraints on lorentz invariance violation. Astrophys. J. **870**, 93 (2019). https://doi.org/10.3847/1538-4357/aaf1c4
2. A.A. Abdo et al., Gamma-ray flaring activity from the gravitationally lensed blazar PKS 1830-211 observed by Fermi LAT. Astrophys. J. **799**, 143 (2015). https://doi.org/10.1088/0004-637X/799/2/143
3. G. Asanov, *Finsler Geometry, Relativity and Gauge Theories* (Reidel, Dordrecht, 1985)
4. L. Barcaroli, L.K. Brunkhorst, G. Gubitosi, N. Loret, C. Pfeifer, Curved spacetimes with local κ -Poincaré dispersion relation. Phys. Rev. D **96**(8), 084010 (2017). https://doi.org/10.1103/PhysRevD.96.084010
5. A. Barnacka, J.F. Glicenstein, Y. Moudden, First evidence of a gravitational lensing-induced echo in gamma rays with Fermi LAT. Astron. Astrophys. **528**, L3 (2011). https://doi.org/10.1051/0004-6361/201016175
6. A. Barnacka, M.J. Geller, L.P. Dell'Antonio, W. Benbow, Resolving the high-energy universe with strong gravitational lensing: the case of PKS 1830-211. Astrophys. J. **809**(1), 100 (2015). https://doi.org/10.1088/0004-637X/809/1/100
7. A. Barnacka, M.J. Geller, L.P. Dell'Antonio, A. Zitrin, The structure of the strongly lensed gamma-ray source B2 0218+35. Astrophys. J. **821**, 58 (2016). https://doi.org/10.3847/0004-637X/821/1/58
8. J. Beem, Indefinite Finsler spaces and timelike spaces. Can. J. Math. **22**, 1035 (1970)
9. A. Bernal, M.A. Javaloyes, M. Sánchez, Foundations of finsler spacetimes from the observers' viewpoint. Universe **6**, 55 (2020). https://doi.org/10.3390/universe6040055
10. V.B. Bezerra, H.R. Christiansen, M.S. Cunha, C.R. Muniz, Exact solutions and phenomenological constraints from massive scalars in a gravity's rainbow spacetime. Phys. Rev. D **96**, 024018 (2017). https://doi.org/10.1103/PhysRevD.96.024018
11. M. Biesiada, A. Piórkowska, Gravitational lensing time delays as a tool for testing Lorentz-invariance violation. Month. Not. R. Astron. Soc. **396**, 946–950 (2009). https://doi.org/10.1111/j.1365-2966.2009.14748.x
12. A.D. Biggs, I.W.A. Browne, A revised lens time delay for JVAS B0218+357 from a reanalysis of VLA monitoring data. Month. Not. R. Astron. Soc. **476**, 5393–5407 (2018). https://doi.org/10.1093/mnras/sty565
13. C.C. Cheung et al., Fermi large area telescope detection of gravitational lens delayed γ-ray flares from blazar B0218+357. Astrophys. J. Lett. **782**, L14 (2014). https://doi.org/10.1088/2041-8205/782/2/L14
14. X.M. Deng, Y. Xie, Gravitational time advancement under gravity's rainbow. Phys. Lett. B **772**, 152–158 (2017). https://doi.org/10.1016/j.physletb.2017.06.036

15. P. Egorov, M. Guzinin, H. Hakobyan, S. Troitsky, Constraining new fundamental physics with multiwavelength astrometry. Month. Not. R Astron. Soc. **437**, L90–L94 (2014). https://doi.org/10.1093/mnrasl/slt148
16. A. Farag Ali, M.M. Khalil, A proposal for testing gravity's rainbow. Europhys. Lett. **110**, 20009 (2015). https://doi.org/10.1209/0295-5075/110/20009
17. R. Gallego Torromé, P. Piccione, On the Lie group structure of pseudo-Finsler isometries. Houston J. Math. **41**, 513 (2015)
18. J.F. Glicenstein, Gravitational lensing time delays with massive photons. Astrophys. J. **850**, 102 (2017). https://doi.org/10.3847/1538-4357/aa9439
19. J.F. Glicenstein, An experimental test of gravity at high energy. J. Cosmol. Astroparticle Phys. **2019**(4), 010 (2019). https://doi.org/10.1088/1475-7516/2019/04/010
20. A.F. Grillo, E. Luzio, F. Méndez, F. Torres, Gravitational lensing in AN energy-dependent spacetime metric. Int. J. Mod. Phys. D **21**, 1250007-1-1250007-18 (2012). https://doi.org/10.1142/S0218271812500071
21. W. Hasse, V. Perlick, Redshift in finsler spacetimes. Phys. Rev. D **100**, 0234033 (2019). https://doi.org/10.1103/PhysRevD.100.024033
22. M. Hohmann, C. Pfeifer, Geodesics and the magnitude-redshift relation on cosmologically symmetric Finsler spacetimes. Phys. Rev. D **95**, 104021 (2017). https://doi.org/10.1103/PhysRevD.95.104021
23. M. Hohmann, C. Pfeifer, N. Voicu, Finsler gravity action from variational completion. Phys. Rev. D **100**, 064035 (2019). https://doi.org/10.1103/PhysRevD.100.064035
24. M.A. Javaloyes, M. Sánchez, Finsler metrics and relativistic spacetimes. Int. J. Geom. Methods Mod. Phys. **11**, 1460032–230 (2014). https://doi.org/10.1142/S0219887814600329
25. M. Knebelman, Collineations and motions in generalized spaces. Am. J. Math. **51**, 527 (1929)
26. I. Kovner, Fermat principle in gravitational fields. Astrophys. J. **351**, 114 (1990). https://doi.org/10.1086/168450
27. C. Laemmerzahl, V. Perlick, W. Hasse, Observable effects in a class of spherically symmetric static. Phys. Rev. D **86**, 104042 (2012). https://doi.org/10.1103/PhysRevD.86.104042
28. J.E.J. Lovell, D.L. Jauncey, J.E. Reynolds, M.H. Wieringa, E.A. King, A.K. Tzioumis, P.M. McCulloch, P.G. Edwards, The time delay in the gravitational lens PKS 1830-211. Astrophys. J. Lett. **508**(1), L51–L54 (1998). https://doi.org/10.1086/311723
29. J. Magueijo, L. Smolin, Gravity's rainbow. Class. Quant. Gravity **21**, 1725–1736 (2004). https://doi.org/10.1088/0264-9381/21/7/001
30. P. McCarthy, S. Rutz, The general four-dimensional spherically symmetric Finsler space. Gen. Relat. Gravit. **25**, 589 (1993). https://doi.org/10.1007/BF00757070
31. P. McCarthy, S. Rutz, Symmetry in Finsler spaces, in *Finsler Geometry*. Contemporary Mathematics, vol. 196 (American Mathematical Society, Providence, 1996), p. 289
32. E. Minguzzi, Light cones in Finsler spacetime. Commun. Math. Phys. **334**, 1529 (2015). https://doi.org/10.1007/s00220-014-2215-6
33. E. Minguzzi, The conformal transformation of the night sky. Class. Quant. Gravity **33**, 235009 (2016). https://doi.org/10.1088/0264-9381/33/23/235009
34. S. Muller, S. Jaswanth, C. Horellou, L. Martí-Vidal, All good things come in threes: the third image of the lensed quasar PKS 1830-211. Astron. Astrophys. **641**, L2 (2020). https://doi.org/10.1051/0004-6361/202038978
35. V. Perlick, On Fermat's principle in general relativity: I. The general case. Class. Quant. Gravity **7**, 1319 (1990). https://doi.org/10.1088/0264-9381/7/8/011
36. V. Perlick, *Ray Optics, Fermat's Principle, and Applications to General Relativity*, vol. 61 (Springer, Berlin, 2000)
37. V. Perlick, Fermat principle in Finsler spacetimes. Gen. Relat. Gravit. **38**, 365 (2006). https://doi.org/10.1007/s10714-005-0225-6
38. C. Pfeifer, Radar orthogonality and radar length in Finsler and metric spacetime geometry. Phys. Rev. D **90**, 064052 (2014). https://doi.org/10.1103/PhysRevD.90.064052
39. C. Pfeifer, Finsler spacetime geometry in physics. Int. J. Geom. Methods Mod. Phys. **16**, 1941004-193 (2019). https://doi.org/10.1142/S0219887819410044

40. C. Pfeifer, M. Wohlfarth, Causal structure and electrodynamics on Finsler spacetimes. Phys. Rev. D **84**, 044039 (2011). https://doi.org/10.1103/PhysRevD.84.044039
41. C. Pfeifer, M. Wohlfarth, Finsler geometric extension of Einstein gravity. Phys. Rev. D **85**, 064009 (2012). https://doi.org/10.1103/PhysRevD.85.064009
42. J. Plebański, A. Krasiński, *An Introduction to General Relativity and Cosmology*. (Cambridge University Press, Cambridge, 2006)
43. S. Rutz, A Finsler generalisation of Einstein's vacuum field equations. Gen. Relat. Gravit. **25**, 1139 (1993)
44. P. Schneider, J. Ehlers, E.E. Falco, *Gravitational Lenses* (Springer, Berlin, 1992). https://doi.org/10.1007/978-3-662-03758-4
45. G. Temple, New systems of normal co-ordinates for relativistic optics. Proc. R. Soc. Lond. A **168**, 122 (1938)
46. R.M. Wald, *General Relativity* (University of Chicago Press, Chicago, 1984)
47. T. Wiklind, F. Combes, Time delay of PKS 1830-211 using molecular absorption lines, in *Gravitational Lensing: Recent Progress and Future Go*, ed. by T.G. Brainerd, C.S. Kochanek. Astronomical Society of the Pacific Conference Series, vol. 237 (Astronomical Society of the Pacific, San Francisco, 2001), p. 155

Part II

Observational Effects Beyond Special and General Relativity: From Cosmic Scales, via Compact Objects to the Lab

Leaving the more fundamental models from Part I behind, we enter the discussion of the imprint of modified and quantum gravity on cosmic messengers, astrophysical systems and laboratory sized experiments.

Since the detection of gravitational waves in 2016 multi-messenger astronomy consists of combining the data of neutrino-, cosmic-ray-, gamma-ray- and gravitational wave telescopes to investigate the properties of the gravitational interaction. In Chap. 6 the search for Lorentz invariance violation from the individual messengers and multi-messenger analyses is discussed.

One of the most exciting laboratories revealing properties of the gravitational interaction are Neutron stars, since the predictions of their characteristics are sensitive to the underlying model of gravity, as we will see in Chap. 7. In particular the existence and form of the universal relations and the quasi-normal mode spectrum is one major source of information.

Chapter 8 turns the discussion towards black holes, whose existence has been directly demonstrated in gravitational wave observations and with the direct images provided by the Event Horizon Telescope in 2019 and 2022. The question discussed is whether the obtained data lead to the conclusion that the black holes are Kerr Black Holes predicted by general relativity, or possibly differ from these, as predicted by modified theories of gravity. In Chap. 9 further details, on how information about the characteristics of a black hole are extracted from gravitational wave signals are presented.

Are the just mentioned observations really pointing towards a black hole, or are they compatible with matter configurations which nearly look like black holes from far? This question is discussed in Chap. 10. Objects which are candidates for such astrophysical systems are Boson stars. Their main difference to a black hole is that they do not possess a horizon, however, from far, it might be difficult to distinguish them from black holes. Moreover, Boson stars are natural counterparts to Neutron stars, and could serve as dark matter candidates.

In Chap. 11 we leave the strong gravity regime and the impact of modified gravity on the evolution of stars is discussed. Here, mostly, corrections to the Newtonian potential play a role. However, this role is significant and provides clear tests of modified gravity.

Another way to test modified gravity is to observe pulsars, as is discussed in Chap. 12. These serve as best clocks in space and in strong gravity. Deviation of their orbits from the ones predicted by general relativity can be detected to high accuracy. Pulsar timing allows for a precise determination of the so called Post Kepplerian Parameters, which can be predicted from the theory under consideration and then be constraint.

Possible deviations from the Newtonian potential become relevant in short distance experiments in laboratories. A particular sensitive way to search for such weak field deviations employs the Casimir effect. In Chap. 13, the details on how the measurement of the Casimir effect can be used to learn about the properties of the gravitational interaction at small distances is presented.

In the IIIrd part of the book, the topic will be the interaction between quantum matter and gravity.

Cosmic Searches for Lorentz Invariance Violation

6

Carlos Pérez de los Heros and Tomislav Terzić

Abstract

Cosmic messengers (gamma rays, cosmic rays, neutrinos and gravitational waves) provide a powerful complementary way to search for Lorentz invariance violating effects to laboratory-based experiments. The long baselines and high energies involved make Cherenkov telescopes, air-shower arrays, neutrino telescopes and gravitational wave detectors unique tools to probe the expected tiny effects that the breaking of Lorentz invariance would cause in the propagation of these messengers, in comparison with the standard scenario. In this chapter we explain the expected effects that the mentioned detectors can measure and summarize current results of searches for Lorentz violation.

6.1 Introduction

The invariance of physical laws under Lorentz transformations is a fundamental requirement that has successfully guided the development of our current theory describing the basic constituents of matter and their interactions, the Standard Model of Particle Physics. Its meaning, that experimental results are independent of the orientation and velocity of the frame of reference, seems also a reasonable criterion for any physical theory. Therefore any deviation from this, in principle, hard-wired requirement of the Standard Model would indicate the emergence of new physics

C. Pérez de los Heros (✉)
Department of Physics and Astronomy, Uppsala University, Uppsala, Sweden
e-mail: cph@physics.uu.se

T. Terzić
University of Rijeka, Faculty of Physics, Rijeka, Croatia
e-mail: tterzic@phy.uniri.hr

C. Pfeifer, C. Lämmerzahl (eds.), *Modified and Quantum Gravity*, Lecture Notes in Physics 1017, https://doi.org/10.1007/978-3-031-31520-6_6

at some given energy scale where the theory would break down, see Chap. 1. We do know that the Standard Model is not the ultimate description of nature, even if only because it does not describe Gravity alongside the other fundamental forces. But there are other reasons why the Standard Model needs to be extended, like the existence of neutrino masses, the strong CP problem or the uncomfortably, to some, number of free parameters in the model that an extended theory could account for. Besides, astrophysical observations, like the existence of dark matter or dark energy, need ultimately to be explained from a fundamental physics point of view and, whatever that explanation might turn out to be, it can not come from within the Standard Model. Not all the extensions of the Standard Model require or predict Lorentz invariance violation (LIV) but, contrarily, the observation of LIV effects would undoubtedly point to physics beyond the Standard Model.

Breaking or deformations of Lorentz invariance are expected in models of quantum gravity with a minimal fundamental length that must be an invariant for all reference frames. This leads to a breakdown of Lorentz invariance at energies near the Planck scale. A consequence is that ultrarelativistic particles (small Compton wavelengths) propagating in vacuum between two points (i.e., source and detector) will perceive this quantum structure of space-time and cover the distance in a different time compared to propagation through classical, continuous, space-time. But LIV is not necessarily connected to the existence of a discrete space-time. Lorentz invariance breaking operators can be added in an effective field theory formulated in standard space-time.

From an experimental point of view, searches for LIV tend to be carried out in a model independent way, just parameterizing the effect of LIV in power expansions of the free particle Lagrangian as we discuss below, and setting limits on the parameters from the lack of observation of any LIV effect. In particular the effect of LIV in the propagation of a particle can be parameterized quite generically as a modification of the dispersion relation, which acquires corrections at a given high-energy scale denoted below by E_{QG}. The usual starting point of experimental searches for Lorentz invariance violation (LIV) is therefore the modified dispersion relation

$$E_{\mathrm{i}}^2 = m_{\mathrm{i}}^2 c^4 + p_{\mathrm{i}}^2 c^2 \left[1 + \sum_{n=1}^{\infty} \eta_n^{(\mathrm{i})} \left(\frac{p_{\mathrm{i}} c}{E_{\mathrm{QG},n}^{(\mathrm{i})}} \right)^n \right], \tag{6.1}$$

where the index i represents the particle type (e.g. photon, electron, proton, neutrino, pion, etc.), c is the standard special-relativistic invariant speed of light, E_{i}, p_{i}, and m_{i} are the particle's energy, momentum, and rest mass, respectively (n.b. for photons $m_\gamma = 0$). The series expansion in the square brackets represents a modification to the standard Lorentz invariant dispersion relation, presumably caused by quantum gravity. Each term is characterised by $E_{\mathrm{QG},n}^{(\mathrm{i})}$, the energy level at which that particular contribution becomes relevant. Typically, these are expected to be on the scale of the Planck energy, but it does not necessarily need to be so.

The parameter $\eta_n^{(i)}$ is either $+1$ or -1. It determines the sign of the modifying contribution, corresponding to superluminal or subluminal effects respectively.

It is important to stress that the modified dispersion relation is not a consequence of any particular model of quantum gravity. Equation (6.1) is a rather simple parameterization, a model used to test LIV. A possible detection of an effect will not necessarily prove any particular theory of quantum gravity. However, it will trigger research to determine the cause and nature of the violation. LIV can be formally incorporated in extensions of the Standard Model in the form of effective theories that explicitly include LIV terms in a consistent manner. The most popular of these approaches is the Standard Model Extension (SME) [1, 2], whose connection to String theory was discussed in Chap. 1. The SME is a renormalizable quantum field theory that inherits the $SU(3) \times SU(2) \times U(1)$ group structure of the SM but extends its Hamiltonian to include general Lorentz violating terms. These terms introduce new effects like energy-dependent velocity of masless particles, modified interactions, modified neutrino oscillations, a possible direction dependence of the photon polarisation, photon instability, CPT violation or the possibility of $\nu - \bar{\nu}$ mixing or neutrino bremsstrahlung, giving rise to a rich phenomenology [3–6].

Although we will not go deep in discussing the underlying theories in this chapter (we refer the interested reader to Chaps. 1 and 2 for discussions on theory and phenomenology), it is worth mentioning that there are other scenarios than the SME in which the Lorentz symmetry is not strictly preserved as we know it, as in doubly special relativity (DSR) models [7, 8]. The underlying principles of the SME and DSR are fundamentally different. As mentioned above, the SME is an effective theory that adds Lorentz violating terms to the Standard Model. On the contrary, DSR, keeps the speed of light c as an observer invariant and a limiting propagation speed for low-energy photons, but adds an additional observer invariant in the form of a fundamental energy E_{QG}. Typically, E_{QG} is taken as the Planck energy but, in some models, it can be orders of magnitude below it. In order to accommodate this new invariant, Lorentz symmetry needs to be deformed, which in turn leads to deformed expressions for conservation laws, resulting in rather different predictions on the propagation of particles. In addition, some phenomena expected in LIV are not allowed in DSR. We will point these out as we go along.

Although no experimental evidence of violation of Lorentz invariance has been found yet, experimental searches on many fronts are being carried out, from laboratory experiments [9–12] to astrophysical probes, the focus of this chapter. In what follows we describe the effects to be expected from LIV using cosmic messengers as probes, the experimental techniques used and the current experimental limits obtained with Cherenkov telescope arrays, neutrino telescopes, air shower arrays, and gravitational wave detectors.

6.2 Cosmic Messengers as a Probe of LIV

The cosmic messengers carrying information about processes in the universe are Gamma Rays, Neutrinos, Cosmic Rays and Gravitational waves. In what follows, we will discuss all of these, and how they are used to detect deviations from local Lorentz invariance.

6.2.1 Gamma Rays

Gamma radiation constitutes a significant portion of the electromagnetic spectrum. Astrophysicists classify every photon with energy above 100 keV as a gamma ray, regardless of the process in which it was created. Considering that the most energetic photon detected up to date was a 1.4 PeV (1.4×10^{15} eV) [13], the gamma-ray band spans over more than 10 orders of magnitude in energy.

Compared to some other cosmic messengers (e.g. neutrinos in Sect. 6.2.2), gamma rays are relatively easily detected and, since they are electrically neutral, they propagate on straight lines (compared to cosmic rays, see Sect. 6.2.3), meaning that their direction of arrival points towards their source. These characteristics, combined with their high energies, make gamma rays superb probes of the most energetic processes in the universe, particularly convenient for searches of LIV.

6.2.1.1 Gamma Ray Detectors

Gamma rays do not penetrate the atmosphere. Therefore, gamma-ray detectors either need to be placed on satellites in orbits above the Earth's atmosphere, or the atmosphere needs to be used as a target in the detection process. In either case, the underlying process is fundamentally the same, though at different scales. When a gamma ray enters the atmosphere, it interacts with atomic nuclei in the air. It is absorbed and an electron–positron pair is created. Both of them lose energy through interactions with other nuclei, emitting additional gamma rays in the process through *bremsstrahlung*. These secondary gamma rays are also absorbed, and new e^-–e^+ pairs are created. A cascade of particles, known as extensive air shower (EAS), develops for as long as bremsstrahlung is the preferred way for electrons to lose energy. Charged particles in the cascade can be energetic enough to propagate faster than light through the medium, instantaneously polarising molecules in the air as they pass through. As the molecules coherently depolarise, they emit flashes of so-called Cherenkov radiation, which peaks in the ultraviolet band. Photomultiplier tubes in Cherenkov detectors record the Cherenkov light emitted as a final consequence of a gamma ray penetrating the atmosphere. Shower images are usually ellipse-shaped, with the longer axis pointing towards the position of the gamma-ray source in the camera, while the image size is connected to the energy of the primary particle. Using imaging techniques, energies and directions of primary gamma rays are estimated.

Showers induced by gamma rays contain predominantly electrons, positrons and photons, and their developments is governed through electromagnetic interaction. For that reason, they are often referred to as electromagnetic showers. Cosmic rays also induce extensive air showers, with several subtle but important differences. Unlike gamma rays, primary cosmic rays survive interactions with atomic nuclei in the medium. More importantly, in addition to the electromagnetic, the weak and strong interactions play important roles in the shower development. As a consequence, the so-called *hadronic* showers are composed of secondary baryons, mesons (such as kaons or pions), and leptons (electrons, muons, neutrinos, etc.). Hadronic showers often contain electromagnetic sub-showers.

Hadronic showers are less homogeneous than gamma showers, which can be used to classify the type of the primary particle. As far as gamma-ray observations are concerned, cosmic rays constitute background, and their rate is in most cases much higher. That means that the region in the detector where the signal is expected will be contaminated with background events. The background flux is usually estimated from another region in the detector where no sources of gamma rays are expected. It is important to note that while the number of background events in the signal region can be estimated, as of yet there is no way of determining which specific event belongs to the signal and which to the background. This is quite a drawback in research which is performed on single events, which is the case in some LIV studies.

The most important present-day Cherenkov telescopes are the High Energy Stereoscopic System H.E.S.S. (https://www.mpi-hd.mpg.de/hfm/HESS, [14, 15]), the Major Atmospheric Gamma Imaging Cherenkov (MAGIC) (https://magic.mpp.mpg.de, [16, 17]) and the Very Energetic Radiation Imaging Telescope Array System (VERITAS) (https://veritas.sao.arizona.edu, [18]), while the next generation Cherenkov experiment, the Cherenkov Telescope Array (CTA) (https://www.cta-observatory.org, [19]) is currently being constructed. More details on the extensive air showers and Cherenkov effect one can find in [20, 21], while for a detailed review of data analysis techniques in Cherenkov telescopes we refer interested reader to [22].

We already mentioned that the shower size is correlated to the energy of the primary particle. Therefore, the detector design will strongly depend on the targeted energy range. Starting with the lowest energies, satellite detectors are used to detect gamma rays with energies up to few hundred GeV. For example the Fermi Large Area Telescope (*Fermi*-LAT) (https://fermi.gsfc.nasa.gov/science/instruments/lat.html, [23]) is composed from two main parts. The first part consists of layers of conversion foils and tracking detectors, where a gamma ray is converted to an e^-–e^+ pair, and their trajectories are tracked to determine the direction of primary gamma ray. Showers are initiated in the calorimeter placed just below, where the energies of all shower constituents are summed to measure the energy of the primary particle. *Fermi*-LAT is sensitive in the energy range 20 MeV–300 GeV. Lower energies can be accessed if Compton scattering is used in the detector instead of pair creation. Some future space detectors, e.g. enhanced.ASTROGAM (e-ASTROGAM) [24], or the All-sky Medium Energy Gamma-ray Observatory (AMEGO) (https://asd.gsfc.nasa.

gov/amego/index.html, [25]), will combine these processes to access lower energies, as well as measure the gamma-ray polarisation. Another deciding factor for detector design selection is the rate of gamma rays, which, as a general rule, falls off as a power-law of energy, requiring larger detector collection areas to access higher energies. The collection area of *Fermi*-LAT of $\lesssim 1\,\mathrm{m}^2$ is too small to detect a relevant number of gamma rays with energies above $\sim 300\,\mathrm{GeV}$. In addition, the showers at these energies are too big to fit in the calorimeter, making gamma-ray energy reconstruction unfeasible.

Cherenkov telescopes on the ground are, on the other hand, sensitive to gamma rays with energies from $\sim$10 GeV to $\gtrsim$100 TeV. Showers below the low energy threshold produce Cherenkov radiation too weak to be detected and can not be properly reconstructed with these instruments. As we go to higher energies, gamma rays become too rare, and air showers too extended to be detected with Cherenkov telescopes. However, charged particles within the showers reach the ground and can be detected by water Cherenkov detectors. Experiments such as the High Altitude Water Cherenkov Observatory (HAWC) (https://www.hawc-observatory.org, [26]) employ arrays of water containers with photomultipliers, which record flashes of Cherenkov light in the water, not unlike water neutrino detectors (see Sect. 6.2.2). There are also hybrid detectors, as the Large High Altitude Air Shower Observatory (LHAASO) (http://english.ihep.cas.cn/lhaaso, [27]), which combine several detection techniques to detect the highest energies and to suppress the background. It is important to notice the significant overlap in energies covered by detectors of different types, which is essential for cross-instrumental calibration and collaboration.

While the gamma-ray detection technique is fundamentally the same in all these instruments, the implementations, and therefore the observation strategies, are quite different. Instruments onboard the *Fermi* satellite have a wide field of view. The *Fermi*-LAT field of view covers almost 20% of the sky, and mostly observes in the sky survey mode. It scans the entire sky in 3 hours, with any given point remaining in the field of view continuously for at least 30 minutes. Cherenkov telescopes, on the other hand, observe only small portions of the sky at the time, with fields of view of up to $\sim$5 deg. Therefore, they perform pointed observations, with usually only one source in the field of view. Considering that Cherenkov telescopes rely on detecting flashes of optical and UV light in the air, they cannot be used during the day, or in bad weather. Satellite detectors, do not suffer of such restrictions. Water Cherenkov detectors have a field of view comparable to *Fermi*-LAT (e.g. HAWC instantaneously covers 15% of the sky) and also perform sky surveys. However, unlike satellite-borne detectors or Cherenkov telescopes, they cannot be pointed nor repositioned. Water Cherenkov detectors also use photomultipliers, but these are enclosed in light-tight water tanks, so their duty cycle is close to 100%, as in satellite detectors.

6.2.1.2 Effects of Lorentz Invariance Violation

Effects of modified photon dispersion relation can be most generally classified in two categories: (i) modification of the propagation speed, and (ii) modifications of photon interactions.

Energy Dependent Group Velocity

Assuming that the group velocity still corresponds to the derivative of energy with respect to momentum, one can easily derive from Eq. (6.1)

$$v_\gamma = \frac{\partial E_\gamma}{\partial p_\gamma} \simeq c\left[1 + \sum_{n=1}^{\infty} \eta_n \frac{n+1}{2}\left(\frac{E}{E_{\mathrm{QG},n}}\right)^n\right]. \tag{6.2}$$

Obviously, the modification introduced in the photon dispersion relation makes the modified photon group velocity energy dependent, and different from c. Depending on the value of η_n, the group velocity can be greater or less than the standard speed of light. These two behaviours are known as *superluminal* ($\eta_n = +1$) and *subluminal* ($\eta_n = -1$), respectively.

Assuming that the photon speed is not constant but that it depends on energy, two photons of different energies will have different times of flight (often abbreviated: ToF). Therefore, by comparing times of flight of photons of different energies, we can calculate the difference in their speeds. Based on the derivation by Jacob & Piran in [28], the delay in the arrival times of two photons of different energies, emitted at the same time from the same source at redshift z_{s} is given by:

$$\Delta t = t_2 - t_1 \simeq -\eta_n \frac{n+1}{2} \frac{E_2^n - E_1^n}{E_{\mathrm{QG},n}^n} D_n(z_{\mathrm{s}}). \tag{6.3}$$

where t_1 and t_2 designate the times of flight of photons of energies E_1 and E_2, respectively.

? Exercise

6.1. Starting from the dispersion relation for photons (Eq. (6.1)), calculate modified photon group velocity (Eq. (6.2)). Derive the expression for the arrival time delay between two photons emitted at the same time.
Hint: follow the procedure from [28].

Notice that the time delay can be either positive or negative. E.g., for superluminal behaviour, and $E_2 > E_1$, t_2' will be shorter than t_1', and $\Delta t'$ will be negative. $D_n(z_{\mathrm{s}})$ accounts for the distance of the source. The expression most commonly used in experimental tests was proposed in [28]:

$$D_n(z_{\mathrm{s}}) = \frac{1}{H_0} \int_0^{z_{\mathrm{s}}} \frac{(1+z)^n}{\sqrt{\Omega_{\mathrm{m}}(1+z)^3 + \Omega_\Lambda}} dz. \tag{6.4}$$

H_0, Ω_m, and Ω_Λ are the Hubble constant, the matter density parameter, and the dark-energy density parameter. This expression was derived from LIV. DSR can result in different distance contributions, depending on the assumptions under which it was obtained. One of the results in the DSR framework and the LIV contribution, as well as the sensitivity of experimental tests to each of them were compared in [29]. It is the only study performed so far in which different phenomenological models are compared on experimental data.

DSR with a specific choice of parameters can lead to expression (6.4). Details of dependence on the redshift notwithstanding, $D_n(z_s)$ serves as a natural amplifier for time delay.

Vacuum Birefringence

Certain aspects of LIV allow photon group velocity to depend not only on the photon energy, but also on the polarisation. This effect is known as vacuum birefringence. In this scenario, the modified photon dispersion relation will be given as

$$E_{\pm}^2 = p^2c^2\left(1 \pm \frac{pc}{E_{\mathrm{QG},1}}\right), \tag{6.5}$$

where $\pm$ represent different circular polarisation states (see, e.g., [30]). This will lead to different propagation velocities for different polarisation states, which will finally result in a rotation of the polarisation vector of a linearly polarised wave. The rotation angle will depend on the energy of the photon and the distance to the source

$$\Delta\theta \simeq \frac{E^2}{E_{\mathrm{QG},1}} D_1(z_s), \tag{6.6}$$

where E denotes the measured energy of the photon, and $D_1(z_s)$ is given in Eq. (6.4) for $n = 1$. Photon birefringence arises in the Standard Model extension from mass dimension five operators, which corresponds to $n = 1$ modification of the photon dispersion relation. For $n = 2$, both polarisation states will have the dispersion relation modified in the same way.

Anomalous Gamma-Ray Absorption

Modifying a particle dispersion relation can affect its interactions, either through modifying the kinematics or dynamics of a process, or both. This can lead to different reaction thresholds and process rates (i.e. decay widths and cross sections) compared to the ones predicted within the Standard Model. Such processes, most relevant for astrophysics, are synchrotron radiation, Compton scattering, Breit–Wheeler and Bethe–Heitler process, etc. Other phenomena, such as vacuum Cherenkov radiation, photon decay, or photon splitting, are kinematically forbidden under standard special relativity, but become possible if Lorentz symmetry is broken.

As we already argued in Sect. 6.1, not all these effects will be present in DSR. Take photon stability as an example: if there is no preferred frame of reference (as is the case in DSR), photons should decay in all frames or none. Otherwise, one observer would see the process take place, while an observer in another equivalent frame would see it as forbidden. This situation is contradictory to requirement of equivalent inertial frames. In addition, DSR is still in its development phase, and the formality is still not fully understood. Therefore, at this stage, one still cannot calculate cross sections.

Another important aspect to note is that photons are not the only particles involved in these processes. Most commonly they will also involve electrons, which poses the question of whether all particles are equally affected by LIV.

Finally, while these effects refer to different processes, they all have a similar net result, which is changing a number of emitted or detected gamma rays at different energies. Therefore, measuring these effects boils down to measuring spectra of astrophysical sources and searching for anomalous behaviour. Let us discuss details with an example.

The universe is filled with low-energy background electromagnetic radiation, usually classified as radio background (RB), cosmic microwave background (CMB), and extragalactic background light (EBL). Gamma rays of very high energies can interact with the photons from the background fields to create electron–positron pairs and are therefore absorbed in these processes. As a rule of thumb in the special-relativistic scenario, the higher energy of a gamma ray, the more likely it is to be absorbed, which results in the softening of the spectrum. The optical depth of the universe to the gamma rays is given as:

$$\tau(E, z_s) = \int_0^{z_s} \frac{dl}{dz} dz \int_{-1}^{1} \frac{1 - \cos\theta'}{2} d\cos\theta' \int_{\epsilon'_{th}}^{\infty} \sigma_{\gamma\gamma}(s)\, n\left(\epsilon', z\right) d\epsilon'. \tag{6.7}$$

It depends on the comoving number density of background photons per unit energy, here marked as $n\left(\epsilon', z\right)$, and the cross section given in standard quantum electrodynamics as

$$\sigma_{\gamma\gamma}(s) = \frac{2\pi\alpha^2}{3m_e^2}\left(1 - \beta^2\right)\left[2\beta\left(\beta^2 - 2\right) + \left(3 - \beta^4\right)\ln\left(\frac{1+\beta}{1-\beta}\right)\right], \tag{6.8}$$

with

$$\beta(s) = \left[-\frac{4m_e^2c^4}{s}\right]^{1/2}, \tag{6.9}$$

and s being the invariant mass. The integrals run (right to left) over all background photon energies, ϵ', scattering angles, θ', and the thickness of the medium between the source and the observer. Primed variables are given in the interaction comoving frame. Therefore, ϵ'_{th} denotes reaction threshold for the Breit–Wheeler process in the comoving frame. The expression for dl/dz depends on the adopted cosmological model. For the ΛCDM cosmology, it is given as:

$$\frac{dl}{dz} = \frac{c}{H_0\,(1+z)\,\sqrt{\Omega_\mathrm{m}\,(1+z)^3 + \Omega_\Lambda}}. \tag{6.10}$$

Violation of Lorentz symmetry can result in the modification of the kinematics or dynamics of the process. It can be relatively easy to show that the reaction threshold changes to [31]

$$\epsilon'_\mathrm{th} = \frac{2m_\mathrm{e}^2c^4}{E'(1-\cos\theta')} - \frac{\eta_n}{2(1-\cos\theta')}\left(\frac{E'}{E_{\mathrm{QG},n}}\right)^n E', \tag{6.11}$$

while the invariant mass becomes

$$s = 2E'\epsilon'(1-\cos\theta') + \eta_n\left(\frac{E'}{E_{\mathrm{QG},n}}\right)^n E'^2. \tag{6.12}$$

? Exercise

6.2. Work out the reaction threshold and the invariant mass for the Breit–Wheeler process in standard special relativity, and in LIV. Compare your results to Eqs. (6.11) and (6.12). Hint: Start by calculating the reaction threshold and invariant mass in the special-relativistic scenario. Then repeat the exercise using modified dispersion relation of the gamma ray. Note that in the latter case, because of the violation of the Lorentz symmetry, the calculation has to be done in the laboratory frame.

In both cases, the first term is the same as in the special relativity, while the second term is a consequence of the modification of the photon dispersion relation. The net result of these modifications is that, depending on the value of η_n, the gamma-ray absorption by the background fields will be weaker (for $\eta_n = -1$, i.e. subluminal behaviour), or stronger (for $\eta_n = +1$, i.e. superluminal behaviour) than in standard special-relativistic case. In other words, the universe will be most transparent to gamma rays in subluminal scenario, and least transparent in the superluminal scenario. Modified gamma-ray absorption is shown in Fig. 6.1. This phenomenon of modified universe transparency will be reflected on the observed spectra of astrophysical sources. To be specific, superluminal scenario will result in the observed spectra to appear even softer than the standard absorption. There is no clear way of resolving the effect of LIV in the superluminal scenario from what is expected in standard physics. Subluminal behaviour, on the other hand, will lead to harder spectra. On top of that, at high enough gamma-ray energies, the absorption

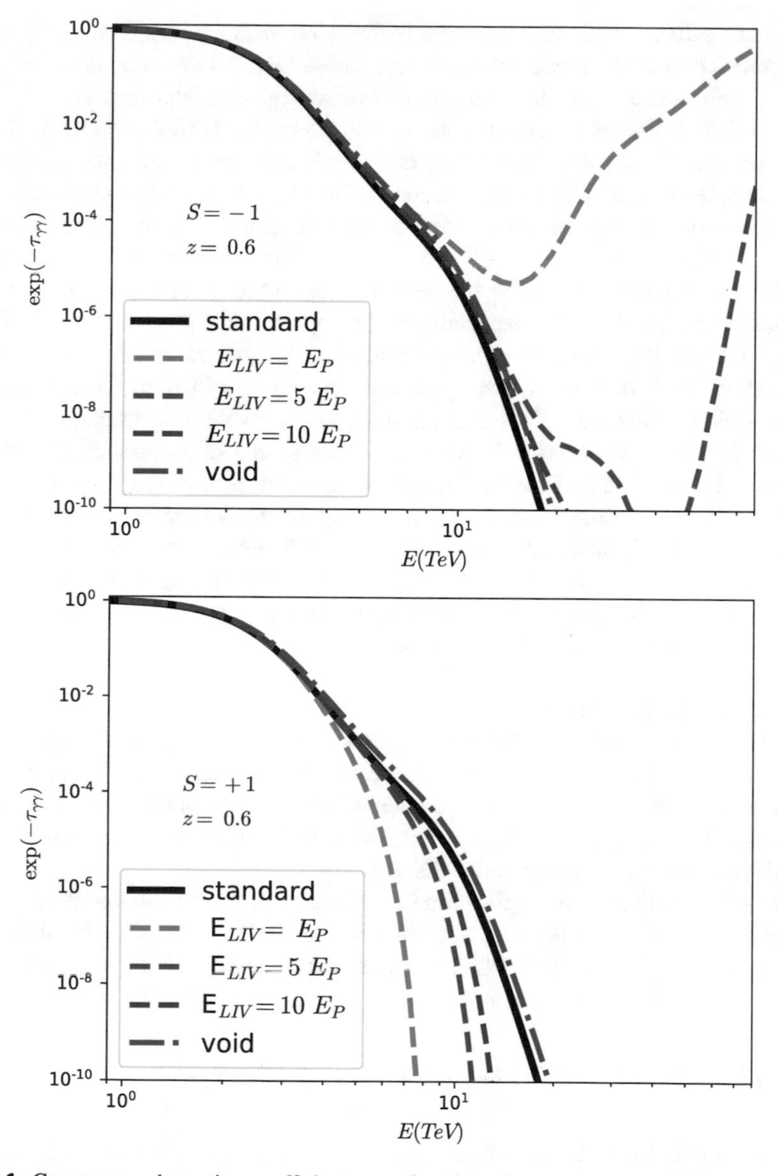

Fig. 6.1 Gamma-ray absorption coefficients as a function of gamma-ray energy for a source at redshift $z_s = 0.6$. The subluminal (here marked by $S = -1$) and superluminal (here marked by $S = +1$) scenarios are represented in the upper and lower plots, respectively. The standard special-relativistic scenario is represented by solid black lines in bot plots, while dashed lines represent modified absorption for different values of quantum gravity energy scale. In all cases $n = 1$ modification was considered. The dot-dashed blue line represents another effect investigated by the authors not connected to LIV. Figure adopted from [32]. ©AAS. Reproduced with permissions. All rights reserved

will cease entirely, and provided there is emission at those energies, the spectrum will correspond to the source intrinsic one. These features could be detected in the spectra, provided the source emits gamma rays at high enough energies.

We will not discuss here other possible consequences of LIV in so many details. However, we will say that, if real, they have similar effects on the observed spectra. For example, in the superluminal scenario the photon can become unstable and decay into an electron–positron pair, or split in several lower-energy photons, where splitting in three photons is the most dominant channel. As a result, not all emitted gamma rays will reach the detector. The effect is similar to the gamma-ray absorption by the EBL, especially in the superluminal case. However, photon decay and splitting would be seen as a cutoff in the spectra, rather than a gradual attenuation. Just as Breit–Wheeler process can be modified in LIV, so can the Bethe–Heitler, which describes gamma-ray interactions with an atomic nucleus. To be specific, the cross section in the subluminal scenario is significantly smaller, while in the superluminal scenario it remains quite similar to the standard special-relativistic one. In either case the number of gamma rays reaching Earth will not change with respect to standard special relativity. However, a smaller cross section means that particle showers will develop deeper in the atmosphere. Consequently, fewer showers will be detected, resulting in the observed spectra resembling the scenario of superluminal EBL absorption.

6.2.1.3 Analysis Methods

Several analysis methods have been proposed to test energy-dependent photon group velocity. Here we will focus on the maximum likelihood (ML) estimation method, most often employed by experimentalists. It is a powerful statistical method applicable to various problems. Alternative analysis methods can be found in e.g. [33–39] (see [40] for a comparative discussion).

We will first discuss the application in the time of flight measurements, for which this method was developed and first introduced in [41]. We start by defining a probability density function (PDF) for a photon of energy E to be detected at a time t, under certain conditions:

$$f^{(s)}(E, t) = \int_0^\infty G(E, E_{\rm true})\; A_{\rm eff}(E_{\rm true}, t)\; \Phi(t', E_{\rm true})\, dE_{\rm true}. \tag{6.13}$$

All our knowledge and assumptions about the emission process and the measurement technique are contained in this expression. In particular:

- **Energy resolution and bias**, $G(E, E_{\rm true})$, takes into account imperfections of the instrument. It is the probability for a gamma ray of energy $E_{\rm true}$ to be reconstructed as E. For this reason, the integral goes over all possible values of $E_{\rm true}$.
- **Instrument acceptance**, $A_{\rm eff}(E_{\rm true}, t)$, sometimes also called *effective area* or *collection area*, is the probability for the instrument to detect a gamma ray of energy $E_{\rm true}$ at a moment t. This function will be null for energies outside

of the instrument sensitivity range. Dependence on time represents changing observation conditions.

- **Emission and propagation effects** are contained in $\Phi(E_{\text{true}}, t')$. This includes the energy and temporal distribution of gamma rays at the emission, as well as propagation effects. The latter in turn include the redshift of the gamma-ray energy, gamma-ray absorption on the background radiation, or effects induced by LIV. The emission time t' is related to the detection time t in the following way:

$$t' = t + \xi_n E_{\text{true}}^n, \tag{6.14}$$

where the parameter ξ_n represents the strength of the LIV effect. It is introduced in the following way to facilitate numerical computations:

$$\xi_n = -\eta_n \frac{n+1}{2} \frac{1}{E_{\text{QG},n}^n} D_n(z_{\text{s}}). \tag{6.15}$$

Note that Eq. (6.14) contains only the LIV-induced modification to the time of flight. An additive constant, corresponding to the energy-independent time of flight between the source and detector, which is the same for all photons, does not affect the final result, and is usually ignored.

Obviously, the probability of detecting a photon of a certain energy at a given time will depend on the assumed distribution of gamma rays at the emission, the so-called emission template. While gamma-ray detectors are capable of precise measurements of arrival times for each photon, our ability to estimate the emission time for each photon emitted from an astronomical object is limited at best. Therefore, in order to make precise measurements on LIV, the emission time needs to be constrained in some way. Unlike in laboratory experiments, astrophysicists cannot control their sources. But we can choose the most adequate ones for a particular study. In this case, sources with strong and fast changes of flux make a good choice. Sources of highly variable gamma-ray flux are pulsars, gamma-ray bursts (GRB), and active galactic nuclei (AGN) in high-emission states, usually called flares.

Pulsars, with their millisecond pulses, provide a very strong constraint on the emission time, and are very reliable. On the other hand, only a handful of pulsars have been detected at energies of a few 100 GeV and above, and all of them inside the Milky Way, thus the LIV effects, if any, are less amplified by the distance. Gamma-ray bursts are violent explosions, usually associated with collapses of massive stars into black holes or mergers of binary neutron stars. These transient events are energetic enough to be seen in other galaxies and at large redshifts, and their short duration and variability provide strong constraints on the emission time. However they are entirely unpredictable. Instruments onboard the *Fermi* satellite are capable of detecting numerous GRBs because of their wide field of view. Cherenkov telescopes, on the other hand, observe only small portions of the sky at the time, and rely on other instruments to alert them of ongoing GRBs. However, repointing takes time, making GRBs notoriously difficult to catch. In addition, the signal above $\sim$100 GeV from GRBs at redshifts above $z_s = 1$ is attenuated because of gamma-ray absorption on extragalactic background light, posing an additional difficulty. So far, only four GRBs were significantly detected with Cherenkov telescopes, and only one was used for LIV study. We will take a closer look into that case shortly. Active galactic nuclei are persistent strong sources of gamma rays at distances from very small to large redshifts. However, steady emission is not particularly useful for time of flight studies. Flaring episodes, on the other hand, are characterised by fast changes of flux, which constrain the emission time and also provide richer data samples of gamma rays. Although the flux variability in AGN flares is not as fast as in pulsars or GRBs, these states of enhanced emission last longer than GRBs, can sometimes be predicted from observations in lower-energy bands, and are in general easier to catch.

? Exercise

6.3. Generate a data sample of 1000 gamma rays. Each event should be characterised with emission time t, given in seconds, and energy E, given in GeV. Let the distribution of emission times follow normal distribution, and distribution of energies follow a power law $\Phi(E) \propto E^{-\gamma}$, with $\gamma = 2.5$. Use Eq. (6.14) to calculate the detection time for each gamma ray assuming they were emitted from a source at redshift 0.5 and $E_{QG} = E_{Pl} = 1.22 \times 10^{19}$ GeV. Plot the distribution of arrival times and compare it to the distribution of the emission times. Do the same both for $n = 1$ and $n = 2$. Repeat the exercise for different values of E_{QG} and different redshifts.

Note that the same exercise could be performed for neutrinos, as well as cosmic rays. What is the problem when using cosmic rays for such analysis?

Now, let us consider an example of LIV study performed on gamma-ray burst GRB 190114C observed by the MAGIC telescopes [42], which was the first such study done on a gamma-ray data observed by Cherenkov telescopes. The MAGIC telescopes detected GRB 190114C above 1 TeV in energy [43]. The observations started 62 seconds after the burst, with the light curve showing smooth decay of the flux. The MAGIC observations results are shown in Fig. 6.2 with black points. Note that the scales are logarithmic, meaning that the flux changes very quickly. However, the decay is a monotone power law. One can easily demonstrate that adding an energy-dependent time delay to individual gamma rays will not change the overall shape of the light curve, therefore making it virtually impossible to resolve source-intrinsic effects from the ones induced by LIV. Fortunately, experts in gamma-ray bursts managed to create model of the gamma-ray emission based on observations in lower energy bands combined with theoretical inferences [44]. The result is represented with a full black line in Fig. 6.2. Using this model as a template for the temporal distribution of events provides a very strong handle on

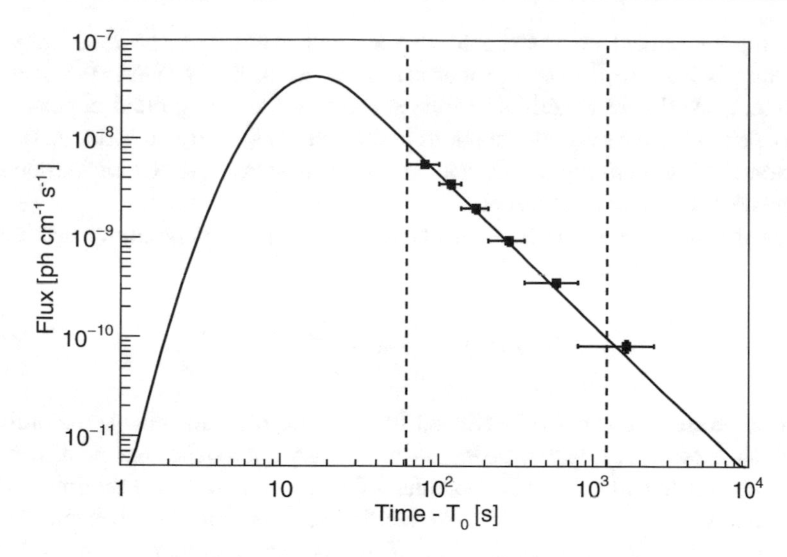

Fig. 6.2 Light curve of GRB 190114C above 300 GeV. Black points represent measured flux by the MAGIC telescopes. The full black line represents the emission template as reported in [44]. Reprinted from [42].

energy-dependent time delay. A very sharp peak in flux will strongly dominate the probability density function (Eq. 6.13). In this case, no change of spectrum with time was detected, so the entire emission template was obtained by folding the light curve with the observed spectrum.

Once we have the probability density function for each event in the data sample, we combine them in the likelihood function. It is the combined probability that a given set of data would be produced from some value of a parameter of the chosen statistical model (see, e.g., [45]). In our case, the data are N_{ON} events in the signal region, and the parameter of interest is the LIV parameter ξ_n (see Eq. 6.15). We construct the likelihood function by multiplying the complete probability density functions of all events in the signal region

$$\mathcal{L}(\xi_n) = \prod_{i=1}^{N_{\mathrm{ON}}} \left(p_i^{(\mathrm{s})} \frac{f^{(\mathrm{s})}(E_i, t_i)}{\int_{E_{\mathrm{min}}}^{E_{\mathrm{max}}} dE \int_{t_{\mathrm{min}}}^{t_{\mathrm{max}}} f^{(\mathrm{s})}(E, t) dt} + p_i^{(\mathrm{b})} \frac{f^{(\mathrm{b})}(E_i, t_i)}{\int_{E_{\mathrm{min}}}^{E_{\mathrm{max}}} dE \int_{t_{\mathrm{min}}}^{t_{\mathrm{max}}} f^{(\mathrm{b})}(E, t) dt} \right). \tag{6.16}$$

The product goes over all events in the signal region. As noted in Sect. 6.2.1.1, the signal region contains also some background events, which also need to be taken into account. The superscripts "s" and "b" stand for signal and background, respectively. $p_i^{(\mathrm{s})}$ is the probability that event i belongs to the signal, and respectively for background. The probability density function for background ($f^{(\mathrm{b})}(E_i, t_i)$) is similar to the one for signal, given in Eq. (6.13). The energy and temporal distributions of the background events is not necessarily the same as for the signal, while the instrument effects should be the same. However, the most important difference is that we do not consider a time delay in background. While it may be present, we do not exactly know the sources of the background effects, so we cannot determine the event distributions at the source; we only consider them at the detector. Putting it simply, we do not care what happens with the background, we just acknowledge that it is present.

Now that we have all the components of the likelihood function, we maximise it for ξ_n. To be more precise, we define test statistics (TS) as

$$L(\xi_n) = -2 \ln \frac{\mathcal{L}_0(\xi_n)}{\mathcal{L}(\xi_n)}, \tag{6.17}$$

where $\mathcal{L}_0$ corresponds to the likelihood function for the null model. According to the Wilks' theorem, the test statistics asymptotically approaches the χ^2 distribution under the null hypothesis [46]. The value of ξ_n for which L has minimum is the estimation of the true value that we are looking for. This procedure enables us also to determine the statistical significance of the measured value of ξ_n, as well as to produce confidence intervals. The strongest constraints on the energy-dependent photon group velocity for the order $n = 1$ were set based on the observation of GRB 090510 with *Fermi*-LAT to $E_{\mathrm{QG},1}^{(-)} > 2.2 \times 10^{19}$ GeV and $E_{\mathrm{QG},1}^{(+)} > 3.9 \times 10^{19}$ GeV, for subluminal and superluminal scenario, respectively [35]. As discussed in [40], the sensitivity to quantum gravity energy scales is some what different for orders $n = 1$ and $n = 2$ (see Table 6.1).

Thanks to the very fast change of flux and relatively large redshift of GRB 090510 the sensitivity to the 1st order modification was very high. When it comes to the 2nd order modification, the maximal energy in the sample carries more weight compared to other two parameters. That is where Cherenkov telescopes have the advantage over space detectors, and why the strongest constraints for the order $n = 2$ were set based on the observation of the blazar Mrk 501 with the H.E.S.S. telescopes: $E_{\mathrm{QG},2}^{(-)} > 8.5 \times 10^{10}$ GeV and $E_{\mathrm{QG},2}^{(+)} > 7.3 \times 10^{10}$ GeV, for subluminal and superluminal scenario, respectively [47]. The GRB 190114C had all three requirements met to a high degree, and while the LIV constraints were not the most constraining in either scenario, they were very close and supporting the results from the studies performed on GRB 090510 and Mrk 501.

From the experimental point of view, the maximum likelihood method is a very natural tool because it allows to directly include the information on the detector response in the analysis. Moreover, any other unknown can be introduced as a

Table 6.1 Sensitivity to $E_{\mathrm{QG},n}$ considering characteristics of the source and the sample. $E_{\max}$ is the highest gamma-ray energy in the sample, t_{var} is the shortest variability timescale in the light curve, and z_{s} is the redshift of the source

$E_{\mathrm{QG},1}$	$\propto$	$E_{\max}$	t_{var}^{-1}	$z_{\mathrm{s}}^{\sim 1}$
$E_{\mathrm{QG},2}$	$\propto$	$E_{\max}$	$t_{\mathrm{var}}^{-1/2}$	$z_{\mathrm{s}}^{\sim 2/3}$

nuisance parameter. On the other hand, one obvious downside of this approach is that it requires making certain assumptions about the emission processes and propagation of gamma rays. Our present knowledge of astrophysical sources is not nearly good enough to predict the exact emission time of each particular photon. Moreover, it is quite possible that the source-intrinsic processes are energy correlated, which could mimic or disguise effects of LIV (see, e.g., [48, 49]). It will still be quite some time before our understanding of sources emission mechanisms becomes precise enough to describe correlations. For the time being, a clever workaround is to combine sources at different distances. While the time delay coming from LIV clearly depends on the distance between the source and the detector, the source intrinsic processes are not expected to be distance dependent. Furthermore, different types of sources have different emission mechanisms, so by combining them in a single analysis the effect of possible correlations in source-intrinsic processes is decreased. This method was explored in [29]. So far, only a proof of concept was demonstrated on Monte Carlo simulations, with a study on real data promised to follow soon.

Probability density, likelihood, and test statistics, as statistical methods, are also used to search for other effects of LIV. Many astrophysical sources emit polarised photons. Provided that the polarisation of emitted light is known, one can measure the birefringence effect by comparing polarisation of the detected photons to the expected polarisation at the emission. However, even if there is no way of establishing the angle of polarisation at the emission, some processes, such as synchrotron radiation, emit strongly polarised photons. If the angle of rotation of the polarisation depends on the photon energy, after crossing astrophysical distances, the signal will be depolarised. Therefore, one can measure the degree of polarisation in the signal against expected degree of polarisation when no LIV is assumed. Such study was performed in [50], using broadband optical polarimetry of 1278 AGN and GRBs. Unfortunately, it is virtually impossible to measure gamma-ray polarisation with detection techniques which rely on particle showers. The highest energies used to measure vacuum birefringence were a few hundred keV. As modest as this may be compared to the PeV energies accessible to us nowadays, depolarisation measurements placed very strong constraints on the energy-dependent polarisation hypothesis.

As explained in Sect. 6.2.1.2, subluminal scenario can induce measurable features in the observed spectra of gamma-ray sources. Such features have not been detected so far, but strong constrains have been set to the quantum gravity energy scales. As in the time of flight studies, we do not know exactly the intrinsic spectra of sources that we observe. However, we do have certain expectations based on our knowledge of our sources. In universe transparency studies, as in the case of active galactic nucleus Mrk 501 observed with the H.E.S.S. telescopes shown in Fig. 6.3, one starts by assuming a source-intrinsic spectrum. In this case, a simple power law was assumed. Then, an absorbed spectrum is calculated using a certain EBL model (in this case, the authors decided on the one from [51]) and assuming no LIV, and fitted to the data. The resulting spectrum is shown with a black line in Fig. 6.3. Then

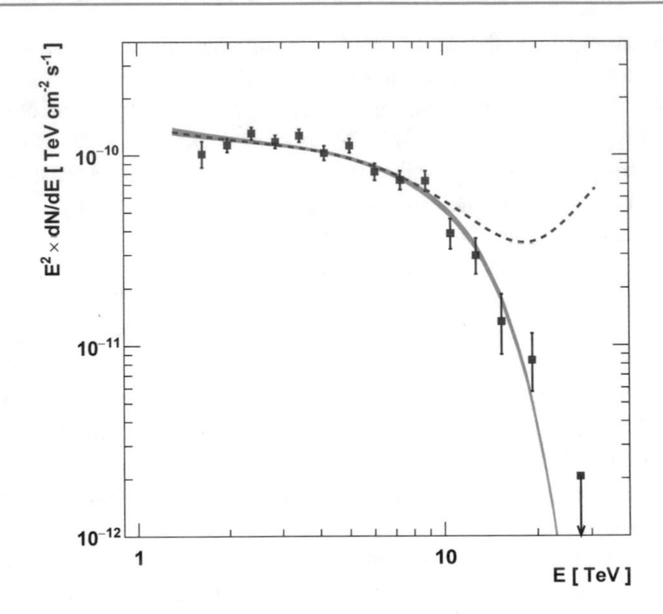

Fig. 6.3 Spectral energy distribution of blazar Mrk 501 observed with the H.E.S.S. telescopes. The black points and full black line represent the fit to the EBL-attenuated spectrum. A simple power law was assumed for the intrinsic spectrum. The expected spectrum observed for the same intrinsic shape but considering subluminal scenario with $n = 1$ and the quantum gravity energy scale corresponding to the Planck energy is represented by the dashed red line. Reprinted from [47]. ©AAS. Reproduced with permissions. All rights reserved

the procedure is repeated for different values of $E_{\rm QG}$ to test hypotheses of different levels of LIV. For each value of $E_{\rm QG}$ test statistics is computed according to

$$TS = \chi^2(E_{\rm QG}) - \chi^2(E_{\rm QG} \to \infty), \tag{6.18}$$

where $E_{\rm QG} \to \infty$ corresponds to no LIV. Putting all test statistics together creates a test statistics profile, which, in a similar manner as in the time of flight studies, allows estimating confidence intervals and constraining the quantum gravity energy scales.

In the study described here, it was assumed that electron dispersion relation is not modified, and the only effect of LIV is modified kinematics of the Breit–Wheeler process. How solid are these assumptions? If we had to take into account modifications of dispersion relations of several particles, and consider different LIV effects on top of that, it would be virtually impossible to obtain a very significant result. Fortunately, there are other types of studies which constrain these other effects, allowing us to neglect them. Let us start with the modification of electron dispersion relation. The modifying term in the electron dispersion relation for $n = 1$ was constrained based on the synchrotron radiation from the Crab nebula to seven orders of magnitude above the Planck energy [52]. Compton scattering is one of the most important processes for production of gamma rays in AGN. LIV could also result in anomalous Compton scattering, which would compete with other processes and effects, making them extremely difficult to resolve. However, based on several combined arguments, it was shown that possible LIV effects on the Compton scattering were unlikely to be relevant in realistic astrophysical environments [32]. Photon decay and splitting is only possible in the superluminal scenario, with no equivalent process in the subluminal scenario. As we argued earlier, gamma-ray absorption in the superluminal scenario becomes even stronger, and very difficult to resolve from source-intrinsic effects, which is why anomalous absorption is usually tested in the subluminal scenario. In that case, there is no danger of photon instability contaminating the results. However, if electromagnetic shower development in the atmosphere was modified in subluminal LIV scenario, it would have the opposite effect to the subluminal LIV gamma-ray absorption, and the two effects would compete against each other. It was, therefore, essential to test for one of these effects in an environment where the other was not present. The effect on the shower development was tested on the gamma rays from the Crab Nebula [53]. The Crab Nebula is located 2.0 ± 0.5 kpc from Earth, which is a too short distance for gamma-ray absorption on EBL to be significant, providing a clean data sample to constrain anomalous shower development. Influence of LIV on the extensive air shower development was constrained to $E_{QG,2} > 2.1 \times 10^{11}$ GeV (note that only order $n = 2$ was constrained). Considering all these constraints, gives us ground to test anomalous gamma-ray absorption independently. As in the time of flight studies, multiple sources can also be used to test the anomalous gamma-ray absorption. Eighteen spectra from six different AGN were used in [54] to set the strongest constraints on universe transparency to date: $E_{QG,1} > 6.9 \times 10^{19}$ GeV and $E_{QG,2} > 1.6 \times 10^{12}$ GeV.

6.2.2 Neutrinos

Neutrinos are excellent cosmic messengers since they are electrically neutral and interact only weakly (and gravitationally). They can therefore reach us without deflection or appreciable absorption directly from their sources. But they are very difficult to detect and that poses an experimental challenge. The ingenuity of experimentalists has made it possible to build large neutrino telescopes that can collect enough events to extract statistically meaningful measurements. Neutrino telescopes are large-volume detectors using open transparent media like water or ice both as a target and as Cherenkov medium. They detect the Cherenkov radiation emitted by particles produced in high-energy neutrino interactions and are able to reconstruct the original direction and energy of the neutrino from this information [55]. They are deployed at great depths to reduce the copious flux of muons produced in cosmic ray interactions in the atmosphere that leave tracks in the detectors. Although the main reason to build neutrino telescopes is to study the universe with high-energy neutrinos, were high-energy refers to neutrino energies above a few tens of GeV and up to several PeV, they have proved to be highly versatile detectors capable of addressing many topics related to fundamental physics in a competitive way to accelerator experiments [56–59].

The detector sizes, typically $O(\text{km}^3)$, are determined by the weak astrophysical neutrino flux they are built to study and by the tiny neutrino cross section with matter. There are currently three large neutrino telescopes in operation, IceCube (https://icecube.wisc.edu, [60]) at the geographic South Pole, KM3NET (https:www.km3net.org, [61]) in the Mediterranean sea and Baikal in lake Baikal (https://baikalgvd.jinr.ru, [62]). An additional project, P-ONE (https://www.pacific-neutrino.org, [63]) is under R&D in the Pacific ocean, off the coast of Canada, while ANTARES (https://antares.in2p3.fr, [64]), the predecessor of KM3NET off the coast of Toulon, in the Mediterranean, has recently been decommissioned after sixteen years of operation. Smaller underground neutrino detectors with an energy threshold of a few MeV, like Super-Kamiokande (https://www-sk.icrr.u-tokyo.ac.jp/sk/index-e.html, [65]), are complementary to their high-energy brothers in energy reach and can also competitively address fundamental physics topics.

The common design feature of all these detectors is an array of optical modules monitoring a volume of a transparent medium. The exact design of the optical modules depends on the detector, but in essence an optical module consists of one, or several, photomultiplier tubes with associated electronics to timestamp and digitize the detected signal. Charged-current neutrino interactions of any flavour on a nucleon N, $\nu_l + N \rightarrow X + l$, will produce the corresponding lepton, l, along with a cascade of particles X at the interaction vertex from the hadronization and decays of the interaction products, see Fig. 6.4. The muon produced in charged-current ν_μ interactions can travel several kilometers in matter, depending on material and energy. At the energies of importance in neutrino telescopes, muons lose energy mainly through bremsstrahlung, direct electron-positron production and photonuclear interactions when they traverse the detector medium. These processes

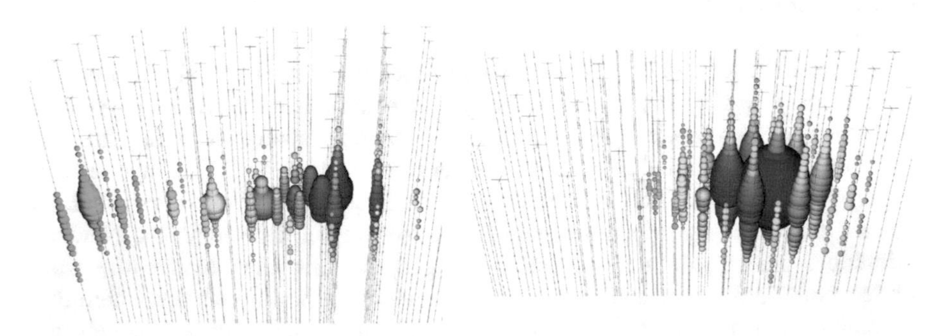

Fig. 6.4 Typical event patterns in a large-scale neutrino telescope. **Left:** A muon track from a ν_μ interaction. **Right:** The signature of the cascade of particles produced in a ν_e or ν_τ charged-current interaction or any flavour neutral-current. Each coloured dot represents a hit optical module. The size of the dot is proportional to the amount of light detected and the colour code is related to the relative timing of light detection: red denotes earlier hits, blue corresponds to later hits. Figure from [56], © Springer Nature. Reproduced under CC-BY-4.0 license

increase with energy but the energy loss scales as $1/m_\mu^2$ or $1/m_\mu$ for bremsstrahlung and pair production respectively. So high energetic muons will still penetrate long distances in media like water or ice, leaving a track in the detector which follows the original direction of the neutrino. Electrons from charged-current ν_e interactions lose energy mainly to bremsstrahlung due to their lower mass, creating gamma rays that induce an electromagnetic shower in the close vicinity of the interaction. Taus from ν_τ interactions are too short lived to leave a track and they also produce a cascade at the production point (which is indistinguishable from the decay point) that is very difficult to tell apart from an electromagnetic cascade of a charged-current ν_e interaction. Only taus of energies larger than 10s of PeV (not detected so far) can travel a few tens of meters before decaying, leaving a track between the production and decay points. Neutral-current interactions, $\nu_l + N \rightarrow X + \nu_l$, where the outgoing neutrino is not detected, will also leave a cascade-type signature in the detector at the interaction point. The absolute time resolution of the optical modules must be of the order of a ns to allow precise event reconstruction, and the detector is calibrated so that the amount of light detected can be translated to the total energy of the neutrino. Additionally, the global timing of the detector is kept synchronized to a master GPS also to the ns level to be able to do astronomy.

These measurements, neutrino arrival time, pointing accuracy and energy resolution, define the performance of a neutrino telescope. Typically the energy reconstruction for cascades is better than for events producing a muon (if this one is not contained, i.e., is produced and decays, in the detector volume). The typical energy resolution for cascade events in a neutrino telescope, depending on design, is 10% or better, while for through-going tracks it can be more than double that. Note that the geometric instrumented volume of a neutrino telescope does not coincide with its effective detection volume. The latter depends on the signal searched for and the event energy, so the effective volume of a neutrino telescope is analysis dependent and it can be both larger or smaller than the geometric volume.

There are several ways to use high energy neutrinos to search for LIV effects and there are excellent reviews on the subject [4, 6, 66–69] but we will concentrate here on two techniques that are especially relevant for the topic of this book: measuring the relative arrival times of the different messengers in a multi-messenger observation of a transient event, like a supernova, GRB, or flaring AGN, or to use neutrino oscillations, an interference phenomenon, which can provide a very sensitive probe of new physics.

Multi-Messenger Observations

In this approach the LIV signature is a deviation in the speed of propagation of energetic neutrinos with respect to electromagnetic radiation in the form of relative time of arrival. The time difference with respect to the standard (no LIV) scenario can be both positive (delay) or negative (pre-arrival) of high energy neutrinos with respect to photons. As for studies performed on gamma-rays alone, GRBs and flaring AGN are excellent candidates for these multi-messenger studies as well, see Sect. 6.2.1.2. The effect on the propagation of high energy neutrinos can be parameterized quite generically as a modification of the dispersion relation, as was done in Eq. (6.1). Note that at the energies of relevance, O(TeV) and above, neutrinos can be considered as massless particles in these studies. In practice there are two problems with this approach. First, the propagation time difference is redshift-dependent (see Eq. 6.3), and only about 20% of the well localized GRBs have an accurately measured redshift. The second problem is related to the reliability of the clock, i.e., to what extent neutrinos are emitted at the same time as, or close enough to, the burst of gamma rays. The evolution of GRBs pre, during and post the gamma-ray emission that is considered the time of the burst is still rather poorly known, and neutrinos could be emitted before, contemporaneously or after the gamma rays [70]. Indeed the fireball model of GRBs [71, 72] predicts that neutrinos can be emitted in different energy ranges during the precursor, burst and afterglow phases of a GRB. Searches for neutrino emission from GRBs with neutrino telescopes are therefore sometimes performed within a time window of up to several days around the detected time of the burst in gamma rays [73]. It will be difficult to assign a difference between the arrival time of neutrinos and electromagnetic radiation from a single GRB to LIV effects if indeed neutrino emission from a GRB is detected in the future. There have been several proposals to try to ameliorate this problem by using similar populations of GRBs. Assuming that the production of neutrinos and photons is similar in similar GRBs, the intrinsic relative delay between neutrinos and photons due to production processes at the source should be independent of distance, while it should depend on distance if the delay is due to LIV effects during propagation [74]. A similar experimental approach is to consider that if the arrival time correlation between gamma rays and neutrinos from GRBs follows a systematic pattern, e.g., neutrinos that can be associated with the position in the sky of an ensemble of similar GRBs, are detected systematically later or earlier than the electromagnetic radiation, that could indicate the effect of LIV [75]. However, it would still be difficult to unambiguously allocate such effect to LIV, instead of a common production mechanism of neutrinos in GRBs. We need to advance our knowledge on the sequence of physical processes that lead and follow a GRB event (a relativistically expanding environment, still poorly known) in order to further distinguish with confidence new physics effects in such objects. Note also that these methods depend on an accurate measurement of the distance (redshift) of the GRBs.

Even if the detection of an anomalous effect in the propagation of neutrinos would need a quite detailed knowledge of the time evolution of the source in order to unambiguously assign it to LIV effects, the non-observation of an unusual propagation can be used to limit the LIV scale without such knowledge of the source. Indeed the only two extragalactic objects that have been identified as neutrino sources so far, the supernova SN1987A and the blazar TXS 0506-056, have been used to such effect.

The supernova SN1987A was the first celestial object to be detected in neutrinos, along with optical, X-ray and radio observations [76]. The other object is the blazar TXS 0506-056, which has been identified as a neutrino source as well as an optical, gamma ray and radio source [77]. These two objects provide an actual example on how the ideas mentioned in the previous paragraph can be adapted to set strong limits on the scale of LIV effects. SN1987A was a supernova in the Large Magellanic Cloud, at a distance of 51.4 kpc, observed on February 23rd 1987. A total of 25 neutrinos were detected by three neutrino detectors in operation at the time (Kamiokande [78], IMB [79] and Baksan [80]. See also [81]) during a short time interval of about 13 seconds, three hours before the visible light was detected. The total amount and energy of the detected neutrinos ($\mathcal{O}(10)$ MeV) and the duration of the burst was found to be in accordance with the predictions of how a type-II supernova forms [82]. A supernova explosion is a complex process where neutrinos are expected to be produced during different stages of the explosion, through different process (β decays of star material, e^+ and e^- capture by nucleons or e^+e^- and $\nu\bar{\nu}$ annihilations) while they are thermalized through elastic scattering with nucleons, electrons and positrons. Although all neutrino flavours escape within a few seconds during the collapse, if one could follow the different phases of the explosion with subsecond resolution, the development of the relative neutrino flux per flavour can be predicted as the collapsing star goes through distinct density phases. Different neutrino flavours feel different opacities in the medium due to their specific flavour-dependent interactions, and free streaming from the system occurs at different times. However none of the detectors mentioned above had such timing resolution, neither sensitivity to all neutrino flavours. The neutrinos detected from SN1987A were probably $\bar{\nu}_e$, through the process $\bar{\nu}_e + p \rightarrow n + e^+$, which has a dominant cross section over the process $\nu_e + e^- \rightarrow \nu_e + e^-$ at MeV energies (the Kamiokande collaboration estimated that only one neutrino among the 12 detected originated from the latter process). But neither a detailed picture of neutrino emission, nor of the relative time of neutrino emission with respect to photon release is necessary to set a limit on LIV using the neutrino signal from Kamiokande, IMB and Baksan. LIV effects on the propagation of neutrinos would result in an energy-dependent arrival time (see Eq. (6.3) in the case of known redshifts z_s), so the measured time span of the burst in the detectors, 13 seconds, can be used to set a limit on the strength of LIV effects, without any further knowledge of the neutrino emission sequence at the source.

Crude as it is, this method gave at the time a much better limit on the scale of LIV than any obtained from neutrino beams in accelerators. If LIV effects on the propagation of neutrinos are parametrized at first order as $v/c = [1 \pm (E/E_{\mathrm{QG}})]$ (compare with Eq. 6.2 for n=1), a limit on E_{QG} can be set requiring that the time spread in the arrival of the neutrinos is not larger than the observed window. This gives a 95% confidence level limit $E_{\mathrm{QG}} > 2.7(2.5) \times 10^{10}$ GeV for subluminal(superluminal) neutrinos (and $E_{\mathrm{QG}} > 4.6(4.1) \times 10^4$ GeV if the dependence on energy is quadratic, i.e., $(E/E_{\mathrm{QG}})^2$) [38]. These are limits that are about five orders of magnitude more constraining than similar ones obtained from an analysis of MINOS data [83].

The second object that has been used to extract limits on LIV is the blazar TXS 0506-056. This blazar has a redshift $z = 0.3365 \pm 0.0010$ and it is known to present flaring episodes, as Fig. 6.5 illustrates for the case of gamma rays. On September 22nd 2017 IceCube detected a neutrino with an estimated energy of 290 TeV from the direction of this object (vertical red dashed line in Fig. 6.5) in

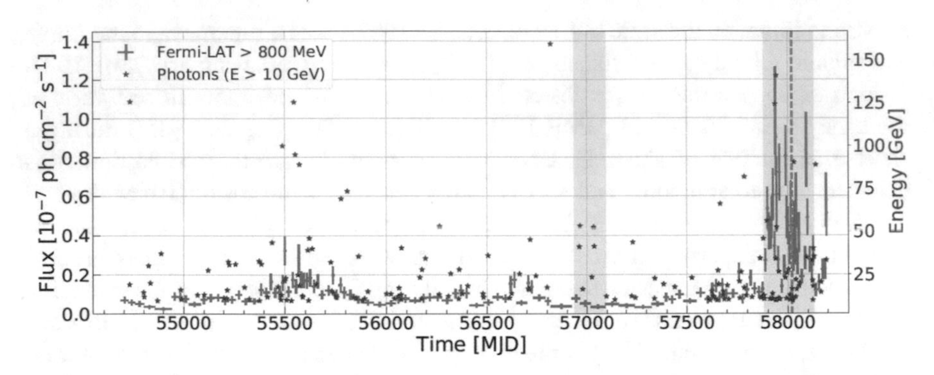

Fig. 6.5 Gamma-ray flux of TXS 0506+056 versus time, integrated above 800 MeV. The green vertical band denotes the gamma-ray flare of 2017/2018. The vertical red dashed line marks the IceCube neutrino alert IceCube-170922A and the vertical yellow band the neutrino flare detected by IceCube in archived data from 2014. Reprinted from [84]. ©AAS. Reproduced with permissions. All rights reserved

coincidence with a gamma-ray flaring period of several weeks. Even if the neutrino emission time can not be assigned to a given specific gamma-ray flare to be able to assign a common clock between photons and neutrinos, the fact that neutrinos are detected within about 10 days of the photons can be already used to extract limits on the difference between their respective velocities in vacuum in a similar way as discussed above for SN1987A [85, 86]. A difference in velocity between neutrinos and electromagnetic radiation induces a difference in arrival time of $\Delta t = \Delta v\, D$, where D is the distance to the object. Assuming a LIV effect linearly proportional to energy, $\Delta v = -E/E_{\rm QG}$ where $E_{\rm QG}$ has the same interpretation as above, a limit of 10 days in Δt translates into a limit of $E_{\rm QG} \gtrsim 3 \times 10^{16}$ GeV. This is still below the Plank mass, where quantum gravity Lorentz violating effects are assumed to surface. But it is a much more restrictive limit that the one obtained from SN1987A, illustrating the strength of using cosmological distances in these kind of studies.

From the experimental point of view, there is a distinctive difference between using supernovae neutrinos with energy of a few ten MeV or high energy astrophysical neutrinos of TeV energies and above. MeV neutrinos do not provide a precise pointing in neutrino detectors and the association of a MeV neutrino burst with a supernova is done purely on timing. The electrons and positrons produced in the reactions $\nu_e + e^- \rightarrow \nu_e + e^-$ and $\bar{\nu}_e + p \rightarrow n + e^+$ quickly scatter (and the positron annihilates), losing information from the direction of the incoming neutrino (the angular distribution of the SuperK and IMB SN1983A events is practically isotropic with respect to the position of the source). For larger, km^3-size, neutrino telescopes the situation is not better due to the large distance between optical sensors, up to $O(100)$ m, in comparison with the short electron (and positron) tracks of less than a meter length (the mean free path of a 10 MeV electron in water/ice is about 5.6 cm [87]). However, the copious neutrino flux expected from a supernova explosion (between $10^{57} - 10^{58}$ neutrinos in total) induces an increase of events in the detectors during a few seconds, which constitutes the signal. In the case of neutrino telescopes a supernova signal would consist of a coherent increase in optical sensor noise across the whole array for the few seconds of the burst, but without the possibility of identifying individual electrons or positrons from the neutrino interactions, neither with the possibility of pointing to the source, e.g. [87].

On the other hand, TeV-PeV neutrinos accompanying GRBs or AGN flares are extremely rare and good pointing is needed to associate the neutrino with the object, which is only achieved with the long muon tracks produced in charged-current ν_μ interactions. Therefore, of the three neutrino flavours, only one is usable in practice if a precise association with the position of the source is required. The advantage is that the expected background from atmospheric neutrinos from the same direction and within the chosen time window can be negligible, less than 0.01 events in a kilometer cube detector during the time of a few seconds burst, or a few events if the time window is of the order of several days (of course larger time windows allow for a higher probability of background sneaking in, weakening the usefulness of the method).

There is another way of using multimessenger probes to study LIV effects without the need of relative timing by taking into account the fact that LIV allows processes like "vacuum neutrino bremsstrahlung",[1] $\nu \rightarrow \nu\gamma$ [88]. In this scenario, the astrophysical neutrino flux would be a source of gamma rays, contributing to the total diffuse flux that has been measured by different experiments, e.g. [89]. By requiring that neutrino bremsstrahlung due to LIV should not contribute to the diffuse gamma-ray flux in a way that contradicts the measurement, a limit on the strength of LIV can be established. Another consequence of the above mentioned process, along with $\nu \rightarrow \nu e^+ e^-$, which is also allowed in LIV models [90], would be a new "GZK-like" effect (see Sect. 6.2.3) for high energy neutrinos. Astrophysical neutrinos would lose energy during their propagation, appearing at lower energies in the spectrum and showing a cutoff at an energy scale that depends on the LIV strength. The observation of astrophysical neutrinos up to a given energy can thus be used to set limits on LIV [69, 91–95]. These two methods rely on a precise measurement of the astrophysical neutrino flux and knowledge of whether it presents a cutoff or not (and at what energy if that is the case), something that is not established at this point. Still, the sheer observation of the first PeV neutrinos by IceCube, implying that any cutoff in the neutrino spectrum lies above such energy, allowed to improve limits on several LIV-coefficients of the SME by up to 20 orders of magnitude (!) with respect to previous limits at the time [96].

Flavour Interferometry

This technique makes use of the two classes of "beam" that a neutrino telescope is subject to: the copious, but relatively short-baseline (at most Earth-diameter), atmo-

[1] This process is usually known as vacuum Cherenkov emission although it does not really resemble Cherenkov radiation since the usual Cherenkov radiation is emitted by the media which the relativistic particle traverses, not by direct radiation from the particle itself.

spheric neutrino flux or the much weaker, but cosmological-baseline, astrophysical flux. Both have their advantages and disadvantages.

The advantage of using atmospheric neutrinos to address new physics with neutrino telescopes is that the flux is relatively well known. "Conventional" atmospheric neutrinos arise mainly from the decay of pions and kaons produced in cosmic ray interactions and their spectrum follows a power-law as $\Phi(E) \propto E^{-\gamma}$, with γ typically quoted as -3.7, although the actual index depends on the energy range under consideration. This flux has been measured by several experiments to a good agreement with theoretical calculations [97]. A "prompt" component arises from the decay of heavier D_s and B_s mesons containing a charm or bottom quark. This prompt flux can become comparable to the conventional atmospheric neutrino spectrum at energies above several hundred TeV. Incidentally, this is the energy where the astrophysical neutrino flux becomes also stronger than the conventional atmospheric neutrino flux, so the prompt flux remains undetected so far. IceCube measurements of the neutrino flux above 100 TeV provide only upper limits on the contribution of the prompt flux, with a lower limit compatible with zero [98].

Without entering into the known details of neutrino oscillation theory (see, e.g., [99] for a review) let us just mention a few facts that will be useful to understand how they can be used to detect LIV effects. Assuming a two-flavour scenario for illustration purposes, a neutrino state of flavour α, $|\nu_\alpha\rangle$, can be expressed as a superposition of the mass states 1 and 2, with masses m_1 and m_2 respectively,

$$|\nu_\alpha(t)\rangle = -\sin\theta e^{-iH_1t}|\nu_1\rangle + \cos\theta e^{-iH_2t}|\nu_2\rangle\,, \tag{6.19}$$

where the Hamiltonian for free-propagating neutrinos, $H_{1,2}$, is based on the dispersion relation $E_i^2 = p_i^2 + m_i^2$ $(i = 1, 2)$ and θ is the "mixing angle", used to parametrize the mass composition of the flavour state while keeping unitarity. The neutrino flavour at the detector, $|\nu_\alpha\rangle$, is related to the flavour at production, $|\nu_\beta\rangle$, by the known relation

$$|\nu_\alpha\rangle = \sum_{\beta=e,\mu,\tau} P\left(\nu_\beta \to \nu_\alpha\right)|\nu_\beta\rangle_S\,, \tag{6.20}$$

where S stands for "source" and the transition probability $P\left(\nu_\beta \to \nu_\alpha\right)$ represents the standard flavor oscillations in the absence of new physics, which is linearly proportional to the square mass difference of the mass eigenstates and the propagation length, and inversely proportional to the neutrino energy,

$$P\left(\nu_\beta \to \nu_\alpha\right) = \sin^2 2\theta \sin^2\left(\frac{\Delta m_{1,2}^2 L}{4E}\right) \tag{6.21}$$

where $\Delta m_{1,2}^2 = m_2^2 - m_1^2$, measured in eV^2, E is measured in GeV and L in km, and where it has been assumed that neutrinos are relativistic and one can set $t \sim L$ in Eq. (6.19), the distance traveled between production and detection.

? Exercise

6.4. Derive the neutrino flavour transition probability for a two-flavour scenario (see Eq. 6.21). Then repeat the derivation assuming modified neutrino dispersion relation to obtain Eq. 6.22.
Hint: follow the procedure from [100].

Figure 6.6 illustrates the idea behind using atmospheric neutrinos to measure neutrino oscillations. If the neutrino energy can be measured to a suitable precision, and it can in neutrino telescopes, then the path-length from production to the detector is related to the arrival direction of the neutrino, which in local detector

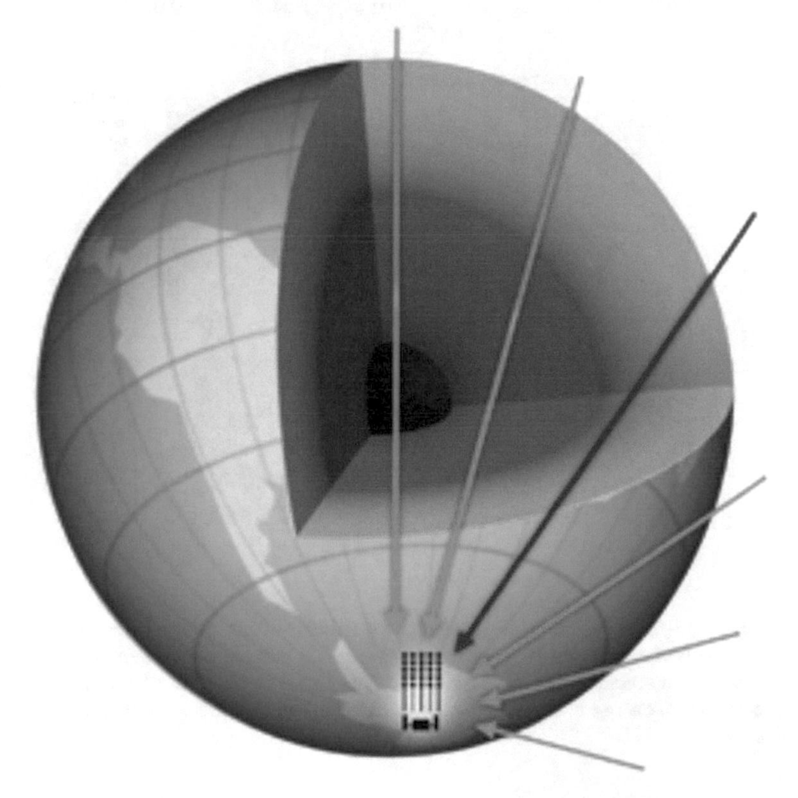

Fig. 6.6 Measuring neutrino oscillations with atmospheric neutrinos. Since the flavour oscillation probability depends on the path-length from production to detector, a measurement of such path-length for atmospheric neutrinos is the zenith angle they arrive to the detector. A measurement of zenith angle and energy shows the typical flavour oscillation pattern, as shown in Fig. 6.7. Illustration courtesy of IceCube collaboration.

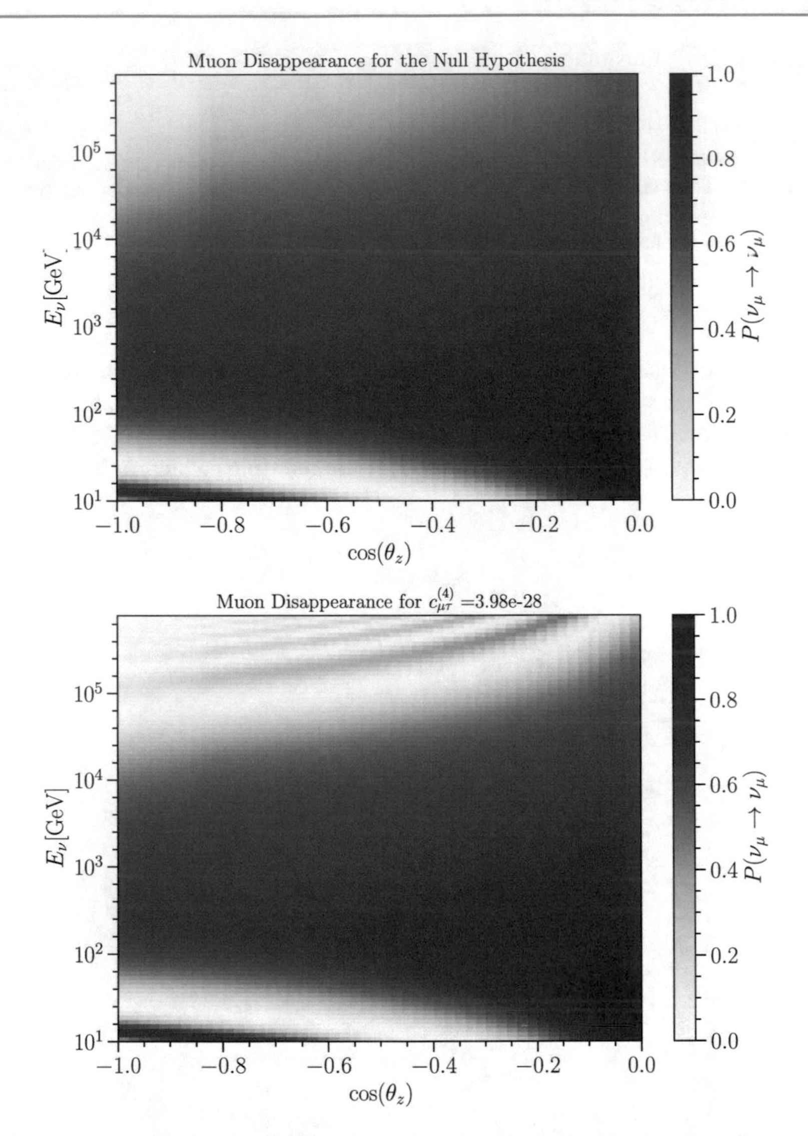

Fig. 6.7 $\nu_\mu \rightarrow \nu_\mu$ survival probability, denoted as the color code, as a function of energy and arrival direction between vertically up-going ($\cos\theta_z = -1$) and horizontal ($\cos\theta_z = 0$). **Upper:** Standard oscillation scenario. **Lower:** Oscillations under the assumption of a Lorentz-violating term proportional to energy with strength set to the current limit. Plots courtesy of B. Skrzypek.

coordinates is the zenith angle. So a measurement of L/E is possible in Eq. (6.21). The left plot of Fig. 6.7 shows the expected atmospheric ν_μ disappearance probability under the assumption of standard oscillations as a function of neutrino energy and arrival direction.

As a simple illustration on how anomalous flavour-changing effects due to Lorentz invariance violation can modify the energy and zenith angle distributions of atmospheric neutrinos let us consider the following simple example. As mentioned above, one of the effects of LIV can be the modification of the dispersion relation for massive particles (Eq. 6.1). Let us assume here for generality that the parameter η is eigenstate dependent, η_i (here η can still be positive or negative, but is not restricted to values ± 1 any more). Under this assumption, the Hamiltonian in Eq. (6.19) is different from the free-propagation case, and a similar calculation that yields Eq. (6.21) now gives (e.g. [100]),

$$P\left(\nu_\beta \rightarrow \nu_\alpha\right) = sin^2 2\theta sin^2 \left(\frac{\Delta m_{1,2}^2 L}{4E} + \frac{\Delta\eta E^{n+1} L}{4E_{\mathrm{P}}^n}\right) \tag{6.22}$$

where $\Delta\eta = \eta_2 - \eta_1$. If $\Delta\eta = 0$, i.e., there is no difference in the propagation of the two neutrino mass eigenstates due to LIV effects, we recover the standard oscillation formula of Eq. (6.21). Note that in this scenario neutrino oscillations are only sensitive to differences in the strength of LIV effects on mass eigenstates, and not on the individual ηs, in a similar way that standard oscillations are sensitive to the difference of the masses squared and not to individual masses. The previous parameterization of LIV effects on neutrino propagation provides a useful way to perform experimental searches, since the parameters $\Delta\eta$ and n can be probed by measuring L and E in neutrino telescopes and comparing the result to the expected oscillation pattern without LIV. The most common way to extract limits on LIV effects is by using oscillograms, two-dimensional plots of the oscillation probability versus E and $\cos(\theta_z)$, as the ones shown in Fig. 6.7. By comparing the data with the expected pattern of a LIV model, the parameters of the model in question can be constrained [100, 101]. Note that in the example shown in the right plot of Fig. 6.7 the effects of LIV appear at high energies, where neutrino telescopes are specially sensitive with respect to smaller neutrino detectors or accelerator experiments. Indeed Super-Kamiokande data were readily used already in 1999 to search for an anomalous oscillation pattern due to new physics, among it LIV, and limits set at the level of $|1 - \beta_\nu| < 10^{-24}$, where β_ν represents the neutrino velocity in units of c [102].

A rather straightforward way to search for LIV effects with atmospheric neutrinos is to exploit the fact that the relatively short distance of travel for horizontal neutrinos ($\cos(\theta_z) = 0$) is not enough to develop any appreciable spectral distortion due to LIV, even at high energies. However, the effect becomes maximal for up-going neutrinos crossing the Earth ($\cos(\theta_z) = -1.0$), as can be seen on the right plot of Fig. 6.7. So the ratio of transition probabilities of vertical events to horizontal events can be used to determine, or set limits to, LIV parameters within a given model. Indeed the Super-Kamiokande, ANTARES and IceCube collaborations have performed searches for LIV using this technique [100, 101, 103]. The left plot of Fig. 6.8 shows such ratio from a search for LIV effects with atmospheric neutrinos by the IceCube collaboration. An example of how specific parameters of the Lorentz

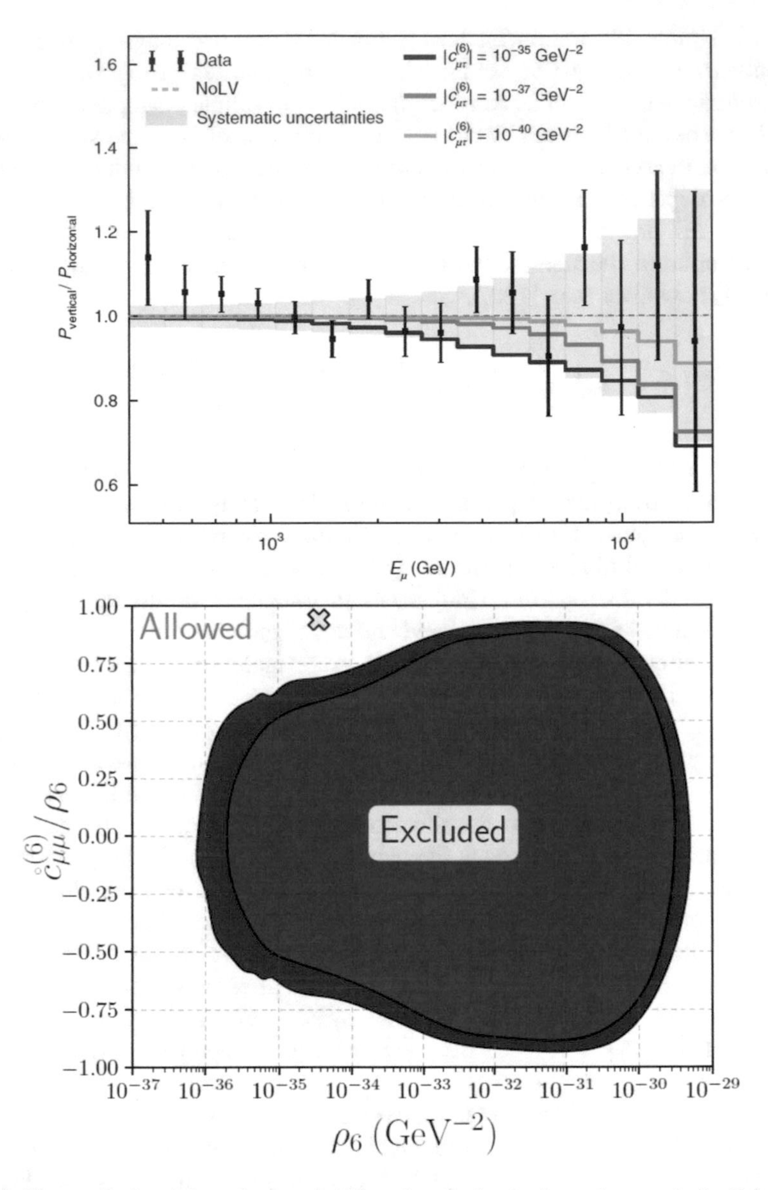

Fig. 6.8 **Upper:** Ratio of the arrival probability of vertical to horizontal events in IceCube, as a function of energy. The coloured lines show the prediction for different strength of LIV effects, defined as different values of the relevant parameters in the SME Hamiltonian. The null hypothesis (expected event ratio under standard oscillations) is shown as the dashed, straight horizontal line. Data is shown as the black dots with accompanying statistical error bars. **Lower:** Example of the excluded parameter space for one of the SME coefficients responsible for LIV. The parameter $c^{(6)}_{\mu\mu}$ represents the contribution to Lorentz violation that can be extracted from muon disappearance measurements. The parameter ρ_6 represents the total strength of the Lorentz violation effect. The super/sub-script 6 indicates the dimension of the operator in the SME Hamiltonian. The best-fit point is shown by the yellow cross and the blue (red) region is excluded at 99% (90%) confidence level respectively. Both figures reproduced from [103]. ©Springer Nature. Reproduced with permissions. All rights reserved

invariance terms in the Lagrangian can be constrained is shown in the right plot of Fig. 6.8, with more examples available in the additional material of reference [103]. This is a simple and powerful approach to access LIV effects in the neutrino sector, thanks to the high sensitivity that the oscillation patterns shows to deviations from the standard scenario.

An intriguing possibility of breaking Lorentz invariance is the introduction of an anisotropy in the space-time (equivalently, a preferred spatial direction) due to the coupling of the Lorentz violation terms with the neutrino propagation vector $\mathbf{p}$. This will induce an oscillation probability that is dependent on the arrival direction of the atmospheric neutrinos which, since the detector is fixed to the Earth, means a time-dependent event rate at the detector in the form of a sinusoidal neutrino disappearance signal. Searches for such an effect have been performed with IceCube data by measuring the event rate as a function of right ascension and, even with a limited data sample [104], competitive constraints were set to the coefficients of the SME Hamiltonian responsible for the predicted anisotropy. This is an effect that can also be searched for with accelerator neutrino beams at a complementary energy regime as has been reported in [105–107]. An equivalent analysis was performed by the SNO collaboration, but using solar neutrinos [108]. In this case the pathlength is the radius of the orbit of the Earth and the expectation is a modulation on the oscillation probability of electron neutrinos over the course of a year, as the Earth moves in the frame of the Sun. The results of the SNO analysis set limits for the first time on 38 previously unconstrained parameters of the SME and confirmed that LIV effects do not occur below an energy scale of about 10^{17} GeV.

However, the most sensitive search for LIV effects is achieved with astrophysical neutrinos. The extremely long path-lengths can compensate the weak expected effects and the results of searches can probe up to the Planck regime. In principle this approach relies on an assumption about the original neutrino flavour ratio at the source, something unknown, and on the shape of the neutrino energy spectrum, something currently measured with limited precision [98]. If neutrinos in astrophysical sources are produced in pion and kaon decays, the same flavour ratio at the source is expected as in atmospheric neutrinos, $\nu_e : \nu_\mu : \nu_\tau = 1 : 2 : 0$. But that might not be the case in scenarios with strong magnetic fields, where muons can lose a significant fraction of their energy before decaying, and the high energy neutrinos produced arise only from the direct pion decay. In this case the expected flavour composition at the source is $\nu_e : \nu_\mu : \nu_\tau = 0 : 1 : 0$ [109]. On the other hand, neutron-rich sources can emit neutrinos through neutron decay producing a flavour composition at the source of $\nu_e : \nu_\mu : \nu_\tau = 1 : 0 : 0$ (although this scenario seems to be disfavoured by current IceCube data [110]). It would seem therefore that an exact knowledge of the initial flavour composition at the source is key to be able to study LIV effects arising during cosmological propagation of neutrinos. In practice, however, the dependence on the source flavour composition is ameliorated since the flavour composition at Earth, after the neutrinos travel over cosmological distances, is reduced to a small region around $\nu_e : \nu_\mu : \nu_\tau = 1 : 1 : 1$ in the flavour parameter space, practically independent of the original flavour composition at the source. The left plot of Fig. 6.9 illustrates this effect [111]. The three axes of the

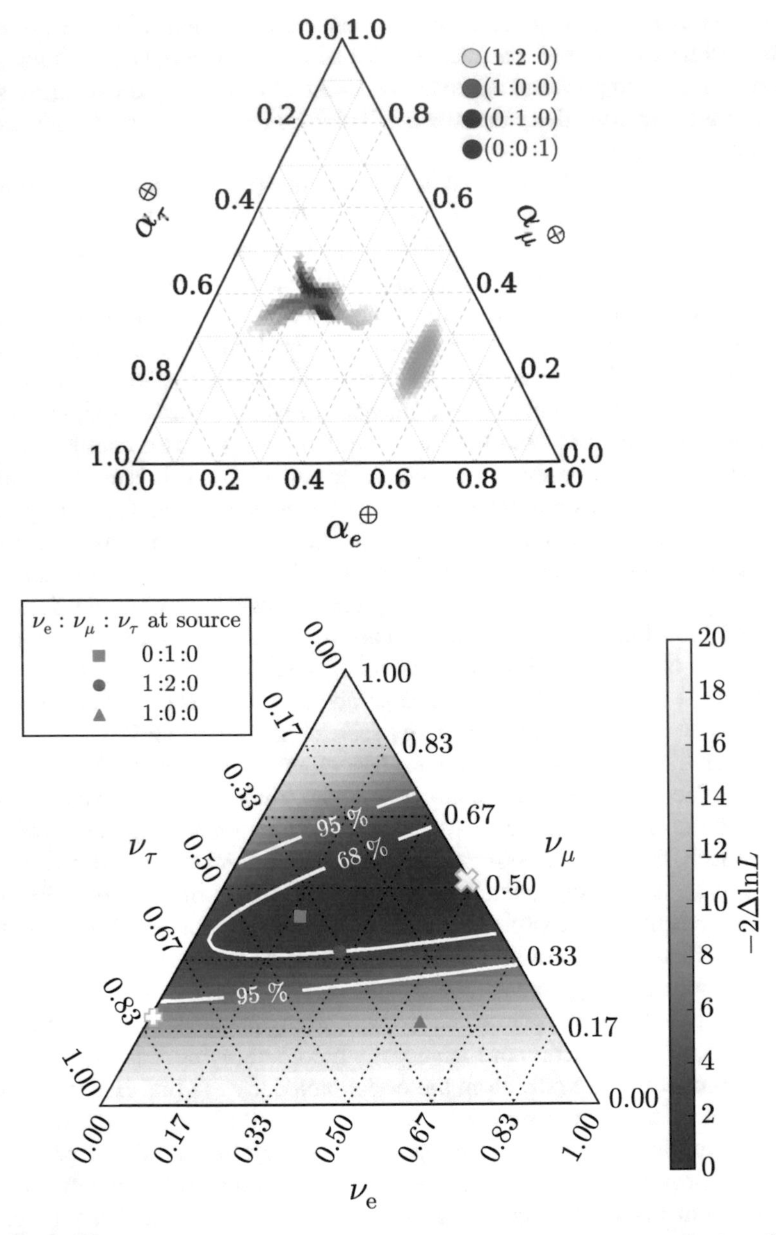

Fig. 6.9 **Upper:** Allowed neutrino flavour composition at Earth (colored areas) when different flavour compositions at the source are assumed (corresponding colour point). The original flavour composition is washed out by standard oscillations over cosmological distances. The three axes of the plot show the fraction of each neutrino flavour, with pure ν_e, pure ν_μ and pure ν_τ at the respective vertices. Current neutrino oscillation parameters have been used. Any measured flavour ratio of astrophysical neutrinos outside the coloured areas would point to new physics, rather independently of any source composition. Figure from [111]. © APS. Reproduced with permission. All rights reserved. **Lower:** Current measurement of the flavour triangle with

plot show the fraction $\alpha_{(e,\mu,\tau)}$ of each neutrino flavour, with pure ν_e, pure ν_μ and pure ν_τ at the respective vertices. The fraction of each flavour at any point in the triangle is obtained by projecting the point onto each axis in the order (e, μ, τ) and in the direction that points closer to the origin of each axis, so that the sum of the projections is one. The coloured regions show the allowed flavour composition at Earth starting from a flavour composition at the source indicated by the color code, assuming standard oscillations and no new physics. Although there is some "memory" of the source flavour ratio, the allowed flavour compositions at Earth concentrate around well delimited regions. A measured flavour ratio outside these regions points to new physics. This is a very powerful method to access new physics effects, not only due to LIV [112], in a way that does not depend on relative arrival timing between different messengers and it is quite independent of the neutrino production process at origin. The right plot of Fig. 6.9 shows results from the IceCube collaboration obtained with the high-energy astrophysical neutrino flux [113]. The white cross in the plot marks the best-fit point from the analysis and the allowed 68 and 95% confidence level regions on the flavour composition of astrophysical neutrinos are marked with the white lines. The plot shows that current data is still compatible with the expectation of standard oscillations and typical source flavour compositions, marked by the orange, red and green points in the upper-left caption of the figure with the same color. More data is needed to reduce the allowed regions and make a more precise measurement of the flavour ratio to be able to establish if any deviation from the null hypothesis (no LIV) is present.

6.2.3 Cosmic Rays

While any particle arriving at the Earth from outer space can be considered a cosmic ray, in this section we will just focus on protons and heavier nuclei of ultrahigh energies (10^{18} eV and above).

The interaction of a cosmic ray with an atom in the higher atmosphere, typically occurring at an altitude of about 25–30 km, produces a cascade of particles that develops through the atmosphere reaching a maximum particle density at a depth that depends on the interaction energy. Stable particles can reach the ground over an area that can cover several km^2. The direction and energy of the primary cosmic ray

Fig. 6.9 (continued) astrophysical neutrinos by the IceCube collaboration. The best-fit point is marked with a white cross ("x") and the 68% and 95% contours are indicated with the white lines. The orange, red and green points mark the expected composition at the detector from the assumptions at source given in the upper-left caption of the figure with the same color. Although a neutron-rich source composition is disfavoured, the results are compatible with standard oscillations and a source flavour composition of both $\nu_e : \nu_\mu : \nu_\tau = 1:2:0$ or $\nu_e : \nu_\mu : \nu_\tau = 0:1:0$. Figure from [113].

Fig. 6.10 Illustration of the hybrid detection of an air shower in the Pierre Auger Observatory. The white dots represent the surface water Cherenkov tanks while the semicircles show the location and field of view of the fluorescence telescopes. The red line represents the original direction of the primary cosmic ray. The blue curve shows the shower particle density as a function of atmospheric depth, directly accessible through the measurement of fluorescence light by the four telescopes. Muons and electrons that reach the surface trigger several of the water Cherenkov tanks, represented by the coloured dots. Reproduced from [116]. ©APS/Carin Cain. Reproduced with permissions. All rights reserved

are reconstructed with the help of large surface arrays of particle detectors.[2] Due to the impossibility to tightly instrument such large surface areas, air shower arrays are segmented detectors, consisting of "stations" separated by a given distance and specialized in detecting electrons and muons. We already discussed water Cherenkov tanks and/or scintillators in Sect. 6.2.1. They are commonly used to detect the electromagnetic component of the shower at ground level, but the array can be supplemented with telescopes that detect the fluorescence light produced by the excitation of nitrogen atoms in the atmosphere during the development of the particle shower, see Fig. 6.10. The largest air shower arrays in operation are the Pierre Auger Observatory in Argentina (https://www.auger.org, [117]), Large

[2] Cosmic rays can also be detected directly by placing particle detectors in space, e.g. [114, 115], although the necessarily small detection area of these kind of detectors limit their energy reach and detection rate.

High Altitude Air Shower Observatory (LHAASO) in Tibet [118] and the Telescope Array in Utah (http://telescopearray.org, [119]), with primary energy thresholds of 10^{17}, 10^{12}, and $10^{16.5}$ eV respectively. The development and ground footprint of an air shower started by a high energetic gamma, a proton or heavy nucleus are slightly different, and they can be distinguished. In this section we discuss only the cosmic ray detection capabilities of air shower arrays since photon detection has been discussed in Sect. 6.2.1.

? Exercise

6.5. A cosmic ray (proton) propagates towards Earth with the kinetic energy T and scatters on a stationary proton in the atmosphere. Calculate the kinetic energies of these two protons in the center of mass system. What is the cosmic ray energy that would be equivalent to proton–proton collision at the LHC at 14 TeV?

The original direction of the cosmic ray primary can be obtained by measuring the relative arrival time to each detector unit of the particles that reach the ground, while the amount of particles detected at the ground is a proxy for the primary energy. Both the measurement of primary direction and energy is aided by the measurement of the fluorescence light during the development of the shower in the atmosphere. Note that the estimation of the energy of the primary relies in detailed Monte Carlo simulations of particle production and decay during the development of the particle shower, which in turn relies on physical quantities like cross sections that have been measured in the laboratory, but not at the center of mass energies reached in cosmic ray interactions (the threshold energy of the Pierre Auger array in the CM frame, a 10^{17} eV proton hitting a stationary nucleon in the atmosphere, is comparable with the 14 TeV CM energy at the LHC). Particle production in air showers also involve processes that can not be perturbatively calculated in QCD, so there is a degree of model dependency in the translation of measured particle multiplicity at ground level and primary energy.

The main difference between cosmic rays and the messengers considered in the previous sections is that cosmic rays are electrically charged and their propagation over astrophysical distances is affected by intergalactic and galactic magnetic fields. Given the poorly known structure and distribution of intergalactic magnetic fields, it is difficult to precisely know above which energy a cosmic ray can point to its source, but it seems clear that one needs energies above 10^{18} eV to achieve pointing of order a degree resolution [120]. This means that cosmic ray sources can not be located as precisely as in the case of gammas or neutrinos, neither can cosmic rays be used in coincidence with electromagnetic or neutrino emission to measure relative arrival times.

However, there are still tests of LIV that can be performed with the observed cosmic ray flux without the need to identify the sources. And there are two levels at which this can be done: either by looking into LIV effects during the propagation of the cosmic rays, or by looking for LIV effects in the development of the particle shower once the cosmic ray has interacted in the atmosphere. Note that the predictions on how a particle shower evolves (production cross sections of different particle species, lifetimes and particle propagation and energy losses in a medium) are based on Lorentz invariant Standard Model processes. Modified particle kinematics duc to LIV will modify the expected relative particle composition and spectrum of air showers at the surface, so any deviation from the expected signal in an air shower detector can be interpreted as new physics. This is easier said than done, since the underlying assumption is then that the development of a particle shower is precisely understood in terms of standard processes and their uncertainties, which is currently far from obvious at energies beyond those available at accelerators or beyond the regime where perturbative QCD can be used.

Calculations on particle shower development assuming LIV effects can of course still be done to evaluate what signatures experimentalists should be aware of and at what strength level they should appear, given current constraints on LIV. But an atmospheric particle shower is a complex system and if individual LIV parameters are introduced, for example, as a modified dispersion relation as in Eq. (6.1) for each particle type, the amount of new parameters makes it difficult to make concrete predictions. So simplifications are usually done where LIV is introduced for some particle type but not for others (Fig. 6.11).

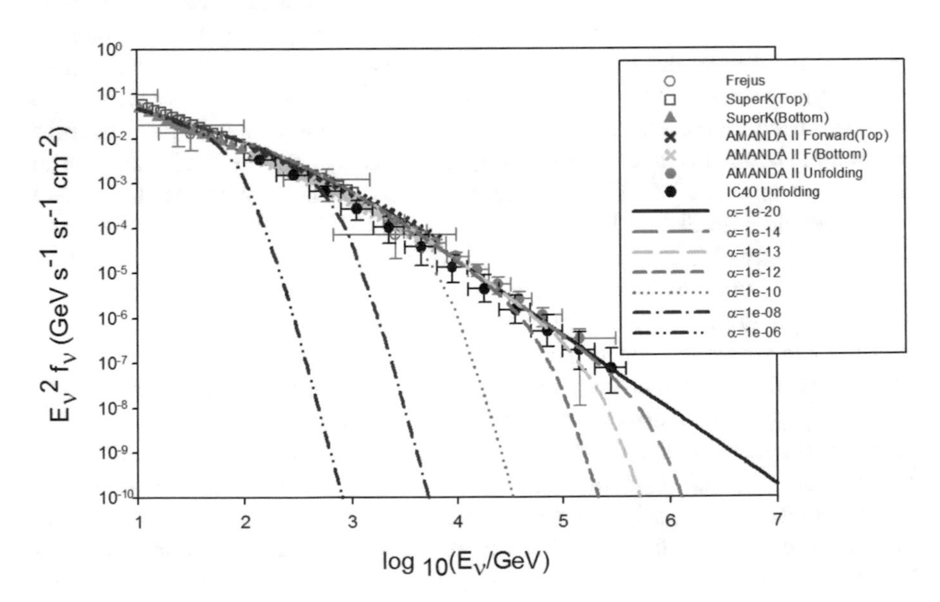

Fig. 6.11 Attenuation length of cosmic rays due to photo-pion production as a function of energy for different LIV coefficients [123]. $\delta_{\mathrm{had},0} = 0$ represents the standard non-LIV case, while stronger LIV effects make the universe more transparent for high energy cosmic rays, challenging the standard interpretation of the GZK cutoff. Figure from [129], reproduced under the CC BY 4.0 license

For example if one focuses on muon production in cosmic ray showers, LIV processes as $\mu \to e+\gamma$ or $\nu_\mu \to \nu_e+\gamma$ will change the muon density of the shower at ground level. Additionally, the injection of e's and γ's during the development of the shower due to the above processes will induce electromagnetic cascades within the main shower. This anomalous electromagnetic component of the shower can be detected by fluorescence detectors and it will also change the electron density at ground level. The effect will be more pronounced for inclined showers due to the longer shower development. The effect of LIV on the muon decay probability can be parameterized as

$$\Gamma = \frac{1}{\gamma \tau_0} + \eta \frac{\gamma^3}{\tau_0}, \tag{6.23}$$

where τ_0 is the standard muon lifetime, γ is the usual Lorentz factor and η is the parameter describing the strength of LIV [121]. A measurement of the electron and muon composition of cosmic ray showers at ground level can thus be used to set a limit on η. Even with the relatively limited data on inclined air showers existing at the time, the authors in [121] could set a limit on $\eta < 10^{-25}$, which was competitive at the time with limits derived from atomic physics or the absence of a Greisen–Zatsepin–Kuzmin (GZK) cutoff (see below). A similar argument can be developed considering the effect of LIV on atmospheric neutrinos, which distorts the standard pion decay kinematics and therefore the atmospheric neutrino spectrum. When considering the measured atmospheric muon-neutrino spectrum, the dispersion relation (Eq. (6.1)) can be simplified to

$$E^2 = m^2 + p^2(1+\eta), \tag{6.24}$$

where now η is the parameter describing the strength of LIV, and is not limited to values ± 1. Using this probe, a limit on the strength of LIV effects can be set at a level of 10^{-13}, see Fig. 6.12, although under certain model assumptions [122].

One of the predicted distinctive features of the cosmic ray energy spectrum is the existence of a cutoff at an energy above $E_{\mathrm{GZK}} \approx 6\times 10^{19}$ eV due to the interaction of protons with the cosmic microwave background and the extra-galactic background light. Although protons are rare with respect to heavier nuclei at ultrahigh energies, let us consider a proton with an energy larger than E_{GZK}. In that case, the process $p + \gamma_{\mathrm{CMB}} \to \Delta_{(1232)} \to p + \pi^0$ (or $\to n + \pi^+$ with the subsequent decay of the neutron into $p + e^- + \bar{\nu}_e$), will result in a proton of a lower energy, where γ_{CMB} is a photon from the all-permeating cosmic microwave background with a mean energy in the current epoch of 6×10^{-4} eV. The above process has an energy threshold of $E_{\mathrm{GZK}} \geq (m_\Delta^2 - m_p^2)/2E_{\gamma_{\mathrm{CMB}}}^2$, so a proton with $E > E_{\mathrm{GZK}}$ will rapidly lose energy through pion photoproduction until its energy falls below E_{GZK}. A similar effect happens for heavier nuclei, which are broken into lighter nuclei of lower energy by photo-dissociation. The GZK cutoff is therefore a universal limitation for the propagation of cosmic rays. It means that if a cosmic ray with an energy $E > E_{\mathrm{GZK}}$ is detected, it either comes from quite close in the universe (the mean free path for

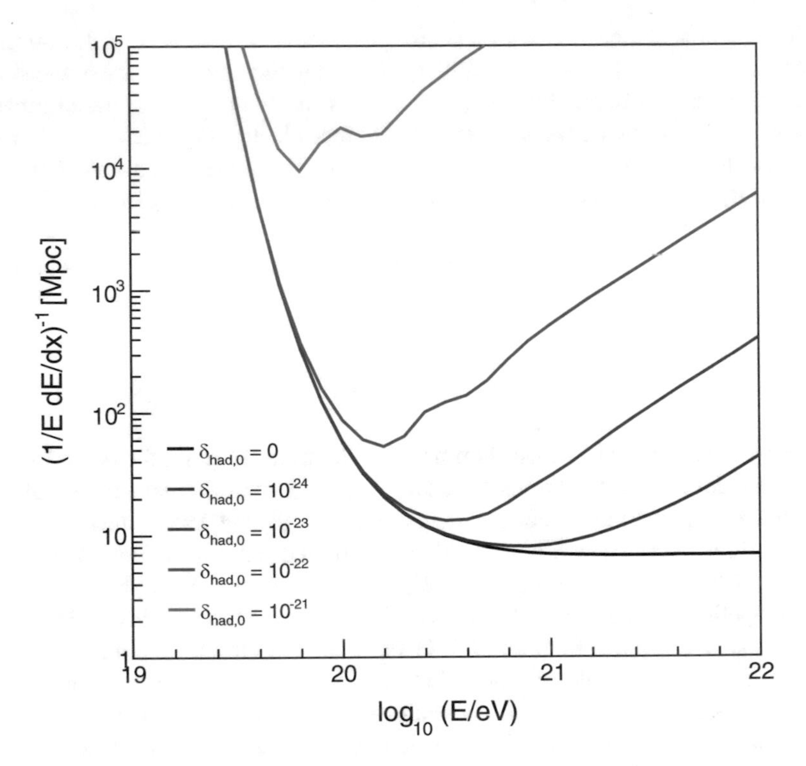

Fig. 6.12 Expected atmospheric muon neutrino energy spectrum for different values of LIV strength, denoted by the parameter α (dashed, dotted and dash-dotted curves) compared with experimental measurements by different detectors. α has the same interpretation as η in Eq. (6.24). Figure reprinted from [122].

a proton with $E > E_{\mathrm{GZK}}$ is about 10 Mpc, quite independent of the initial energy), or some non-standard physics like LIV is at play in its propagation. Turning the argument around, the observation of the GZK cutoff [124, 125] sets a limit on the strength of LIV effects.

? Exercise

6.6. Calculate the threshold energy for the GZK cutoff.

There is one caveat in this argument and it is that the position of the GZK cutoff we observe is the entangled effect of production at the source and propagation through cosmic distances. If LIV effects become important above a given energy that is achieved during the acceleration or production of the cosmic rays, as well as during propagation, we do not have a way to disentangle these two sources of LIV. Usually, though, it is effects during propagation, through the assumption of non-standard particle dispersion relations or anomalous thresholds for photo-production, that are used to probe LIV with cosmic rays [126–128]. Figure 6.11 illustrates the effect that different strengths of LIV (parameterized by the coefficients δ, see Eq. (6.26) below) have on the attenuation length of cosmic rays. Simulations of the propagation of ultrahigh energy cosmic rays including LIV show that they will interact less with the background photons and they can propagate further than they would do in the absence of LIV. This is the basis for using the observed cosmic ray spectrum to set limits on the δ parameters. Given that LIV effects should appear at high energies (we do not have any evidence for LIV at the energies reached at accelerators), the modifying term in Eq. (6.1) is necessarily small, and redefining

$$\delta_{i,n} = \frac{\eta_{i,n}}{E_{\rm QG}^n} \tag{6.25}$$

where n represents the order of LIV and i denotes a particle species, Eq. (6.1) can be recast as a power expansion in energy,

$$E_i^2 = p_i^2 + m_i^2 + \Sigma_{n=0}^N \delta_{i,n} E_i^{2+n} \tag{6.26}$$

Measurements of the energy spectrum and cosmic ray flux composition at ultrahigh energies by the Pierre Auger Observatory, see Fig. 6.13, leave little room

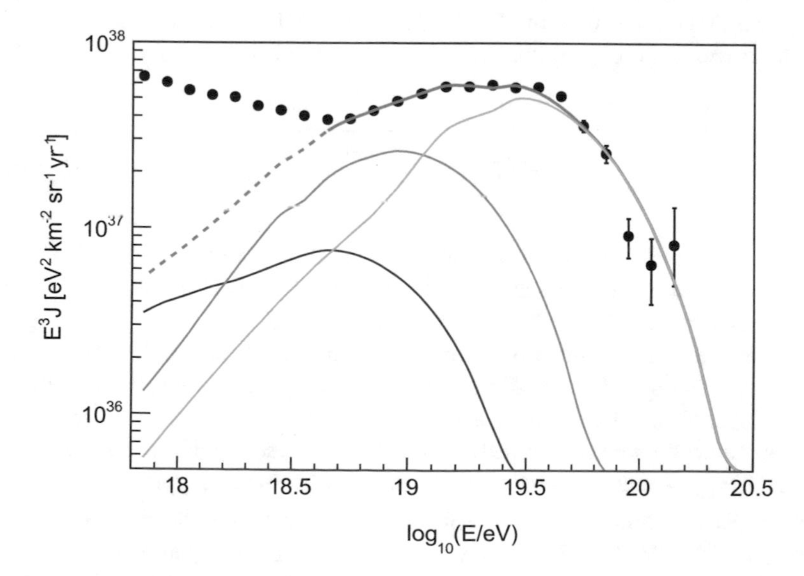

Fig. 6.13 Energy spectrum for standard (non-LIV) cosmic ray propagation (brown curve) compared to the Pierre Auger Observatory data (black dots). Different colors denote the contribution from different mass numbers (A=1, 2<A<4 and 5<A<22 in red, grey and green, respectively). Figure from [129], reproduced under the CC BY 4.0 license

for LIV effects in the propagation of ultrahigh energy cosmic rays and allow to set very stringent limits on several δ parameters for hadrons: $\delta_{\text{had},0} < 10^{-19}$, $\delta_{\text{had},1} < 10^{-38}\ eV^{-1}$ and $\delta_{\text{had},2} < 10^{-57}\ eV^{-2}$ at 5σ confidence level [129]. The difficulty in this kind of analysis is the modeling of the "standard" propagation scenario against which data are compared. On the astrophysical side, the distribution of the diffuse extra-galactic background light is difficult to asses due to the large local backgrounds. On the particle physics side, photo-nuclear cross sections need to be extrapolated over several orders of magnitude in energy with respect to measured values. These unknowns add uncertainties to the simulations of extensive air showers and therefore on the extracted mass composition through the measured X_{max}, the depth at which the shower contains the maximum number of particles. Meaningful limits on LIV coefficients can still be set as long as the characteristics of the observed cosmic ray spectrum and composition are compatible with expectations within uncertainties. But a detailed understanding of the standard model physics in the propagation and interaction of ultrahigh energy cosmic rays is needed if ever a deviation from expectations is to be explained by new physics.

There is another process linked to the existence of LIV that can also lead to a suppression of ultrahigh energy cosmic rays: "gravitational Cherenkov radiation", an effect that has been argued that could take place when the speed of a particle exceeds the speed of propagation of gravity. In the presence of LIV the radiation of a graviton by cosmic rays with speeds larger than the speed of gravitational waves would result in an unconventional energy loss of the particle, which could become observable over cosmological travel distances as a lack of cosmic rays above certain energy. Therefore, the observation of ultrahigh energy cosmic rays can be used to set stringent limits on the strength of LIV. Using the SME as a benchmark, the authors in [130] arrive at an expression for the limit on the coefficients s^dresponsible for LIV as

$$s^d(\hat{p}) < \sqrt{\frac{\mathcal{F}(d)}{G_N E^{2d-5} L}} \tag{6.27}$$

where d is the mass dimension of the corresponding operator in the SME, $\mathcal{F}$ is a dimensionless numerical factor that depends on d but also on the particle species under consideration, G_N is Newton's constant and E and L are the energy and travel distance of the cosmic ray which arrives at Earth from the direction $\hat{p}$. Since LIV effects can depend on the direction of arrival of cosmic rays, the s coefficients further depend on orientation in a solar-centered coordinate system through the quantum number j when expressed as an expansion in spherical harmonics, $s^d(\hat{p}) = \sum Y_{j,m}(\hat{p}) s^d_{j,m}$. Assuming a value of the order of Mpc for L, typical distance to the closest AGN, constraints on s^d for different dimensions (order of $s^d < 10^{-14}$, 10^{-30} GeV^{-2}, 10^{-46} GeV^{-4} for $d = 4, 6, 8$ respectively, slightly dependent on orientation through the quantum number j) have been set from the measured energies of cosmic rays [130]. These are quite competitive constraints from a quite simple analysis, which shows again the advantage of using cosmic messengers to probe tiny new physics effects.

6.2.4 Gravitational Waves

September 14th, 2015 marked the beginning of gravitational-wave astronomy with the direct detection of the event GW150914 [131] by the LIGO (https://www.ligo.caltech.edu) and Virgo (https://www.virgo-gw.eu, [132]) collaborations, an event compatible with the signal predicted by general relativity for the spiral and merger of two black holes of 36 and 29 solar masses. Since then, several black hole and neutron star mergers have been detected. Gravitational waves, along with cosmic rays and neutrinos, complete our non-electromagnetic probes of the universe and, as those other messengers, also provide a glimpse into fundamental physics.

Gravitational effects can be used to search for LIV effects either in the gravitational sector alone (through the dependence of the gravitational wave speed on frequency and/or arrival direction) or in relation to the propagation of other messengers (using the collapse of binary systems to provide a clock for the event and comparing arrival times with other messengers as far as their emission can be assumed to be simultaneous). These are the same kind of tests that we have mentioned with other messengers above, but provide a useful complementary probe into new physics at the Planck scale.

If we concentrate on the specific tests that can be done with information from the propagation of gravitational waves alone,[3] the handles that can be used are the deformation of the waveform due to different propagation speeds for different frequencies (dispersion) and changes in the polarisation modes (birefringence) with respect to the expected signal without LIV. The SME can also be used here as a generic tool to test LIV effects in an effective way by writing the Lagrangian as an expansion of terms of mass dimension d in a linearized metric, $g_{\mu\nu} = \eta_{\mu\nu} + h_{\mu\nu}$, where $h_{\mu\nu}$ is understood as a perturbation to the flat Minkowski metric $\eta_{\mu\nu}$ [133]. In the case of strong dispersion or birefringence effects the gravitational wave event might be "diluted" and fall under the detector threshold, not producing the typical chirp signature. As with other messengers, multiple observations from different directions in the sky are necessary to study LIV effects with gravitational waves since Lorentz violation can depend on direction: the above mentioned effect can render observations of gravitational waves from a given direction difficult if indeed LIV is at play.

The event GW150914 alone was already used in [134] to set limits on LIV coefficients with dimension 4 and 5 from the absence of a splitting in the signal polarisation modes. This sets constraints in the dispersion relation of gravitational waves when written as a function of Lorentz violating coefficients. However the

[3] Tests using relative timing with respect to other messengers or direction-dependent observables are similar to what was discussed in Sect. 6.2.2

strongest limits on modifications of the propagation of gravitational waves currently come from analyses of the LIGO/Virgo collaborations themselves using the events collected in the GWTC-1 catalog [135, 136]. Taking a pure phenomenological approach, a Lorentz-violating dispersion relation for GWs can be parameterized as

$$E^2 = p^2c^2 + A_\alpha p^\alpha c^\alpha. \tag{6.28}$$

Limits on the coefficients A_α can be set by comparing the characteristics of the detected signals with the predicted waveforms obtained from general relativity (see Fig. 6.14).

As just argued, cataclysmic events produce gravitational waves over a short period of time that provide a timing for the emission that, in conjunction with other messengers, can be used to probe general relativity and deviations thereof.

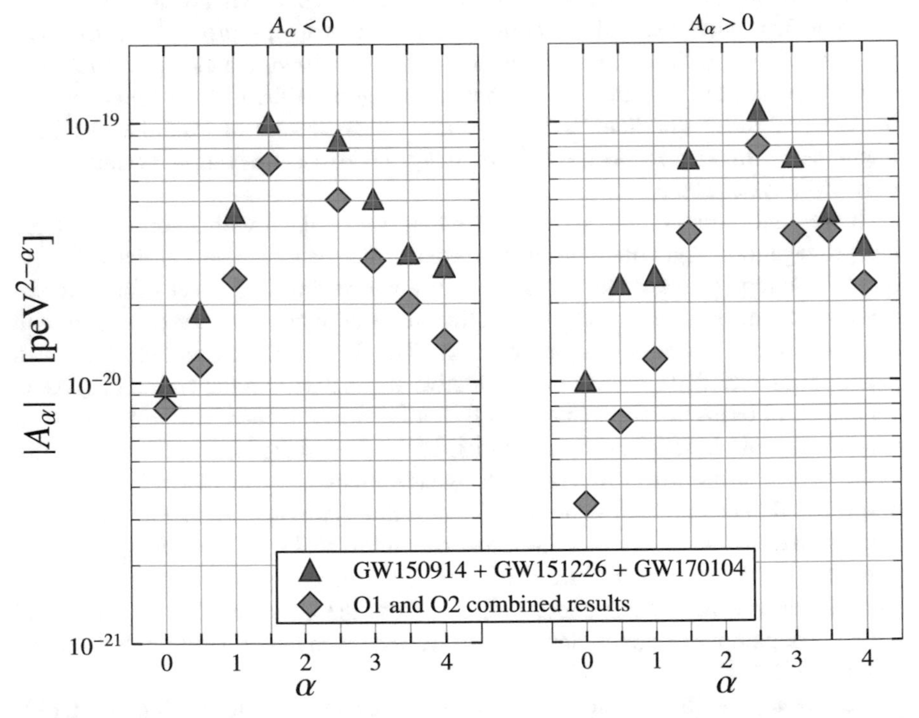

Fig. 6.14 90% upper limits on the parameter A_α. Blue triangles denote the results obtained on GW150914, GW151226, and GW170104, which are updated with respect to the results previously published in [137]. Grey diamonds represent results obtained jointly on GW150914, GW151012, GW151226, GW170104, GW170608, GW170729, GW170809, GW170814, GW170818, and GW170823. The scale of PeV was chosen because it is equivalent to 250 Hz, which is close to the frequencies in which gravitational wave detectors LIGO and Virgo are the most sensitive. Figure reprinted from [136]. Copyright 2022 by the American Physical Society. All rights reserved

A further source of gravitational waves are non-spherically symmetric rotating neutron stars (pulsars). Due to the strong gravity on their surface, it is difficult for a neutron star to maintain mass distribution irregularities or asymmetric deformations that would make it non spherical. But the distribution of the magnetic field inside the star, internal convection or accretion from a close companion can cause a small ellipticity ϵ on rotating neutron stars of the order of $\epsilon < 10^{-8} - 10^{-6}$ [138]. Since pulsars can be found relatively close and well spatially resolved, a detection of continuous, practically monochromatic gravitational waves from a pulsar would provide a unique additional way to test general relativity, but deviations of it as well, i.e. LIV effects [139,140]. The existence of a preferred direction in space, a source of LIV, can modify the rotation of free neutron stars by producing a torque that forces the angular momentum of the star to precess around an axis aligned with the spatial preferred direction. This process results in the continuous emission of gravitational waves with a different spectrum than those emitted by non-spheroidal stars just rotating in isotropic space. Searches for continuous emission of gravitational waves from rotating pulsars have been carried out, although without success so far [141] since the expected frequency of the gravitational waves from rotating neutron stars (about 1 Hz) lies near the lower sensitivity limit of current detectors.

Third-generation gravitational wave detectors currently in R&D, like the Einstein Telescope [142], will be able to increase the detection sensitivity to frequencies as low as 1 Hz (the current lower threshold of LIGO is about 10 Hz) while space-based interferometers like LISA [143] will have a peak sensitivity at 10^{-2} Hz, sufficient to explore gravitational wave emission from binaries in our Galaxy and vastly increase the sensitivity to deviations from general relativity [144].

6.3 Outlook

Cosmic searches for Lorentz invariance violation are an attempt at probing effects of quantum gravity. The formulation of a quantum theory of gravity proved to be quite a challenge so far. An experimental detection of a possible effect of quantum gravity, hinting at what the quantum nature of gravity might be, could be decisive for reaching this paramount scientific goal. However, quantum gravity is expected to manifest at energies on the order of Planck scale, far beyond the reach of accelerator experiments. That is where experiments with cosmic messengers come into play. The energies of astrophysical gamma rays, neutrinos, and cosmic rays can be orders of magnitude higher than energies attainable in accelerators, and more likely to probe quantum gravity. In addition, these messengers traverse enormous distances, over which tiny effects of quantum gravity are expected to accumulate enough to be detected.

The highest measured energies of cosmic messengers are still orders of magnitude below the Planck scale, so what is really tested are effective theories which emulate consequences of quantum gravity at lower energies. LIV as one possible effect of quantum gravity can be introduced by adding terms in the Standard Model Extension, while modifying dispersion relation as in Eq. (6.1) is a very simple way of modelling physics outside of Standard Model. Modifying the dispersion relations of particles leads to a plethora of possible new effects. Some of these, such as energy-dependent group velocity of massless particles, photon instability, vacuum bremsstrahlung or Cherenkov radiation, vacuum birefringence, etc. are forbidden in the standard special-relativistic scenario. On the other hand, propagation speed of massive particles, particle interaction thresholds (e.g. gamma-ray absorption on the background photon fields, GZK cutoff), neutrino flavour oscillations, etc. are modified with respect to the same phenomena predicted by the standard physics.

Detectors of cosmic messengers already yielded some important results in the searches for traces of Lorentz invariance violation. So far, no such effects were found. However, strong constraints have been set on some of LIV parameters. The energy scale at which LIV effects could manifest for some of them has been bounded to above the Planck energy; in some cases, even several orders of magnitude above. A census of experimental tests and limits on LIV can be found in the QG-MM Catalogue [145],[4] created and maintained by the COST Action 18108.

Experimental studies will surely become more sensitive and keep probing new regions of parameter space, especially with the introduction of new experiments such as CTA [146, 147], IceCube-Gen2, KM3NET and the third generation of gravitational wave detectors. However, the vast majority of these test were performed using single messenger data sets. In fact, only a handful astrophysical events were confirmed as multi-messenger until now. With more sensitive instruments, we expect to detect multi-messenger events at a greater rate, enabling more frequent and more sensitive multi-messenger tests of LIV, leading us closer to be able to paint the whole picture of what quantum gravity might have in stock.

6.4 Further Reading

It is impossible in such a short review to give credit to all the relevant work published in the subject by theorists and experimentalists. Since we have tried to keep the references to the point, we list here a few additional papers that, even without the aim of being comprehensive, can be useful for a reader interested in the subject of this chapter:

- The COST Action CA18108 "Quantum gravity phenomenology in the multi-messenger approach" has published a comprehensive review of theory, phenomenology, and experimental searches for effects of quantum gravity in astrophysical observations "Quantum gravity phenomenology at the dawn of the multi-messenger era—A review" [127].
- A detailed description of LIV studies performed with Cherenkov telescopes can be found in [40] and references therein. Various effects, analysis methods, and results are mutually compared and discussed.

[4] https://qg-mm.unizar.es/wiki/.

- We already mentioned some these, but it does not hurt to repeat: there are excellent reviews on the phenomenology and signatures of LIV in high energy neutrinos [4, 6, 66–69, 91, 148].
- The ingredients needed for searches for LIV with cosmic rays are well covered in [149],
- The search for fundamental physics with gravitational waves is covered in [150], [144], and [133].

References

1. D. Colladay, V.A. Kostelecky, CPT violation and the standard model. Phys. Rev. D **55**, 6760–6774 (1997)
2. D. Colladay, V.A. Kostelecky, Lorentz violating extension of the standard model. Phys. Rev. D **58**, 116002 (1998)
3. V.A. Kostelecky, M. Mewes, Lorentz and CPT violation in neutrinos. Phys. Rev. D **69**, 016005 (2004)
4. D. Hooper, D. Morgan, E. Winstanley, Lorentz and CPT invariance violation in high-energy neutrinos. Phys. Rev. D **72**, 065009 (2005)
5. A. Kostelecky, M. Mewes, Neutrinos with Lorentz-violating operators of arbitrary dimension. Phys. Rev. D **85**, 096005 (2012)
6. J.S. Diaz, Neutrinos as probes of Lorentz invariance. Adv. High Energy Phys. **2014**, 962410 (2014)
7. G. Amelino-Camelia, Doubly special relativity. Nature **418**, 34–35 (2002)
8. G. Amelino-Camelia, Doubly-special relativity: facts, myths and some key open issues. Symmetry **2**, 230–271 (2010)
9. E.G. Adelberger, J.H. Gundlach, B.R. Heckel, S. Hoedl, S. Schlamminger, Torsion balance experiments: a low-energy frontier of particle physics. Prog. Part. Nucl. Phys. **62**, 102–134 (2009)
10. M.S. Safronova, D. Budker, D. DeMille, D.F.J. Kimball, A. Derevianko, C.W. Clark, Search for new physics with atoms and molecules. Rev. Mod. Phys. **90**(2), 025008 (2018)
11. A.N. Ivanov, M. Wellenzohn, H. Abele, Probing of violation of Lorentz invariance by ultracold neutrons in the standard model extension. Phys. Lett. B **797**, 134819 (2019)
12. T. Zhang, J. Bi, Y. Zhi, J. Peng, L. Li, L. Chen, Test of Lorentz invariance using rotating ultra-stable optical cavities. Phys. Lett. A **416**, 127666 (2021)
13. Z. Cao et al., Ultrahigh-energy photons up to 1.4 petaelectronvolts from 12 γ-ray Galactic sources. Nature **594**, 33–36 (2021)
14. M. Holler et al., Observations of the Crab Nebula with H.E.S.S. phase II. PoS **ICRC2015**, 847 (2016)
15. M. Holler, A. Balzer, R. Chalmé-Calvet, M.D. Naurois, D. Zaborov, Photon reconstruction for H.E.S.S. using a semi-analytical model. PoS **ICRC2015**, 980 (2016)
16. J. Aleksić et al., The major upgrade of the MAGIC telescopes, Part I: the hardware improvements and the commissioning of the system. Astropart. Phys. **72**, 61–75 (2016)
17. J. Aleksić et al., The major upgrade of the MAGIC telescopes, Part II: a performance study using observations of the Crab Nebula. Astropart. Phys. **72**, 76–94 (2016)
18. N. Park, Performance of the VERITAS experiment. PoS **ICRC2015**, 771 (2016)
19. B.S. Acharya et al., *Science with the Cherenkov Telescope Array* (World Scientific Publishing, Singapore, 2018)
20. M. Spurio, *Indirect Cosmic Rays Detection: Particle Showers in the Atmosphere* (Springer, Cham, 2015)
21. A. De Angelis, M. Pimenta, R. Conceição, *Particle and Astroparticle Physics: Problems and Solutions* (Springer, Cham, 2021)

22. G. D'Amico, Statistical tools for imaging atmospheric Cherenkov telescopes. Universe **8**(2), 90 (2022)
23. W.B. Atwood et al., The large area telescope on the fermi gamma-ray space telescope mission. Astrophys. J. **697**, 1071–1102 (2009)
24. M. Tavani et al., Science with e-ASTROGAM: a space mission for MeV–GeV gamma-ray astrophysics. J. High Energy Astrophys **19**, 1–106 (2018)
25. R. Caputo et al., All-sky medium energy gamma-ray observatory: exploring the extreme multimessenger universe (2019). https://arxiv.org/abs/1907.07558
26. A.U. Abeysekara et al., Observation of the Crab Nebula with the HAWC gamma-ray observatory. Astrophys. J. **843**(1), 39 (2017)
27. A. Addazi et al., The large high altitude air shower observatory (LHAASO) science book (2021 Edition). Chin. Phys. C **46**, 035001–035007 (2022)
28. U. Jacob, T. Piran, Lorentz-violation-induced arrival delays of cosmological particles. J. Cosmol. Astroparticle Phys. **1**, 031 (2008)
29. J. Bolmont et al., First combined study on lorentz invariance violation from observations of energy-dependent time delays from multiple-type gamma-ray sources. I. Motivation, method description, and validation through simulations of H.E.S.S., MAGIC, and VERITAS data sets. Astrophys. J. **930**(1), 75 (2022)
30. K. Toma, S. Mukohyama, D. Yonetoku, T. Murakami, S. Gunji, T. Mihara, Y. Morihara, T. Sakashita, T. Takahashi, Y. Wakashima, et al., Strict limit on CPT violation from polarization of gamma-ray burst. Phys. Rev. Lett. **109**, 241104 (2012)
31. O. Blanch, J. Lopez, M. Martinez, Testing the effective scale of quantum gravity with the next generation of gamma-ray telescopes. Astropart. Phys. **19**, 245–252 (2003)
32. H. Abdalla, M. Böttcher, Lorentz invariance violation effects on gamma–gamma absorption and compton scattering. Astrophys. J. **865**(2), 159 (2018)
33. S.D. Biller et al., Limits to quantum gravity effects from observations of TeV flares in active galaxies. Phys. Rev. Lett. **83**, 2108 (1999)
34. J. Albert et al. [MAGIC and Other Contributors], Probing quantum gravity using photons from a flare of the active galactic nucleus Markarian 501 observed by the MAGIC telescope. Phys. Lett. B **668**, 253 (2008)
35. V. Vasileiou, A. Jacholkowska, F. Piron, J. Bolmont, C. Couturier, J. Granot, F.W. Stecker, J. Cohen-Tanugi, F. Longo, Constraints on Lorentz invariance violation from fermi-large area telescope observations of gamma-ray bursts. Phys. Rev. D **87**(12), 122001 (2013)
36. U. Barres de Almeida, M.K. Daniel, A simple method to test for energy-dependent dispersion in high energy light-curves of astrophysical sources. Astropart. Phys. **35**, 850 (2012)
37. J.D. Scargle, J.P. Norris, J.T. Bonnell, An algorithm for detecting quantum-gravity photon dispersion in gamma-ray bursts: DISCAN. Astrophys. J. **673**(2), 972 (2008)
38. J.R. Ellis, N. Harries, A. Meregaglia, A. Rubbia, A. Sakharov, Probes of Lorentz violation in neutrino propagation. Phys. Rev. D **78**, 033013 (2008)
39. J. Ellis, R. Konoplich, N.E. Mavromatos, L. Nguyen, A.S. Sakharov, E.K. Sarkisyan-Grinbaum, Robust constraint on lorentz violation using Fermi-LAT gamma-ray burst data. Phys. Rev. D **99**(8), 083009 (2019)
40. T. Terzić, D. Kerszberg, J. Strišković, Probing quantum gravity with imaging atmospheric cherenkov telescopes. Universe **7**, 345 (2021)
41. M. Martinez, M. Errando, A new method to study energy-dependent arrival delays on photons from astrophysical sources. Astropart. Phys. **31**, 226–232 (2009)
42. V.A. Acciari et al., Bounds on lorentz invariance violation from MAGIC observation of GRB 190114C. Phys. Rev. Lett. **125**, 021301 (2020)
43. V.A. Acciari et al., Teraelectronvolt emission from the γ-ray burst GRB 190114C. Nature **575**(7783), 455–458 (2019)
44. V.A. Acciari et al., Observation of inverse Compton emission from a long γ-ray burst. Nature **575**(7783), 459–463 (2019)
45. Barlow, R.J., *Statistics: A Guide to the Use of Statistical Methods in the Physical Sciences*. Manchester Physics Series (Wiley Blackwell, 1989)

46. S.S. Wilks, The large-sample distribution of the likelihood ratio for testing composite hypotheses. Ann. Math. Stat. **9**(1), 60–62 (1938)
47. H. Abdalla et al., The 2014 TeV γ-ray flare of Mrk 501 seen with H.E.S.S.: temporal and spectral constraints on Lorentz invariance violation. Astrophys. J. **870**, 93 (2019)
48. M. Ajello et al., [Fermi-LAT], The first Fermi LAT gamma-ray burst catalog. Astrophys. J. Suppl. **209**, 11 (2013)
49. G. Castignani, D. Guetta, E. Pian, L. Amati, S. Puccetti, S. Dichiara, Time delays between Fermi-LAT and GBM light curves of gamma-ray bursts. Astron. Astrophys. **565**, A60 (2014)
50. A.S. Friedman et al., Improved constraints on anisotropic birefringent lorentz invariance and *cpt* violation from broadband optical polarimetry of high redshift galaxies. Phys. Rev. D **102**, 043008 (2020)
51. A. Franceschini, G. Rodighiero, M. Vaccari, Extragalactic optical-infrared background radiation, its time evolution and the cosmic photon-photon opacity. Astron. Astrophys. **487**, 837–852 (2008)
52. T. Jacobson, S. Liberati, D. Mattingly, A strong astrophysical constraint on the violation of special relativity by quantum gravity. Nature **424**, 1019–1021 (2003)
53. G. Rubtsov, P. Satunin, S. Sibiryakov, Constraints on violation of Lorentz invariance from atmospheric showers initiated by multi-TeV photons. J. Cosmol. Astroparticle Phys. **5**, 049 (2017)
54. R.G. Lang, H. Martínez-Huerta, V. de Souza, Improved limits on Lorentz invariance violation from astrophysical gamma-ray sources. Phys. Rev. D **99**(4), 043015 (2019)
55. G. Anton, Neutrino telescopes, in *Probing Particle Physics with Neutrino Telescopes*, ed. by P. de los Heros, chap. 2 (World Scientific Publishing, Singapore, 2020), pp. 11–32
56. M. Ahlers, K. Helbing, C. Pérez de los Heros, Probing particle physics with IceCube. Eur. Phys. J. C **78**(11), 924 (2018)
57. A. Margiotta, Searches for exotica and dark matter with neutrino telescopes. Phil. Trans. Roy. Soc. Lond. A **377**(2161), 20190084 (2019)
58. M. Ackermann et al., Fundamental physics with high-energy cosmic neutrinos. Bull. Am. Astron. Soc. **51**, 215 (2019)
59. C. Pérez de los Heros (Ed.), *Probing Particle Physics with Neutrino Telescopes* (World Scientific Publishing, Singapore, 2020)
60. A. Achterberg et al., First year performance of the IceCube neutrino telescope. Astropart. Phys. **26**, 155–173 (2006)
61. S. Adrian-Martinez et al., Letter of intent for KM3NeT 2.0. J. Phys. G **43**(8), 084001 (2016)
62. A. Avrorin et al., The Baikal neutrino experiment. Nucl. Instrum. Methods A **626–627**, S13–S18 (2011)
63. M. Agostini et al., The Pacific ocean neutrino experiment. Nat. Astron. **4**(10), 913–915 (2020)
64. M. Ageron et al., ANTARES: the first undersea neutrino telescope. Nucl. Instrum. Methods A **656**, 11–38 (2011)
65. Y. Fukuda et al., The super-kamiokande detector. Nucl. Instrum. Methods A **501**, 418–462 (2003)
66. A. Roberts, Astrophysical neutrinos in testing Lorentz symmetry.Galaxies **9**(3), 47 (2021)
67. M.D.C. Torri, Neutrino oscillations and lorentz invariance violation. Universe **6**(3), 37 (2020)
68. F.W. Stecker, Testing Lorentz symmetry using high energy astrophysics observations. Symmetry **9**(10), 201 (2017)
69. F.W. Stecker, S.T. Scully, S. Liberati, D. Mattingly, Searching for traces of planck-scale physics with high energy neutrinos. Phys. Rev. D **91**(4), 045009 (2015)
70. T. Pitik, I. Tamborra, M. Petropoulou, Neutrino signal dependence on gamma-ray burst emission mechanism. J. Cosmol. Astroparticle Phys. **5**, 034 (2021)
71. T. Piran, Gamma-ray bursts and the fireball model.Phys. Rep. **314**, 575–667 (1999)
72. B. Zhang, P. Meszaros, Gamma-ray bursts: progress, problems & prospects. Int. J. Mod. Phys. A **19**, 2385–2472 (2004)
73. R. Abbasi et al., Searches for neutrinos from precursors and afterglows of gamma-ray bursts using the IceCube neutrino observatory. PoS **ICRC2021**, 1118 (2021)

74. U. Jacob, T. Piran, Neutrinos from gamma-ray bursts as a tool to explore quantum-gravity-induced Lorentz violation. Nat. Phys. 3, 87–90 (2007)
75. G. Amelino-Camelia, D. Guetta, T. Piran, Icecube neutrinos and Lorentz invariance violation. Astrophys. J. **806**(2), 269 (2015)
76. W. Hillebrandt, P. Hoflich, The supernova SN1987A in the large magellanic cloud. Rep. Prog. Phys. **52**, 1421–1473 (1989)
77. M.G. Aartsen et al., Multimessenger observations of a flaring blazar coincident with high-energy neutrino IceCube-170922A. Science **361**(6398), eaat1378 (2018)
78. K.S. Hirata et al., Observation in the Kamiokande-II detector of the neutrino burst from supernova SN 1987a. Phys. Rev. D **38**, 448–458 (1988)
79. T. Haines et al., Neutrinos from SN1987A in the IMB detector. Nucl. Instrum. Meth. A **264**, 28–31 (1988)
80. E.N. Alekseev, L.N. Alekseeva, I.V. Krivosheina, V.I. Volchenko, Detection of the neutrino signal from SN1987A in the LMC using the InrBaksan underground scintillation telescope. Phys. Lett. B **205**,209–214 (1988)
81. M. Aglietta et al., On the event observed in the Mont Blanc underground neutrino observatory during the occurrence of Supernova 1987a. Europhys. Lett. **3**, 1315–1320 (1987)
82. H.T. Janka, Neutrino emission from supernovae, in *Handbook of Supernovae*, ed. by A. Alsabti, P. Murdin (Springer, Berlin, 2017)
83. P. Adamson et al., Measurement of neutrino velocity with the MINOS detectors and NuMI neutrino beam. Phys. Rev. D **76**, 072005 (2007)
84. S. Garrappa et al., Investigation of two Fermi-LAT gamma-ray blazars coincident with high-energy neutrinos detected by IceCube. Astrophys. J. **880**(2), 880:103 (2019)
85. J. Ellis, N.E. Mavromatos, A.S. Sakharov, E.K. Sarkisyan-Grinbaum, Limits on neutrino lorentz violation from multimessenger observations of TXS 0506+056. Phys. Lett. B **789**, 352–355 (2019)
86. K. Wang, S.Q. Xi, L. Shao, R.Y. Liu, Z. Li, Z.K. Zhang, Limiting superluminal neutrino velocity and lorentz invariance violation by neutrino emission from the blazar TXS 0506+056. Phys. Rev. D **102**(6), 063027 (2020)
87. L. Köpke, Improved detection of supernovae with the IceCube observatory. J. Phys. Conf. Ser. **1029**(1), 012001 (2018)
88. T. Jacobson, S. Liberati, D. Mattingly, Threshold effects and planck scale lorentz violation: combined constraints from high-energy astrophysics. Phys. Rev. D **67**, 124011 (2003)
89. M. Ackermann et al., Unresolved gamma-ray sky through its angular power spectrum. Phys. Rev. Lett. **121**(24), 241101 (2018)
90. G. Somogyi, I. Nándori, U.D. Jentschura, Neutrino splitting for lorentz-violating neutrinos: detailed analysis. Phys. Rev. D **100**(3), 035036 (2019)
91. J. Christian, Testing quantum gravity via cosmogenic neutrino oscillations. Phys. Rev. D **71**, 024012 (2005)
92. D.M. Mattingly, L. Maccione, M. Galaverni, S. Liberati, G. Sigl, Possible cosmogenic neutrino constraints on planck-scale Lorentz violation. J. Cosmol. Astroparticle Phys. **2**, 007 (2010)
93. E. Borriello, S. Chakraborty, A. Mirizzi, P.D. Serpico, Stringent constraint on neutrino Lorentz-invariance violation from the two IceCube PeV neutrinos. Phys. Rev. D **87**(11), 116009 (2013)
94. F.W. Stecker, S.T. Scully, Propagation of superluminal PeV IceCube neutrinos: a high energy spectral cutoff or new constraints on Lorentz invariance violation. Phys. Rev. D **90**(4), 043012 (2014)
95. G. Tomar, S. Mohanty, S. Pakvasa, Lorentz invariance violation and IceCube neutrino events. J. High Energy Phys. **11**, 022 (2015)
96. J.S. Diaz, A. Kostelecky, M. Mewes, Testing relativity with high-energy astrophysical neutrinos. Phys. Rev. D **89**(4), 043005 (2014)
97. T.K. Gaisser, Atmospheric neutrinos, in *Probing Particle Physics with Neutrino Telescopes*, ed. by P. de los Heros, chap. 3 (World Scientific Publishing, Singapore, 2020), pp. 33–74

98. R. Abbasi et al., The IceCube high-energy starting event sample: description and flux characterization with 7.5 years of data. Phys. Rev. D **104**, 022002 (2021)
99. G. Fantini, A. Gallo Rosso, F. Vissani, V. Zema, Introduction to the formalism of neutrino oscillations. Adv. Ser. Direct. High Energy Phys. **28**, 37–119 (2018)
100. D. Morgan, E. Winstanley, L.F. Thompson, J. Brunner, L.F. Thompson, Neutrino telescope modelling of Lorentz invariance violation in oscillations of atmospheric neutrinos. Astropart. Phys. **29**, 345–354 (2008)
101. K. Abe et al., Test of Lorentz invariance with atmospheric neutrinos. Phys. Rev. D **91**(5), 052003 (2015)
102. G.L. Fogli, E. Lisi, A. Marrone, G. Scioscia, Testing violations of special and general relativity through the energy dependence of muon-neutrino <-> tau-neutrino oscillations in the Super-Kamiokande atmospheric neutrino experiment. Phys. Rev. D **60**, 053006 (1999)
103. M.G. Aartsen et al., Neutrino interferometry for high-precision tests of Lorentz symmetry with IceCube. Nat. Phys. **14**(9), 961–966 (2018)
104. R. Abbasi et al., Search for a Lorentz-violating sidereal signal with atmospheric neutrinos in IceCube. Phys. Rev. D **82**, 112003 (2010)
105. A.A. Aguilar-Arevalo et al., Test of Lorentz and CPT violation with short baseline neutrino oscillation excesses. Phys. Lett. B **718**, 1303–1308 (2013)
106. K. Abe et al., Search for Lorentz and CPT violation using sidereal time dependence of neutrino flavor transitions over a short baseline. Phys. Rev. D **95**(11), 111101 (2017)
107. P. Adamson et al., Search for Lorentz invariance and CPT violation with muon antineutrinos in the MINOS near detector. Phys. Rev. D **85**, 031101 (2012)
108. B. Aharmim et al., Tests of Lorentz invariance at the sudbury neutrino observatory. Phys. Rev. D **98**(11), 112013 (2018)
109. S. Hummer, M. Maltoni, W. Winter, C. Yaguna, Energy dependent neutrino flavor ratios from cosmic accelerators on the Hillas plot. Astropart. Phys. **34**, 205–224 (2010)
110. M. Bustamante, M. Ahlers, Inferring the flavor of high-energy astrophysical neutrinos at their sources. Phys. Rev. Lett. **122**(24), 241101 (2019)
111. C.A. Argüelles, T. Katori, J. Salvado, New physics in astrophysical neutrino flavor. Phys. Rev. Lett. **115**, 161303 (2015)
112. N. Song, S.W. Li, C.A. Argüelles, M. Bustamante, A.C. Vincent, The future of high-energy astrophysical neutrino flavor measurements. J. Cosmol. Astroparticle Phys. **4**, 054 (2021)
113. M.G. Aartsen et al., A combined maximum-likelihood analysis of the high-energy astrophysical neutrino flux measured with IceCube. Astrophys. J. **809**(1), 98 (2015)
114. M. Boezio, R. Munini, P. Picozza, Cosmic ray detection in space. Prog. Part. Nucl. Phys. **112**, 103765 (2020)
115. R. Battiston, Spaceborne experiments, in *Particle Physics Reference Library. Volume 2: Detectors for Particles and Radiation*, ed. by C.W. Fabjan, H. Schopper (Springer, Berlin, 2020), pp. 823–870
116. T.K. Gaisser, Cosmic-ray showers reveal muon mystery. Physics **9**, 125 (2016)
117. A. Aab et al., The pierre auger cosmic ray observatory. Nucl. Instrum. Meth. A **798**, 172–213 (2015)
118. X.H. Ma et al., Chapter 1 LHAASO instruments and detector technology*. Chin. Phys. C **46**(1), 030001 (2022)
119. H. Kawai et al., Telescope array experiment. Nucl. Phys. B Proc. Suppl. **175–176**, 221–226 (2008)
120. R. Alves Batista, E.M. de Gouveia Dal Pino, K. Dolag, S. Hussain, Cosmic-ray propagation in the turbulent intergalactic medium, in *30th General Assembly of the International Astronomical Union* (2018)
121. R. Cowsik, B.V. Sreekantan, A bound on violations of Lorentz invariance. Phys. Lett. B **449**, 219–222 (1999)
122. R. Cowsik, T. Madziwa-Nussinov, S. Nussinov, U. Sarkar, Testing violations of Lorentz invariance with cosmic-rays. Phys. Rev. D **86**, 045024 (2012)

123. S.T. Scully, F.W. Stecker, Lorentz invariance violation and the observed spectrum of ultrahigh energy cosmic rays. Astropart. Phys. **31**, 220 (2009)
124. R.U. Abbasi et al., First observation of the Greisen-Zatsepin-Kuzmin suppression. Phys. Rev. Lett. **100**, 101101 (2008)
125. J. Abraham et al., Observation of the suppression of the flux of cosmic rays above 4×10^{19}eV. Phys. Rev. Lett. **101**, 061101 (2008)
126. S.R. Coleman, S.L. Glashow, Cosmic ray and neutrino tests of special relativity. Phys. Lett. B **405**, 249–252 (1997)
127. D. Mattingly, Modern tests of Lorentz invariance. Living Rev. Rel. **8**, 5 (2005)
128. H. Martínez-Huerta, A. Pérez-Lorenzana, Restrictions from Lorentz invariance violation on cosmic ray propagation. Phys. Rev. D **95**(6), 063001 (2017)
129. P. Abreu et al., Testing effects of Lorentz invariance violation in the propagation of astroparticles with the pierre auger observatory. J. Cosmol. Astropart. Phys. **1**(1), 023 (2022)
130. V.A. Kostelecký, J.D. Tasson, Constraints on Lorentz violation from gravitational Čerenkov radiation. Phys. Lett. B 749, 551–559 (2015)
131. B.P. Abbott et al., Observation of gravitational waves from a binary black hole merger. Phys. Rev. Lett. **116**(6), 061102 (2016)
132. F. Acernese et al., Virgo detector characterization and data quality during the O3 run. (2022). https://arxiv.org/abs/2210.15633
133. M. Mewes, Signals for Lorentz violation in gravitational waves. Phys. Rev. D **99**(10), 104062 (2019)
134. V.A. Kostelecký, M. Mewes, Testing local Lorentz invariance with gravitational waves. Phys. Lett. B **757**, 510–514 (2016)
135. B.P. Abbott et al., GWTC-1: a gravitational-wave transient catalog of compact binary mergers observed by LIGO and virgo during the first and second observing runs. Phys. Rev. X **9**(3), 031040 (2019)
136. B.P. Abbott et al., Tests of general relativity with the binary black hole signals from the LIGO-virgo catalog GWTC-1. Phys. Rev. D **100**(10), 104036 (2019)
137. B.P. Abbott et al., GW170104: observation of a 50-solar-mass binary black hole coalescence at redshift 0.2. Phys. Rev. Lett. **118**(22), 221101 (2017). [Erratum: Phys.Rev.Lett. 121, 129901 (2018)]
138. P.D. Lasky, Gravitational waves from neutron stars: a review. Publ. Astron. Soc. Austral. **32**, e034 (2015)
139. M. Zimmermann, Gravitational waves from rotating and precessing rigid bodies. 1. General solutions and computationally useful formulas. Phys. Rev. D **21**, 891–898 (1980)
140. R. Xu, Y. Gao, L. Shao, Precession of spheroids under Lorentz violation and observational consequences for neutron stars. Phys. Rev. D **103**(8), 084028 (2021)
141. B.P. Abbott et al., First search for gravitational waves from known pulsars with Advanced LIGO. Astrophys. J. **839**(1), 12 (2017). [Erratum: Astrophys.J. 851, 71 (2017)]
142. S. Di Pace et al., Research facilities for Europe's next generation gravitational-wave detector einstein telescope. Galaxies **10**(3), 65 (2022)
143. P. Amaro-Seoane et al., Low-frequency gravitational-wave science with eLISA/NGO. Class. Quant. Grav. **29**, 124016 (2012)
144. S.E. Perkins, N. Yunes, E. Berti, Probing fundamental physics with gravitational waves: the next generation. Phys. Rev. D 103(4), 044024 (2021)
145. COST Action 18108, QG-mm catalogue (2021). https://qg-mm.unizar.es/wiki
146. H. Abdalla, et al., Sensitivity of the Cherenkov telescope array for probing cosmology and fundamental physics with gamma-ray propagation. J. Cosmol. Astropart. Phys. **02**, 048 (2021)
147. M. Fairbairn, A. Nilsson, J. Ellis, J. Hinton, R. White, The CTA sensitivity to Lorentz-violating effects on the gamma-ray horizon. J. Cosmol. Astropart. Phys. **06**, 005 (2014)

148. F.W. Stecker, Testing Lorentz invariance with neutrinos, in *'Neutrino Physics and Astrophysics', "Encyclopedia of Cosmology Series II"*, ed. by G.G. Fazio (World Scientific Publishing Company, Singapore, 2022)
149. W. Bietenholz, Cosmic rays and the search for a Lorentz invariance violation. Phys. Rep. **505**, 145–185 (2011)
150. C.M. Will, The confrontation between general relativity and experiment. Living Rev. Rel. **17**, 4, (2014)

Neutron Stars

7

Jutta Kunz

Abstract

Neutron stars are highly compact astrophysical objects and therefore of utmost relevance to learn about theories of gravity. Whereas the proper equation of state of the nuclear matter inside neutron stars is not yet known, and a wide range of equations of state is still compatible with observations, this uncertainty can be overcome to a large extent, when dimensionless neutron star properties are considered. In this case *universal relations* between neutron star properties and for the gravitational radiation in the form of quasi-normal modes arise. These *universal relations* can be rather distinct for alternative theories of gravity as compared to General Relativity. Moreover, the presence of new degrees of freedom in alternative theories of gravity leads to further types of gravitational radiation, that may be revealed by pulsar observations and gravitational wave detectors. Here an introduction to neutron stars, their properties and *universal relations* is presented, followed by two examples of alternative theories of gravity featuring interesting effects for neutrons stars.

7.1 Introduction

When massive stars with initial mass $M \gtrsim 8M_\odot$ have burnt the nuclear fuel in their core gravitational collapse results, leaving behind a highly compact remnant, a neutron star or a black hole. (The latter will be discussed in the next Chaps. 8 and 9.) While predicted shortly after the discovery of the neutron [1], neutron stars were only observed in the late 60s, when very regular radio pulses appeared in the data

J. Kunz (✉)
Institute of Physics, University of Oldenburg, Oldenburg, Germany
e-mail: jutta.kunz@uni-oldenburg.de

C. Pfeifer, C. Lämmerzahl (eds.), *Modified and Quantum Gravity*, Lecture Notes in Physics 1017, https://doi.org/10.1007/978-3-031-31520-6_7

taken by Jocelyn Bell [2]. The radio pulses were emitted by a pulsar, now known as PSR B1919+21, a rapidly rotating neutron star with misaligned magnetic field. Ever since numerous pulsars including a double pulsar have been discovered [3–7]. The extreme regularity of these pulses allows for high precision tests of General Relativity and severe constraints for various alternative theories of gravity (see e.g. [8,9]).

On the theoretical side, Tolman, Oppenheimer and Volkoff (TOV) considered in the 30s already the description of neutron stars, deriving and solving the TOV equations for a simple equation of state (EOS) of the nuclear matter, namely a cold Fermi gas [10, 11]. This work had profound implications, since it showed that neutron stars can be supported against the gravitational pull only up to a maximum mass, while beyond this mass the collapse of the stellar core will continue and lead to a black hole. The value of the maximum mass depends of course on the EOS for the nuclear matter. The proper EOS for nuclear matter under such extreme conditions as present in neutron stars is still unknown, though [12–14]).

In recent years much progress has been made based on the discovery of gravitational waves and the advent of multi-messenger astronomy [15–17]. In particular, the observation of GW170817, where the merger of a neutron star binary was reported and analyzed led to new constraints on the neutron star EOS, since it allowed to put constraints on the tidal effects experienced by the coalescing bodies and on the neutron star radii [18]. Further analysis also hinted at a new value (range) for the maximum mass of neutron stars [19,20]. Previous observations of pulsars had already revealed the existence of neutron stars with masses of about 2 solar masses [21–23].

In the following we will focus mainly on static neutron stars. We will start with the derivation of the TOV equations, and then address a set of important neutron star properties. Besides their mass and radius, we will consider their moment of inertia, their rotational quadrupole moment and their tidal deformability. Subsequently we will address seismology of neutron stars. Thus we will consider the normal modes and quasi-normal modes (QNMs) of neutron stars, representing their reaction to perturbations. The uncertainty of the EOS reflected in the neutron star properties and QNMs will then be largely reduced with the help of universal relations (see e.g. [24, 25]). Our final concern will be the consideration of neutron stars in a set of alternative gravity theories featuring an additional scalar degree of freedom, where we will highlight some interesting new aspects as compared to General Relativity.

7.2 Static Neutron Stars

7.2.1 Tolman-Oppenheimer-Volkoff Equations

In General Relativity (GR) static neutron stars are obtained by solving the Tolman-Oppenheimer-Volkoff (TOV) equations for a given equation of state (EOS) of the nuclear matter. We will now derive this set of equations.

To this end we start from the Einstein equations

$$G_{\mu\nu} = \mathcal{R}_{\mu\nu} - \frac{1}{2} g_{\mu\nu} \mathcal{R} = 8\pi G T_{\mu\nu}, \tag{7.1}$$

and employ the stress-energy tensor of an isotropic perfect fluid describing the nuclear matter

$$T_{\mu\nu} = (\rho + P)\, u_\mu u_\nu + g_{\mu\nu} P, \tag{7.2}$$

whose four-velocity in the rest system is $u^\mu = \left(u^t, 0, 0, 0\right)$. In mixed co- and contravariant components the stress-energy tensor then reads $T^\mu_\nu = \text{diag}(-\rho,\ P,\ P,\ P\ )$ with energy density ρ and pressure P.

A convenient metric ansatz for static spherically symmetric neutron stars is given by

$$ds^2 = g_{\mu\nu} dx^\mu dx^\nu = -e^{2\Phi(r)} dt^2 + e^{2\Lambda(r)} dr^2 + r^2 \left(d\theta^2 + \sin^2\theta d\phi^2\right), \tag{7.3}$$

which contains two unknown functions, $\Phi(r)$ and $\Lambda(r)$, where the latter can be expressed in terms of the mass function $m(r)$

$$e^{2\Lambda(r)} = \frac{1}{1 - \frac{2m(r)}{r}}. \tag{7.4}$$

? Exercise

7.1. Show that the Einstein Tensor becomes with this ansatz

$$G_{00} = e^{2\Phi} \frac{2m'}{r^2}, \tag{7.5}$$

$$G_{rr} = \frac{2}{r} \left(\Phi' - \frac{m}{r^2} \left(1 - \frac{2m}{r} \right)^{-1} \right), \tag{7.6}$$

$$G_{\theta\theta} = r^2 \left[\left(\Phi'' + \Phi'^2 \right) \left(1 - \frac{2m}{r} \right) + \frac{\Phi'}{r} \left(1 - m' - \frac{m}{r} \right) - \frac{1}{r^2} \left(m' - \frac{m}{r} \right) \right], \tag{7.7}$$

$$G_{\varphi\varphi} = \sin^2\theta G_{\theta\theta}. \tag{7.8}$$

From the Einstein equations $G_0^0 = \kappa T_0^0$ and $G_r^r = \kappa T_r^r$ ($\kappa = 8\pi G$) we find

$$m' = \frac{\kappa}{2}\rho r^2, \tag{7.9}$$

$$\Phi' = \frac{\frac{\kappa}{2}r^3 P + m}{\left(1 - \frac{2m}{r}\right)r^2}. \tag{7.10}$$

Employing these two equations in the Einstein equation $G_\theta^\theta = \kappa T_\theta^\theta$ using

$$\Phi'' = \frac{d}{dr}\Phi' = \frac{d}{dr}\left(\frac{\frac{\kappa}{2}r^3 P + m}{\left(1 - \frac{2m}{r}\right)r^2}\right) \tag{7.11}$$

we obtain the equation for pressure P, where we can identify the Newtonian part (underlined) and the relativistic corrections

$$\underline{P' = -\frac{m\rho}{r^2}\left(1 + \frac{P}{\rho}\right)}\left(1 + \frac{\kappa}{2}\frac{P}{m}r^3\right)\left(1 - \frac{2m}{r}\right)^{-1}. \tag{7.12}$$

The system of equations (7.9), (7.10) and (7.12) are the **TOV equations**, representing three equations for four unknowns. Therefore we have to provide an EOS $\rho(P)$ in order to solve the equations.

A relatively simple EOS is the so-called polytropic EOS

$$\rho = \frac{P}{\Gamma - 1} + \left(\frac{P}{K}\right)^\Gamma, \tag{7.13}$$

where Γ is the adiabatic index and K the polytropic constant. Many realistic EOSs can be parametrized as piecewise polytropic EOSs (see e.g. [26]).

Neutron stars are compact objects with a given radius R. Outside this radius the pressure and the density vanish. Therefore, the exterior is simply described by the Schwarzschild spacetime. Asymptotic flatness requires that the function Φ satisfies $\Phi(\infty) = 0$. The mass function $m(r)$ assumes its asymptotic value M at the surface of the star, where M corresponds to the mass of the neutron star in geometric units. At the center regularity requires that $m(0) = 0$. The density and the pressure at the center are $\rho_c(P_c)$ and $P_c = P(0)$, respectively. P_c is a free parameter.

By varying the central pressure a family of neutron star solutions for a given EOS is obtained. The mass-radius relation of numerous such families of neutron stars is shown in Fig. 7.1. Clearly, there is a strong EOS dependence of the mass-radius relation. Observations of high mass pulsars constrain the EOSs, however, since their maximum mass should allow for the measured mass values [21–23].

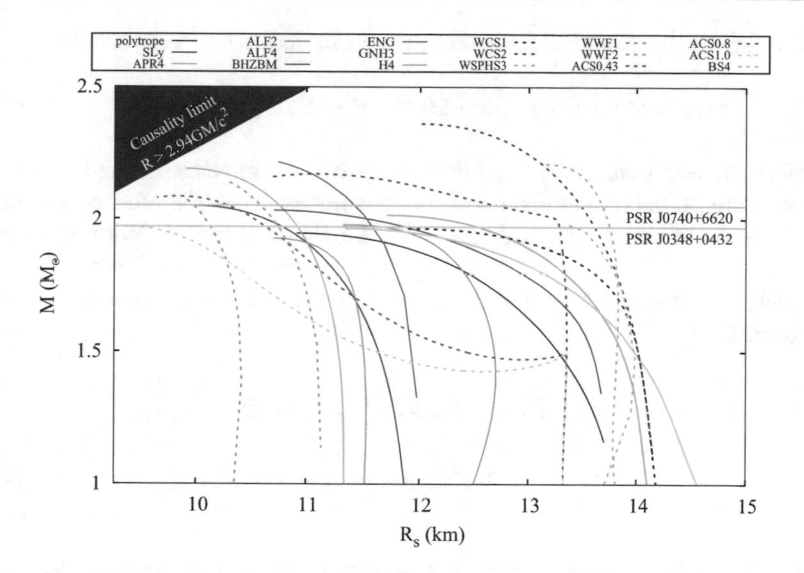

Fig. 7.1 Mass-radius relation of neutron stars in GR: mass M (in solar masses $M_\odot$) vs radius R (in km) for numerous EOSs. The horizontal lines indicate high mass pulsars, the data was obtained from [21] ©2010 Macmillan Publishers Limited. All rights reserved; [23] ©the Author(s), under exclusive license to Springer Nature Limited 2019; and [27] ©2021 by Annual Reviews under CC BY 4.0 license. The upper left corner marks the causality limit obtained in [28] ©2022 The Authors under CC BY 4.0 license. Data reprinted with permission

7.2.2 Properties

While the mass and radius of a neutron star are easily obtained, once an equation of state and a central pressure are specified, further properties of interest typically involve perturbation theory around the TOV background solution. In lowest order perturbation theory the moment of inertia I is obtained. To this end, we consider a slowly rotating neutron star, that rotates with uniform angular velocity Ω around the axis $\theta = 0\ (\pi)$. The metric then acquires a non-diagonal component

$$\delta g_{t\phi} = -\epsilon\omega r^2 \sin^2\theta, \tag{7.14}$$

where ϵ is a perturbation bookkeeping parameter, and the new metric function ω arising from the rotation needs to be determined. All other rotational effects in the metric are of higher order. This also holds for the effects on the density and pressure, which are even functions under time reversal. The fluid velocity receives a contribution

$$\delta U^\mu = \left(0, 0, 0, \epsilon\Omega U^t\right), \tag{7.15}$$

where $U^t = e^{-\Phi}$ is the time-component in the non-rotating frame.

The slow rotation induces a new component in the stress-energy tensor

$$\delta T_{t\phi} = r^2 (\rho + P) \epsilon (\omega - \Omega) \sin^2\theta - P\epsilon\omega r^2 \sin^2\theta. \tag{7.16}$$

A priori, the new function ω could depend on two coordinates, r and θ. Moreover, the resulting partial differential equation does not separate, therefore an expansion of ω in terms of vector spherical harmonics should be made [29, 30]. Inspection of the boundary conditions w.r.t. regularity and asymptotic flatness shows, however, that only a single l can contribute, $l = 1$, leaving ω as a function of r only, determined by

$$\omega'' + \left(\frac{4}{r} - \Phi' - \Lambda'\right)\omega' - 2\left[\Phi'' + (\Phi' - \Lambda')\left(\Phi' + \frac{1}{r}\right)\right]\omega$$
$$+ 2\kappa e^{2\Lambda}\left[P\omega + (\rho + P)(\Omega - \omega)\right] = 0. \tag{7.17}$$

? Exercise

7.2. Assume $\omega = \omega(r)$ and derive the Einstein tensor for the metric

$$ds^2 = -e^{2\Phi(r)}dt^2 + e^{2\Lambda(r)}dr^2 + r^2\left(d\theta^2 + \sin^2\theta d\phi^2\right) - 2\epsilon\omega r^2 \sin^2\theta dt d\phi, \tag{7.18}$$

to first order in ϵ. Consider the energy-momentum tensor component (7.16), and show that the $t\phi$ component of the Einstein equations is given by (7.17).

Expansion at infinity allows to extract the angular momentum J

$$\omega(r) = \frac{2J}{r^3} + O(\frac{1}{r^5}), \tag{7.19}$$

and the moment of inertia I, since $J = I\Omega$. When calculating the moment of inertia for various EOSs, one obtains a large variation of its value for neutron stars with the same mass, as expected from the large variation of the radii, shown in Fig. 7.1 (see e.g. [31]). This is illustrated in Fig. 7.2, where the moment of inertia is shown versus the mass (Fig. 7.2a) and the radius (Fig. 7.2b) for several EOSs.

Of considerable interest is also the quadrupole moment Q that is induced by rotation. However, to extract the quadrupole moment one has to go to second order in Ω. As shown by Hartle and Thorne [29, 32, 33] the appropriate parametrization of

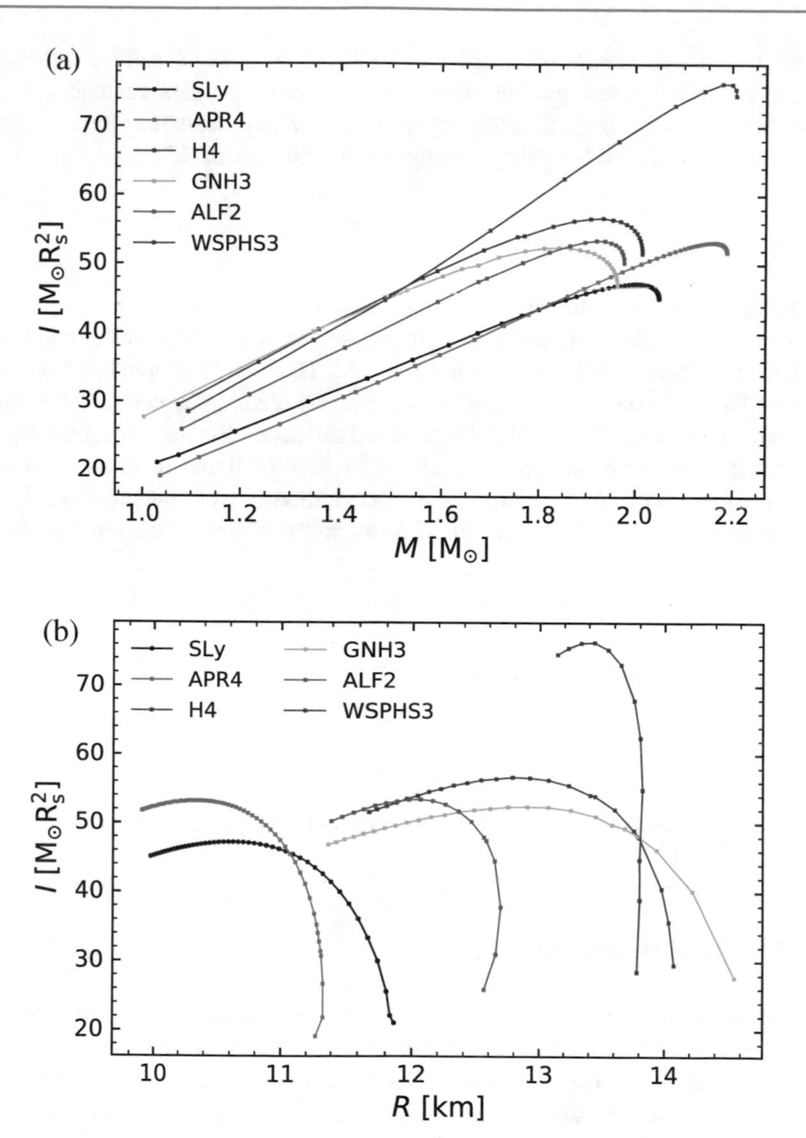

Fig. 7.2 Moment of inertia I of neutron stars in GR: (**a**, upper) I (in solar masses $M_\odot$ times the squared solar gravitational radius R_S) vs mass M (in solar masses $M_\odot$); (**b**, lower) I vs radius R (in km) for several EOSs

the metric is given by

$$
\begin{aligned}
ds^2 = &- e^{2\Phi}\left[1 + 2\epsilon^2 (h_0 + h_2 P_2)\right] dt^2 + e^{2\Lambda}\left[1 + 2e^{2\Lambda}\epsilon^2 (m_0 + m_2 P_2)/r\right] dr^2 \\
&+ r^2\left[1 + 2\epsilon^2 (v_2 - h_2) P_2\right]\left[d\theta^2 + \sin^2\theta\, (d\phi - \epsilon\omega dt)^2\right],
\end{aligned}
\tag{7.20}
$$

where P_2 is the Legendre polynomial $P_2 = \left(3\cos^2\theta - 1\right)/2$, and h_0, h_2, m_0, m_2 and v_2 are radial functions. The density and pressure possess analogous second order terms. After solving the resulting set of differential equations the quadrupole moment Q can be read from the asymptotic behavior [29, 32, 33]

$$h_2(r) \to \frac{Q}{r^3}. \tag{7.21}$$

A further important property of neutron stars is their tidal deformability [34]. In this case one considers a binary system, where the tidal forces of the companion compact object deform the neutron star [33–35]. The tidal Love number λ is related to the tidal quadrupole moment and is obtained by placing the neutron star into an external quadrupolar tidal field. The appropriate ansatz for the perturbations then consists of a subset of the previous ansatz for the rotational quadrupole moment with $\omega = h_0 = m_0 = 0$ [33]. The boundary conditions are of course different, since an external quadrupole field is present. The asymptotic form of the function h_2,

$$h_2 \to a_{-2}r^2 + a_{-1}r + \frac{a_3}{r^3}, \tag{7.22}$$

then provides the tidal Love number λ

$$\lambda = \frac{a_3}{3a_{-2}}. \tag{7.23}$$

In a similar manner one can also obtain the higher multipole moments and the higher Love numbers [24].

7.2.3 Quasi-Normal Modes

Asteroseismology allows to extract important information on the stability and ringdown of neutron stars, when perturbed (see e.g. [36, 37]). Neutron stars possess a rich spectrum of modes, associated with the nuclear matter and the gravitational field. Since General Relativity features gravitational waves starting with quadrupole ($l = 2$) radiation, QNMs arise when $l \geq 2$. These possess a complex eigenvalue ω whose real part is the characteristic frequency ω_R of the mode, while the imaginary part ω_I represents its decay rate.

The linear perturbations of the metric Ansatz and the fluid read [38]

$$g_{\mu\nu} = g^{(0)}_{\mu\nu}(r) + \epsilon h_{\mu\nu}(t, r, \theta, \varphi) \,, \tag{7.24}$$

$$\rho = \rho_0(r) + \epsilon\delta\rho(t, r, \theta, \varphi) \,, \tag{7.25}$$

$$p = p_0(r) + \epsilon\delta p(t, r, \theta, \varphi) \,, \tag{7.26}$$

$$u_\mu = u^{(0)}_\mu(r) + \epsilon\delta u_\mu(t, r, \theta, \varphi) \,, \tag{7.27}$$

where the superscript (0) denotes the static and spherically symmetric background solutions. The perturbations, in contrast, depend on all four coordinates.

To proceed one then expands the perturbations in tensorial spherical harmonics characterized by multipole numbers l and m [39]. The high symmetry of the background solutions then leads to a split of the perturbations into two separate classes: axial perturbations and polar perturbations. Axial perturbations transform as $(-1)^{l+1}$ under parity, and therefore do not couple to the fluid. They are pure space-time modes. Polar modes on the other hand are parity-even and transform as $(-1)^l$. These include the perturbations of the pressure and energy density of the fluid.

Expansion and Fourier decomposition of the axial perturbations of the metric yields

$$h_{\mu\nu}^{(axial)} = \sum_{l,m} \int \begin{bmatrix} 0 & 0 & -h_0 \frac{1}{\sin\theta}\frac{\partial}{\partial\phi} Y_{lm} & h_0 \sin\theta \frac{\partial}{\partial\theta} Y_{lm} \\ 0 & 0 & -h_1 \frac{1}{\sin\theta}\frac{\partial}{\partial\phi} Y_{lm} & h_1 \sin\theta \frac{\partial}{\partial\theta} Y_{lm} \\ -h_0 \frac{1}{\sin\theta}\frac{\partial}{\partial\phi} Y_{lm} & -h_1 \frac{1}{\sin\theta}\frac{\partial}{\partial\phi} Y_{lm} & 0 & 0 \\ h_0 \sin\theta \frac{\partial}{\partial\theta} Y_{lm} & h_1 \sin\theta \frac{\partial}{\partial\theta} Y_{lm} & 0 & 0 \end{bmatrix} e^{-i\omega t} d\omega, \tag{7.28}$$

whereas for polar perturbations one finds

$$h_{\mu\nu}^{(\mathrm{polar})} = \sum_{l,m} \int \begin{bmatrix} r^l e^{2\nu} H_0 Y_{lm} & -i\omega r^{l+1} H_1 Y_{lm} & 0 & 0 \\ -i\omega r^{l+1} H_1 Y_{lm} & r^l e^{2\lambda} H_2 Y_{lm} & 0 & 0 \\ 0 & 0 & r^{l+2} K Y_{lm} & 0 \\ 0 & 0 & 0 & r^{l+2} \sin^2\theta K Y_{lm} \end{bmatrix} e^{-i\omega t} d\omega\,, \tag{7.29}$$

(using (t, r, θ, φ) order for the matrix). The corresponding decomposition of the density and the pressure of the fluid inside the star is

$$\delta\rho = \sum_{l,m} \int r^l E_1 Y_{lm} e^{-i\omega t} d\omega\,, \quad \delta p = \sum_{l,m} \int r^l \Pi_1 Y_{lm} e^{-i\omega t} d\omega\,, \tag{7.30}$$

and the perturbation of the velocity is

$$\delta u_\mu = \sum_{l,m} \int \begin{bmatrix} \frac{1}{2} r^l e^\nu H_0 Y_{lm} \\ r^l i\omega e^{-\nu} \left(e^\lambda W/r - r H_1\right) Y_{lm} \\ -i\omega r^l e^{-\nu} V \partial_\theta Y_{lm} \\ -i\omega r^l e^{-\nu} V \partial_\phi Y_{lm} \end{bmatrix} e^{-i\omega t} d\omega \,, \tag{7.31}$$

while outside the star there is no fluid, of course. All perturbation functions depend only on the radial coordinate r, the multipole numbers l, m, and the complex eigenvalue ω. (In contrast to the previous section, here ω is just a complex number, not a function.) The resulting systems of ordinary differential equations must then be simplified by specific choices of gauge and solved subject to an appropriate set of boundary conditions. These boundary conditions require regularity at the center of the star and a purely outgoing wave behavior at infinity. Moreover, they require continuity of the metric perturbation functions and their derivatives at the border of the star, where the pressure and the energy density vanish. Together all these requirements then select a discrete set of values for the complex eigenvalue ω for a given l, representing the fundamental frequency and its overtones (see e.g. [40, 41] for further details). In Sect. 9.3.1, further details on QNMs will also be discussed in the context of BHs.

The fundamental f mode ($l = 2$) in GR is illustrated in Fig. 7.3, where the frequency ω_R (Fig. 7.3a) and the decay time $\tau = 1/\omega_I$ (Fig. 7.3b) are shown versus the mass of the neutron stars for several EOSs. The figure reveals clearly the significant dependence of the modes on the EOS.

7.3 Universal Relations

As discussed above, dimensionful neutron star properties depend significantly on the employed EOS. If, however, properly scaled dimensionless quantities are considered instead, an important set of *universal relations* arises in GR, which exhibit only little EOS dependence.

7.3.1 I-Love-Q Relations

In geometric units the so-called compactness C is a simple dimensionless quantity. It represents the ratio of the mass M and the radius R of a neutron star, $C = M/R$. The compactness of neutron stars ranges typically in the interval $0.1 < C < 0.3$, while a Schwarzschild black hole has a compactness of $C = 0.5$, since its horizon radius is given by $R = 2M$. Clearly, compactness is a relevant physical property, and being dimensionless, it is a suitable candidate to feature in *universal relations*.

A first *universal relation* can thus be envisaged that exhibits a suitably scaled moment of inertia $\bar{I}$ versus the compactness C. Since $J = \Omega I$, a dimensionless

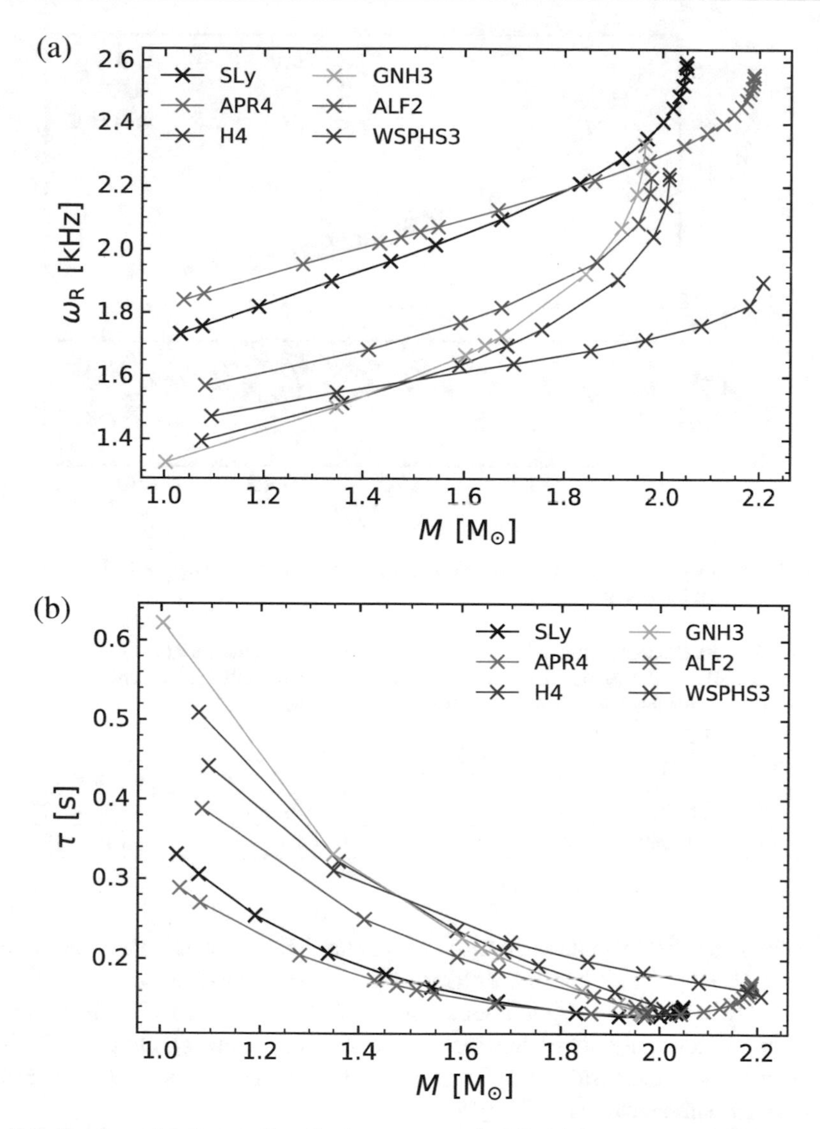

Fig. 7.3 Fundamental f mode ($l = 2$) of neutron stars in GR: (**a**) frequency ω_R (in kHz) vs mass M (in solar masses $M_\odot$); (**b**) decay time $\tau = 1/\omega_I$ (in s) vs mass M (in solar masses $M_\odot$) for several EOSs

moment of inertia is obtained in geometric units in terms of $\bar{I} = I/M^3$. (Recall, that J/M^2 is dimensionless.) This $\bar{I}$-C relation is demonstrated in Fig. 7.4 for several EOSs. While dependence on the EOS has been reduced considerably for these dimensionless quantities as compared to the dimensionful quantities I, M and R shown in Fig. 7.2, this relation is less impressive than the I-Love-Q relations discussed in the following.

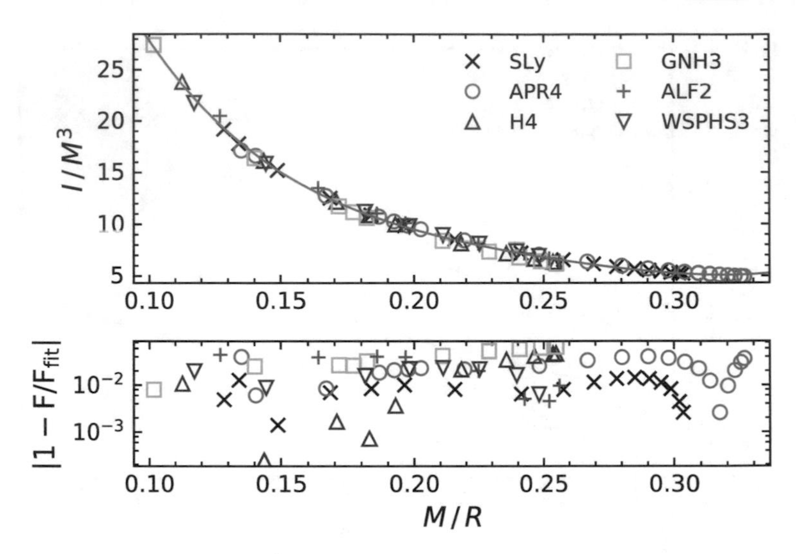

Fig. 7.4 Upper figure: Universal $\bar{I}$-C relation for several EOSs. Lower figure: Relative deviation of a value F from the best fit F_{fit}

Table 7.1 Best fit coefficients in (7.32) for the relations between the variables $\bar{I}$, $\bar{Q}$ and $\bar{\lambda}$. Data taken from [24] ©2017 Elsevier B.V. All rights reserved. Data reprinted with permissions. Earlier approaches towards this latest data have been already published in [42]

y_i	x_i	a_i	b_i	c_i	d_i	e_i
$\bar{I}$	$\bar{\lambda}$	1.496	0.05951	0.02238	-6.953×10^{-4}	8.345×10^{-6}
$\bar{I}$	$\bar{Q}$	1.393	0.5471	0.03028	0.01926	4.434×10^{-4}
$\bar{Q}$	$\bar{\lambda}$	0.1940	0.09163	0.04812	-4.283×10^{-3}	1.245×10^{-4}

Besides the dimensionless moment of inertia $\bar{I}$ the dimensionless quadrupole moment $\bar{Q} = QM/J^2$ and the dimensionless Love number $\bar{\lambda} = \lambda/M^5$ feature prominently in the I-Love-Q relations. These relations do not involve the compactness, but consider only the dimensionless quantities $\bar{I}$, $\bar{\lambda}$ and $\bar{Q}$. Obtained by Yagi and Yunes [24, 42], the truly remarkable I-Love, I-Q, and Q-Love relations can be expressed as simple curves of the type

$$\ln y_i = a_i + b_i \ln x_i + c_i (\ln x_i)^2 + d_i (\ln x_i)^3 + e_i (\ln x_i)^4, \tag{7.32}$$

where y_i represent the first and x_i the second dimensionless quantity, as seen in Tables 7.1 and 7.2, where the coefficients a_i to e_i yield an excellent fit to the data of a very large number of EOSs with very different properties of the matter of the star [24, 42]. In fact, the deviations of the data from the best fit shown are below 1%.

Analogous relations can be considered for the higher multipole moments and higher Love numbers. In the usual nomenclature the mass corresponds to the lowest mass moment $M = M_0$ and the angular momentum to the lowest current moment

Table 7.2 Best fit coefficients in (7.32) for the relations between the variables $\bar{S}_3$, $\bar{M}_4$ and $\bar{Q}$. Data taken from [24] ©2017 Elsevier B.V and [48] ©2014. The American Astronomical Society. All rights reserved. Data reprinted with permissions

y_i	x_i	a_i	b_i	c_i	d_i	e_i
$\bar{S}_3$	$\bar{Q}$	3.131×10^{-3}	2.071	−0.7152	0.2458	−0.03309
$\bar{M}_4$	$\bar{Q}$	−0.02287	3.849	−1.540	0.5863	-8.337×10^{-2}

$J = S_1$. Higher mass moments possess even index, M_{2l}, and higher current moments odd index, S_{2l+1} [39, 43–46]. The quadrupole moment then corresponds to $Q = M_2$. Higher tidal mass and current moments are referred to as λ_n [47]. Examples of *universal relations* for higher moments are exhibited in Table II [24,48]. These represent the $\bar{S}_3$-$\bar{Q}$ and $\bar{M}_4$-$\bar{Q}$ relations, that possess larger deviations (4% and 10%) than the I-Love-Q relations.

In this connection the expression *three hair relations* was coined [24,48]. This is a generalization of the *no-hair* (*two hair relation*), highlighting that Kerr black holes are fully determined by only two quantities (hairs), their mass and their angular momentum. In the *three hair relations* of neutron stars the additional quantity besides the mass and the angular momentum is the quadrupole moment [24, 48]. In contrast to the *two hair relation* of black holes, the *three hair relations* of neutron stars are only approximate relations. Their validity for neutron stars has been associated with an approximate symmetry that emerges at high compactness: the self-similarity of isodensity surfaces [24, 49].

7.3.2 Quasi-Normal Modes

Universal relations arise also in the study of quasi-normal modes. As illustrated above, quasi-normal modes feature a considerable dependence on the EOS. However, Anderson and Kokkotas pointed out rather early that *universal relations* may reduce this EOS dependence significantly [36,50], as confirmed in numerous further studies, e.g., [51–58].

A set of *universal relations* for the fundamental f mode of neutron stars is illustrated in Fig. 7.5, where Fig. 7.5a and b exhibit the dimensionless scaled frequency $M\omega_R/c$ and the dimensionless scaled decay rate $M\omega_I/c$, respectively, versus the compactness $C = M/R$ for several EOSs. Analogous relations are shown in Fig. 7.5c and d, where instead of the compactness C the so-called effective compactness $\eta = \sqrt{M^3/I} = \bar{I}^{-1/2}$ was used, that is based on the dimensionless moment of inertia $\bar{I}$. It was introduced in [55], where also the following best fit was provided for the f mode

$$M\omega_R = -0.0047 + 0.133\eta + 0.575\eta^2, \quad M\omega_I = 0.00694\eta^4 - 0.0256\eta^6.$$

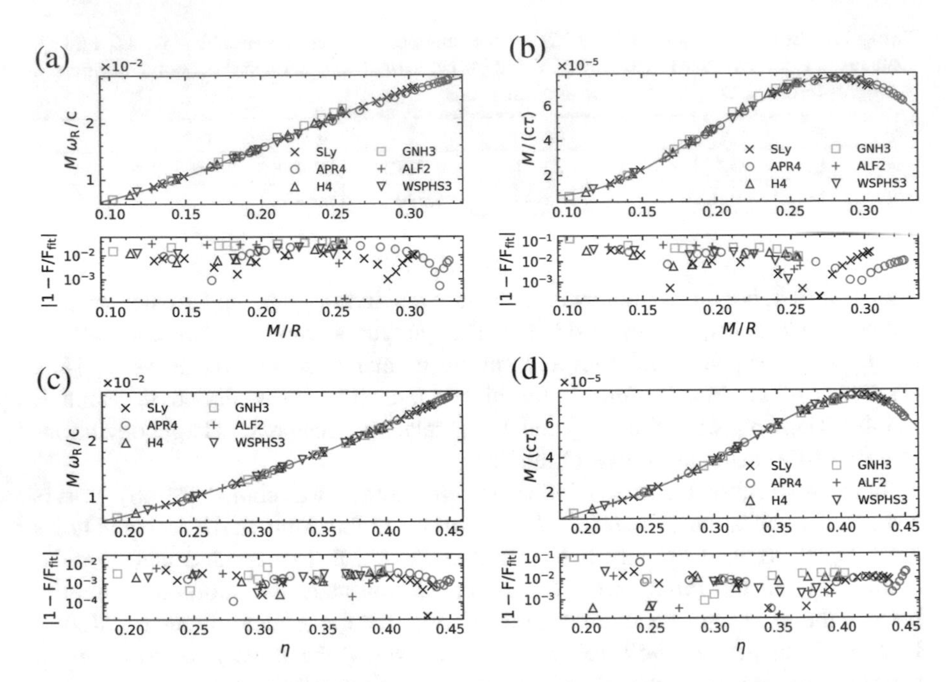

Fig. 7.5 Universal relation for the fundamental f mode ($l = 2$) of neutron stars in GR: (**a**, upper left) scaled frequency $M\omega_R/c$ vs compactness $C = M/R$; (**b**, upper right) scaled decay rate $M\omega_I/c$ vs compactness $C = M/R$ for several EOSs; (**c**, lower left) and (**d**, lower right) analogous, but vs the effective compactness $\eta = \sqrt{M^3/I}$

The parametrization in terms of the effective compactness reduces the errors (as compared to the compactness) and is therefore preferable. Various further *universal relations* for the quasi-normal modes have been found, among them for instance a relation between the scaled frequency of a mode and the scaled damping rate [56, 57].

Universal relations can be of use in many different circumstances [24]. First of all, they can be employed to extract further information on neutron star properties not yet known from explicit measurements. In case of the lowest moments, for instance, the I-Love-Q relations would allow to obtain any two of the three quantities, once the third one would be measured [24], while the seven lowest moments could be obtained from measurements of the mass, rotation period and moment of inertia with the help of the *three-hair-relations* [48]. On the other hand, in the case of the quasi-normal mode measurement of an axial or polar mode would allow the determination of the mass M and the moment of inertia I of a star by invoking the $M\omega_R$-η and $M\omega_I$-η relations [55]. Moreover, the radius R might be extracted and conclusions with respect to the EOS might be possible. Currently *universal relations* are already employed to reduce degeneracies in the analysis of gravitational waves [24]. Last but not least, as discussed next *universal relations* also provide a means to test alternative gravity theories.

7.4 Neutron Stars in Alternative Gravity Theories

Studies of alternative gravity theories are motivated largely by the quest for a theory of quantum gravity and by cosmological issues like dark matter and dark energy, see also the discussions in the previous Chaps. 1–5. Such theories typically involve new degrees of freedom, with the simplest being a real scalar field (see Sects. 4.4.4 and 4.4.5 for torsion-scalar gravity theories for example). If indeed such additional degrees of freedom would be present, their consequences might not only resolve the issues intended, but they might also have observable consequences that could be tested by observations in the solar system or observations of black holes and neutron stars and gravitational waves emitted by these compact objects [59–62]. In the following neutron stars will be discussed for two widely employed types of alternative gravity theories.

7.4.1 Scalar-Tensor-Theories

Scalar-tensor theories introduce in addition to the gravitational metric tensor field a gravitational scalar field (see e.g. [63–66]). A generic action for such scalar-tensor theories is given by

$$S = \frac{1}{16\pi G}\int d^4x\sqrt{-\tilde{g}}\left[F(\Phi)\tilde{\mathcal{R}} - Z(\Phi)\tilde{g}^{\mu\nu}\partial_\mu\Phi\partial_\nu\Phi - 2U(\Phi)\right] + S_m\left[\Psi_m;\tilde{g}_{\mu\nu}\right], \tag{7.33}$$

where the tilde indicates that the respective quantities are in the so-called Jordan frame, Φ is the gravitational scalar field, and S_m denotes any additional matter fields Ψ_m. In the Jordan frame the gravitational scalar field does not couple directly to the matter fields and the weak equivalence principle is retained. The functions $F(\Phi)$ and $Z(\Phi)$ cannot be chosen arbitrary, but need to meet some physical restrictions [67].

While neutron stars can be studied directly in the Jordan frame, it is typically more convenient to transform to the so-called Einstein frame, which can be achieved by means of a conformal transformation of the metric $g_{\mu\nu} = F(\Phi)\tilde{g}_{\mu\nu}$, and an associated transformation of the gravitational scalar field denoted by φ now [64–66].

$$\left(\frac{d\varphi}{d\Phi}\right)^2 = \frac{3}{4}\left(\frac{d\ln(F(\Phi))}{d\Phi}\right)^2 + \frac{Z(\Phi)}{2F(\Phi)}. \tag{7.34}$$

In the Einstein frame, the action then reads

$$S = \frac{1}{16\pi G}\int d^4x\sqrt{-g}\left[\mathcal{R} - 2g^{\mu\nu}\partial_\mu\varphi\partial_\nu\varphi - 4V(\varphi)\right] + S_m[\Psi_m; \mathrm{A}^2(\varphi)g_{\mu\nu}], \tag{7.35}$$

where the Einstein frame quantities are denoted without tilde, and the following relations hold

$$A(\varphi) = F^{-1/2}(\Phi), \; 2V(\varphi) = U(\Phi)F^{-2}(\Phi). \tag{7.36}$$

In the simplest case the scalar potential is chosen to vanish, $U(\Phi) = 0 = V(\varphi)$. The gravitational scalar field is then massless and has no self-interactions.

? Exercise

7.3. Show that variation of the action (7.35) leads to the Einstein equations

$$\mathcal{R}_{\mu\nu} - \frac{1}{2}g_{\mu\nu}\mathcal{R} = 2\partial_\mu\varphi\partial_\nu\varphi - g_{\mu\nu}g^{\alpha\beta}\partial_\alpha\varphi\partial_\beta\varphi + 8\pi T_{\mu\nu} \tag{7.37}$$

and gravitational scalar field equation

$$\nabla^\mu\nabla_\mu\varphi = -4\pi k(\varphi)T, \tag{7.38}$$

where $T = T_\mu^\mu$, and $k(\varphi) = \frac{d\ln(A(\varphi))}{d\varphi}$.

The function $A(\varphi)$ determines the coupling between the scalar field and the matter. The stress-energy tensor $\tilde{T}_{\mu\nu}$ is provided in the physical Jordan frame and then transformed into the Einstein frame

$$T_{\mu\nu} = A^2\tilde{T}_{\mu\nu}, \tag{7.39}$$

where the Bianchi identities yield

$$\nabla_\mu T^\mu{}_\nu = k(\varphi)T\partial_\nu\varphi. \tag{7.40}$$

The freedom in the choice of coupling function $A(\varphi)$ leads to different types of scalar-tensor theories, and thus different consequences for neutron stars in these theories. At the same time it leads to a variety of physical effects, that can be compared to observations and thus result in more or less stringent constraints from observations. Brans-Dicke theory, for instance, is obtained for the simple parametrization $A = e^{\kappa\varphi}$, i.e., $k(\varphi) = \kappa$ with constant κ, addressed further below [63]. Here another coupling function is considered,

$$A(\varphi) = e^{\frac{1}{2}\beta\varphi^2}, \quad k(\varphi) = \beta\varphi, \tag{7.41}$$

that leads to the interesting phenomenon of spontaneous scalarization of neutron stars, discovered by Damour and Esposito-Farèse [68].

Spontaneous scalarization in neutron stars is matter induced. It can arise in theories with coupling functions, that possess a quadratic dependence on the gravitational scalar field such that it satisfies a Klein-Gordon type equation with an effective mass, i.e.,

$$\nabla^{\mu}\nabla_{\mu}\varphi = m_{\rm eff}^{2}\varphi. \tag{7.42}$$

In that case the GR neutron star solutions remain solutions of the scalar-tensor theory, since for vanishing scalar field the equations reduce to the GR equations. However, in addition to the GR solutions new solutions with a gravitational scalar field may arise, when the neutron matter represents a sufficiently strong source to induce a tachyonic instability, $m_{\rm eff}^2 = -4\pi G\beta T < 0$. While typical neutron stars possess $T = 3p - \rho < 0$, so $\beta < 0$ must be chosen for spontaneous scalarization to occur, both T and β could also be positive, but in this case the neutron stars would need a pressure dominated core [69, 70].

When evaluating a family of GR neutron stars by increasing the central pressure, at some point a neutron star with a zero mode arises. Beyond this point scalarization sets in, and a branch of scalarized neutron stars is present in addition to the GR neutron stars. In fact, GR neutron stars then possess an unstable mode, whereas the scalarized neutron stars become the physically preferred stable configurations (see, e.g., [68, 71]). The scalar field at the center of the star and the scalar charge are largely independent of the EOS, and thus basically universal, only depending on the gravitational potential at the center of the neutron star [31].

Pulsar observations have by now virtually excluded the possibility of spontaneous scalarization of neutron stars for the simplest case of a massless scalar field [72]. These conclusions are based on the expected effects of dipolar and thus scalar radiation on the orbits of the compact objects. However, the inclusion of a genuine mass term with a sufficiently large mass for the scalar field allows to circumvent these observational constraints, since the dipolar radiation becomes rather negligible when the orbital separation of a binary star system is much larger than the scalar field Compton wavelength [73]. Evaluation of the properties and quasi-normal modes of scalarized neutron stars with a massive gravitational scalar field also leads to *universal relations* (see e.g., [74–76]). Depending on the strength of the scalarization, they may differ significantly from those of neutron stars in GR.

7.4.2 $f(\mathcal{R})$ Theories

In $f(\mathcal{R})$ theories the gravitational action is no longer given by the curvature scalar $\mathcal{R}$, but by some function of the curvature scalar, $f(\mathcal{R})$ [77–79]. A particular well-motivated such theory is based on

$$f(\mathcal{R}) = \mathcal{R} + a\mathcal{R}^2\,, \tag{7.43}$$

since this model (called the Starobinsky model) is capable to predict the inflationary phase of the early universe consistent with observations. $f(\mathcal{R})$ theories can also be transformed to the Einstein frame, since from a mathematical point of view they are equivalent to scalar-tensor theories. As shown in [80, 81] such a transformation then leads to a scalar field with a Brans-Dicke type coupling function and a scalar field potential

$$A(\varphi) = e^{-\frac{1}{\sqrt{3}}\varphi}, \quad V(\varphi) = \frac{3m_\varphi^2}{2}\left(1 - e^{-\frac{2\varphi}{\sqrt{3}}}\right)^2 . \tag{7.44}$$

The scalar field mass m_φ is identified in the transformation from the coefficient of the φ^2 term of the potential $V(\varphi)$, that arises in the transformation, and is thus a function of the coupling constant a of the $f(R)$ theory considered, $m_\varphi = 1/\sqrt{6a}$. The parameter a therefore determines the mass of the scalar field, and can be chosen well within the current observational window [73].

? Exercise

7.4. Transform (7.43) into a scalar tensor theory. Follow the steps which were outlined in Sect. 4.4.5 for teleparallel $f(G)$ theories of gravity.

Besides leading to distinct I-Love-Q relations (see e.g., [82, 83]), this $f(\mathcal{R})$ theory has a distinct spectrum of quasi-normal modes [28, 84–86]. In particular, in contrast to GR, monopole ($l = 0$) and dipole ($l = 1$) radiation arises due to the additional degree of freedom. In GR neutron stars possess only $l = 0$ normal modes. But in such an $f(\mathcal{R})$ theory, these modes become propagating modes. Interestingly, these modes feature a very small decay rate ω_I, which means that they are ultra long-lived [85]. Moreover, the scale of the frequency ω_R is determined by the neutron star size for small Compton wavelength $L_\varphi = 1/m_\varphi$ of the scalar field, while for large L_φ the frequency follows L_φ.

The *universal relations* for quasi-normal modes exhibit distinct features, as well, and therefore might be exploited to put further bounds on such a theory [28, 84]. Figure 7.6 illustrates a set of *universal relations* for the fundamental f mode ($l = 2$) for two values of the scalar mass m_φ for this $f(\mathcal{R})$ theory, analogous to Fig. 7.5 for GR. Due to the presence of the new degree of freedom, however, this f mode is not the only polar quadrupole ($l = 2$) mode. There is an additional scalar $l = 2$ mode present, and there are also the scalar dipole and monopole modes, all of them exhibiting *universal relations* [28].

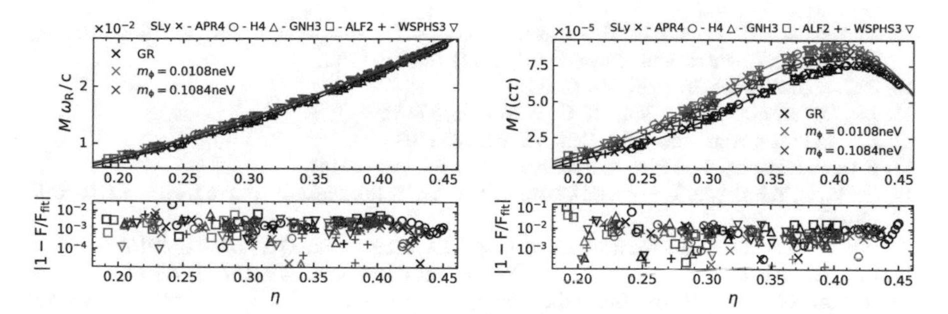

Fig. 7.6 Universal relation for the fundamental f mode ($l = 2$) of neutron stars in $f(\mathcal{R}) = \mathcal{R}+a\mathcal{R}^2$ theory with $m_\varphi = 0.0108$ neV and 0.1084 neV: (left) scaled frequency $M\omega_R/c$ vs effective compactness $\eta = \sqrt{M^3/I}$ (right) scaled decay rate $M\omega_I/c$ vs effective compactness $\eta = \sqrt{M^3/I}$ for several EOSs. For comparison also the GR relations are shown

7.5 Conclusion

Since neutron stars are highly compact objects, they represent ideal astrophysical objects to learn about gravity. While the current lack of knowledge of the physical EOS of nuclear matter under these extreme conditions leads to larger ranges of possible values of their physical properties and their emitted radiation, the presence of *universal relations*, which are rather independent of the EOS, reduces these uncertainties to a large extent. Moreover, *universal relations* may be rather different for GR and for alternative theories of gravity, thus allowing to put bounds on such theories once the corresponding measurements will have been achieved with sufficient accuracy.

Acknowledgments I would like to thank the organizers for the invitation to the interesting meeting *Signatures and experimental searches for modified and quantum gravity*. I would also like to thank my collaborators and here, in particular, Jose Luis Blázquez-Salcedo and Vincent Preut, for providing the above figures. Furthermore, I would like to gratefully acknowledge support by the DFG Research Training Group 1620 *Models of Gravity* and the COST Actions CA15117 and CA16104.

References

1. W. Baade, F. Zwicky, Proc. Natl. Acad. Sci. **20**(5), 254 (1934)
2. A. Hewish, S.J. Bell, J.D.H. Pilkington, P.F. Scott, R.A. Collins, Nature **217**, 709 (1968)
3. J.H. Taylor, R.N. Manchester, A.G. Lyne, Astrophys. J. Suppl. **88**, 529 (1993)
4. M. Kramer, J.F. Bell, R.N. Manchester, A.G. Lyne, F. Camilo, I.H. Stairs, N. D'Amico, V.M. Kaspi, G. Hobbs, D.J. Morris et al., Mon. Not. Roy. Astron. Soc. **342**, 1299 (2003)
5. M. Burgay, N. D'Amico, A. Possenti, R.N. Manchester, A.G. Lyne, B.C. Joshi, M.A. McLaughlin, M. Kramer, J.M. Sarkissian, F. Camilo et al., Nature **426**, 531 (2003)
6. A.A. Abdo et al., [Fermi-LAT], Astrophys. J. Suppl. **208**, 17 (2013)
7. The EPN Database of Pulsar Profiles: http://www.epta.eu.org/epndb/
8. I.H. Stairs, Living Rev. Rel. **6**, 5 (2003)

9. M. Kramer, I.H. Stairs, R.N. Manchester, N. Wex, A.T. Deller, W.A. Coles, M. Ali, M. Burgay, F. Camilo, I. Cognard et al., Phys. Rev. X **11**(4), 041050 (2021)
10. R.C. Tolman, Phys. Rev. **55**, 364 (1939)
11. J.R. Oppenheimer, G.M. Volkoff, Phys. Rev. **55**, 374 (1939)
12. J.M. Lattimer, Ann. Rev. Nucl. Part. Sci. **62**, 485 (2012)
13. F. Özel, P. Freire, Ann. Rev. Astron. Astrophys. **54**, 401 (2016)
14. G. Baym, T. Hatsuda, T. Kojo, P.D. Powell, Y. Song, T. Takatsuka, Rep. Prog. Phys. **81**, 056902 (2018)
15. B.P. Abbott et al., [LIGO Scientific and Virgo], Phys. Rev. Lett. **116**, 061102 (2016)
16. B.P. Abbott et al., [LIGO Scientific and Virgo], Phys. Rev. Lett. **119**, 161101 (2017)
17. B.P. Abbott et al., [LIGO Scientific, Virgo, Fermi GBM, INTEGRAL, IceCube, AstroSat Cadmium Zinc Telluride Imager Team, IPN, Insight-Hxmt, ANTARES, Swift, AGILE Team, 1M2H Team, Dark Energy Camera GW-EM, DES, DLT40, GRAWITA, Fermi-LAT, ATCA, ASKAP, Las Cumbres Observatory Group, OzGrav, DWF (Deeper Wider Faster Program), AST3, CAASTRO, VINROUGE, MASTER, J-GEM, GROWTH, JAGWAR, CaltechNRAO, TTU-NRAO, NuSTAR, Pan-STARRS, MAXI Team, TZAC Consortium, KU, Nordic Optical Telescope, ePESSTO, GROND, Texas Tech University, SALT Group, TOROS, BOOTES, MWA, CALET, IKI-GW Follow-up, H.E.S.S., LOFAR, LWA, HAWC, Pierre Auger, ALMA, Euro VLBI Team, Pi of Sky, Chandra Team at McGill University, DFN, ATLAS Telescopes, High Time Resolution Universe Survey, RIMAS, RATIR and SKA South Africa/MeerKAT], Astrophys. J. Lett. **848**, L12 (2017)
18. B.P. Abbott et al., [LIGO Scientific and Virgo], Phys. Rev. Lett. **121**, 161101 (2018)
19. J. Alsing, H.O. Silva, E. Berti, Mon. Not. Roy. Astron. Soc. **478**, 1377 (2018)
20. L. Rezzolla, E.R. Most, L.R. Weih, Astrophys. J. Lett. **852**, L25 (2018)
21. P. Demorest, T. Pennucci, S. Ransom, M. Roberts, J. Hessels, Nature **467**, 1081 (2010)
22. J. Antoniadis, P.C.C. Freire, N. Wex, T.M. Tauris, R.S. Lynch, M.H. van Kerkwijk, M. Kramer, C. Bassa, V.S. Dhillon, T. Driebe et al., Science **340**, 6131 (2013)
23. H.T. Cromartie et al., [NANOGrav], Nat. Astron. **4**, 72 (2019)
24. K. Yagi, N. Yunes, Phys. Rep. **681**, 1 (2017)
25. D.D. Doneva, G. Pappas, Astrophys. Space Sci. Libr. **457**, 737 (2018)
26. J.S. Read, B.D. Lackey, B.J. Owen, J.L. Friedman, Phys. Rev. D **79**, 124032 (2009)
27. J.M. Lattimer, Ann. Rev. Nucl. Part. Sci. **71**, 433 (2021)
28. J.L. Blázquez-Salcedo, B. Kleihaus, J. Kunz, Universe **8**, 153 (2022)
29. J.B. Hartle, Astrophys. J. **150**, 1005 (1967)
30. H. Sotani, Phys. Rev. D **86**, 124036 (2012)
31. Z. Altaha Motahar, J.L. Blázquez-Salcedo, B. Kleihaus, J. Kunz, Phys. Rev. D **96**, 064046 (2017)
32. J.B. Hartle, K.S. Thorne, Astrophys. J. **153**, 807 (1968)
33. P. Pani, E. Berti, Phys. Rev. D **90**, 024025 (2014)
34. T. Hinderer, Astrophys. J. **677**, 1216 (2008)
35. K. Yagi, N. Yunes, Phys. Rev. D **88**, 023009 (2013)
36. N. Andersson, K.D. Kokkotas, Mon. Not. Roy. Astron. Soc. **299**, 1059 (1998)
37. K.D. Kokkotas, B.G. Schmidt, Living Rev. Rel. **2**, 2 (1999)
38. K.S. Thorne, A. Campolattaro, Astrophys. J. **149**, 591 (1967). Erratum: Astrophys. J. **152**, 673 (1968)
39. K.S. Thorne, Rev. Mod. Phys. **52**, 299 (1980)
40. L. Lindblom, S.L. Detweiler, Astrophys. J. Suppl. **53**, 73 (1983)
41. S.L. Detweiler, L. Lindblom, Astrophys. J. **292**, 12 (1985)
42. K. Yagi, N. Yunes, Science **341**, 365 (2013)
43. R.P. Geroch, J. Math. Phys. **11**, 2580 (1970)
44. R.O. Hansen, J. Math. Phys. **15**, 46 (1974)
45. C. Hoenselaers, Z. Perjes, Class. Quant. Grav. **10**, 375 (1993)
46. T.P. Sotiriou, T.A. Apostolatos, Class. Quant. Grav. **21**, 5727 (2004)
47. T. Damour, A. Nagar, Phys. Rev. D **80**, 084035 (2009)

48. L.C. Stein, K. Yagi, N. Yunes, Astrophys. J. **788**, 15 (2014)
49. K. Yagi, L.C. Stein, G. Pappas, N. Yunes, T.A. Apostolatos, Phys. Rev. D **90**, 063010 (2014)
50. N. Andersson, K.D. Kokkotas, Phys. Rev. Lett. **77**, 4134 (1996)
51. K.D. Kokkotas, T.A. Apostolatos, N. Andersson, Mon. Not. Roy. Astron. Soc. **320**, 307 (2001)
52. O. Benhar, E. Berti, V. Ferrari, Mon. Not. Roy. Astron. Soc. **310**, 797 (1999)
53. O. Benhar, V. Ferrari, L. Gualtieri, Phys. Rev. D **70**, 124015 (2004)
54. L.K. Tsui, P.T. Leung, Mon. Not. Roy. Astron. Soc. **357**, 1029 (2005)
55. H.K. Lau, P.T. Leung, L.M. Lin, Astrophys. J. **714**, 1234 (2010)
56. J.L. Blazquez-Salcedo, L.M. Gonzalez-Romero, F. Navarro-Lerida, Phys. Rev. D **87**, 104042 (2013)
57. J.L. Blázquez-Salcedo, L.M. González-Romero, F. Navarro-Lérida, Phys. Rev. D **89**, 044006 (2014)
58. C. Chirenti, G.H. de Souza, W. Kastaun, Phys. Rev. D **91**, 044034 (2015)
59. C.M. Will, Living Rev. Rel. **9**, 3 (2006)
60. V. Faraoni, S. Capozziello, Fundam. Theor. Phys. **170** (2010)
61. E. Berti et al., Class. Quant. Grav. **32**, 243001 (2015)
62. E.N. Saridakis et al., *Modified Gravity and Cosmology - An Update by the CANTATA Network*, ed. by E.N. Saridakis, R. Lazkoz, V. Salzano, P.V. Moniz, S. Capozziello, J.B. Jiménez, M. De Laurentis, G.J. Olmo. Springer (2021). ISBN: 978-3-030-83714-3. https://link.springer.com/book/10.1007/978-3-030-83715-0
63. C. Brans, R.H. Dicke, Phys. Rev. **124**, 925 (1961)
64. T. Damour, G. Esposito-Farese, Class. Quant. Grav. **9**, 2093 (1992)
65. T. Damour, G. Esposito-Farese, Phys. Rev. D **54**, 1474 (1996)
66. Y. Fujii, K. Maeda, *The Scalar-Tensor Theory of Gravitation* (Cambridge University Press, Cambridge, 2007)
67. G. Esposito-Farese, D. Polarski, Phys. Rev. D **63**, 063504 (2001)
68. T. Damour, G. Esposito-Farese, Phys. Rev. Lett. **70**, 2220 (1993)
69. R.F.P. Mendes, Phys. Rev. D **91**, 064024 (2015)
70. R.F.P. Mendes, N. Ortiz, Phys. Rev. D **93**, 124035 (2016)
71. R.F.P. Mendes, N. Ortiz, Phys. Rev. Lett. **120**, 201104 (2018)
72. J. Zhao, P.C.C. Freire, M. Kramer, L. Shao, N. Wex, Class. Quant. Grav. **39**, 11LT01 (2022)
73. F.M. Ramazanoğlu, F. Pretorius, Phys. Rev. D **93**, 064005 (2016)
74. S.S. Yazadjiev, D.D. Doneva, D. Popchev, Phys. Rev. D **93**, 084038 (2016)
75. D.D. Doneva, S.S. Yazadjiev, J. Cosmol. Astropart. Phys. **11**, 019 (2016)
76. Z. Altaha Motahar, J.L. Blázquez-Salcedo, D.D. Doneva, J. Kunz, S.S. Yazadjiev, Phys. Rev. D **99**, 104006 (2019)
77. T.P. Sotiriou, V. Faraoni, Rev. Mod. Phys. **82**, 451 (2010)
78. A. De Felice, S. Tsujikawa, Living Rev. Rel. **13**, 3 (2010)
79. S. Capozziello, M. De Laurentis, Phys. Rep. **509**, 167 (2011)
80. S.S. Yazadjiev, D.D. Doneva, K.D. Kokkotas, K.V. Staykov, J. Cosmol. Astropart. Phys. **06**, 003 (2014)
81. K.V. Staykov, D.D. Doneva, S.S. Yazadjiev, K.D. Kokkotas, J. Cosmol. Astropart. Phys. **10**, 006 (2014)
82. S.S. Yazadjiev, D.D. Doneva, K.D. Kokkotas, Phys. Rev. D **91**, 084018 (2015)
83. D.D. Doneva, S.S. Yazadjiev, K.D. Kokkotas, Phys. Rev. D **92**, 064015 (2015)
84. J.L. Blázquez-Salcedo, D.D. Doneva, J. Kunz, K.V. Staykov, S.S. Yazadjiev, Phys. Rev. D **98**, 104047 (2018)
85. J.L. Blázquez-Salcedo, F.S. Khoo, J. Kunz, Europhys. Lett. **130**, 50002 (2020)
86. J.L. Blázquez-Salcedo, F.S. Khoo, J. Kunz, V. Preut, Front. Phys. **9**, 741427 (2021)

Black Holes: On the Universality of the Kerr Hypothesis

8

Carlos A. R. Herdeiro

Abstract

To what extent are *all* astrophysical, dark, compact objects both black holes (BHs) and described by the Kerr geometry? We embark on the exercise of defying the *universality* of this remarkable idea, often called the "Kerr hypothesis". After establishing its rationale and timeliness, we define a minimal set of reasonability criteria for alternative models of dark compact objects. Then, as proof of principle, we discuss concrete, dynamically robust non-Kerr BHs and horizonless imitators, that (1) pass the basic theoretical, and in particular dynamical, tests, (2) match (some of the) state of the art astrophysical observables and (3) only emerge at some (macroscopic) scales. These examples illustrate how the universality (at all macroscopic scales) of the Kerr hypothesis can be challenged.

8.1 The Kerr Hypothesis

The elegant Kerr metric [1,2],

$$ds^2 = -\frac{\Delta}{\Sigma}\left(dt - a\sin^2\theta d\phi\right)^2 + \frac{\Sigma}{\Delta}dr^2 + \Sigma d\theta^2 + \frac{\sin^2\theta}{\Sigma}[adt - (r^2+a^2)d\phi]^2\,, \tag{8.1}$$

where $\Sigma \equiv r^2 + a^2\cos^2\theta$, $\Delta \equiv r^2 - 2Mr + a^2$, is the currently accepted model to describe the phenomenology of *all* astrophysical black hole (BH) candidates.

C. A. R. Herdeiro (✉)
Departamento de Matemática da Universidade de Aveiro, Aveiro, Portugal
e-mail: herdeiro@ua.pt

C. Pfeifer, C. Lämmerzahl (eds.), *Modified and Quantum Gravity*, Lecture Notes in Physics 1017, https://doi.org/10.1007/978-3-031-31520-6_8

Since (8.1) is only described by two macroscopic parameters, the mass M and angular momentum Ma of the BH, this is clearly an economical scenario. The very same (almost featureless) theoretical model describes astrophysical objects ranging, at least, 10 orders of magnitude in mass. There is evidence for BHs in the stellar mass range, from $\sim M_\odot$ to $\sim 100\ M_\odot$, obtained from X-ray binaries [3] and gravitational wave (GW) detections [4–6] and in the supermassive range, from $\sim 10^5$ to $10^{10}\ M_\odot$, as radio sources [3], from Very Large Base Line Interferometry [7] and infrared observations in our own galactic centre [8, 9]. According to the "Kerr hypothesis" these BHs have exactly the same spacetime structure, simply rescaled by the mass, and with only one extra macroscopic degree of freedom, their spin. If true, this is remarkable. As we have seen in the previous Chap. 7, Neutron stars for example lead to a much larger variety of spacetime structures.

The goal of this chapter is to discuss, with concrete proof of concept models, the possibility if (and under which circumstances) astrophysical BH candidates could be described by something else rather than the Kerr geometry in some *range of (macroscopic) scales*,[1] both in view of theoretical consistency, in particular dynamics, as well as in view of the current observational developments.

8.2 Non-Kerrness: Guiding Principles and Testing Grounds

Non-Kerr models for astrophysical BH candidates must obey theoretical reasonability criteria. Here is a possible minimal list [10]:

1. to appear in well motivated and consistent physical theories. In the case of Kerr, this is vacuum General Relativity (GR), which, albeit an incomplete theory, due e.g. to singularities, within its regime of validity fulfills this criterion;
2. to have a dynamical formation mechanism, which for Kerr is gravitational collapse [11], together with accretion and mergers;
3. to be sufficiently stable, meaning it can play a role in astrophysical or cosmological time scales. For Kerr in GR, mode stability has been established long ago [12].

The two last criteria establish *dynamical robustness*. This is one of the unifying principles of the models discussed below: there should be a route for forming them and they should be sufficiently stable against the unavoidable perturbations in any realistic environment. This is a restrictive criterion. Some models of alternative BHs or horizonless compact objects, e.g. wormholes, have no established formation mechanism.

[1] It is widely accepted that sufficiently microscopic BHs will require quantum gravity effects. This is not the range of scales we shall be interested in, but rather macroscopic scales for which there is astrophysical evidence for BHs.

Being a good theoretical model in the sense of the previous paragraph is not, however, enough to describe Nature. The model must give rise to all the correct phenomenology attributed to astrophysical BHs, both in the electromagnetic and GWs channels. At the moment, there is not (yet?) clear tension between observations and the Kerr model. But the limitations of GR (e.g. the unavoidable singularities behind BH horizons [13]) and of the standard model (SM) of particle physics (which fails to explain, say, dark matter) leverage us to consider non-Kerr models and how much we can distinguish them from the paradigm, with state of the art observables. Moreover, we should bear in mind that, at the moment, we are not testing simultaneously (i.e. for the same mass range) all these observables. Thus, degeneracy in any single observable (even if not in all), for some model, may be of interest.

The first class of state of the art strong gravity observables, to test against, comes from GWs. Amongst the many detections, a most interesting (and intriguing) one is GW190521 [14]. Key novel aspects of this event include: (i) the very massive progenitors, of about 85 and 66 $M_\odot$, meaning that, within the uncertainties, at least one is in the so called "pair instability supernova gap", a gap in the mass spectrum starting somewhere between 45 and 55 $M_\odot$ wherein BHs cannot form from the collapse of massive stars according to standard stellar evolution [15]. So, how did the progenitors of this event come to be? (ii) it is a low frequency event, implying there is no inspiral in the observed signal, leaving room to speculate about the true nature of the event. Thus, GW190521 (and similar events) promises to be a fertile ground for theoretical modelling, in particular for testing the Kerr hypothesis. A detailed discussion about how one can obtain information about the structure of BHs from GW observations, is exemplified in the next Chap. 9.

A second remarkable observable is the first image of a BH—M87*—resolving horizon scale structure [7]. Since this observation is probing the strong gravity region of a few Schwarzschild radii, even though it has only been done for a single BH and even though there may be a non negligible impact of the astrophysical environment, which is not fully under control, it is worth understanding how it can be used to confirm the Kerr hypothesis or constrain deviations thereof, by comparing the shadow [16, 17] and emission ring in the M87* image with non-Kerr models.

8.3 Non-Kerrness: Two Families of Examples

The Carter-Robinson uniqueness theorem [18–20] establishes that physically reasonable BHs in vacuum GR fall into the family (8.1). Hence they are rather featureless; they have *no-hair* [21]. Thus, to find non-Kerr BHs, one must either include matter in GR or consider modified gravity theories (such as for example the ones which have been discussed in Chaps. 3 and 4). This is necessary but not sufficient. For instance, including minimally coupled matter fields, with standard kinetic terms and obeying (say) the dominant energy condition is rather restrictive; it prevents non-Kerr BHs in many models. This conclusion has been established by model-specific no-hair theorems [22, 23]. Some of these theorems are powerful since they do not require much, a paradigmatic example being Bekenstein's theorem ruling out BH "hair" of a real scalar field minimally coupled to Einstein's gravity, possibly with a mass term or even some classes of self interactions [24]. Still, theorems have assumptions, and dropping some of them one may (sometimes) find reasonable scenarios where new BHs emerge.

Here, we shall be interested both in examples in GR with minimally coupled scalar (or massive vector) fields, where the usual theorems are circumvented in a subtle way, by a property called "symmetry non-inheritance", and in models beyond GR where the theorems are circumvented by the use of non-minimal couplings. In either case, the theoretical foundation will invoke new physics which introduces a new scale. Moreover, in both cases, the non-Kerr BHs co-exist with Kerr as solutions of the model. At some scales, however, the former emerge dynamically from the latter. In other words, the Kerr solution becomes *unstable* due to some new physics and a new preferred state emerges.

The first family is in GR but with matter beyond the SM, in fact ultralight bosonic particles, that have been proposed as dark matter candidates [25–27]. The two basic members of this family are described by the action ($G = 1 = c$)

$$S^{(s)} = \int d^4x\sqrt{-g}\left[\frac{R}{16\pi} + \mathcal{L}^{(s)}\right], \tag{8.2}$$

where $\mathcal{L}^{(0)} = -\bar{\Phi}_{,\alpha}\Phi^{,\alpha} - \mu^2\bar{\Phi}\Phi$ and $\mathcal{L}^{(1)} = -\bar{F}_{\alpha\beta}F^{\alpha\beta}/4 - \mu^2\bar{A}_\alpha A^\alpha/2$ for the scalar ($s = 0$) and vector ($s = 1$) cases, respectively. Φ and A_μ are a complex scalar and vector field, $F = dA$ and overbar denotes complex conjugate. The field mass introduces a new scale, μ, which is an inverse length, and which, for the scenario herein, is taken to be *ultralight*, with mass between 10^{-10} to 10^{-20} eV. Kerr BHs with a horizon scale comparable to the Compton wavelength of this new particle, become (efficiently) unstable against a process called *superradiance* [28], transferring part of their energy and angular momentum to a scalar cloud around the BH, which becomes a "BH with synchronised (scalar or Proca) hair" [29, 30], and different from Kerr. These BHs were first reported in [31] and [32] for the scalar and Proca case, respectively.

? Exercise

8.1. Derive the field equations for action (8.2) for $s = 0$ and $s = 1$. Are the field equations for $s = 1$ gauge invariant under the transformation $A_\mu \rightarrow A'_\mu = A_\mu + \partial_\mu f$?

The second family is beyond GR, in modified gravity. A member of this family is the extended scalar-tensor Gauss-Bonnet (eSTGB) model, described by the action

$$S = \frac{1}{16\pi}\int d^4x\sqrt{-g}\Big[R - 2\partial_\mu\phi\partial^\mu\phi + \lambda^2 f(\phi)(R^{\mu\nu\alpha\beta}R_{\mu\nu\alpha\beta} - 4R_{\mu\nu}R^{\mu\nu} + R^2)\Big], \tag{8.3}$$

where ϕ is a real scalar field and $f(\phi)$ an yet unspecified coupling function. The Gauss-Bonnet (GB) coupling introduces a new scale λ, which has units of length. Then, BHs with a horizon scale comparable to this new scale, can undergo a strong gravity phase transition, *spontaneous scalarisation* [33, 34].[2] This leads, dynamically, to new types of BHs, dubbed "scalarised BHs".

Model (8.2) also accommodates horizonless compact objects known as "bosonic stars", scalar [36–38] or vector [39–41]. (See also the detailed discussion about Boson stars in Chap. 10.) The latter are also known as "Proca stars". Some of these solutions are dynamically robust [42, 43] and have a formation mechanism, via a process called "gravitational cooling" [43–46]. When the new scale, μ is in the aforementioned range, between 10^{-10} to 10^{-20} eV, the maximal mass of these bosonic stars is in the astrophysical range of $1 - 10^{10}\ M_\odot$, and these objects mimic the mass of astrophysical BHs. Scalar and vector bosonic stars have some interesting differences which will be emphasised in the examples below concerning the *(BH) imitation game*.

8.4 Non-Kerr BHs: The Example of Synchronisation

8.4.1 Dynamical Considerations

Dynamical synchronisation occurs in many systems, both in biology, e.g. communities of fireflies or crickets, and in physics, e.g. sets of metronomes or pendulums [47]. In these systems, individual cycles converge dynamically to the same phase due to appropriate interactions, yielding a configuration that would otherwise look fine-tuned.

In Newtonian gravitational dynamics, synchronisation occurs in binary systems of extended objects, such as planets or stars [48]. Tidal effects tend to synchronise and lock orbital and rotational periods. This effect led the moon to always show the same face towards the Earth, and the Earth is (very slowly) tending to show the same face towards the moon. This is a ubiquitous behaviour observed in all planets-moons of the solar system.

In the context of relativistic gravity, the aforementioned BHs with synchronised bosonic hair can be interpreted as synchronised configurations (hence the name).

[2] Spontaneous scalarisation was first proposed for neutron stars, in a different model [35], see also the discussion in Sect. 7.4.1.

The synchronisation condition reads $\Omega_H = \omega/m$ [31, 49, 50], where Ω_H is the horizon angular velocity, and ω/m is the phase angular velocity of the bosonic field, which has a dependence $\Psi \sim e^{-i(\omega t - m\varphi)}$, where Ψ represents either Φ or A_μ; t, φ are the time and azimuthal coordinates of the stationary and axisymmetric spacetime; ω is the frequency of the bosonic field's harmonic time dependence and m is an integer azimuthal harmonic index. This condition is stating that the phase of the field is co-rotating in synchrony with the horizon. Are these synchronous hairy BHs attained dynamically, as in the case of the fireflies or pendulums?

There is indeed one dynamical channel that leads to synchronisation: the process of superradiance. Fully non-linear numerical simulations have been successfully performed when the bosonic field is the Proca one [29]. They have shown that the process of superradiant rotational energy extraction spins down the BH, saturates and a new equilibrium state is attained, precisely when the BH and the dominant superradiant mode obey $\Omega_H = \omega/m$. These simulations obtained a maximum of $\sim$9% of energy transfer from the BH into the bosonic cloud. This is close to the maximum expected in the evolution of the dominant superradiant mode, which is $\sim$10%, also for the scalar case [51]. Using the data from these simulations the new equilibrium state was identified with a BH with synchronised Proca hair [30].

We thus have a process creating a new sort of BH, which takes some time scale. But the time scale depends crucially on a resonance between the Compton wavelength of the fundamental boson $1/\mu$ and the Schwarzschild radius of the BH, $\sim M$. Maximal efficiency occurs when they are similar, $M\mu \sim 1$ [52]. Otherwise the time scale quickly grows and becomes larger than the Cosmological time sufficiently far from the sweet spot. This therefore selects a mass scale of BHs: depending on the sort of mass of the fundamental boson, BHs in a certain (narrow) mass range become hairy, but outside this range they (effectively) do not [10]. The punch line is that, under the assumption a single ultralight boson with some mass μ exists, non-Kerrness would manifest itself only for BHs in some narrow mass range around $1/\mu$. Observational evidence for such non-Kerrness would therefore identify μ.

There may be other formation channels for these synchronised BHs, in particular, from mergers of bosonic stars, both scalar and vector [53]. There is a key difference in this latter channel: superradiance forms a synchronised configuration by spinning down the BH. In this new channel, mergers of bosonic stars lead to a synchronised system by spinning up the BH that results from the merger, and which accretes part of the field remnant. In the latter case, however, fine tuning seems necessary [53].

8.4.2 Comparison with Observations

Can we constrain BHs with synchronised hair with current observations? Concerning GWs, both the perturbation theory (for the ringdown) and fully non-linear dynamical evolutions (for the inspiral and merger) remain essentially unexplored (but see [54]). Concerning shadows, on the other hand, more progress has been achieved, e.g. [55, 56].

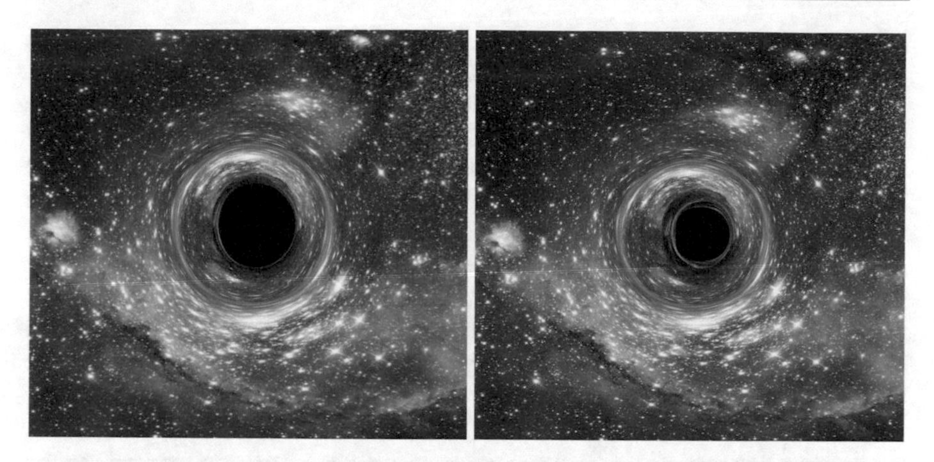

Fig. 8.1 Shadow and lensing of a Kerr BH (left) and a BH with synchronised scalar hair (right) with the same ADM mass and angular momentum, using a background image of Infant Stars in the Small Magellanic cloud from the Hubble space telescope. Adapted from the results in [55].

This family of BHs with synchronised hair interpolates between vacuum Kerr BHs and bosonic stars. Consequently, close to the Kerr limit phenomenological differences are as small as desired (compared to Kerr) due to the continuity of the solutions; but sufficiently far away they are large and can become huge in the solution space region close to bosonic stars. Figure 8.1 shows lensing images obtained by ray tracing for these BHs and comparing with Kerr [57], using an aesthetically appealing starry sky as the light source, rather than a realistic astrophysical environment. The hairy BH shown (righ panel) has the same mass and angular momentum as the Kerr one (left panel), but 75% of the mass and 85% of the angular momentum are stored in the scalar field (rather than the horizon). One can see that due to the transfer of part of the energy and spin of the BH to the scalar "hair" the actual shadow is considerably smaller than that of a Kerr BH with the same total mass and spin and comparable observation conditions. Actually it is about 25% smaller in terms of the average radius, which seems to exclude this particular example of hairy BH as a good model for M87* with current EHT observations. This illustrates how sufficiently large departures from Kerr can be falsified even with current observations. However, how are the differences in the dynamically viable region, assuming formation from superradiance?

In the dynamically most interesting region of the parameter space, where the hairy BHs can form from Kerr in an astrophysical time scale and are stable (against higher superradiant modes) for at least a Hubble time [10], the differences are generically small. Recall that only about $\sim$10% of the Kerr BH mass can be transferred to the bosonic cloud. Yet... this is a non-negligible amount of mass for a 6 billion solar masses BH like M87*. Analysing in this region of the parameter space if the M87* image could then distinguish the hairy BH from Kerr, one concludes that all dynamically viable BHs could be mistaken for a Kerr BH within the current error bars [56]. This conclusion seems quite natural, since roughly, the error in the measurement of the emission ring diameter is about 10%.

The conclusion is the following. Consider that an ultralight (dark matter) particle with a mass of $\sim 10^{-20}$ eV exists in Nature, so that $\mu M_{M87*} \sim 1$. Then, as M87* grew throughout its history to its current mass, superradiance became efficient around its current mass, and endowed it with synchronised hair, a process that took a few million years. The (current) hairy M87* BH is effectively stable against higher superradiant modes, for a Hubble time. In this scenario, M87* is not a Kerr BH, but rather a BH with synchronised hair. However, the current Event Horizon Telescope precision would not be able to tell the difference. Observe, moreover, that all other BHs with a mass smaller than $\lesssim 10^9 M_\odot$ would not become hairy, as superradiance is not efficient. They would be Kerr BHs.

8.5 Non-Kerr BHs: The Example of Scalarisation

We now consider a different scenario wherein non-Kerr BHs also arise dynamically, for some scales, but via a different instability of Kerr BHs: spontaneous scalarisation.

In the landscape of modified gravity models there are question marks on the well-posedness and theoretical consistency of many of them. A fairly well motivated class of models with higher curvature corrections is the class of eSTGB models (8.3). Here, the model is defined by the choice of the coupling between the GB and the scalar field, $f(\phi)$. Choosing a dilatonic coupling, $f(\phi) \sim e^{\beta\phi}$, which is motivated, say, by string theory or dimensional reduction (in a somewhat similar way to the emergence of the Chern-Simons gravity from string theory, which was discussed in Chap. 1), Schwarzschild/Kerr BHs are not solutions; there are new BHs which are stable (in some regime) and which have a qualitatively new feature, namely a minimal BH size [58, 59]. This can be interpreted as due to a repulsive effect, sourced by the GB term, that destabilises the horizon when the GB term becomes dominant, i.e. for sufficiently small BHs.

Changing the scalar-curvature coupling into a more general $f(\phi)$, there are various interesting cousin models. For example, one can consider shift-symmetric models, which are close to the previous dilatonic model but with this additional shift-symmetry. BHs in this model have been constructed [60, 61] and shown to exhibit dynamical formation [62]. On the other hand, the condition $df/d\phi(\phi = 0) = 0$, yields a class of models admitting *both* vacuum GR and scalarised BHs [33, 34, 63]. Moreover, depending on the sign of $d^2 f/d\phi^2(\phi = 0)$, the GR BHs may be unstable in some scales, i.e. when M/λ is smaller than some threshold, which suggests the non-GR BHs could emerge dynamically, via spontaneous scalarisation. Entropic considerations support this possibility in some models, e.g. [33, 64].

In the case of these models, the fully dynamical process has not been established, but there are dynamical results showing the exponential growth and saturation of a self-interacting scalar field in the decoupling limit [65]. Moreover, fully non-linear numerical evolutions could be performed of the spontaneous scalarisation of a Reissner-Nordström BH in a cousin model, showing the phenomenon occurs dynamically and leads to a perturbatively stable scalarised BH [66].

The scalarised Kerr BHs that would be the endpoint of the scalarisation process were first constructed in [64] (see also [67]), in model (8.3) with $f(\phi) = (1 - e^{-\beta\phi^2})/(2\beta)$. Choosing this coupling is illustrative. Some properties (but not all) of the scalarised solutions are universal for all couplings allowing scalarisation, e.g. the threshold between Kerr and scalarised BHs. All solutions (as in the case of the BHs with synchronised hair) are numerical. Consider first static, spherical BHs, taking λ as fixed (new) scale. For sufficiently large BHs as compared to λ (greater than ~ 0.6) there are only Schwarzschild BHs. No scalarised BHs exist. Decreasing the mass of the BH, Schwarzschild BHs becomes unstable against scalarisation and the scalarised BHs appear. These are stable, at least against radial perturbations up to some smaller size, where the scalarised BHs cease to be stable [68]. So, there is a window of sizes where scalarised BHs are stable and dynamically preferred to Schwarzschild.

The effect of the angular momentum is interesting: spin suppresses the effects of scalarisation [64]. This spin suppression of the non-Kerrness can be seen using the diagnosis of the associated shadows—Fig. 8.2. Using the same background image as before, one sees that for non-spinning BHs (top panels), the distinction is visible with a naked eye: the Schwarzschild BH shadow (left panel) and that of a scalarised BH with the same mass and under similar observation conditions (right panel) are clearly different. The latter is in the perturbatively stable region. But as the spin parameter increases (bottom panels) the difference is suppressed and even for still fairly low spins, say $j \sim 0.5$, they become negligible.

In [64] a comparison with the Event Horizon Telescope data was performed. Taking into account that very little is known about the spin of M87*, however, this comparison is not very informative. Even in the most optimistic scenario (for this model) where the spin is low, one can only put a rather weak constraint on the new scale that the model introduces.

Let us close this discussion on spontaneous scalarisation with a different class of models that introduces a new twist. In the models we have just discussed, the Kretschmann scalar of the vacuum BH is providing the instability, endowing scalar perturbations with a tachyonic mass. This Kretschmann scalar is positive for Schwarzschild. For Kerr it starts to become negative (around the poles) for the dimensionless spin, $j > 0.5$. Thus, if we consider models with the opposite coupling sign, only sufficiently fast spinning BHs will scalarise. This has been called "spin induced scalarisation" [69].

In [70] (see also [71]) the domain of existence of the corresponding BHs was explored, for the same illustrative coupling as before. Whereas in the model before, spin quenched the Kerr deviations, now spin enhances the Kerr deviations; in fact it is mandatory. So, this illustrates how there are models in which only some BHs, either with small or with large spin can differ from Kerr, which moreover only occurs for some mass scales. In the case of spin induced scalarisation a detailed

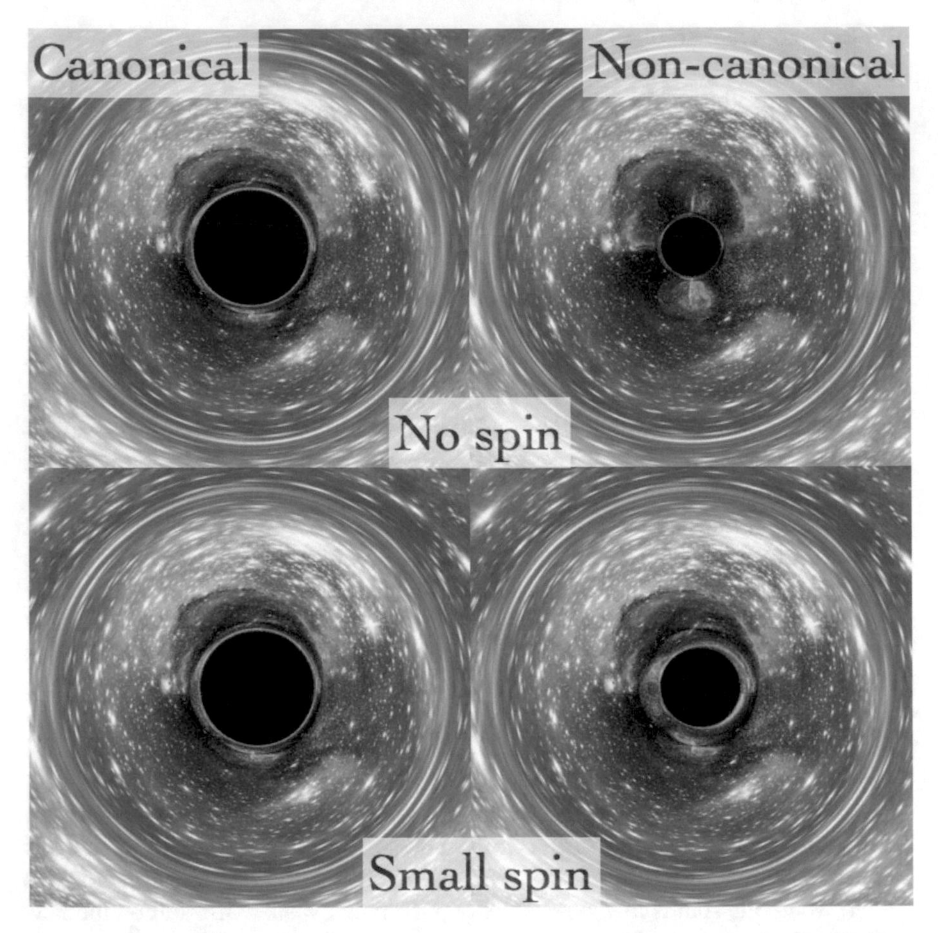

Fig. 8.2 Shadow and lensing of scalarised BHs (right panels) and Schwarzschild/Kerr BHs (right panels) with the same ADM mass and angular momentum, using the same background image as in Fig. 8.1. Adapted from [64]. Figure credit: ©Americal Physical Society. Reproduced with permissions. All rights reserved

phenomenological study of the solutions, namely of the shadows, has not yet been reported.

8.6 The Imitation Game: Non-BHs Mimicking BH Observables

Let us now discuss "the imitation game", or how non-BHs can mimic some (even if not all) BH observables.

There are several motivations to consider BH mimickers, e.g., the singularity problem of BHs. This has led to many models of horizonless compact objects that could behave as BH imitators. It has been pointed out, however, that an imitator needs to have a *light ring* (LR) to mimic the BH ringdown [72]. LRs also play a key role in determining the edge of the shadow [73]. Thus, it seems that for compact object to mimic the most strong field current observations they need to possess LRs (and are then called "ultra-compact").

There may be, however, a generic possible issue with horizonless ultracompact objects. A theorem established in [74] states that for generic equilibrium ultracompact objects, resulting from a smooth, incomplete gravitational collapse, (thus, for which there is plausible formation mechanism) LRs come in pairs and one is stable. A stable LR has been suggested to trigger a (non-linear) spacetime instability [75], as massless perturbations can pile up in its vicinity. Not much is known about this instability or its timescale, but it raises a shadow of doubt about the dynamical robustness of horizonless, ultracompact objects.

Let us therefore ask the question if real data, like a true GW event or the Event Horizon Telescope M87* observation could be imitated by a compact object mimicker that does not have LRs. Interestingly, as the following case study examples indicate the answer is yes in both cases.

8.6.1 Mimicking a GW Event

Let us first address the imitation of a GW event. In GR coupled to ultralight bosonic fields, there has been, for many years, developments in evolving scalar and vector boson stars dynamically, using numerical relativity techniques [42]. In 2019 an unexpected difference between spinning scalar and vector bosonic stars was found: the most fundamental spinning scalar boson stars develop a non-axisymmetric instability [43]. When this instability kicks in, these stars collapse into a BH. On the other hand, this instability is absent in the cousin Proca model. Indeed, even if one perturbs considerably a spinning Proca star, no instability is seen. So, spinning Proca stars, without self-interactions, are dynamically robust, unlike their scalar cousins.

Given the dynamically robustness of spinning Proca Stars, recently we considered simulations of mergers of spinning Proca stars and compared them with real data, in particular with the intriguing event GW190521 commented on earlier. We have found, through a Bayesian analysis, that a collision of two spinning Proca stars actually fits slightly better the data than the vanilla binary BH model that was used by the LIGO-Virgo collaboration [76].

It follows that if, even just as a proof of concept, one takes seriously the Proca model, we can use the data to infer the mass of the ultralight bosonic (in this case vector) particle. We have obtained a mass of around 8×10^{-13} eV. More similar events are needed to confirm this possibility. In the most likely case scenario this is just a proof of concept showing (1) how there can be degeneracy in real data between two very different models and (2) how one could extract physical information about a new fundamental, dark matter particle from GW data (in this case the mass of the boson). Note that the colliding Proca stars are compact but not ultracompact, so they do not have LRs. Of course, in a much more exciting possibility, a potential confirmation of the bosonic star scenario would be a first hint for the long sought dark matter nature.

Under this rationale, one may ask why could this event be such a Proca star collision and not the other events detected so far? The point is again mass selection. The mass of the ultralight bosonic particle determines the maximal mass of the corresponding stars. For the above quoted ultralight boson, this mass turns out to be about $\sim 173\ M_{\odot}$. All other events correspond to a smaller final mass. Since we have observed BH formation, due to the ringdown, then these cannot be Proca star collisions. For the Proca star collisions to form a BH they have to overshoot this limit, which requires events as massive as GW190521.

8.6.2 Mimicking a BH Shadow Without LRs

Let us now address the imitation of a BH shadow by a mimicker without LRs.

In a generic stationary and axi-symmetric BH spacetime, one can associate the edge of the BH shadow to a set of photon bound orbits, which we refer to as "fundamental photon orbits" [73]. In the Schwarzschild case, all of these are planar LRs. In the Kerr case they are called "spherical photon orbits", since they have a constant radial Boyer-Lindquist coordinate.

In a real astrophysical environment, however, an effective shadow seen may depend on the details of the light source. This was nicely illustrated in [77], where GR magnetohydrodynamic (GRMHD) simulations were performed on the background of BHs and of some models of static scalar boson stars. None of these stars have LRs (or a horizon); yet, some could produce an effective shadow, where others did not. In all boson stars models considered, they admitted stable timelike circular orbits until their very centre; but in the case where an effective shadow was seen the angular velocity of the timelike circular orbits attains a maximum at some non-zero areal radius R_{Ω}. This new scale is observed to determine the inner edge of the accretion disk in the simulations, under some assumptions, including that the loss of angular momentum of the orbiting matter is driven by the magneto-rotational instability and that the radiation relevant for the BH shadow observations is mostly due to synchroton emission. There are, however, two caveats for the models in [77] that produce an effective shadow, to be seriously considered as imitators. First, the imitation is not perfect, since the effective shadow is considerably smaller than that of a comparable Scwharzschild BH with the same mass. Secondly, the boson stars producing an effective shadow are perturbatively unstable.

It turns out, as discussed in [78], that this imitation game works better for Proca stars (rather than scalar boson stars). Within the stable branch of spherical, fundamental Proca stars, there are solutions that display the necessary new scale, that is, for which the timelike circular orbit with the maximal angular velocity is at some radius $R_{\Omega} \neq 0$. One can even choose a particular solution for which this new scale equals the location of the Innermost Stable Circular Orbit (ISCO) of a Schwarzschild BH with the same mass. So, there is a dynamically robust solution

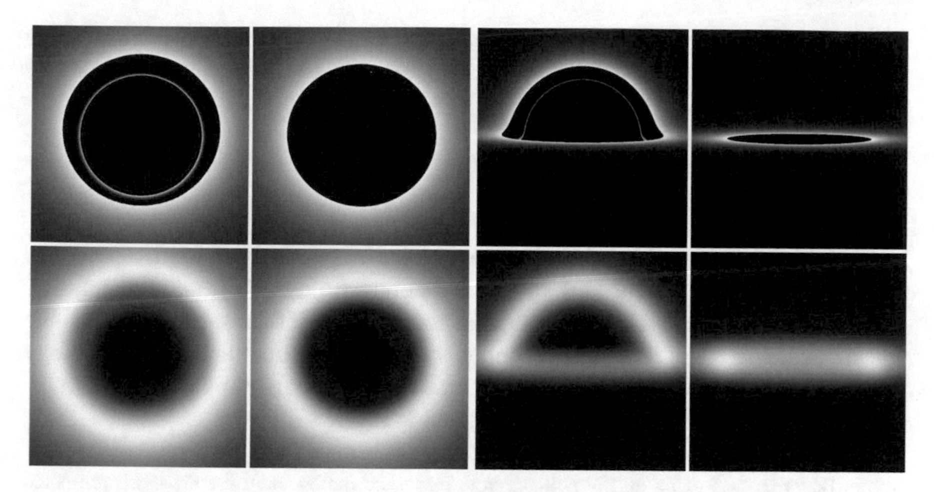

Fig. 8.3 Shadow and lensing of the Proca star discussed in the text and of a comparable Schwarzschild BH, both illuminated by a thin, equatorial accretion disk, at two different observation angles. The bottom images are blurred to mimic current observational limitations. Adapted from [78]. Figure credit: ©IOP Science. Reproduced with permissions. All rights reserved

that, under some accretion models yields an accretion disk morphology mimicking that of a Schwarzschild BH of the same mass.

To check the potential shadow degeneracy, in [78] ray tracing images were produced considering a simplified astrophysical setup, wherein the only radiation source is an opaque and thin accretion disk located on the equatorial plane around the central compact object. The disk has an inner edge with an areal radius $R_\Omega = 6M$ in both spacetimes. The most interesting case for degeneracy occurs for an observer close to the poles (to match the estimated angle at which M87*, was observed from Earth)—Fig. 8.3. For this angle, the images of the Schwarzschild (top left panel) and Proca star (top middle left panel) look similar, although some finer additional lensing features are still visible in the Schwarzschild case. But these subtle differences are washed away if one applies a Gaussian blurring filter to the images, to mimic the current observations limited angular resolution, of the order of the compact object itself. The images so obtained are shown below the corresponding unblurred image, and are essentially indistinguishable.

On the other hand, a near equatorial observation leads to fairly different images. The Schwarzschild one (top middle right panel) resembles the BH shape displayed in the Hollywood movie "Interstellar", whereas the Proca star simply looks like a flat accretion disk with a hole in it, as seen from the side (top right panel). This is because the gravitational potential well of the Proca star is shallow and so the bending of light it produces is weak. Consequently, the accretion disk has an almost flat spacetime appearance. Applying the same blurring as before, even with limited resolution, the two objects could be distinguished. Let us stress that full GRMHD analysis and ray-tracing in the background of this Proca star are required to fully settle the question: to what degree can it imitate a BH observation? The case built herein, nonetheless, clearly confirms a potential degeneracy, but only under some observation conditions. And these Proca stars do not have LRs and are dynamically stable.

8.7 Concluding Remarks

Let us close with two following final remarks. Firstly, all these models we have discussed have caveats; but they illustrate theoretical possibilities of dynamically robust non-Kerr BHs, or BH imitators, that could manifest themselves only at some specific scales of mass and spin. Secondly, that producing detailed phenomenology will constrain the model and the corresponding (admittedly exotic) physics or, in the best case scenario, provide a smoking gun to this new physics.

Acknowledgments I thank all my collaborators that were fundamental for the work presented here. Special thanks, by virtue of our long and fruitful collaboration go to P. Cunha, E. Radu and N. Sanchis-Gual. This work is supported by the Center for Research and Development in Mathematics and Applications (CIDMA) through the Portuguese Foundation for Science and Technology (FCT—Fundação para a Ciência e a Tecnologia), references UIDB/04106/2020, UIDP/04106/2020 and the projects PTDC/FIS-OUT/28407/2017, CERN/FIS-PAR/0027/2019, PTDC/FIS-AST/3041/2020 and CERN/FIS-PAR/0024/2021. This work has further been supported by the European Union's Horizon 2020 research and innovation (RISE) programme H2020-MSCA-RISE-2017 Grant No. FunFiCO-777740.

References

1. R.P. Kerr, Gravitational field of a spinning mass as an example of algebraically special metrics. Phys. Rev. Lett. **11**, 237–238 (1963)
2. R.H. Boyer, R.W. Lindquist, Maximal analytic extension of the Kerr metric. J. Math. Phys. **8**, 265 (1967)
3. R. Narayan, J.E. McClintock, Observational evidence for black holes (Dec 2013)
4. B.P. Abbott et al., GWTC-1: a gravitational-wave transient catalog of compact binary mergers observed by LIGO and virgo during the first and second observing runs. Phys. Rev. X **9**(3), 031040 (2019)
5. R. Abbott et al., GWTC-2: compact binary coalescences observed by LIGO and virgo during the first half of the third observing run. Phys. Rev. X **11**, 021053 (2021)
6. R. Abbott et al., GWTC-3: compact binary coalescences observed by LIGO and virgo during the second part of the third observing run (Nov 2021)
7. K. Akiyama et al., First M87 event horizon telescope results. I. The shadow of the supermassive black hole. Astrophys. J. Lett. **875**, L1 (2019)
8. S. Gillessen, F. Eisenhauer, S. Trippe, T. Alexander, R. Genzel, F. Martins, T. Ott, Monitoring stellar orbits around the massive black hole in the galactic center. Astrophys. J. **692**, 1075–1109 (2009)
9. A.M. Ghez et al., Measuring distance and properties of the Milky Way's central supermassive black hole with stellar orbits. Astrophys. J. **689**, 1044–1062 (2008)
10. J.C. Degollado, C.A.R. Herdeiro, E. Radu, Effective stability against superradiance of Kerr black holes with synchronised hair. Phys. Lett. B **781**, 651–655 (2018)
11. T. Nakamura, K. Oohara, Y. Kojima, General relativistic collapse to black holes and gravitational waves from black holes. Prog. Theor. Phys. Suppl. **90**, 1–218 (1987)

12. B.F. Whiting, Mode stability of the Kerr black hole. J. Math. Phys. **30**, 1301 (1989)
13. R. Penrose, Gravitational collapse and space-time singularities. Phys. Rev. Lett. **14**, 57–59 (1965)
14. R. Abbott et al., GW190521: a binary black hole merger with a total mass of $150M_{\odot}$. Phys. Rev. Lett. **125**(10), 101102 (2020)
15. R. Abbott et al., Properties and astrophysical implications of the 150 $M_{\odot}$ binary black hole merger GW190521. Astrophys. J. Lett. **900**(1), L13 (2020)
16. H. Falcke, F. Melia, E. Agol, Viewing the shadow of the black hole at the galactic center. Astrophys. J. Lett. **528**, L13 (2000)
17. P.V.P. Cunha, C.A.R. Herdeiro, Shadows and strong gravitational lensing: a brief review. Gen. Rel. Grav. **50**(4), 42 (2018)
18. B. Carter, Axisymmetric black hole has only two degrees of freedom. Phys. Rev. Lett. **26**, 331–333 (1971)
19. D.C. Robinson, Uniqueness of the Kerr black hole. Phys. Rev. Lett. **34**, 905–906 (1975)
20. P.T. Chrusciel, J. Lopes Costa, M. Heusler, Stationary black holes: uniqueness and beyond. Living Rev. Rel. **15**, 7 (2012)
21. R. Ruffini, J.A. Wheeler, Introducing the black hole. Phys. Today **24**(1), 30 (1971)
22. C.A.R. Herdeiro, E. Radu, Asymptotically flat black holes with scalar hair: a review. Int. J. Mod. Phys. D **24**(09), 1542014 (2015)
23. M.S. Volkov, Hairy black holes in the XX-th and XXI-st centuries, in *14th Marcel Grossmann Meeting on Recent Developments in Theoretical and Experimental General Relativity, Astrophysics, and Relativistic Field Theories*, vol. 2 (2017), pp. 1779–1798
24. J.D. Bekenstein, Transcendence of the law of baryon-number conservation in black hole physics. Phys. Rev. Lett. **28**, 452–455 (1972)
25. L. Hui, J.P. Ostriker, S. Tremaine, E. Witten, Ultralight scalars as cosmological dark matter. Phys. Rev. D **95**(4), 043541 (2017)
26. A. Suárez, V.H. Robles, T. Matos, A review on the scalar field/Bose-Einstein condensate dark matter model. Astrophys. Space Sci. Proc. **38**, 107–142 (2014)
27. F.F. Freitas, C.A.R. Herdeiro, A.P. Morais, A. Onofre, R. Pasechnik, E. Radu, N. Sanchis-Gual, R. Santos, Ultralight bosons for strong gravity applications from simple Standard Model extensions. J. Cosmol. Astropart. Phys. **12**(12), 047 (2021)
28. R. Brito, V. Cardoso, P. Pani, Superradiance: new frontiers in black hole physics. Lect. Notes Phys. **906**, 1–237 (2015)
29. W.E. East, F. Pretorius, Superradiant instability and backreaction of massive vector fields around Kerr Black Holes. Phys. Rev. Lett. **119**(4), 041101 (2017)
30. C.A.R. Herdeiro, E. Radu, Dynamical formation of Kerr Black Holes with synchronized hair: an analytic model. Phys. Rev. Lett. **119**(26), 261101 (2017)
31. C.A.R. Herdeiro, E. Radu, Kerr black holes with scalar hair. Phys. Rev. Lett. **112**, 221101 (2014)
32. C. Herdeiro, E. Radu, H. Rúnarsson, Kerr black holes with Proca hair. Class. Quant. Grav. **33**(15), 154001 (2016)
33. D.D. Doneva, S.S. Yazadjiev, New Gauss-Bonnet black holes with curvature-induced scalarization in extended scalar-tensor theories. Phys. Rev. Lett. **120**(13), 131103 (2018)
34. H.O. Silva, J. Sakstein, L. Gualtieri, T.P. Sotiriou, E. Berti, Spontaneous scalarization of black holes and compact stars from a Gauss-Bonnet coupling. Phys. Rev. Lett. **120**(13), 131104 (2018)
35. T. Damour, G. Esposito-Farese, Nonperturbative strong field effects in tensor - scalar theories of gravitation. Phys. Rev. Lett. **70**, 2220–2223 (1993)
36. D.J. Kaup, Klein-Gordon Geon. Phys. Rev. **172**, 1331–1342 (1968)
37. R. Ruffini, S. Bonazzola, Systems of selfgravitating particles in general relativity and the concept of an equation of state. Phys. Rev. **187**, 1767–1783 (1969)
38. F.E. Schunck, E.W. Mielke, General relativistic boson stars. Class. Quant. Grav. **20**, R301–R356 (2003)

39. R. Brito, V. Cardoso, C.A.R. Herdeiro, E. Radu, Proca stars: gravitating Bose–Einstein condensates of massive spin 1 particles. Phys. Lett. B **752**, 291–295 (2016)
40. C.A.R. Herdeiro, A.M. Pombo, E. Radu, Asymptotically flat scalar, Dirac and Proca stars: discrete vs. continuous families of solutions. Phys. Lett. B **773**, 654–662 (2017)
41. C. Herdeiro, I. Perapechka, E. Radu, Ya. Shnir, Asymptotically flat spinning scalar, Dirac and Proca stars. Phys. Lett. B **797**, 134845 (2019)
42. S.L. Liebling, C. Palenzuela, Dynamical Boson stars. Living Rev. Rel. **15**, 6 (2012)
43. N. Sanchis-Gual, F. Di Giovanni, M. Zilhão, C. Herdeiro, P. Cerdá-Durán, J.A. Font, E. Radu, Nonlinear dynamics of spinning bosonic stars: formation and stability. Phys. Rev. Lett. **123**(22), 221101 (2019)
44. E. Seidel, W.-M. Suen, Formation of solitonic stars through gravitational cooling. Phys. Rev. Lett. **72**, 2516–2519 (1994)
45. F. Siddhartha Guzman, L.A. Urena-Lopez, Gravitational cooling of self-gravitating Bose-Condensates. Astrophys. J. **645**, 814–819 (2006)
46. F. Di Giovanni, N. Sanchis-Gual, C.A.R. Herdeiro, J.A. Font, Dynamical formation of Proca stars and quasistationary solitonic objects. Phys. Rev. D **98**(6), 064044 (2018)
47. S. Strogatz, *Sync: How Order Emerges from Chaos in the Universe, Nature, and Daily Life* (Hachette Books, New York, 2004)
48. P. Hut, Tidal evolution in close binary systems. Astron. Astrophys. **99**, 126–140 (1981)
49. S. Hod, Stationary scalar clouds around rotating black holes. Phys. Rev. D **86**, 104026 (2012) [Erratum: Phys. Rev. D **86**, 129902 (2012)]
50. O.J.C. Dias, G.T. Horowitz, J.E. Santos, Black holes with only one killing field. J. High Energy Phys. **07**, 115 (2011)
51. C.A.R. Herdeiro, E. Radu, N.M. Santos, A bound on energy extraction (and hairiness) from superradiance. Phys. Lett. B **824**, 136835 (2022)
52. S.R. Dolan, Superradiant instabilities of rotating black holes in the time domain. Phys. Rev. D **87**(12), 124026 (2013)
53. N. Sanchis-Gual, M. Zilhão, C. Herdeiro, F. Di Giovanni, J.A. Font, E. Radu, Synchronized gravitational atoms from mergers of bosonic stars. Phys. Rev. D **102**(10), 101504 (2020)
54. L.G. Collodel, D.D. Doneva, S.S. Yazadjiev, Equatorial EMRIs in KBHsSH Spacetimes (Aug 2021)
55. P.V.P. Cunha, C.A.R. Herdeiro, E. Radu, H.F. Runarsson, Shadows of Kerr black holes with scalar hair. Phys. Rev. Lett. **115**(21), 211102 (2015)
56. P.V.P. Cunha, C.A.R. Herdeiro, E. Radu, EHT constraint on the ultralight scalar hair of the M87 supermassive black hole. Universe **5**(12), 220 (2019)
57. P.V.P. Cunha, C.A.R. Herdeiro, E. Radu, H.F. Runarsson, Shadows of Kerr black holes with and without scalar hair. Int. J. Mod. Phys. D **25**(09), 1641021 (2016)
58. P. Kanti, N.E. Mavromatos, J. Rizos, K. Tamvakis, E. Winstanley, Dilatonic black holes in higher curvature string gravity. Phys. Rev. D **54**, 5049–5058 (1996)
59. B. Kleihaus, J. Kunz, S. Mojica, E. Radu, Spinning black holes in Einstein–Gauss-Bonnet–dilaton theory: nonperturbative solutions. Phys. Rev. D **93**(4), 044047 (2016)
60. T.P. Sotiriou, S.-Y. Zhou, Black hole hair in generalized scalar-tensor gravity: an explicit example. Phys. Rev. D **90**, 124063 (2014)
61. J.F.M. Delgado, C.A.R. Herdeiro, E. Radu, Spinning black holes in shift-symmetric Horndeski theory. J. High Energy Phys. **04**, 180 (2020)
62. R. Benkel, T.P. Sotiriou, H. Witek, Dynamical scalar hair formation around a Schwarzschild black hole. Phys. Rev. D **94**(12), 121503 (2016)
63. G. Antoniou, A. Bakopoulos, P. Kanti, Evasion of no-hair theorems and novel black-hole solutions in Gauss-Bonnet theories. Phys. Rev. Lett. **120**(13), 131102 (2018)
64. P.V.P. Cunha, C.A.R. Herdeiro, E. Radu, Spontaneously scalarized Kerr black holes in extended scalar-tensor–Gauss-Bonnet gravity. Phys. Rev. Lett. **123**(1), 011101 (2019)
65. D.D. Doneva, S.S. Yazadjiev, Dynamics of the nonrotating and rotating black hole scalarization. Phys. Rev. D **103**(6), 064024 (2021)

66. C.A.R. Herdeiro, E. Radu, N. Sanchis-Gual, J.A. Font, Spontaneous scalarization of charged black holes. Phys. Rev. Lett. **121**(10), 101102 (2018)
67. L.G. Collodel, B. Kleihaus, J. Kunz, E. Berti, Spinning and excited black holes in Einstein-scalar-Gauss–Bonnet theory. Class. Quant. Grav. **37**(7), 075018 (2020)
68. J. Luis Blázquez-Salcedo, D.D. Doneva, J. Kunz, S.S. Yazadjiev, Radial perturbations of the scalarized Einstein-Gauss-Bonnet black holes. Phys. Rev. D **98**(8), 084011 (2018)
69. A. Dima, E. Barausse, N. Franchini, T.P. Sotiriou, Spin-induced black hole spontaneous scalarization. Phys. Rev. Lett. **125**(23), 231101 (2020)
70. C.A.R. Herdeiro, E. Radu, H.O. Silva, T.P. Sotiriou, N. Yunes, Spin-induced scalarized black holes. Phys. Rev. Lett. **126**(1), 011103 (2021)
71. E. Berti, L.G. Collodel, B. Kleihaus, J. Kunz, Spin-induced black-hole scalarization in Einstein-scalar-Gauss-Bonnet theory. Phys. Rev. Lett. **126**(1), 011104 (2021)
72. V. Cardoso, E. Franzin, P. Pani, Is the gravitational-wave ringdown a probe of the event horizon? Phys. Rev. Lett. **116**(17), 171101 (2016) [Erratum: Phys. Rev. Lett. **117**, 089902 (2016)]
73. P.V.P. Cunha, C.A.R. Herdeiro, E. Radu, Fundamental photon orbits: black hole shadows and spacetime instabilities. Phys. Rev. D **96**(2), 024039 (2017)
74. P.V.P. Cunha, E. Berti, C.A.R. Herdeiro, Light-ring stability for ultracompact objects. Phys. Rev. Lett. **119**(25), 251102 (2017)
75. J. Keir, Slowly decaying waves on spherically symmetric spacetimes and ultracompact neutron stars. Class. Quant. Grav. **33**(13), 135009 (2016)
76. J. Calderón Bustillo, N. Sanchis-Gual, A. Torres-Forné, J.A. Font, A. Vajpeyi, R. Smith, C. Herdeiro, E. Radu, S.H.W. Leong, GW190521 as a merger of Proca stars: a potential new vector Boson of 8.7×10^{-13} eV. Phys. Rev. Lett. **126**(8), 081101 (2021)
77. H. Olivares, Z. Younsi, C.M. Fromm, M. De Laurentis, O. Porth, Y. Mizuno, H. Falcke, M. Kramer, L. Rezzolla, How to tell an accreting boson star from a black hole. Mon. Not. Roy. Astron. Soc. **497**(1), 521–535 (2020)
78. C.A.R. Herdeiro, A.M. Pombo, E. Radu, P.V.P. Cunha, N. Sanchis-Gual, The imitation game: Proca stars that can mimic the Schwarzschild shadow. J. Cosmol. Astropart. Phys. **04**, 051 (2021)

Probing the Horizon of Black Holes with Gravitational Waves

9

Elisa Maggio

Abstract

Gravitational waves open the possibility to investigate the nature of compact objects and probe the horizons of black holes. Some models of modified gravity predict the presence of horizonless and singularity-free compact objects. Such dark compact objects would emit a gravitational-wave signal which differs from the standard black hole scenario. In this chapter, we overview the phenomenology of dark compact objects by analysing their characteristic frequencies in the ringdown and the emission of gravitational-wave echoes in the postmerger signal. We show that future gravitational-wave detectors will allow us to perform model-independent tests of the black hole paradigm.

9.1 Tests of the Black Hole Paradigm

Black holes (BHs) are the end result of the gravitational collapse and the most compact objects in the Universe. According to the no-hair theorems of general relativity (GR), any compact object heavier than a few solar masses is well described by the Kerr geometry [1, 2]. Kerr BHs are determined uniquely by two parameters, i.e., their mass M and angular momentum J defined through the dimensionless spin parameter $\chi \equiv J/M^2$ [3]. Therefore, any observation of deviation from the properties of Kerr BHs would be an indication of a departure from GR, see also the discussion in the previous Chap. 8.

Gravitational waves (GWs) provide a unique channel for probing the nature of astrophysical sources. The GW signal emitted by the coalescence of compact

E. Maggio (✉)
Max Planck Institute for Gravitational Physics (Albert Einstein Institute), Potsdam, Germany
e-mail: elisa.maggio@aei.mpg.de

C. Pfeifer, C. Lämmerzahl (eds.), *Modified and Quantum Gravity*, Lecture Notes in Physics 1017, https://doi.org/10.1007/978-3-031-31520-6_9

binaries is characterized by three main stages: the *inspiral*, when the two bodies spiral in towards each other as they loose energy into gravitational radiation; the *merger*, when the two bodies coalesce; and the *ringdown*, when the final remnant relaxes to an equilibrium solution. In particular, the analysis of the ringdown would allow us to infer the properties of the compact remnants.

The ringdown is dominated by the complex characteristic frequencies of the remnant, the so-called *quasi-normal modes* (QNMs) (which are derived similarly as it was explained for Neutron Stars in Sect. 7.2.3), which describe the response of the compact object to a perturbation [4], i.e.

$$\omega_{\ell mn} = \omega_{R,\ell mn} + i\omega_{I,\ell mn} \, , \tag{9.1}$$

where $\omega_{R/I,\ell mn} \in \mathrm{Re}$. Each mode is described by three integers, namely the angular number of the perturbation ℓ (where $\ell \geq 0$), the azimuthal number of the perturbation m (such that $|m| \leq \ell$), and the overtone number n (where $n \geq 0$). The fundamental mode with $n = 0$ corresponds to the mode with the smallest imaginary part. The ringdown is modeled as a sum of exponentially damped sinusoids whose frequencies $f_{\ell mn}$ (damping times $\tau_{\ell mn}$) are related to the real (imaginary) part of the QNMs of the remnant via

$$f_{\ell mn} = \omega_{R,\ell mn}/(2\pi) \, , \tag{9.2}$$

$$\tau_{\ell mn} = -1/\omega_{I,\ell mn} \, . \tag{9.3}$$

Therefore, from the detection of the ringdown signal it is possible to infer the QNMs of the remnant and understand the nature of the latter.

The fundamental QNM has been observed in the ringdown of several GW events [5]. The ringdown detections are compatible with Kerr BH remnants, however the characterization of the remnant requires further analyses. Indeed, the measurement of one complex QNM allows us only to estimate the mass and the spin of the remnant. A test of the BH paradigm would require the identification of at least two QNMs in the ringdown. Next generation detectors, e.g. the space-based interferometer LISA, will allow for tests of the BH paradigm with unprecedented precision [6].

9.2 Horizonless Compact Objects

On the theoretical side, the presence of horizons in Kerr BHs poses some issues. In particular, the horizon hides a curvature singularity with infinite tidal forces where the Einstein equations break down. Moreover, the spacetime within the horizon can contain closed time-like hypersurfaces that violate causality.

Several attempts to regularize the BH solution predict the existence of horizonless and singularity-free compact objects [7]. Some models are solutions to quantum-gravity extensions of GR, e.g. the fuzzball in string theory as an ensemble of a large number of regular and horizonless microstate geometries with the same asymptotic charges of a BH [8]. Other models of horizonless compact objects are solutions to GR in the presence of dark matter or exotic fields, e.g. boson stars as self-gravitating solutions formed by massive bosonic fields which are coupled minimally to GR [9], see the discussion in Chaps. 8 and 10.

Horizonless compact objects can mimic BHs in terms of electromagnetic observations since they can be as compact as BHs [10]. For example, the observation of the supermassive object at the center of the galaxy M87 by the Event Horizon Telescope constrained weakly some models of horizonless compact objects [11]. Moreover, horizonless compact objects can be used to study GW events in the mass gap between neutron stars and BHs and due to pair-instability supernova processes [12, 13].

In this context, horizonless compact objects allow us to quantify the existence of horizons in astrophysical sources. We analyse a generic model of *dark compact object* which deviate from a BH for two parameters [14]:

- the compactness, which is defined as the inverse of the effective radius of the object in units of mass, i.e. $C = M/r_0$, where

$$r_0 = r_+(1 + \epsilon) \tag{9.4}$$

  is the location of the effective radius of the object and $r_+ = M\left(1 + \sqrt{1 - \chi^2}\right)$ is the horizon of a Kerr BH. Depending on their compactness, two categories of horizonless compact objects can be distinguished: compact objects whose effective radius is comparable with the light ring of BHs, i.e. $\epsilon \approx 0.1, 1$; and ultracompact objects with Planckian corrections at the horizon scale due to quantum fluctuations, i.e. $\epsilon \approx 10^{-40}$. The two categories of horizonless compact objects give rise to different fingerprints in the GW signal. In particular, a merger remnant with $\epsilon \approx 0.1, 1$ would emit a ringdown signal which differs from the BH ringdown at early stages, whereas an ultracompact horizonless object would emit a modulated train of GW echoes at late times, as discussed in Sect. 9.3.2;
- the "darkness", which is related to the reflectivity of the compact object $\mathcal{R}(\omega)$ at its effective radius. The BH is a totally absorbing object with $\mathcal{R} = 0$ at the horizon, whereas a horizonless compact object can have $0 \leq |\mathcal{R}(\omega)|^2 \leq 1$ depending on its interior structure. The $|\mathcal{R}(\omega)|^2 = 1$ case describes a perfectly reflecting object of perturbations moving towards the object. This is the case, for example, of neutron stars where the absorption of radiation through viscosity is negligible. Intermediate values of $\mathcal{R}(\omega)$ describe partially absorbing compact objects due to dissipation, viscosity, fluid mode excitations, nonlinear effects, etc.

9.3 Phenomenology

Let us derive the GW signatures of horizonless compact objects in the postmerger phase of compact binary coalescences. In this section, we overview the quasi-normal mode spectrum and the GW signal in the time domain at variance with the BH case.

9.3.1 Quasi-Normal Mode Spectrum

For simplicity, let us analyse a static and spherically symmetric horizonless compact object. Let us assume that GR is a reliable approximation outside the radius of the object and some modifications appear at the horizon scale. Owing to the Birkhoff theorem, the exterior spacetime is described by the Schwarzschild metric

$$ds^2 = -f(r)dt^2 + \frac{1}{f(r)}dr^2 + r^2\left(d\theta^2 + \sin^2\theta d\phi^2\right), \tag{9.5}$$

where (t, r, θ, ϕ) are the Boyer-Lindquist coordinates and $f(r) = 1 - 2M/r$. The radius of the compact object is located as in Eq. (9.4), where $r_+ = 2M$ is the horizon of a Schwarzschild BH. In order to derive the QNM spectrum of the horizonless compact object, let us perturb the background geometry with a gravitational perturbation. The radial component of the gravitational perturbation is governed by a second-order differential equation [15, 16]

$$\frac{d^2\psi(r)}{dr_*^2} + \left[\omega^2 - V(r)\right]\psi(r) = 0, \tag{9.6}$$

where r_* is the tortoise coordinate defined such that $dr_*/dr = 1/f(r)$ with $f(r) = 1 - 2M/r$, and the effective potential reads

$$V_{\text{axial}}(r) = f(r)\left[\frac{\ell(\ell+1)}{r^2} - \frac{6M}{r^3}\right], \tag{9.7}$$

$$V_{\text{polar}}(r) = 2f(r)\left[\frac{q^2(q+1)r^3 + 3q^2Mr^2 + 9M^2(qr+M)}{r^3(qr+3M)^2}\right], \tag{9.8}$$

for axial and polar perturbations, respectively, with parity $(-1)^{\ell+1}$ and $(-1)^{\ell}$, where $q = (\ell-1)(\ell+2)/2$. Figure 9.1 shows the effective potential as a function of the tortoise coordinate for a BH (top panel) and a horizonless compact object (bottom panel). The effective potentials display a peak approximately at the light ring, $r \approx 3M$, which is the unstable circular orbit of photons around the compact object. In the BH case, the perturbation is purely ingoing towards the horizon; whereas in the case of a horizonless compact object, the absence of the horizon implies the existence of a cavity between the radius of the object and the light ring. The cavity can support

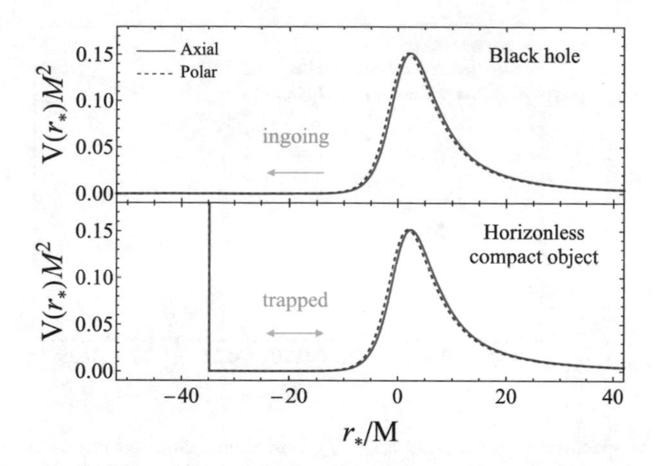

Fig. 9.1 Effective potential as a function of the tortoise coordinate of a Schwarzschild BH (top panel) and a static horizonless compact object with radius $r_0 = 2M(1+\epsilon)$ (bottom panel), for axial (continuous line) and polar (dashed line) $\ell = 2$ gravitational perturbations. The effective potential has a peak approximately at the light ring, $r \approx 3M$. In the case of a horizonless compact object, the effective potential features a cavity between the radius of the object and the light ring. Adapted from [7].

trapped modes that are responsible for a completely different QNM spectrum with respect to the BH case.

By adding two boundary conditions to Eq. (9.6), the system defines an eigenvalue problem whose complex eigenvalues are the QNMs of the object. At infinity, we impose that the perturbation is a purely outgoing wave, i.e.

$$\psi(r) \sim e^{i\omega r_*}, \quad \text{as } r_* \to +\infty. \tag{9.9}$$

In the case of a horizonless ultracompact object ($\epsilon \ll 1$), the perturbation can be decomposed a superposition of ingoing and outgoing waves at the radius of the object, i.e.

$$\psi(r) \sim C_{\text{in}}(\omega)e^{-i\omega r_*} + C_{\text{out}}(\omega)e^{i\omega r_*}, \quad \text{as } r_* \to r_*^0, \tag{9.10}$$

where the reflectivity of the compact object is defined as [17]

$$\mathcal{R}(\omega) = \frac{C_{\text{out}}(\omega)}{C_{\text{in}}(\omega)} e^{2i\omega r_*^0}. \tag{9.11}$$

Let us derive the fundamental ($n = 0$) $\ell = 2$ QNM which is the mode with the longest damping time (in the static and spherically symmetric case, the QNMs do not depend on the azimuthal number m). Figure 9.2 shows the QNM spectrum of a horizonless ultracompact object with a perfectly reflecting surface $\left(|\mathcal{R}(\omega)|^2 = 1\right)$

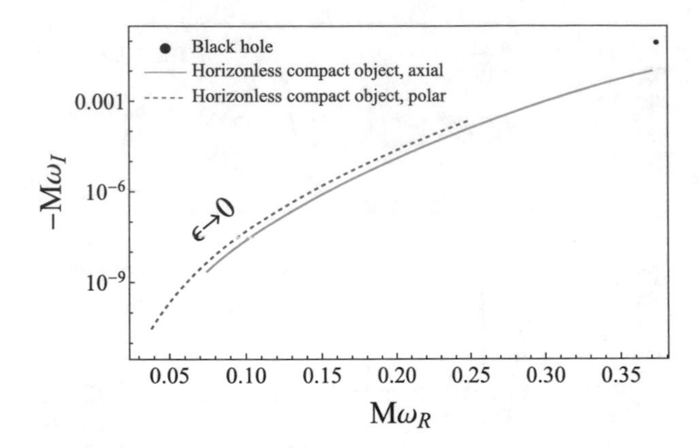

Fig. 9.2 QNM spectrum of a perfectly reflecting horizonless compact object with radius $r_0 = 2M(1+\epsilon)$ and $\epsilon \in (10^{-10}, 10^{-2})$ compared to the fundamental $\ell = 2$ QNM of a Schwarzschild BH. Axial and polar modes are not isospectral as in the BH case. As $\epsilon \to 0$, the QNM spectrum is low-frequencies and long-lived. Modified from [18] and [14].

and $\epsilon \in \left(10^{-10}, 10^{-2}\right)$ from the left to the right of the plot compared to the fundamental $\ell = 2$ QNM of a Schwarzschild BH, i.e.

$$M\omega_{\text{BH}} = 0.3737 - i0.08896\,. \tag{9.12}$$

A first important signature of horizonless compact object is the breaking of isospectrality between axial and polar modes differently from BHs in GR. Indeed, Schwarzschild BHs have a unique QNM spectrum [4] despite the effective potentials for axial and polar perturbations differ from each other (see Eqs. (9.7), (9.8)). Conversely, the radius of horizonless compact objects is responsible for the appearance of a mode doublet for axial and polar QNMs.

? Exercise

9.1. The isospectrality of axial and polar modes in BHs can be demonstrated from the Darboux transformation between the Regge-Wheeler and Zerilli wave functions governing axial and polar modes, respectively, both satisfying Eq. (9.6), i.e.

$$\psi_{\text{RW}} = A\frac{d\psi_{\text{Z}}}{dr_*} + B(r)\psi_{\text{Z}}\,, \tag{9.13}$$

where

$$A = -M\left[i\omega M + \frac{1}{3}q(q+1)\right]^{-1}, \tag{9.14}$$

$$B(r) = \frac{q(q+1)(qr+3M)r^2 + 9M^2(r-2M)}{r^2(qr+3M)[q(q+1)+3i\omega M]}\,. \tag{9.15}$$

Demonstrate that the BH boundary condition $\psi = C_{\text{in}(\omega)} e^{-i\omega r_*}$ as $r \to 2M$ for both Regge-Wheeler and Zerilli wave functions satisfies the Darboux transformation in Eq. (9.13). Conversely, demonstrate that the boundary condition of a horizonless ultracompact object in Eq. (9.10) does not satisfy the Darboux transformation in Eq. (9.13).

Furthermore, a relevant feature of horizonless compact objects is that the QNM spectrum is low-frequency and long-lived in the limit $\epsilon \to 0$. For example, the fundamental $\ell = 2$ QNMs of a perfectly reflecting compact object with $\epsilon = 10^{-10}$ are:

$$M\omega_{\text{axial}} = 0.07470 - i2.299 \times 10^{-9}\,, \tag{9.16}$$

$$M\omega_{\text{polar}} = 0.03791 - i2.739 \times 10^{-11}\,. \tag{9.17}$$

This finding might seem surprising since, in the limit of a compactness close to the BH case, the QNM spectrum of a horizonless compact object deviates significantly from the BH QNM spectrum. A key role is played by the boundary condition in Eq. (9.10), particularly by the fact that the reflective properties of a horizonless compact object differ generically from the totally absorbing BH case.

Low-frequency QNMs can be understood in terms of the trapped modes between the radius of the compact object and the light ring, as shown in Fig. 9.1. The real part of the QNMs depends on the width of the cavity in the effective potential, whereas the imaginary part of the QNMs depends on the amplification factor of the modes in the cavity and the reflectivity at the radius of the compact object. For $\epsilon \ll 1$, the QNMs can be derived analytically in the low-frequency regime as [7, 19–21]

$$\omega_R \sim -\frac{\pi}{2|r_*^0|}\,(p+1)\,, \tag{9.18}$$

$$\omega_I \sim -\frac{\beta_{2\ell}}{|r_*^0|}\,(2M\omega_R)^{2\ell+2}\,, \tag{9.19}$$

where $\sqrt{\beta_{2\ell}} = \frac{(\ell-2)!(\ell+2)!}{(2\ell)!(2\ell+1)!!}$ and p is a positive odd (even) integer for polar (axial) modes. The real part of the QNMs scales with the compactness of the object as $\omega_R \sim |\log\epsilon|^{-1}$, whereas the imaginary part of the QNMs scales as $\omega_I \sim -|\log\epsilon|^{-(2\ell+3)}$.

Let us notice that the boundary condition in Eq. (9.10) can be imposed at the radius of the compact object when $\epsilon \ll 1$ and the effective potential is vanishing. To derive the QNMs of horizonless compact objects with any compactness, we can make use of the membrane paradigm. The original BHs membrane paradigm states that a static observer outside the BH horizon can replace the interior of the perturbed BH by a *fictitious* membrane located at the horizon [22, 23]. The generalisation of the membrane paradigm to horizonless compact objects allows us to describe any compact object with a Schwarzschild exterior where no specific model is assumed for the object's interior. The compactness of the horizonless object is generic and the reflectivity of the object is mapped in terms of the properties of the fictitious membrane.

The Israel-Darmois junction conditions fix the properties of the fictious membrane relating the exterior and the interior spacetime to the radius of the compact object, i.e. [24, 25]

$$[[K_{ab} - K h_{ab}]] = -8\pi T_{ab}\,, \qquad [[h_{ab}]] = 0\,, \tag{9.20}$$

where h_{ab} is the induced metric on the membrane, K_{ab} is the extrinsic curvature, $K = K_{ab}h^{ab}$, T_{ab} is the membrane stress-energy tensor, and [[...]] is the jump of a quantity across the membrane (detailed definitions of the above quantities are in Ref. [26]). For the membrane paradigm, the fictitious membrane is such that the extrinsic curvature of the interior spacetime vanishes. As a consequence, Eq. (9.20) impose that the fictitious membrane is a *viscous* fluid with stress-energy tensor

$$T_{ab} = \rho u_a u_b + (p - \zeta\Theta)\gamma_{ab} - 2\eta\sigma_{ab}\,, \tag{9.21}$$

where η and ζ are the shear and bulk viscosities of the fluid, ρ, p and u_a are the density, pressure and 3-velocity of the fluid, $\Theta = u^a_{\ ;a}$ is the expansion, σ_{ab} is the shear tensor, and the semicolon is the covariant derivative compatible with the induced metric. BHs are described by the following values of the shear and bulk viscosities of the membrane:

$$\eta_{\rm BH} = \frac{1}{16\pi}\,, \qquad \zeta_{\rm BH} = -\frac{1}{16\pi}\,; \tag{9.22}$$

whereas horizonless compact objects have values of the shear and bulk viscosities which are generically complex and frequency dependent. For a specific model for the interior of the compact object, the shear and the bulk viscosities are uniquely determined. The junction conditions in Eq. (9.20) with the stress-energy tensor in Eq. (9.21) allow us to derive the boundary conditions at the radius of the horizonless compact object, i.e. [26]

$$\frac{d\psi(r_0)/dr_*}{\psi(r_0)} = -\frac{i\omega}{16\pi\eta} - \frac{r_0^2 V_{\rm axial}(r_0)}{2(r_0 - 3M)}\,, \qquad \text{axial}\,, \tag{9.23}$$

$$\frac{d\psi(r_0)/dr_*}{\psi(r_0)} = -16\pi i\eta\omega + G(r_0, \omega, \eta, \zeta)\,, \quad \text{polar}\,, \tag{9.24}$$

where $G(r_0, \omega, \eta, \zeta)$ is a cumbersome function given in Ref. [26]. The boundary conditions in Eqs. (9.23), (9.24) describe a horizonless object with any compactness whose reflective properties are mapped in terms of the shear and bulk viscosities of the fictitious membrane.

? Exercise

9.2.

1. 1. Demonstrate that, in the limit ($r_0 \to 2M$), the axial boundary condition in Eq. (9.23) reduces to a purely ingoing wave when the condition in Eq. (9.22) is satisfied.
2. 2. For $\epsilon \ll 1$, the axial boundary condition in Eq. (9.23) reduces to the boundary condition in Eq. (9.10) for horizonless ultracompact objects. Derive that the relation between the reflectivity of the compact object and the shear viscosity of the membrane is in the large-frequency limit:

$$|\mathcal{R}|^2 = \left(\frac{1 - \eta/\eta_{\rm BH}}{1 + \eta/\eta_{\rm BH}}\right)^2 . \tag{9.25}$$

This shows that a compact object is a perfect absorber of high-frequency waves ($|\mathcal{R}|^2 = 0$) if $\eta = \eta_{\rm BH}$, whereas it is a perfect reflector of high-frequency waves ($|\mathcal{R}|^2 = 1$) when either $\eta = 0$ or $\eta \to \infty$.

Figure 9.3 shows the ratio of the real (left panel) and imaginary (right panel) part of the QNMs of a horizonless compact object to the fundamental $\ell = 2$ QNM of a Schwarzschild BH as a function of the compactness. Let us notice that as $\epsilon \to 0$, the QNM spectrum of the horizonless compact object coincides with the BH spectrum. This is because a horizonless compact object with the shear and bulk viscosities as in Eq. (9.22) has the same reflective properties of a BH. For relatively large values of ϵ, the compactness of the object decreases and the QNMs deviate from the BH QNM. The highlighted regions are the maximum allowed deviation (with 90% credibility) for the least-damped QNM in the event GW150914, and correspond to

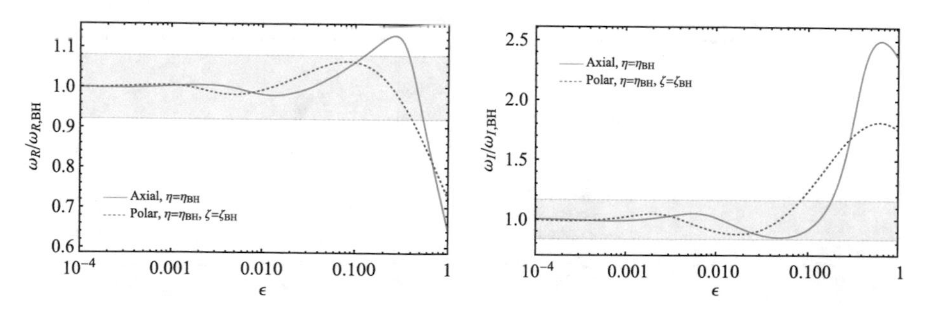

Fig. 9.3 Real (left panel) and imaginary (right panel) part of the QNMs of a horizonless compact object described by a fictitious fluid with shear viscosity $\eta = \eta_{\rm BH}$ and bulk viscosity $\zeta = \zeta_{\rm BH}$ compared to the fundamental $\ell = 2$ QNM of a Schwarzschild BH, as a function of ϵ where the radius of the object is located at $r_0 = 2M(1+\epsilon)$. The highlighted region is the maximum deviation (with 90% credibility) for the least-damped QNM in the event GW150914 [27]. Horizonless compact objects with $\epsilon \lesssim 0.1$ are compatible with current measurement accuracies. Adapted from [26].

$\sim 16\%$ and $\sim 33\%$ for the real and imaginary part of the QNM, respectively [27]. Figure 9.3 shows that horizonless compact objects with $\epsilon \lesssim 0.1$ are compatible with current measurement accuracies. Next-generation detectors would allow us to set more stringent constraints on the radius of compact objects.

9.3.2 Gravitational-Waves Echoes

In this section, we shall analyse the modifications that would appear in the post-merger GW signal if the remnant of a compact binary coalescence is a horizonless compact object. The phenomenology depends strongly on the compactness of the object. In particular, if the remnant is a horizonless ultracompact object ($\epsilon \ll 1$) the prompt ringdown would be nearly indistinguishable from the BH ringdown since it is due to the excitation of the light ring that occurs approximately at the same location as shown in Fig. 9.1. Afterwards, some trapped modes travel within the cavity of the effective potential and are reflected back at the radius of the compact object. After the interaction with the light ring, an additional GW signal is emitted at infinity in the form a GW echo. Multiple reflections of the trapped modes in the cavity can give rise to a train of GW echoes.

The left panel of Fig. 9.4 shows the GW signal that would be emitted in the case of a horizonless compact object compared to the BH case. The delay time between subsequent GW echoes is fixed and depends on the width of the cavity, i.e. the compactness of the object. The delay time is computed as the round-trip time of the radiation to travel in the cavity between the light ring and the radius of the compact object. In the static and spherically symmetric case [18],

$$\tau_{\text{echo}} = 2 \int_{r_0}^{3M} \frac{dr}{f(r)} \sim 2M \left[1 - 2\epsilon - 2 \log (2\epsilon)\right] . \tag{9.26}$$

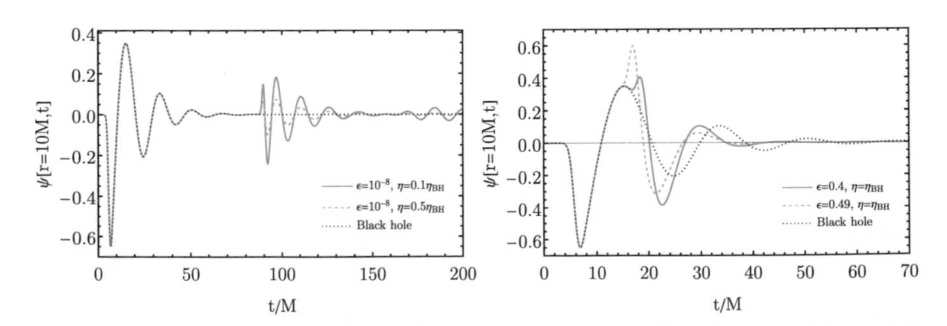

Fig. 9.4 Left panel: GW echoes emitted in the postmerger signal by an ultracompact horizonless object ($\epsilon \ll 1$) with different reflective properties parametrised by the shear viscosity η of the membrane. Right panel: Ringdown of an horizonless compact object with small compactness ($\epsilon \gtrsim 0.01$) and the same reflective properties of a BH ($\eta = \eta_{\text{BH}}$). The ringdown signal is modified due to the interference of the first GW echo with the prompt ringdown. Adapted from [26]. Copyright by the American Physical Society, all rights reserved

The logarithmic dependence in Eq. (9.26) allows us to detect even Planckian corrections at the horizon scale ($\epsilon \sim l_{\text{Planck}}/M$) few ms after the merger with a remnant of $M \sim 10M_{\odot}$. The amplitude of the GW echoes depends on the reflective properties of the compact object, as shown in the left panel of Fig. 9.4 for several values of the shear viscosity of the fictitious membrane. Furthermore, the light ring acts as a frequency-dependent high-pass filter, i.e. each GW echo has a lower frequency content than the previous one. At late times, the GW signal is dominated by the low-frequency QNMs of the horizonless compact object shown in Fig. 9.2.

If the remnant of a binary coalescence is a horizonless compact object with small compactness ($\epsilon \gtrsim 0.01$), the GW phenomenology in the postmerger signal would be different. In particular, the delay time of the first GW echo in Eq. (9.26) would be comparable with the decay time of the prompt ringdown, i.e. $\tau_{\text{ringdown}} = -1/\omega_{I,\text{BH}} \approx 10M$. Therefore, the first GW echo would interfere with the prompt ringdown as shown in the right panel of Fig. 9.4. Finally, subsequent GW echoes are suppressed because the cavity between the light ring and the radius of the compact object is so small that does not trap the modes efficiently.

9.4 Detectability

Several searches for GW echoes have been performed based on matched-filter techniques and unmodeled searches [7,28]. In the time domain, some phenomenological templates are based on inspiral-merger-ringdown templates in GR with additional parameters related to the morphology of GW echoes [29] and the superposition of sine-Gaussians with free parameters [30]. In the frequency domain, some waveform templates depend explicitly on the physical parameters of the horizonless compact object, i.e., its compactness and reflectivity [31–33]. Moreover, some unmodeled searches have been performed based on the superposition of generalized wavelets [34] and with Fourier windows [35].

Tentative evidence for GW echoes has been reported in the events of the first and second observing runs of LIGO and Virgo [29, 35], followed by independent searches arguing that the statistical significance of GW echoes is consistent with noise [36–39]. Furthermore, no evidene for GW echoes has been reported in the third observing run of the LIGO, Virgo, KAGRA collaboration [5].

The next generation detectors have promising prospects of testing the BH paradigm. The ground-based observatories Einstein Telescope [40] and Cosmic Explorer [41] will observe GWs with an overall improvement of the signal-to-noise ratio by an order of magnitude than current detectors. Moreover, the future space-based interferometer LISA [42] will detect GWs in the $10^{-4} - 1$ Hz frequency band from a variety of astrophysical sources. The sensitivity of the detectors will allow us to resolve the QNMs at percent level and perform multiple tests of the BH paradigm with the detection of higher modes.

Figure 9.5 shows the relative percentage difference between the fundamental $\ell = 2$ QNM of a Schwarzschild BH and the fundamental $\ell = 2$ QNMs of a horizonless compact object with radius $r_0 = 2M(1 + \epsilon)$ and reflectivity defined

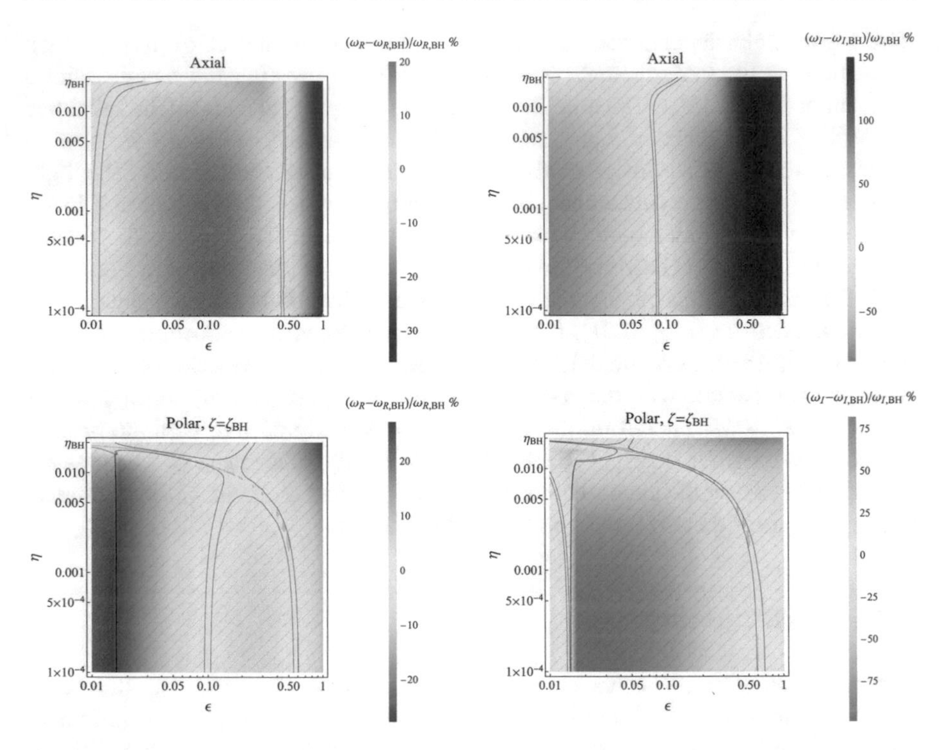

Fig. 9.5 Relative percentage difference of the real (left panels) and imaginary (right panels) part of the QNMs of a horizonless compact object to the fundamental QNM of a Schwarzschild BH for axial (top panels) and polar (bottom panels) perturbations. The dashed areas are the regions that would be excluded by individual measurements of the real and imaginary part of the QNMs by next-generation detectors. The plot shows that next-generation detectors will allow us to constraint the whole region of the (ϵ, η) parameter space shown in the diagram. Adapted from [26]. Copyright by the American Physical Society, all rights reserved

by the shear viscosity of the fictitious membrane. The QNM spectrum is a function of the parameter ϵ (x-axis) and the shear viscosity of the fictitious membrane $0 \leq \eta \leq \eta_{\mathrm{BH}}$ (y-axis) where $\eta = 0$ describes a perfectly reflecting compact object and $\eta = \eta_{\mathrm{BH}}$ describes a totally absorbing compact object. The left (right) panels show the relative percentage difference of the real (imaginary) part of the QNMs for axial and polar perturbations in the top and bottom panels, respectively. The dashed areas are the regions of the (ϵ, η) parameter space that would be excluded by individual measurements of the real and imaginary part of the fundamental QNM with next-generation detectors whose accuracy is assumed to be an order of magnitude better than current detectors [27]. Figure 9.5 shows that almost the whole region of the (ϵ, η) parameter space would be constrained. Therefore, next-generation detectors will allow us to set very stringent constraints on the radius and the reflective properties of compact objects.

Acknowledgments EM acknowledges funding from the Deutsche Forschungsgemeinschaft (DFG)—project number: 386119226.

References

1. B. Carter, Axisymmetric black hole has only two degrees of freedom. Phys. Rev. Lett. **26**, 331–333 (1971)
2. D.C. Robinson, Uniqueness of the Kerr black hole. Phys. Rev. Lett. **34**, 905–906 (1975)
3. R.P. Kerr, Gravitational field of a spinning mass as an example of algebraically special metrics. Phys. Rev. Lett. **11**, 237–238 (1963)
4. S. Chandrasekhar, S.L. Detweiler, The quasi-normal modes of the Schwarzschild black hole. Proc. Roy. Soc. Lond. A **344**, 441–452 (1975)
5. R. Abbott, et al., [LIGO Scientific, VIRGO and KAGRA], Tests of General Relativity with GWTC-3. arXiv:2112.06861 [gr-qc]
6. E. Berti, V. Cardoso, C.M. Will, On gravitational-wave spectroscopy of massive black holes with the space interferometer LISA. Phys. Rev. D **73**, 064030 (2006)
7. V. Cardoso, P. Pani, Testing the nature of dark compact objects: a status report. Living Rev. Rel. **22**(1), 4 (2019)
8. S.D. Mathur, The Fuzzball proposal for black holes: an elementary review. Fortsch. Phys. **53**, 793–827 (2005)
9. S.L. Liebling, C. Palenzuela, Dynamical boson stars. Living Rev. Rel. **15**, 6 (2012)
10. M.A. Abramowicz, W. Kluzniak, J.P. Lasota, No observational proof of the black hole event-horizon. Astron. Astrophys. **396**, L31–L34 (2002)
11. K. Akiyama, et al., [Event Horizon Telescope], First M87 event horizon telescope results. V. physical origin of the asymmetric ring. Astrophys. J. Lett. **875**(1), L5 (2019)
12. J.C. Bustillo, N. Sanchis-Gual, A. Torres-Forné, J.A. Font, A. Vajpeyi, R. Smith, C. Herdeiro, E. Radu, S.H.W. Leong, GW190521 as a merger of proca stars: a potential new vector boson of 8.7×10^{-13} eV. Phys. Rev. Lett. **126**(8), 081101 (2021)
13. R. Abbott, et al., [LIGO Scientific and Virgo], GW190814: gravitational waves from the coalescence of a 23 solar mass black hole with a 2.6 solar mass compact object. Astrophys. J. Lett. **896**(2), L44 (2020)
14. E. Maggio, P. Pani, G. Raposo, Testing the nature of dark compact objects with gravitational waves. arXiv:2105.06410 [gr-qc]
15. T. Regge, J.A. Wheeler, Stability of a schwarzschild singularity. Phys. Rev. **108**, 1063–1069 (1957)
16. F.J. Zerilli, Effective potential for even parity Regge-Wheeler gravitational perturbation equations. Phys. Rev. Lett. **24**, 737–738 (1970)
17. E. Maggio, P. Pani, V. Ferrari, Exotic compact objects and how to quench their ergoregion instability. Phys. Rev. D **96**(10), 104047 (2017)
18. V. Cardoso, E. Franzin, P. Pani, Is the gravitational-wave ringdown a probe of the event horizon? Phys. Rev. Lett. **116**(17), 171101 (2016). Erratum: Phys. Rev. Lett. **117**(8), 089902 (2016)
19. E. Maggio, V. Cardoso, S.R. Dolan, P. Pani, Ergoregion instability of exotic compact objects: electromagnetic and gravitational perturbations and the role of absorption. Phys. Rev. D **99**(6), 064007 (2019)
20. A. Vilenkin, Exponential amplification of waves in the gravitational field of ultrarelativistic rotating body. Phys. Lett. B **78**, 301–303 (1978)
21. A.A. Starobinskil, S.M. Churilov, Amplification of electromagnetic and gravitational waves scattered by a rotating black hole. Sov. Phys. JETP **65**(1), 1–5 (1974)
22. T. Damour, Surface effects in black-hole physics. in *Proceedings of the Second Marcel Grossmann Meeting of General Relativity*, ed. by R. Ruffini (North Holland, Amsterdam, 1982), pp. 587–608

23. K.S. Thorne, R. Price, D. Macdonald, Black Holes: The Membrane Paradigm. Yale University Press (1986)
24. G. Darmois, Les équations de la gravitation einsteinienne. Mémorial Sci. Math. **fascicule 25**, 1–48 (1927)
25. W. Israel, Singular hypersurfaces and thin shells in general relativity. Nuovo Cim. B **44S10**, 1 (1966). erratum: Nuovo Cim. B **48**, 463 (1967)
26. E. Maggio, L. Buoninfante, A. Mazumdar, P. Pani, How does a dark compact object ringdown? Phys. Rev. D **102**(6), 064053 (2020)
27. A. Ghosh, R. Brito, A. Buonanno, Constraints on quasinormal-mode frequencies with LIGO-Virgo binary–black-hole observations. Phys. Rev. D **103**(12), 124041 (2021)
28. J. Abedi, N. Afshordi, N. Oshita, Q. Wang, Quantum black holes in the sky. Universe **6**(3), 43 (2020)
29. J. Abedi, H. Dykaar, N. Afshordi, Echoes from the abyss: tentative evidence for planck-scale structure at black hole horizons. Phys. Rev. D **96**(8), 082004 (2017)
30. A. Maselli, S.H. Völkel, K.D. Kokkotas, Parameter estimation of gravitational wave echoes from exotic compact objects. Phys. Rev. D **96**(6), 064045 (2017)
31. Z. Mark, A. Zimmerman, S.M. Du, Y. Chen, A recipe for echoes from exotic compact objects. Phys. Rev. D **96**(8), 084002 (2017)
32. A. Testa, P. Pani, Analytical template for gravitational-wave echoes: signal characterization and prospects of detection with current and future interferometers. Phys. Rev. D **98**(4), 044018 (2018)
33. E. Maggio, A. Testa, S. Bhagwat, P. Pani, Analytical model for gravitational-wave echoes from spinning remnants. Phys. Rev. D **100**(6), 064056 (2019)
34. K.W. Tsang, M. Rollier, A. Ghosh, A. Samajdar, M. Agathos, K. Chatziioannou, V. Cardoso, G. Khanna C. Van Den Broeck, A morphology-independent data analysis method for detecting and characterizing gravitational wave echoes. Phys. Rev. D **98**(2), 024023 (2018)
35. R.S. Conklin, B. Holdom, J. Ren, Gravitational wave echoes through new windows. Phys. Rev. D **98**(4), 044021 (2018)
36. J. Westerweck, A. Nielsen, O. Fischer-Birnholtz, M. Cabero, C. Capano, T. Dent, B. Krishnan, G. Meadors, A.H. Nitz, Low significance of evidence for black hole echoes in gravitational wave data. Phys. Rev. D **97**(12), 124037 (2018)
37. A.B. Nielsen, C.D. Capano, O. Birnholtz, J. Westerweck, Parameter estimation and statistical significance of echoes following black hole signals in the first Advanced LIGO observing run. Phys. Rev. D **99**(10), 104012 (2019)
38. K.W. Tsang, A. Ghosh, A. Samajdar, K. Chatziioannou, S. Mastrogiovanni, M. Agathos, C. Van Den Broeck, A morphology-independent search for gravitational wave echoes in data from the first and second observing runs of Advanced LIGO and Advanced Virgo. Phys. Rev. D **101**(6), 064012 (2020)
39. R.K.L. Lo, T.G.F. Li, A.J. Weinstein, Template-based gravitational-wave echoes search using Bayesian model selection. Phys. Rev. D **99**(8), 084052 (2019)
40. M. Punturo, M. Abernathy, F. Acernese, B. Allen, N. Andersson, K. Arun, F. Barone, B. Barr, M. Barsuglia, M. Beker, et al., The Einstein telescope: a third-generation gravitational wave observatory. Class. Quant. Grav. **27**, 194002 (2010)
41. D. Reitze, R.X. Adhikari, S. Ballmer, B. Barish, L. Barsotti, G. Billingsley, D.A. Brown, Y. Chen, D. Coyne, R. Eisenstein, et al., Cosmic explorer: the U.S. contribution to gravitational-wave astronomy beyond LIGO. Bull. Am. Astron. Soc. **51**(7), 035 (2019)
42. P. Amaro-Seoane, et al., [LISA], Laser interferometer space antenna. arXiv:1702.00786 [astro-ph.IM]

Boson Stars

10

Yakov Shnir

Abstract

We review particle-like configurations of complex scalar field, localized by gravity, so-called boson stars. In the simplest case, these solutions posses spherical symmetry, they may arise in the massive Einstein-Klein-Gordon theory with global $U(1)$ symmetry, as gravitationally bounded lumps of scalar condensate. Further, there are spinning axially symmetric boson stars which possess non-zero angular momentum, and a variety of non-trivial multipolar stationary configurations without any continuous symmetries. In this short overview we discuss important dynamic properties of the boson stars, concentrating on recent results on the construction of multicomponent constellations of boson stars.

10.1 Q-Balls and Boson Stars

One of the most interesting directions in modern theoretical physics is related with investigation of spatially localized field configurations with finite energy bounded by gravity, see i.e. [1–4] for detailed review. The idea that the gravitational attraction may stabilize a fundamental matter field, was pioneered by Wheeler [5], who considered classical self-gravitating lumps of electromagnetic field, so-called *geons*. The geon is a localized regular solution of the coupled system of the field equations of the Einstein-Maxwell theory. Notably, Wheeler emphasized the unstability of geons with respect to linear perturbations of the fields.

From a modern perspective, the geons represent a *soliton*, a field configuration which may exist in diverse non-linear models in a wide variety of physical

---

Y. Shnir (✉)
BLTP, JINR, Dubna, Russia
e-mail: shnir@theor.jinr.ru

C. Pfeifer, C. Lämmerzahl (eds.), *Modified and Quantum Gravity*, Lecture Notes in Physics 1017, https://doi.org/10.1007/978-3-031-31520-6_10

contexts. Roughly speaking, the solitons can be can be divided into two groups, the topological and non-topological solitons, see e.g. [6, 7]. Topological solitons, like kinks, vortices, monopoles or skyrmions, are characterized by a conserved topological charge. This is not a property of non-topological solitons which occur in various non-linear systems with an unbroken global symmetry. A typical example in Minkovski spacetime are Q-balls, they represent time-dependent lumps of a complex scalar field with a stationary oscillating phase [8–10].

It was pointed out by Kaup [11], Feinblum and McKinley [12], and subsequently by Ruffini and Bonazzola [13], that stable localized soliton-type configurations, now dubbed as *boson stars* (BSs) may arise as the complex scalar field becomes coupled to gravity. In the simplest case, spherically symmetric boson star represent a particle-like self-gravitating asymptotically flat stationary solution of the $(3 + 1)$-dimensional Einstein-Klein-Gordon (EKG) theory. In this model the scalar field possess a mass term only, without self-interaction. The corresponding configurations can be considered as lumps of the scalar condensate, macroscopic quantum state, which is prevented from gravitationally collapsing by Heisenberg′s uncertainty principle. These mini-boson stars do not have a regular flat spacetime limit. On the contrary, the BSs in the models with polynomial potentials [14, 15], or in the two-component Einstein-Friedberg-Lee-Sirlin model [16], are linked to the corresponding flat space Q-balls. The BSs in the model with a repulsive self-interaction [17] are more massive than the mini-boson stars in the EKG model, further, inclusion of a sextic potential [14–16] allows for existence of very massive and highly compact objects, near of the threshold of gravitational collapse [18, 19]. Clearly, these configurations resemble neutron stars (which were discussed in Chap. 7), further astrophysical applications of BSs include consideration of hypothetical weakly-interacting ultralight component of cosmological dark matter [20, 21], axions [22, 23], and black hole mimickers [24–26], see also the discussion in Chap. 8. Bosons stars attracted a lot of attention in study of their evolution in binaries and in search for gravitational-wave signals produced by collision of BSs [27–29], see also Chap. 9.

Both Q-balls and BSs have a harmonic time dependence with a constant angular frequency ω, they carry a Noether charge Q associated with an unbroken continuous global $U(1)$ symmetry. This charge is proportional to the frequency ω and represents the boson particle number of the configurations. Further, there are charged Q-balls in gauged models with local $U(1)$ symmetry [30–38]. The presence of the electromagnetic interaction affects the properties of the gauged Q-balls, in particular, they may exist for a restricted range of values of the gauge coupling. Charged BSs arise in extended Einstein-Maxwell-scalar theories, these solutions were studied in [39–44]. Besides, BSs exist in the asymptotically anti-de Sitter spacetime [45, 46].

In Minkowski spacetime, Q-balls exist only within a restricted interval of values of the angular frequency ω: there is a maximal value ω_{max}, which corresponds to the mass of the scalar excitations, and some minimal value ω_{min}, that depends on the form of the potential. Notably, $\omega_{min} = 0$ in the two-component Friedberg-Lee-Sirlin (FLS) model [9, 47, 48]. Both the mass M and the charge Q diverge, as the frequency ω approaches the limiting values. Typically, there are two branches of flat space Q-balls, merging and ending at the minimal values of charge and mass. This bifurcation corresponds to some critical value of the frequency $\omega_{cr} \in [\omega_{min}, \omega_{max}]$, from where they increase monotonically towards both limiting values of ω.

The situation is different for BSs: coupling of the scalar field to gravity modifies the critical behavior pattern of the configurations. The fundamental branch of the solutions starts off from the perturbative excitations at $\omega \sim \omega_{max}$, at which both the mass and the charge trivialize (rather than diverge). Then, the BSs exhibit a spiral-like frequency dependence of the charge and the mass, where both quantities tend to some finite limiting values at the centers of the corresponding spirals [16]. Qualitatively, the appearance of the frequency-mass spiral may be related to oscillations in the force balance between the repulsive scalar interaction and the gravitational attraction in equilibria [49]. This spiraling behavior is reminiscent of the mass radius relation of neutron stars beyond the maximum mass star.

Simplest BSs are spherically symmetric, for each fundamental solutions there exist a tower of radially excited states, which possess some number of nodes in profile of the scalar field [1, 16, 50]. The mass of these excited solution is higher, than the mass of the corresponding fundamental boson star with the same angular frequency ω, however the properties of the spherically symmetric excited BSs are not very different from those of the nodeless boson stars. Also multi-state BSs have been studied, these configurations represent spherically symmetric superposition of the fundamental and the first excited solutions [51]. The radial pulsations and radiation of BSs were studied in numerical relativity [19, 52, 53], the solutions are shown to be stable on the first branch.

Rotating BSs are axially symmetric, they possess non-zero angular momentum J which is quantized in terms of the charge, $J = nQ$ [14, 15, 54, 55]. In other words, the BSs do not admit slow rotating limit. Rotating BSs possess some peculiar geometrical features, in particular, ergo-regions may arise for such solutions [15, 56]. Interestingly, radially excited rotating BSs do not exhibit a spiraling behavior; instead, the second branch extends back to the upper critical value of the frequency ω_{max}, forming a loop [57].

Both axially-symmetric spinning Q-balls in Minkowski spacetime and the rotating BSs may be either symmetric with respect to reflections in the equatorial plane, $\theta \to \pi - \theta$, or antisymmetric. The solutions of the first type are referred to as parity-even, while the configurations of the second type are termed parity-odd [14, 15, 58–60], for each value of integer winding number n, there should be two types of spinning solutions possessing different parity.

Notably, the character of the scalar interaction between Q-balls and BSs depends on their relative phase [61, 62], If the solitons are in phase, the scalar interaction is attractive, if they are out of phase, there is a repulsive scalar force between them. Thus, a pair of boson stars may exist as a saddle point solution of the EKG model [63–65]. Furthermore, scalar repulsion can be balanced by the gravitational attraction in various multicomponent bounded systems of BSs [64, 65].

Below we briefly review the basic properties of boson stars and discuss multi-component BS configurations constructed recently in [64, 65].

10.2 The Model: Action, Field Equations, and Global Charges

We consider a massive complex scalar field Φ, which is minimally coupled to Einstein's gravity in an asymptotically flat (3 + 1)-dimensional space-time. The corresponding action of the system is

$$\mathcal{S} = \int d^4x \sqrt{-g} \left[\frac{R}{16\pi G} - \frac{1}{2} g^{\mu\nu} \left(\Phi^*_{,\mu} \Phi_{,\nu} + \Phi^*_{,\nu} \Phi_{,\mu} \right) - U(|\Phi|^2) \right], \quad (10.1)$$

where R is the Ricci scalar curvature, G is Newton's constant, the asterisk denotes complex conjugation, U denotes the scalar field potential and we employ the usual compact notation $\Phi_{,\mu} \equiv \partial_\mu \Phi$.

Variation of the action (10.1) with respect to the metric leads to the Einstein equations

$$E_{\mu\nu} \equiv R_{\mu\nu} - \frac{1}{2} g_{\mu\nu} R - 8\pi G \, T_{\mu\nu} = 0 \,, \quad (10.2)$$

where

$$T_{\mu\nu} \equiv \Phi^*_{,\mu} \Phi_{,\nu} + \Phi^*_{,\nu} \Phi_{,\mu} - g_{\mu\nu} \left[\frac{1}{2} g^{\sigma\tau} (\Phi^*_{,\sigma} \Phi_{,\tau} + \Phi^*_{,\tau} \Phi_{,\sigma}) + U(|\Phi|^2) \right], \quad (10.3)$$

is the stress-energy tensor of the scalar field.

The corresponding equation of motion of the scalar field is the non-linear Klein-Gordon equation

$$\left(\Box - \frac{dU}{d|\Phi|^2} \right) \Phi = 0 \,, \quad (10.4)$$

where $\Box$ represents the covariant d'Alembert operator.

? Exercise

10.1. Verify (10.2) to (10.4).

The solutions considered below have a static line-element (with a timelike Killing vector field $\xi = \partial_t$), being topologically trivial and globally regular, *i.e.* without an event horizon or conical singularities, while the scalar field is finite and smooth everywhere. Also, they approach asymptotically the Minkowski spacetime background. Their mass M can be obtained from the respective Komar expressions [66],

$$M = 2\int_{\Sigma} R_{\mu\nu}n^{\mu}\xi^{\nu}dV = 2\int_{\Sigma}\left(T_{\mu\nu} - \frac{1}{2}g_{\mu\nu}T_{\gamma}^{\ \gamma}\right)n^{\mu}\xi^{\nu}dV. \tag{10.5}$$

Here Σ denotes a spacelike hypersurface (with the volume element dV), while n^{μ} is a time-like vector normal to Σ, $n_{\mu}n^{\mu} = -1$.

The axially symmetric spinning boson stars are characterized by the mass M and by the angular momentum

$$J = -\int_{\Sigma} R_{\mu\nu}n^{\mu}\eta^{\nu}dV = -\int_{\Sigma}\left(T_{\mu\nu} - \frac{1}{2}g_{\mu\nu}T_{\gamma}^{\ \gamma}\right)n^{\mu}\eta^{\nu}dV \tag{10.6}$$

where the second commuting Killing vector field is $\eta = \partial_{\varphi}$.

The action (10.1) is invariant with respect to the global U(1) transformations of the complex scalar field, $\phi \to \phi e^{i\chi}$, where χ is a constant. The following Noether 4-current is associated with this symmetry

$$j_{\mu} = -i(\Phi\partial_{\mu}\Phi^{*} - \Phi^{*}\partial_{\mu}\Phi)\,. \tag{10.7}$$

It follows that integrating the timelike component of this 4-current in a spacelike slice Σ yields a second conserved quantity—the *Noether charge*:

$$Q = \int_{\Sigma} j^{\mu}n_{\mu}dV\,. \tag{10.8}$$

Semiclassically, the charge Q can be interpreted as a measure of the number of scalar quanta condensed in the BS. There is the quantization relation for the angular momentum of the scalar field J (10.6), $J = nQ$ [55].

10.2.1 Potential

In the simplest case of the non-self interacting EKG model, the potential contains just the mass term, $U = \mu^2|\Phi|^2$, where parameter μ yields the mass of the scalar field. The corresponding mini-BSs represent a gravitationally bound system of globally regular massive interacting bosons, it does not possess the flat space limit. It should be noted that in the EKG model the natural units are set by the mass parameter μ and by the effective gravitational coupling $\alpha^2 = 4\pi G$. They can be rescaled away via transformations of the coordinates and the field, $x_{\mu} \to x_{\mu}/\mu$, $\Phi \to \Phi/\alpha$. Note that the scalar field frequency changes accordingly, $\omega \to \omega/\mu$.

The quartic self-interaction potential

$$U = \lambda|\Phi|^4 + \mu^2|\Phi|^2\,, \tag{10.9}$$

was considered in many works, see e.g. [17, 67, 68]. Such potential can stabilize excited BSs, however the corresponding solutions do not posses the flat space limit.

The non-renormalizable self-interacting sixtic potential, originally proposed in [69, 70]

$$U = \nu|\Phi|^6 - \lambda|\Phi|^4 + \mu^2|\Phi|^2 \tag{10.10}$$

allows for the existence of very massive BSs, they are linked to the corresponding Q-balls on a Minkowski spacetime background [14, 16, 59, 60]. Similar to the case of the EKG model, two of the parameters of the model (10.1), (10.10) can be absorbed into a redefinition of the coordinates together with a rescaling of the scalar field,

$$x_\mu \to \frac{a}{\mu} x_\mu \,, \qquad \Phi \to \frac{\sqrt{\mu}}{\nu^{1/4}\sqrt{a}}\Phi \,,$$

where a is an arbitrary constant. Thus, the potential of the rescaled model becomes

$$U = |\Phi|^6 - \tilde{\lambda}|\Phi|^4 + a^2|\Phi|^2$$

with the usual choice $\tilde{\lambda} = \frac{a\lambda}{\mu\sqrt{\nu}} = 2$ and $a^2 = 1.1$.

? Exercise

10.2. Derive the action for the rescaled model, by employing the rescaling in (10.1) with (10.10).

Then the dimensionless effective gravitational coupling becomes $\alpha^2 = \frac{4\pi G\mu}{a\sqrt{\nu}}$. Evidently, for large values of the gravitational coupling, the nonlinearity of the potential (10.10) becomes suppressed and the system approaches the EKG model with its corresponding mBS solutions. However, as the gravitational attraction remains relatively weak, the scalar interaction becomes more important, it allows for existence of very large massive BSs.

The sixtic potential (10.10) can be considered as a limiting form of the periodic axion potential which describes a real quantized scalar field Φ,

$$U = m_a f_a \left(1 - \cos(\Phi/f_a)\right)$$

where f_a is the axion decay constant and m_a is the mass of the axion [22, 23].

Certainly, there are many other possible choices of a potential term for the boson stars. In particular, there is a class of flat potentials arising in the models with gauge-

and gravity-mediated supersymmetry breaking mechanism [71, 72]. Such potentials may be of the logarithmic or the exponential form, for example [71, 73]

$$U = \mu^2\eta^2\left[1 - \exp\left(-\frac{\Phi^2}{\eta^2}\right)\right]$$

where μ is the mass of the scalar field Φ and the parameter η is defines the mass scale below which supersymmetry is broken.

Domain of existence of the BSs is determined by the form of the potential. The maximal value ω_{max} corresponds to the mass of the scalar excitations $\mu^2 = \frac{dU}{d|\Phi|^2}$, the minimal value ω_{min} depends on explicit form of the potential and on the strength of the gravitational coupling α. Hereafter we assume that $\mu = 1$, without loss of generality, hence in the EKG model $\omega_{max} = 1$. Since the Planck mass is defined as $M_{Pl} = 1/\sqrt{G}$, the EKG BSs can be interpreted as *macroscopic quantum states*, they are prevented from gravitational collapse by the uncertainty principle. The critical mass of the EKG BSs is $M \approx M_{Pl}^2/\mu$ [11], more massive BSs become unstable w.r.t. linear fluctuations [74, 75]. In the models with non-linear potentials, like (10.9), (10.10), the BSs may have larger mass, they represent lumps of a macroscopic self-gravitating Bose-Einstein condensate. In the discussion below we mainly focus on the microscopic BSs in the EKG model and fix the value of the gravitational coupling $\alpha = 0.5$.

10.2.2 The Ansatz and the Field Equations

For the stationary spinning scalar field we can adopt a general Ansatz with a harmonic time dependence:

$$\Phi = f(r, \theta, \varphi)e^{-i(\omega t + n\varphi)}, \tag{10.11}$$

where r, θ, φ are the usual spherical coordinates, $\omega \geq 0$ is the angular frequency, $n \in \mathbb{Z}$ is the azimuthal winding number, and $f(r, \theta, \varphi)$ is a real spatial profile function. Notably, harmonic time dependency of the scalar field does not affect the physical quantities, like the stress-energy tensor (10.3). On the other hand, it allows us to evade scaling arguments of the Derrick's theorem [76], which does not support existence of static scalar soliton solutions in three spatial dimensions.

Allowing an angular dependence for the profile function of the BSs requires considering a metric Ansatz with sufficient generality. In particular, considering configurations with $n = 0$, which carry no angular momentum, we can make use of the line element without any spatial isometries

$$ds^2 = -F_0 dt^2 + F_1 dr^2 + F_2(rd\theta + S_1 dr)^2 + F_3(r\sin\theta d\varphi + S_2 dr + S_3 r d\theta)^2 \tag{10.12}$$

where seven metric functions F_0, F_1, F_2, F_3 and S_1, S_2, S_3 depend on spherical coordinates r, θ, φ [65].

By substituting the ansatz (10.11) into the scalar field Eq. (10.4) we obtain

$$\frac{1}{\sqrt{-g}}\frac{\partial}{\partial r}\left(g^{rr}\sqrt{-g}\frac{\partial f}{\partial r}\right)+\frac{1}{\sqrt{-g}}\frac{\partial}{\partial \theta}\left(g^{\theta\theta}\sqrt{-g}\frac{\partial f}{\partial \theta}\right)+\frac{1}{\sqrt{-g}}\frac{\partial}{\partial \varphi}\left(g^{\varphi\varphi}\sqrt{-g}\frac{\partial f}{\partial r}\right)$$
$$-(n^2g^{\varphi\varphi}-2g^{\varphi t}+\omega^2g^{tt})f=\frac{dU}{d|\phi|^2}f \tag{10.13}$$

Note that on the spatial asymptotic the metric approaches the Minkowski spacetime, then the field Eq. (10.13) tends to the usual Klein-Gordon equation with general solution for the scalar field $f \sim \sum_{l,n} R_l(r)Y_{ln}(\theta, \varphi)$. Here the radial part is

$$R_l(r) \sim \frac{1}{\sqrt{r}}K_{l+\frac{1}{2}}(r, \sqrt{\mu^2-\omega^2}) \tag{10.14}$$

where $K_{l+\frac{1}{2}}$ is the modified Bessel function of the first kind of order l and $Y_{ln}(\theta, \varphi)$ are the real spherical harmonics, which form a complete basis on the sphere S^2 and integers $l \geq n$ are the usual quantum numbers. Because of central character of gravitational interaction, this basis remains for any scalar multipole configuration of the BSs. Furthermore, for each particular set of values of the quantum numbers l, n, there are two types of the solutions, the parity even for even l and the parity-odd for odd l. They are symmetric and anti-symmetric, respectively, under a reflection along the equatorial plane. The spherical harmonics $Y_{ln}(\theta, \varphi)$ possess $2n$ φ-zeros, each describing a nodal longitude line and $l - n$ θ-zeros, each yielding a nodal latitude line. These nodal distributions define a multipolar configuration of BSs [65] briefly discussed below.

Simplest BSs are spherically symmetric [11–13], in such a case $l = n = 0$ and the profile function depends on the radial coordinate only, $f = f(r)$. The corresponding line element can be reduced to the Schwarzschild type metric, it can be written as

$$ds^2 = -N(r)\sigma^2(r)dt^2 + \frac{dr^2}{N(r)} + r^2(d\theta^2 + r^2\sin^2\theta\, d\varphi^2) \tag{10.15}$$

with $N(r) = 1 - 2m(r)/r$. Here $m(r)$ is so-called mass function, the Arnowitz-Deser-Misner (ADM) mass of the BS is $M = \lim_{r\to\infty} m(r)$. Clearly, the angular momentum of such BS is zero.

The resulting system of coupled ordinary differential equations on three radial functions $f(r), \sigma(r)$ and $m(r)$ can be solved numerically, using, for example a shooting method [77]. Along with the fundamental nodeless mode, there is an infinite tower of radial excitations of the BSs [1, 16], they are classified according to the number of nodes k of the scalar profile function $f(r)$, see Fig. 10.1.

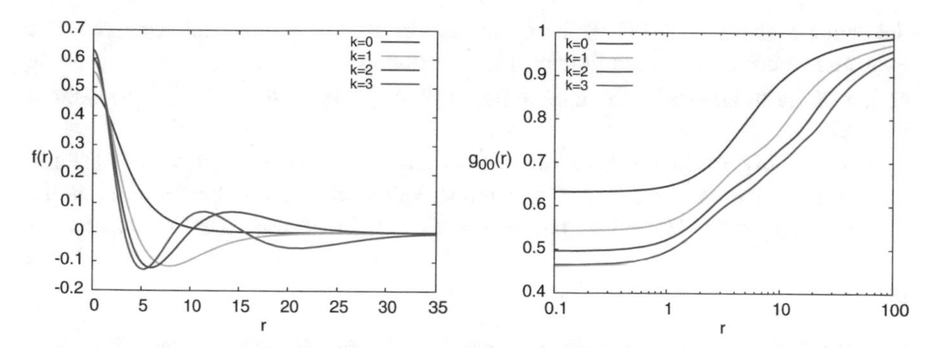

Fig. 10.1 The profile functions of the scalar field (left) and the metric component g_{00} (right) of the non-rotating $n = 0$ fundamental Einstein-Klein-Gordon boson star $k = 0$ and its first three radial excitations are displayed on the first branch of solutions at $\omega = 0.90$ as functions of the radial coordinate

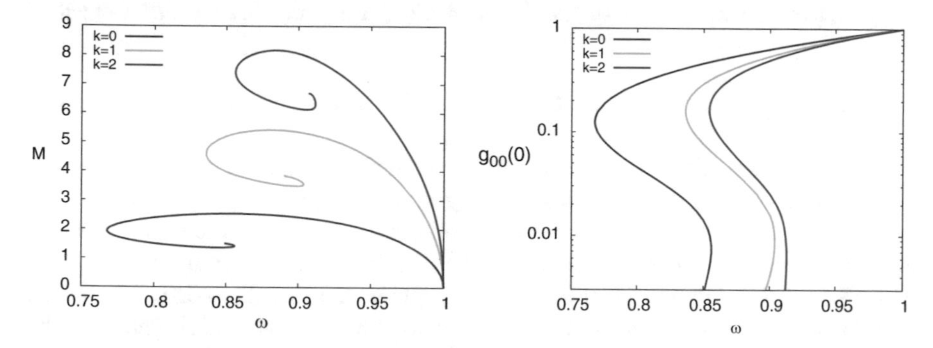

Fig. 10.2 Non-rotating $n = 0$ fundamental ($k = 0$) and radially excited ($k = 1, 2$) Einstein-Klein-Gordon boson stars. The mass of the solutions (left plot) and the minimal values of the metric component $g_{00}(0)$ (right plot) are displayed as functions of the angular frequency ω

The fundamental nodeless ground state solution is an analog of the 1s hydrogen orbital. This branch of BSs emerges from the vacuum fluctuations with angular part Y_{00} at the maximal frequency ω_{max}, given by the boson mass. Notably, unlike the case of Q-balls in flat space, where mass and charge diverge, these quantities vanish in this limit. Decreasing the frequency yields the fundamental branch of solutions which terminates at the first backbending of the curve, at which point it moves toward larger frequencies, as seen in Fig. 10.2. These solutions are stable with respect to linear perturbations [50]. The curve then follows a spiraling/oscillating pattern, with successive backbendings, while the minimum of the metric component $g_{00}(0)$ and the maximum of the scalar profile function $f(0)$ show damped oscillations [16]. Both mass and charge tend to some finite limiting values at the centers of the corresponding spirals, see Fig. 10.2. Qualitatively, the appearance of the frequency-mass spiral may be related to oscillations in the alternating force balance between the repulsive scalar interaction and the gravitational attraction in equilibria. There is an infinite set of branches, leading towards a critical solution at the center of the spiral. Plotting the Q (instead of M) also yields similar curves. The extremal values of the scalar field profile function and the metric function g_{00} at the center of the star do not seem to be finite, with f_{max} diverging and $g_{00}^{(min)}$ vanishing in this limit.

The radially excited spherically symmetric BSs also exhibit such spiraling behavior, as seen in Fig. 10.2. They emerge similarly from the vacuum at the maximal frequency. These BSs posses higher mass, increase of the nodal number k leads to increase of the minimal critical frequency ω_{min}, as seen in Fig. 10.2, left plot.

? Exercise

10.3. Using the Ansatz (10.11) with $n = 0$ and the Schwarzschild type metric (10.15) derive the system of coupled ordinary differential equations on three radial functions $f(r), \sigma(r)$ and $m(r)$ for the spherically symmetric Q-ball in the Einstein-Klein-Gordon theory.

Stationary spinning BSs are axially symmetric, their angular momentum is quantized in units of the azimuthal winding number, $J = nQ$ [14, 15, 59]. Similar to the case of non-rotating spherically symmetric BSs, they exhibit an analogous spiralling frequency/mass dependence. The rotating BSs exist in the EKG model [54,55,58,78] and in the model with solitonic potential (10.10) [14,15] as well as in other systems. The mass and the charge of the rotating BSs are much higher than the fundamental spherically symmetric counterparts, as seen in Fig. 10.3. The energy density distribution of these solutions is torus-like, the scalar field is vanishing at the origin, it possess a maximal value in the equatorial plane.

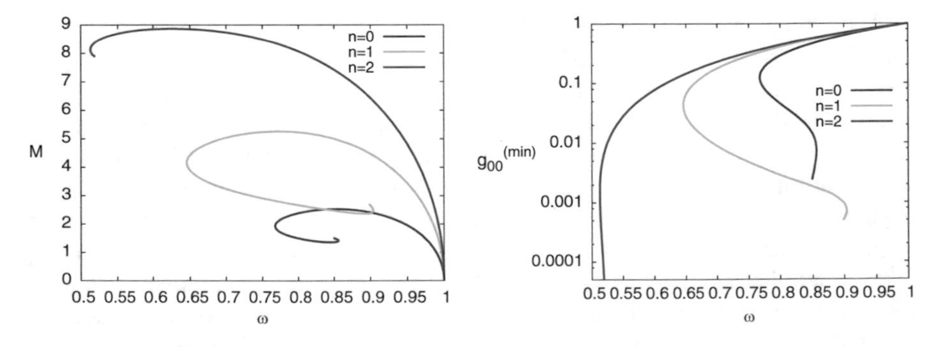

Fig. 10.3 Rotating Einstein-Klein-Gordon boson stars. The mass of the solutions (left plot) and the minimal values of the metric component g_{00} (right plot) are displayed as functions of the angular frequency ω

Remarkably, rapidly rotating BSs develop an ergoregion where the Killing vector field $\xi = \partial_t$ becomes spacelike [15, 78], or equally, $g_{tt} < 0$. Topologically, this region represent a torus. The existence of ergoregions is typical for a Kerr black hole, for the BSs it is an indication of instability of the configuration. The instability mechanism is related to the rotational superradiance [79], an excited relativistic BSs decays into less energetic state via emission of scalar quanta and gravitational waves. On the other hand, for the Kerr black hole with synchronized scalar hair [80, 81] the superradiance mechanism may induce transitions from the $n = 0$ state to higher n solutions [82].

The frequency dependence of rotating nodeless axially symmetric BSs is similar to that of the fundamental $n = 0$ solutions, the mass (and the angular momentum) form a spiral, as ω varies, while the minimum of the metric component g_{00} and the maximum of the scalar function f shows damped oscillations, see Fig. 10.3. The minimal value of the angular frequency ω_{min} is decreasing as the winding number n increases. Further, for each value of integer winding number n, there are two types of spinning BSs possessing different parity, so called parity-even and parity-odd rotating hairy BHs [14, 15, 58–60]. These configurations are symmetric or anti-symmetric, respectively, with respect to a reflection through the equatorial plane, i.e. under $\theta \to \pi - \theta$. In other words, the scalar field of the parity-odd BSs posses an angular node at $\theta = \pi/2$.

The energy density distribution of rotating BSs with positive parity forms a torus, while the energy density of rotating parity-odd BSs corresponds to a double torus, see see Fig. 10.4. More generally, there is a sequence of angularly excited BSs with some number of nodes of the scalar field in θ-direction [83], which are closely related to the real spherical harmonics $Y_{lm}(\theta, \varphi)$. For example, the angular part of the $n = 1$ spinning parity-even BSs corresponds to the harmonic Y_{11} while the angular part of the corresponding parity-odd BSs corresponds to the harmonic Y_{21},

Fig. 10.4 Einstein-Klein-Gordon boson stars. Surfaces of constant energy density of the (i) fundamental $n = 0$ solution; (ii) parity-even $n = 1$ rotating boson star; (iii) parity-odd $n = 1$ rotating boson star; and (iv) angularly excited parity-even $n = 1$ boson star, from left to right, all configurations at $\omega = 0.92$ on the first branch at $\alpha = 0.5$

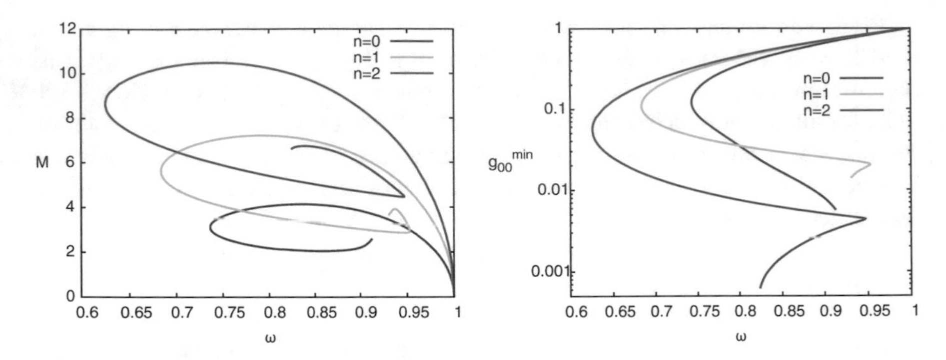

Fig. 10.5 Rotating parity-odd Einstein-Klein-Gordon boson stars. The mass of the solutions (left plot) and the minimal values of the metric component g_{00} (right plot) are displayed as functions of the angular frequency ω

the triple torus configuration, displayed in the right plot of Fig. 10.4, corresponds to the harmonic Y_{31}, etc.

The mass and the charge of both parity-even and parity-odd EKG BSs exhibit similar spiraling behavior, cf Figs. 10.3 and 10.5. However, the situation becomes different for the rotating radially excited axially symmetric BSs with non-zero angular momentum [57]. In such a case, there are two branches of solutions, merging and ending at the minimal values of the charge and the mass of the configurations, the second branch extends all the way back to the upper critical value of the frequency ω_{max}, forming a loop.

Notably, the gravitational interaction stabilizes the parity-odd BSs even in the limit $n = 0$. This axially-symmetric configuration with zero angular momentum represents a pair of boson stars, a saddle point solution of the EKG model [63–65, 84]. Its existence is related to a delicate force balance between the repulsive scalar interaction and gravity. Indeed, if the flat space Q-balls are in phase, they attract each other, if they are out of phase, there is a repulsive scalar force between them [61, 62]. The inversion of the sign of the scalar field function Φ under reflections $\theta \to \pi - \theta$ corresponds to the shift of the phase $\omega \to \omega + \pi$. Hence, the static pair of BSs with a single node of the scalar field on the symmetry axis, can be thought of as the limit of negative parity spinning configurations considered in [15].

The curves of the mass/frequency dependency of the pair of BSs are different from the case of a single spherical BS [64]. Instead of the paradigmatic spiraling curve one finds a truncated scenario with only two branches, ending at a limiting solution with finite values of ADM mass and Noether charge.

Furthermore, scalar repulsion can be balanced by the gravitational attraction in various multicomponent bounded systems of BSs [64, 65]. Figure 10.6 displays an overview of a selection of multipolar EKG BSs with various structure of nodes [65]. Constructing these solutions we do not impose any restrictions of symmetry, they all arise as corresponding linearized perturbations of the scalar field in the asymptotic region, as ω approaches the mass threshold. Gravitational attraction

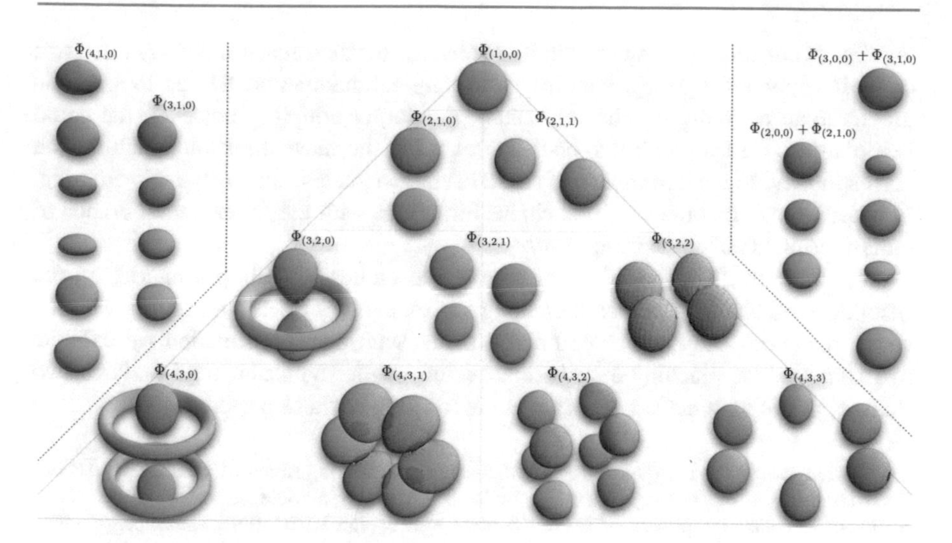

Fig. 10.6 Surfaces of constant energy density for a selection of multicomponent BSs in the EKG model. Reprinted (without modification) from [65]. © 2021 The Authors of [65], under the CC BY 4.0 license

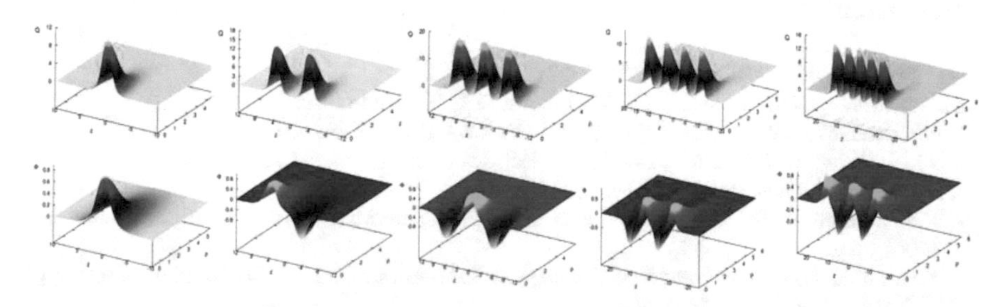

Fig. 10.7 Chains of BSs with one to five constituents on the first branch for $\alpha = 0.25$ at $\omega/\mu = 0.80$: $3d$ plots of the $U(1)$ scalar charge distributions (upper row) and the scalar field Φ (bottom row) versus the coordinates $\rho = r \sin\theta$ and $z = r \cos\theta$

stabilizes the excitations with nodal structure of the $\Phi(r, \theta, \varphi) \sim R_k(r)Y_{ln}(\theta, \varphi))$ wavefunctions. As the angular frequency decreases, the mass and the charge of the multicomponent configurations increase, however, the nodal structure remains unaffected [65]. Similar to the fundamental spherically symmetric solution, the fundamental branch of the multicomponent BSs ends in a spiraling/oscillating pattern. Clearly, all these solutions do not exist in Minkowsky space-time.

Analogous multipolar configurations with zero angular momentum exist in models with various potentials. For example, the chains of BSs in the system with sixtic potential (10.10) were discussed in [64]. Figure 10.7 exhibits a few examples of these chains.

In such a case the pattern of dynamical evolution of the multicomponent BSs becomes different from the above-discussed EKG systems. Chains with an odd number of constituents show a spiraling behavior for their mass and charge in terms of their angular frequency, similarly to a single fundamental BS, as long as the gravitational coupling is relatively small. For larger coupling however, the spiral is replaced by a lace with two ends approaching the mass threshold, each branch corresponding to the dominance of either of the two states, and with a self-crossing. In other words, the branch of odd chains bifurcates with the fundamental branch of radially excited spherical boson stars.

For the even chains we do not observe the endless spiraling scenario, on the second, or on the third branch the configuration evolves toward a limiting solution which retain basically two central constituents, whose metric function g_{00} exhibits two sharp peaks, reaching a very small value, while the scalar field features two sharp opposite extrema located right at the location of these peaks [64].

Acknowledgments Ya.S. gratefully acknowledges the networking support by the COST Actions CA16104. This work was supported in part by the Ministry of Science and High Education of Russian Federation, project FEWF-2020–0003 and by the BLTP JINR Heisenberg-Landau program 2020.

References

1. P. Jetzer, Phys. Rep. **220**, 163–227 (1992)
2. A.R. Liddle, M.S. Madsen, Int. J. Mod. Phys. D **1**, 101–144 (1992)
3. F.E. Schunck, E.W. Mielke, Class. Quant. Grav. **20**, R301–R356 (2003)
4. S.L. Liebling, C. Palenzuela, Living Rev. Rel. **15**, 6 (2012)
5. J.A. Wheeler, Phys. Rev. **97**, 511–536 (1955)
6. N.S. Manton, P. Sutcliffe, *Topological Solitons* (Cambridge University Press, Cambridge, 2004)
7. Y.M. Shnir, *Topological and Non-Topological Solitons in Scalar Field Theories*. (Cambridge University Press, Cambridge, 2018)
8. G. Rosen, J. Math. Phys. **9**, 996–999 (1968)
9. R. Friedberg, T.D. Lee, A. Sirlin, Phys. Rev. D **13**, 2739 (1976)
10. S.R. Coleman, Nucl. Phys. B **262**, 263 (1985). Erratum: Nucl. Phys. B **269**, 744 (1986)
11. D.J. Kaup, Phys. Rev. **172**, 1331 (1968)
12. D.A. Feinblum, W.A. McKinley, Phys. Rev. **168**(5), 1445 (1968)
13. R. Ruffini, S. Bonazzola, Phys. Rev. **187**, 1767 (1969)
14. B. Kleihaus, J. Kunz, M. List, Phys. Rev. D **72**, 064002 (2005)
15. B. Kleihaus, J. Kunz, M. List, I. Schaffer, Phys. Rev. D **77**, 064025 (2008)
16. R. Friedberg, T.D. Lee, Y. Pang, Phys. Rev. D **35**, 3658 (1987)
17. M. Colpi, S.L. Shapiro, I. Wasserman, Phys. Rev. Lett. **57**, 2485–2488 (1986)
18. T.D. Lee, Phys. Rev. D **35**, 3637 (1987)
19. S.H. Hawley, M.W. Choptuik, Phys. Rev. D **62**, 104024 (2000)
20. A. Suárez, V.H. Robles, T. Matos, Astrophys. Space Sci. Proc. **38**, 107–142 (2014)
21. L. Hui, J.P. Ostriker, S. Tremaine, E. Witten, Phys. Rev. D **95**(4), 043541 (2017)
22. D. Guerra, C.F.B. Macedo, P. Pani, J. Cosmol. Astroparticle Phys. **09**(09), 061 (2019)
23. J.F.M. Delgado, C.A.R. Herdeiro, E. Radu, J. Cosmol. Astroparticle Phys. **06**, 037 (2020)
24. V. Cardoso, P. Pani, Living Rev. Rel. **22**(1), 4 (2019)
25. K. Glampedakis, G. Pappas, Phys. Rev. D **97**(4), 041502 (2018)
26. C.A.R. Herdeiro, A.M. Pombo, E. Radu, P.V.P. Cunha, N. Sanchis-Gual, J. Cosmol. Astroparticle Phys. **04**, 051 (2021)

27. C. Palenzuela, L. Lehner, S.L. Liebling, Phys. Rev. D **77**, 044036 (2008)
28. M. Bezares, C. Palenzuela, C. Bona, Phys. Rev. D **95**(12), 124005 (2017)
29. C. Palenzuela, P. Pani, M. Bezares, V. Cardoso, L. Lehner, S. Liebling, Phys. Rev. D **96**(10), 104058 (2017)
30. K.M. Lee, J.A. Stein-Schabes, R. Watkins, L.M. Widrow, Phys. Rev. D **39**, 1665 (1989)
31. C.H. Lee, S.U. Yoon, Mod. Phys. Lett. A **6**, 1479 (1991)
32. A. Kusenko, M.E. Shaposhnikov, P.G. Tinyakov, Pisma Zh. Eksp. Teor. Fiz. **67**, 229 (1998). JETP Lett. **67**, 247 (1998)
33. K.N. Anagnostopoulos, M. Axenides, E.G. Floratos, N. Tetradis, Phys. Rev. D **64**, 125006 (2001)
34. I.E. Gulamov, et al., Phys. Rev. D **92**(4), 045011 (2015)
35. I.E. Gulamov, E.Y. Nugaev, M.N. Smolyakov, Phys. Rev. D **89**(8), 085006 (2014)
36. A.G. Panin, M.N. Smolyakov, Phys. Rev. D **95**(6), 065006 (2017)
37. E.Y. Nugaev, A.V. Shkerin, J. Exp. Theor. Phys. **130**(2), 301–320 (2020)
38. V. Loiko, Y. Shnir, Phys. Lett. B **797**, 134810 (2019)
39. P. Jetzer, J.J. van der Bij, Phys. Lett. B **227**, 341–346 (1989)
40. P. Jetzer, Phys. Lett. B **231**, 433–438 (1989)
41. P. Jetzer, P. Liljenberg, B.S. Skagerstam, Astropart. Phys. **1**, 429–448 (1993)
42. D. Pugliese, H. Quevedo, J.A. Rueda, R. Ruffini, Phys. Rev. D **88**, 024053 (2013)
43. B. Kleihaus, J. Kunz, C. Lammerzahl, M. List, Phys. Lett. B **675**, 102–115 (2009)
44. S. Kumar, U. Kulshreshtha, D. Shankar Kulshreshtha, Class. Quant. Grav. **31**, 167001 (2014)
45. D. Astefanesei, E. Radu, Nucl. Phys. B **665**, 594–622 (2003)
46. O. Kichakova, J. Kunz, E. Radu, Phys. Lett. B **728**, 328–335 (2014)
47. A. Levin, V. Rubakov, Mod. Phys. Lett. A **26**, 409 (2011)
48. V. Loiko, I. Perapechka, Y. Shnir, Phys. Rev. D **98**(4), 045018 (2018)
49. R. Friedberg, T.D. Lee, Y. Pang, Phys. Rev. D **35**, 3640 (1987)
50. E. Seidel, W.M. Suen, Phys. Rev. D **42**, 384–403 (1990)
51. A. Bernal, J. Barranco, D. Alic, C. Palenzuela, Phys. Rev. D **81**, 044031 (2010)
52. M. Gleiser, R. Watkins, Nucl. Phys. B **319**, 733–746 (1989)
53. B. Kain, Phys. Rev. D **103**(12), 123003 (2021)
54. V. Silveira, C.M.G. de Sousa, Phys. Rev. D **52**, 5724–5728 (1995)
55. F.E. Schunck, E.W. Mielke, Phys. Lett. A **249**, 389–394 (1998)
56. V. Cardoso, P. Pani, M. Cadoni, M. Cavaglia, Phys. Rev. D **77**, 124044 (2008)
57. L.G. Collodel, B. Kleihaus, J. Kunz, Phys. Rev. D **96**(8), 084066 (2017)
58. S. Yoshida, Y. Eriguchi, Phys. Rev. D **56**, 762–771 (1997)
59. M.S. Volkov, E. Wohnert, Phys. Rev. D **66**, 085003 (2002)
60. E. Radu, M.S. Volkov, Phys. Rept. **468**, 101 (2008)
61. R. Battye, P. Sutcliffe, Nucl. Phys. B **590**, 329–363 (2000)
62. P. Bowcock, D. Foster, P. Sutcliffe, J. Phys. A **42**, 085403 (2009)
63. S. Yoshida, Y. Eriguchi, Phys. Rev. D **55**, 1994–2001 (1997)
64. C.A.R. Herdeiro, J. Kunz, I. Perapechka, E. Radu, Y. Shnir, Phys. Rev. D **103**(6), 065009 (2021)
65. C.A.R. Herdeiro, J. Kunz, I. Perapechka, E. Radu, Y. Shnir, Phys. Lett. B **812**, 136027 (2021)
66. R.M. Wald, *General Relativity* (University of Chicago Press, Chicago, 1984)
67. C.A.R. Herdeiro, E. Radu, H. Rúnarsson, Phys. Rev. D **92**(8), 084059 (2015)
68. N. Sanchis-Gual, C. Herdeiro, E. Radu, arXiv:2110.03000 [gr-qc]
69. W. Deppert, E.W. Mielke, Phys. Rev. D **20**, 1303–1312 (1979)
70. E.W. Mielke, R. Scherzer, Phys. Rev. D **24**, 2111 (1981)
71. E.J. Copeland, M.I. Tsumagari, Phys. Rev. D **80**, 02501 (2009)
72. B. Hartmann, J. Riedel, Phys. Rev. D **87**(4), 044003 (2013)
73. L. Campanelli, M. Ruggieri, Phys. Rev. D **77**, 043504 (2008)
74. T.D. Lee, Y. Pang, Nucl. Phys. B **315**, 477 (1989)
75. M. Gleiser, Phys. Rev. D **38**, 2376 (1988). Erratum: Phys. Rev. D **39**(4), 1257 (1989)
76. G.H. Derrick, J. Math. Phys. **5**, 1252 (1964)

77. Ó.J.C. Dias, J.E. Santos, B. Way, Class. Quant. Grav. **33**(13), 133001 (2016)
78. P. Grandclement, C. Somé, E. Gourgoulhon, Phys. Rev. D **90**(2), 024068 (2014)
79. R. Brito, V. Cardoso, P. Pani, Lect. Notes Phys. **906**, 1–237 (2015)
80. S. Hod, Phys. Rev. D **86**, 104026 (2012). Erratum: Phys. Rev. D **86**, 129902 (2012)
81. C.A.R. Herdeiro, E. Radu, Phys. Rev. Lett. **112**, 221101 (2014)
82. J.F.M. Delgado, C.A.R. Herdeiro, E. Radu, Phys. Lett. B **792**, 436–444 (2019)
83. Y. Brihaye, B. Hartmann, Phys. Rev. D **79**, 064013 (2009)
84. C. Palenzuela, I. Olabarrieta, L. Lehner, S.L. Liebling, Phys. Rev. D **75**, 064005 (2007)

11 Stellar and Substellar Objects in Modified Gravity

Aneta Wojnar

Abstract

The last findings on stellar and substellar objects in modified gravity are presented, allowing a reader to quickly jump into this topic. Early stellar evolution of low-mass stars, cooling models of brown dwarfs and giant gaseous exoplanets as well as internal structure of terrestrial planets are discussed. Moreover, possible test of models of gravity with the use of the discussed objects are proposed.

11.1 Basic Equations

There are modifications to the Einstein's gravity which turn out to survive, depending on the features of a given theory of gravity, in the non-relativistic limit derived from their fully relativistic equations. That is, some of those proposals modify Newtonian gravity, which is commonly used to describe stellar objects, such as the Sun and other stars of the Main Sequence. Those equations are also used to study the substellar family, starting with brown dwarf stars, giant gaseous planets, and even those more similar to the Earth. Therefore, there has appeared a need to explore non-relativistic objects not only for the consistency in describing different astrophysical bodies and gravitational phenomena with the use of the *same* theory

A. Wojnar (✉)
Institute of Physics, University of Tartu, Tartu, Estonia

Department of Theoretical Physics and IPARCOS, Faculty of Physical Sciences, Complutense University of Madrid, Madrid, Spain
e-mail: aneta.magdalena.wojnar@ut.ee; awojnar@ucm.es

C. Pfeifer, C. Lämmerzahl (eds.), *Modified and Quantum Gravity*, Lecture Notes in Physics 1017, https://doi.org/10.1007/978-3-031-31520-6_11

of gravity[1] but this fact is also an opportunity to understand the nature of the theory, since we better understand the density regimes of such objects. Moreover, since data sets of the discussed stars and exoplanets as well as the accuracy of the observations are still growing, the objects described by non-relativistic equations can be used to constrain some of the gravitational proposals, as presented in the further part of this chapter.

Before discussing the recent findings regarding the topic of non-relativistic objects in modified gravity, we will go through a suitable formalism needed to study low-mass stars and other objects living in the cold and dark edge of the Hertzsprung-Russell diagram (see the Fig. 11.1 and basic literature [1–4]).

As a working theory we will consider Palatini $f(\bar{R})$ gravity for the Starobinsky model

$$f(\bar{R}) = \bar{R} + \beta\bar{R}^2, \tag{11.1}$$

where β is the theory parameter,[2] but similar results as the ones presented here are expected to happen in any theory of gravity which alters Newtonian limit. To read more about Palatini gravity, see [5], because we will now focus directly on the modified hydrostatic equilibrium equation without its derivation [6–12]. Therefore, we will consider a toy-model of a star or planet, that is, a spherical-symmetric low-mass object without taking into account nonsphericity, magnetic fields, and time-dependency, described by the non-relativistic hydrostatic equilibrium equation with modifications given by the Palatini $f(\bar{R})$ gravity

$$p' = -g\rho(1 + \kappa c^2\beta[r\rho' - 3\rho])\,, \tag{11.2}$$

where prime denotes the derivative with respect to the radius coordinate r, $\kappa = -8\pi G/c^4$, G and c are Newtonian constant and speed of light, respectively. The quantity g is the surface gravity, approximated on the object's atmosphere as a constant value ($r_{atmosphere} \approx R$, where R is the radius of the object):

$$g \equiv \frac{Gm(r)}{r^2} \sim \frac{GM}{R^2} = \text{constant}, \tag{11.3}$$

where $M = m(R)$. We will consider only the usual definition for the mass function (however, see the discussion in [12, 13] on modified gravity issues)

$$m'(r) = 4\pi r^2\rho(r). \tag{11.4}$$

[1] However the "which one?" is a question which many physicists try to answer.

[2] In the further part, we will introduce the rescaled model parameter α to simplify some expression; see the discussion after (11.12).

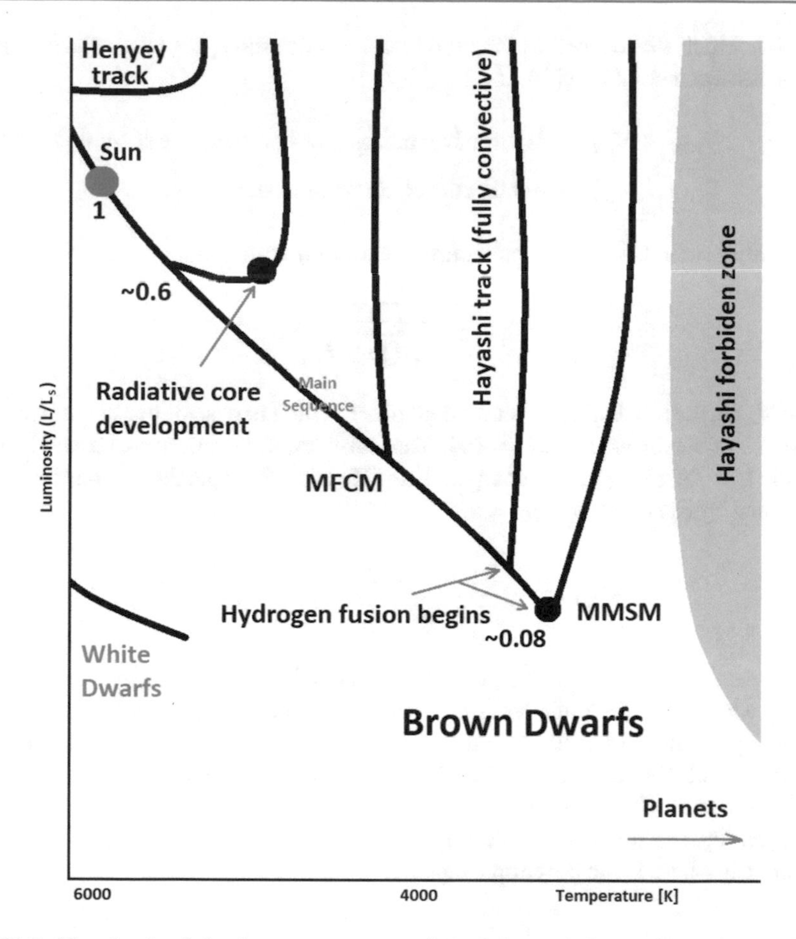

Fig. 11.1 The sketch of the low-temperature region of the evolutionary Hertzsprung-Russell diagram for astrophysical objects discussed in this chapter (the proportions of the evolution and scales are not preserved). A baby star is travelling along the Hayashi track till it reaches the Main Sequence, possibly burning lithium and deuterium. Depending on the star's mass, the object can reach the Main Sequence (MMSM—Minimum Main Sequence Mass indicated as stars with masses $\sim 0.08 M_\odot$ for hydrogen burning) as a fully convective star (MFCM—Maximal Fully Convective Mass marked), or it can develop a radiative core (it happens for stars with masses $\sim 0.6 M_\odot$) and then move along the Henyey track. The Hayashi forbidden zone as well as region occupied by brown dwarfs are also indicated. Giant gaseous planets can be found in the colder and dimmer region of the diagram

Using (11.4) and (11.3), the Eq. (11.2) can be written as

$$p' = -g\rho\left(1 + 8\beta\frac{g}{c^2 r}\right). \tag{11.5}$$

One of the most important elements in the star's or planet's modelling is the heat transport through object's interior and its atmosphere. A simple and common criterion which determines which kind of the energy transport takes place is given by the Schwarzschild one [14, 15]:

$$\nabla_{rad} \leq \nabla_{ad} \quad \text{pure diffusive radiative or conductive transport} \tag{11.6}$$

$$\nabla_{rad} > \nabla_{ad} \quad \text{adiabatic convection is present locally.} \tag{11.7}$$

The gradient stands for the temperature T variation with depth

$$\nabla_{rad} := \left(\frac{d\ln T}{d\ln p}\right)_{rad}, \tag{11.8}$$

while ∇_{ad} is the adiabatic temperature gradient, which in case of perfect, monatomic gas has a constant value $\nabla_{ad} = 0.4$. The Schwarzschild criterion turns out to be modified in Palatini gravity compared to GR [9], (the additional contribution is multiplied by the parameter β),

$$\nabla_{rad} = \frac{3\kappa_{rc} l p}{16\pi acGmT^4}\left(1 + 8\beta\frac{Gm}{c^2r^3}\right)^{-1}, \tag{11.9}$$

with l being the local luminosity, the constant $a = 7.57 \times 10^{-15}\frac{erg}{cm^3K^4}$ the radiation density while κ_{rc} is the radiative and/or conductive opacity. The additional $\beta-$term, depending on the sign of the parameter, has a stabilizing or destabilizing effect. On the other hand, the adiabatic gradient ∇_{ad} is a constant value for particular cases, as we will see in the further part.

Regarding the microscopic description of matter, an approximation which we will be using here is the polytropic equation of state (EoS):

$$p = K\rho^{1+\frac{1}{n}}, \tag{11.10}$$

It is good enough for our purposes, particularly taking into account the fact that K, since it depends on the composition of the fluid, carries information about the interactions between particles, the effects of electron degeneracy, and phase transitions,...[16]. We will use at least 3 different polytropic EoS, depending on the physical situation. On the other hand, the value of the polytropic index n is related to the class of the astrophysical objects we study [17]. The simplest case we will deal with is a fully convective objects with the interior modelled by non-relativistic degenerate electron gas for which $n = 3/2$ while K is given by [1]:

$$K = \frac{1}{20}\left(\frac{3}{\pi}\right)^{\frac{2}{3}}\frac{h^2}{m_e}\frac{1}{(\mu_e m_u)^{\frac{5}{3}}}. \tag{11.11}$$

It is always useful in the case of analytic EoS to write it in the polytropic form (11.10) since there exists a very convenient approach, called the Lane-Emden (LE) formalism, allowing to rewrite all relevant equations in the dimensionless form. It can be shown that for our particular model of gravity the Eq. (11.5) transforms into the modified Lane-Emden equation [6]

$$\frac{1}{\xi}\frac{d^2}{d\xi^2}\left[\sqrt{\Phi}\xi\left(\theta-\frac{2\alpha}{n+1}\theta^{n+1}\right)\right]=-\frac{(\Phi+\frac{1}{2}\xi\frac{d\Phi}{d\xi})^2}{\sqrt{\Phi}}\theta^n, \tag{11.12}$$

where $\Phi = 1 + 2\alpha\theta^n$ and the rescaled modifed gravity parameter (from now on multiplying the contribution beyond GR to star evolution) is

$$\alpha = \kappa c^2 \beta \rho_c.$$

The dimensionless θ and ξ are defined in the following way

$$r = r_c\xi, \quad \rho = \rho_c\theta^n, \quad p = p_c\theta^{n+1}, \quad r_c^2 = \frac{(n+1)p_c}{4\pi G\rho_c^2}, \tag{11.13}$$

with p_c and ρ_c being the core values of pressure and density, respectively.

? Exercise

11.1. Derive the Lane-Emden equation (11.12) with the use of the definitions (11.13))

The Eq. (11.12) can be solved numerically, and its solution θ provides star's mass, radius, central density, and temperature:

$$M = 4\pi r_c^3\rho_c\omega_n, \quad R = \gamma_n\left(\frac{K}{G}\right)^{\frac{n}{3-n}} M^{\frac{1-n}{n-3}}, \tag{11.14}$$

$$\rho_c = \delta_n\left(\frac{3M}{4\pi R^3}\right), \quad T = \frac{K\mu}{k_B}\rho_c^{\frac{1}{n}}\theta_n, \tag{11.15}$$

where k_B is Boltzmann's constant, μ the mean molecular weight while ξ_R is the dimensionless radius for which $\theta(\xi_R) = 0$. In the case of the model of gravity used here the constants (11.16) and (11.18) appearing in the above equations also include modifications [7] but it is not a common feature of modified gravity (see the case of Horndeski gravity, for instance [18], or in Eddington-inspired Born-Infeld gravity, [19]):

$$\omega_n = -\frac{\xi^2\Phi^{\frac{3}{2}}}{1+\frac{1}{2}\xi\frac{\Phi_\xi}{\Phi}}\frac{d\theta}{d\xi}\Big|_{\xi=\xi_R}, \tag{11.16}$$

$$\gamma_n = (4\pi)^{\frac{1}{n-3}}(n+1)^{\frac{n}{3-n}}\omega_n^{\frac{n-1}{3-n}}\xi_R, \tag{11.17}$$

$$\delta_n = -\frac{\xi_R}{3\frac{\Phi^{-\frac{1}{2}}}{1+\frac{1}{2}\xi\frac{\Phi_\xi}{\Phi}}\frac{d\theta}{d\xi}\Big|_{\xi=\xi_R}}. \tag{11.18}$$

? Exercise

11.2. Use the LE formalism to rewrite (11.5) and (11.9) as

$$p' = -g\rho\left(1 - \frac{4\alpha}{3\delta}\right), \quad \nabla_{rad} = \frac{3\kappa_{rc} l p}{16\pi acGmT^4}\left(1 - \frac{4\alpha}{3\delta}\right)^{-1}. \tag{11.19}$$

Notice that index n in the parameter δ (11.18) has been skipped.

Some of the objects we will consider in those notes are massive enough to burn light elements in their core; it can be either hydrogen, deuterium, or lithium. The product of any of those energy generation processes is luminosity, which can be obtained by the integration of the below expression:

$$\frac{dL_{burning}}{dr} = 4\pi r^2 \dot{\epsilon}\rho. \tag{11.20}$$

The energy generation rate $\dot{\epsilon}$ is a function of energy density, temperature, and stellar composition, however it can be approximated as a power-low function of the two first [20]. The energy produced in the core is radiated through the surface and can be expressed by the Stefan-Boltzmann law (L is luminosity)

$$L = 4\pi f\sigma T_{eff}^4 R^2, \tag{11.21}$$

where σ is the Stefan-Boltzmann constant. We have added the factor $f \leq 1$ which allows to include planets, which obviously radiate less than the black-body with the same effective temperature T_{eff}. This particular temperature (as well as other parts of atmosphere modelling) is usually difficult to determine and can carry significant uncertainties. Notwithstanding, there is a tool which we will often use when we look for some characteristics of the atmosphere. It is the optical depth τ, averaged over the object's atmosphere, see e.g. [1, 2]):

$$\tau(r) = \bar{\kappa}\int_r^\infty \rho dr, \tag{11.22}$$

where $\bar{\kappa}$ is a mean opacity. In the further part, since we will mainly work with the objects whose atmospheres have low temperatures, we will use Rosseland mean opacities which are given by the simple Kramers' law

$$\bar{\kappa} = \kappa_0 p^u T^w, \tag{11.23}$$

where κ_0, u and w are values depending on different opacity regimes [3, 21]. We will also assume that the atmosphere is made of particles satisfying the ideal gas relation (N_A is the Avogardo constant)

$$\rho = \frac{\mu p}{N_A k_B T}. \tag{11.24}$$

? Exercise

11.3. Use the polytropic EoS (11.10) to rewrite above as

$$p = \tilde{K} T^{1+n}, \quad \tilde{K} = \left(\frac{N_A k_B}{\mu} \right)^{1+n} K^{-n}. \tag{11.25}$$

Notice thate K can be shown to be a function of solutions of the modified Lane-Emden equation, an therefore it depends on the theory of gravity [9].

11.2 Pre-Main Sequence Phase

In the following section we will discuss some of the processes related to the early stellar evolution. Before reaching the Main Sequence, a baby star being on the so-called Hayashi track still contracts, decreasing its luminosity but not changing too much its surface temperature. Often the conditions present in the core are sufficient to burn light elements such as deuterium and lithium for instance, however in order to burn hydrogen, the temperature in the star's core must be much higher than in the lithium's case. Moreover, during its journey down along the Hayashi track the pre-Main Sequence star is fully convective apart from its radiative atmosphere. As already mentioned, because of the gravitational contractions the physical conditions in the core are changing and it may happen that the convective core will become radiative. In such a situation, the star will follow subsequently the Henyey track. This phase is much shorten than Hayashi one, it is followed by more massive stars (see the Fig. 11.1), and will not be discussed here. The radiative core development, hydrogen burning, and other processes related to the early stellar evolution not only depend on the star's mass but also on a theory of gravity, as we will see in the following subsections.

11.2.1 Hayashi Track

The photosphere is defined at the radius for which the optical depth (11.22) with mean opacity κ is equaled to 2/3. Using this relation in order to integrate the hydrostatic equilibrium Eq. (11.5) with $r = R$ and $M = m(R)$, and applying the absorption law (11.23) for a stellar atmosphere dominated by H^- in the temperature range $3000 < T < 6000\,\mathrm{K}$ with $\kappa_0 \approx 1.371 \times 10^{-33} Z\mu^{\frac{1}{2}}$ and $u = \frac{1}{2}$, $w = 8.5$, where $Z = 0.02$ is solar metallicity [2], one gets

$$p_{ph} = 8.12 \times 10^{14} \left(\frac{M\left(1 - \frac{4\alpha}{3\delta}\right)}{L T_{ph}^{4.5} Z \mu^{\frac{1}{2}}} \right)^{\frac{2}{3}}, \tag{11.26}$$

in which the Stefan-Boltzmann law $L = 4\pi\sigma R^2 T_{ph}^4$ with $T_{eff}|_{r=R} \equiv T_{ph}$ was already used. On the other hand, from (11.25) taken on the photosphere with $n = 3/2$ and applying the Stefan-Boltzmann law again, we have

$$T_{ph} = 9.196 \times 10^{-6} \left(\frac{L^{\frac{3}{2}} M p_{ph}^2 \mu^5}{-\theta' \xi_R^5} \right)^{\frac{1}{11}}. \tag{11.27}$$

The pressure appearing above is the pressure of the atmosphere; therefore, using (11.26) and rescaling mass and luminosity to the solar values $M_\odot$ and $L_\odot$, respectively, we can finally write

$$T_{ph} = 2487.77 \mu^{\frac{13}{51}} \left(\frac{L}{L_\odot} \right)^{\frac{1}{102}} \left(\frac{M}{M_\odot} \right)^{\frac{7}{51}} \left(\frac{\left(\frac{1 - \frac{4\alpha}{3\delta}}{Z} \right)^{\frac{4}{3}}}{\xi_R^5 \sqrt{-\theta'}} \right)^{\frac{1}{17}} \mathrm{K}. \tag{11.28}$$

? Exercise

11.4. Derive the above Hayashi track and see how it differs with respect to various values of the parameter α and the Lane-Emden functions.

The obtained formula relates the effective temperature and luminosity of the pre-main sequence star for a given mass M and mean molecular weight μ. That is, it provides an evolutionary track called Hayashi track [22]. Those tracks, being almost vertical lines on the right-hand side of the H-R diagram, are followed by the baby stars until they develop the radiative core, or they reach the Main Sequence. Immediately we observe that the effective temperature is nearly constant; but also notice that the temperature coefficient is too low—it is caused by our toy-model assumptions, mainly related to the atmosphere modelling. However, this simplified analysis allows us to agree that indeed modified gravity shifts the curves (see the figure 2 in [9]), leading to the possibility of constraining models of gravity by studying T Tauri stars [23] positioned nearby the Hayashi forbidden zone (see the Fig. 11.1).

11.2.2 Lithium Burning

In the fully convective stars (in such a case we may assume that the star is well-mixed) with mass M and hydrogen fraction X, the depletion rate is given by the expression

$$M\frac{df}{dt} = -\frac{Xf}{m_H}\int_0^M \rho\langle\sigma v\rangle dM, \tag{11.29}$$

where f is the lithium-to-hydrogen ratio. The non-resonant reaction rate for the temperature $T < 6 \times 10^6$ K is

$$N_A\langle\sigma v\rangle = Sf_{scr}T_{c6}^{-2/3}\exp\left[-aT_{c6}^{-\frac{1}{3}}\right]\frac{\text{cm}^3}{\text{sg}}, \tag{11.30}$$

where $T_{c6} \equiv T_c/10^6$ K and f_{scr} is the screening correction factor, while $S = 7.2 \times 10^{10}$ and $a = 84.72$ are dimensionless parameters in the fit to the reaction rate $^7\text{Li}(p,\alpha)\,^4\text{He}$ [24–26]. The Lane-Emden formalism for Palatini gravity provides the expressions for the central temperature T_c and central density ρ_c (11.15). However, instead of the simplest polytropic model (11.11), we need to take into account an arbitrary electron degeneracy degree Ψ and mean molecular weight μ_{eff}, and thus the radius is

$$\frac{R}{R_\odot} \approx \frac{7.1 \times 10^{-2}\gamma}{\mu_{eff}\mu_e^{\frac{2}{3}}F_{1/2}^{\frac{2}{3}}(\Psi)}\left(\frac{0.1M_\odot}{M}\right)^{\frac{1}{3}}, \tag{11.31}$$

where $F_n(\Psi)$ is the nth order Fermi-Dirac function. Inserting the quantities T_c, ρ_c, and R given by the Lane-Emden formalism, changing the variables to the spatial ones, and assuming that the burning process is restricted to the central region of the star (so then we can use the near center solution of LE) the depletion rate (11.29) can be written as [10]

$$\frac{d}{dt}\ln f = -6.54\left(\frac{X}{0.7}\right)\left(\frac{0.6}{\mu_{eff}}\right)^3\left(\frac{0.1M_\odot}{M}\right)^2 \times Sf_{scr}a^7u^{-\frac{17}{2}}e^{-u}\left(1+\frac{7}{u}\right)^{-\frac{3}{2}}\xi_R^2(-\theta'(\xi_R)), \tag{11.32}$$

where $u \equiv aT_6^{-1/3}$.

? Exercise

11.5. Find the central density and central density from the Lane-Emden formalism. Substitute them to Eq. (11.29) to get the above expression. For the near center solution in Palatini gravity, see the formula (11.43).

In order to proceed further, we need to find the dependence on time of the central temperature parameter u, which can be obtained from the Stefan-Boltzman equation together with the virial theorem

$$L = 4\pi R^2 T_{eff}^4 = -\frac{3}{7}\Omega\frac{GM^2}{R^2}\frac{dR}{dt}. \tag{11.33}$$

The factor Ω stands for modified gravity effects on the equation (in Palatini quadratic model $\Omega = 1$ for $n = 3/2$).

? Exercise

11.6. Find that the above relations provides the radius and luminosity as functions of time during the contraction phase

$$\frac{R}{R_\odot} = 0.85\Omega^{\frac{1}{3}}\left(\frac{M}{0.1M_\odot}\right)^{\frac{2}{3}}\left(\frac{3000\text{K}}{T_{eff}}\right)^{\frac{4}{3}}\left(\frac{\text{Myr}}{t}\right)^{\frac{1}{3}} \tag{11.34}$$

$$\frac{L}{L_\odot} = 5.25\times 10^{-2}\Omega\left(\frac{M}{0.1M_\odot}\right)^{\frac{4}{3}}\left(\frac{T_{eff}}{3000\text{K}}\right)^{\frac{4}{3}}\left(\frac{\text{Myr}}{t}\right)^{\frac{2}{3}}, \tag{11.35}$$

with the contraction time given as

$$t_{cont} \equiv -\frac{R}{dR/dt} \approx 841.91\left(\frac{3000\text{K}}{T_{eff}}\right)^4\left(\frac{0.1M_\odot}{M}\right) \times\left(\frac{0.6}{\mu_{eff}}\right)^3\left(\frac{T_c}{3\times 10^6\text{K}}\right)^3\frac{\xi_R^2(-\theta'(\xi_R))\Omega}{\delta^2}\text{Myr}. \tag{11.36}$$

Using Eqs. (11.34) and (11.31) it is possible to express the central temperature T_c with the time during the contraction epoch, which results as

$$\frac{u}{a} = 1.15\left(\frac{M}{0.1M_\odot}\right)^{2/9}\left(\frac{\mu_e F_{1/2}(\eta)}{t_6 T_{3eff}^4}\right)^{2/9}\times\left(\frac{\xi_R^5\Omega^{2/3}(-\theta'(\xi_R))^{2/3}}{\gamma\delta^{2/3}}\right)^{1/3}, \tag{11.37}$$

where $T_{3eff} \equiv T_{eff}/3000\,\text{K}$ and $t_6 \equiv t/10^6$.

Let us focus now on stars with masses $M < 0.2M_\odot$ such that the degeneracy effects are insignificant and $\dot{\mu}_{eff}$ can be neglected when compared to $\dot{R}$. Then, we can write the depletion rate as

$$\frac{\mathrm{d}\ln f}{\mathrm{d}u} = 1.15 \times 10^{13}\, T_{3eff}^{-4} \left(\frac{X}{0.7}\right)\left(\frac{0.6}{\mu_{eff}}\right)^6 \left(\frac{M_\odot}{M}\right)^3 \times\; S f_{scr} a^{16} u^{-\frac{37}{2}} e^{-u}\left(1-\frac{21}{2u}\right)\frac{\xi_R^4(-\theta'(\xi_R))^2\Omega}{\delta^2}. \tag{11.38}$$

The above equation can be integrated from $u_0 = \infty$ to u ($\mathcal{F} \equiv \ln\frac{f_0}{f}$):

$$\mathcal{F} = 1.15 \times 10^{13} \frac{X}{0.7}\left(\frac{0.6}{\mu_{eff}}\right)^6 \left(\frac{M_\odot}{M}\right)^3 \frac{S f_{scr} a^{16} g(u)}{T_{3eff}} \frac{\xi_R^4(-\theta'(\xi_R))^2\Omega}{\delta^2}, \tag{11.39}$$

where $g(u) = u^{-37/2}e^{-u} - 29\Gamma(-37/2, u)$ while $\Gamma(-37/2, u)$ is an upper incomplete gamma function. Notice that this expression depends on the functions ξ_R, θ', δ, and the central temperature, given by the Eq. (11.37), which are derived from the solution of the modified Lane-Emden equation (11.12). Therefore, they differ with respect to their GR values making that the ^{7}Li abundance depends on the gravity model.

One obtains the central temperature T_c from $u(\mathcal{F})$ for a given depletion $\mathcal{F}$. The star's age, radius, and luminosity are given by the Eqs. (11.36), (11.34), and (11.35). Let us emphasize that all these values depend on the model of gravity, clearly altering the pre-Main Sequence stage of the stellar evolution. Moreover, age determination techniques which are based on lithium abundance measurements are not model-independent: they do depend on a model of gravity used, as presented above (see details in [10]).

11.2.3 Approaching the Main Sequence: Hydrogen Burning

The process of becoming a true star is related to the stable hydrogen burning. It means that the energy produced in this reaction is radiated away through the star's atmosphere, and that the pressure appearing there because of the energy transport balances the gravitational contraction. When a star contracts, the central temperature increases and when it reaches the values $\sim 3 \times 10^6$ K in the core, the thermonuclear ignition of hydrogen starts. There are three reactions responsible for this process: $p + p \rightarrow d + e^+ + \nu_e$, $p + e^- + p \rightarrow d + \nu_e$, $p + d \rightarrow {}^3H_e + \gamma$, where the first one is slow and a bottle-neck for the lower-mass objects; that is, it stands behind the Minimum Main Sequence Mass (MMSM) term. It was demonstrated that the energy generation rate per unit mass for the hydrogen ignition process can be well described by the power law form [20, 27]

$$\dot{\epsilon}_{pp} = \dot{\epsilon}_c \left(\frac{T}{T_c}\right)^s \left(\frac{\rho}{\rho_c}\right)^{u-1}, \quad \dot{\epsilon}_c = \epsilon_0 T_c^s \rho_c^{u-1}, \tag{11.40}$$

where the two exponents can be approximated as $s \approx 6.31$ and $u \approx 2.28$, while $\dot{\epsilon}_0 \approx 3.4 \times 10^{-9}$ ergs g^{-1}s^{-1}. For a baby star with the hydrogen fraction $X = 0.75$ the number of baryons per electron in low-mass stars is $\mu_e \approx 1.143$.

Using the energy generation rate (11.40) and luminosity (11.20) formulae, we can integrate the latter over the stellar volume ($M_{-1} = M/(0.1M_\odot)$):

$$\frac{L_{HB}}{L_\odot} = 4\pi r_c^3 \rho_c \dot{\epsilon}_c \int_0^{\xi_R} \xi^2 \theta^{n(u+\frac{2}{3}s)} d\xi = \frac{1.53 \times 10^7 \Psi^{10.15}}{(\Psi + \alpha_d)^{16.46}} \frac{\delta_{3/2}^{5.487} M_{-1}^{11.977}}{\omega_{3/2} \gamma_{3/2}^{16.46}}, \tag{11.41}$$

where we used the Lane-Emden formalism with

$$K = \frac{(3\pi^2)^{2/3}\hbar}{5 m_e m_H^{5/3} \mu_e^{5/3}} \left(1 + \frac{\alpha_d}{\Psi}\right), \tag{11.42}$$

and the near center solution of the LE equation which is $\theta(\xi \approx 0) = 1 - \frac{\xi^2}{6} \sim \exp\left(-\frac{\xi^2}{6}\right)$ for Palatini $f(R)$ gravity. Here, $\alpha_d \equiv 5\mu_e/2\mu \approx 4.82$.

? Exercise

11.7. Show that for the considered model of gravity, the near center solution of the Lane-Emden equation is

$$\theta(\xi \approx 0) = 1 - \frac{\xi^2}{6} \sim \exp\left(-\frac{\xi^2}{6}\right) \tag{11.43}$$

Now we will focus on finding the photospheric luminosity which must be equaled to (11.41) in order to have a star as a stable system. Therefore, the surface gravity (11.3) needs to be rewritten wrt the Lane-Emden variables:

$$g = \frac{3.15 \times 10^6}{\gamma_{3/2}^2} M_{-1}^{5/3} \left(1 + \frac{\alpha_d}{\Psi}\right)^{-2} \text{cm/s}^2. \tag{11.44}$$

The most tricky part is to find the photospheric temperature. Usually it is obtained from matching the specific entropy of the gas and metallic phases of the H-He mixture [27] (here without the phase transition points [16])

$$T_{ph} = 1.8 \times 10^6 \frac{\rho_{ph}^{0.42}}{\Psi^{1.545}} \mathrm{K}\,. \tag{11.45}$$

Applying these two results into (11.5) and (11.24) one writes the photospheric energy density as ($\kappa_{-2} = \kappa_R/(10^{-2}\,\mathrm{cm^2 g^{-1}})$, κ_R is Rosseland's mean opacity):

$$\frac{\rho_{ph}}{\mathrm{g/cm^3}} = 5.28 \times 10^{-5} M_{-1}^{1.17} \left(\frac{1+8\beta\frac{g}{c^2 R}}{\kappa_{-2}}\right)^{0.7} \frac{\Psi^{1.09}}{\gamma_{3/2}^{1.41}} \left(1+\frac{\alpha_d}{\Psi}\right)^{-1.41}. \tag{11.46}$$

Notice that here we are using again the Starobisky parameter β, which introduces the modification term to the GR equation (that is, for $\beta = 0$, we recover the GR case). Inserting it into T_{ph} and using the stellar luminosity (11.21) we find

$$L_{ph} = 28.18 L_\odot \frac{M_{-1}^{1.305}}{\gamma_{3/2}^{2.366}\Psi^{4.351}} \times \left(\frac{1+8\beta\frac{g}{c^2 R}}{\kappa_{-2}}\right)^{1.183} \left(1+\frac{\alpha_d}{\Psi}\right)^{-0.366}. \tag{11.47}$$

Finally, writing $L_{HB} = L_{ph}$ and performing non-complicated algebra:

$$M_{-1}^{MMSM} = 0.290 \frac{\gamma_{3/2}^{1.32}\omega_{3/2}^{0.09}}{\delta_{3/2}^{0.51}} \frac{(\alpha_d+\Psi)^{1.509}}{\Psi^{1.325}} \left(1 - 1.31\alpha \frac{\left(\frac{\alpha_d+\Psi}{\Psi}\right)^4}{\delta_{3/2}\kappa_{-2}}\right)^{0.111} \tag{11.48}$$

we have derived the MMSM.[3] It is clearly modified by our model of gravity not only by the parameter α, but also by the solutions of the LE Eq. (11.12), manifested by γ, ω, and δ.

11.3 Low-Mass Main Sequence Stars

In every stellar modelling one needs to determine which kind of the energy transport mechanism is present in each particular layer of the given star. It is usually given by the Schwarzschild criterion (11.9) which is also altered by the model of gravity [9]. Using that result we will demonstrate that the mass limit of fully convective stars on the Main Sequence is shifted and can have a significant effect on how we model the

[3] We have used the relation between the Starobinsky parameter β and rescaled parameter α again.

stars from this mass range. Newtonian-based models predict that Main Sequence stars' interiors with masses smaller than $\sim 0.6 M_\odot$ are fully convective.

Since the star's luminosity decreases when it contracts following the Hayashi track, it may happen that there appears a radiative zone in the star's interior, and then the star will start following the Henyey track [28–30]. In the case of the low-mass stars, however, the fully convective baby star may also reach the Main Sequence without developing a radiative core. In order to deal with such a situation, the decreasing luminosity in the Schwarschild condition for the radiative core development condition (it happens when $\nabla_{rad} = \nabla_{ad}$, where in our simplified model $\nabla_{ad} = 0.4$) cannot be lower than the luminosity of H burning (11.41). Therefore, the modified Schwarzschild criterion, after inserting (11.21) and (11.15) with homology contaction argument, provides the minimum luminosity for the radiative core development:

$$L_{min} = 9.89 \times 10^7 L_\odot \frac{\delta_{3/2}^{1.064}(\frac{3}{4}\delta_{3/2} - \alpha)}{\xi^{8.67}(-\theta')^{1.73}} \left(\frac{T_{eff}}{\kappa_0}\right)^{0.8} M_{-1}^{4.4}, \tag{11.49}$$

where we have used the Kramer's absorption law (11.23) with $u = 1$ and $w = -4.5$. Thus, a star on the onset of the radiative core development will reach the Main Sequence when $L_{min} = L_{HB}$; so the mass of the maximal fully convective star on the Main Sequence is given by the following expression:

$$M_{-1} = 1.7 \frac{\mu^{0.9} T_{eff}^{0.11} (\alpha_d + \Psi)^{2.173}}{\Psi^{1.34} \kappa_0^{0.11}} \frac{\gamma^{2.173} \omega^{0.132}}{\delta_{3/2}^{0.58} \xi^{1.14} (-\theta')^{0.23}}. \tag{11.50}$$

Let us firstly focus on the GR case, that is, when $\alpha = 0$. Considering a star with $\alpha_d = 4.82$, the degree of the degeneracy electron pressure as $\Psi = 9.4$, and the mean molecular weight $\mu = 0.618$ with $T_{eff} = 4000\,\text{K}$, the maximal mass of the fully convective star on the Main Sequence is:

$$M = 4.86 M_\odot \kappa_0^{-0.11}. \tag{11.51}$$

We notice immediately that the final value does depend on the opacity. Considering two Kramers' opacities: the total bound-free and free-free estimated to be (in cm^2g^{-1}), [2]

$$\kappa_0^{bf} \approx 4 \times 10^{25} \mu \frac{Z(1+X)}{N_A k_B}, \quad \kappa_0^{ff} \approx 4 \times 10^{22} \mu \frac{(X+Y)(1+X)}{N_A k_B}, \tag{11.52}$$

the corresponding masses, for $X = 0.75$ and $Z = 0.02$, are

$$M_{bf} = 0.099 M_\odot, \quad M_{ff} = 0.135 M_\odot, \tag{11.53}$$

respectively. The obtained masses, as we expected, are too low—it is a result of our simplified analysis, mainly related to the atmosphere's description and gas behaviour in the considered pressure and temperature regimes. However, we may use the obtained values as reference to compare the result arriving from modified gravity: depending on the parameter's value, the masses can even differ around 50% [9].

11.4 Aborted Stars: Brown Dwarfs

Let us discuss a family of objects which do not satisfy necessary conditions in their core to ignite hydrogen[4] and subsequently to enter the Main Sequence phase. Such an object will radiate away all stored energy, being a result of gravitational contraction and eventual light elements burning in the early stage's evolution. It will stop contracting when the electron degeneracy pressure balances the gravitational pulling, and consequently it will be cooling down with time. In order to study a simple but accurate cooling model of brown dwarfs, we need to consider a more realistic description of matter, as the brown dwarf stars are composed of the mixture of degenerate and ideal gas states at finite temperature. It turns out however that such an EoS can be rewritten in the polytropic form for $n = 3/2$ [16], but with more complicated polytropic function with $K = C\mu_e^{-\frac{5}{3}}(1 + b + a\eta)$, where the constant $C = 10^{13}\,\mathrm{cm}^4 g^{-2/3} s^{-2}$, $a = \frac{5}{2}\mu_e\mu_1^{-1}$, while the number of baryons per electron is represented by μ_e. Here, we use $\eta = \Psi^{-1}$ as the electron degeneracy parameter, while μ_1 takes into account ionization, and it is defined as

$$\frac{1}{\mu_1} = (1 + x_{H^+})X + \frac{Y}{4}, \tag{11.54}$$

where x_{H^+} is the ionization fraction of hydrogen X (Y stand for helium one) and depends on the phase transitions points [31]. Besides, the quantity b is

$$b = -\frac{5}{16}\eta\ln(1 + e^{-1/\eta}) + \frac{15}{8}\eta^2\left(\frac{\pi^2}{3} + \mathrm{Li}_2[-e^{-1/\eta}]\right), \tag{11.55}$$

where Li_2 denotes the second order polylogarithm function and the degeneracy parameter is given as $\eta = \frac{k_B T}{\mu_F}$. Therefore, we can still use the LE formalism for our purposes, that is, we can express the star's central pressure, radius, central density, and temperature $T_c = \frac{K\mu}{k_B}\rho_c^{\frac{1}{n}}$ as functions of the above parameters; we will see soon that the degeneracy parameter depends on time because of the still ongoing gravitational contraction.

[4] some massive brown dwarfs do burn hydrogen, however the process is not stable ($L_{HB} \neq L_{ph}$) and since although there is some energy production, the object radiates more than produces, therefore it is cooling down and following the BDs' evolution.

As already commented, the most uncertain part of our calculations is related to the photospheric values of, for instance, effective temperature. In brown dwarfs' case one usually uses the entropy method, that is, matching the entropy of non-ionized molecular mixture of H and He at the atmosphere to the interior one, composed mainly of degenerate electron gas [16, 27]:

$$S_{interior} = \frac{3}{2}\frac{k_B N_A}{\mu_{1mod}}(\ln\eta + 12.7065) + C_1, \tag{11.56}$$

where C_1 is an integration constant of the first law of thermodynamics and μ_{1mod} is modified μ_1 at the photosphere (see the details and its form in [11, 16]. The matching provides the effective temperature as

$$T_{eff} = b_1 \times 10^6 \rho_{ph}^{0.4} \eta^{\nu} \text{ K}, \tag{11.57}$$

where the parameters b_1 and ν depend on the specific model describing the phase transition between a metallic H and He state in the BD's interior and the photosphere composed of molecular ones [31]. Following the analogous steps as in the Sect. (11.2.3), one gets the photospheric temperature as

$$T_{eff} = \frac{2.558\times 10^4\text{ K}}{\kappa_R^{0.286}\gamma^{0.572}}\left(\frac{M}{M_\odot}\right)^{0.4764}\frac{\eta^{0.714\nu}b_1^{0.714}}{(1+b+a\eta)^{0.571}}\left(1-1.33\frac{\alpha}{\delta}\right)^{0.286}, \tag{11.58}$$

where $\mu_e = 1.143$ was used. That allows to find the luminosity of the brown dwarf; hence using the Stefan-Boltzman equation one gets:

$$L = \frac{0.0721 L_\odot}{\kappa_R^{1.1424}\gamma^{0.286}}\left(\frac{M}{M_\odot}\right)^{1.239}\frac{\eta^{2.856\nu}b_1^{2.856}}{(1+b+a\eta)^{0.2848}}\left(1-1.33\frac{\alpha}{\delta}\right)^{1.143}. \tag{11.59}$$

The above luminosity depends on time since the electron degeneracy η does. To find such a relation for the latter one [27, 32], let us consider the pace of cooling and contraction given by the first and the second law of thermodynamics

$$\frac{\text{dE}}{\text{dt}} + p\frac{\text{dV}}{\text{dt}} = T\frac{\text{dS}}{\text{dt}} = \dot{\epsilon} - \frac{\partial L}{\partial M}, \tag{11.60}$$

in which the energy generation term $\dot{\epsilon}$ is negligible in brown dwarfs. We can integrate the above equation over mass to find

$$\frac{\text{d}\sigma}{\text{dt}}\left[\int N_A k_B T \text{dM}\right] = -L, \tag{11.61}$$

where L is a surface luminosity and we have defined $\sigma = S/k_B N_A$. The LE polytropic relations allow to get rid of T and ρ and write down

$$\frac{\mathrm{d}\sigma}{\mathrm{dt}}\frac{N_A A\mu_e\eta}{C(1+b+a\eta)}\int p\mathrm{dV} = -L, \tag{11.62}$$

where $A = (3\pi\hbar^3 N_A)^{\frac{2}{3}}/(2m_e) \approx 4.166\times 10^{-11}$. The integral in the above equation can be simply found to be $\int p\mathrm{dV} = \frac{2}{7}\Omega G\frac{M^2}{R}$ with $\Omega = 1$ for $n = 3/2$ in Palatini gravity [7, 10].

? Exercise

11.8. Show that using the entropy formula (11.56) one can easily get the entropy rate as (let us recall that $\sigma = S/k_B N_A$):

$$\frac{\mathrm{d}\sigma}{\mathrm{dt}} = \frac{1.5}{\mu_{1mod}}\frac{1}{\eta}\frac{\mathrm{d}\eta}{\mathrm{dt}}. \tag{11.63}$$

Inserting the above expression into (11.62) together with the luminosity (11.59) gives us the evolutionary equation for the degeneracy parameter η

$$\begin{aligned}\frac{\mathrm{d}\eta}{\mathrm{dt}} = &-\frac{1.1634\times 10^{-18} b_1^{2.856}\mu_{1mod}}{\kappa_R^{1.1424}\mu_e^{8/3}}\left(\frac{M_\odot}{M}\right)^{1.094}\\ &\times\eta^{2.856\upsilon}(1+b+a\eta)^{1.715}\frac{\gamma^{0.7143}}{\Omega}\left(1-1.33\frac{\alpha}{\delta}\right)^{1.143}.\end{aligned} \tag{11.64}$$

This equation, together with the luminosity Eq. (11.59) and initial conditions $\eta = 1$ at $t = 0$, provides the cooling process model for a brown dwarf star in Palatini $f(\bar{R})$ gravity. To see how modified gravity affects such an evolution after solving these equations numerically,[5] see [11].

11.5 (Exo)-Planets

As we will see, some theories of gravity can change the giant planets' evolution, and may also affect the internal structure of gaseous and terrestrial ones. This fact can change our understanding of the Solar System's formation, as well as it can be used to constrain different gravitational proposals when observational and experimental data with high accuracy are at our disposal. Missions such as ESA's Cosmic Visions[6]

[5] https://github.com/mariabenitocst/brown_dwarfs_palatini

[6] https://www.esa.int/Science_Exploration/Space_Science/Voyage_2050_sets_sail_ESA_chooses_future_science_mission_themes

will bring soon more data on the physical properties of Jupiter-like planets, while improved seismic experiments [33], as well as those performed in laboratories [34], or with the use of the new generation of the neutrinos' telescopes [35] will provide more information about the matter behaviour in the Earth's core and its more exact composition.

11.5.1 Jovian Planets

Giant gaseous planets, although their formation processes differs significantly from the one followed by stars and brown dwarfs [3, 4], do also contract and cool down until it reaches the thermal equilibrium, that is, when the received energy from its parent star is equalled to the energy radiated away from the surface of the planet. Their inner description is quite similar to the one of brown dwarfs'; however, the main difference in the cooling process between these two substellar object is that the jovian planets possess an additional source of energy provided by the parent star which cannot be ignored. When a planet with the radius R_p and in the distance R_{sp} from its parent star is in the mentioned thermal equilibrium, it means that its equilibrium temperature

$$(1 - A_p)\left(\frac{R_p}{2R_{sp}}\right)^2 L_s = 4\pi f \sigma T_{eq}^4 R_p^2, \tag{11.65}$$

where A_p is an albedo of the planet while L_s the star's luminosity, is equalled to its effective one. However, when we are dealing with some additional energy sources such as for instance gravitational contraction, Ohmic heating, or tidal forces, it is not so since the planets radiates more than it receives. Therefore, we need a relation between these two temperatures; it is derived from the radiative transport equation with the use of Eddington's approximation [2]:

$$4T^4 = 3\tau(T_{eff}^4 - T_{eq}^4) + 2(T_{eff}^4 + T_{eq}^4), \tag{11.66}$$

where T is the stratification temperature in the atmosphere while τ is the optical depth. This will allow, when we integrate the Eq. (11.19) with (11.23), to write down the atmospheric pressure as (see [36] for $w = 4$):

$$p_{w\neq 4}^{u+1} = \frac{4^{\frac{w}{4}} g}{3\kappa_0} \frac{u+1}{1-\frac{w}{4}} \left(1 - \frac{4\alpha}{3\delta}\right) T_-^{-1}\left((3\tau T_- + 2T_+)^{1-\frac{w}{4}} - (2T_+)^{1-\frac{w}{4}}\right), \tag{11.67}$$

where we have defined $T_- := T_{eff}^4 - T_{eq}^4$ and $T_+ := T_{eff}^4 + T_{eq}^4$. The atmosphere is radiative so there must exist a region in which the convective transport of energy in the planet's interior becomes radiative. In order to find this boundary, we will use the Schwarzschild criterion (11.9) to find the critical depth in which the radiative process is replaced with the convective one:

$$\tau_c = \frac{2}{3}\frac{T_+}{T_-}\left(\left(1+\frac{8}{5}\left(\frac{\frac{w}{4}-1}{u+1}\right)\right)^{\frac{1}{\frac{w}{4}-1}}-1\right), \quad w\neq 4 \tag{11.68}$$

Substituting those expressions into (11.67) and (11.66) we may write the formulas for the boundary pressure and temperature

$$p_{conv}^{u+1} = \frac{8g}{15\kappa_0}\frac{4^{\frac{w}{4}}\left(1-\frac{4\alpha}{3\delta}\right)}{T_-(2T_+)^{w-1}}\left(\frac{5(u+1)}{5u+8\frac{w}{4}-3}\right), \tag{11.69}$$

$$T_{conv}^4 = \frac{T_+}{2}\left(\frac{5u+8\frac{w}{4}-3}{5(u+1)}\right)^{\frac{w}{4}-1}, \quad w\neq 4. \tag{11.70}$$

On the other hand, to describe the planet's convective interior, let us consider a combination of pressures [37]

$$p = p_1 + p_2, \tag{11.71}$$

where p_1 is pressure arising from electron degeneracy, given by the polytropic EoS (11.10) with $n = 3/2$, while p_2 is pressure of ideal gas (11.24).

? Exercise

11.9. Show that such a mixture can be again written as a polytrope [32].

Matching the above interior pressure with (11.69) provides a relation between the effective temperature T_{eff} with the radius of the planet R_p which depends on modified gravity:

$$\begin{aligned} T_+^{\frac{5}{8}u+\frac{w}{4}-\frac{3}{8}}T_- &= CG^{-u}M_p^{\frac{1}{3}(2-u)}R_p^{-(u+3)}\mu^{\frac{5}{2}(u+1)}k_B^{-\frac{5}{2}(u+1)} \\ &\times \gamma^{u+1}(G\gamma^{-1}M_p^{\frac{1}{3}}R_p-K)^{\frac{5}{2}(u+1)}\left(1-\frac{4\alpha}{3\delta}\right) \end{aligned} \tag{11.72}$$

where C is a constant depending on the opacity constants u and w:

$$C_{w\neq 4} = \frac{16}{15\kappa_0}2^{\frac{5}{8}(1+u)+\frac{w}{4}}\left(\frac{5u+8\frac{w}{4}-3}{5(u+1)}\right)^{1+\frac{5}{8}(1+u)(\frac{w}{4}-1)}. \tag{11.73}$$

Since the contraction of the planet is a quasi-equilibrium process, the planet's luminosity is a sum of the total energy absorbed by the planet and the internal energy such that for a polytrope with $n = 3/2$ [11] we may write

$$L_p = (1 - A_p) \left(\frac{R_p}{2R_{sp}} \right)^2 L_s - \frac{3}{7} \frac{GM_p^2}{R_p^2} \frac{dR_p}{dt}. \tag{11.74}$$

Using (11.21), (11.23) and integrating it from an initial radius R_0 to the final one R_F, and inserting (11.72) to get rid of T_- we can derive the cooling equation for jovian planets:

$$t = -\frac{3}{7} \frac{GM_p^{\frac{4}{3}} k_B^{\frac{5}{2}(u+1)} \kappa_0}{\pi a c \gamma \mu^{\frac{5}{2}(u+1)} K^{\frac{3}{2}u+\frac{5}{2}} C} \left(1 - \frac{4\alpha}{3\delta}\right)^{-1} \int_{x_0}^{x_p} \frac{(T_{eff}^4 + T_{eq}^4)^{\frac{5}{8}u+\frac{w}{4}-\frac{3}{8}} dx}{x^{1-u}(x-1)^{\frac{5}{2}(u+1)}}.$$

This, together with (11.72) providing the effective temperature for a given radius allows to find the age of the planet which clearly differ from the values given by Newtonian physics (see the figure 2 and tables 1–2 in [36]).

11.5.2 Terrestrial Planets

In this section we will just comment some findings regarding the rocky planets, such as for example the Earth and Mars. Although the numerical analysis demonstrates that we should not expect a large degeneracy in the mass-radius plots for the Earth-sized and smaller planets[7] [13]—however have a look on a more realistic approach in [39,40]—it turns out that there is a considerable difference in the density profiles $\rho(r)$, which could be used to constrain and test models of gravity. Knowing what is the density profile in a given planet allows to obtain the polar moment of inertia C (R_p is the planet's radius)

$$C = \frac{8\pi}{3} \int_0^{R_p} \rho(r) r^4 dr. \tag{11.75}$$

The density profiles provide information on the number of layers composed of different materials (that is, EoS), and their boundaries. The inner structure of the Earth is given by the PREM model [41–44] being a result of the seismic data analysis, while the martian interior will be known soon, when the Seismic Experiment for Interior Structure from NASA's MARS InSight Mission's seismometer[8] provides the required data.

[7] in the case of larger terrestrial planets we observe a significant difference, making the exoplanet's composition more difficult to determine [13,38].

[8] https://mars.nasa.gov/insight/spacecraft/instruments/seis/

Since density profiles (central and boundary values of density/pressure, and layers' thickness) are slightly different in modified gravity than those obtained from Newtonian gravity, it means that this fact has an influence on the polar moment of inertia (11.75), yielding different results for different models of gravity. Such a phenomenon can be compare with the observational value C provided by precession rate $d\eta/dt$ being caused by gravitational torques from the Sun [45]:

$$\frac{d\eta}{dt} = -\frac{3}{2}J_2 \cos\epsilon(1-e^2)\frac{n^2}{\omega}\frac{MR^2}{C} \tag{11.76}$$

where the orbital eccentricity e, obliquity ϵ, the rotation rate ω, the effective mean motion n and the gravitational harmonic coefficient J_2 are well-known with high accuracy for the Solar System planets, especially for the Earth [46] and Mars [47–49]. Therefore, the computed polar moment of inertia from a given model of gravity must agree with the observational one provided by (11.76). That procedure, when the theoretical modelling improved, can be a powerful tool to test theories of gravity which alters Newtonian equations.

11.6 Summary

Many theories of gravity happen to introduce additional terms to the hydrostatic equilibrium equation. This fact however, as we could see in this brief review of the current research, has a non-trivial effect on many processes occurring in the stellar and substellar interiors. Therefore, modelling evolutionary phases and compositions of those objects must be undertaken carefully, rather expecting changes on each step than assuming that because of the weaker gravitational field some processes will undergo in the same way as in GR or Newtonian case. Although most of the results presented here concern toy-model approach, it provided a necessary insight into a kind of modifications one has to take into account during modelling more realistic objects.

Acknowledgments This work was supported by the EU through the European Regional Development Fund CoE program TK133 "The Dark Side of the Universe".

References

1. R. Kippenhahn, A. Weigert, A. Weiss, *Stellar Structure and Evolution*, vol. 192, 2nd edn. (Springer-Verlag, Berlin, 1990)
2. C.J. Hansen, S.D. Kawaler, V. Trimble, *Stellar Interiors: Physical Principles, Structure, and Evolution* (Springer Science & Business Media, New York, 2012)
3. P.J. Armitage, *Astrophysics of Planet Formation* (Cambridge University Press, Cambridge, 2010)
4. P. Irwin, *Giant Planets of Our Solar System: Atmospheres, Composition, and Structure* (Springer Science & Business Media, New York, 2009)

5. A. De Felice, S. Tsujikawa, Living Rev. Rel. **13**, 3 (2010)
6. A. Wojnar, Eur. Phys. J. C **79**(1), 51 (2019)
7. A. Sergyeyev, A. Wojnar, Eur. Phys. J. C **80**(4), 313 (2020)
8. G. Olmo, D. Rubiera-García, A. Wojnar, Phys. Rev. D **100**(4), 044020 (2019)
9. A. Wojnar, Phys. Rev. D **102**(12), 124045 (2020)
10. A. Wojnar, Phys. Rev. D **103**(4), 044037 (2021)
11. M. Benito, A. Wojnar, Phys. Rev. D **103**(6), 064032 (2021)
12. A. Kozak, A. Wojnar, Phys. Rev. D **104**(8), 084097 (2021)
13. A. Kozak, A. Wojnar, Eur. Phys. J. C **81**(6), 492 (2021)
14. K. Schwarzschild, Nachrichten Göttingen. Math.-phys. Klasse **195**, 41–53 (1906)
15. M. Schwarzschild, *Structure and Evolution of Stars* (Princeton University Press, Princeton, 2015)
16. S. Auddy, S. Basu, S.R. Valluri, Adv. Astron. **2016**, 574327 (2016)
17. G. Horedt, *Polytropes: Applications in Astrophysics and Related Fields*, vol. 306 (Springer Science & Business Media, New York, 2004)
18. K. Koyama, J. Sakstein, Phys. Rev. D **91**(12), 124066 (2015)
19. M. Guerrero, D. Rubiera-García, A. Wojnar, Eur. Phys. J. C **82**(8), 707 (2022)
20. W.A. Fowler, G.R. Caughlan, B.A. Zimmerman, Annu. Rev. Astron. Astrophys. **13**, 69 (1975)
21. T.W.A. Müller, W. Kley, Astron. Astrophys. **539**, A18 (2012)
22. C. Hayashi, Publ. Astron. Soc. Jpn. **13**, 450 (1961)
23. C. Bertout, Ann. Rev. Astron. Astrophys. **27**, 351 (1989)
24. G. Ushomirsky, et al., Astrophys. J. **497**(1), 253 (1998)
25. G.R. Caughlan, W.A. Fowler, At. Data Nucl. Data Tables **40**, 283 (1998)
26. G. Raimann, Z. Phys. A Hadrons Nucl. **347**(1), 73–74 (1993)
27. A. Burrows, J. Liebert, Rev. Mod. Phys. **65**, 301 (1993)
28. L. Henyey, R. Lelevier, R.D. Levee, Publ. Astron. Soc. Pac. **67**, 154, 396 (1955)
29. L. Henyey, J.E. Forbes, N.L. Gould, Astrophys. J. **139**, 306 (1964)
30. L. Henyey, M.S. Vardya, P. Bodenheimer, Astrophys. J. **142**, 841 (1965)
31. G. Chabrier, et al., Astrophys. J. **391**, 817–826 (1992)
32. D.J. Stevenson, Ann. Rev. Astron. Astrophys. **29**(1), 163–193 (1991)
33. R. Butler, S. Tsuboi, Phys. Earth Planet. Inter. **321**, 106802 (2021)
34. S. Merkel, et al., Phys. Rev. Lett. **127**(20), 205501 (2021)
35. A. Donini, S. Palomares-Ruiz, J. Salvado, Nature Phys. **15**(1), 37–40 (2019)
36. A. Wojnar, Phys. Rev. D **104**(10), 104058 (2021)
37. J.R. Donnison, I.P. Williams, Astrophys. Space Sci. **29**(2), 387–396 (1974)
38. S. Seager et al., Astrophys. J. **669**, 1279 (2007)
39. A. Kozak, A. Wojnar, Int. J. Geom. Meth. Mod. Phys. **19**(Suppl. 01), 2250157 (2022)
40. A. Kozak, A. Wojnar, Universe **8**(1), 3 (2021)
41. A.M. Dziewonski, D.L. Anderson, Phys. Earth Planet. Inter. **25**(4), 297–356 (1981)
42. B. Kustowski, et al., J. Geophys. Res. Solid Earth **113**(B6) (2008)
43. B.L.N. Kennett, E.R. Engdahl, Geophys. J. Int. **105**(2), 429–465 (1991)
44. B.L.N. Kennett, E.R. Engdahl, R. Buland, Geophys. J. Int. **122**(1), 108–124 (1995)
45. W.M. Kaula, *An Introduction to Planetary Physics: The Terrestrial Planets* (Wiley, New York, 1968)
46. J.G. Williams, Astron. J. **108**, 711 (1994)
47. A.S. Konopliv, W.L. Sjogren, Publication 95–3 (Jet Propulsion Laboratory/California Institute of Technology, Pasadena, 1995)
48. D.E. Smith, et al., J. Geophys. Res. **98**, 20871 (1995)
49. W.M. Folkner, et al., J. Geophys. Res. **102**, 4057 (1997)

12 Radio Pulsars as a Laboratory for Strong-Field Gravity Tests

Lijing Shao

Abstract

General relativity offers a classical description to gravitation and spacetime, and is a cornerstone for modern physics. It has passed a number of empirical tests with flying colours, mostly in the weak-gravity regimes, but nowadays also in the strong-gravity regimes. Radio pulsars provide one of the earliest extrasolar laboratories for gravity tests. They, in possession of strongly self-gravitating bodies, i.e. neutron stars, are playing a unique role in the studies of strong-field gravity. Radio timing of binary pulsars enables very precise measurements of system parameters, and the pulsar timing technology is extremely sensitive to various types of changes in the orbital dynamics. If an alternative gravity theory causes modifications to binary orbital evolution with respect to general relativity, the theory prediction can be confronted with timing results. In this chapter, we review the basic concepts in using radio pulsars for strong-field gravity tests, with the aid of some recent examples in this regard, including tests of gravitational dipolar radiation, massive gravity theories, and the strong equivalence principle. With more sensitive radio telescopes coming online, pulsars are to provide even more dedicated tests of strong gravity in the near future.

L. Shao (✉)
Kavli Institute for Astronomy and Astrophysics, Peking University, Beijing, China

National Astronomical Observatories, Chinese Academy of Sciences, Beijing, China
e-mail: lshao@pku.edu.cn

C. Pfeifer, C. Lämmerzahl (eds.), *Modified and Quantum Gravity*, Lecture Notes in Physics 1017, https://doi.org/10.1007/978-3-031-31520-6_12

12.1 Introduction

Pulsars are rotating magnetized neutron stars. On the one hand, due to their large moment of inertia ($I \sim 10^{38}\,\mathrm{kg\,m^2}$) and usually small external torque, their rotation is extremely stable. If a pulsar sweeps a radiating beam in the direction of the Earth, a radio pulse could be recorded using large-area telescopes for each rotation. As fundamentally known in physics, such a periodic signal can be viewed as a *clock*. Therefore, pulsars are famously recognized as astrophysical clocks in astronomy. Even better, thanks to a sophisticated technique called pulsar timing [58], pulsar astronomers can *accurately* record a number of periodic pulse signals. These pulses' times of arrival are compared with atomic clocks at the telescope sites. Some of these observations can be carried out and last for decades. From a large number of times of arrival of these pulse signals, the physical properties of pulsar systems are inferred to a great precision [39]. For example, a recent study with 16 years of timing data of the Double Pulsar,[1] PSR J0737−3039A/B, gives the rotational frequency of pulsar A in the binary system [37],

$$\nu = 44.05406864196281(17)\,\mathrm{Hz}\,. \tag{12.1}$$

It has sixteen significant digits, and the numbers in the parenthesis give the uncertainty of the last-two digits. Such a precision rivals the precision of atomic clocks on the Earth [30], and also it possibly calls for an extension of the usual use of floating numbers in computer numerics for future precision pulsar timing experiments. Pulsars are truly *precision clocks*.

? Exercise

12.1. During the 16 years of observation, how many cycles have PSR J0737−3039A rotated?

On the other hand, neutron stars are the densest objects known that are made of standard-model materials. For such a compact object, gravity plays a vital role in shaping its internal structure and affecting its external dynamics. As explicitly demonstrated by Damour and Esposito-Farèse [16], if gravity is described by an alternative theory to the general relativity—in their case, a class of scalar-tensor gravity theories—nonperturbative phase-transition-like behaviours might happen for neutron stars, resulting in large deviations from general relativity in the strong field of neutron stars [17,25,46]. These large deviations will manifest in the timing

[1] Currently, PSR J0737−3039A/B is the only discovered double neutron star system whose two neutron stars were both detected as pulsars [14,41,49], known as Pulsar A and Pulsar B.

data of pulsars in some way (cf. Sect. 12.2), and they could provide smoking-gun signals for gravity theories regarding the strong-field properties. Combining the strong-field nature of neutron stars and the precision measurements of times of arrival, radio pulsars are truly ideal to test alternative theories of gravity [51, 63, 64], augmenting what have been done in the weak field of the Solar System [65], and complementing what are recently being performed with gravitational waves [3–5] and black hole shadows [6–8, 50].

Currently, more than three thousands of radio pulsars are discovered[2] [43]. The most useful subset of pulsars in testing alternative gravity theories are millisecond pulsars in *clean* binaries.[3] Their times of arrival at telescopes are imprinted with information from the following sources:

1. the Solar system dynamics which affect the motion of radio telescopes;
2. the binary dynamics which are resulted from the mutual gravitational interaction between the two binary components; and
3. the interstellar medium which affects the propagation of radio waves in a frequency-dependent way, in terms of dispersion, scattering, and so on.

A formalism, which includes the above effects and connects the proper time of the pulse signals in the pulsar frame to the observed coordinate time at the telescopes, is called a *pulsar timing model*. One of the widely used timing models for binary pulsars is the Damour-Deruelle timing model [15]. It is a phenomenological model that applies to a large set of alternative gravity theories which are possibly being the underlying theory for the binary's orbital motion.

In the Damour-Deruelle timing model, a handful of parameterized post-Keplerian (PPK) parameters are introduced for generic Lorentz-invariant extensions of gravity theories [19]. The values of PPK parameters differ in different gravity theories. Therefore, measurements of these PPK parameters can be converted into constraints on parameters in the alternative gravity theories. The most frequently used PPK parameters include $\dot{\omega}$, $\dot{P}_b$, γ, r, and s. The PPK parameter $\dot{\omega}$ describes the periastron advance of the binary orbit, the PPK parameter $\dot{P}_b$ describes the orbital period decay caused by the radiation of gravitational waves, the PPK parameter γ describes combined effects from the Doppler time delay and gravitational time delay, and the PPK parameters (r, s) describe the Shapiro time delay imprinted by the spacetime curvature of the companion star. The values of these five PPK parameters in the

[2] https://www.atnf.csiro.au/people/pulsar/psrcat/.

[3] In one case, a pulsar in a triple system, PSR J0337+1715, provides the best limit on the strong equivalence principle [9, 47, 59].

general relativity are given in Damour and Deruelle [15] and Lorimer and Kramer [39],

$$\dot{\omega} = 3\left(\frac{P_b}{2\pi}\right)^{-5/3}(T_\odot M)^{2/3}\left(1-e^2\right)^{-1}, \tag{12.2}$$

$$\dot{P}_b = -\frac{192\pi}{5}\left(\frac{P_b}{2\pi}\right)^{-5/3}\left(1+\frac{73}{24}e^2+\frac{37}{96}e^4\right)\left(1-e^2\right)^{-7/2}T_\odot^{5/3}m_A m_B M^{-1/3}, \tag{12.3}$$

$$\gamma = e\left(\frac{P_b}{2\pi}\right)^{1/3}T_\odot^{2/3}M^{-4/3}m_B(m_A+2m_B), \tag{12.4}$$

$$r = T_\odot m_B, \tag{12.5}$$

$$s = x\left(\frac{P_b}{2\pi}\right)^{-2/3}T_\odot^{-1/3}M^{2/3}m_B^{-1}, \tag{12.6}$$

where P_b and e are respectively the orbital period and orbital eccentricity, m_A and m_B are the masses of the pulsar and its companion in unit of the Solar mass ($M_\odot$), the total mass $M \equiv m_A + m_B$, and $T_\odot \equiv GM_\odot/c^3 = 4.925490947\ \mu s$. Equations (12.2)–(12.6) take different forms in alternative gravity theories, often with dependence on the extra charges of the binary components in the theory, e.g., these PPK parameters depend on scalar charges of the pulsar and its companion in the scalar-tensor theory [17]. In pulsar-timing observation, each PPK parameter is *independently* measured. Eventually, for a gravity theory to pass the tests from pulsar timing, it should give consistent predictions to *all* the measured values of PPK parameters with a *unique* set of physical parameters of the binary system. These consistency checks are often illustrated in the mass-mass diagram. For an example, in Fig. 12.1 the measurements of three PPK parameters, $\dot{\omega}$, γ, and $\dot{P}_b$, from the Hulse-Taylor pulsar PSR B1913+16, give consistent component masses when the general relativistic Eqs. (12.2)–(12.4) are used [61]. Therefore, general relativity passes the tests posed by the Hulse-Taylor pulsar [61].

? Exercise

12.2. For the Hulse-Taylor pulsar PSR B1913+16, the following parameters are measured directly via pulsar timing: $P_b = 0.322997448918(3)$ d, $e = 0.6171340(4)$, $\dot{\omega} = 4.226585(4)$ deg yr^{-1}, and $\gamma = 0.004307(4)$ s [61]. Assuming general relativity, please derive the two component masses for this binary system.

In this following, we will give a few more concrete and recent examples where binary pulsars play a key role in limiting alternative gravity theories, including the gravitational dipolar radiation in the scalar-tensor gravity (Sect. 12.2), two

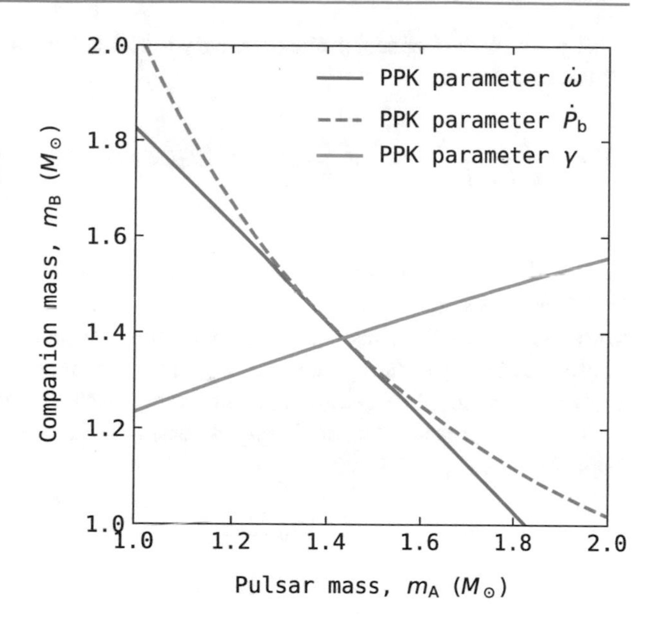

Fig. 12.1 Consistency of general relativity in describing three measured PPK parameters ($\dot{\omega}$, γ, and $\dot{P}_b$) from PSR B1913+16 in the mass-mass diagram [61]

classes of massive gravity theories (Sect. 12.3), and the strong equivalence principle (Sect. 12.4). These examples are by no means complete, and certainly reflect the somehow biased topics that the author is interested in. A short perspective discussion is given in Sect. 12.5. For more extensive reviews on using radio pulsars for gravity tests, readers are referred to Refs. [35, 42, 51, 57, 63, 64].

12.2 Strong-Field Effects and Gravitational Dipolar Radiation

Scalar-tensor gravity theories represent a well posed, healthy extension of Einstein's general relativity by including a nonminimally coupled scalar field in the Lagrangian of gravity [10, 13, 65]; see also Sects. 4.4.4 and 7.4.1 for the discussion of gravity theories involving additional scalar fields as mediator of the gravitational interaction. Shortly after the first discovery of the Hulse-Taylor binary pulsar, Eardley [24] pointed out that a gravitational dipolar radiation could be used as a discriminant for such a class of gravity theories. An extra dipolar radiation term can be tested with the PPK parameter $\dot{P}_b$. Investigation along this line was boosted by the theoretical discovery that in a slightly extended version of the original scalar-tensor gravity, nonperturbative effects develop for certain neutron stars [16, 17]. The so-called *spontaneous scalarization* (see also Sect. 7.4.1 for more details) introduces a much enhanced gravitational dipolar radiation for a scalarized neutron star in a binary. The dipolar radiation in principle can even dominate over the quadrupolar radiation predicted by the general relativity in binary pulsar observations [cf. Eq. (12.3)], but still keeping all weak-field gravity tests satisfied. This enters the regime of *strong-field* gravity tests, where weak-field tests have a rather limited power.

A general class of scalar-tensor gravity theories have the following action in the Einstein frame,

$$S = \frac{c^4}{16\pi G_*} \int \frac{d^4x}{c} \sqrt{-g_*} \left[R_* - 2g_*^{\mu\nu} \partial_\mu \varphi \partial_\nu \varphi - V(\varphi)\right] + S_m \left[\psi_m; A^2(\varphi) g^*_{\mu\nu}\right], \tag{12.7}$$

where $g_*^{\mu\nu}$ and R_* are the metric tensor and Ricci scalar respectively, ψ_m collectively denotes standard-model matter fields, φ is an extra scalar field, and quantities with stars are in the Einstein frame. The novel aspect lies in the fact that it is a conformal metric $A^2(\varphi)g^*_{\mu\nu}$ instead of $g^*_{\mu\nu}$ itself that couples to matter fields. Such a *nonminimal* coupling is important for the discussions below.

The class of scalar-tensor gravity theories carefully examined by Damour and Esposito-Farèse [16, 17] has

$$V(\varphi) = 0, \tag{12.8}$$

$$A(\varphi) = \exp\left(\beta_0 \varphi^2/2\right), \tag{12.9}$$

$$\alpha_0 = \beta_0 \varphi_0, \tag{12.10}$$

where φ_0 is the asymptotic value of φ at infinity, and α_0 and β_0 are two theory parameters. This is the class of scalar-tensor theories, sometimes denoted as $T_1(\alpha_0, \beta_0)$ and called the *Damour-Esposito-Farèse theory*, that are most widely confronted with pulsar observations [28, 53, 63, 71].

? Exercise

12.3. Derive field equations for the Damour-Esposito-Farèse theory.

? Exercise

12.4. Based on the field equations, derive the modified Tolman-Oppenheimer-Volkoff equations for the Damour-Esposito-Farèse theory, for a spherically symmetric neutron star.

By integrating the modified Tolman-Oppenheimer-Volkoff equations derived from theory (12.7), one gets a boost in a neutron star's scalar charge when its mass reaches a critical point. This phenomenon is understood from the viewpoint of Landau's phase transition theory when a tachyonic instability kicks in and

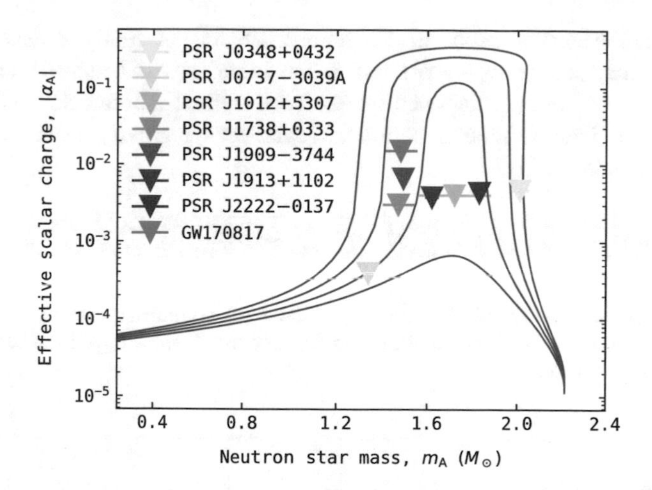

Fig. 12.2 Blue curves show the effective scalar charge in the Damour-Esposito-Farèse scalar-tensor gravity theory with $|\alpha_0| = 10^{-5}$ and, from top to bottom, $\beta_0 = -4.8, -4.6, -4.4, -4.2$. The AP4 equation of state is assumed in the calculation. Triangles show the observational bounds from binary pulsars [53, 71] and gravitational waves [1, 2] at the 90% confidence level. The mass uncertainty for these neutron stars is indicated at the 68% confidence level

a new branch of neutron star solutions with scalar charges are energetically favored [25, 34, 46]. We define the *effective scalar charge* of a neutron star [16],

$$\alpha_A \equiv \frac{\partial \ln m_A}{\partial \varphi_0}, \tag{12.11}$$

which is a representative quantity characterizing the strength of deviation from general relativity. In Fig. 12.2, example curves for the effective scalar charge as a function of neutron star mass are given in blue lines from top to bottom for $\beta_0 = -4.8, -4.6, -4.4, -4.2$, assuming the AP4 equation of state and $|\alpha_0| = 10^{-5}$. As we can easily seen, indeed that for certain mass range of neutron stars, $|\alpha_A|$ can be very large while keeping its value very small in weak-gravity fields.

The emission of gravitational dipolar radiation in a binary pulsar is proportional to the difference in the effective scalar couplings of the two binary components A and B, and to the leading order, it contributes to an additional decay rate of orbital period via [17],

$$\dot{P}_b^{\rm dipole} = -\frac{2\pi G_*}{c^3}\left(1+\frac{e^2}{2}\right)\left(1-e^2\right)^{-5/2}\left(\frac{2\pi}{P_b}\right)\frac{m_A m_B}{M}\left(\alpha_A-\alpha_B\right)^2 . \tag{12.12}$$

While neutron stars have significant scalar charges, white dwarfs, being weak-field objects, are hardly different from their counterparts in general relativity with a vanishingly small scalar charge $\alpha_B \simeq \alpha_0 \to 0$, where α_0 is well constrained by Solar System weak-field tests [65]. Therefore, neutron-star white-dwarf binaries turn out to be the most sensitive probe in this regard [28, 53]. Recently, a new study [71] shows explicitly that neutron-star neutron-star binaries with a significant difference in the masses of binary components are also excellent laboratories. Therefore, to test the gravitational dipolar radiation in scalar-tensor gravity, asymmetric binary pulsars are needed[4] [63].

? Exercise

12.5. When the dipolar-radiation-induced orbital decay is comparable to the quadrupolar-radiation-induced orbital decay for the Hulse-Taylor pulsar? Derive the critical value for the effective scalar charge.

Some illustration for a specific equation of state, `AP4`, is given in Fig. 12.2, along with constraints on the gravitational dipolar radiation from seven binary pulsars [71]: five neutron-star white-dwarf binaries (PSRs J0348+0432, J1012+5307, J1738+0333, J1909−3744, and J2222−0137) and two asymmetric neutron-star neutron-star binaries (PSRs J0737−3039A and J1913+1102). For comparison, we also show a constraint from the first binary neutron star merger observed via gravitational waves [2]. In principle, the uncertainty in the superanuclear neutron-star matter is entangled with strong-field gravity tests [48]. Nevertheless, nowadays we have enough well-measured binary pulsar systems to populate the whole mass range for neutron stars, and a combined study [53, 71] has verified that for each reasonable equation of state, the possibility for spontaneous scalarization in the Damour-Esposito-Farèse scalar-tensor gravity theory is very low. Following the method developed by Shao et al. [53], a dedicated Bayesian parameter-estimation study combining the above-mentioned seven pulsar systems has basically closed the possibility of developing spontaneous scalarization for an effective scalar coupling larger than 10^{-2} for the theory given by Eqs. (12.7)–(12.10), no matter of the underlying yet-uncertain equation of state for supranuclear neutron-star matters.

It is worth to mention that, when performing Markov-chain Monte Carlo Bayesian parameter estimation, the integration of the modified Tolman-Oppenheimer-Volkoff equations needs to be carried out by more than millions of times on the fly thus computationally expensive. Recently, reduced-order surrogate models, which extract dominating features to represent accurate enough integration results, were bulit to aid the speedup of the calculation [29, 70]. The codes of these reduced-order surrogate models are publicly available at https://github.com/BenjaminDbb/pySTGROM and https://github.com/mh-guo/pySTGROMX for community use.

---

[4] Unfortunately, we have not detected yet suitable neutron-star black-hole binaries for this test, which are also potentially very good testbeds [38].

Although the original Damour-Esposito-Farèse scalar-tensor gravity theory is disfavored by binary pulsar timing results, in further extended, generic scalar-tensor gravity theories, neutron stars can still be scalarized. This is particularly true for a massive scalar-tensor theory with $V(\varphi) \sim m^2\varphi^2$ when the Compton wavelength of the scalar field is smaller than the orbital separation of the binary [45, 66, 69]. Basically the modification with respect to the general relativity in the orbital dynamics is suppressed exponentially in a Yukawa fashion. Fortuitously, without giving much details, such kind of massive scalar-tensor theories can be efficiently probed via the tidal deformability measurement in gravitational waves [1, 31, 32]. In this sense, a combination of pulsar timing data and gravitational wave data is called for to probe a larger parameter space for scalar-tensor gravity theories [53].

In the past few years, other variants of scalar-tensor gravity theories triggered great enthusiasm. Some of them not only give scalarized neutron stars, but also scalarized black holes, in contrast to the *no-hair theorem*. A particularly interesting class of such theory includes a topological Gauss-Bonnet term,

$$\mathcal{G} = R_{\alpha\beta\gamma\delta}R^{\alpha\beta\gamma\delta} - 4R_{\alpha\beta}R^{\alpha\beta} + R^2 , \tag{12.13}$$

coupling to the scalar field [23, 56, 67]. In Eq. (12.13), $R_{\alpha\beta\gamma\delta}$ and $R_{\alpha\beta}$ are the Riemann tensor and Ricci tensor respectively. Preliminary constraints on the scalar-Gauss-Bonnet gravity from binary pulsars are presented by Danchev et al. [20]. This is a new field where observations of compact objects including neutron stars and black holes are crucial to reveal the strong-field information of gravitation.

12.3 Radiative Effects in Massive Gravity Theories

Radiative tests from binary pulsars are powerful, as the related PPK parameter, $\dot{P}_b$, can be very well measured from a long-term timing project on suitable pulsars [19]. This parameter improves with observational time span $T_{\rm obs}$ quite fast, as $T_{\rm obs}^{-5/2}$. The orbital decay rate $\dot{P}_b$ is not only useful for constraining the dipolar gravitational wave emission, but also in other radiative aspects of gravitation, for example, in constraining the extra radiation caused by a certain model of the breaking down of the Lorentz symmetry [68],[5] or a nonzero mass of gravitons [21, 26, 44, 55]. Here we give a brief introduction to the latter.

In general relativity, the hypothetical quantum particle for gravity, graviton, is a massless spin-2 particle. However, massive gravity theories are found to provide interesting phenomena related to the evolution of the Universe, e.g. the accelerated expansion and dark energy [21]. Therefore, probing the upper bounds of the graviton mass is fundamentally important to field theories and cosmology studies, and it is one of the central topics in gravitational physics.

[5] Recall that there are numerous different models of Lorentz invariance violation or doubly special relativity, see Chaps. 1 or 2.

One of the early study of using binary pulsars to test the graviton mass was performed by Finn and Sutton in 2002 [26]. They investigated a linearized gravity with a massive graviton with the action,

$$S = \frac{1}{64\pi} \int d^4x \left[\partial_\lambda h_{\mu\nu} \partial^\lambda h^{\mu\nu} - 2\partial^\nu h_{\mu\nu} \partial_\lambda h^{\mu\lambda} + 2\partial^\nu h_{\mu\nu} \partial^\mu h - \partial^\mu h \partial_\mu h - 32\pi h_{\mu\nu} T^{\mu\nu} + m_g^2 \left(h_{\mu\nu} h^{\mu\nu} - \frac{1}{2} h^2 \right) \right], \tag{12.14}$$

where the last term gives a unique graviton mass under certain conditions[6] [26] while the others are just linearized expansions from the Einstein-Hilbert action with $h_{\mu\nu} \equiv g_{\mu\nu} - \eta_{\mu\nu}$ and $h \equiv h^\mu{}_\mu$. It was shown that extra gravitational wave radiation exists in theory (12.14), which results in a fractional change in the orbital decay rate, by Finn and Sutton [26]

$$\frac{\dot{P}_b - \dot{P}_b^{GR}}{\dot{P}_b^{GR}} = \frac{5}{24} \frac{\left(1-e^2\right)^3}{1 + \frac{73}{24}e^2 + \frac{37}{96}e^4} \left(\frac{P_b}{2\pi\hbar} \right)^2 m_g^2 . \tag{12.16}$$

Here $\dot{P}_b^{GR}$ is the value predicted by the general relativity in Eq. (12.3). Notice that the fractional change is proportional to $\propto P_b^2 m_g^2$. Therefore, if the precision of $\dot{P}_b$ is given, binary pulsars with larger orbits have a larger figure of merit for the test. However, usually, the precision of $\dot{P}_b$ crucially depends on the orbital size, and it turns out that, still, binary pulsars with smaller orbits have a larger figure of merit.

The most recent constraint in this Finn-Sutton framework was provided by a combination of multiple best-timed binary pulsars with a Bayesian statistical treatment. A collection of nine best-timed binary pulsars (PSRs J0348+0432, J0737−3039, J1012+5307, B1534+12, J1713+0747, J1738+0333, J1909−3744, B1913+16, and J2222−0137) provide a tight bound on the graviton mass,

$$m_g < 5.2 \times 10^{-21}\ \mathrm{eV}/c^2, \quad (90\%\ \mathrm{C.L.}), \tag{12.17}$$

using a uniform prior in $\ln m_g$ [44]. This limit is not the strongest limit on the graviton mass [22]. However, from a theoretical point of view, it is a bound from binary orbital dynamics, complementary to, e.g. the kinematic dispersion-relation

[6] The conditions are that (i) the wave equation takes a standard form for the trace-reversed metric perturbation $\bar{h}_{\mu\nu}$

$$\left(\Box - m_g^2\right) \bar{h}_{\mu\nu} + 16\pi T_{\mu\nu} = 0, \tag{12.15}$$

and the theory recovers the general relativity in the limit when $m_g \to 0$, namely, there is no van Dam-Veltman-Zakharov discontinuity [26].

tests from the LIGO/Virgo/KAGRA observation of gravitational waves [5]. It is worth mentioning that the theory (12.14) has some drawbacks including ghosts and instability [22, 26], and here it is only used as a strawman target for illustration.

? Exercise

12.6. Derive the lower limit for the Compton wavelength of gravitons from Eq. (12.17).

It is interesting to note, that in different massive gravity theories, the dependence of the extra radiation on the graviton mass is in general different. It depends on the specifics of the illustrated gravity theory. This is due to the deep fundamental principles in the designs of a number of variants of massive gravity theories. For example, in a cosmologically motivated massive gravity theory, known as the *cubic Galileon theory* with the action [21],

$$S = \int \mathrm{d}^4x \left[-\frac{1}{4} h^{\mu\nu} (\mathcal{E}h)_{\mu\nu} + \frac{h^{\mu\nu} T_{\mu\nu}}{2M_{\mathrm{Pl}}} - \frac{3}{4} (\partial\varphi)^2 \left(1 + \frac{1}{3m_{\mathrm{g}}^2 M_{\mathrm{Pl}}} \Box\varphi \right) + \frac{\varphi T}{2M_{\mathrm{Pl}}} \right], \tag{12.18}$$

the specific way of the addition of the scalar field φ introduces the so-called *screening mechanism*, thus avoids the stringent constraints from the Solar System, yet provides important changes to the cosmological evolution. In the action (12.18), φ is the Galileon scalar field, $T_{\mu\nu}$ is the matter energy-momentum tensor, $T \equiv T^{\mu}{}_{\mu}$, M_{Pl} is the Planck mass, and

$$(\mathcal{E}h)_{\mu\nu} \equiv -\frac{1}{2} \Box h_{\mu\nu} + \cdots \tag{12.19}$$

is the Lichnerowicz operator. For a central massive body with mass M, the *screening radius* is $r_\star = \left(M/16 m_{\mathrm{g}}^2 M_{\mathrm{Pl}}^2 \right)^{1/3}$, within which, the theory exhibits strong couplings and it reduces to the canonical gravity.

? Exercise

12.7. By knowing that the Earth is within the screening radius of the Sun, derive the upper limit of graviton mass in the cubic Galileon theory.

According to de Rham et al. [21], though with a screening mechanism to suppress modification at the high density region within $r_\star$, this cubic Galileon theory predicts a different scaling behaviour for the gravitational radiation. For a system with a typical length scale L, the *fifth-force* suppression factor is $\sim (L/r_\star)^{3/2}$, and the suppression factor for the gravitational radiation is $\sim (P_b/r_\star)^{3/2}$. As for a binary system, one has $L \sim vP_b$ where v is a characteristic velocity. Therefore, the gravitational radiation is, compared with the fifth force, *less* suppressed by a factor of $v^{3/2}$, and it provides a valuable window to look for evidence of this theory via radiative channels, for example, in binary pulsar systems.

Analytic radiative powers were worked out by de Rham et al. [21], and the extra radiative channels include monopolar radiation, dipolar radiation, and quadrupolar radiation. For binary pulsar systems with different orbital periods and orbital eccentricities, the dominate radiation channel can be different [55]. For the current set of binary pulsars, the quadrupole radiation is the dominating factor among the extra channels [21, 55].

The most up-to-date constraint from binary pulsars is

$$m_g < 2 \times 10^{-28}\,\mathrm{eV}/c^2, \quad (95\%\ \mathrm{C.L.})\,, \tag{12.20}$$

for the cubic Galileon theory, and the cumulative probability distributions of the graviton mass are given in Fig. 12.3 for two different priors [55]. Such a tight constraint was obtained from the combination of fourteen best-timed binary pulsar systems, including PSRs J0348+0432, J0437−4715, J0613−0200, J0737−3039, J1012+5307, J1022+1001, J1141−6545, B1534+12, J1713+0747, J1738+0333, J1756−2251, J1909−3744, B1913+16, and J2222−0137. One should keep in mind that, the limit (12.20) is theory specific, and in this situation, only applies to the cubic Galileon theory given in Eq. (12.18). Nonetheless, it provides an interesting example that for a gravity theory designed for cosmological purposes at corresponding lengthscales, binary pulsar systems with astronomical lengthscales still provide intriguing and useful bounds. It is an illustration of using binary pulsars

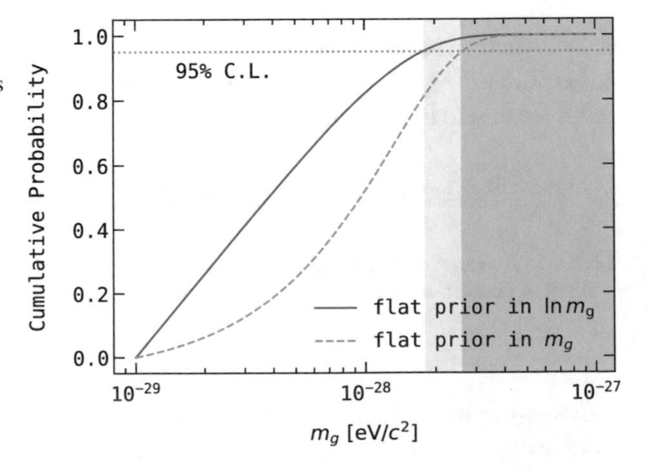

Fig. 12.3 Cumulative probability for the graviton mass with two different priors in the cubic Galileon theory [55]. Shaded regions show the excluded graviton mass values at the 95% confidence level

in the studies of cosmology by examining the modification to binary orbits brought by a cosmologically-motivated modified gravity.

12.4 Strong Equivalence Principle and Dark Matters

Binary pulsars are not only useful for the radiative tests introduced in the above sections, they also provide superb limiting power in the conservative aspects of gravitational dynamics for orbital evolutions. Below we introduce an example of examining the strong equivalence principle via the conservative dynamics of binary pulsars [18, 72], and its extension to test certain interesting properties of dark matters [54, 60].

As discovered by Damour and Schäfer [18], a perturbed binary orbit with an equivalence-principle-violating abnormal acceleration has a characteristic evolution in its orbital elements. The notable change is the appearance of a *vectorized superposition* of two eccentricity vectors for the real orbital eccentricity. It provides a graphical understanding of the underlying dynamics for a binary in presence of equivalence principle violations. The *real* orbital eccentricity vector, $\boldsymbol{e}(t)$, is an addition of a rotating normal eccentricity vector, $\boldsymbol{e}_{\mathrm{R}}(t)$, in its post-Newtonian fashion, and an extra abnormal eccentricity vector, $\boldsymbol{e}_{\Delta}$, which is time independent and whose length is proportional to the Eötvös parameter, Δ, describing the violation of the equivalence principle. If $\Delta = 0$, the abnormal eccentricity vector $\boldsymbol{e}_{\Delta} = 0$ and it returns to the precessing case in the general relavitity. A graphical illustration is given in Fig. 12.4. As we discussed in Sect. 12.1, the pulsar timing

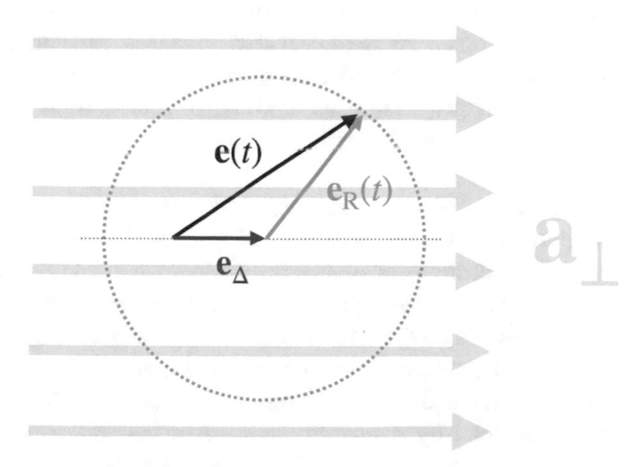

Fig. 12.4 Graphical illustration of the time-varying orbital eccentricity vector, $\boldsymbol{e}(t)$, for a binary pulsar, in the presence of strong equivalence principle violation [18]. The orbital eccentricity vector evolves according to $\boldsymbol{e}(t) = \boldsymbol{e}_{\Delta} + \boldsymbol{e}_{\mathrm{R}}(t)$, where $\boldsymbol{e}_{\mathrm{R}}(t)$ is the usual precessing eccentricity vector in the general relativity, and the *constant* abnormal eccentricity is in the direction of $\boldsymbol{a}_{\perp}$, which is the projection of the external Galactic acceleration in the orbital plane

technique is very sensitive to tiny changes in the orbit, and such a change can be captured in pulsar timing data [18].

At the beginning, such a scenario was applied to a few binary pulsars in a statistical sense by marginalizing over some unknown angles to obtain constraints on the violation of the equivalence principle [18]. Later it was implemented to a handful of binary pulsars with an improved statistical methodology to better account for the movements of binary pulsars in the Milky Way [63]. Then, with better data and more information about binary pulsar systems, a direct method was developed [27]. The direct method not only can constrain the equivalence principle violation, but in principle can detect it if it exists.

The most stringent limit using binary pulsars comes from a precisely timed long-orbital-period binary pulsar, PSR J1713+0747 [72], as larger orbits have higher figures of merit in such a test [18]. Using the improved direct method, the limit on the Eötvös parameter from PSR J1713+0747 is [72],

$$|\Delta| < 2 \times 10^{-3}\,, \quad (95\%\ \text{C.L.})\,. \tag{12.21}$$

Though it is much less limiting than the earlier constraint obtained from the Solar System [60, 65], the limit (12.21) encodes strong-field effects. For example, in the case of the aforementioned scalar-tensor gravity, the strong-field version of Eötvös parameter will be very different from its weak-field counterpart [27]. Therefore, such a limit from neutron stars is a *standalone* bound and applicable to the strong version of equivalence principle [51, 63].

The limit (12.21) is not only interesting to gravitational physics, it also has its value when we look at it from a different angle. As we now know, the binary pulsar is actually immersed in the ocean of dark matters in the Milky Way. As we have not really understood what the very nature of dark matter is, the above method for testing the equivalence principle provides a non-traditional probe to dark matter's properties. Shao et al. [54] proposed a method where such a limit, with a proper handle, can be converted to the interaction properties between dark matters and ordinary matters.

If there is a *long-range fifth force* between dark matter particles and ordinary matter fields, as many field theories will suggest [60], it is likely to introduce an *apparent* violation of the strong equivalence principle if we have not taken the fifth force into account in our standard assumptions. The role of the Galactic acceleration in Fig. 12.4, whose projection on the orbital plane is $\boldsymbol{a}_\perp$, is replaced by the attraction of dark matters to the binary system. The difference in the acceleration to two binary components (a neutron star and a white dwarf in the case of PSR J1713+0747), described by Δ, is replaced by a quantity related to the long-range fifth-force between dark matters and ordinary standard-model matters [54].

Detailed analysis of PSR J1713+0747 [54] took into consideration of the Galactic distribution of dark matters, and gave a very different bound in nature that could be obtained from terrestrial experiments [60]. The current observational data of PSR J1713+0747 already imply that, if there is such a long-range fifth force between dark matters and ordinary matters, its magnitude should be no more than 1% of the gravitational force between them. Such a limit provides a useful complement to other types of dark-matter experiments, which are usually looking for *short-range* forces between the hypothesized dark-matter particles and the standard-model particles [60], including the searches in underground laboratories, particle colliders, and X-ray/γ-ray observations via high-energy satellites.

12.5 Summary

In this chapter, we present some basic concepts of using binary pulsars as *fundamental clocks* in a curved spacetime to probe various types of modifications to the binary orbits. These modifications could have been caused by a modified gravity theory or some other new physics like a long-range fifth force between dark matters and ordinary matters. As pulsar timing provides us with very *accurate* measurements, it puts constraints on tiny changes caused by an alternative gravity theory other than the general relativity. Moreover, neutron stars are intrinsically *strong-gravity* objects, and nonperturbative aspects of the strong-field gravity can also be studied via radio pulsar experiments. Actually, quite many strong-field limits are still best provided by pulsar timing experiments, even nowadays in presence of new types of observations like gravitational waves and black hole shadows. A careful study shows that the limits from pulsar timing are actually complementary to those from gravitational wave detections and black hole shadows [8, 53]. Proper combinations of these strong-gravity experiments could provide a more complete landscape to gravitation in the strong-field .

Solely focusing on the radio pulsar side, the timing experiments can be carried out for decades, in particular for some interesting systems like the Hulse-Taylor pulsar PSR B1913+16 [61] and the Double Pulsar PSR J0737−3039A/B [37]. Long-term observations improve the precision of PPK parameters with the observational time span $T_{\rm obs}$. For examples, the precision in the orbital decay parameter, $\dot{P}_{\rm b}$, improves very fast, as $T_{\rm obs}^{-5/2}$, and the precision in the periastron advance rate, $\dot{\omega}$, improves as $T_{\rm obs}^{-3/2}$. Furthermore, the sensitivity of radio telescopes is also improving, notably with the Five-hundred-meter Aperture Spherical Telescope in China [33, 40] and the Square Kilometre Array in South Africa and Australia [52, 62]. The former has already been operating for a couple of years, while the latter has also entered the construction phase recently. The improvement in the sensitivity of radio telescopes directly converts to improvements in the timing precision. Therefore, the real improvement for PPK parameters is faster than the theoretical power law predictions. Last but not the least, radio telescopes are also continuously discovering new pulsar systems, and some of these systems with suitable system properties will contribute to strong-field gravity tests. We are even looking forward to discovering yet-undetected binary pulsar systems like neutron-star black-hole binaries with short orbital periods $P_{\rm b} \lesssim 1$ day or pulsars around the Sgr A* black hole with orbital periods $P_{\rm b} \lesssim 10$ years [11, 12, 38], which will provide completely new gravity tests in the strong-field regimes [36].

Acknowledgments We are grateful to the 740th WE-Heraeus-Seminar "Experimental Tests and Signatures of Modified and Quantum Gravity", organized by Christian Pfeifer and Claus Lämmerzahl. LS was supported by the National SKA Program of China (2020SKA0120300), the National Natural Science Foundation of China (11975027, 11991053, 11721303), and the Max Planck Partner Group Program funded by the Max Planck Society.

References

1. B.P. Abbott et al., GW170817: observation of gravitational waves from a binary beutron star inspiral. Phys. Rev. Lett. **119**(16), 161101 (2017)
2. B.P. Abbott et al., Tests of General Relativity with GW170817. Phys. Rev. Lett. **123**(1), 011102 (2019)
3. B.P. Abbott et al., Tests of General Relativity with the Binary Black Hole Signals from the LIGO-Virgo Catalog GWTC-1. Phys. Rev. D **100**(10), 104036 (2019)
4. R. Abbott et al., Tests of general relativity with binary black holes from the second LIGO-Virgo gravitational-wave transient catalog. Phys. Rev. D **103**(12), 122002 (2021)
5. R. Abbott et al., Tests of general relativity with GWTC-3 (2021). arXiv:2112.06861
6. K. Akiyama et al., First M87 event horizon telescope results. I. the shadow of the supermassive black hole. Astrophys. J. Lett. **875**L1 (2019)
7. K. Akiyama et al., First sagittarius A* event horizon telescope results. I. the shadow of the supermassive black hole in the center of the milky way. Astrophys. J. Lett. **930**(2), L12 (2022)
8. K. Akiyama et al., First sagittarius A* event horizon telescope results. VI. testing the black hole metric. Astrophys. J. Lett. **930**(2), L17 (2022)
9. A.M. Archibald, N.V. Gusinskaia, J.W.T. Hessels, A.T. Deller, D.L. Kaplan, D.R. Lorimer, R.S. Lynch, S.M. Ransom, I.H. Stairs, Universality of free fall from the orbital motion of a pulsar in a stellar triple system. Nature **559**(7712), 73–76 (2018)
10. E. Berti, et al., Testing general relativity with present and future astrophysical observations. Class. Quant. Grav. **32**, 243001 (2015)
11. G.C. Bower, et al., Galactic Center Pulsars with the ngVLA. ASP Conf. Ser. **517**, 793 (2018)
12. G.C. Bower, et al., Fundamental Physics with Galactic Center Pulsars. Bull. Am. Astron. Soc. **51**(3), 438 (2019)
13. C. Brans, R.H. Dicke, Mach's principle and a relativistic theory of gravitation. Phys. Rev. **124**, 925–935 (1961)
14. M. Burgay, et al., An increased estimate of the merger rate of double neutron stars from observations of a highly relativistic system. Nature **426**, 531–533 (2003)
15. T. Damour, N. Deruelle, General relativistic celestial mechanics of binary systems. II. The post-Newtonian timing formula. Ann. Inst. Henri Poincaré Phys. Théor. **44**, 263–292 (1986)
16. T. Damour, G. Esposito-Farèse, Nonperturbative strong field effects in tensor-scalar theories of gravitation. Phys. Rev. Lett. **70**, 2220–2223 (1993)
17. T. Damour, G. Esposito-Farèse, Tensor-scalar gravity and binary pulsar experiments. Phys. Rev. D **54**, 1474–1491 (1996)
18. T. Damour, G. Schaefer, New tests of the strong equivalence principle using binary pulsar data. Phys. Rev. Lett. **66**, 2549–2552 (1991)
19. T. Damour, J.H. Taylor, Strong field tests of relativistic gravity and binary pulsars. Phys. Rev. D **45**, 1840–1868 (1992)
20. V.I. Danchev, D.D. Doneva, S.S. Yazadjiev, Constraining scalarization in scalar-Gauss-Bonnet gravity through binary pulsars. Phys. Rev. D **106**, 124001 (2022)
21. C. de Rham, A.J. Tolley, D.H. Wesley, Vainshtein mechanism in binary pulsars. Phys. Rev. D **87**(4), 044025 (2013)
22. C. de Rham, J.T. Deskins, A.J. Tolley, S.-Y. Zhou, Graviton mass bounds. Rev. Mod. Phys. **89**(2), 025004 (2017)

23. D.D. Doneva, S.S. Yazadjiev, New Gauss-Bonnet black holes with curvature-induced scalarization in extended scalar-tensor theories. Phys. Rev. Lett. **120**(13), 131103 (2018)
24. D.M. Eardley, Observable effects of a scalar gravitational field in a binary pulsar. Astrophys. J. **196**, L59–L62 (1975)
25. G. Esposito-Farèse, Tests of scalar-tensor gravity. AIP Conf. Proc. **736**(1), 35–52 (2004)
26. L.S. Finn, P.J. Sutton, Bounding the mass of the graviton using binary pulsar observations. Phys. Rev. D **65**, 044022 (2002)
27. P.C.C. Freire, M. Kramer, N. Wex, Tests of the universality of free fall for strongly self-gravitating bodies with radio pulsars. Class. Quant. Grav. **29**, 184007 (2012)
28. P.C.C. Freire, N. Wex, G. Esposito-Farèse, J.P.W. Verbiest, M. Bailes, B.A. Jacoby, M. Kramer, I.H. Stairs, J. Antoniadis, G.H. Janssen, The relativistic pulsar-white dwarf binary PSR J1738+0333 II. The most stringent test of scalar-tensor gravity. Mon. Not. R. Astron. Soc. **423**, 3328 (2012)
29. M. Guo, J. Zhao, L. Shao, Extended reduced-order surrogate models for scalar-tensor gravity in the strong field and applications to binary pulsars and gravitational waves. Phys. Rev. D **104**(10), 104065 (2021)
30. G. Hobbs, et al., Development of a pulsar-based timescale. Mon. Not. R. Astron. Soc. **427**, 2780–2787 (2012)
31. H. Hu, M. Kramer, N. Wex, D.J. Champion, M.S. Kehl, Constraining the dense matter equation-of-state with radio pulsars. Mon. Not. R. Astron. Soc. **497**(3), 3118–3130 (2020)
32. Z. Hu, Y. Gao, R. Xu, L. Shao, Scalarized neutron stars in massive scalar-tensor gravity: X-ray pulsars and tidal deformability. Phys. Rev. D **104**(10), 104014 (2021)
33. P. Jiang, et al., Commissioning progress of the FAST. Sci. China Phys. Mech. Astron. **62**(5), 959502 (2019)
34. M. Khalil, R.F.P. Mendes, N. Ortiz, J. Steinhoff, Effective-action model for dynamical scalarization beyond the adiabatic approximation. Phys. Rev. D **106**, 104016 (2022)
35. M. Kramer, Pulsars as probes of gravity and fundamental physics. Int. J. Mod. Phys. D **25**(14), 1630029 (2016)
36. M. Kramer, D.C. Backer, J.M. Cordes, T.J.W. Lazio, B.W. Stappers, S. Johnston, Strong-field tests of gravity using pulsars and black holes. New Astron. Rev. **48**, 993–1002 (2004)
37. M. Kramer, et al., Strong-field gravity tests with the double pulsar. Phys. Rev. X **11**(4), 041050 (2021)
38. K. Liu, R.P. Eatough, N. Wex, M. Kramer, Pulsar–black hole binaries: prospects for new gravity tests with future radio telescopes. Mon. Not. R. Astron. Soc. **445**(3), 3115–3132 (2014)
39. D.R. Lorimer, M. Kramer, Handbook of Pulsar Astronomy. Cambridge University Press, Cambridge (2005)
40. J. Lu, K. Lee, R. Xu, Advancing Pulsar Science with the FAST. Sci. China Phys. Mech. Astron. **63**(2), 229531 (2020)
41. A.G. Lyne, et al., A double-pulsar system: a rare laboratory for relativistic gravity and plasma physics. Science **303**, 1153–1157 (2004)
42. R.N. Manchester, Pulsars and gravity. Int. J. Mod. Phys. D **24**(06), 1530018 (2015)
43. R.N. Manchester, G.B. Hobbs, A. Teoh, M. Hobbs, The Australia Telescope National Facility pulsar catalogue. Astron. J. **129**, 1993 (2005)
44. X. Miao, L. Shao, B.-Q. Ma, Bounding the mass of graviton in a dynamic regime with binary pulsars. Phys. Rev. D **99**(12), 123015 (2019)
45. F.M. Ramazanoğlu, F. Pretorius, Spontaneous scalarization with massive fields. Phys. Rev. D **93**(6), 064005 (2016)
46. N. Sennett, L. Shao, J. Steinhoff, Effective action model of dynamically scalarizing binary neutron stars. Phys. Rev. D **96**(8), 084019 (2017)
47. L. Shao, Testing the strong equivalence principle with the triple pulsar PSR J0337+1715. Phys. Rev. D **93**(8), 084023 (2016)
48. L. Shao, Degeneracy in Studying the Supranuclear Equation of State and Modified Gravity with Neutron Stars. AIP Conf. Proc. **2127**(1), 020016 (2019)
49. L. Shao, General relativity withstands double pulsar's scrutiny. APS Phys. **14**, 173 (2021)

50. L. Shao, Imaging supermassive black hole shadows with a global very long baseline interferometry array. Front. Phys. (Beijing) **17**(4), 44601 (2022)
51. L. Shao, N. Wex, Tests of gravitational symmetries with radio pulsars. Sci. China Phys. Mech. Astron. **59**(9), 699501 (2016)
52. L. Shao, et al., Testing gravity with pulsars in the SKA era, in *Advancing Astrophysics with the Square Kilometre Array*, vol. AASKA14. Proceedings of Science (2015), p. 042
53. L. Shao, N. Sennett, A. Buonanno, M. Kramer, N. Wex, Constraining nonperturbative strong-field effects in scalar-tensor gravity by combining pulsar timing and laser-interferometer gravitational-wave detectors. Phys. Rev. X **7**(4), 041025 (2017)
54. L. Shao, N. Wex, M. Kramer, Testing the universality of free fall towards dark matter with radio pulsars. Phys. Rev. Lett. **120**(24), 241104 (2018)
55. L. Shao, N. Wex, S.-Y. Zhou, New graviton mass bound from binary pulsars. Phys. Rev. D **102**(2), 024069 (2020)
56. H.O. Silva, J. Sakstein, L. Gualtieri, T.P. Sotiriou, E. Berti, Spontaneous scalarization of black holes and compact stars from a Gauss-Bonnet coupling. Phys. Rev. Lett. **120**(13), 131104 (2018)
57. I.H. Stairs, Testing general relativity with pulsar timing. Living Rev. Rel. **6**, 5 (2003)
58. J.H. Taylor, Pulsar timing and relativistic gravity. Phil. Trans. A. Math. Phys. Eng. Sci. **341**(1660), 117–134 (1992)
59. G. Voisin, I. Cognard, P.C.C. Freire, N. Wex, L. Guillemot, G. Desvignes, M. Kramer, G. Theureau, An improved test of the strong equivalence principle with the pulsar in a triple star system. Astron. Astrophys. **638**, A24 (2020)
60. T.A. Wagner, S. Schlamminger, J.H. Gundlach, E.G. Adelberger, Torsion-balance tests of the weak equivalence principle. Class. Quant. Grav. **29**, 184002 (2012)
61. J.M. Weisberg, Y. Huang, Relativistic measurements from timing the binary pulsar PSR B1913+16. Astrophys. J. **829**(1), 55 (2016)
62. A. Weltman, et al., Fundamental Physics with the Square Kilometre Array. Publ. Astron. Soc. Austral. **37**, e002 (2020)
63. N. Wex, Testing relativistic gravity with radio pulsars, in *Frontiers in Relativistic Celestial Mechanics: Applications and Experiments*, ed. by S.M. Kopeikin, vol. 2 (Walter de Gruyter GmbH, Berlin/Boston, 2014), p. 39
64. N. Wex, M. Kramer, Gravity tests with radio pulsars. Universe **6**(9), 156 (2020)
65. C.M. Will. The confrontation between general relativity and experiment. Living Rev. Rel. **17**, 4 (2014)
66. R. Xu, Y. Gao, L. Shao, Strong-field effects in massive scalar-tensor gravity for slowly spinning neutron stars and application to X-ray pulsar pulse profiles. Phys. Rev. D **102**(6), 064057 (2020)
67. R. Xu, Y. Gao, L. Shao, Neutron stars in massive scalar-Gauss-Bonnet gravity: Spherical structure and time-independent perturbations. Phys. Rev. D **105**(2), 024003 (2022)
68. K. Yagi, D. Blas, E. Barausse, N. Yunes, Constraints on Einstein-Æther theory and Hořava gravity from binary pulsar observations. Phys. Rev. D **89**(8), 084067 (2014). Erratum: Phys.Rev.D 90, 069902 (2014), Erratum: Phys.Rev.D 90, 069901 (2014)
69. S.S. Yazadjiev, D.D. Doneva, D. Popchev, Slowly rotating neutron stars in scalar-tensor theories with a massive scalar field. Phys. Rev. D **93**(8), 084038 (2016)
70. J. Zhao, L. Shao, Z. Cao, B.-Q. Ma, Reduced-order surrogate models for scalar-tensor gravity in the strong field regime and applications to binary pulsars and GW170817. Phys. Rev. D **100**(6), 064034 (2019)
71. J. Zhao, P.C.C. Freire, M. Kramer, L. Shao, N. Wex, Closing a spontaneous-scalarization window with binary pulsars. Class. Quant. Grav. **39**(11), 11LT01 (2022)
72. W.W. Zhu, et al., Tests of gravitational symmetries with pulsar binary J1713+0747. Mon. Not. R. Astron. Soc. **482**(3), 3249–3260 (2019)

Testing Gravity and Predictions Beyond the Standard Model at Short Distances: The Casimir Effect

13

Galina L. Klimchitskaya and Vladimir M. Mostepanenko

Abstract

The Standard Model of elementary particles and their interactions does not include the gravitational interaction and faces problems in understanding dark matter, dark energy, strong CP violation etc. To solve these problems, many predictions of new light elementary particles and hypothetical interactions have been made. These predictions can be constrained by many means including measuring the Casimir force caused by the zero-point and thermal fluctuations. After discussing the theory of the Casimir effect, the strongest constraints on the power-type and Yukawa-type corrections to Newtonian gravity, following from measuring the Casimir force are considered. Next, the problems of dark matter, dark energy and their probable constituents are discussed. This is followed by an analysis of constraints on the dark matter particles, including axions and axion-like particles, obtained from the Casimir effect. The question of whether the Casimir effect can be used for constraining the spin-dependent interactions is considered. Then the constraints on the dark energy particles, like chameleons and symmetrons, are examined. In all cases we discuss not only measurements of the Casimir force but some other relevant table-top experiments as well.

G. L. Klimchitskaya
Central Astronomical Observatory at Pulkovo of the Russian Academy of Sciences, Saint Petersburg, Russia

Peter the Great Saint Petersburg Polytechnic University, Saint Petersburg, Russia

V. M. Mostepanenko (✉)
Central Astronomical Observatory at Pulkovo of the Russian Academy of Sciences, Saint Petersburg, Russia

Peter the Great Saint Petersburg Polytechnic University, Saint Petersburg, Russia

Kazan Federal University, Kazan, Russia

C. Pfeifer, C. Lämmerzahl (eds.), *Modified and Quantum Gravity*, Lecture Notes in Physics 1017, https://doi.org/10.1007/978-3-031-31520-6_13

In conclusion, the prospects of the Casimir effect for constraining theoretical predictions beyond the Standard Model are summarized.

13.1 Introduction: Gravity, the Standard Model and Beyond

The gravitational force is familiar to everybody from the day-to-day experience. If some body is released, it falls to earth under the influence of gravitational attraction. The laws of free fall were experimentally discovered by Galileo Galilei who found that in a vacuum the bodies of different weight fall with a uniform acceleration and reach the earth concurrently. This great (and somewhat counterintuitive) result was later derived theoretically by Newton from his second law and law of gravity under a fundamental assumption that the inertial and gravitational masses are equal (the equivalence principle).

It is common knowledge that according to Newton's law of gravity two point masses m_1 and m_2 separated by a distance r attract each other with the force

$$F_{gr}(r) = -\frac{dV_{gr}(r)}{dr} = -G\frac{m_1 m_2}{r^2}, \tag{13.1}$$

where G is the gravitational constant and V_{gr} is the gravitational interaction energy

$$V_{gr}(r) = -G\frac{m_1 m_2}{r}. \tag{13.2}$$

Einstein's general relativity theory [1] changed seriously the conceptual pattern of gravity. According to this theory, gravity is a curved space-time whose geometrical properties are determined not only by the masses of material bodies but by all components of their stress-energy tensor. It is important, however, that corrections to (13.1) and (13.2) predicted by the general relativity theory for mass and separation scales characteristic of a physical laboratory are negligibly small [2]. The gravitational force ensures the stability of planets, solar system, galaxies, and determines the structure and evolution of the whole Universe.

Another force which manifests itself in day-to-day life is the electromagnetic one. The classical theory of this force was created by Maxwell and is known as classical electrodynamics [3]. Unlike the gravitational force which is universal and acts between all material bodies, the electromagnetic force acts only between bodies possessing electric charges. The Maxwell theory describes only the classical aspects of electromagnetic interaction, whereas the full picture is given by quantum electrodynamics [4] created in the middle of the last century by Feynman, Schwinger, Tonomaga, and Dyson. Electromagnetic forces bind nuclei and electrons into atoms, create chemical bonds which make possible the existence of molecules. They are responsible for the structure of crystal lattices and are heavily used in electronics and all modern technologies.

Two other types of fundamental forces existing in nature, weak and strong interaction, are entirely quantum. They are not visible to the naked eye. The weak interaction is responsible for a decay of many elementary particles whereas the strong interaction binds protons and neutrons into atomic nuclei. In the middle of sixties of the last century, Weinberg, Salam and Glashow developed the unified theory of weak and electromagnetic interactions [5]. For electromagnetic interaction, the intermediate particles between electrically charged particles (the so-called force carriers) are the massless photons. Photons are the specific case of gauge bosons, i.e., particles of spin one which mediate different interactions. The force carriers for a weak interaction between particles are the three massive bosons, two of which, W^+ and W^-, are electrically charged and one, Z_0, is neutral.

By the middle of seventies, owing to works by Nambu, Gross, Wilczek, Politzer and other scientists, the theory of strong interactions had been elaborated. According to this theory, strongly interacting particles, e.g., protons and neutrons, consist of quarks possessing spin 1/2 and the new type of charge called color. Unlike the electric charge, which may be either positive or negative, the color charge has three different values (and respective anticolors). The force carriers for a strong interaction between quarks are eight massless gauge bosons called gluons which bear the color charges. Due to this the theory of strong interactions was called quantum chromodynamics [6].

The Standard Model is a unified theory of the three fundamental interactions—electromagnetic, weak, and strong [7]. According to the Standard Model, there are three pairs (generations) of quarks possessing the color charge and three pairs of spin 1/2 particles called leptons (the most familiar of them are electrons and respective electronic neutrinos). There are also as many antiquarks and antileptons as quarks and leptons. Next, the Standard Model includes the force carriers of electromagnetic, weak and strong interactions, i.e., photons, three massive bosons, W^+, W^-, Z^0, and eight gluons. Finally, an important element of the Standard Model is the heavy particle of zero spin called the Higgs boson predicted by Higgs in 1964, which is responsible for a generation of masses of other elementary particles. By now all the above elements of the Standard Model were observed in the accelerator experiments and many other theoretical predictions made on this basis found their experimental confirmation.

Great successes of the Standard Model in particle physics do not mean, however, that we already have the theory of everything. The major problem is that the general relativity theory remains to be isolated from the Standard Model, and the gravitational force avoids unification with other three fundamental interactions in spite of persistent efforts undertaken during several decades. Both Newtonian gravity and Einstein's general relativity are the entirely classical theories. However, for a description of physical phenomena happening in the close proximity of space-time singularities predicted by the general relativity theory, one evidently needs some theory which takes into account the quantum effects. There are also unresolved problems of dark matter and dark energy which are observed only indirectly through their gravitational interactions but are not explained in the context of the Standard Model. It should be mentioned also that at short distances below a micrometer the Newton law (13.1) lacks of experimental confirmation and leaves a room for modifications at the cost of different quantum effects.

There are also other serious problems of the Standard Model. Among them one should mention the hierarchy problem, i.e., an unanswered question of why there is a difference by the factor of 10^{24} between the strength of weak and gravitational interactions, the problem of neutrino mass, which was zero in the original formulation of the model but turns out to be nonzero according to precise measurements, and the problem of an asymmetry between matter and antimatter. An important problem is also the strong CP violation (i.e., the violation of invariance relative to the charge conjugation accompanied by the parity transformation) which is admitted by the formalism of quantum chromodynamics but is not observed in experiments involving only the strong interaction.

All these problems are widely discussed in the literature, and many theoretical approaches to their resolution are proposed in the framework of extended standard model, supersymmetry, supergravity [8], and string theory [9] (see also the discussions in Part I of this book) The mentioned approaches go beyond the Standard Model and introduce additional particles, interactions, and symmetries leading to some theoretical predictions which can be verified experimentally using the powerful high energy accelerators, astrophysical observations, and laboratory experiments.

Below we consider some of these predictions which can be verified in experiments on measuring the Casimir force arising between two closely spaced uncharged material bodies due to the zero-point and thermal fluctuations of the electromagnetic field. As it is shown below, these relatively cheap and compact laboratory experiments can compete with huge accelerators in testing some important theoretical predictions beyond the Standard Model.

13.2 Electromagnetic Casimir Force and the Quantum Vacuum

In 1948, Casimir [10] considered two parallel, uncharged ideal metal planes in vacuum at zero temperature spaced at a distance a and calculated the zero-point energy of the electromagnetic field in the presence and in the absence of these planes, i.e., in free space. The case with the planes differs in that the tangential component of electric field and the normal component of magnetic induction vanish on their surfaces. Casimir considered a difference between the zero-point energies per unit area in the presence and in the absence of planes

$$E(a) = \hbar \int_0^\infty \frac{k_\perp dk_\perp}{2\pi} \left({\sum_{l=0}^{\infty}}' \omega_{k_\perp,l} - \frac{a}{\pi} \int_0^\infty dk_z \omega_k \right), \tag{13.3}$$

where $\boldsymbol{k} = (k_x, k_y, k_z)$ is the wave vector, $k_\perp = \sqrt{k_x^2 + k_y^2}$ is the magnitude of the wave vector projection on the planes, the prime on the summation sign divides the term with $l = 0$ by 2, and the frequencies of the zero-point oscillations are given by the following expressions:

$$\omega_{k_\perp,l} = c\sqrt{k_\perp^2 + \left(\frac{\pi l}{a}\right)^2}, \qquad \omega_k = c\sqrt{k_\perp^2 + k_z^2}. \tag{13.4}$$

? Exercise

13.1. Verify the expression for $\omega_{k_\perp,l}$ in Eq. 13.4. Consult [11] for details.

Although both terms on the right-hand side of (13.3) are infinitely large, their difference is finite. Using the Abel-Plana formula for a difference between the sum and the integral [12], one obtains

$$E(a) = -\frac{\pi^2\hbar c}{a^3}\int_0^\infty y\,dy \int_y^\infty \frac{\sqrt{t^2 - y^2}}{e^{2\pi t} - 1}dt. \tag{13.5}$$

Then, calculating the integrals in (13.5), one arrives at the famous Casimir result

$$E(a) = -\frac{\pi^2}{720}\frac{\hbar c}{a^3} \tag{13.6}$$

and at respective expression for the Casimir force per unit area of the plates

$$P(a) = -\frac{dE(a)}{da} = -\frac{\pi^2}{240}\frac{\hbar c}{a^4}, \tag{13.7}$$

i.e., the Casimir pressure. This pressure is some kind of a macroscopic quantum effect determined entirely by the zero-point oscillations of quantized electromagnetic field. Thus, for two ideal metal planes separated by a distance $a = 1\,\mu\text{m}$ we obtain from (13.7) an attractive pressure $P(a) = -1.3\,\text{mPa}$.

In a physical laboratory we deal not with ideal metals but with real material bodies made of metallic, dielectric or semiconductor materials. In 1955, Lifshitz [13] created the general theory describing the free energy and force arising between two thick material plates (semispaces) spaced at a separation a in thermal equilibrium with the environment at temperature T. The material properties in this theory were characterized by the dielectric permittivities $\varepsilon^{(n)}(\omega)$ of the first and second plates ($n = 1, 2$). Later the Lifshitz results were generalized for the plates possessing magnetic properties characterized by the magnetic permeabilities $\mu^{(n)}(\omega)$ [14].

In the framework of the Lifshitz theory, the free energy of interaction caused by the zero-point and thermal fluctuations of the electromagnetic field per unit area of the plates is given by Lifshitz [13] and Bordag et al. [11]

$$\mathcal{F}(a,T) = \frac{k_B T}{2\pi} \sum_{l=0}^{\infty}{}' \int_0^\infty k_\perp dk_\perp \sum_\alpha \ln\left[1 - r_\alpha^{(1)}(i\xi_l, k_\perp) r_\alpha^{(2)}(i\xi_l, k_\perp) e^{-2aq_l}\right], \tag{13.8}$$

where k_B is the Boltzmann constant, $\xi_l = 2\pi k_B T l/\hbar$ are the Matsubara frequencies, $q_l = \sqrt{k_\perp^2 + \xi_l^2/c^2}$, and the reflection coefficients for two independent polarizations of the electromagnetic field, transverse magnetic (α = TM) and transverse electric (α = TE), are given by

$$\begin{aligned} r_{\mathrm{TM}}^{(n)}(i\xi_l, k_\perp) &= \frac{\varepsilon^{(n)}(i\xi_l) q_l - k^{(n)}(i\xi_l, k_\perp)}{\varepsilon^{(n)}(i\xi_l) q_l + k^{(n)}(i\xi_l, k_\perp)}, \\ r_{\mathrm{TE}}^{(n)}(i\xi_l, k_\perp) &= \frac{\mu^{(n)}(i\xi_l) q_l - k^{(n)}(i\xi_l, k_\perp)}{\mu^{(n)}(i\xi_l) q_l + k^{(n)}(i\xi_l, k_\perp)}, \end{aligned} \tag{13.9}$$

where

$$k^{(n)}(i\xi_l, k_\perp) = \sqrt{k_\perp^2 + \varepsilon^{(n)}(i\xi_l)\mu^{(n)}(i\xi_l)\frac{\xi_l^2}{c^2}}. \tag{13.10}$$

In a similar way, the Casimir force per unit area of real material plates is expressed as

$$\begin{aligned} P(a,T) = -\frac{\partial \mathcal{F}(a,T)}{\partial a} &= -\frac{k_B T}{\pi} \sum_{l=0}^{\infty}{}' \int_0^\infty q_l k_\perp dk_\perp \\ &\times \sum_\alpha \left[\frac{e^{2aq_l}}{r_\alpha^{(1)}(i\xi_l, k_\perp) r_\alpha^{(2)}(i\xi_l, k_\perp)} - 1\right]^{-1}. \end{aligned} \tag{13.11}$$

Taking into account that ideal metal is the perfect reflector, so that

$$r_{\mathrm{TM}}^{(n)}(i\xi_l, k_\perp) = -r_{\mathrm{TE}}^{(n)}(i\xi_l, k_\perp) = 1 \tag{13.12}$$

at all ξ_l, at $T = 0$ one obtains from (13.8) and (13.11) the Casimir results (13.6) and (13.7). In the limiting case of small separations, (13.8) and (13.11) describe the familiar van der Waals force which depends on $\hbar$ but does not depend on the speed of light c. In the opposite limiting case of large separations, the resulting free energy and force do not depend either on $\hbar$ or c. This is the so-called classical regime where the Casimir interaction depends only on T.

Precise measurements of the Casimir force allowing quantitative comparison between experiment and theory were performed by means of an atomic force microscope, whose sharp tip was replaced with a relatively large sphere, and a micromechanical torsional oscillator (see [11, 15] for a review). All these experiments measured the Casimir force not between two parallel plates but between a sphere and a plate. The Casimir force between a sphere of radius R and a plate $F^{SP}(a,T)$ can be calculated in the framework of the Lifshitz theory using the proximity force approximation [11, 15]

$$F^{SP}(a,T) = 2\pi R\mathcal{F}(a,T), \tag{13.13}$$

where the Casimir free energy between two parallel plates $\mathcal{F}$ is given by the Lifshitz formula (13.8). Exact calculations of the Casimir force in sphere-plate geometry using the scattering approach [16–19] and the gradient expansion [20–24] have shown that the errors introduced by (13.13) are less than a/R, i.e., less than a fraction of a percent in the most of experimental configurations.

By calculating the derivative of (13.13) with respect to separation, one can express another quantity measured in many experiments, i.e., the gradient of the Casimir force in sphere-plate geometry via the Casimir force (13.11) per unit area of two parallel plates

$$\frac{\partial}{\partial a}F^{SP}(a,T) = -2\pi RP(a,T). \tag{13.14}$$

For comparison of theoretical predictions with the measurement data of precise experiments, one should compute the Casimir free energy (13.8) and the Casimir pressure (13.11) with sufficient precision. To do so, one needs to have the values of dielectric permittivities of plate materials at sufficiently large number of pure imaginary Matsubara frequencies. This is usually achieved by means of the Kramers-Kronig relation using the measured optical data for the complex indices of refraction of plate materials. In doing so the terms of the Lifshitz formulas (13.8) and (13.11) with $l = 0$ play an important role in obtaining the physically correct results.

Unfortunately, the optical data are available at only sufficiently high frequencies $\omega \geqslant \omega_{\min}$. Because of this, the obtained dielectric permittivity is usually extrapolated down to zero frequency using some theoretical model. For experiments with metallic test bodies, which are used below for testing the predictions beyond the Standard Model, the most reasonable extrapolation seems to be by means of the well tested Drude model. In this case, the dielectric permittivities of plate materials take the form

$$\varepsilon_D^{(n)}(i\xi_l) = \varepsilon_c^{(n)}(i\xi_l) + \frac{\omega_{p,n}^2}{\xi_l(\xi_l+\gamma_n)}, \tag{13.15}$$

where $\varepsilon_c^{(n)}(i\xi_l)$ is a contribution due to core electrons determined by the optical data, $\omega_{p,n}$ is the plasma frequency and γ_n is the relaxation parameter.

It turned out, however, that the measurement data of all precise experiments with nonmagnetic (Au) metals [25–33] and magnetic (Ni) metals [34–37] exclude the theoretical predictions of the Lifshitz theory using the dielectric functions (13.15). Specifically, for two test bodies made of Ni a disagreement between experiment and theory in measurements of the differential Casimir force is up to a factor of 1000 [37]. If, however, one makes an extrapolation by means of the plasma model, i.e., puts $\gamma_n = 0$ in (13.15),

$$\varepsilon_p^{(n)}(i\xi_l) = \varepsilon_c^{(n)}(i\xi_l) + \frac{\omega_{p,n}^2}{\xi_l^2}, \tag{13.16}$$

the predictions of the Lifshitz theory come to a very good agreement with the measurement data of all precise experiments [25–37].

This situations calls for some clarification because at low frequencies conduction electrons really possess relaxation properties described by the phenomenological parameter γ_n. It is then unclear why one should put $\gamma_n = 0$ in computations of the Casimir force. Although the ultimate answer to this question is not found yet, theory suggests some plausible explanation. First of all, it was proven [38–41] that for metals with perfect crystal lattices the Casimir entropy calculated using the dielectric permittivity (13.15) violates the third law of thermodynamics, the Nernst heat theorem, but satisfies it if the permittivity (13.16) is used.

Next, for graphene, which is a novel 2D material [42], the dielectric permittivity is not of a model character. At low energies characteristic for the Casimir effect, the dielectric properties of graphene can be calculated on the basis of first principles of quantum electrodynamics at nonzero temperature using the polarization tensor in (2+1)-dimensional space-time [43, 44]. It was found that graphene is described by two spatially nonlocal dielectric permittivities, i.e., depending on both the frequency ω and the 2D wave vector $\boldsymbol{k}$ [45, 46]. The Lifshitz theory using these permittivities turned out to be in perfect agreement with measurements of the Casimir force from graphene [47–50] and with the Nernst heat theorem [51–55].

This suggests that the model dielectric permittivity (13.15), which is well-checked for the propagating electromagnetic waves on the mass shell in vacuum, may be inapplicable to the evanescent (off-the-mass-shell) waves. The latter contribute essentially to the Casimir free energy and force (13.8) and (13.11) caused by the electromagnetic fluctuations. First steps on the road to justification of this conjecture were made by the recently proposed spatially nonlocal dielectric permittivities which describe nearly the same response, as does the Drude model, to the propagating waves but an alternative response to the evanescent ones [56, 57]. The Lifshitz theory employing these permittivities is in as good agreement with measurements of the Casimir force between nonmagnetic metals and with the Nernst heat theorem as when it uses the plasma model (13.16) [56–58]. Recently it was also shown that it agrees equally well with measurements of the Casimir force between magnetic metals [57, 59].

By and large one can conclude that although there is a continuing discussion in the literature on theoretical description of the Casimir interaction between real material bodies (see [60] for a review), the predictions of the Lifshitz theory are now found in good agreement with the measurement data of all precise experiments and the measure of this agreement can be used for constraining the hypothetical forces of nonelectromagnetic origin.

13.3 Testing the Power-Type Corrections to Newtonian Gravity from the Casimir Effect

From the point of view of quantum field theory, the gravitational interaction energy (13.2) can be considered as originating from an exchange of one massless particle between two massive particles m_1 and m_2. Exactly in this way the Coulomb potential is derived in quantum electrodynamics by considering an exchange of one photon between two charged particles.

The Standard Model does not contain massless particles in a free state except of photons (gluons are confined inside of barions). There are, however, massless particles predicted by some extensions of the Standard Model. For instance, theory of electroweak interactions with an extended Higgs sector predicts pseudoscalar massless particles called arions [61]. An exchange of one arion between electrons belonging to atoms of two neighboring test bodies leads to the spin-dependent effective potential which averages to zero when integrating over their volumes. The spin-independent effective potential decreasing with separation as r^{-3} arises from the process of two-arions exchange [62].

In a similar way, the effective potential decreasing with separation as r^{-5} arises from an exchange of neutrino-antineutrino pair between two neutrons [63, 64]. The power-type potentials result also from an exchange of even numbers of goldstinos which are the massless fermions introduced in the theoretical schemes with a spontaneously broken supersymmetry [65] and other predicted particles.

Taking into account that the power-type interactions with different powers coexist with the gravitational potential, the resulting interaction energy is usually represented as

$$V_l(r) = -\frac{Gm_1m_2}{r}\left[1 + \Lambda_l\left(\frac{r_0}{r}\right)^{l-1}\right], \tag{13.17}$$

where Λ_l is the dimensionless interaction constant, $l = 1, 2, 3, \ldots$, and r_0 with the dimension of length is introduced to preserve the dimension of energy for $V_l(r)$. Following many authors, we put $r_0 = 1\text{ F} = 10^{-15}$ m. For $l = 1$, the quantity $1+\Lambda_1$ has the meaning of a factor connecting the values of inertial and gravitational masses, for $l = 3$ the second term in (13.17) presents a correction to the Newtonian potential due to an exchange of two arions, and for $l = 5$—due to an exchange of neutrino-antineutrino pair.

The power-type corrections to Newton's law arise not only due to an exchange of massless hypothetical particles but in extensions of the Standard Model which exploit the extra-dimensional unification schemes with noncompact but warped extra dimensions. In this case, the modified gravitational interaction energy at separations $r \gg K_w$ takes the form [66, 67]

$$V_3(r) = -\frac{Gm_1m_2}{r}\left(1 + \frac{2}{3K_w^2r^2}\right), \tag{13.18}$$

where K_w is the warping scale. This is the potential of the form of (13.17) with $\Lambda_3 = 2/(3K_w^2r_0^2)$.

Constraints on the values of interaction constant Λ_l with different l can be obtained from the gravitational experiments of Eötvos and Cavendish type. In the Eötvos-type experiments one verifies a validity of the equivalence principle, i.e., places limits on possible deviations between the inertial and gravitational masses. Using (13.17), these limits can be recalculated in the constraints on Λ_1. Thus, from the most precise short-range Eötvos-type experiments [68, 69] the constraint $|\Lambda_1| \leqslant 1 \times 10^{-9}$ was obtained.

In the Cavendish-type experiments, one measures probable deviations of the force acting between two bodies from the Newton law (13.1). From the power-type interaction energy (13.17) one finds the respective force

$$F_l(r) = -\frac{dV_l(r)}{dr} = -\frac{Gm_1m_2}{r^2}\left[1 + l\Lambda_l\left(\frac{r_0}{r}\right)^{l-1}\right]. \tag{13.19}$$

Then the constraints on Λ_l can be found from the measured limits on the dimensionless quantity

$$\varepsilon_l = \frac{1}{rF_l(r)}\frac{d}{dr}\left[r^2F_l(r)\right], \tag{13.20}$$

which is equal to zero if $\Lambda_l = 0$, i.e., no power-type interaction in addition to gravity is present. Using this approach, from the Cavendish-type experiment [70] the following constraints on Λ_l were obtained [71]: $|\Lambda_2| \leqslant 4.5\times 10^8$, $|\Lambda_3| \leqslant 1.3\times 10^{20}$, $|\Lambda_4| \leqslant 4.9 \times 10^{31}$, $|\Lambda_5| \leqslant 1.5 \times 10^{43}$.

In [62, 72] it was suggested to obtain constraints on the power-type interactions from measurements of the Casimir force. The Casimir force $F^{LP}(a)$ between a spherical lens of centimeter-size radius and a plate both made of quartz was measured at distances $a \leqslant 1\,\mu$m in [73] with a relative error $\Delta F/F^{LP} \approx 10\%$, where ΔF is the absolute error. In the limits of this error, the measurement data were found to be in agreement with theoretical predictions of the Lifshitz theory.

Any hypothetical interaction energy of power type between an atom of the lens at a point $\boldsymbol{r}_1$ and an atom of the plate at a point $\boldsymbol{r}_2$ is given by (13.17) where $r =$

$|\boldsymbol{r}_1 - \boldsymbol{r}_2|$. Then, the total interaction force between the experimental test bodies (the lens and the plate) is given by the integration over their volumes V_1 and V_2 with subsequent negative differentiation with respect to the distance a of their closest approach

$$F_l^{LP}(a) = -n_1 n_2 \frac{\partial}{\partial a} \int_{V_1} d^3 r_1 \int_{V_2} d^3 r_2 V_l(|\boldsymbol{r}_1 - \boldsymbol{r}_2|), \tag{13.21}$$

where n_1 and n_2 are the numbers of atoms per unit volume of the first and second test bodies. In doing so, one can neglect by the Newtonian contribution on the right-hand side of (13.21) because it is negligibly small as compared to the experimental error in the micrometer separation range.

Taking into account that no additional interaction was observed within the limits of measurement errors, the constraints on Λ_l with $l = 1, 2, 3, 4,$ and 5 were obtained from the inequality [62, 72]

$$|F_l^{LP}(a)| \leqslant \Delta F(a). \tag{13.22}$$

Among these constraints, that ones on Λ_2 and Λ_3 turned out to be stronger as compared with constraints found from older Cavendish-type experiments available in 1987 [74].

It would be interesting to estimate potentialities of modern measurements of the Casimir force for constraining the power-type interactions. For this purpose we consider the most recent experiment [33] on measuring the Casimir force between an Au-coted sphere of $R = 149.7\,\mu$m radius and an Au-coated plate in the micrometer separation range. The sphere is spaced at a height a above the plate. To estimate the strongest constraints that could be obtained from the experiments of this kind, we consider both the sphere and the plate as all-gold (in real experiment the sapphire sphere and silicon plate were coated with Au films of 250 and 150 nm thicknesses, respectively). The plate can be considered as infinitely large becausc its size was much larger then the sphere radius.

Let the plate top be in the plane $z = 0$ and an atom of the sphere has the coordinates $\boldsymbol{r}_1 = (0, 0, z)$. For all powers $l \geqslant 3$ in (13.17) the plate can be considered as infinitely thick. The atom-plate force arising due to the second contribution on the right-hand side of (13.17) is given by

$$\begin{aligned} F_l^{AP}(a) &= G m_1 m_2 n_2 \Lambda_l r_0^{l-1} \frac{\partial}{\partial z} \int_{V_2} d^3 r_2 \frac{1}{|\boldsymbol{r}_1 - \boldsymbol{r}_2|^{l-1}} \\ &= -\frac{2\pi}{l-2} G \rho_2 m_1 \Lambda_l r_0^{l-1} \frac{1}{z^{l-2}}, \end{aligned} \tag{13.23}$$

where $\rho_2 = m_2 n_2$ is the mass density of the plate material (Au).

? Exercise

13.2. Derive (13.23) by performing integration in the cylindrical coordinate system.

Now we integrate (13.23) over the volume of a sphere. The density of atoms at a height $z \geqslant a$ in thin horizontal layer of the sphere is given by

$$\pi n_1 \left[2R(z-a)-(z-a)^2\right]. \tag{13.24}$$

Then, the sphere-plate force is found by integrating (13.23) with the weight (13.24)

$$F_l^{SP}(a) = -\frac{2\pi^2}{l-2} G\rho_1\rho_2\Lambda_l r_0^{l-1} \int_a^{2R+a} \frac{2R(z-a)-(z-a)^2}{z^{l-2}}\,dz, \tag{13.25}$$

where $\rho_1 = m_1 n_1$ is the mass density of the sphere material (in our case also Au).

Introducing the new integration variable $t = z-a$, we rewrite (13.25) in the form

$$F_l^{SP}(a) = -\frac{2\pi^2}{l-2} G\rho_1\rho_2\Lambda_l r_0^{l-1} \int_0^{2R} \frac{2Rt-t^2}{(a+t)^{l-2}}\,dt. \tag{13.26}$$

Finally, calculating the integral in (13.26), one arrives at

$$F_l^{SP}(a) = -\frac{2\pi^2}{l-2} G\Lambda_l\rho_1\rho_2 \frac{r_0^{l-1}R^3}{a^{l-2}}\, {}_2F_1(l-2,2;3;-2Ra^{-1}), \tag{13.27}$$

where ${}_2F_1(a,b;c;z)$ is the hypergeometric function.

Substituting (13.27) in place of F_l^{LP} in (13.22), one finds the strongest constraints on Λ_l obtainable from the experiment [33] if it would be performed with the all-gold test bodies. The numerical analysis shows that the most strong constraints follow at $a = 3\,\mu$m where $\Delta F(a) = 2.2\,$fN [33]. The obtained constraints are: $|\Lambda_3| \leqslant 1.3 \times 10^{23}$, $|\Lambda_4| \leqslant 1.8 \times 10^{34}$, and $|\Lambda_5| \leqslant 5.6 \times 10^{44}$. It is seen that these constraints are weaker than those mentioned above following from the Cavendish-type experiment [70, 71].

The case of $l = 2$ should be considered separately. In this case, it is necessary to take into account the finite thickness of the plate $D = 50\,\mu$m because for $l = 2$ an integral over the plate of infinitely large thickness (i.e., over the semispace) diverges.

Table 13.1 The strongest constraints on the constants of power-type hypothetical interaction following from the Eötvos-type (line 1) and Cavendish-type (lines 2–5) experiments

l	$\|\Lambda_l\|_{max}$		$\|\Lambda_l\|_{max}$	
1	1×10^{-9}	[68]	1×10^{-9}	[68]
2	4.5×10^{8}	[71]	3.7×10^{8}	[75]
3	1.3×10^{20}	[71]	7.5×10^{19}	[75]
4	4.9×10^{31}	[71]	2.2×10^{31}	[75]
5	1.5×10^{43}	[71]	6.7×10^{42}	[75]

By performing calculations in the same way as above, one obtains

$$F_2^{SP}(a) = -\frac{2\pi^2}{3}\rho_1\rho_2 G\Lambda_2 r_0 \left[2RD(2a+2R+D) + (a+3R)a^2 \ln\frac{a+2R}{a} - (a+D)^2(a+3R+D)\ln\frac{a+2R+D}{a+D} + 4R^3 \ln\frac{a+2R+D}{a+2R}\right]. \tag{13.28}$$

Substituting (13.28) in (13.22) in place of F_l^{LP} and using $\Delta F(a) = 2.2\,\text{fN}$, one finds $|\Lambda_2| \leqslant 2.85 \times 10^{12}$. This is again a much weaker constraint than that obtained in [71] based on the Cavendish-type experiment [70]. One can conclude that the short-separation Cavendish-type experiments are more prospective for constraining the power-type hypothetical interactions than measurements of the Casimir force.

This conclusion finds further confirmation from the recently performed Cavendish-type experiment which presents an improved test of Newton's gravitational law at short separations [75]. The constraints on Λ_l with $l = 2, 3, 4$, and 5 obtained in this work are somewhat stronger than those cited above [70, 71]. In Table 13.1 (line 1) we present the strongest constraint on Λ_1 following from the Eötvos-type experiment [68]. In columns 2 and 3 (lines 2–5), the strongest constraints Λ_l with $l = 2, 3, 4$, and 5 following from the Cavendish-type experiments [71] and [75], respectively, are presented. As is seen in Table 13.1, the strength of constraints quickly drops with increase of the interaction power.

13.4 Testing the Yukawa-Type Corrections to Newtonian Gravity from the Casimir Effect

The interaction energy of Yukawa type between two pointlike particles (atoms or molecules) separated by a distance r arises due to an exchange of one light scalar particle. The Standard Model considered in Sect. 13.1 contains only one scalar particle, the Higgs boson, which is very heavy and cannot serve as an exchange boson in long-range interactions. The extensions of the Standard Model predict, however, a number of light scalar particles such as moduli [76], which arise in supersymmetric theories, dilaton [77], which appears in extra-dimensional models with the varying volume of compactified dimensions, scalar axion [78], which is a superpartner of an axion, etc. (see Sect. 13.5 and also the discussions on scalar-teleparallel and scalar-tensor theories in Chaps. 4, 7 and 10).

Similar to the power-type interactions, the interaction of Yukawa type between two particles of masses m_1 and m_2 coexists with gravity and is usually parametrized as

$$V_{\mathrm{Yu}}(r) = -\frac{Gm_1m_2}{r}\left(1+\alpha e^{-r/\lambda}\right), \tag{13.29}$$

where α is the dimensionless interaction constant and λ is the interaction range having the meaning of the Compton wavelength of exchange scalar particle of mass m_s: $\lambda = \hbar/(m_s c)$.

Another prediction of the Yukawa-type correction to Newton's gravitational law comes from the extra-dimensional models with compact extra dimensions and low-energy compactification scale [79, 80]. In the framework of this approach beyond the Standard Model, the space-time has $D = 4 + N$ dimensions where N extra dimensions are compactified at relatively low Planck energy scale in D dimensions

$$E_{\mathrm{Pl}}^{(D)} = \left(\frac{\hbar^{1+N}c^{5+N}}{G_D}\right)^{\frac{1}{2+N}} \sim 1\ \mathrm{TeV}. \tag{13.30}$$

Here, G_D is the gravitational constant in the extended D-dimensional space-time $G_D = G\Omega_N$ and $\Omega_N \sim R_*^N$, R_* being the size of compact manifold.

In fact the approach under consideration was suggested as a possible solution of hierarchy problem discussed in Sect. 13.1 since due to (13.30) the characteristic energy scales of the gravitational and gauge interactions of the Standard Model coincide. In doing so the size of compact manifold is given by Arkani-Hamed et al. [80]

$$R_* \sim \frac{\hbar c}{E_{\mathrm{Pl}}^{(D)}}\left[\frac{E_{\mathrm{Pl}}}{E_{\mathrm{Pl}}^{(D)}}\right]^{\frac{2}{N}} \sim 10^{\frac{32-17N}{N}}, \tag{13.31}$$

where the usual Planck energy $E_{\mathrm{Pl}} = (\hbar c^5/G)^{1/2} \sim 10^{19}$ GeV.

According to the developed approach, the standard Newton law (13.1) and (13.2) is not valid in D-dimensional space-time. It was shown [81, 82] that at separations $r \gg R_*$ the gravitational interaction energy takes the form (13.29) with $\lambda \sim R_*$. Although for one extra dimension ($N = 1$) Eq. (13.31) leads to too large $R_* \sim 10^{15}$ cm, which is excluded by the tests of Newton's law in the solar system [83], for $N = 2$ and 3 (13.31) leads to the more realistic results $R_* \sim 1$ mm and $R_* \sim 5$ nm, respectively. This means that a search for deviations from the Newton law at short distances is not only a quest for hypothetical particles, but for extra dimensions as well.

Constraints on the parameters of Yukawa-type interaction α and λ can be obtained from the Cavendish-type experiments. The potential energy (13.29) results in the force acting between two particles m_1 and m_2

$$F_{\mathrm{Yu}}(r) = -\frac{dV_{\mathrm{Yu}}(r)}{dr} = -\frac{Gm_1m_2}{r^2}\left[1+\alpha e^{-r/\lambda}\left(1+\frac{r}{\lambda}\right)\right]. \tag{13.32}$$

Then the quantity

$$\varepsilon_{\mathrm{Yu}} = \frac{1}{rF_{\mathrm{Yu}}(r)}\frac{d}{dr}\left[r^2F_{\mathrm{Yu}}(r)\right] \tag{13.33}$$

is not equal to zero due to a nonzero strength of the Yukawa force α. The deviation of this quantity from zero (if any) could be determined from the results of Cavendish-type experiments. Depending on the range of λ, different Cavendish-type experiments lead to the strongest constraints on α. For $8\,\mu\mathrm{m} < \lambda < 9\,\mu\mathrm{m}$ the most strong constraints follow from the short-range test of Newtonian gravity at 20 micrometers [84]. The Cavendish-type experiment [70], already discussed in Sect. 13.3 in the context of power-type interactions, leads to the strongest constraints on α within the wide interaction range $9\,\mu\mathrm{m} < \lambda < 4\,\mathrm{mm}$ [71]. It should be noted, however, that in the part of this interval $40\,\mu\mathrm{m} < \lambda < 0.35\,\mathrm{mm}$ the obtained results have been strengthened by up to a factor of 3 in the refined experiment [75] which was also mentioned in Sect. 13.3. Finally, an older Cavendish-type experiment [85] performed at larger separations allows obtaining the strongest constraints on α in the range $4\,\mathrm{mm} < \lambda < 1\,\mathrm{cm}$. In the range of even larger λ, unrelated to the Casimir force, the strongest constraints on α follow from the Eötvos-type experiments [69, 86].

In Fig. 13.1, we present the constraints on α obtained from different gravitational experiments by the line labeled gr. Only the range of λ below $66\,\mu\mathrm{m}$ is included neighboring to the region considered below where the strongest constraints on α follow from experiments performed in the Casimir regime. The values of parameters of the Yukawa-type interaction belonging to the area of (λ, α) plane above the line are excluded by the results of Cavendish-type experiments mentioned above, whereas the area of the same plane below the line is allowed. At $\lambda = 8\,\mu\mathrm{m}$ the line gr intersects with the end of the line Casimir-less which is discussed below in this section.

As is seen in Fig. 13.1, the strength of constraints obtained from gravitational experiments quickly drops with decreasing λ. As an example, for $r = \lambda = 10\,\mu\mathrm{m}$ the Cavendish-type experiments do not exclude an existence of the Yukawa-type force between two particles which exceeds the Newtonian gravitational force by the factor of 10^4. This means that the Newtonian law of gravitation lacks of sufficient experimental confirmation at short separations which prevents obtaining strong constraints on some other forces from gravitational experiments.

In fact at separations of the order of micrometer and less the main background force between two material bodies far exceeding the gravitational interaction is the

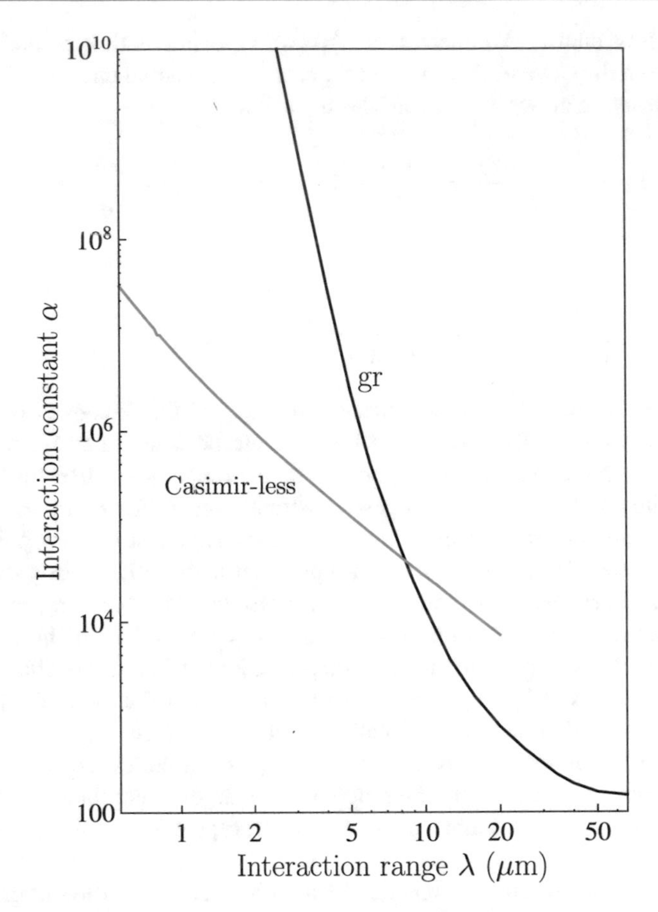

Fig. 13.1 Constraints on the interaction constant α of Yukawa-type interaction are shown as functions of the interaction range λ by the lines labeled gr and Casimir-less obtained from the gravitational and Casimir-less experiments, respectively. The regions of (λ, α) plane above each line are excluded and below are allowed

Casimir force considered in Sect. 13.2. In [87] it was suggested to constrain the hypothetical Yukawa-type interaction from experiments on measuring the van der Waals and Casimir forces.

Similar to the case of power-type interactions, the Yukawa-type force acting between two test bodies spaced at a closest separation a can be obtained by an integration of the interaction energy (13.29) over their volumes with subsequent negative differentiation with respect to a

$$F_{\rm Yu}^{V_1V_2}(a) = -n_1n_2\frac{\partial}{\partial a}\int_{V_1} d^3r_1 \int_{V_2} d^3r_2 V_{\rm Yu}(|\boldsymbol{r}_1 - \boldsymbol{r}_2|). \tag{13.34}$$

Taking into account that within the experimental error $\Delta F(a)$ the measured Casimir force was found to be in agreement with theoretical predictions, the constraints on F_{Yu} can be found from the inequality

$$|F_{\text{Yu}}^{V_1V_2}(a)| \leqslant \Delta F(a). \tag{13.35}$$

Following this approach, the first constraints on the Yukawa-type interaction with $\lambda < 20\,\text{cm}$ were obtained [87] from two experiments [73, 88] performed long ago.

During the last 20 years many experiments on measuring the Casimir interaction have been performed (some of them are mentioned in Sect. 13.2). All these experiments use the configuration of a sphere above a plate which surfaces may be coated by some additional material layers or covered with sinusoidal corrugations. We start with the simplest configuration of a smooth sphere of radius R at the closest separation a above a large smooth plate of thickness D. We again consider an atom m_1 of the sphere at a height z above the plate and integrate the Yukawa interaction energy (13.29) over the plate volume V_2 with subsequent negative differentiation according to (13.34). As explained above, the contribution of gravitational interaction can be neglected. Then, similar to (13.23), for the atom-plate force one obtains

$$F_{\text{Yu}}^{AP}(z) = -2\pi G\rho_2 m_1\alpha\lambda e^{-z/\lambda}\left(1 - e^{-D/\lambda}\right). \tag{13.36}$$

? Exercise

13.3. Derive (13.36) using the cylindrical coordinate system.

Now we integrate (13.36) over the sphere volume using (13.24) and obtain the Yukawa-type force acting between a sphere and a plate [89]

$$\begin{aligned} F_{\text{Yu}}^{AP}(a) &= -2\pi^2 G\rho_1\rho_2\alpha\lambda\left(1 - e^{-D/\lambda}\right)\int_a^{2R+a} dz\left[2R(z-a) - (z-a)^2\right]e^{-z/\lambda} \\ &= -4\pi^2 G\rho_1\rho_2\alpha\lambda^3\left(1 - e^{-D/\lambda}\right)e^{-a/\lambda}\Phi(R.\lambda), \end{aligned} \tag{13.37}$$

where the following notation is introduced

$$\Phi(r, \lambda) = r - \lambda + (r + \lambda)e^{-2r/\lambda}. \tag{13.38}$$

The strongest current constraints on the Yukawa-type force in the wide interaction range from 10 nm to 8 μm follow from four experiments of the Casimir physics.

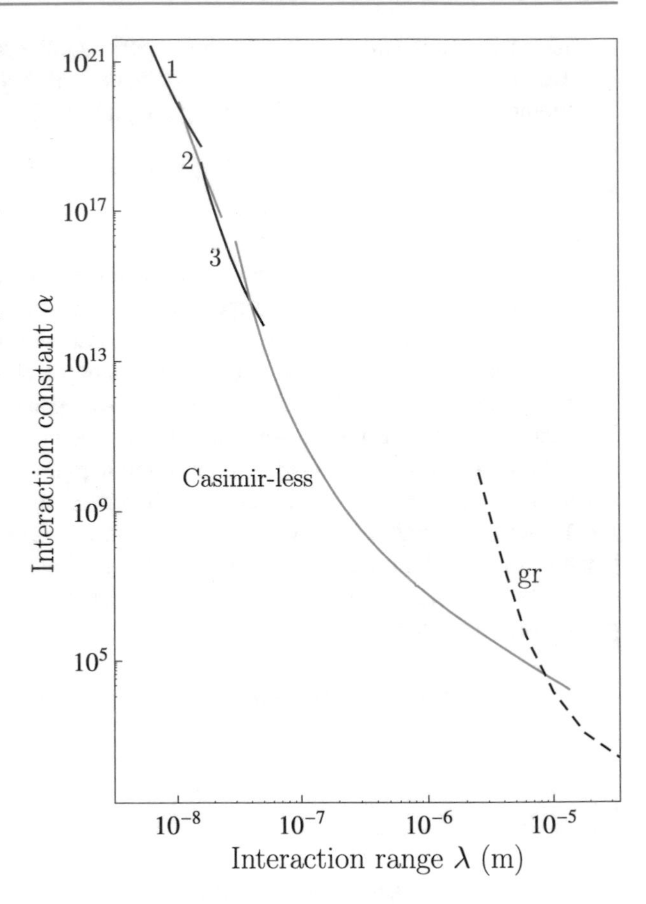

Fig. 13.2 Constraints on the interaction constant α of Yukawa-type interaction are shown as functions of the interaction range λ by the lines labeled 1, 2, 3, Casimir-less, and gr obtained from measuring the lateral and normal Casimir forces between the sinusoidally corrugated surfaces, effective Casimir pressure, from the Casimir-less experiment, and gravitational experiments, respectively. The regions of (λ, α) plane above each line are excluded and below are allowed

The first of them is devoted to measurements of the lateral Casimir force between the surfaces of a sphere and a plate covered with coaxial longitudinal sinusoidal corrugations and coated with an Au film [90, 91]. This experiment was performed by means of an atomic force microscope.

The respective constraints were obtained in [92]. For this purpose, the interaction energy of Yukawa type between the corrugated test bodies was calculated by integrating (13.29) over their volumes, and the lateral force was found by the negative differentiation of the obtained result with respect to the phase shift between corrugations (see [92] for details). The obtained constraints cover a wide interaction range but currently they are the strongest ones only in the narrow region $10\,\mathrm{nm} < \lambda < 11.6\,\mathrm{nm}$ (see the line labeled 1 in Fig. 13.2).

The second experiment, also performed by using an atomic force microscope, is on measuring the usual (normal) Casimir force between the sinusoidally corrugated Au-coated surfaces of a sphere and a plate under some angle between the corrugation axes [93, 94]. The constraints on the Yukawa parameters α and λ, following from this experiment, were obtained in [95]. Currently, these constraints remain the strongest ones in the region $11.6\,\text{nm} < \lambda < 17.2\,\text{nm}$. They are shown by the line labeled 2 in Fig. 13.2.

In the next, third, experiment the effective Casimir force per unit area of two Au-coated plates (i.e., the effective Casimir pressure) was determined by means of a micromechanical torsional oscillator [27, 28]. In fact it was recalculated from the directly measured gradient of the Casimir force, F'_{sp}, between a sphere and a plate using (13.14). In the same way, calculating the gradient of (13.37) one finds from (13.14) the Yukawa-type pressure between two parallel plates

$$P_{\text{Yu}}(a) = -2\pi G\rho_1\rho_2\alpha\lambda^2\left(1 - e^{-D/\lambda}\right)e^{-a/\lambda}. \tag{13.39}$$

Here, following [27, 28], we took into account that $\lambda \ll R$ leading to $\Phi(R, \lambda) \approx R$.

In this experiment, the test bodies were not homogeneous. A sapphire sphere of density $\rho_s = 4.1\,\text{g/cm}^3$ was coated with the first layer of Cr with density $\rho_{\text{Cr}} = 7.14\,\text{g/cm}^3$ and thickness $\Delta_1 = 10\,\text{nm}$, and then with the second, external, layer of Au of density $\rho_{\text{Au}} = 19.28\,\text{g/cm}^3$ and thickness $\Delta_2 = 180\,\text{nm}$. The plate was made of Si with density $\rho_{\text{Si}} = 2.33\,\text{g/cm}^3$ and coated with a layer of Cr of thickness $\Delta_1 = 10\,\text{nm}$ and external layer of Au of thickness $\widetilde{\Delta}_2 = 210\,\text{nm}$. Taking into account that the layers contribute to the Casimir pressure additively, we obtain from (13.39) the following expression valid in the experimental configuration

$$\begin{aligned} P_{\text{Yu}}(a) = &-2\pi G\alpha\lambda^2 e^{-a/\lambda}\left[\rho_{\text{Au}} - (\rho_{\text{Au}} - \rho_{\text{Cr}})e^{-\Delta_2/\lambda} - (\rho_{\text{Cr}} - \rho_s)e^{-(\Delta_2+\Delta_1)/\lambda}\right] \\ &\times\left[\rho_{\text{Au}} - (\rho_{\text{Au}} - \rho_{\text{Cr}})e^{-\widetilde{\Delta}_2/\lambda} - (\rho_{\text{Cr}} - \rho_{\text{Si}})e^{-(\widetilde{\Delta}_2+\Delta_1)/\lambda}\right]. \end{aligned} \tag{13.40}$$

The constraints on the Yukawa parameters α and λ were obtained from the inequality

$$|P_{\text{Yu}}(a)| \leqslant \Delta P(a), \tag{13.41}$$

where the experimental error $\Delta P(a)$ in measuring the effective Casimir pressure, with which the theoretical predictions of the Lifshitz theory were confirmed, was determined at the 95% confidence level. Currently, the constraints obtained from this experiment are the strongest ones over the interaction range $17.2\,\text{nm} < \lambda < 39\,\text{nm}$. They are found from (13.41) at $a = 180\,\text{nm}$ where $\Delta P(a) = 4.8\,\text{mPa}$ [27, 28] and shown by the line labeled 3 in Fig. 13.2.

The last, fourth, experiment leading to the strongest constraints on the Yukawa-type interaction over a wide interaction range $39\,\text{nm} < \lambda < 8\,\mu\text{m}$ is performed in such a way, that the contribution of the Casimir force to the measured signal is nullified [96]. This was achieved by measuring the differential force between a sphere of $R = 149.3\,\mu\text{m}$ radius and an especially structured plate using a micromechanical torsional oscillator. The sphere made of sapphire was coated with a $\Delta_1 = 10\,\text{nm}$ layer of Cr and $\Delta_2 = 250\,\text{nm}$ layer of Au. The plate consisted of Si and Au parts of $D = 2.1\,\mu$m thickness both coated with a Cr and Au overlayers of thicknesses $\Delta_1 = 10\,$nm and $\widetilde{\Delta}_2 = 150\,$nm, respectively.

The Casimir forces between a sphere and two halves of the patterned Au-Si plate are equal because the thickness of an Au overlayer is sufficiently large in order it could be considered as a semispace [11]. As a result, when the sphere is moved back and forth above the patterned plate, the measured differential force is equal to a difference of the Yukawa-type forces between a sphere and two halves of the plate. Using (13.37) with $\Phi(R, \lambda) \approx R$ and taking into account the Au and Cr layers covering the test bodies, the differential Yukawa-type force takes the form

$$F_{\rm Yu,Au}^{SP}(a) - F_{\rm Yu,Si}^{SP}(a) = -4\pi^2 G\alpha\lambda^3 R(\rho_{\rm Au} - \rho_{\rm Si}) e^{-(a+\widetilde{\Delta}_2+\Delta_1)/\lambda}\left(1 - e^{-D/\lambda}\right)$$
$$\times\left[\rho_{\rm Au} - (\rho_{\rm Au} - \rho_{\rm Cr})e^{-\Delta_2/\lambda} - (\rho_{\rm Cr} - \rho_s)e^{-(\Delta_2+\Delta_1)/\lambda}\right]. \qquad (13.42)$$

The constraints have been obtained from the inequality

$$|F_{\rm Yu,Au}^{SP}(a) - F_{\rm Yu,Si}^{SP}(a)| \leqslant \Xi(a), \qquad (13.43)$$

where a sensitivity of the setup to force differences $\Xi(a)$ is equal to a fraction of 1 fN. Note that both the residual electric and Newtonian gravitational forces contribute well below this sensitivity [96]. The strongest current constraints of α and λ obtained from (13.43) extend over a wide interaction range $40\,\text{nm} < \lambda < 8\,\mu\text{m}$ (see the line labeled Casimir-less in Fig. 13.2).

Thus, Fig. 13.2 presents the strongest constraints on the Yukawa-type interaction obtained from Casimir physics. Almost all these constraints with except of the region from 10 to 39 nm are obtained in [96] which is the fourth experiment discussed above. At $\lambda = 8\,\mu$m the constraints found from the differential force measurements (which are also called the Casimir-less experiment) are of the same strength as the constraints found from the Cavendish-type experiments. At larger λ the strongest constraints are shown by a beginning of the line labeled gr reproduced from Fig. 13.1.

In Fig. 13.3 we reproduce the beginning of the line labeled Casimir-less in Fig. 13.2 on an enlarged scale in order to better demonstrate the constraints obtained in the range $10\,\text{nm} < \lambda < 39\,\text{nm}$ [27, 28, 92, 95] from the experiments [27, 28, 90, 91, 93, 94] (i.e., from the first, second and third experiments discussed above). In this figure, the lines labeled n also indicate the strongest constraints on the Yukawa-type interaction obtained at $\lambda < 10\,\text{nm}$ from the experiments on neutron scattering [97, 98] (in the region $0.03\,\text{nm} < \lambda < 0.1\,\text{nm}$ the strongest constraints follow from the experiment using a pulsed neutron beam [99]).

There are also many other papers in the scientific literature devoted to constraining the Yukawa-type corrections to Newtonian gravity from the Casimir effect (see, e.g., [100–104]). The constraints obtained there are, however, somewhat weaker than the current strongest constraints presented in Figs. 13.2 and 13.3. It should be mentioned that in the range of extremely small λ the constraints on α have been

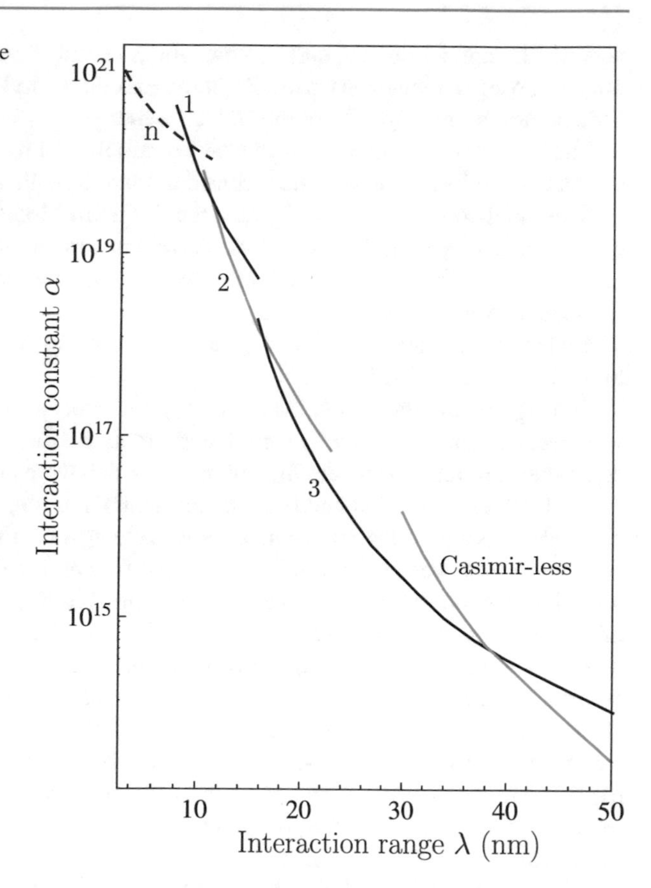

Fig. 13.3 Constraints on the interaction constant α of Yukawa-type interaction are shown as functions of the interaction range λ by the lines labeled n, 1, 2, 3, and Casimir-less, obtained from the experiments on neutron scattering, measuring the lateral and normal Casimir force between the sinusoidally corrugated surfaces, effective Casimir pressure, and from the Casimir-less experiment, respectively. The regions of (λ, α) plane above each line are excluded and below are allowed

obtained from spectroscopic measurements in simple atomic systems like hydrogen and deuterium whose spectra can be calculated and measured with high precision. Thus, it was found that in the range $2 \times 10^{-4}\,\text{nm} < \lambda < 20\,\text{nm}$ the maximum strength of Yukawa interaction varies from 2×10^{27} to 2×10^{25} [105].

13.5 Dark Matter, Dark Energy and Their Hypothetical Constituents

According to astrophysical observations, the visible matter in the form of stars, galaxies, planets and radiation constitutes only about 5% of the total mass of the Universe. By studying stellar motion in the neighborhood of our galaxy 90 years ago, Oort found [106] that the galaxy mass must be much larger than the mass of all stars belonging to it. At the same time, an application of the virial theorem to the Coma cluster of galaxies by Zwicky [107] resulted in a much larger mass than that found by summing up the masses of all observed galaxies belonging to this cluster. In succeeding years, these results received ample recognition. Presently it is generally agreed that the dark matter, which reveals itself only gravitationally, is not composed of elementary particles of the Standard Model listed in Sect. 13.1 and adds up approximately 27% of the Universe energy.

The problem of what dark matter is remains unresolved. There are many approaches to its resolution which consider some hypothetical particles introduced in different theoretical schemes beyond the Standard Model as possible constituents of dark matter. Among these particles are axions, arions, massive neutrinos, weakly interacting massive particles (WIMP) etc. The possibility to explain the observational data by a modification of the gravitational theory in place of compensating for a deficiency in matter is also investigated. All these approaches are widely discussed in the literature [108–112].

During the last few years, the major support from astrophysics and cosmology was received by the model of cold dark matter. This model suggests that the constituents of dark matter are light particles which were produced at the first stages of the Universe evolution and became nonrelativistic long ago. The best candidate of this kind is a pseudoscalar Nambu-Goldstone boson called an axion.

This particle was introduced [78, 113, 114] for solving the problem of strong CP violation in quantum chromodynamics mentioned in Sect. 13.1, i.e., independently of the problem of dark matter. The point is that all the experimental data show that strong interactions are CP invariant and the electric dipole moment of a neutron is equal to zero. In contrast to these facts, the vacuum state of quantum chromodynamics depends on an angle θ which violates the CP invariance and allows a nonzero electric dipole moment of a neutron. To resolve this contradiction between experiment and theory, Peccei and Quinn [78] introduced the new symmetry which received their names. In doing so the emergence of axions is a direct consequence of the violation of this symmetry [113, 114].

Later it was understood that axions and other axionlike particles arise in many extensions of the Standard Model. They can interact with particles of the Standard Model, e.g., with photons, electrons and nucleons, and lead to a number of processes which could be observed both in the laboratory experiments and in astrophysics and cosmology (see [108–112, 115–121] for a review). The question arises whether it is possible to constrain the parameters of axions from the Casimir effect. This question is considered below in Sects. 13.6 and 13.7.

Another unresolved problem of modern physics is the problem of dark energy. In the end of twentieth century, the observations of supernovae demonstrated that an expansion of the Universe is accelerating [122]. This fact is in some contradiction with expectations based on the general relativity theory and the properties of matter described by the Standard Model because the gravitational interaction of usual matter is attractive and should make the Universe expansion slower.

The concept of dark energy, i.e., some new kind of invisible matter which causes a repulsion, was introduced in numerous discussions of this problem. One approach to its resolution goes back to Einstein's cosmological constant which is closely connected with the problem of the quantum vacuum. According to the observational data, the dark energy constitutes as much as approximately 68% of the Universe energy. This corresponds to some background medium (physical vacuum) possessing the energy density of $\epsilon \approx 10^{-9}$ J/m^3. In order that this medium could accelerate the Universe expansion, it should possess the equation of state $P = -\epsilon < 0$, i.e., the negative pressure.

The cosmological term Λg_{ik}, where g_{ik} is the metrical tensor, when added to Einstein's equations of general relativity theory, results in just this equation of state. In doing so, the value of Λ is determined by the above value of ϵ determined from the observed acceleration of the Universe expansion

$$\Lambda = 8\pi G\epsilon \approx 2 \times 10^{-52}\text{m}^{-2}. \tag{13.44}$$

It was argued, however, that quantum field theory using a cutoff at the Planck momentum $p_{Pl} = E_{Pl}/c$ leads to quite a different value of the vacuum energy density $\epsilon_{vac} \approx 10^{111}$ J/m^3 which is different from ϵ determined from observations by the factor of 10^{120} [123, 124]. If to take into account that the vacuum energy density admits an interpretation in term of the cosmological constant [125], it becomes clear why this discrepancy by the factor of 10^{120} was called the vacuum catastrophe [126].

Another approach to the understanding of dark energy attempts to model it by the fields and respective particles with unusual physical properties. One of the models of this kind introduces the real self-interacting scalar field ϕ with a variable mass called chameleon [127]. The distinctive feature of chameleon particles is that they become heavier in more dense environments and lighter in free space.

Another model similar in spirit suggests that the interaction constant of the self-interacting real scalar field with usual matter depends on the density in the environment. The fields and particles of this kind are called symmetrons [128–130]. Symmetrons interact with usual matter described by the Standard Model weaker if the density of the environment is higher.

There are also other hypothetical particles which could lead to the negative pressure and help to understand the accelerating expansion of the Universe. For instance, the negative pressure originates from the Maxwell stress-energy tensor of massive photons in the Maxwell-Proca electrodynamics [131].

If the exotic particles, such as chameleons, symmetrons, massive photons etc., exist in nature, this should lead to some additional forces between the closely spaced macrobodies. In Sect. 13.8 the possibility of constraining these forces from measurements of the Casimir force is discussed.

13.6 Constraining Dark Matter Particles from the Casimir Effect

As was mentioned in previous section, the main candidate for the role of a dark matter particle is light pseudoscalar particle called an axion which can interact with photons, electrons, and nucleons. It can be easily seen that the interaction of axions with photons and electrons does not lead to sufficiently large forces between the closely spaced bodies which could be constrained from measurements of the Casimir force. These interactions of axions are investigated by other means. For example, the conversion process of photons into axions in strong magnetic field (the so-called Primakoff process) is used for an axion search in astrophysics [132] (see also reviews [117–120] for already obtained constraints on interactions of axions with photons and electrons).

Here, we concentrate our attention on the interaction of axions with nucleons (neutrons and protons) which could lead to some noticeable additional force between two neighboring bodies. The interaction Lagrangian density between the originally introduced axion field $a(x)$ and the fermionic field $\psi(x)$ is given by [115, 118]

$$\mathcal{L}_{pv}(x) = \frac{g}{2m_a}\hbar^2\bar{\psi}(x)\gamma_5\gamma_\mu\psi a(x)\partial^\mu a(x), \tag{13.45}$$

where g is the dimensionless interaction constant, m_a is the axion mass, γ_μ with $\mu = 0, 1, 2, 3$ and γ_5 are the Dirac matrices. The Lagrangian density (13.45) is called pseudovector. It describes the interaction of fermions with pseudo Nambu-Goldstone bosons.

Various extensions of the Standard Model called the Grand Unified Theories (GUT) introdice the axionlike particles which interact with fermions through the pseudoscalar Lagrangian density [115, 118, 133]

$$\mathcal{L}_{ps}(x) = -ig\hbar c\bar{\psi}(x)\gamma_5\psi(x)a(x). \tag{13.46}$$

Unlike (13.45), which contains a dimensional effective interaction constant g/m_a, the Lagrangian density (13.46) results in a renormalizable field theory.

When one considers an exchange of a single axion between two nucleons of mass m belonging to the closely spaced test bodies, both Lagrangian densities (13.45) and (13.46) lead to the common effective potential energy [134, 135]

$$\begin{aligned} V_{an}(r; \boldsymbol{\sigma}_1, \boldsymbol{\sigma}_2) = \frac{g^2\hbar^3}{16\pi m^2 c}\Bigg[& (\boldsymbol{\sigma}_1\cdot\boldsymbol{n})(\boldsymbol{\sigma}_2\cdot\boldsymbol{n})\left(\frac{m_a^2c^2}{\hbar^2 r} + \frac{3m_a c}{\hbar r^2} + \frac{3}{r^3}\right) \\ & - (\boldsymbol{\sigma}_1\cdot\boldsymbol{\sigma}_2)\left(\frac{m_a c}{\hbar r^2} + \frac{1}{r^3}\right)\Bigg], \end{aligned} \tag{13.47}$$

where $r = |\boldsymbol{r}_1 - \boldsymbol{r}_2|$ is a distance between nucleons, $\boldsymbol{\sigma}_1$, $\boldsymbol{\sigma}_2$ are their spins, and $\boldsymbol{n} = (\boldsymbol{r}_1 - \boldsymbol{r}_2)/r$ is the unit vector along the line connecting these nucleons.

The effective interaction energy (13.47) depends on the spins of nucleons and the respective force averages to zero after a summation over the volumes of unpolarized test bodies. Because of this, using (13.47), the parameters of axion g and m_a can not be constrained from experiments on measuring the Casimir force discussed in Sect. 13.2 (the possibilities of constraining the spin-dependent interactions are considered in the next section).

There is, however, the possibility to obtain the spin-independent interaction energy between two nucleons by considering the process of two-axion exchange. If the Lagrangian density (13.46) is used, the effective interaction energy is given by [83, 136, 137]

$$V_{aan}(r) = -\frac{g^4\hbar^2}{32\pi^3 m^2}\frac{m_a}{r^2}K_1\left(\frac{2m_a cr}{\hbar}\right), \quad (13.48)$$

where $K_1(z)$ is the modified Bessel function of the second kind.

In the case of Lagrangian density (13.45), the respective field theory is nonrenormalizable. As a result, the effective interaction energy between nucleons due to an exchange of two axions is not yet available (see [138] for more details). This means that measurements of the Casimir force can be used for constraining only the parameters of GUT axions described by the pseudoscalar Lagrangian density (13.46).

Similar to the cases of power-type and Yukawa-type interactions in (13.21) and (13.34), the hypothetical force between two experimental test bodies due to two-axion exchange of their nucleons is given by

$$F_{aan}(a) = -n_1 n_2 \frac{\partial}{\partial a}\int_{V_1} d^3 r_1 \int_{V_2} d^3 r_2 V_{aan}(|\boldsymbol{r}_1 - \boldsymbol{r}_2|), \quad (13.49)$$

where a is the closest distance between these bodies and n_1, n_2 are the numbers of nucleons per unit volume of their materials.

We consider first a homogeneous Au sphere above a homogeneous Si plate of thickness D which is assumed to be infinitely large. Substituting (13.48) in (13.49) and using the integral representation [139]

$$\frac{K_1(z)}{z} = \int_1^\infty du\sqrt{u^2-1}e^{-zu}, \quad (13.50)$$

one obtains

$$F_{aan}^{SP}(a) = -\frac{\pi m_a \hbar^2}{m^2 m_{\mathrm{H}}^2} C_1 C_2 \int_1^\infty du \frac{\sqrt{u^2-1}}{u}\left(1 - e^{-2m_a cuD/\hbar}\right)$$
$$\times \int_a^{2R+a}\left[2R(z-a) - (z-a)^2\right] e^{-2m_a cuz/\hbar} dz. \quad (13.51)$$

Here, the coefficients C_1 and C_2 are defined for a sphere and a plate materials, respectively, in the following way:

$$C_{1,2} = \rho_{1,2}\frac{g_{an}^2}{4\pi}\left(\frac{Z_{1,2}}{\mu_{1,2}} + \frac{N_{1,2}}{\mu_{1,2}}\right), \quad (13.52)$$

where $Z_{1,2}$ and $N_{1,2}$ are the numbers of protons and the mean number of neutrons in the sphere and plate atoms, respectively, and $\mu_{1,2} = m_{1,2}/m_{\mathrm{H}}$ are defined as the mean masses of a sphere and a plate atoms divided by the mass of atomic hydrogen.

By integrating in (13.51) with respect to z, one obtains

$$F_{aan}^{SP}(a) = -\frac{\pi\hbar^4}{2m_a m^2 m_{\mathrm{H}}^2 c^2} C_1 C_2 \int_1^\infty du \frac{\sqrt{u^2-1}}{u^3} e^{-2m_a cua/\hbar} \times \left(1 - e^{-2m_a cuD/\hbar}\right) \chi\left(R, \frac{m_a cu}{\hbar}\right), \tag{13.53}$$

where the function $\chi(r, z)$ similar to $\Phi(r, \lambda)$ in (13.38) is defined as

$$\chi(r, z) = r - \frac{1}{2z} + \left(r + \frac{1}{2z}\right) e^{-4rz}. \tag{13.54}$$

The strongest constraints on the parameters of axionlike particles were obtained [140] from the differential measurements where the contribution of the Casimir force was nullified [96]. This experiment was already discussed in Sect. 13.4. Taking into account the structure of the plate consisting of Au and Si halves, as well as additional Cr and Au layers (see Sect. 13.4), and using (13.53), the differential force in the experimental configuration takes the form

$$F_{aan,\mathrm{Au}}^{SP}(a) - F_{aan,\mathrm{Si}}^{SP}(a) = -\frac{\pi\hbar^4}{2m_a m^2 m_{\mathrm{H}}^2 c^2} (C_{\mathrm{Au}} - C_{\mathrm{Si}}) \int_1^\infty du \frac{\sqrt{u^2-1}}{u^3} \times e^{-2m_a cu(a+\widetilde{\Delta}_2+\Delta_1)/\hbar} \left(1 - e^{-2m_a cuD/\hbar}\right) X\left(\frac{m_a cu}{\hbar}\right), \tag{13.55}$$

where the following notation is introduced

$$\begin{aligned} X(z) = {} & C_{\mathrm{Au}} \left[\chi(R, z) - e^{-2z\Delta_2} \chi(R - \Delta_2, z)\right] \\ & + C_{\mathrm{Cr}} e^{-2z\Delta_2} \left[\chi(R - \Delta_2, z) - e^{-2z\Delta_1} \chi(R - \Delta_2 - \Delta_1, z)\right] \\ & + C_s e^{-2z(\Delta_2+\Delta_1)} \chi(R - \Delta_2 - \Delta_1, z) \end{aligned} \tag{13.56}$$

and the values of all coefficients C for Au, Cr, Si, and sapphire can be calculated by using (13.52) and numerical data for all involved quantities presented in [83].

The constraints on the parameters of hypothetical forces due to two-axion exchange between nucleons follow from the inequality

$$|F_{aan,\mathrm{Au}}^{SP}(a) - F_{aan,\mathrm{Si}}^{SP}(a)| \leqslant \Xi(a), \tag{13.57}$$

where $\Xi(a)$ is the setup sensitivity to force differences in the experiment [96]. This inequality is similar to (13.43) used in constraining the interaction of Yukawa type. The strongest current constraints on the coupling constant of axions to nucleons g follow from (13.57) in the region of axion masses 4.9 meV $< m_a c^2 <$ 0.5 eV.

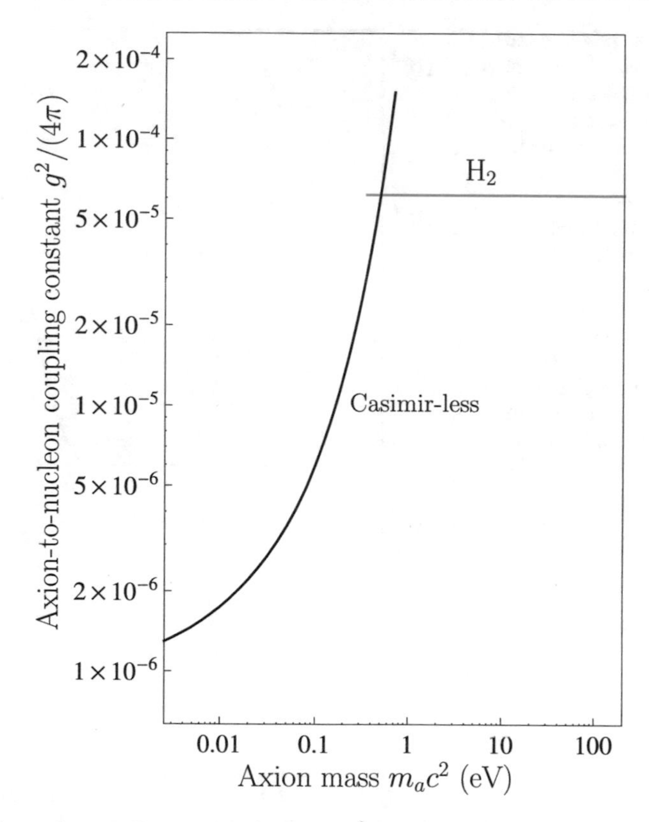

Fig. 13.4 Constraints on the coupling constant $g^2/(4\pi)$ of axions to nucleons are shown as functions of the axion mass $m_a c^2$ by the lines labeled Casimir-less and H_2 obtained from the Casimir-less experiment and from measuring dipole-dipole forces between protons in the beam of molecular hydrogen, respectively. The regions of $[m_a c^2, g^2/(4\pi)]$ plane above each line are excluded and below are allowed

In Fig. 13.4 the obtained constraints are shown by the line labeled Casimir-less. Similar to all previous figures, the values of axion parameters belonging to the area of $[m_a c^2, g^2/(4\pi)]$ plane above the line are excluded by the results of differential force measurements whereas the plane area below the line is allowed.

For $m_a c^2 > 0.5\,\text{eV}$ the strongest current constraints on g are obtained by comparing with theory the measurement results for the dipole-dipole forces between two protons in the beam of molecular hydrogen [141, 142]. They are shown by the line labeled H_2 in Fig. 13.4. In this experiment, the additional force between protons arises due to an exchange of one axion and is described by the spin-dependent interaction energy (13.47). As a result, for sufficiently large m_a the obtained constraints on g are much stronger than those found from the differential force measurements. What is more, the constraints of line H_2 are valid both for the originally introduced axions whose interaction with nucleons is described by the

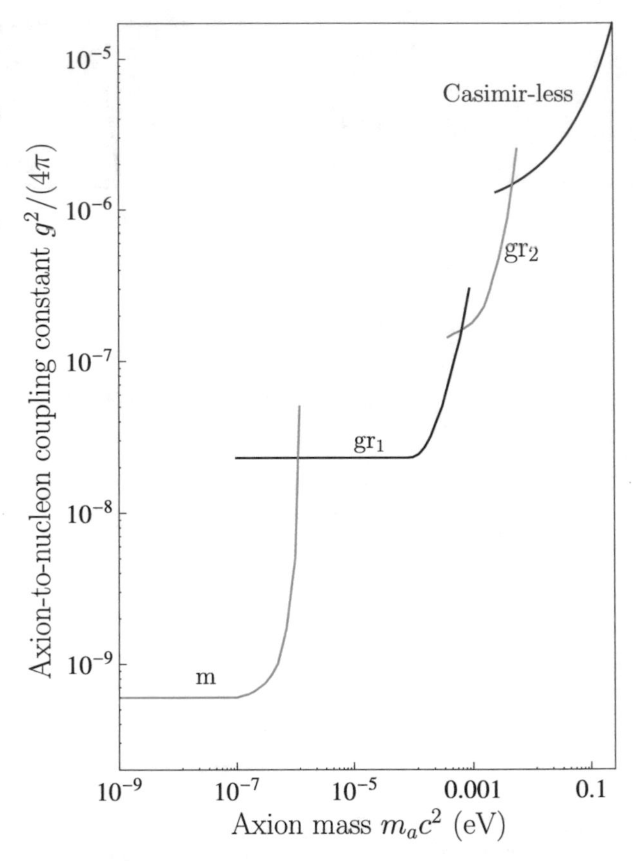

Fig. 13.5 Constraints on the coupling constant $g^2/(4\pi)$ of axions to nucleons are shown as functions of the axion mass $m_a c^2$ by the lines labeled m, gr_1, gr_2, and Casimir-less obtained from the magnetometer measurements, Cavendish-type experiment, measuring the minimum force of gravitational strength, and from the Casimir-less experiment, respectively. The regions of $[m_a c^2, g^2/(4\pi)]$ plane above each line are excluded and below are allowed

Lagrangian density (13.45) and for axionlike particles with respective Lagrangian density (13.46).

In the region of axion masses $m_a c^2 < 4.9\,\mathrm{meV}$ the strongest constraints of g follow from gravitational experiments. In Fig. 13.5, the line labeled gr_1 shows the constraints on g found [71] from the Cavendish-type experiment [70]. These constraints are the strongest ones in the region of axion masses $1\,\mu\mathrm{eV} < m_a c^2 < 0.676\,\mathrm{meV}$. Within the relatively short range of axion masses $0.676\,\mathrm{meV} < m_a c^2 < 4.9\,\mathrm{meV}$ the strongest constraints were obtained [138] by using the planar torsional oscillator for measuring the minimum force of gravitational strength [143, 144]. The respective constraints are shown by the line labeled gr_2 in Fig. 13.5. The gravitational constraints shown by the lines gr_1 and gr_2 are valid for only the axionlike particles whose interaction with nucleons is described by the Lagrangian density (13.46) because in the gravitational experiments the test bodies used are unpolarized.

For the smallest axion masses $m_a c^2 < 1\,\mu\mathrm{eV}$, the strongest constraints on g were again obtained from considering the spin-dependent forces which arise due to a one-axion exchange in the comagnetometer measurements using the spin-polarized

K and ^{3}He atoms and the ^{3}He spin source [145]. These constraints are shown by the line labeled m in Fig. 13.5. They are valid for all types of axions and axionlike particles.

The competitive constraints on the parameters of axionlike particles were obtained also from several other experiments on measuring the Casimir interaction (see, e.g., [103, 104, 146–151]). They are, however, weaker than those shown by the line labeled Casimir-less and H_2 in Fig. 13.4.

13.7 Could the Casimir Effect be Used For Testing Spin-Dependent Interactions?

As explained in the previous section, the process of one-axion exchange between two nucleons described by the pseudovector Lagrangian density (13.45) results in the spin-dependent interaction energy (13.47) which does not lead to any additional force between two unpolarized test bodies. The parameters of originally introduced axions described by this interaction energy were constrained using, e.g., the magnetometer measurements or from measuring dipole-dipole forces between protons (see Sect. 13.6).

There are also other predictions of spin-dependent interactions beyond the Standard Model. In fact the coupling constant of axions to nucleons considered above describes either the pseudoscalar or pseudovector interactions. This can be notated as $g \equiv g_P$. In addition to the one-axion exchange, it is possible to consider an exchange of one light vector particle between two nucleons with a vector and axial vector couplings. This corresponds to the following Lagrangian density:

$$\mathcal{L}_{VA}(x) = \hbar c \bar{\psi}(x)\gamma^{\mu}(g_V + g_A\gamma_5)\psi(x)A_{\mu}(x). \tag{13.58}$$

This Lagrangian density results in the effective spin-dependent interaction energies between two nucleons [145]

$$V_1(r) = \frac{g_A^2}{4\pi r}\hbar(\boldsymbol{\sigma}_1 \cdot \boldsymbol{\sigma}_2)e^{-m_A cr/\hbar} \tag{13.59}$$

or

$$V_2(r) = -\frac{g_A g_V}{4\pi m}\hbar^2([\boldsymbol{\sigma}_1 \times \boldsymbol{\sigma}_2] \cdot \boldsymbol{n})\left(\frac{m_A c}{\hbar r} + \frac{1}{r^2}\right) e^{-m_A cr/\hbar}, \tag{13.60}$$

where m_A is the mass of a vector field A_{μ}. The parameters of the interaction energies (13.59) and (13.60) were constrained by the same comagnetometer measurements [145] which have already been used in Sect. 13.6 for constraining the interaction energy (13.47).

Below we discuss the possibility of constraining the spin-dependent interaction energy (13.47) from measuring the effective Casimir pressure. For this purpose it was proposed [152] to use the Casimir plates made of silicon carbide (SiC) with aligned nuclear spins. It has been known that the nuclear spin of ^{29}Si is equal to 1/2 owing to the presence of one neutron with an uncompensated spin. In native Si there is only 4.68% of the isotope ^{29}Si. In nanotechnology, however, the special procedures are elaborated for growing the isotopically controlled bulk Si [153].

In [152] it was assumed that a fraction of Si atoms κ in both plates is polarized in some definite direction due to the polarization of their nuclear spins (it was shown that an additional force due to the electronic polarization does not permit to obtain competitive constraints on the coupling constant of axions to electrons). In order to obtain the nonzero additional force between plates due to one-axion exchange, the atomic polarization should be perpendicular to the plates and directed either in one direction or in the opposite directions [152].

Under these conditions, by integrating the interaction energy (13.47) over the volumes of two parallel plates of density ρ and thickness D, for the force per unit area of the plates (i.e., pressure) one obtains [152]

$$P_{an}(a) = \pm g^2 \frac{\kappa^2 \rho^2 \hbar^3}{8m^2 m_{\mathrm{H}}^2 c} e^{-m_a c a/\hbar} \left(1 - e^{-m_a c D/\hbar}\right)^2. \tag{13.61}$$

The force (13.61) could be constrained from the experiments [27, 28] on measuring the effective Casimir pressure using a micromechanical torsional oscillator if the test bodies were made of SiC with aligned nuclear spins. Similar to (13.41), in this case the constraints are obtained from the inequality

$$|P_{an}(a)| \leqslant \Delta P(a). \tag{13.62}$$

For the pressure (13.61), the strongest constraints follow at $a = 300$ nm, where $\Delta P(a) = 0.22$ mPa [27, 28], under the conditions that $\kappa = 1$ and $D \gg \hbar/(m_a c)$, i.e., the plates are sufficiently thick. The density of SiC is $\rho = 3.21$ g/cm^3. The most strong constraint that can be placed in this way on axions and axionlike particles with $m_a = 0.0126$ eV is $g^2/(4\pi) \leqslant 4.43 \times 10^{-5}$ [152], which is weaker than that found from the Casimir-less experiment (see Fig. 13.4) but is applicable to all kinds of axions.

An experiment on measuring the Casimir pressure between parallel plates with aligned nuclear spins can be also used for constraining the interaction of axions with nucleons from a simultaneous account of one- and two-axion exchange. In this case the constraints are obtained from the inequality

$$|P_{an}(a) + P_{aan}(a)| \leqslant \Delta P(a). \tag{13.63}$$

Here the additional pressure between two thick plates due to one-axion exchange is given by (13.61), where the last factor on the right-hand side is replaced with unity, and P_{aan} is obtained by integration of the interaction energy (13.48) over the volumes of both plates

$$P_{aan}(a) = -\frac{C_{\mathrm{SiC}}^2 \hbar^3}{2m^2 m_{\mathrm{H}}^2 c} \int_1^\infty du \frac{\sqrt{u^2 - 1}}{u^2} e^{-2m_a cau/\hbar}. \tag{13.64}$$

The constant C_{SiC} is defined as in (13.52) using the numerical data presented in [83].

It has been shown [152] that using (13.63) in the region of axion masses below 1 eV one could obtain up to an order of magnitude stronger constraints on the coupling of axions with nucleons than from (13.62). These constraints, however, would be valid only for the GUT axions which interaction with nucleons is described by the pseudoscalar Lagrangian density (13.46).

13.8 Constraining Dark Energy Particles from the Casimir Effect

Axions considered above as the most probable constituents of dark matter are the Nambu-Goldstone bosons, which appear in the formalism of quantum field theory when some symmetry (in this case the Peccei-Quinn symmetry) is broken both spontaneously (i.e., the vacuum is not invariant) and dynamically (i.e., in the Lagrangian). Although these particles are not the part of the Standarn Model, they can be considered as its natural supplement. The axionlike particles introduced later also fall into the standard pattern of quantum field theory.

The particles proposed as the possible constituents of dark energy (chameleons, symmetrons, etc., see Sect. 13.5) are quite different. Unlike all conventional elementary particles, the properties of these particles depend on the environmental conditions.

We begin with chameleons whose mass is larger, i.e., the interaction range is shorter, in environments with higher energy density (see Sect. 13.5). Mathematically chameleons are described by the real self-interacting scalar field Φ possessing a variable mass. In the static case, this field satisfies the simplest equation of the form [127, 154]:

$$\Delta\Phi = \frac{1}{(\hbar c)^4} \frac{\partial V(\Phi)}{\partial \Phi} + \frac{\rho}{M} e^{\hbar\Phi/(Mc)}, \tag{13.65}$$

where M is the typical mass of conventional particles forming the background matter of density ρ and $V(\Phi)$ is the self-interaction which decreases monotonically with increasing Φ.

An interaction between chameleons and background matter with density ρ in (13.65) implies that the effective interaction potential describing the chameleon field is given by

$$V_{\mathrm{eff}}(\Phi) = V(\Phi) + \rho\hbar^3 c^5 e^{\hbar\Phi/(Mc)}. \tag{13.66}$$

Although the self-interaction V is assumed to be monotonic, this effective potential takes the minimum value for Φ_0 satisfying the condition

$$\frac{\partial V(\Phi_0)}{\partial \Phi} + \frac{\rho(\hbar c)^4}{M} e^{\hbar \Phi_0/(Mc)} = 0. \tag{13.67}$$

Then the mass of the field Φ_0 is given by

$$m_{\Phi_0}^2 \equiv \frac{1}{\hbar^2 c^6} \frac{\partial^2 V_{\text{eff}}(\Phi_0)}{\partial \Phi^2} = \frac{1}{\hbar^2 c^6} \frac{\partial^2 V(\Phi_0)}{\partial \Phi^2} + \frac{\rho}{M^2} \left(\frac{\hbar}{c}\right)^3 e^{\hbar \Phi_0/(Mc)} \tag{13.68}$$

and depends on the background mass density ρ.

According to the above assumption, V is a decreasing function of Φ. Then, $\partial V/\partial \Phi$ is negative and monotonously increasing whereas $\partial^2 V/\partial \Phi^2$ is positive and decreasing. According to (13.68), this means that we have larger m_{Φ_0} and smaller Φ_0 for larger values of the background mass density ρ [127].

There are different possible forms of the chameleon self-interaction suggested in the literature [127, 154, 155], e.g.,

$$V(\Phi) = E^4 \left(\frac{E}{\hbar c \Phi}\right)^n, \qquad V(\Phi) = E_0^4 \, e^{E^n/(\hbar c \Phi)^n} \tag{13.69}$$

or

$$V(\Phi) = E_0^4 \left[1 + \left(\frac{E}{\hbar c \Phi}\right)^n\right], \tag{13.70}$$

where E_0 and E are the quantities with a dimension of energy and $n = 1, 2, 3$, etc. According to [154], if the chameleon field is responsible for the presently observed acceleration of the Universe, it should be $E_0 \approx 2.4 \times 10^{-12}$ GeV.

Similar to axions, an exchange of chameleons between two constituent particles of two closely spaced test bodies results in some additional force. The constraints on this force can be obtained from experiments on measuring the Casimir force [154]. In this case, however, both the additional force and constraints on it strongly depend not only on the specific experimental setup but also on the form of chameleon self-interaction and other related parameters (see [154–156] for some specific results obtained in the configuration of two parallel plates and a sphere above a plate with different models of self-interaction).

Another hypothetical particle mentioned in Sect. 13.5 as a possible constituent of dark energy is a symmetron whose interaction with usual matter becomes weaker with increasing mass density of the environment [128, 129]. In the static case the symmetron field Φ_s satisfies the equation [128–130]

$$\Delta\Phi_s = \frac{1}{(\hbar c)^4}\frac{\partial V^s(\Phi_s)}{\partial \Phi_s} + \left[\frac{\hbar}{c}\frac{\rho}{M^2} - \left(\frac{\mu c}{\hbar}\right)^2\right]\Phi_s, \tag{13.71}$$

where V^s is the symmetron self-interaction and μ is the symmetron mass.

The respective effective potential leading to the right-hand side of (13.71) is given by Hinterbichler et al. [130]

$$V^s_{\text{eff}}(\Phi_s) = V^s(\Phi_s) + \frac{1}{2}(\hbar c)^4\left[\frac{\hbar}{c}\frac{\rho}{M^2} - \left(\frac{\mu c}{\hbar}\right)^2\right]\Phi_s^2, \tag{13.72}$$

where the self-interaction of a symmetron field takes the standard form

$$V^s(\Phi_s) = \frac{1}{4}\lambda\Phi_s^4, \tag{13.73}$$

and λ is a dimensionless constant.

The effective potential (13.72), (13.73) takes the minimum value for $\Phi_s = \Phi_{s,0}$ satisfying the condition

$$\lambda\Phi_{s,0}^3 + (\hbar c)^4\left[\frac{\hbar}{c}\frac{\rho}{M^2} - \left(\frac{\mu c}{\hbar}\right)^2\right]\Phi_{s,0} = 0. \tag{13.74}$$

If the density of background matter $\rho < \rho_0 = c^3M^2\mu^2/\hbar^3$, the minimum value of V^s_{eff} is attained at

$$\Phi_{s,0} = \frac{(\hbar c)^2}{\sqrt{\lambda}}\left[\left(\frac{\mu c}{\hbar}\right)^2 - \frac{\hbar}{c}\frac{\rho}{M^2}\right]^{1/2}. \tag{13.75}$$

Under the opposite condition $\rho > \rho_0$, the minimum value of V^s_{eff} is at $\Phi_{s,0} = 0$. Thus, if $\rho < \rho_0$ the reflection symmetry is broken and the vacuum expectation value of $\Phi_{s,0}$ takes a nonzero value. By contrast, in the regions of high density of background matter $\rho > \rho_0$, the vacuum expectation value of $\Phi_{s,0}$ turns into zero.

The exchange of symmetrons between two closely spaced material bodies results in some additional force which was calculated in [157] for the experimental configurations of two parallel plates and a sphere above a plate. According to the results of [157], strong constraints on the parameters of a symmetron can be obtained from measurements of the Casimir force in these configurations. These measurements, however, are not yet performed. Prospects in constraining various hypothetical interactions beyond the Standard Model and some other laboratory experiments are discussed in the next section.

13.9 Outlook

Many experiments on measuring the Casimir interaction mentioned above were performed entirely in an effort to investigate the Casimir effect. This means that the constraints on corrections to Newtonian gravity and axionlike particles discussed above were obtained as some kind of by-product. In [158] some improvements in the configurations of experiments employing both smooth and sinusoidally corrugated surfaces of a sphere and a plate were suggested which allow obtaining up to an order of magnitude stronger constraints. Specifically, for the configurations with corrugated surfaces this could be reached by using smaller corrugation periods and larger corrugation amplitudes [158].

There are many proposals of new Casimir experiments aimed for testing gravity and predictions beyond the Standard Model at short distances. Thus, it is suggested to measure the Casimir pressure between two parallel plates at separations up to 10–20 μm (Casimir and Non-Newtonian Force Experiment called CANNEX) [159–163]. This experiment promises obtaining stronger constraints not only on non-Newtonian gravity and axionlike particles, but also on chameleon, symmetron and some other theoretical predictions beyond the Standard Model.

The Casimir-Polder interaction between two atoms or an atom and a cavity wall can also be used for constraining the hypothetical interactions. The constraints on an axion to nucleon coupling constant obtained in this way [146] were mentioned in Sect. 13.6. In [164] it was suggested to measure the Casimir-Polder force between a Rb atom and a movable Si plate screened with an Au film. This makes it possible to strengthen constraints on the Yukawa interaction constant α in the interaction range around 1 μm. According to [165], the measured deviations of the Casimir-Polder force between two polarized particles, arising for photons of nonzero mass, from the standard one calculated for massless photons can be used for constraining the extradimensional unification models.

An interesting method for detecting the interaction of axion with nucleons by means of a levitated optomechanical system was suggested in [166]. In fact this is a version of the Casimir-less experiment [96] where the contribution of the Casimir force is nullified (see Sects. 13.4 and 13.6). The suggested method could further strengthen the already obtained constraints on the coupling constant of axions to nucleons and on the Yukawa-type corrections to Newtonian gravity.

There are also many proposed laboratory experiments which are not closely related to the Casimir physics but could lead to constraints on the hypothetical interactions in the same or neighboring regions of parameters as the Casimir effect. Some of them are discussed below.

Thus, the neutron interferometry already used for constraining the Yukawa-type forces (see Sect. 13.4) has large potential for improving the obtained constraints. Several experiments of this kind have been performed and suggested pursuing this goal (see, e.g., [167–171]).

There is a continuing interest in the literature to constraining the power-type, Yukawa-type and other hypothetical interactions by means of atomic and molecular spectroscopy. A few experiments of this kind await for their realization [172–174].

It has been known that the levitated nanoparticle sensors are sensitive to the static forces down to 10^{-17} N. In [175] it was suggested to use such sensors for obtaining constraints on the Yukawa-type corrections to Newtonian gravity. The optomechanical methods exploiting the levitated sensors were proposed also for constraining the hypothetical interaction of Yukawa type [176].

Recent literature also contains information on already performed experiment constraining the exotic interaction between moving polarized electrons and unpolarized nucleons by means of a magnetic force microscope [177], on the general scheme allowing an extraction of constraints on any specific model from different experiments [178], and on a compressed ultrafast photography system using temporal lensing for probing short-range gravity [179].

Interest in all these topics has quickened in the past few years. One may expect that measurements of the Casimir force and related table-top laboratory experiments will furnish insights into the nature of some theoretical predictions beyond the Standard Model and their relationship to reality.

Acknowledgments G.L.K. was partially funded by the Ministry of Science and Higher Education of Russian Federation ("The World-Class Research Center: Advanced Digital Technologies," contract No. 075-15-2022-311 dated April 20, 2022). The research of V.M.M. was partially carried out in accordance with the Strategic Academic Leadership Program "Priority 2030" of the Kazan Federal University.

References

1. R.M. Wald, *General Relativity* (University of Chicago Press, Chicago, 2010)
2. S. Weinberg, *Gravitation and Cosmology: Principles and Applications of the General Theory of Relativity* (Wiley, New York, 1972)
3. J.D. Jackson, *Classical Electrodynamics* (Wiley, New York, 1998)
4. R.P. Feynman, *Quantum Electrodynamics* (Westview Press, Boulder, 1998)
5. P. Renton, *Electroweak Interactions. An Introduction to the Physics of Quarks and Leptons* (Cambridge University Press, Cambridge, 1990)
6. W. Greiner, S. Schramm, E. Stein, *Quantum Chromodynamics* (Springer, Berlin, 2007)
7. C. Burgess, G. Moore, *The Standard Model* (Cambridge University Press, Cambridge, 2011)
8. J. Wess, J. Bagger, *Supersymmetry and Supergravity* (Prinston University Press, Prinston, 1992)
9. B. Zwiebach, *A First Course in String Theory* (Cambridge University Press, Cambridge, 2006)
10. H.B.G. Casimir, On the attraction between two perfectly conducting plates. Proc. K. Ned. Akad. Wet. B **51**, 793–795 (1948)
11. M. Bordag, G.L. Klimchitskaya, U. Mohideen, V.M. Mostepanenko, *Advances in the Casimir Effect*. (Oxford University Press, Oxford, 2015)
12. A. Erdélyi, W. Magnus, F.G. Oberhettinger, *Higher Transcendental Functions*, vol. 1 (Kriger, New York, 1981)
13. E.M. Lifshitz, The theory of molecular attractive forces between solids. Zh. Eksp. Teor. Fiz. **29**, 94–110 (1955); Sov. Phys. JETP **2**, 73–83 (1956)

14. P. Richmond, B.W. Ninham, A note on the extension of the Lifshitz theory of van der Waals forces to magnetic media. J. Phys. C: Solid State Phys. **4**, 1988–1993 (1971)
15. G.L. Klimchitskaya, U. Mohideen, V.M. Mostepanenko, The Casimir force between real materials: experiment and theory. Rev. Mod. Phys. **81**, 1827–1885 (2009)
16. A. Canaguier-Durand, P.A. Maia Neto, I. Cavero-Pelaez, A. Lambrecht, S. Reynaud, Casimir interaction between plane and spherical metallic surfaces. Phys. Rev. Lett. **102**, 230404 (2009)
17. M. Hartmann, G.-L. Ingold, P.A. Maia Neto, Plasma versus drude modeling of the casimir force: beyond the proximity force approximation. Phys. Rev. Lett. **119**, 043901 (2017)
18. B. Spreng, M. Hartmann, V. Henning, P.A. Maia Neto, G.-L. Ingold, Proximity force approximation and specular reflection: application of the WKB limit of Mie scattering to the Casimir effect. Phys. Rev. A **97**, 062504 (2018)
19. M. Hartmann, G.-L. Ingold, P.A. Maia Neto, Advancing numerics for the Casimir effect to experimentally relevant aspect ratios. Phys. Scr. **93**, 114003 (2018)
20. C.D. Fosco, F.C. Lombardo, F.D. Mazzitelli, Proximity force approximation for the Casimir energy as a derivative expansion. Phys. Rev. D **84**, 105031 (2011)
21. G. Bimonte, T. Emig, R.L. Jaffe, M. Kardar, Casimir forces beyond the proximity force approximation. Europhys. Lett. **97**, 50001 (2012)
22. G. Bimonte, T. Emig, M. Kardar, Material dependence of Casimir force: gradient expansion beyond proximity. Appl. Phys. Lett. **100**, 074110 (2012)
23. L.P. Teo, Material dependence of Casimir interaction between a sphere and a plate: first analytic correction beyond proximity force approximation. Phys. Rev. D **88**, 045019 (2013)
24. G. Bimonte: Going beyond PFA: a precise formula for the sphere-plate Casimir force. Europhys. Lett. **118**, 20002 (2017)
25. R.S. Decca, E. Fischbach, G.L. Klimchitskaya, D.E. Krause, D. L'opez, V.M. Mostepanenko, Improved tests of extra-dimensional physics and thermal quantum field theory from new Casimir force measurements. Phys. Rev. D **68**, 116003 (2003)
26. R.S. Decca, D. L'opez, E. Fischbach, G.L. Klimchitskaya, D.E. Krause, V.M. Mostepanenko, Precise comparison of theory and new experiment for the Casimir force leads to stronger constraints on thermal quantum effects and long-range interactions. Ann. Phys. (N.Y.) **318**, 37–80 (2005)
27. R.S. Decca, D. L'opez, E. Fischbach, G.L. Klimchitskaya, D.E. Krause, V.M. Mostepanenko, Tests of new physics from precise measurements of the Casimir pressure between two gold-coated plates. Phys. Rev. D **75**, 077101 (2007)
28. R.S. Decca, D. L'opez, E. Fischbach, G.L. Klimchitskaya, D.E. Krause, V.M. Mostepanenko, Novel constraints on light elementary particles and extra-dimensional physics from the Casimir effect. Eur. Phys. J. C **51**, 963–975 (2007)
29. C.-C. Chang, A.A. Banishev, R. Castillo-Garza, G.L. Klimchitskaya, V.M. Mostepanenko, U. Mohideen, Gradient of the Casimir force between Au surfaces of a sphere and a plate measured using an atomic force microscope in a frequency-shift technique. Phys. Rev. B **85**, 165443 (2012)
30. J. Xu, G.L. Klimchitskaya, V.M. Mostepanenko, U. Mohideen, Reducing detrimental electrostatic effects in Casimir-force measurements and Casimir-force-based microdevices. Phys. Rev. A **97**, 032501 (2018)
31. M. Liu, J. Xu, G.L. Klimchitskaya, V.M. Mostepanenko, U. Mohideen, Examining the Casimir puzzle with an upgraded AFM-based technique and advanced surface cleaning. Phys. Rev. B **100**, 081406(R) (2019)
32. M. Liu, J. Xu, G.L. Klimchitskaya, V.M. Mostepanenko, U. Mohideen, Precision measurements of the gradient of the Casimir force between ultraclean metallic surfaces at larger separations. Phys. Rev. A **100**, 052511 (2019)
33. G. Bimonte, B. Spreng, P.A. Maia Neto, G.-L. Ingold, G.L. Klimchitskaya, V.M. Mostepanenko, R.S. Decca, Measurement of the Casimir Force between 0.2 and 8 μm: experimental Procedures and Comparison with Theory. Universe **7**, 93 (2021)

34. A.A. Banishev, C.-C. Chang, G.L. Klimchitskaya, V.M. Mostepanenko, U. Mohideen, Measurement of the gradient of the Casimir force between a nonmagnetic gold sphere and a magnetic nickel plate. Phys. Rev. B **85**, 195422 (2012)
35. A.A. Banishev, G.L. Klimchitskaya, V.M. Mostepanenko, U. Mohideen, Demonstration of the Casimir force between ferromagnetic surfaces of a ni-coated sphere and a ni-coated plate. Phys. Rev. Lett. **110**, 137401 (2013)
36. A.A. Banishev, G.L. Klimchitskaya, V.M. Mostepanenko, U. Mohideen, Casimir interaction between two magnetic metals in comparison with nonmagnetic test bodies. Phys. Rev. B **88**, 155410 (2013)
37. G. Bimonte, D. L'opez, R.S. Decca, Isoelectronic determination of the thermal Casimir force. Phys. Rev. B **93**, 184434 (2016)
38. V.B. Bezerra, G.L. Klimchitskaya, V.M. Mostepanenko, C. Romero, Violation of the Nernst heat theorem in the theory of thermal Casimir force between Drude metals. Phys. Rev. A **69**, 022119 (2004)
39. M. Bordag, I. Pirozhenko, Casimir entropy for a ball in front of a plane. Phys. Rev. D **82**, 125016 (2010)
40. G.L. Klimchitskaya, V.M. Mostepanenko, Low-temperature behavior of the Casimir free energy and entropy of metallic films. Phys. Rev. A **95**, 012130 (2017)
41. G.L. Klimchitskaya, C.C. Korikov, Analytic results for the Casimir free energy between ferromagnetic metals. Phys. Rev. A **91**, 032119 (2015)
42. A.H. Castro Neto, F. Guinea, N.M.R. Peres, K.S. Novoselov, A.K. Geim, The electronic properties of graphene. Rev. Mod. Phys. **81**, 109–162 (2009)
43. M. Bordag, G.L. Klimchitskaya, V.M. Mostepanenko, V.M. Petrov, Quantum field theoretical description for the reflectivity of graphene. Phys. Rev. D **91**, 045037 (2015); Erratum in **93**, 089907 (2016)
44. M. Bordag, I. Fialkovskiy, D. Vassilevich, Enhanced Casimir effect for doped graphene. Phys. Rev. B **93**, 075414 (2016); Erratum in **95**, 119905 (2017)
45. G.L. Klimchitskaya, V.M. Mostepanenko, Bo E. Sernelius, Two approaches for describing the Casimir interaction with graphene: Density-density correlation function versus polarization tensor. Phys. Rev. B **89**, 125407 (2014)
46. G.L. Klimchitskaya, V.M. Mostepanenko, Conductivity of pure graphene: theoretical approach using the polarization tensor. Phys. Rev. B **93**, 245419 (2016)
47. A.A. Banishev, H. Wen, J. Xu, R.K. Kawakami, G.L. Klimchitskaya, V.M. Mostepanenko, U. Mohideen, Measuring the Casimir force gradient from graphene on a SiO_2 substrate. Phys. Rev. B **87**, 205433 (2013)
48. G.L. Klimchitskaya, U. Mohideen, V.M. Mostepanenko, Theory of the Casimir interaction for graphene-coated substrates using the polarization tensor and comparison with experiment. Phys. Rev. B **89**, 115419 (2014)
49. M. Liu, Y. Zhang, G.L. Klimchitskaya, V.M. Mostepanenko, U. Mohideen, Demonstration of an unusual thermal effect in the casimir force from graphene. Phys. Rev. Lett. **126**, 206802 (2021)
50. M. Liu, Y. Zhang, G.L. Klimchitskaya, V.M. Mostepanenko, U. Mohideen, Experimental and theoretical investigation of the thermal effect in the Casimir interaction from graphene. Phys. Rev. B **104**, 085436 (2021)
51. V.B. Bezerra, G.L. Klimchitskaya, V.M. Mostepanenko, C. Romero, Nernst heat theorem for the thermal Casimir interaction between two graphene sheets. Phys. Rev. A **94**, 042501 (2016)
52. G.L. Klimchitskaya, V.M. Mostepanenko, Low-temperature behavior of the Casimir-Polder free energy and entropy for an atom interacting with graphene. Phys. Rev. A **98**, 032506 (2018)
53. G.L. Klimchitskaya, V.M. Mostepanenko, Nernst heat theorem for an atom interacting with graphene: Dirac model with nonzero energy gap and chemical potential. Phys. Rev. D **101**, 116003 (2020)

54. G.L. Klimchitskaya, V.M. Mostepanenko, Quantum field theoretical description of the Casimir effect between two real graphene sheets and thermodynamics. Phys. Rev. D **102**, 016006 (2020)
55. G.L. Klimchitskaya, V.M. Mostepanenko, Casimir and Casimir-Polder forces in graphene systems: quantum field theoretical description and thermodynamics. Universe **6**, 150 (2020)
56. G.L. Klimchitskaya, V.M. Mostepanenko, An alternative response to the off-shell quantum fluctuations: a step forward in resolution of the Casimir puzzle. Eur. Phys. J. C **80**, 900 (2020)
57. G.L. Klimchitskaya, V.M. Mostepanenko, Theory-experiment comparison for the Casimir force between metallic test bodies: a spatially nonlocal dielectric response. Phys. Rev. A **105**, 012805 (2022)
58. G.L. Klimchitskaya, V.M. Mostepanenko, Casimir entropy and nonlocal response functions to the off-shell quantum fluctuations. Phys. Rev. D **103**, 096007 (2021)
59. G.L. Klimchitskaya, V.M. Mostepanenko, Casimir effect for magnetic media: Spatially nonlocal response to the off-shell quantum fluctuations. Phys. Rev. D **104**, 085001 (2021)
60. V.M. Mostepanenko: Casimir puzzle and casimir conundrum: discovery and search for resolution. Universe **7**, 84 (2021)
61. A.A. Anselm, N.G. Uraltsev, A second massless axion? Phys. Lett. B **114**, 39–41 (1982)
62. V.M. Mostepanenko, I.Y. Sokolov, Restrictions on long-range forces following from the Casimir effect. Yadern. Fiz. **46**, 1174–1180 (1987); Sov. J. Nucl. Phys. **46**, 685–688 (1987)
63. G. Feinberg, J. Sucher, Long-range forces from neutrino-pair exchange. Phys. Rev. **166**, 1638–1644 (1968)
64. E. Fischbach, Long-range forces and neutrino mass. Ann. Phys. (N.Y.) **247**, 213–291 (1996)
65. S. Deser, B. Zumino, Broken Supersymmetry and Supergravity. Phys. Rev. Lett. **38**, 1433–1436 (1977)
66. L. Randall, R. Sundrum, Large mass hierarchy from a small extra dimension. Phys. Rev. Lett. **83**, 3370–3373 (1999)
67. L. Randall, R. Sundrum, An Alternative to Compactification. Phys. Rev. Lett. **83**, 4690–4693 (1999)
68. J.H. Gundlach, G.L. Smith, E.G. Adelberger, B.R. Heckel, H.E. Swanson, Short-range test of the equivalence principle. Phys. Rev. Lett. **78**, 2523–2526 (1997)
69. G.L. Smith, C.D. Hoyle, J.H. Gundlach, E.G. Adelberger, B.R. Heckel, H.E. Swanson, Short-range tests of the equivalence principle. Phys. Rev. D **61**, 022001 (2000)
70. D.J. Kapner, T.S. Cook, E.G. Adelberger, J.H. Gundlach, B.R. Heckel, C.D. Hoyle, H.E. Swanson, Tests of the gravitational inverse-square law below the dark-energy length scale. Phys. Rev. Lett. **98**, 021101 (2007)
71. E.G. Adelberger, B.R. Heckel, S. Hoedl, C.D. Hoyle, D.J. Kapner, A. Upadhye, Particle-physics implications of a recent test of the gravitational inverse-square law. Phys. Rev. Lett. **98**, 131104 (2007)
72. V.M. Mostepanenko, I.Y. Sokolov, The Casimir effect leads to new restrictions on long-range force constants. Phys. Lett. A **125**, 405–408 (1987)
73. B.V. Derjaguin, I.I. Abrikosova, E.M. Lifshitz, Direct measurement of molecular attraction between solids separated by a narrow gap. Quat. Rev. **10**, 295–329 (1956)
74. G. Feinberg, J. Sucher, Is there a strong van der Waals between hadrons? Phys. Rev. D **20**, 1717–1724 (1979)
75. W.-H. Tan, A.-B. Du, W.-C. Dong, S.-Q. Yang, C.-G. Shao, S.-G. Guan, Q.-L. Wang, B.-F. Zhan, P.-S. Luo, L.-C. Tu, J. Luo, Improvement for testing the gravitational inverse-square law at the submillimeter range. Phys. Rev. Lett. **124**, 051301 (2020)
76. S. Dimopoulos, G.F. Guidice, Macroscopic forces from supersymmetry. Phys. Lett. B **379**, 105–114 (1996)
77. Y. Fujii, The theoretical background of the fifth force. Int. J. Mod. Phys. A **6**, 3505–3557 (1991)
78. R.D. Peccei, H.R. Quinn, CP conservation in the presence of pseudoparticles. Phys. Rev. Lett. **38**, 1440–1143 (1977)

79. I. Antoniadis, N. Arkani-Hamed, S. Dimopoulos, G. Dvali, New dimensions at a millimeter to a fermi and superstrings at a TeV. Phys. Lett. B **436**, 257–263 (1998)
80. N. Arkani-Hamed, S. Dimopoulos, G. Dvali, Phenomenology, astrophysics, and cosmology of theories with millimeter dimensions and TeV scale quantum gravity. Phys. Rev. D **59**, 086004 (1999)
81. E.G. Floratos, G.K. Leontaris, Low scale unification, Newton's law and extra dimensions. Phys. Lett. B **465**, 95–100 (1999)
82. A. Kehagias, K. Sfetsos, Deviations from $1/r^2$ Newton law due to extra dimensions. Phys. Lett. B **472**, 39–44 (2000)
83. E. Fischbach, C.L. Talmadge, *The Search for Non-Newtonian Gravity* (Springer-Verlag, New York, 1999)
84. S.J. Smullin, A.A. Geraci, D.M. Weld, J. Chiaverini, S. Holmes, A. Kapitulnik, Constraints on Yukawa-type deviations from Newtonian gravity at 20 microns. Phys. Rev. D **72**, 122001 (2005)
85. J.K. Hoskins, R.D. Newman, R. Spero, J. Schultz, Experimental tests of the gravitational inverse-square law for mass separations from 2 to 105 cm. Phys. Rev. D **32**, 3084–3095 (1985)
86. S. Schlamminger, K.-J. Choi, T.A. Wagner, J.H. Gundlach, E.G. Adelberger, Test of the equivalence principle using a rotating torsion balance. Phys. Rev. Lett. **100**, 041101 (2008)
87. V.A. Kuzmin, I.I. Tkachev, M.E. Kaposhnikov, Restrictions imposed on light scalar particles by measurements of van der Waals forces. Pis'ma v Zh. Eksp. Teor. Fiz. **36**, 49–52 (1982); JETP Lett. **36**, 59–62 (1982)
88. D. Tabor, R.H.S. Winterton, Surface forces: direct measurement of normal and retarded van der Waals forces. Nature **219**, 1120–1121 (1968)
89. R.S. Decca, E. Fischbach, G.L. Klimchitskaya, D.E. Krause, D. L'opez, V.M. Mostepanenko, Application of the proximity force approximation to gravitational and Yukawa-type forces. Phys. Rev. D **79**, 124021 (2009)
90. H.C. Chiu, G.L. Klimchitskaya, V.N. Marachevsky, V.M. Mostepanenko, U. Mohideen, Demonstration of the asymmetric lateral Casimir force between corrugated surfaces in the nonadditive regime. Phys. Rev. B **80**, 121402(R) (2009)
91. H.C. Chiu, G.L. Klimchitskaya, V.N. Marachevsky, V.M. Mostepanenko, U. Mohideen, Lateral Casimir force between sinusoidally corrugated surfaces: Asymmetric profiles, deviations from the proximity force approximation, and comparison with exact theory. Phys. Rev. B **81**, 115417 (2010)
92. V.B. Bezerra, G.L. Klimchitskaya, V.M. Mostepanenko, C. Romero, Advance and prospects in constraining the Yukawa-type corrections to Newtonian gravity from the Casimir effect. Phys. Rev. D **81**, 055003 (2010)
93. A.A. Banishev, J. Wagner, T. Emig, R. Zandi, U. Mohideen, Demonstration of Angle-Dependent Casimir Force between Corrugations. Phys. Rev. Lett. **110**, 250403 (2013)
94. A.A. Banishev, J. Wagner, T. Emig, R. Zandi, U. Mohideen, Experimental and theoretical investigation of the angular dependence of the Casimir force between sinusoidally corrugated surfaces. Phys. Rev. B **89**, 235436 (2014)
95. G.L. Klimchitskaya, U. Mohideen, V.M. Mostepanenko, Constraints on corrections to Newtonian gravity from two recent measurements of the Casimir interaction between metallic surfaces. Phys. Rev. D **87**, 125031 (2013)
96. Y.J. Chen, W.K. Tham, D.E. Krause, D. L'opez, E. Fischbach, R.S. Decca, Stronger limits on hypothetical Yukawa interactions in the 30–8000 nm range. Phys. Rev. Lett. **116**, 221102 (2016)
97. V.V. Nesvizhevsky, G. Pignol, K.V. Protasov, Neutron scattering and extra short range interactions. Phys. Rev. D **77**, 034020 (2008)
98. Y. Kamiya, K. Itagami, M. Tani, G.N. Kim, S. Komamiya, Constraints on new gravitylike forces in the nanometer range. Phys. Rev. Lett. **114**, 161101 (2015)
99. C.C. Haddock, N. Oi, K. Hirota, T. Ino, M. Kitaguchi, S. Matsumoto, K. Mishima, T. Shima, H.M. Shimizu, W.M. Snow, T. Yoshioka, Search for deviations from the inverse square law of gravity at nm range using a pulsed neutron beam. Phys. Rev. D **97**, 062002 (2018)

100. V.M. Mostepanenko, M. Novello, Constraints on non-Newtonian gravity from the Casimir force measurements between two crossed cylinder. Phys. Rev. D **63**, 115003 (2001)
101. M. Masuda, M. Sasaki, limits on nonstandard forces in the submicrometer range. Phys. Rev. Lett. **102**, 171101 (2009)
102. G.L. Klimchitskaya, U. Mohideen, V.M. Mostepanenko, Constraints on non-Newtonian gravity and light elementary particles from measurements of the Casimir force by means of a dynamic atomic microscope. Phys. Rev. D **86**, 065025 (2012)
103. G.L. Klimchitskaya, P. Kuusk, V.M. Mostepanenko, Constraints on non-Newtonian gravity and axionlike particles from measuring the Casimir force in nanometer separation range. Phys. Rev. D **101**, 056013 (2020)
104. G.L. Klimchitskaya, V.M. Mostepanenko, Dark matter axions, non-newtonian gravity and constraints on them from recent measurements of the Casimir force in the micrometer separation range. Universe **7**, 343 (2021)
105. S.G. Karshenboim, Constraints on a long-range spin-independent interaction from precision atomic physics. Phys. Rev. D **82**, 073003 (2010)
106. J.H. Oort, The force exerted by the stellar system in the direction perpendicular to the galactic plane and some related problems. Bull. Astron. Inst. Neth. **6**, 249–287 (1932)
107. F. Zwicky, Die Rotverschiebung von extragalaktischen Nebeln. Helv. Phys. Acta **6**, 110–127 (1933)
108. G. Bertone, D. Hooper, Hystory of dark matter. Rev. Mod. Phys. **90**, 045002 (2018)
109. J.M. Overduin, P.S. Wesson, Dark matter and background light. Phys. Rep. **402**, 267–406 (2004)
110. G. Bertone, D. Hooper, J. Silk, Particle dark matter: evidence, candidates and constraints. Phys. Rep. **405**, 279–390 (2005)
111. R.H. Sanders, *The Dark Matter Problem: A Historical Perspective* (Cambridge University Press, Cambridge, 2010)
112. S. Matarrese, M. Colpi, V. Gorini, U. Moshella (eds.), *Dark Matter and Dark Energy* (Springer, Dordrecht, 2011)
113. S. Weinberg, A new light boson? Phys. Rev. Lett. **40**, 223–226 (1978)
114. F. Wilczek, Problem of strong P and T invariance in the presence of instantons. Phys. Rev. Lett. **40**, 279–283 (1978)
115. J.E. Kim, Light pseudoscalars, particle physics and cosmology. Phys. Rep. **150**, 1–177 (1987)
116. E.G. Adelberger, B.R. Heckel, C.W. Stubbs, W.F. Rogers, Searches for new macroscopic forces. Annu. Rev. Nucl. Part. Sci. **41**, 269–320 (1991)
117. L.J. Rosenberg, K.A. van Bibber, Searches for invisible axions. Phys. Rep. **325**, 1–39 (2000)
118. G.G. Raffelt, Axions—motivation, limits and searches. J. Phys. A Math. Theor. **40**, 6607–6620 (2007)
119. M. Kawasaki, K. Nakayama, Axions: theory and cosmological role. Annu. Rev. Nucl. Part. Sci. **63**, 69–95 (2013)
120. I.G. Ivastorza, J. Redondo, New experimental approaches in the search for axion-like particles. Progr. Part. Nucl. Phys. **102**, 89–159 (2018)
121. G.L. Klimchitskaya, Constraints on theoretical predictions beyond the standard model from the casimir effect and some other tabletop physics. Universe **7**, 47 (2021)
122. P.J.E. Peebles, B. Ratra, The cosmological constant and dark energy. Rev. Mod. Phys. **75**, 559–606 (2003)
123. S. Weinberg, The cosmological constant problem. Rev. Mod. Phys. **61**, 1–23 (1989)
124. J.A. Frieman, M.S. Turner, D. Huterer, Dark energy and the accelerating universe. Ann. Rev. Astron. Astrophys. **46**, 385–432 (2008)
125. Y.B. Zel'dovich, The cosmological constant and the theory of elementary particles. Uspekhi Fiz. Nauk **95**, 209–230 (1968); Sov. Phys. Usp. **11**, 381–393 (1968)
126. R.J. Adler, B. Casey, O.C. Jacob, Vacuum catastrophe: an elementary exposition of the cosmological constant problem. Am. J. Phys. **63**, 620–626 (1995)
127. J. Khoury, A. Weltman, Chameleon cosmology. Phys. Rev. D **69**, 044026 (2004)

128. K.A. Olive, M. Pospelov: Environmental dependence of masses and coupling constants. Phys. Rev. D **77**, 043524 (2008)
129. K. Hinterbichler, J. Khoury, Screening long-range forces through local symmetry restoration. Phys. Rev. Lett. **104**, 231301 (2010)
130. K. Hinterbichler, J. Khoury, A. Levy, A. Matas, Symmetron cosmology. Phys. Rev. D **84**, 103521 (2011)
131. D.D. Ryutov, D. Budker, V.V. Flambaum, A hypothetical effect of the maxwell-proca electromagnetic stresses on galaxy rotation curve. Astrophys. J. **871**, 218 (2019)
132. Y.N. Gnedin, S.V. Krasnikov, Polarimetric effects associated with the detection of Goldstone bosons in stars and galaxies. Sov. Phys. JETP **75**, 933–937 (1992); Zh. Eksp. Teor. Fiz. **102**, 1729–1738 (1992)
133. J.E. Moody, F. Wilczek, New macroscopic forces? Phys. Rev. D **30**, 130–139 (1984)
134. A. Bohr, B.R. Mottelson, *Nuclear Structure*, vol. 1 (Benjamin, New York, 1969)
135. E.G. Adelberger, E. Fischbach, D.E. Krause, R.D. Newman, Constraining the couplings of massive pseudoscalars using gravity and optical experiments. Phys. Rev. D **68**, 062002 (2003)
136. F. Ferrer, M. Nowakowski, Higgs- and Goldstone-boson-mediated long range forces. Phys. Rev. D **59**, 075009 (1999)
137. S.D. Drell, K. Huang, Many-body forces and nuclear saturation. Phys. Rev. **91**, 1527–1543 (1953)
138. S. Aldaihan, D.E. Krause, J.C. Long, W.M. Snow, Calculations of the dominant long-range, spin-independent contributions to the interaction energy between two nonrelativistic Dirac fermions from double-boson exchange of spin-0 and spin-1 bosons with spin-dependent couplings. Phys. Rev. D **95**, 096005 (2017)
139. I.S. Gradshtein, I.M. Ryzhik, *Table of Integrals, Series and Products* (Academic Press, New York, 1980)
140. G.L. Klimchitskaya, V.M. Mostepanenko, Improved constraints on the coupling constants of axion-like particles to nucleons from recent Casimir-less experiment. Eur. Phys. J. C **75**, 164 (2015)
141. N.F. Ramsey, The tensor force between two protons at long range. Phys. A **96**, 285–289 (1979)
142. M.P. Ledbetter, M.V. Romalis, D.F. Jackson Kimball, Constraints on short-range spin-dependent interactions from scalar spin-spin coupling in deuterated molecular hydrogen. Phys. Rev. Lett. **110**, 040402 (2013)
143. J.C. Long, H.W. Chan, A.B. Churnside, E.A. Gulbis, M.C.M. Varney, J.C. Price, Upper limits to submillimetre-range forces from extra space-time dimensions. Nature **421**, 922–925 (2003)
144. J.C. Long, V.A. Kosteleck'y, Search for Lorentz violation in short-range gravity. Phys. Rev. D **91**, 092003 (2015)
145. G. Vasilakis, J.M. Brown, T.R. Kornak, M.V. Romalis, Limits on new long range nuclear spin-dependent forces set with a K-^{3}He comagnetometer. Phys. Rev. Lett. **103**, 261801 (2009)
146. V.B. Bezerra, G.L. Klimchitskaya, V.M. Mostepanenko, C. Romero, Constraints on the parameters of an axion from measurements of the thermal Casimir-Polder force. Phys. Rev. D **89**, 035010 (2014)
147. V.B. Bezerra, G.L. Klimchitskaya, V.M. Mostepanenko, C. Romero, Stronger constraints on an axion from measuring the Casimir interaction by means of a dynamic atomic force microscope. Phys. Rev. D **89**, 075002 (2014)
148. V.B. Bezerra, G.L. Klimchitskaya, V.M. Mostepanenko, C. Romero, Constraining axion-nucleon coupling constants from measurements of effective Casimir pressure by means of micromachined oscillator. Eur. Phys. J. C **74**, 2859 (2014)
149. V.B. Bezerra, G.L. Klimchitskaya, V.M. Mostepanenko, C. Romero, Constraints on axion-nucleon coupling constants from measuring the Casimir force between corrugated surfaces. Phys. Rev. D **90**, 055013 (2014)
150. G.L. Klimchitskaya, V.M. Mostepanenko, Constraints on axionlike particles and non-Newtonian gravity from measuring the difference of Casimir forces. Phys. Rev. D **95**, 123013 (2017)

151. V.M. Mostepanenko, G.L. Klimchitskaya, The state of the art in constraining axion-to-nucleon coupling and non-newtonian gravity from laboratory experiments. Universe **6**, 147 (2020)
152. V.B. Bezerra, G.L. Klimchitskaya, V.M. Mostepanenko, C. Romero, Constraining axion coupling constants from measuring the Casimir interaction between polarized test bodies. Phys. Rev. D **94**, 035011 (2016)
153. K. Takyu, K.M. Itoh, K. Oka, N. Saito, V.I. Ozhogin, Growth and characterization of the isotopically enriched ^{28}Si bulk single crystal. Jpn. J. Appl. Phys. **38**, L1493 (1999)
154. P. Brax, C. van de Bruck, A.-C. Davis, D.F. Mota, D. Shaw, Detecting chameleons through Casimir force measurements. Phys. Rev. D **76**, 124034 (2007)
155. D.F. Mota, D.J. Shaw, Evading equivalence principle violations, cosmological, and experimental constraints in scalar field theories with a strong coupling to matter. Phys. Rev. D **75**, 063501 (2007)
156. C. Burrage, J. Sakstein, Tests of chameleon gravity. Living Rev. Relativ. **21**, 1 (2018)
157. B. Elder, V. Vardanyan, Y. Arkami, P. Brax, A.-C. Davis, R.S. Decca, Classical symmetron force in Casimir experiments. Phys. Rev. D **101**, 064065 (2020)
158. G.L. Klimchitskaya, Recent breakthrough and outlook in constraining the non-Newtonian gravity and axion-like particles from Casimir physics. Eur. Phys. J. C **77**, 315 (2017)
159. A. Almasi, P. Brax, D. Iannuzzi, R.I.P. Sedmik, Force sensor for chameleon and Casimir force experiments with parallel-plate configuration. Phys. Rev. D **91**, 102002 (2015)
160. R. Sedmik, P. Brax, Status report and first light from Cannex: Casimir force measurements between flat parallel plates. J. Phys. Conf. Ser. **1138**, 012014 (2018)
161. G.L. Klimchitskaya, V.M. Mostepanenko, R.I.P. Sedmik, H. Abele, Prospects for searching thermal effects, non-newtonian gravity and axion-like particles: CANNEX test of the quantum vacuum. Symmetry **11**, 407 (2019)
162. G.L. Klimchitskaya, V.M. Mostepanenko, R.I.P. Sedmik, Casimir pressure between metallic plates out of thermal equilibrium: proposed test for the relaxation properties of free electrons. Phys. Rev. A **100**, 022511 (2019)
163. R.I.P. Sedmik, Casimir and non-newtonian force experiment (CANNEX): review, status, and outlook. Int. J. Mod. Phys. A **35**, 2040008 (2020)
164. R. Bennett, D.H.J. O'Dell, Revealing short-range non-Newtonian gravity through Casimir-Polder shielding. New J. Phys. **21**, 033032 (2019)
165. L. Mattioli, A.M. Frassino, O. Panella, Casimir-Polder interactions with massive photons: implications for BSM physics. Phys. Rev. D **100**, 116023 (2019)
166. L. Chen, J. Liu, K. Zhuy, Constraining the axion-nucleon coupling and non-Newtonian gravity with a levitated optomechanical device. Phys. Rev. D **106**, 095007 (2022)
167. C.C. Haddock, N. Oi, K. Hirota, T. Ino, M. Kitaguchi, S. Matsumoto, K. Mishima, T. Shima, H.M. Shimizu, W.M. Snow, T. Yoshioka, Search for deviations from the inverse square law of gravity at nm range using a pulsed neutron beam. Phys. Rev. D **97**, 062002 (2018)
168. P. Brax, S. Fichet, G. Pignol, Bounding quantum dark forces. Phys. Rev. D **97**, 115034 (2018)
169. S. Sponar, R.I.P. Sedmik, M. Pitschmann, H. Abele, Y. Hasegawa, Tests of fundamental quantum mechanics and dark interactions with low-energy neutrons. Nature Rev. Phys. **3**, 309–327 (2021)
170. B. Heacock, T. Fujiie, R.W. Haun, A. Henins, K. Hirota, T. Hosobata, M.G. Huber, M. Kitaguchi, D.A. Pushin, H. Shimizu, M. Takeda, R. Valdillez, Y. Yamagata, A. Young, Pendell"osung interferometry probes the neutron charge radius, lattice dynamics, and fifth forces. Science **373**, 1239–1243 (2021)
171. J.M. Rocha, F. Dahia, Neutron interferometry and tests of short-range modifications of gravity. Phys. Rev. D **103**, 124014 (2021)
172. M. Borkowski, A.A. Buchachenko, R. Ciuryło, P.S. Julienne, H. Yamada, Y. Kikuchi, Y. Takasu, Y. Takahashi, Weakly bound molecules as sensors of new gravitylike forces. Sci. Rep. **9**, 14807 (2019)
173. W.G. Hollik, M. Linster, T. Tabet, A study of new physics searches with tritium and similar molecules. Eur. Phys. J. C **80**, 661 (2020)

174. A.S. Lemos, Submillimeter constraints for non-Newtonian gravity from spectroscopy. Europhys. Lett. **135**, 11001 (2021)
175. E. Hebestreit, M. Frimmer, R. Reimann, L. Novotny, Sensing static forces with free-falling nanoparticles. Phys. Rev. Lett. **121**, 063602 (2018)
176. J. Liu, K.-D. Zhu, Detecting large extra dimensions with optomechanical levitated sensors. Eur. Phys. J. C **79**, 18 (2019)
177. X. Ren, J. Wang, R. Luo, L. Yin, J. Ding, G. Zeng, P. Luo, Search for an exotic parity-odd spin- and velocity-dependent interaction using a magnetic force microscope. Phys. Rev. D **104**, 032008 (2021)
178. H. Banks, M. McCullough, Charting the fifth force landscape. Phys. Rev. D **103**, 075018 (2021)
179. M. Faizal, H. Patel, Probing short distance gravity using temporal lensing. Int. J. Mod. Phys. A **36**, 2150115 (2021)

Part III
Quantum Systems and Gravity

Form the observational consequences of modified and quantum, gravity in cosmic, astrophysical and laboratory sized setups, as they were discussed in Part II, we move on to the interaction of quantum matter with gravity. The laboratory sized setups involving quantum effects test the properties of the weak gravitational field to very high precession.

Chapter 14 discusses quantum test of gravity, i.e., tests of gravity with quantum states of matter like optical cavities, cold atom interferometry and Bose-Einstein condensates. Their advantage, compared to classical tests, lies in a high accuracy and the insights about the coupling between quantum matter and gravity. They mainly test deviations from the Newtonian potential due to additional scalar fields emerging from modified gravity or low energy limits of String theory.

The question about the coupling between quantum matter and gravity is picked up again in Chap. 15, where the impact of the gravitational field of a plane electromagnetic wave on different neutral fields, like scalars, spinors and vectors is presented. In this way the indirect coupling between neutral fields, via gravity, to the electromagnetic field can be predicted, and the universality of the gravity matter coupling can be tested.

Last but not least Chap. 16 discusses the Newtonian weak field limit of quantum field theory on curved spacetimes, including the backreaction of quantum fields on the gravitational field, and thus the non-relativistic limit of the coupling of relativistic quantum fields to gravity. It is demonstrated that a systematic mathematical treatment ensures that no coupling terms are missed in the non-relativistic c^{-1} expansion of the field equations and what consequences can be deduced from the semi-classical Einstein equations.

Quantum Tests of Gravity

14

Sven Herrmann and Dennis Rätzel

Abstract

We give an introduction to quantum tests of gravity including simple exercises. The first part comprises a brief review of the foundations and the status of laboratory-based tests of gravity. In the second part, we discuss a specific platform for quantum tests of gravity represented by cold atom interferometry (CAI) in detail and introduce the non-expert reader to its basic operation principles. The motivation for quantum tests of gravity has two aspects that will both be addressed. The first is technical and based on the fact that quantum technology devices such as CAI are intrinsically very sensitive and allow for highly accurate measurements. As such they may simply help to perform otherwise classical test experiments at improved accuracy. The second motivation roots in the quantum nature of the applied test masses which opens avenues to test gravity theory in ways that are not available with classical systems. We briefly discuss some of the proposed ideas and point the reader to the related literature for further reading.

14.1 Introduction

Experimental tests of gravitation have been on the agenda of science since the early days of Galilei and Newton. However, due to the weakness of the gravitational force, the first proper gravitational experiment in a laboratory was performed much later by Cavendish who used it to precisely measure the gravitational constant. In general, one can define laboratory tests or local tests of gravity as those where one has control over the complete measurement system directly affected by the gravitational field

S. Herrmann (✉) · D. Rätzel
ZARM, University Bremen, Bremen, Germany
e-mail: sven.herrmann@zarm.uni-bremen.de; dennis.raetzel@zarm.uni-bremen.de

C. Pfeifer, C. Lämmerzahl (eds.), *Modified and Quantum Gravity*, Lecture Notes in Physics 1017, https://doi.org/10.1007/978-3-031-31520-6_14

and can perform repeated preparations and measurements. Following the famous experiment by Cavendish, there have been many similar laboratory experiments in the last 200 years measuring the gravitational force with increasing precision on scales down to micrometer distances and millimeter-sized masses [1–4], (as has also been discussed in Chap. 13, in context of the Casimir effect). However, tests of gravity are not restricted to force measurements. Since general relativity has extended the realm of gravitation to the very fabric of space and time, laboratory tests of gravity comprise tests of fundamental principles like the equivalence principle.

In this work, we focus on a particular branch of laboratory tests, that is, tests that employ quantum systems as sensors. Instead of providing a comprehensive review, we want to give a concise and self-contained introduction to the topic for beginners. To this end, in the first part, we give a short non-comprehensive wrap-up of the main features of gravity that can be tested in experiments and mention a few examples of tests with quantum sensors. In the second part, we provide details about a specific experimental platform for quantum tests of gravity, cold atom interferometry (CAI).

14.2 Some Lab-Testable Features of Gravity

As mentioned above, gravity's manifestations are manifold and as such are potential laboratory tests. We start right away with the most obvious.

14.2.1 The Gravitational Force

Indeed, the first feature of gravity that has been tested many times and is still a central element of tests of gravity is its manifestation as a force between masses. The first proper laboratory test of this kind was the one performed by Cavendish. It consists of a torsion balance: essentially, two spherical masses connected by a rod which is suspended from a wire that is attached to its center (see Fig. 14.1). Two more spherical masses are then brought close to those of the torsion balance to measure the resulting gravitational force between the masses. The torsion balance is still an important element of experimental gravity research today [1]. A particularly important feature and basis for its success is that it is fairly insensitive to time dependent gravitational acceleration generated by changes in the surrounding mass distribution.

A millimeter-scale implementation of the torsion balance concept has recently been used for the measurement of the gravitational force between gold spheres of less than 100 mg [4]. This experiment was based on a light beam that is reflected from a mirror attached to the rod of the torsion balance, and thus, is part of the wider field of optomechanics. Some experimental platforms in this field have already entered the quantum regime [6]. With decreasing size and increasing control, torsion balances may eventually also enter this regime, where the ultimate precision is dominated by quantum fluctuations [7]. Very small and precise sensor systems are

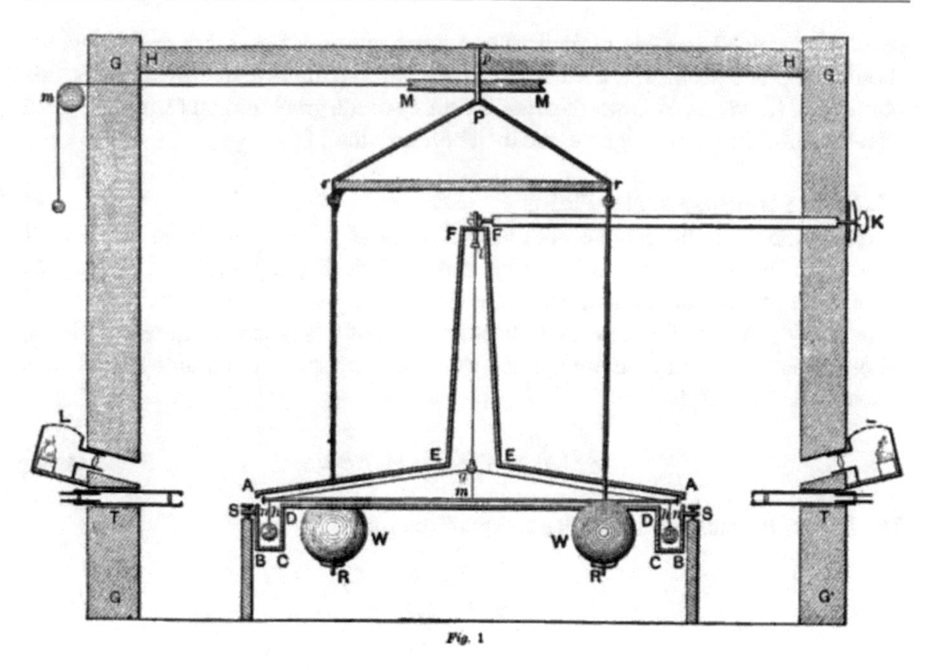

Fig. 14.1 Original drawing of the torsion balance used by Henry Cavendish for his experiment [5]

particularly interesting for tests of gravity at short distances, which we will discuss in the following.

14.2.1.1 Gravity at Short Distances

At short distances, one can test for deviations from Newton's gravitational force law that are strongly distance dependent. A well motivated form of a deformed gravitational force law is that corresponding to a Yukawa potential acting on a spherical probe mass m in the presence of a spherical source mass M_S

$$V(r) = -G\frac{M_S m}{r}\left[1 + \alpha e^{-r/\lambda}\right], \tag{14.1}$$

where α describes a modified strength and λ a modified range of the interaction. For $\alpha \approx 1$, that is an interaction of comparable strength to gravity, experiments down to $\lambda \approx 10\,\mu\text{m}$ have been performed using torsion pendula or microcantilevers [1]. For more details on tests of Yukawa corrections to Newton's law with help of force measurements at short distances, see Sect. 13.4.

A particular motivation for modifications of the Yukawa type comes from scalar tensor theories of gravity where a scalar field is added to the metric tensor of space-time to explain the observed acceleration of the universe that is phenomenologically described as "dark energy" [8]. As the nature of dark energy is one of the most pressing unsolved puzzles in physics and there are only few ideas as to where it should originate from, the community has a strong interest in investigating this possibility. However, Yukawa-like modifications of the gravitational force have been already constrained to a high degree by laboratory tests [1–3].

14.2.1.2 Screened Scalar Fields

A modification of the simple addition of a scalar field is the inclusion of self interaction that can lead to a screening mechanism that suppresses effects of the scalar field in regions of significant mass densities thus evading its straightforward detection [8]. A specific class of candidate models of this type are chameleon fields, whose equations of motion are defined by a Lagrangian that contains an effective potential density of the form

$$V_{\mathrm{eff}}(\phi) = V(\phi) + V_{int}(\phi). \tag{14.2}$$

The first term describes a self-interaction of the form

$$V_{(\phi)} = \frac{\Lambda^{4+n}}{\phi^n} \tag{14.3}$$

(in natural units) where Λ needs to be on the order of the observed value of the cosmological constant $\Lambda_0 = 2.4\,\mathrm{meV}$ in order to link such a Chameleon field to dark energy and accelerated cosmic expansion. The second term describes interaction of the field with ordinary matter, and for the simplest case of universal coupling can be written as

$$V_{int} = \frac{\rho}{M}\phi. \tag{14.4}$$

The parameter M has the dimension of mass and is expected to be on the order respectively below the level of the Planck mass M_{Pl}. Thus there are essentially two parameters that need to be constrained in order to rule out the existence of a Chameleon field of given n (Fig. 14.2).

The goal of an experiment would then be to measure the anomalous gravitational force next to a source mass. Given a spherical source mass M_S and a spherical probe mass m, this anomalous force is given as [11]

$$F_C = \frac{GM_S m}{r^2}\left[1 + \alpha_C\left(1 + \frac{r}{\lambda_C}\right)e^{-r/\lambda_C} f(R_P/\lambda_C, r/\lambda_c)\right], \tag{14.5}$$

where α_C and λ_C are the modified strength and modified range in the sense of the Yukawa potentials above , respectively, which depend on M, Λ as well as the radius and the density of the source mass. The function $f(R_P/\lambda_C, r/\lambda_C)$ is a form factor that depends on the mass m and radius R_P of the probe sphere and converges to 1 for $R_P/\lambda_C \to 0$. The modified strength can be written as $\alpha_C = 2\zeta_S\zeta_P(M_{Pl}/M)^2$, where ζ_S and ζ_P are screening factors of the source sphere and the probe sphere

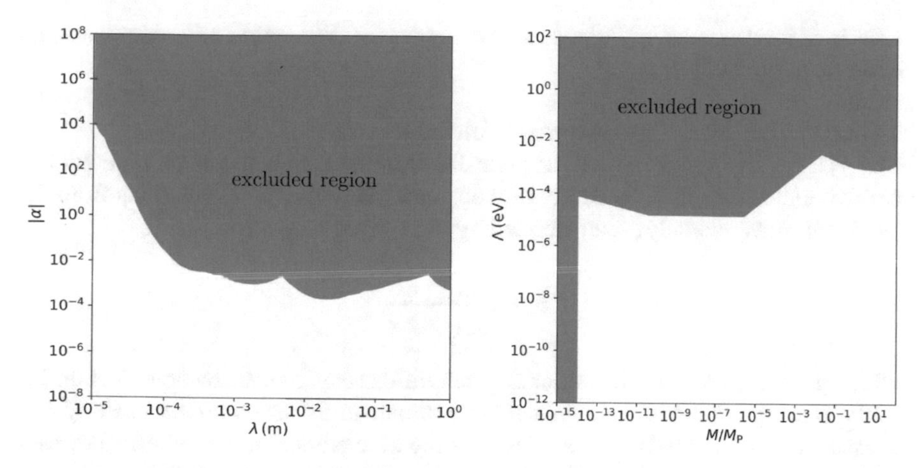

Fig. 14.2 Left plot: experimental bounds on Yukawa parameters α and λ. Redrawn from Fig. 8 of [9] ©2015 IOP Publishing Ltd, all rights reserved, and recent results presented in figure 6 of [10] ©2020 American Physical Society. All rights reserved. Right plot: bounds on mass and energy-scale chameleon field parameters M and Λ for $n = 1$ as compiled in Ref [8] ©The Authors 2018. Distributed under CC-BY 4.0 license

respectively. ζ_S, ζ_P and λ_C decrease for increasing background density. This is where the screening mechanism takes place, that is, the anomalous contribution to F_C only becomes visible in high vacuum, for example, in an experiment with a vacuum chamber. Furthermore, the screening factors ζ_S and ζ_P decrease with increasing density and radius of the source sphere and the probe sphere, respectively. A way to maximize ζ_P would be to use smaller and smaller probe masses, which is not trivial with the usual torsion balance experiments. Instead one can use atoms as probe masses as in the CAI experiments that will be discussed in more detail in Sect. 14.3. Another type of quantum systems that have been used as test masses is neutrons [8].

? Exercise

14.1. Calculate the gravitational force due to the Yukawa potential in Eq. (14.1) and compare with Eq. (14.5) for $R_P/\lambda_C \to 0$. Show that in the short distance limit $r/\lambda \to 0$ only a modification of Newton's gravitational constant remains.

14.2.2 The Equivalence Principle(s)

The experimental basis of General Relativity is set in the Einstein Equivalence Principle (EEP), which following [12] comprises the Weak Equivalence Principle (WEP), Local Lorentz Invariance (LLI) and the Local Position Invariance (LPI). The

EEP is of fundamental importance as it supports the description of gravity based on a metric tensor [12].

14.2.2.1 The Weak Equivalence Principle

The WEP is the statement that all point-like masses follow the same trajectory in a gravitational field independently of their material properties. Deviations from the WEP are conventionally parameterized by the Eötvös parameter

$$\eta = 2\frac{g_A - g_B}{g_A + g_B}, \tag{14.6}$$

where g_A and g_A are the gravitational accelerations experienced by objects A and B in the same gravitational field. When the anomalous chameleon force in Eq. (14.5) is associated with gravity, it is obvious that the acceleration of two spherical masses of the same mass but different density will, in general, be different. This would be a case of violating the WEP.

In the context of general relativity, the WEP is formulated as universality of propagation of free-falling test particles along time-like geodesics. That is, if the WEP is fulfilled, the trajectory γ of a point-like mass will always be governed by the geodesic equation

$$\ddot{\gamma}^{\mu} = -\Gamma^{\mu}_{\nu\rho}\dot{\gamma}^{\nu}\dot{\gamma}^{\rho}, \tag{14.7}$$

where $\Gamma^{\mu}_{\nu\rho}$ are the Christoffel symbols defined through the metric tensor of the spacetime, $\dot{\gamma}^{\nu}$ is the tangent to the curve γ and the derivative is taken with respect to the curve's affine parameter.

Obviously, the WEP can only be tested approximately as no real point-like mass can be used. In a realistic experiment, a test could be based on, for example, the free fall of masses with different compositions in the gravitational field of the earth or force measurements between a source mass and test masses like in torsion balance experiments.

14.2.2.2 Local Lorentz Invariance

The principle of LLI states that the outcome of any local non-gravitational experiment is independent of the velocity of the freely-falling reference frame in which it is performed. This implies, for example, that the speed of light is the same in all reference systems. Corresponding tests date back even to the pre-relativity times with the Michelson-Morley interferometer experiment [13]. A detailed review of the current state of tests of LLI can be found in [14], and see also the discussion about breaking or deforming LLI in Chaps. 1 and 2 as well as the constraints from cosmic messengers in Chap. 6.

To quantify the precision at which LLI is confirmed by experiments, tests may be based on the Robertson–Mansouri–Sexl framework [15, 16] or on an extension of the standard model of particle physics (SME) that allows for violation of Lorentz invariance [17, 18]. The latter uses general, Lorentz invariance violating scalar expressions formed from fields and dimensionless tensors to extend the fundamental Lagrangian underlying each sector of the SME. From this the dynamics and modified equations of motion are derived and the observable consequences of broken Lorentz symmetry in precision experiments are modeled. In [19] Bailey and Kostelecky have extended this concept of SME to include also a possible violation of Lorentz invariance for gravitational experiments at the post-Newtonian level. One consequence of Lorentz invariance violation found there, is an anisotropy contained in the elements of a dimensionless tensor $\bar{s}$ introduced into the Lagrangian that describes gravitational interaction of two point-like masses m and M at a separation of $\mathbf{r}$:

$$\mathcal{L} = \frac{1}{2}mv^2 + G\frac{Mm}{2r}\left(2 + 3\bar{s}^{00} + \bar{s}^{jk}\hat{r}^j\hat{r}^k - 3\bar{s}^{0j}v^j - \bar{s}^{0j}\hat{r}^j v^k\hat{r}^k\right), \tag{14.8}$$

where $\hat{r} = \mathbf{r}/r$. Another approach to LLI violation in gravity has been formulated by Nordtvedt much earlier in an anisotropic PPN formalism with a Lagrangian of similar form, the parameters of which find corresponding counterparts in the various elements of the SME tensor $\bar{s}$ [20].

14.2.2.3 Local Position Invariance

The principle of LPI states that the outcome of any local non-gravitational experiment is independent of where and when in the universe it is performed. One important consequence of LPI is the constancy of the fundamental constants that is tested in various ways by direct measurements of these constants (see [21] for a detailed review). Another consequence is the Universality of Gravitational Redshift (UGR) [22], which states that rates of clocks at different positions are related by the gravitational redshift.

Gravitational redshift has been tested in various ways starting from the first famous experiment by Pound and Rebka [23] utilizing the Mössbauer effect to obtain extremely narrow resonance lines of a crystal and sending photons up in a tower. The most obvious test principle is to employ highly precise clocks placed at different heights and a simple clock comparison [24, 25]. The corresponding test theory to quantify deviations of the predictions of general relativity is a one-parametric deviation of the gravitational redshift formula $z = (1 + \alpha)\Delta V/c^2$. The parameter α has been constrained to the level 10^{-5} by tests with clocks in space [24] and in a tower [25].

Although all of the described tests employ quantum systems, their quantumness is not of fundamental importance for the test strategy. For example, the quantum properties of optical clocks are only used to obtain very high precision. In Sect. 14.3 and in the following, we will discuss a few examples, where tests are based on quantum properties of the sensors.

14.2.3 Quantum Features of Gravity

Quantum properties of the sensing systems are essential when quantum features of gravity are to be tested in laboratory experiments. Tests of two particular features have been widely discussed: the Heisenberg algebra of position and momentum of point particles and the degree at which gravity is a quantum coherent mediator. Deformations of the Heisenberg algebra—often associated with Generalized Uncertainty Principles (GUP)—can be motivated from a minimal length scale associated with quantized gravity [26] and may be tested, for example, by precision measurement of the Lamb shift, Landau levels, and the tunneling current in a scanning tunneling microscope [27]. Note that deformations of the Heisenberg algebra can also have an imprint in the dynamics of macroscopic systems (as implied by the correspondence principle, see e.g. [28]), and thus, may be restricted, for example, by experiments with macroscopic oscillators (see e.g. [29, 30]) or tests of the WEP [31]. If gravity may be described by a fully quantized theory, it should act as a quantum coherent mediator. That is, given a gravitating system in a coherent superposition of states, where each would lead to a different gravitational field, the resulting effect on a quantum test particle should be a coherent superposition as well. This would then lead, for example, to the creation of entanglement between the source system and test particle, which may be measured in an experiment [32, 33]. Let us assume that two spheres of mass m that are freely falling are each brought into a spatial superposition state such that they are either closer together or further apart as depicted in Fig. 14.3. Neglecting the dispersion of the wave packets of the spheres and the change of momentum induced by the gravitational interaction, the only remaining element of the time evolution of the joint state of the two spheres is the accumulation of distance dependent phases. Then, the initial state evolves as

$$
\begin{aligned}
(|L\rangle_1 + |R\rangle_1)(|L\rangle_2 + |R\rangle_2)/2 \rightarrow \quad & e^{i\frac{Gm^2 t}{\hbar d}}|L\rangle_1|L\rangle_2/2 + e^{i\frac{Gm^2 t}{\hbar(d-\Delta x)}}|R\rangle_1|L\rangle_2/2 \qquad (14.9)\\
& + e^{i\frac{Gm^2 t}{\hbar(d+\Delta x)}}|L\rangle_1|R\rangle_2/2 + e^{i\frac{Gm^2 t}{\hbar d}}|R\rangle_1|R\rangle_2/2\,,
\end{aligned}
$$

m m

$|L\rangle + |R\rangle \quad |L\rangle + |R\rangle$

r

Δx Δx

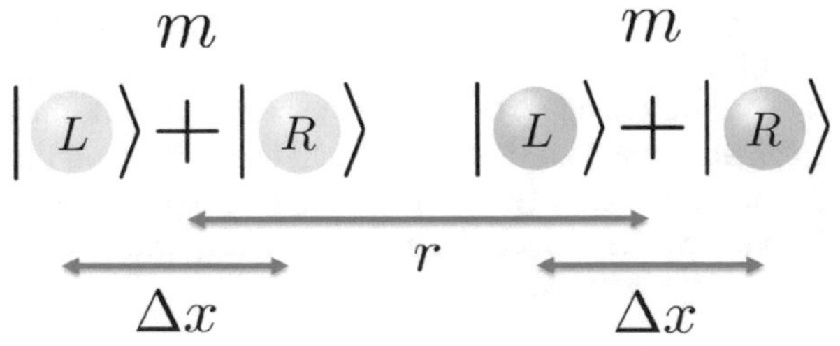

Fig. 14.3 Gravitational interaction of two masses in coherent superposition states of positions. If gravity is fundamentally quantum, the interaction should lead to the creation of entanglement [32, 33]

where d is the initial/average distance between the spheres and Δx is the distance between the two positions of the superposition state of each sphere. For general t, this state cannot be written as a product of separated states of each sphere, and thus, the state is entangled. The entanglement may then be certified by correlated measurements on the two spheres [32]. A similar experiment can be envisioned based on gravitationally interacting optomechanical systems [34–36].

Then, it is the question what one can learn about gravity from quantum coherent mediation beyond that gravity is compatible with the superposition principle. For example, one can argue that gravity needs to be quantized to ensure that relativistic causality holds [37, 38]. The very least one can do is to exclude some frameworks that describe the coupling of quantum matter via classical gravity such as semiclassical gravity and Classical Channel Gravity [39].

? Exercise

14.2. Derive Eq. (14.9) for the setup in Fig. 14.3 from the Schrödinger equation as described in the text.

14.3. The entanglement of the state on the right hand side of Eq. (14.9) can be estimated with the linear entropy $S_L = 1 - \mathrm{Tr}_1[\hat{\rho}_1^2]$, where $\hat{\rho}_1 = \mathrm{Tr}_2[\hat{\rho}]$ is the reduced density matrix of mass 1 after the trace has been performed over the Hilbert space of mass 2. Assuming that $|L\rangle$ and $|R\rangle$ are orthogonal states for both masses, show that the linear entropy is $S_L = \sin^2(Gm^2 t\Delta x^2/(d^2(d-\Delta x)))/2$. For $\Delta x = d/3$, $d = 300$ μm and $m = 10^{-14}$ kg find the smallest value for t for which S_L is maximized.

14.3 Tests of Classical Gravity with CAI

After the discussion of potential features of gravity that can be tested in experiments, in this section, we will present details of a specific category of tests performed with quantum systems: matter wave interferometry with cold atoms.

14.3.1 Light-Pulse CAI: Basic Concepts

Matter wave interferometry as discussed in the following builds on very cold, dilute gases of neutral atoms. For such atomic ensembles the quantum-mechanical de Broglie wavelength is a function of temperature T as given by

$$\lambda_{dB}(T) = \frac{h}{\sqrt{2\pi m k_B T}}. \tag{14.10}$$

At lowest temperatures λ_{dB} may even become comparable to the interatomic distances. In this regime the wave characteristics of the ensemble become apparent, and with bosonic atoms even a phase transition to a Bose-Einstein Condensate (BEC) eventually occurs.

To perform interferometry with such atoms requires the matter wave equivalent of a beamsplitter. Here, this is realized by interaction with short laser pulses. Such a beamsplitter creates a spatially delocalized superposition of the atomic wave function which then evolves along different paths of the interferometer and is coherently manipulated by further beamsplitters or mirrors. The partial wave packets eventually interfere in the output ports of a final beamsplitter, and from the observed number of atoms in those ports the relative accumulated phase is determined. This phase depends sensitively on the action of inertial and gravitational forces, which couple to the atoms during free evolution of the superposition.

The measurements we describe below all rely on the phase evolution of freely falling atoms, with no external fields applied for levitation or guiding. A lot of research has been done on guided interferometry as well, but no precision measurements of gravity have been reported from that work yet, thus it is not covered here. Also note that for most purposes the phase evolution throughout the interferometer can be treated semiclassically, i.e. with classical light fields, and the interferometer paths are usually taken to be the classical trajectories of a point particle at the center of mass of the partial atomic wave functions.

14.3.1.1 Matter Wave Source

Two steps are most commonly applied to cool a dilute gas to the required level of μK and below. The first is laser cooling in a magneto-optical trap (MOT) followed by a so called optical molasses. This provides an ensemble of typically up to 10^9 atoms at few μK temperature. Many CAI measurements start from such a molasses, possibly applying some velocity selection of the atoms to provide the necessary coherence for subsequent interferometry.

To obtain even lower temperature, evaporative cooling in a conservative magnetic or optical trapping potential can be applied. This reduces the temperature and particle number to typically few 100 nK and 10^6 atoms or less. If at the same time the ensemble density is maintained sufficiently high, quantum degeneracy and Bose-Einstein condensation (BEC) may be achieved.

Using a BEC rather than a molasses allows for better spatial coherence and reduced spatial spreading of the free falling atomic cloud. With a BEC the interaction driven expansion can be as low as mm/s and less as compared to several cm/s for a thermal cloud of atoms. This helps to reduce systematic errors and to extend the free evolution time in an interferometer measurement. Furthermore, the velocity spread of the BEC is less than the momentum separation imparted at the beamsplitters. Thus, the interferometer paths can be distinctly delocalized and the expanding clouds in the different paths do not overlap. The cost of using a BEC on the other hand is a substantially higher complexity of the setup when combining laser and evaporative cooling, as well as the reduced atom number and thus reduced signal to noise ratio.

Atomic species that are typically used in CAI are those that can be easily laser cooled. In the first place, these are alkali atoms such as Li, Na, Rb, K and

Cs which feature only one valence electron and thus simple hyperfine spectra. In particular ^{87}Rb and ^{133}Cs have been the work horses applied in many CAI precision experiments. For the bosonic isotopes evaporative cooling has allowed to achieve Bose-Einstein condensation with most such alkali atoms. There the evaporative cooling efficiency depends largely on the atoms' collisional scattering properties, which determines how well the ensemble rethermalizes during evaporative cooling. More recent CAI measurements have also employed Earth-Alkali group-II elements such as Sr or Yb. This group of elements is commonly used in optical clocks based on cold atoms. The required cooling techniques are usually more involved due to the more complex energy level structure, but some features of these atoms make them very attractive in precision CAI measurements. For example interferometry with some of these isotopes is much less sensitive to magnetic field variations. Also, the narrow clock transition and thus long lived excited state they provide, allows for single-photon beamsplitting. In CAI applications with very long baselines, this provides an important advantage over two-photon processes (see below) as used for Alkali atoms.

14.3.1.2 Matter Wave Beamsplitters and Mirrors

Coherent matter wave beamsplitters and mirrors for cold atoms can be achieved by interaction with short laser pulses. Most often these are two-photon processes that transfer momentum kicks of two photons to an atom. We briefly describe the most common implementations of these below and discuss their respective benefits and drawbacks.

Stimulated Raman Transition

A stimulated two-photon Raman transition couples two hyperfine levels of the ground state of an atom, $|g_1\rangle$ and $|g_2\rangle$, via an excited state $|e\rangle$, as shown in Fig. 14.4a. Here, the two laser beams of frequencies ω_1 and ω_2 are detuned by Δ with respect to the optical transition thus avoiding spontaneous emission from the intermediate excited state. This forms an effective two-level system and Rabi oscillations can be

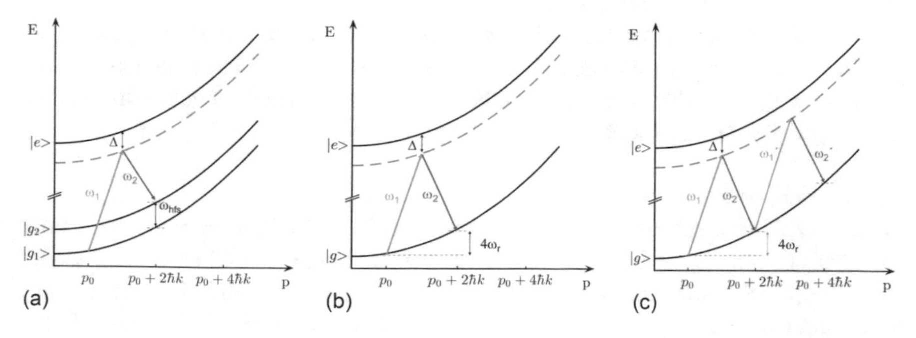

Fig. 14.4 Schematic representation of (**a**) stimulated Raman transition, (**b**) Bragg diffraction and (**c**) Bloch oscillations. For the latter two exemplary subsequent two-photon transitions are shown. The recoil frequency shift is given by $\omega_r = \frac{\hbar k^2}{2m}$, where $k = k_1 \approx k_2$ is the laser wave vector

driven between the coupled states where the effective Rabi frequency is combined from the single-photon Rabi frequencies Ω_1, Ω_2 as $\Omega = \frac{\Omega_1 \Omega_2}{4\Delta}$. With pulses of duration τ such that $\Omega\tau = \pi/2$ a superposition of the two hyperfine states is obtained, while complete state inversion requires $\Omega\tau = \pi$.

Using such a two-photon transition offers two main advantages over a single-photon transition: First, the coupled hyperfine states are long lived and spontaneous emission can be neglected. Second, it is only the difference frequency of the two laser beams that needs to be accurately tuned to resonance. This frequency is typically in the GHz regime and can be controlled much more easily than the optical laser frequency as would be required for a single-photon transition.

In order to obtain a matter wave beamsplitter or mirror from this configuration the two laser beams need to be counter-propagating such that the recoil from absorbed ($\hbar k_1$) and emitted ($\hbar k_2$) photons add up. Then, for a $\pi/2$-pulse a delocalized superposition of atoms in different hyperfine and momentum states is obtained, where the two partial wave packets separate with a relative velocity of two photon recoils $\hbar k_1 + \hbar k_2 \approx 2\hbar k$ on the order of few cm/s.

Bragg Diffraction

Coherent coupling of two momentum states separated by $2\hbar k$ can also be obtained by absorption and stimulated emission of photons within the same internal state as depicted in Fig. 14.4b. In this case, the frequency difference of the exchanged photons is typically on the order of kHz and matches the energy difference of the momentum states. For long pulses i.e. with a Fourier width that is smaller than the momentum separation, this process can be viewed equivalently to Bragg diffraction of the matter wave on a moving optical lattice. Since the atoms remain in the same internal state upon diffraction, several systematic errors due to state dependent phase shifts can be avoided. A drawback though is that the output ports of the interferometer cannot be discriminated by state selective detection any more. Instead, they need to separate after a time of flight for independent spatially resolved detection, which requires that the ensemble expansion is reduced below the recoil velocity. This is either achieved with evaporatively cooled atoms and BEC or with strong velocity selection on a molasses.

As with regular Bragg diffraction, high-order diffraction transferring multiples of photon momenta is also possible. However, the required laser intensity scales strongly with the order of such multi-photon processes, which typically limits their use to lower orders and specific applications.

Bloch Oscillations

Bloch oscillations offer a convenient way to transfer multiple pairs of photon recoils in n subsequent two-photon processes. For this, atoms are adiabatically loaded into an optical lattice and then accelerated along with the lattice by adiabatically sweeping the detuning of the lasers forming the lattice (see Fig. 14.4c). Such Bloch oscillations have been used to transfer thousands of photon recoils, and have allowed to form large area interferometers. For beam splitting, they are usually combined

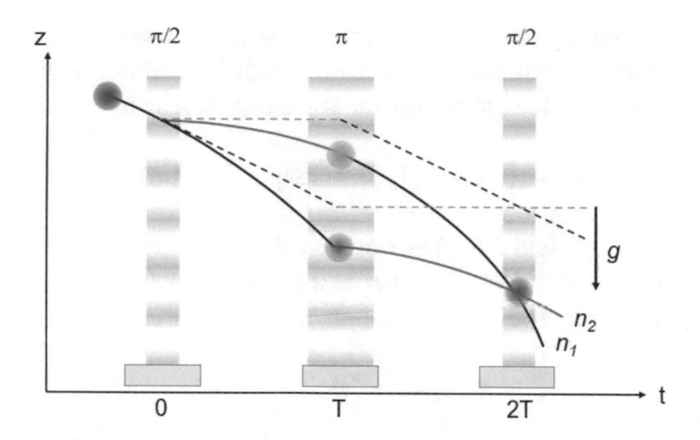

Fig. 14.5 Mach-Zehnder geometry of a cold atom gravimeter. Dashed trajectories are in absence of gravity, solid lines indicate trajectories of atoms in free fall bent by gravitational acceleration. In this depiction beamsplitters and mirrors are formed by laser pulses that are retro-reflected. With two frequency components ω_1 and ω_2 in each pulse a moving lattice with counter propagating wave vectors k_1 and k_2 forms

with an initial splitting process of either Bragg or Raman type, to allow the distinct loading and subsequent acceleration of either interferometer arm into the lattice.

14.3.1.3 Phase Evolution and Interferometry

To illustrate the phase evolution in an atom interferometer, we consider a sequence of $\pi/2 - \pi - \pi/2$ pulses spanning a Mach-Zehnder geometry as depicted in Fig. 14.5. This configuration is the most relevant one for the topic of precision measurements of gravity as discussed here, and is typically used in CAI gravimeters. Other pulse sequences and interferometer topologies are also possible providing different sensitivity or common mode suppression to certain phase contributions.

In any interferometer, the total accumulated phase difference can be decomposed into contributions of different origin as

$$\Phi_{\text{tot}} = \Phi_{\text{prop}} + \Phi_{\text{light}} + \Phi_{\text{sep}}. \tag{14.11}$$

The first contribution is that from the free evolution of the matter wave in either path. Using the path integral approach, the action along the classical paths Γ_i can be calculated from the Lagrangian of the system as in

$$S_{\Gamma_i} = \int_{\Gamma_i} \mathcal{L}_i(r_i(t), v_i(t))dt \tag{14.12}$$

which we use to obtain the phase difference between the upper (u) and lower (l) interferometer paths

$$\Phi_{\text{prop}} = (S_{\Gamma_u} - S_{\Gamma_l})/\hbar. \tag{14.13}$$

The second contribution is due to the interaction of the atoms with the laser light during beamsplitter and mirror pulses. Upon each interaction with the laser light field, a phase factor is added to the matter wave function as in

$$|i\rangle \rightarrow i\exp(k_{\text{eff}}z(t))|f\rangle \quad |f\rangle \rightarrow i\exp(-k_{\text{eff}}z(t))|i\rangle, \tag{14.14}$$

where we use the so called effective wave vector $k_{\text{eff}} = k_1 + k_2 \approx 2k$. All phases thus imparted on the atoms in the upper and the lower path respectively are added up and contribute to the total phase difference as

$$\Phi_{\text{light}} = \sum_{\Gamma_u} \phi_u - \sum_{\Gamma_l} \phi_l. \tag{14.15}$$

Finally, the last term is due to a possible displacement of the interfering portions of the matter wave at the final beamsplitter. Such a displacement could occur if for example a gravity gradient prevents symmetric closure of the interferometer:

$$\Phi_{\text{sep}} = \frac{p_0 \Delta z}{\hbar}. \tag{14.16}$$

Here p_0 is the average momentum of the atoms at the output and Δz the separation of upper and lower path at this position.

An evaluation of all the above contributions for a Mach-Zehnder interferometer as in Fig. 14.5 and for a homogeneous gravitational field results in

$$\Phi_{\text{tot}} = k_{\text{eff}} T^2 g. \tag{14.17}$$

A few remarks on this result are in order:

- The gravitational acceleration is linked to the measurement quantity (phase) by a remarkably simple scale factor: $S = k_{\text{eff}} T^2$. This scaling depends only on the laser frequencies ($k_{\text{eff}} = (\omega_1 + \omega_2)/c$), that can be fixed according to universal atomic properties and on the timing of the pulse sequence. Thus it is not subject to an instrument specific calibration and does not suffer from drifts and ageing as in many classical gravimeter instruments such as spring gravimeters. Also these quantities can be determined very precisely which has allowed CAI gravimeters to achieve the most accurate absolute measurements of local free fall acceleration to date.
- The dynamic range of the interferometer can be adjusted simply by changing the pulse separation time. The latter enters quadratically, thus even very large scale factors on the order of $S > 10^6$ can be realized with $T > 1s$. This enables highly sensitive measurements.
- At this leading order, the above equation comprises only phase contributions from interaction with the laser light. Indeed, the underlying measurement

principle can be thought of as using the laser wavefronts like a ruler to determine the free fall distance over time intervals T and $2T$.

? Exercise

14.4. Retrace the derivation of Eq. (14.17) by evaluating Eqs. (14.12)–(14.15) using the Lagrangian of a point mass in free fall in a homogeneous gravitational field. For the laser phases added to upper and lower path consider $\phi_i = k_{\text{eff}} z_i$ with i indicating the different instances of interaction during the sequence.

14.3.1.4 Interferometer Read-Out

The normalized relative population $P = (n_2 - n_1)/(n_2 + n_1)$ of the two output ports of the interferometer depends on the interferometer phase as

$$P(\Phi_{\text{tot}}) \sim C \times \cos(\Phi_{\text{tot}}). \tag{14.18}$$

Here $C \leq 1$ accounts for the finite contrast of the interferometer. A change in phase leads to a change in relative population that can be detected at the output of the interferometer. This is typically done either by illuminating the atoms with resonant laser light and collecting the resulting fluorescence or by imaging the shadow of the ensemble due to absorption on a camera (the latter typically only with ultra-cold, dense ensembles). From this the total number of atoms in either port can be determined. Ultimately, this detection is limited by shot noise respectively quantum projection noise, that is the signal to noise ratio scales as $1/\sqrt{N}$ where $N = n_2 + n_1$ is the total number of atoms. At the center of the fringe the uncertainty in Φ due to shot noise thus is $\sigma_\Phi = 1/(C\sqrt{N})$. Considering an atom shot noise limited gravimeter with leading order phase shift as given in Eq. (14.17), we can estimate the statistical uncertainty in a measurement of local gravity as

$$\sigma_g = \frac{1}{C\sqrt{N}} \frac{1}{n k_{\text{eff}} T^2}. \tag{14.19}$$

Here n indicates the number of two-photon transitions used for beam splitting. Obviously the sensitivity can be increased either by extending pulse separation time T i.e. extending the free fall time or by using so called Large-Momentum-Transfer (LMT) beamsplitters that transfer n multiples of two-photon momenta, such as high order Bragg diffraction or Bloch oscillations. A third possibility is to overcome the standard quantum limit by using momentum entangled atoms in the interferometer. Here the fundamental detection noise scales as $1/N$ rather than $1/\sqrt{N}$ which could ultimately further boost precision tests of fundamental physics with these instruments.

14.3.2 CAI Precision Tests of Gravity

With the basic idea of a CAI gravimeter established above, it is now easy to see how these instruments are employed in tests of gravity. In particular, the quantum nature of the applied test masses opens new avenues to test gravity theory in ways that are not available with classical tests. We will briefly address some of the proposed ideas and point the reader to the related literature for further reading.

14.3.2.1 Measurements of Local Free Fall Acceleration g and Gravity Gradients

Extremely accurate gravimeters have been realized in atomic fountain setups following the scheme shown in Fig. 14.5. There, interferometer times of $2T$ close to 1 second have been realized resulting in a large scale factor and thus sensitive measurements of local gravitational acceleration at 2×10^{-8} m/s$^2/\sqrt{Hz}$ and an overall absolute uncertainty of few parts in 10^9 [40]. Even mobile devices of this kind are being developed today, which is of high interest in geodesy, particularly in areas where classical gravimeters cannot be employed such as in airborne or shipborne measurement campaigns [41,42]. One current limitation in these devices is due to the thermal spread of the atomic cloud in combination with wavefront distortions of the beamsplitters and efforts using compact BEC gravimeters are currently pursued to overcome this. Besides gravimetry also gravity gradiometry with cold atoms is done, where simply two such gravimeters are operated in close proximity on two atomic ensembles, typically sharing the same beamsplitter lasers possibly even within the same vacuum system. These measurements benefit from common mode suppression of various error sources such as laser phase noise or vibrations and have been able to measure gravity gradients as low as $4 \times 10^{-10}\, g/\mathrm{m}$. An elegant way to extract the differential phase and thus δg in these measurements is to use ellipse fitting, where the fringes are plotted on x and y axis. Since both interferometers use the same scale factor common phase fluctuations map out an ellipse while the differential phase determines the eccentricity of the ellipse and can be determined from a respective fit.

In terms of testing fundamental physics, precision measurements of local gravitational acceleration g have been used to test Lorentz invariance in the gravitational sector of an extension of the standard model of particle physics (SME) that incorporates a violation of Lorentz invariance. This was done by using precision gravimetry data from an atomic fountain recorded over several days and corrected for tidal effects from sun and moon [43]. As the laboratory rotates along with Earth's rotation a corresponding anisotropy in the residual gravitational acceleration might indicate a violation of Lorentz invariance. No such anisotropy could be found and together with similar analysis of lunar laser ranging data [44], these measurements provide stringent bounds on elements of the tensor $\bar{s}$ of Eq. (14.8) down to the level of 10^{-9}.

14.3.2.2 Measurements of Newton's Constant G

Several experiments have used cold atom interferometry to perform a precision measurement of Newton's constant G by measuring the phase shift due to the gravitational force of a nearby source mass on the atomic ensemble. Typically these apply a gravity gradiometer as described above and thus rely on common mode cancellation of several important systematic errors and noise sources. To allow for a measurement of G, they employ a pair of heavy source masses next to the setup and symmetrically alternate their distance with respect to the atoms from measurement run to measurement run. From the observed modulation of the local gravity gradient, the value of G can then be derived.

These experiments have reached a precision at the level of 150 ppm [45–47] and provide an important alternative experimental approach to measuring G as compared to Cavendish type torsion pendulum experiments. Such systematically different experiments are especially important, because different precision measurements of G have resulted in values severely differing by hundreds of ppm, while at the same time each claiming an accuracy at the level of few tens of ppm, and so far there seems to be no convergence towards an agreed upon value at this level of accuracy.

To obtain a measurement of G from CAI, similar to the most sensitive tests with macroscopic test masses, requires a further improvement by about a factor of five in both statistical and systematic uncertainties. The first should be attainable by significantly longer measurement campaigns of weeks rather than days and possibly increased number of atoms. The second will need, among several other improvements, atomic clouds with reduced size and momentum spread as well as source masses of better homogeneity.

14.3.2.3 Gravity at Short Distance

Gravity gradient measurements as just presented also allow to perform sensitive tests of the Newtonian force law in the form of the Yukawa potential of Eq. (14.1). In [48] a horizontal CAI gradiometer has been shown to enable a constraint on α at the 8×10^{-3} level on the length scale of $\lambda = 0.1$ m. While this still falls into the region already excluded by other experiments, an improvement down to the level of $\alpha < 10^{-5}$ appears feasible, thus extending beyond the excluded parameter space shown in the left panel of Fig. 14.2. To achieve that, the atom to source mass distance would need to be reduced from 10 mm to the level of 1 mm at closest proximity. Also the source mass would need to be increased about tenfold to 5000 kg.

14.3.2.4 Screened Scalar Fields

CAI experiments are particularly well suited for the search for screened scalar fields because (1) they are performed in a high vacuum with reduced background mass density and (2) they the provide very high sensitivity with almost point-like test masses. These are important benefits when it comes to detecting chameleon fields, because it suppresses the screening significantly in comparison to tests with macroscopic masses (e.g. with optomechanical systems [11]).

In a CAI search for chameleon fields one would aim to detect an anomalous gravitational acceleration of atoms next to a source mass M_S of the form

$$a = \frac{GM_S}{r^2}\left[1 + 2\zeta_S\zeta_a\left(\frac{M_{Pl}}{M}\right)^2\right], \tag{14.20}$$

as follows from Eq. (14.5) for $R_P/\lambda_C, r/\lambda_C \to 0$ (compare with [49]).

A specific experiment to improve bounds on the parameters of the chameleon field M and Λ using atom interferometry has been presented in [50]. There a CAI accelerometer has been implemented with the atoms in direct vicinity to a gravitational, cylindric source mass. Alternating the position of the source mass from close to distant in subsequent measurement runs, allowed to take differential measurements and isolate the gravitational acceleration of the test mass from that of Earth and other nearby objects. A potential anomaly in gravitational acceleration could be restricted to $(9 \pm 24)\,\mathrm{nm/s^2}$ by this experiment. Considering Eq. (14.5) and the parameter space shown in the right panel of Fig. 14.2, this excludes mass scales up to $M < 10^{-5} M_{Pl}$ at $\Lambda = 2.4\,\mathrm{meV}$ which is the value consistent with the observed cosmological expansion.

14.3.2.5 Tests of the WEP

A selection of WEP tests based on cold atom interferometry and their respective results is presented in Table 14.1. The simplest way to test the WEP with CAI, is to operate a CAI gravimeter with two different atomic species A and B and compare the measured free fall acceleration g with each species to obtain the Eötvös parameter (14.6). This has been done for example by Fray et al. [51] who operated an instrument with interleaving single runs with different isotopes of Rb. More recent such tests prepare a mixture of both atomic species with superimposed centers of mass and then run the two interferometer sequences simultaneously. This way, systematic errors from gravity gradients or temporal variations can be largely suppressed. For a comparison of two different isotopes of the same atomic species, it is even possible to use the same beam splitting lasers with both ensembles. This allows for a good match in scale factors and a strong suppression of laser phase noise in the differential measurement. Comparisons of atomic species of different chemical elements on the other hand typically require different laser systems and are thus more complex. Yet of course, they offer a much larger and possibly more interesting variety of test mass pairs.

The most accurate test based on CAI so far has been reported in [52]. From a comparison of ^{87}Rb and ^{85}Rb they obtained a limit of $\eta = (1.6 \pm 1.8 \pm 3.4) \times 10^{-12}$. This experiment applied a fountain setup with a large baseline of 10 m, as well as evaporatively cooled atoms and sequential Bragg beamsplitters allowing for a momentum separation of $12\,\hbar k$. The main systematic uncertainties were due to differential AC Stark shifts on the two isotopes as well as from the initial kinematics of the ensembles (center of mass displacements and velocity differences). Some mitigation to the latter was achieved by adjusting the laser frequencies during the sequence [53–55]. To further reduce also the effects of AC Stark shifts requires larger laser detuning and thus upgrading the laser system power in the future.

Table 14.1 WEP tests based on cold atoms. (a) Comparison of a CAI vs a FG5 gravimeter i.e. a falling corner cube. (b) Comparison of bosonic vs fermionic isotopes, done by Bloch oscillations in an optical lattice. (c) Comparison of atoms in different magnetic substates $m_F = \pm 1$. (d) Comparison of atoms with different spin i.e.in different hyperfine states. (e) Comparison of atoms in superposition of hyperfine states. The data of this table has been collected from the references given in the last column of the table

Laboratory	Year	Test masses	Eötvös limit	Citation
Stanford, Palo Alto	2001	^{133}Cs / FG5 a)	$(7.0 \pm 7.0) \times 10^{-9}$	[40] ©2001 IOP Publishing Ltd
MPQ, Garching	2004	^{87}Rb/^{85}Rb	$(1.2 \pm 1.7) \times 10^{-7}$	[51] ©2004 American Physical Society
SYRTE, Paris	2010	^{87}Rb/ FG5 a)	$(4.3 \pm 6.4) \times 10^{-9}$	[86] ©2010 BIPM and IOP Publishing Ltd
ONERA, Palaiseau	2013	^{87}Rb/^{85}Rb	$(1.2 \pm 3.2) \times 10^{-7}$	[87] ©2013 American Physical Society
LENS, Florence	2014	^{88}Sr/^{87}Sr b)	$(0.2 \pm 1.6) \times 10^{-7}$	[81] ©2014 American Physical Society
Wuhan	2015	^{87}Rb/^{85}Rb	$(2.8 \pm 3.0) \times 10^{-8}$	[88] ©2015 American Physical Society
Wuhan	2016	^{87}Rb c)	$(0.2 \pm 1.2) \times 10^{-7}$	[80] ©2016 American Physical Society
LENS, Florence	2017	^{87}Rb d) e)	$(3.3 \pm 2.9) \times 10^{-9}$	[84] ©2017 TheAuthors
IQO, Hannover	2020	^{87}Rb/^{39}K	$(-1.9 \pm 3.7) \times 10^{-7}$	[57] ©2020 TheAuthors
Stanford, Palo Alto	2020	^{87}Rb/^{85}Rb	$(1.6 \pm 1.8 \pm 3.4) \times 10^{-12}$	[52] ©2020 American Physical Society
Wuhan	2021	^{87}Rb/^{85}Rb d)	$< 4.1 \times 10^{-10}$	[89] ©2021 American Physical Society

With respect to a comparison of atomic species of different chemical elements, the most accurate test has been reported by Schlippert et al. [56] and Albers et al. [57] with an Eötvös parameter $\eta = (-1.9 \pm 3.2) \times 10^{-7}$. This experiment did a comparison of ^{87}Rb and ^{39}K over a rather short baseline ($T = 41$ ms) and applied laser beamsplitters at different wavelengths of 780 nm and 767 nm. The accuracy was limited by phase uncertainties arising from wavefront aberrations that are sampled by the thermally expanding molasses as well as magnetic field gradients. Further improvements might thus be achieved by implementing evaporative cooling and thus a reduction of the ensemble spreading as well as improvements on the magnetic shielding.

Now, obviously classical tests have already achieved significantly better precision down to $\eta < 2 \times 10^{-14}$. What then should be the interest in such CAI tests of the WEP? Two aspects can be emphasized here:

First, CAI is still a maturing technology and its potential is by far not yet fully explored. We can still expect significant progress overcoming current technical limitations, which will ultimately allow to surpass classical measurements. For example, several long baseline facilities are being or have been set up [60], [61] to employ and investigate the use of significantly enhanced scale factors S. Besides the continuing work of [52], this comprises plans to have a comparison of the free fall of ^{170}Yb and ^{87}Rb with 10^{-13} sensitivity in a 10 m baseline setup in Hannover, Germany [61]. Also LMT beamsplitters are investigated to further boost the sensitivity. These coherently transfer $2n$ photon momenta to increase the scaling, with n on the order of hundreds and ultimately aiming beyond a thousand. Finally, major efforts are currently underway to investigate interferometry with entangled atoms. This will offer a way to overcome the limit set by quantum projection noise as pointed out above. With these developments ongoing, CAI tests of the WEP should achieve significant improvements in the future.

In order to push their sensitivity to the ultimate limits, quantum tests of the WEP will need to eventually overcome the limited free fall time of an Earth-bound setup, similar as done with the classical WEP test of the MICROSCOPE mission. Such a space-based quantum test has been proposed and has been studied by a European consortium in the framework of the ESA Cosmic Vision program in 2014 [62, 63]. There, a comparison of ^{87}Rb and ^{85}Rb at the 10^{-15} level was targeted, but the mission proposal was not selected mostly for lack of space readiness of the required CAI technology. Technology and space qualification has since been further advanced for example with the first BECs prepared in space in 2017 [64] and in orbit in 2018 [65]. At the same time, the hard requirements on source preparation demanding control of center of mass overlap and center of mass velocity at the nm and nm/s level could also be relaxed by orders of magnitude as discussed in [66,67]. This now allows to target space-based tests even at the 10^{-17} level and with different atomic species such as ^{87}Rb and ^{41}K.

The second point to be emphasized is that CAI tests allow to target aspects, which may either be not as easily accessible in classical tests or which are inherently quantum in nature. They also contribute a novel set of test mass combinations to be used complementary to classical tests (Table 14.2). Test mass pairs e.g. in [68] or [69] have been selected in order to maximize a potential WEP violation signal due to

Table 14.2 Test mass pairs and their respective charge differences, with charges coupling to a violation of WEP, as described in [58] ©2012 IOP Publishing Ltd. All rights reserved and [59]. Data reused with permissions

Test mass pair	$\Delta Q_1 \times 10^4$	$\Delta Q_2 \times 10^4$	$\Delta f_{e+p-n} \times 10^2$	$\Delta f_{e+p+n} \times 10^4$
Ti/Pt	19.92	26.65		
Ti/Be	−15.46	−70.20	1.48	−4.16
^{87}Rb/^{85}Rb	0.84	−0.79	−1.01	1.81
^{87}Rb/^{39}K	−6.69	−23.69	−6.31	1.90
^{87}Rb/^{170}Yb	−12.87	−13.92	−1.36	−8.64
^{87}Sr/^{88}Sr	0.42	−0.39	−0.49	2.04
^{87}Sr/^{114}Cd	−2.62	−6.95	−2.3	−2.11

hypothetical charges that couple to gravity and are linked to the respective baryon, proton or neutron numbers, such as a dilaton charge [58,70–72]. Similarly, CAI tests can be done on combinations of atoms with favourable proton/neutron ratios. For example a comparison of Li isotopes has been identified to be particularly promising in this respect [73]. Obviously, a classical test comparing different isotopes of this kind is not as readily available as it is in a CAI measurement.

Classical tests of the universality of free fall (UFF) on the other hand have already applied spin polarized or rotating test masses e.g. in [74]. This is because one possible UFF violation scenario is based on spin-gravity or spin-torsion coupling [75–79]. With atoms in well defined spin states, CAI experiments are particularly suited for these kind of tests. Thus, an experiment has been done that tested the free fall of Rb atoms with different spin orientations [80]. Also the free fall of bosonic ^{87}Sr and fermionic ^{88}Sr has been compared using Bloch oscillations in a vertical optical lattice [81] .

Further, even a quantum version of the equivalence principle could be tested. A formulation of such a principle has been proposed by Zych and Brukner [82] and Orlando et al. [83], requiring the equivalence of inertial and gravitational rest mass energy. In this model, atoms in different internal energy eigenstates represent interesting test mass pairs. Consequently, in [51, 84] free fall comparisons of Rb atoms in different hyperfine ground states have been performed. In particular, also the equivalence of off-diagonal elements of the quantum mechanical mass-energy operator can be probed if the atoms are prepared in a superposition of internal energy eigenstates, which is not feasible in classical tests. Such a measurement was done in a gravity gradiometer configuration in [84], comparing Rb atoms in a superposition of the hyperfine ground states against Rb atoms prepared in a single hyperfine state. From this, off-diagonal elements of the mass-energy operator were limited to $< 5 \times 10^{-8}$.

Finally, an experiment that should significantly extend the concept of a quantum test of the WEP has been devised, aiming to compare the free fall of entangled atom pairs. In [85] they outline a viable experimental procedure for such a measurement and estimate a feasible sensitivity of $\eta < 10^{-7}$.

? Exercise

14.5. Consider Eq. (14.19) and how this translates into sensitivity with respect to the Eötvös parameter (14.6). Use some typical numbers for a Rb atomic fountain ($\lambda = 780$ nm) to estimate the attainable sensitivity in a WEP test, e.g. $T = 500$ ms, $N = 10^6$, $n = 2$, and $C = 0.5$ with Rb isotopes. Which combination of particle number, pulse separation time, photon momenta would be needed to achieve the level of $\eta < 10^{-13}$ in a single measurement run of a future high precision test.

14.3.2.6 Gravitational Redshift and CAI

Many tests of Local Position Invariance probe the universality of the gravitational redshift in highly accurate atomic clocks of different types as they move through the gravitational potential of Earth and sun. While these clocks operate on principles of quantum physics, these tests are conceptually still classical, as simply worldlines of point-like, independent clocks are considered and assigned with the respective general relativistic proper time. In the microscopic quantum mechanical description of these clocks however, time is only a global parameter and the concept of proper time does not apply.

In recent years, the quantum nature of such tests has thus been scrutinized and in particular CAI experiments have been proposed that strive to perform measurements of the gravitational redshift within the quantum realm. Initially the discussion was fueled by claims of a redshift test based on a reanalysis of a CAI gravimeter experiment [90]. This has been generally dismissed as not viable for any interferometer where interfering atoms are prepared in energy eigenstates and travel in a homogeneous gravitational field [91,92]. Further experimental proposals then laid out how general relativistic proper time in a CAI might manifest itself by a decrease of the visibility of interferometer fringes. This was based on the notion that proper time labeling the interferometer arms at different gravitational potential might provide which-way information [93]. Further, clock interferometry schemes have been presented that might allow to directly measure the gravitational redshift within a CAI. For that, atoms in either interferometer arm at different gravitational potential would generally need to be prepared in a superposition state i.e. representing a quantum clock [92]. However, all of these proposed experiments come with tremendous technical challenges. To provide either a measurable contrast loss or even a phase shift from time dilation, they require a combination of seconds of free evolution as well as superposition of states that are separated by an optical frequency, rather than the usual microwave splitting of hyperfine states. Very long baseline interferometry such as in [61] and the use of single-photon transitions in Earth-Alkali atoms such as Sr or Yb might be able to meet these requirements though. Indeed, a recent proposal has shown a realistic path towards an implementation [94]. In particular this proposal also addresses the technical challenges of how to create a superposition of atoms in either path with just one laser and presents a viable solution.

14.3.3 Testing Quantum Features of Gravity and Search for Signatures of Quantum Gravity

Residual effects from a yet unknown quantum gravity might be revealed by low-energy, laboratory experiments of highest precision. CAI experiments in particular could be well suited to perform such quantum gravity phenomenology. A possible such signature could be a violation of the Weak Equivalence Principle as discussed above. Other such deviations from standard physics that CAIs could reveal are fundamental decoherence, modifications to the spreading of wave packets or modified energy-momentum relations. The latter can be probed sensitively in CAI precision measurements of the photon recoil, which typically apply a Ramsey-Bordé geometry (four $\pi/2$ pulses) and have achieved the ppb level of accuracy [95]. In [96] it has been shown how such measurements provide bounds on parameters of modified dispersion relations that can be phenomenologically derived from quantum gravity models such as loop quantum gravity. Other such phenomenological models have argued for fundamental decoherence arising from space-time fluctuations at Planck scales [97] as well as an anomalous spreading of wave packets [98].

? Exercise

14.6. In principle, the methods of light-pulse CAI may also be employed for an experimental test of the coherent mediation of gravity as described in Sect. 14.2.3. An optical dipole transition of a nano-particle, for example, provided by an embedded two-level system, may be addressed for momentum transfer via Bragg diffraction to coherently split the center-of-mass wave function of the nano-particle. Let us assume that the nano-particle has a mass of 10^{-14} kg and a momentum of $2000\hbar k_{\mathrm{eff}}$ is transferred to the particle with $k_{\mathrm{eff}} \approx 2\pi/\lambda$ and $\lambda \sim 800$ nm. Calculate the spread of the particle's wave function after 1 s of free-fall time by calculating the velocity difference from the transferred momentum. Can this experiment be used to test gravity as a quantum coherent mediator in practice?

14.4 Summary

We hope that it has become evident that the general notion of quantum tests of gravity comprises a huge variety of different experimental setups and regimes of gravity from quantum enhanced sensing of gravitational forces and tests of principles that are at the foundations of general relativity to tests of quantum properties of gravity. In the case of force sensing and tests of fundamental principles, advantages of employing quantum systems are, on the one hand, the high level of control over these systems (e.g. ultra-cold atoms, optomechanical sensors at their motional ground state), on the other hand, the possibility to employ non-classical states to overcome the shot noise limit of measurement precision. In the case of tests of quantum properties of gravity, the quantum properties of the employed systems are absolutely essential as we may only be able to study the non-classical response of the gravitational field through the creation of non-classical states of gravitationally interacting quantum systems. The community of researchers working in the field of quantum tests of gravity has grown extensively in recent years and we expect this trend to continue. With the concentrated effort of these many researchers and the increasing experimental control and variety of experimental setups, we expect to see many advances as the century of quantum technologies is unfolding.

References

1. J.H. Gundlach, New J. Phys. **7**(1), 205 (2005)
2. W.H. Tan, A.B. Du, W.C. Dong, S.Q. Yang, C.G. Shao, S.G. Guan, Q.L. Wang, B.F. Zhan, P.S. Luo, L.C. Tu, et al., Phys. Rev. Lett. **124**(5), 051301 (2020)
3. J.G. Lee, E.G. Adelberger, T.S. Cook, S.M. Fleischer, B.R. Heckel, Phys. Rev. Lett. **124**, 101101 (2020). https://doi.org/10.1103/PhysRevLett.124.101101.
4. T. Westphal, H. Hepach, J. Pfaff, M. Aspelmeyer, Nature **591**(7849), 225 (2021)
5. H. Cavendish, Philos. Trans. R. Soc. Lond. (88), 469 (1798)
6. S. Barzanjeh, A. Xuereb, S. Gröblacher, M. Paternostro, C.A. Regal, E.M. Weig, Nature Phys. 1–10 (2021)
7. M. Aspelmeyer, T.J. Kippenberg, F. Marquardt, Rev. Mod. Phys. **86**(4), 1391 (2014)
8. C. Burrage, J. Sakstein, Living Rev. Relat. **21**(1), 1 (2018)
9. J. Murata, S. Tanaka, Class. Quantum Gravity **32**(3), 033001 (2015). https://doi.org/10.1088/0264-9381/32/3/033001
10. W.H. Tan, A.B. Du, W.C. Dong, S.Q. Yang, C.G. Shao, S.G. Guan, Q.L. Wang, B.F. Zhan, P.S. Luo, L.C. Tu, J. Luo, Phys. Rev. Lett. **124**, 051301 (2020). https://doi.org/10.1103/PhysRevLett.124.051301
11. S. Qvarfort, D. Rätzel, S. Stopyra, New J. Phys. **24**(3), 033009 (2022). https://doi.org/10.1088/1367-2630/ac3e1b
12. C.M. Will, arXiv:gr-qc/0510072 (2005)
13. A.A. Michelson, E.W. Morley, Sidereal Messenger **6**, 306 (1887)
14. D. Mattingly, Living Rev. Relat. **8**(1), 1 (2005)
15. H.P. Robertson, Rev. Mod. Phys. **21**(3), 378 (1949)
16. R. Mansouri, R.U. Sexl, General Relat. Gravit. **8**(7), 497 (1977)
17. V.A. Kostelecký, C.D. Lane, Phys. Rev. D **60**(11), 116010 (1999). https://doi.org/10.1103/PhysRevD.60.116010
18. D. Colladay, V.A. Kostelecký, Phys. Rev. D **58**(11), 116002 (1998). https://doi.org/10.1103/PhysRevD.58.116002
19. Q.G. Bailey, V.A. Kostelecký, Phys. Rev. D **74**(4), 045001 (2006). https://doi.org/10.1103/PhysRevD.74.045001
20. K. Nordtvedt, Phys. Rev. D **14**, 1511 (1976). https://doi.org/10.1103/PhysRevD.14.1511
21. J.P. Uzan, Living Rev. Relat. **14**(1), 1 (2011)
22. D. Giulini, in *Quantum Field Theory and Gravity* (Springer, Berlin, 2012), pp. 345–370
23. R.V. Pound, G.A. Rebka, Jr., Phys. Rev. Lett. **3**(9), 439 (1959)
24. P. Delva, N. Puchades, E. Schönemann, F. Dilssner, C. Courde, S. Bertone, F. Gonzalez, A. Hees, C. Le Poncin-Lafitte, F. Meynadier, et al., Phys. Rev. Lett. **121**(23), 231101 (2018)
25. M. Takamoto, I. Ushijima, N. Ohmae, T. Yahagi, K. Kokado, H. Shinkai, H. Katori, Nature Photon. **14**(7), 411 (2020)
26. L.J. Garay, Int. J. Mod. Phys. A **10**(02), 145 (1995)
27. A.F. Ali, S. Das, E.C. Vagenas, Phys. Rev. D **84**, 044013 (2011). https://doi.org/10.1103/PhysRevD.84.044013
28. A.F. Ali, Class. Quantum Gravity **28**(6), 065013 (2011)

29. I. Pikovski, M.R. Vanner, M. Aspelmeyer, M.S. Kim, Ä. Brukner, Nature Phys. **8**(5), 393 (2012). https://doi.org/10.1038/nphys2262
30. M. Bawaj, C. Biancofiore, M. Bonaldi, F. Bonfigli, A. Borrielli, G. Di Giuseppe, L. Marconi, F. Marino, R. Natali, A. Pontin, et al., Nature Commun. **6**(1), 1 (2015)
31. S. Ghosh, Class. Quantum Gravity **31**(2), 025025 (2013)
32. S. Bose, A. Mazumdar, G.W. Morley, H. Ulbricht, M. ToroÅ, M. Paternostro, A.A. Geraci, P.F. Barker, M. Kim, G. Milburn, Phys. Rev. Lett. **119**(24), 240401 (2017). https://doi.org/10.1103/PhysRevLett.119.240401
33. C. Marletto, V. Vedral, Phys. Rev. Lett. **119**(24), 240402 (2017). https://doi.org/10.1103/PhysRevLett.119.240402
34. C. Wan, Quantum superposition on nano-mechanical oscillator. Ph.D. Thesis, Imperial College, London, 2017
35. H. Miao, D. Martynov, H. Yang, A. Datta, Phys. Rev. A **101**(6), 063804 (2020)
36. A. Matsumura, K. Yamamoto, Phys. Rev. D **102**(10), 106021 (2020)
37. A. Belenchia, R.M. Wald, F. Giacomini, E. Castro-Ruiz, Ä. Brukner, M. Aspelmeyer, Phys. Rev. D **98**(12), 126009 (2018). https://doi.org/10.1103/PhysRevD.98.126009
38. D. Carney, Phys. Rev. D **105**(2), 024029 (2022). https://doi.org/10.1103/PhysRevD.105.024029
39. D. Kafri, J.M. Taylor, G.J. Milburn, New J. Phys. **16**(6), 065020 (2014). https://doi.org/10.1088/1367-2630/16/6/065020
40. A. Peters, K.Y. Chung, S. Chu, Metrologia **38**(1), 25 (2001). https://doi.org/10.1088/0026-1394/38/1/4
41. C. Freier, M. Hauth, V. Schkolnik, B. Leykauf, M. Schilling, H. Wziontek, H.G. Scherneck, J. Müller, A. Peters, J. Phys. Conf. Series **723**, 012050 (2016). https://doi.org/10.1088/1742-6596/723/1/012050
42. Y. Bidel, N. Zahzam, C. Blanchard, A. Bonnin, M. Cadoret, A. Bresson, D. Rouxel, M.F. Lequentrec-Lalancette, Nature Commun. **9**(1), 627 (2018). https://doi.org/10.1038/s41467-018-03040-2
43. H. Müller, S.w. Chiow, S. Herrmann, S. Chu, K.Y. Chung, Phys. Rev. Lett. **100**(3), 031101 (2008). https://doi.org/10.1103/PhysRevLett.100.031101
44. J.B.R. Battat, J.F. Chandler, C.W. Stubbs, Phys. Rev. Lett. **99**, 241103 (2007). https://doi.org/10.1103/PhysRevLett.99.241103
45. G. Rosi, F. Sorrentino, L. Cacciapuoti, M. Prevedelli, G.M. Tino, Nature **510**(7506), 518 (2014). https://doi.org/10.1038/nature13433
46. M. Prevedelli, L. Cacciapuoti, G. Rosi, F. Sorrentino, G.M. Tino, Philos. Trans. R. Soc. A Math. Phys. Eng. Sci. **372**(2026), 20140030 (2014). https://doi.org/10.1098/rsta.2014.0030
47. J.B. Fixler, G.T. Foster, J.M. McGuirk, M.A. Kasevich, Science **315**(5808), 74 (2007). https://doi.org/10.1126/science.1135459
48. G.W. Biedermann, X. Wu, L. Deslauriers, S. Roy, C. Mahadeswaraswamy, M.A. Kasevich, Phys. Rev. A **91**, 033629 (2015). https://doi.org/10.1103/PhysRevA.91.033629
49. P. Hamilton, M. Jaffe, P. Haslinger, Q. Simmons, H. Müller, J. Khoury, Science **349**(6250), 849 (2015). https://doi.org/10.1126/science.aaa8883
50. M. Jaffe, P. Haslinger, V. Xu, P. Hamilton, A. Upadhye, B. Elder, J. Khoury, H. Müller, Nature Phys. **13**(10), 938 (2017). https://doi.org/10.1038/nphys4189
51. S. Fray, C.A. Diez, T.W. Hänsch, M. Weitz, Phys. Rev. Lett. **93**(24), 240404 (2004). https://doi.org/10.1103/PhysRevLett.93.240404
52. P. Asenbaum, C. Overstreet, M. Kim, J. Curti, M.A. Kasevich, Phys. Rev. Lett. **125**(19), 191101 (2020). https://doi.org/10.1103/PhysRevLett.125.191101
53. A. Roura, W. Zeller, W.P. Schleich, New J. Phys. **16**(12), 123012 (2014). https://doi.org/10.1088/1367-2630/16/12/123012
54. A. Roura, Phys. Rev. Lett. **118**, 160401 (2017). https://doi.org/10.1103/PhysRevLett.118.160401
55. C. Overstreet, P. Asenbaum, T. Kovachy, R. Notermans, J.M. Hogan, M.A. Kasevich, Phys. Rev. Lett. **120**, 183604 (2018). https://doi.org/10.1103/PhysRevLett.120.183604

56. D. Schlippert, J. Hartwig, H. Albers, L.L. Richardson, C. Schubert, A. Roura, W.P. Schleich, W. Ertmer, E.M. Rasel, Phys. Rev. Lett. **112**(20), 203002 (2014). https://doi.org/10.1103/PhysRevLett.112.203002
57. H. Albers, A. Herbst, L.L. Richardson, H. Heine, D. Nath, J. Hartwig, C. Schubert, C. Vogt, M. Woltmann, C. Lämmerzahl, S. Herrmann, W. Ertmer, E.M. Rasel, D. Schlippert, Eur. Phys. J. D **74**(7), 145 (2020). https://doi.org/10.1140/epjd/e2020-10132-6
58. T. Damour, Class. Quantum Gravity **29**(18), 184001 (2012). https://doi.org/10.1088/0264-9381/29/18/184001
59. M.A. Hohensee, H. Müller, R.B. Wiringa, Phys. Rev. Lett. **111**(15), 151102 (2013). https://doi.org/10.1103/PhysRevLett.111.151102
60. L. Zhou, Z.Y. Xiong, W. Yang, B. Tang, W.C. Peng, K. Hao, R.B. Li, M. Liu, J. Wang, M.S. Zhan, General Relat. Gravit. **43**(7), 1931 (2011). https://doi.org/10.1007/s10714-011-1167-9
61. J. Hartwig, S. Abend, C. Schubert, D. Schlippert, H. Ahlers, K. Posso-Trujillo, N. Gaaloul, W. Ertmer, E.M. Rasel, New J. Phys. **17**(3), 035011 (2015). https://doi.org/10.1088/1367-2630/17/3/035011
62. D.N. Aguilera, H. Ahlers, B. Battelier, A. Bawamia, A. Bertoldi, R. Bondarescu, K. Bongs, P. Bouyer, C. Braxmaier, L. Cacciapuoti, C. Chaloner, M. Chwalla, W. Ertmer, M. Franz, N. Gaaloul, M. Gehler, D. Gerardi, L. Gesa, N. Gürlebeck, J. Hartwig, M. Hauth, O. Hellmig, W. Herr, S. Herrmann, A. Heske, A. Hinton, P. Ireland, P. Jetzer, U. Johann, M. Krutzik, A. Kubelka, C. Lämmerzahl, A. Landragin, I. Lloro, D. Massonnet, I. Mateos, A. Milke, M. Nofrarias, M. Oswald, A. Peters, K. Posso-Trujillo, E. Rasel, E. Rocco, A. Roura, J. Rudolph, W. Schleich, C. Schubert, T. Schuldt, S. Seidel, K. Sengstock, C.F. Sopuerta, F. Sorrentino, D. Summers, G.M. Tino, C. Trenkel, N. Uzunoglu, W.v. Klitzing, R. Walser, T. Wendrich, A. Wenzlawski, P.W. sels, A. Wicht, E. Wille, M. Williams, P. Windpassinger, N. Zahzam, Class. Quantum Gravity **31**(11), 115010 (2014). https://doi.org/10.1088/0264-9381/31/11/115010
63. B. Altschul, Q.G. Bailey, L. Blanchet, K. Bongs, P. Bouyer, L. Cacciapuoti, S. Capozziello, N. Gaaloul, D. Giulini, J. Hartwig, L. Iess, P. Jetzer, A. Landragin, E. Rasel, S. Reynaud, S. Schiller, C. Schubert, F. Sorrentino, U. Sterr, J.D. Tasson, G.M. Tino, P. Tuckey, P. Wolf, Adv. Space Res. **55**(1), 501 (2015). https://doi.org/10.1016/j.asr.2014.07.014
64. D. Becker, M.D. Lachmann, S.T. Seidel, H. Ahlers, A.N. Dinkelaker, J. Grosse, O. Hellmig, H. Müntinga, V. Schkolnik, T. Wendrich, A. Wenzlawski, B. Weps, R. Corgier, T. Franz, N. Gaaloul, W. Herr, D. Lüdtke, M. Popp, S. Amri, H. Duncker, M. Erbe, A. Kohfeldt, A. Kubelka-Lange, C. Braxmaier, E. Charron, W. Ertmer, M. Krutzik, C. Lämmerzahl, A. Peters, W.P. Schleich, K. Sengstock, R. Walser, A. Wicht, P. Windpassinger, E.M. Rasel, Nature **562**(7727), 391 (2018). https://doi.org/10.1038/s41586-018-0605-1
65. D.C. Aveline, J.R. Williams, E.R. Elliott, C. Dutenhoffer, J.R. Kellogg, J.M. Kohel, N.E. Lay, K. Oudrhiri, R.F. Shotwell, N. Yu, R.J. Thompson, Nature **582**(7811), 193 (2020). https://doi.org/10.1038/s41586-020-2346-1
66. S. Loriani, C. Schubert, D. Schlippert, W. Ertmer, F. Pereira Dos Santos, E.M. Rasel, N. Gaaloul, P. Wolf, Phys. Rev. D **102**, 124043 (2020). https://doi.org/10.1103/PhysRevD.102.124043
67. B. Battelier, J. Berg, A. Bertoldi, L. Blanchet, K. Bongs, P. Bouyer, C. Braxmaier, D. Calonico, P. Fayet, N. Gaaloul, C. Guerlin, A. Hees, P. Jetzer, C. Lämmerzahl, S. Lecomte, C.L. Poncin-Lafitte, S. Loriani, G. Mtris, M. Nofrarias, E. Rasel, S. Reynaud, M. Rodrigues, M. Rothacher, A. Roura, C. Salomon, S. Schiller, W.P. Schleich, C. Schubert, C. Sopuerta, F. Sorrentino, T.J. Sumner, G.M. Tino, P. Tuckey, W. von Klitzing, L. Wrner, P. Wolf, M. Zelan. Exploring the foundations of the universe with space tests of the equivalence principle. Exp. Astron. **51**, 1695–1736 (2019)
68. P. Touboul, G. Métris, M. Rodrigues, Y. André, Q. Baghi, J. Bergé, D. Boulanger, S. Bremer, P. Carle, R. Chhun, B. Christophe, V. Cipolla, T. Damour, P. Danto, H. Dittus, P. Fayet, B. Foulon, C. Gageant, P.Y. Guidotti, D. Hagedorn, E. Hardy, P.A. Huynh, H. Inchauspe, P. Kayser, S. Lala, C. Lämmerzahl, V. Lebat, P. Leseur, F. Liorzou, M. List, F. Löffler, I. Panet, B. Pouilloux, P. Prieur, A. Rebray, S. Reynaud, B. Rievers, A. Robert, H. Selig, L. Serron, T. Sumner, N. Tanguy, P. Visser, Phys. Rev. Lett. **119**(23), 231101 (2017). https://doi.org/10.

1103/PhysRevLett.119.231101
69. S. Schlamminger, K.Y. Choi, T.A. Wagner, J.H. Gundlach, E.G. Adelberger, Phys. Rev. Lett. **100**(4), 041101 (2008). https://doi.org/10.1103/PhysRevLett.100.041101
70. T. Damour, A.M. Polyakov, Nucl. Phys. B **423**(2), 532 (1994). https://doi.org/10.1016/0550-3213(94)90143-0
71. T. Damour, Class. Quantum Gravity **13**(11A), A33 (1996). https://doi.org/10.1088/0264-9381/13/11A/005. ArXiv: gr-qc/9606080
72. T. Damour, F. Piazza, G. Veneziano, Phys. Rev. Lett. **89**(8), 081601 (2002). https://doi.org/10.1103/PhysRevLett.89.081601
73. M.A. Hohensee, H. Müller, J. Mod. Opt. **58**(21), 2021 (2011). https://doi.org/10.1080/09500340.2011.606376
74. B.R. Heckel, E.G. Adelberger, C.E. Cramer, T.S. Cook, S. Schlamminger, U. Schmidt, Phys. Rev. D **78**, 092006 (2008). https://doi.org/10.1103/PhysRevD.78.092006
75. A. Peres, Phys. Rev. D **18**, 2739 (1978). https://doi.org/10.1103/PhysRevD.18.2739
76. B. Mashhoon, Class. Quantum Gravity **17**(12), 2399 (2000). https://doi.org/10.1088/0264-9381/17/12/312
77. I.L. Shapiro, Phys. Rep. **357**(2), 113 (2002). https://doi.org/10.1016/S0370-1573(01)00030-8
78. A.J. Silenko, O.V. Teryaev, Phys. Rev. D **76**, 061101 (2007). https://doi.org/10.1103/PhysRevD.76.061101
79. Y.N. Obukhov, A.J. Silenko, O.V. Teryaev, Phys. Rev. D **90**, 124068 (2014). https://doi.org/10.1103/PhysRevD.90.124068
80. X.C. Duan, X. Deng, M.K. Zhou, K. Zhang, W.J. Xu, F. Xiong, Y.Y. Xu, C. Shao, J. Luo, Z.K. Hu, Phys. Rev. Lett. **117** (2016). https://doi.org/10.1103/PhysRevLett.117.023001
81. M.G. Tarallo, T. Mazzoni, N. Poli, D.V. Sutyrin, X. Zhang, G.M. Tino, Phys. Rev. Lett. **113**, 023005 (2014). https://doi.org/10.1103/PhysRevLett.113.023005
82. M. Zych, C. Brukner. Quantum formulation of the Einstein equivalence principle. Nature Phys. **14**(10), 1027–1031 (2015)
83. P.J. Orlando, R.B. Mann, K. Modi, F.A. Pollock, Class. Quantum Gravity **33**(19), 19LT01 (2016). https://doi.org/10.1088/0264-9381/33/19/19lt01
84. G. Rosi, G. D'Amico, L. Cacciapuoti, F. Sorrentino, M. Prevedelli, M. Zych, Č. Brukner, G.M. Tino, Nature Commun. **8**, 15529 (2017). https://doi.org/10.1038/ncomms15529
85. R. Geiger, M. Trupke, Phys. Rev. Lett. **120**, 043602 (2018). https://doi.org/10.1103/PhysRevLett.120.043602
86. S. Merlet, Q. Bodart, N. Malossi, A. Landragin, F.P.D. Santos, O. Gitlein, L. Timmen, Metrologia **47**(4), L9 (2010). https://doi.org/10.1088/0026-1394/47/4/l01
87. A. Bonnin, N. Zahzam, Y. Bidel, A. Bresson, Phys. Rev. A **88**(4), 043615 (2013). https://doi.org/10.1103/PhysRevA.88.043615
88. L. Zhou, S. Long, B. Tang, X. Chen, F. Gao, W. Peng, W. Duan, J. Zhong, Z. Xiong, J. Wang, Y. Zhang, M. Zhan, Phys. Rev. Lett. **115**(1), 013004 (2015). https://doi.org/10.1103/PhysRevLett.115.013004
89. L. Zhou, C. He, S.T. Yan, X. Chen, D.F. Gao, W.T. Duan, Y.H. Ji, R.D. Xu, B. Tang, C. Zhou, S. Barthwal, Q. Wang, Z. Hou, Z.Y. Xiong, Y.Z. Zhang, M. Liu, W.T. Ni, J. Wang, M.S. Zhan, Phys. Rev. A **104**, 022822 (2021). https://doi.org/10.1103/PhysRevA.104.022822
90. H. Müller, A. Peters, S. Chu, Nature **463**(7283), 926 (2010). https://doi.org/10.1038/nature08776
91. P. Wolf, L. Blanchet, C.J. Bordé, S. Reynaud, C. Salomon, C. Cohen-Tannoudji, Class. Quantum Gravity **28**(14), 145017 (2011). https://doi.org/10.1088/0264-9381/28/14/145017
92. A. Roura, Phys. Rev. X **10**, 021014 (2020). https://doi.org/10.1103/PhysRevX.10.021014.
93. M. Zych, F. Costa, I. Pikovski, Ä. Brukner, Nature Commun. **2**(1), 505 (2011). https://doi.org/10.1038/ncomms1498.
94. A. Roura, C. Schubert, D. Schlippert, E.M. Rasel, Phys. Rev. D **104**, 084001 (2021). https://doi.org/10.1103/PhysRevD.104.084001
95. R.H. Parker, C. Yu, W. Zhong, B. Estey, H. Müller, Science **360**(6385), 191 (2018). https://doi.org/10.1126/science.aap7706

96. G. Amelino-Camelia, C. Lämmerzahl, F. Mercati, G.M. Tino, Phys. Rev. Lett. **103**, 171302 (2009). https://doi.org/10.1103/PhysRevLett.103.171302
97. H.P. Breuer, E. Göklü, C. Lämmerzahl, Class. Quantum Gravity **26**(10), 105012 (2009). https://doi.org/10.1088/0264-9381/26/10/105012
98. E. Göklü, C. Lämmerzahl, A. Camacho, A. Macias, Class. Quantum Gravity **26**(22), 225010 (2009). https://doi.org/10.1088/0264-9381/26/22/225010

The Gravity of Light

15

Jan W. van Holten

Abstract

The gravitational field of an idealized plane-wave solution of the Maxwell equations can be described in closed form. After discussing this particular solution of the Einstein-Maxwell equations, the motion of neutral test particles, which are sensitive only to the gravitational background field, is analyzed. This is followed by a corresponding analysis of the dynamics of neutral fields in the particular Einstein-Maxwell background, considering scalars, Majorana spinors and abelian vector fields, respectively.

15.1 Light and Gravity

Light and gravity provide the main tools for studying the universe at large; gravity, as it determines the interactions and paths of celestial bodies, and light as it makes them visible to us and enables us to unravel their properties. The theoretical descriptions of light, as a form of electromagnetism, and gravity have much in common. The classical theories of electromagnetism and gravity are both local relativistic field theories; these fields carry physical degrees of freedom propagating energy, momentum and angular momentum at a finite speed c, commonly referred

J. W. van Holten (✉)
Nikhef, Amsterdam, Netherlands

Lorentz Insitute, Leiden University, Leiden, Netherlands
e-mail: v.holten@nikhef.nl

C. Pfeifer, C. Lämmerzahl (eds.), *Modified and Quantum Gravity*, Lecture Notes in Physics 1017, https://doi.org/10.1007/978-3-031-31520-6_15

to as the speed of light, even though gravity, and also the color charges of subatomic particles, propagate their interactions at the same universal speed as well.[1]

Of course, at the microscopic level electromagnetism is more than a classical field theory, as quantum effects become essential to its propagation and interaction with matter in the form of electrons and other charged particles. A similar change in the way gravity behaves in this domain is expected as well, although experimental confirmation of these ideas has as yet remained out of reach.

Even though the sources of gravity and electromagnetism are different, with gravity coupling to the local density of energy and momentum, and electromagnetism to the local density of electric charges and currents, the corresponding physical degrees of freedom (classical fields in the macroscopic world) do influence each other, but in an asymmetric way. The present chapter is dedicated to a discussion of some aspects of this mutual interaction. Unless specified otherwise (when numerical estimates are required) units are used in which the speed of light is unity: $c = 1$.

15.2 Einstein-Maxwell Theory

General Relativity (GR), the classical theory of gravity, states that space-time is endowed with a geometry, encoded in the metric $g_{\mu\nu}$, determined by the distribution of all combined energy- and momentum-densities. This geometry expresses itself in the motion of matter and light in the universe. For the interaction between gravity and electromagnetic fields this results in an Einstein equation specifying the Ricci curvature in terms of the electro-magnetic energy-momentum tensor:

$$R_{\mu\nu} - \frac{1}{2} g_{\mu\nu} R = -8\pi G T_{\mu\nu}[F], \tag{15.1}$$

with the local energy-momentum density of electromagnetic fields given by

$$T_{\mu\nu}[F] = F_{\mu\lambda} F_{\nu}^{\ \lambda} - \frac{1}{4} g_{\mu\nu} F_{\kappa\lambda} F^{\kappa\lambda}. \tag{15.2}$$

At the same time the classical dynamics of the electromagnetic field is specified by the generalized Maxwell equations in a space-time with given dynamical metric:

$$D_\mu F^{\mu\nu} = \partial_\mu F^{\mu\nu} + \Gamma_{\mu\lambda}^{\ \ \mu} F^{\lambda\nu} + \Gamma_{\mu\lambda}^{\ \ \nu} F^{\mu\lambda} = -j^\nu, \tag{15.3}$$

where j^ν is the electric charge-current density, and $\Gamma_{\mu\lambda}^{\ \ \nu}$ the Riemann-Christoffel connection. In the absence of charges and currents: $j^\nu = 0$, Eqs. (15.1)–(15.3) form a closed system describing gravity interacting with dynamical electromagnetic fields

[1] For a discussion of the role of the universal constant c characterizing the relations between inertial frames, see ref. [1].

in otherwise empty space; this set-up applies in particular to the coupling of gravity with electromagnetic radiation.

? Exercise

15.1. Show that (15.1)–(15.3) can be derived by variational calculus from the action

$$S[g, A] = \int_M d^4x \sqrt{-\det g} \left(-\frac{R}{8\pi G} + \alpha F_{\mu\nu} F^{\mu\nu} + \beta A_\mu j^\nu \right), \quad F_{\mu\nu} = \partial_\mu A_\nu - \partial_\nu A_\mu . \tag{15.4}$$

Determine α and β.

In places where the energy-momentum density of the electromagnetic field is small compared to the Ricci-curvature determined by external sources, such as the sun or compact bodies like neutron stars or black holes, one can to first approximation neglect the contribution of the electromagnetic fields to the curvature and describe the electromagnetic fields outside the external source regions by Maxwell equations in the gravitational background of the external sources. This approach is usually taken in studies of gravitational lensing, which offered one of the first tests of GR: observing the bending of light by the sun [2]; and in a more extreme case the recent observations of a black-hole shadow by the Event Horizon Telescope [3].

However, as Eq. (15.1) indicates, curvature can also be induced by electromagnetic fields themselves, even though this requires rather extreme electromagnetic energy densities. Indeed, according to this equation (temporarily reinstating the speed of light c) the curvature R measured in $1/m^2$ corresponding to an energy flux Φ in W/m^2 is numerically of the order

$$\frac{R}{1/\mathrm{m}^2} \sim \frac{8\pi G}{c^5} \Phi \simeq 2 \times 10^{-52} \frac{\Phi}{1\,\mathrm{W/m}^2}. \tag{15.5}$$

Measuring the curvature due to even intense electromagnetic radiation will therefore be an even more extreme challenge than the curvature due to the collision of very distant black holes and neutron stars [4]. In the following sections a more precise analysis is presented.

15.3 Plane Waves

An complete radiative solution of the Einstein-Maxwell equations is that of a plane electromagnetic wave of infinite width, which is accompanied by a parallel plane gravitational wave of *pp*-type [5–11]. With the wave propagating in the z-direction, it is convenient to use light-cone co-ordinates $u = t - z$ and $v = t + z$; the traveling plane-wave solution of the electromagnetic field is then expressed in terms of a transverse vector potential

$$A_i(u) = \int_{\infty}^{\infty} \frac{dk}{2\pi} \left(a_i(k) \sin ku + b_i(k) \cos ku \right), \tag{15.6}$$

where $i = (1, 2)$ labels the directions in the transverse x-y-plane; the corresponding electric and magnetic field strengths are given by

$$E_i(u) = -\varepsilon_{ij} B_j(u) = F_{ui}(u), \tag{15.7}$$

where $\varepsilon_{12} = -\varepsilon_{21} = +1$ whilst $\varepsilon_{11} = \varepsilon_{22} = 0$. Such solutions can take the form of wave packets of finite lengths carrying a finite energy flux per unit area. Specifically in the absence of external sources of curvature the energy density in the transverse plane is constant and the transverse geometry can be taken to be flat; the energy flux is then given in terms of the energy-momentum tensor by the only non-zero component

$$T_{uu}(u) = F_{ui} F_u{}^i = \frac{1}{2} \left(\mathbf{E}^2 + \mathbf{B}^2 \right)(u), \tag{15.8}$$

provided the metric is of the Brinkmann type

$$ds^2 = -du dv - \Phi(u, x^i) du^2 + dx^{i\,2}, \tag{15.9}$$

which is flat in the x-y-plane as required. In this Brinkmann geometry the only non-zero components of the Riemann curvature and Ricci tensor are

$$R_{uiuj} = -\frac{1}{2} \partial_i \partial_j \Phi, \qquad R_{uu} = -\frac{1}{2} \left(\partial_x^2 + \partial_y^2 \right) \Phi. \tag{15.10}$$

? Exercise

15.2. Show that the Einstein Eq. (15.1) then reduces to a single equation linking the *uu*-components of the Ricci and energy-momentum tensor:

$$\left(\partial_x^2 + \partial_y^2 \right) \Phi = 8\pi G \left(\mathbf{E}^2 + \mathbf{B}^2 \right). \tag{15.11}$$

The general solution of this inhomogeneous linear equation for the gravitational potential, the metric component $\Phi(u, x^i)$, is

$$\Phi = 2\pi G \left(x^2 + y^2\right) \left(\mathbf{E}^2 + \mathbf{B}^2\right) + \Phi_0, \tag{15.12}$$

where $\Phi_0(u, x^i)$ is an arbitrary solution of the homogeneous equation

$$\left(\partial_x^2 + \partial_y^2\right) \Phi_0 = 0. \tag{15.13}$$

These solutions of the homogeneous equation represent pure gravitational waves of pp-type [5]:

$$\Phi_0 = \kappa_+(u) \left(x^2 - y^2\right) + 2\kappa_\times(u) xy. \tag{15.14}$$

Equation (15.10) then implies that the co-efficients $\kappa_{+,\times}(u)$ represent the components of the corresponding Riemann tensor in the transverse plane:

$$R^{(0)}_{uiuj} = - \begin{pmatrix} \kappa_+ & \kappa_\times \\ \kappa_\times & \kappa_+ \end{pmatrix}. \tag{15.15}$$

Under a rotation in the transverse plane over an angle φ they transform as quadrupole components:

$$\begin{pmatrix} \kappa'_+ \\ \kappa'_\times \end{pmatrix} = \begin{pmatrix} \cos 2\varphi & -\sin 2\varphi \\ \sin 2\varphi & \cos 2\varphi \end{pmatrix} \begin{pmatrix} \kappa_+ \\ \kappa_\times \end{pmatrix}. \tag{15.16}$$

In contrast, the special solution (15.12) of the inhomogenous equation proportional to the energy density of the electromagnetic field: $\Phi - \Phi_0 \sim x^2 + y^2$, is of monopole type, being invariant under rotations in the transverse plane. Thus the plane electromagnetic wave is accompanied by a scalar gravitational wave, on which a free gravitational wave of quadrupole type can be superimposed.

15.4 Motion in the Background of a Plane Wave

Classical motion of electrically neutral particles in the background of this specific gravitational wave is described in terms of geodesics [10, 11]. The worldline $X^\mu(\tau) = (U(\tau), V(\tau), X(\tau), Y(\tau))$, in light cone coordinates, of a massive particle parametrized by the proper time τ is restricted by the constraint

$$\dot{U}\dot{V} - \Phi(U, X^i)\dot{U}^2 + \dot{X}^2 + \dot{Y}^2 = 1, \tag{15.17}$$

with the overdot denoting a proper-time derivative. This constraint is one of the integrals of motion of the geodesic equation

$$\ddot{X}^{\mu} + \Gamma_{\lambda\nu}^{\;\;\;\mu}(X)\dot{X}^{\lambda}\dot{X}^{\nu} = 0, \tag{15.18}$$

The existence of a Killing vector of the metric defined by ∂_v implies another constant of motion

$$\dot{U} = \gamma = \text{constant}. \tag{15.19}$$

As by definition of the laboratory velocity $v^a = dX^a/dT$ (where $a = 1, 2, 3$ labels spatial components only) we get

$$\frac{dU}{dT} = 1 - v_z, \tag{15.20}$$

it follows from a rewriting of the constraint (15.17) that

$$\frac{1 - \mathbf{v}^2}{(1 - v_z)^2} + \Phi = \frac{1}{\gamma^2}. \tag{15.21}$$

As $dU = \gamma d\tau$, the geodesic equations in the transverse plane can be written alternatively as

$$\frac{d^2 X^i}{dU^2} + \frac{1}{2}\frac{\partial \Phi}{\partial X^i} = 0. \tag{15.22}$$

In particular for the electromagnetic wave (15.12) these equations simplify to those of a 2-dimensional parametric oscillator:

$$\frac{d^2 X^i}{dU^2} + 2\pi G\left(\mathbf{E}^2 + \mathbf{B}^2\right) X^i = 0. \tag{15.23}$$

In the special case $2\pi G(\mathbf{E}^2 + \mathbf{B}^2) = \mu^2 =$ constant the solutions take the form

$$X^i(U) = X^i_0 \cos \mu(U - U_0). \tag{15.24}$$

In this case the non-zero components of the Riemann tensor are

$$R_{uiuj} = -\mu^2 \delta_{ij}. \tag{15.25}$$

Therefore the remarkable consequence is, that the geodesics oscillate in the transverse plane at a frequency proportional to the square root of the curvature. Scattering of neutral test particles with a wavetrain of finite length has been discussed in this formalism in refs. [10, 11] and references therein.

15.5 Scalar Fields in a Plane-Wave Background

The gravitational field of the light wave can be probed by neutral test particles, as discussed in the previous section, or by electrically neutral fields of scalar, vector or spinor type, which at the classical level are sensitive only to the non-trivial gravitational background. At the quantum level these fields can describe e.g. pions, photons or neutrinos. As a first example we discuss a massive real scalar field S, with action

$$I[S] = -\frac{1}{2}\int d^4x\,\sqrt{-g}\left(g^{\mu\nu}\partial_\mu S\partial_\nu S + m^2 S^2\right)$$
$$= \int d^4x\left(2\partial_u S\partial_v S - 2\Phi(\partial_v S)^2 - \frac{1}{2}(\partial_x S)^2 - \frac{1}{2}(\partial_y S)^2 - \frac{m^2}{2}S^2\right), \tag{15.26}$$

where $\nabla_\perp$ represents the gradient in the transverse plane. The corresponding field equation is

$$\left(4\partial_u\partial_v - 4\Phi\partial_v^2 - \partial_x^2 - \partial_y^2 + m^2\right)S = 0. \tag{15.27}$$

? Exercise

15.3. Verify (15.27).

To solve this equation we introduce the expansion

$$S(u, v, x_i) = \int \frac{dqds}{2\pi}\,\kappa(q, s; x_i)e^{i(qv+su)}, \tag{15.28}$$

with $\kappa^*(q, s; x_i) = \kappa(-q, -s; x_i)$. Note that in terms if standard space-time coordinates we get

$$qv + su = (q+s)t + (q-s)z \equiv Et + pz \quad\Leftrightarrow\quad q = \frac{1}{2}(E+p),$$
$$s = \frac{1}{2}(E-p). \tag{15.29}$$

The field equation then contstrains the amplitudes in the expansion to solutions of

$$\left(-\partial_x^2 - \partial_y^2 + 4q^2\Phi - 4sq + m^2\right)\kappa = 0. \tag{15.30}$$

For the special case (15.25) with constant energy density T_{uu} this equation becomes that of a 2-dimensional harmonic quantum oscillator:

$$\left(-\partial_x^2 - \partial_y^2 + \omega_q^2(x^2 + y^2) - 4sq + m^2\right)\kappa = 0, \tag{15.31}$$

where $\omega_q = 2\mu|q|$. Introducing the notation $\xi_i = \sqrt{\omega_q}\, x_i$ the solutions are of the form

$$\kappa(q, s; x_i) = \sum_{n_1,n_2=0}^{\infty} c_{n_1 n_2}(q, s) H_{n_1}(\xi_1) H_{n_2}(\xi_2) e^{-(\xi_1^2+\xi_2^2)/2}, \tag{15.32}$$

where the $H_n(\xi)$ are standard Hermite polynomials. It follows that the energy and momentum dispersion relation is quantized according to

$$E^2 = p^2 + m^2 + 2(n_1 + n_2 + 1)\omega_q. \tag{15.33}$$

15.6 Spinor Fields in a Plane-Wave Background

Our second example is a Majorana spinor field $\Psi = \Psi^c \equiv C\bar{\Psi}^T$, where C is the charge conjugation operator and T denotes transposition in spinor space (for our conventions on the Dirac algebra including charge conjugation, see the Appendix); such a field can describe e.g. neutrinos of Majorana type.

As spinor fields are primarily defined in Minkowski space, we need to introduce the formalism of translating between the curved space-time manifold and the flat local tangent space-time; this is achieved by the use of vierbein-fields $e^a_{\ \mu}$ such that

$$g_{\mu\nu} = \eta_{ab} e^a_{\ \mu} e^b_{\ \nu}. \tag{15.34}$$

Here a labels vector components in the local Minkowski space, and μ does the same thing in the curved space-time manifold. Using the vierbein fields one can define 1-forms $E^a = e^a_{\ \mu} dx^\mu$, which for the metric (15.9) have the component form

$$E^a = \left(\frac{1}{2}\left(dv + (\Phi + 1)du\right), dx, dy, \frac{1}{2}\left(dv + (\Phi - 1)du\right)\right). \tag{15.35}$$

Defining the inverse vierbein $e^\mu_{\ a}$ by

$$e^\mu_{\ a} e^a_{\ \nu} = \delta^\mu_\nu, \tag{15.36}$$

there is a corresponding gradient operator

$$\nabla_a = e^{\mu}_{\ a}\partial_\mu = \left(\partial_u + (1-\Phi)\partial_v,\, \partial_x,\, \partial_y,\, -\partial_u + (1+\Phi)\partial_v\right). \tag{15.37}$$

? Exercise

15.4. Check the property 15.36.

In order for the metric to be covariantly constant, the vierbein must satisfy the more general condition

$$dE^a = \omega^a_{\ b} \wedge E^b, \tag{15.38}$$

where the antisymmetric-tensor valued 1-form $\omega^{ab} = -\omega^{ba} = \omega_\mu^{\ ab} dx^\mu$ defines the spin connection. For the special vierbein (15.35) it is reduced to the form

$$\omega^a_{\ b} = \omega_{u\ b}^{\ a}\, du, \qquad \omega_{u\ b}^{\ a} = -\frac{1}{2}\begin{pmatrix} 0 & \partial_x\Phi & \partial_y\Phi & 0 \\ \partial_x\Phi & 0 & 0 & -\partial_x\Phi \\ \partial_y\Phi & 0 & 0 & -\partial_y\Phi \\ 0 & \partial_x\Phi & \partial_y\Phi & 0 \end{pmatrix}, \tag{15.39}$$

modulo an arbitrary local Lorentz transformation in tangent space.

Defining $\omega_a^{\ bc} = \omega_\mu^{\ bc} e^{\mu}_{\ a}$, the curved-space Dirac operator now is

$$\gamma \cdot D = \gamma^a \left(\nabla_a - \frac{1}{2}\omega_a^{\ bc}\sigma_{bc}\right). \tag{15.40}$$

For our special metrics (15.9) or vierbeins (15.35) a great simplification is, that the spin-connection term $\omega_a^{\ bc}\sigma_{bc}$ actually vanishes after contraction with γ^a [9]; therefore the Dirac operator simplifies to

$$\gamma \cdot D = \gamma^a \nabla_a = -i\begin{pmatrix} 0 & \sigma_i\nabla_i + \nabla_0 \\ -\sigma_i\nabla_i + \nabla_0 & 0 \end{pmatrix}, \tag{15.41}$$

and in the 2-component notation introduced in the Appendix the Dirac equation becomes:

$$\begin{aligned} 2\partial_v\chi_2 - \left(\partial_x + i\partial_y\right)\chi_1 &= im\chi_1^*, \\ \left(\partial_x - i\partial_y\right)\chi_2 - 2\left(\partial_u - \Phi\partial_v\right)\chi_1 &= im\chi_2^*. \end{aligned} \tag{15.42}$$

The complex conjugate components χ^* can be eliminated by applying the complex conjugate Dirac equation to get

$$\left(4\partial_u\partial_v - 4\Phi\partial_v^2 - \partial_x^2 - \partial_y^2 + m^2\right)\chi_1 = 0,$$
$$\left(4\partial_u\partial_v - 4\Phi\partial_v^2 - \partial_x^2 - \partial_y^2 + m^2\right)\chi_2 = 2\left(\partial_x + i\partial_y\right)\Phi\,\partial_v\chi_1. \tag{15.43}$$

Therefore the general solution for χ_1 is fully analoguous to that for the scalar field S, with the same spectrum of energy and momentum states; in contrast, the general solution for χ_2 consists of a special solution, defined in terms of the solution for χ_1 by the right-hand side of the second Eq. (15.43), plus an arbitrary solution of the homogeneous free Klein-Gordon equation, as for χ_1. Therefore all solutions are found to have a structure similar to the scalar field, except that any non-trivial solution χ_1 is accompanied by a special dependend solution for χ_2 constructed from χ_1 by

$$\left[-\left(\partial_x^2 + \partial_y^2\right) + m^2\right]\chi_2 = 2(\partial_x + i\partial_y)\left(\partial_u - \Phi\partial_v\right)\chi_1 - 2im\left(\partial_u - \Phi\partial_v\right)\chi_1^*. \tag{15.44}$$

15.7 Massless Abelian Vector Fields in a Plane-Wave Background

Finally we describe the propagation of a massless abelian vector field $a = a^\mu\partial_\mu$ in the plane-wave gravitational background. The Maxwell-action for this field in light cone coordinates takes the form

$$I[a] = \int dudvdxdy\ \left[(\partial_u a_v - \partial_v a_u)^2 + (\partial_u a_i - \partial_i a_u)(\partial_v a_i - \partial_i a_v)\right.$$
$$\left. - \Phi(\partial_v a_i - \partial_i a_v)^2 - \tfrac{1}{8}\left(\partial_i a_j - \partial_j a_i\right)^2\right]. \tag{15.45}$$

The resulting field equations are

$$4\partial_u\partial_v a_v - \Delta_\perp a_v - 2\partial_v\left(\partial_u a_v + \partial_v a_u - \frac{1}{2}\partial_i a_i\right) = 0,$$
$$4\partial_u\partial_v a_u - \Delta_\perp a_u - 2\partial_u\left(\partial_u a_v + \partial_v a_u - \frac{1}{2}\partial_i a_i\right) + 2\partial_i\left[\Phi\left(\partial_i a_v - \partial_v a_i\right)\right] = 0,$$
$$-2\partial_u\partial_v a_i + \frac{1}{2}\Delta_\perp a_i + \partial_i\left(\partial_u a_v + \partial_v a_u - \frac{1}{2}\partial_j a_j\right) - 2\partial_v\left[\Phi\left(\partial_i a_v - \partial_v a_i\right)\right] = 0. \tag{15.46}$$

Gauge transformations $a'_\mu = a_\mu + \partial_\mu \Lambda$ can be used to simplify these equations. First note that we can take

$$\partial_u a'_v + \partial_v a'_u - \frac{1}{2}\,\partial_i a'_i = 0, \tag{15.47}$$

by taking Λ as the solution of

$$\left(-4\partial_u\partial_v + \partial_x^2 + \partial_y^2\right)\Lambda = 2\left(\partial_u a_v + \partial_v a_u\right) - \partial_i a_i. \tag{15.48}$$

The remaining field equations are

$$\begin{aligned}
&4\partial_u\partial_v a'_v - \left(\partial_x^2 + \partial_y^2\right) a'_v = 0,\\
&4\partial_u\partial_v a'_u - \left(\partial_x^2 + \partial_y^2\right) a'_u + 2\partial_i\left[\Phi\left(\partial_i a'_v - \partial_v a'_i\right)\right] = 0,\\
&4\partial_u\partial_v a'_i - \left(\partial_x^2 + \partial_y^2\right) a'_i + 4\partial_v\left[\Phi\left(\partial_i a'_v - \partial_v a'_i\right)\right] = 0.
\end{aligned} \tag{15.49}$$

Next we can still make a residual gauge transformation to eliminate a'_v by taking Λ' restricted by

$$\left(-4\partial_u\partial_v + \partial_x^2 + \partial_y^2\right)\Lambda' = 0, \qquad a''_v = a'_v + \partial_v\Lambda' = 0. \tag{15.50}$$

Then the gauge contraint (15.47) reduces to

$$\partial_v a''_u = \frac{1}{2}\,\partial_i a''_i. \tag{15.51}$$

Therefore we are left with

$$\begin{aligned}
&4\partial_u\partial_v a''_u - 4\Phi\,\partial_v^2 a''_u - \left(\partial_x^2 + \partial_y^2\right) a''_u = 2\partial_i\Phi\,\partial_v a''_i,\\
&4\partial_u\partial_v a''_i - 4\Phi\,\partial_v^2 a''_i - \left(\partial_x^2 + \partial_y^2\right) a''_i = 0.
\end{aligned} \tag{15.52}$$

? Exercise

15.5. Verify (15.52).

This set of equations looks similar to that of the Majorana-Dirac equations in the previous section: the transverse components a_i'' are solutions of the scalar Klein-Gordon equation, accompanied by a special fixed solution $\bar{a}_u''$:

$$-\left(\partial_x^2+\partial_y^2\right)\bar{a}_u''=-2\partial_i\left[\left(\partial_u-\Phi\partial_v\right)a_i''\right]. \tag{15.53}$$

But in contrast to the Majorana-Dirac case there is no independent dynamical solution for a_u'', as for vanishing $a_i''=0$ the homogeneous equation for a_u'' implies its vanishing as well:

$$\partial_v a_u''=0 \quad \text{and} \quad \left(\partial_x^2+\partial_y^2\right)a_u''=0. \tag{15.54}$$

These conditions do not allow normalizable solutions for a_u''; in fact, for $a_i''=0$ the longitudinal component a_u'' can be gauged away by a third residual gauge transformation with a gauge function Λ'' satisfying the constraints

$$a_u'''=a_u''+\partial_u\Lambda''=0, \quad \partial_v\Lambda''=0, \quad \left(\partial_x^2+\partial_y^2\right)\Lambda''=0. \tag{15.55}$$

Therefore the transverse components are the only dynamical ones, taking the same form as solutions of the massless scalar wave equation, with the same spectrum of energy and momentum, whilst $a_u''=\bar{a}_u''$ is a dependend field fixed entirely in terms of the transverse components by Eq. (15.53).

15.8 Conclusions

In summary, in the above it has been shown that the equations of motion of neutral test particles, and the field equations of neutral scalar fields, Majorana-Dirac spinor fields and abelian vector fields can all be solved in the background of gravitational *pp*-waves such as those accompanying infinite plane electromagnetic waves. For energy-momentum density of the source field constant in time, such as that of a circularly polarized plane light wave, a distinct signature is, that test particles oscillate in the transverse plane of the wave, whilst the spectrum of transverse momentum of the fields becomes discrete. Because of the small curvature to be expected from such waves, these effects will be difficult to observe; moreover, in realistic conditions beams of electromagnetic wave will be of finite width, introducing modifications to the above conclusions which still have to be considered. Nevertheless, as a matter of principle the scattering of neutral particles by beams of electromagnetic waves will be another test in establishing the universality and dynamics of gravitational interactions.

Appendix: Spinors and the Dirac Algebra

Spinor fields in curved space-time are most easily described in the tangent Minkowski space, using the vierbein formulation to translate the results to the curved space-time manifold. We use the flat-space representation of the Dirac algebra in which γ_5 is diagonal:

$$\gamma_0 = i \begin{pmatrix} 0 & 1 \\ 1 & 0 \end{pmatrix}, \quad \gamma_i = \begin{pmatrix} 0 & -i\sigma_i \\ i\sigma_i & 0 \end{pmatrix}, \quad \gamma_5 = \begin{pmatrix} 1 & 0 \\ 0 & -1 \end{pmatrix}, \tag{15.56}$$

such that for $a = (0, 1, 2, 3)$

$$\{\gamma_a, \gamma_b\} = 2\eta_{ab}\mathbf{1}, \qquad \gamma_5^2 = \mathbf{1}, \qquad \{\gamma_5, \gamma_a\} = 0. \tag{15.57}$$

The generators of the Lorentz transformations on spinors are defined by

$$\sigma_{ab} = \frac{1}{4}\left[\gamma_a, \gamma_b\right], \tag{15.58}$$

with commutation relations

$$[\sigma_{ab}, \sigma_{cd}] = \eta_{ad}\sigma_{bc} - \eta_{ac}\sigma_{bd} - \eta_{bd}\sigma_{ac} + \eta_{bc}\sigma_{ad}. \tag{15.59}$$

Hermitean conjugation is achieved by

$$\gamma_a^\dagger = \gamma_0\gamma_a\gamma_0. \tag{15.60}$$

The charge conjugation operator C is defined by

$$C = C^\dagger = C^{-1} = -C^T = \gamma_2\gamma_0 = \begin{pmatrix} \sigma_2 & 0 \\ 0 & -\sigma_2 \end{pmatrix}, \tag{15.61}$$

such that

$$C^{-1}\gamma_a C = -\gamma_a^T. \tag{15.62}$$

If the spinor Ψ is a solution of the Dirac equation in Minkowski space

$$(\gamma \cdot \partial + m)\, \Psi = 0, \tag{15.63}$$

then this is also true for the charge-conjugate:

$$\Psi^c = C\bar{\Psi}^T = -\gamma_2\Psi^* \quad \Rightarrow \quad (\gamma \cdot \partial + m)\, \Psi^c = 0. \tag{15.64}$$

The Majorana constraint $\Psi^c = \Psi$ reduces the number of independent spinor components from 4 to 2 complex ones. This makes it easy to work in terms of 2-component spinors (χ, η) which are eigenspinors of γ_5, by the decomposition

$$\Psi = \Psi^c = \begin{bmatrix} \chi \\ \eta \end{bmatrix} \qquad \eta = -i\sigma_2\chi^*. \tag{15.65}$$

Acknowledgments Discussions with Ernst Traanberg of the Lorentz Institute in Leiden on solving field equations in the plane-wave background are gratefully acknowledged.

References

1. G. Koekoek, J.W. van Holten and U. Wyder, The path to special relativity. Preprint (2022). https://arxiv.org/abs/2109.11925
2. F. Dyson, A. Eddington, C. Davidson, A determination of the deflection of light by the sun's gravitational field. Phil. Trans. R. Soc. A **220**, 571 (1920)
3. K. Akiyama, et al., (Event Horizon Telescope Collaboration), First M87 event horizon telescope results. I. The shadow of the supermassive black hole. Astrophys. J. Lett. **875**, L1 (2019)
4. B. Abbott, et al., (LIGO Scientific Collaboration and Virgo Collaboration), Observation of gravitational waves from a binary black-hole merger. Phys. Rev. Lett. **116**, 061102 (2016)
5. H.W. Brinkmann, Proc. Natl. Acad. Sci. **9**, 1 (1923)
6. O. Baldwin, G. Jeffery, The relativity theory of plane waves. Proc. R. Soc. A **111**, 95 (1926)
7. W. Rindler, *Essential Relativity* (Springer, Berlin, 1977)
8. H. Stephani, D. Kramer, M. MacCallum, C. Hoenselaers, E. Herlt, *Exact Solutions of Einstein's Field Equations* (Cambridge University Press, Cambridge, 2003)
9. J.W. van Holten, Gravitational waves and masless particle fields, in *Proceedings of the XXXVth Karpacz Winterschool of Theor. Physics*, ed. by J. Kowalski-Glikman (Springer, Berlin, 1999), p. 365
10. J.W. van Holten, The gravitational field of a light wave. Fortschr. Phys. **59**, 284 (2011)
11. J.W. van Holten, The gravity of light-waves. Universe **4**, 110 (2018)

Coupling Quantum Matter and Gravity

16

Domenico Giulini, André Großardt, and Philip K. Schwartz

Abstract

In this chapter we deal with several issues one encounters when trying to couple quantum matter to classical gravitational fields. We start with a general background discussion and then move on to two more technical sections. In the first technical part we consider the question how the Hamiltonian of a composite two-particle system in an external gravitational field can be computed in a systematic post-Newtonian setting without backreaction. This enables us to reliably estimate the consistency and completeness of less systematic and more intuitive approaches that attempt to solve this problem by adding 'relativistic effects' by hand. In the second technical part we consider the question of how quantum matter may act as source for classical gravitational fields via the semiclassical Einstein equations. Statements to the effect that this approach is fundamentally inconsistent are critically reviewed.

D. Giulini
Institute for Theoretical Physics, University of Hannover, Hannover, Germany

Center of Applied Space Technology and Microgravity, University of Bremen, Bremen, Germany
e-mail: giulini@itp.uni-hannover.de

A. Großardt
Institute for Theoretical Physics, Friedrich Schiller University Jena, Jena, Germany
e-mail: andre.grossardt@uni-jena.de

P. K. Schwartz (✉)
Institute for Theoretical Physics, University of Hannover, Hannover, Germany
e-mail: philip.schwartz@itp.uni-hannover.de

C. Pfeifer, C. Lämmerzahl (eds.), *Modified and Quantum Gravity*, Lecture Notes in Physics 1017, https://doi.org/10.1007/978-3-031-31520-6_16

16.1 Introduction and Preliminary Discussion

The central concern of this contribution is the relation between gravity—as described by the classical (i.e. not quantised) theory of General Relativity (henceforth abbreviated by GR)—and the theory of all other 'interactions', which are described by relativistic quantum (field) theories (henceforth abbreviated by RQFT).[1] To this end, we will address several specific technical as well as conceptual issues which we consider important. Some of these issues can be resolved (we believe), whereas others may perhaps require rethinking before any resolution can be proposed.

In this first section we wish to share and discuss a few thoughts concerning the relation of 'gravity' on one side, and 'matter' on the other. This preliminary discussion is not only meant to set the stage for the later, more technical parts of this contribution, but also tries to convey a sense of appreciation for the distinguishing features of the 'gravitational interaction'.

In Sect. 16.2 we first show how to systematically incorporate in form of a post-Newtonian expansion the interaction of an electromagnetically bound model-atom (consisting of two charged point particles) with an external gravitational field. A second part of that section revisits in some detail the question of allegedly 'anomalous' couplings of internal energies to the centre-of-mass motion of composite systems.

Section 16.3 deals with the question of backreaction, i.e. how quantum matter might source a classical gravitational field. Much debated issues concerning consistency and causality are addressed in separate subsections, as well as alternative schemes to that of semiclassical gravity, like collapse models.

16.1.1 Why Care?

We recall Einstein's equations[2]

$$R_{\mu\nu} - \frac{1}{2} R\, g_{\mu\nu} = \frac{8\pi G}{c^4} T_{\mu\nu}\,, \tag{16.1}$$

which relate the spacetime metric g to the matter content, though the matter does not determine the metric. More precisely, the ten components $T_{\mu\nu}$ comprising the matter's energy and momentum densities and flux-densities, determine the 10 Ricci components out of the 20 Riemann curvature components. The metric carries its

[1] Note that we distinguish the general notion of quantum field theory from the specific form it takes in presence of Poincaré invariance, in which case we write RQFT.

[2] Our conventions are the usual ones: $R_{\mu\nu} := R^{\lambda}{}_{\mu\lambda\nu}$ is the Ricci tensor if $R^{\lambda}{}_{\mu\sigma\nu}$ denotes the Riemann tensor, $R := g^{\mu\nu} R_{\mu\nu}$ is the Ricci scalar, $T_{\mu\nu}$ denote the covariant components of the energy-momentum tensor with $T_{00} := T(e_0, e_0)$ the energy-density for the observer characterised by the unit timelike vector e_0. G is Newton's gravitational constant and c the speed of light in vacuum. Our signature convention is 'mostly plus' $(-, +, +, +)$.

own degrees of freedom, over and above those of the matter, which are capable to transport physical quantities like energy and momentum from one material system to another through spacetime regions which are entirely devoid of any matter. This happens, for example, if a distant binary star-system emits gravitational waves to a detector on Earth. According to GR, the mere existence of spacetime is logically independent of the existence of matter. Hence we arrive at the following dichotomy of our fundamental laws:

- Gravity is modelled by *classical* GR. It requires a spacetime consisting of a pair $(\mathcal{M}, g)$, where $\mathcal{M}$ is a 4-dimensional differentiable manifold and g denotes a Lorentzian metric on it. The latter determines a connection, the Levi–Civita connection for g, and hence an inertial structure (see below). The coupling between spacetime and matter—sometimes referred to as *backreaction*—is then given by Einstein's equations (16.1). Classical matter is described by fields *on* $\mathcal{M}$ (i.e. sections in various vector bundles over $\mathcal{M}$) the dynamics of which usually follows from a Lagrange density $\mathcal{L}$ from which we also obtain the energy-momentum tensor $T_{\mu\nu}$ through the functional derivative $\delta\mathcal{L}/\delta g^{\mu\nu}$.
- Fundamentally, matter is modelled by *quantum* field theories. In a Poincaré invariant context, states of quantum fields are commonly defined as elements of the (bosonic or fermionic) Fock space $\mathcal{F}_{\pm}(\mathcal{H})$ over a single-particle Hilbert space $\mathcal{H}$, usually (and loosely speaking) the space of positive-frequency solutions of some (classical) field equation *on flat Minkowski spacetime*. The dynamical laws of these fields are determined by their interactions with each other as defined by the total Lagrange density $\mathcal{L}$—leading to the common picture where forces between fermionic particles are carried by gauge bosons.

Attempts to cure this division at the level of the most fundamental theories by replacing the classical spacetime structure and Einstein's equations by concepts more compatible with the Hilbert space structure of matter are summarised under the label of 'quantum gravity' and have so far not led to any generally accepted scheme. Less pervasive 'escape strategies' are those in which gravity stays classical. Treating gravity and matter with different mathematical principles need not necessarily result in inconsistencies. Two aspects of such strategies should be distinguished:

1. *Ignoring backreaction* means to take a fixed, generally *curved*, spacetime $(\mathcal{M}, g)$ and consider quantum fields on it. At least in the regime in which the energy-momentum density of the fields is small enough compared to that of the *sources* of the background gravitational field, this approach will provide a sensible approximation. The difficult task remains to formulate quantum field theory without Poincaré invariance (or any other maximal symmetry, like de Sitter or anti-de Sitter). In spite of much mathematical progress over the last decades [1–3], which also shows how many of the familiar concepts of RQFT fail to exist in Quantum Field Theory in Curved Spacetime (henceforth abbreviated by QFTCS)

due to the lack of Poincaré invariance, QFTCS is still far from being able to address and answer specific physical problems with the desired certainty.[3]

However, when considering weak gravitational fields, i.e. small deviations from Minkowskian geometry, and small velocities of the systems involved, we may restrict ourselves to fixed-particle sectors. The systems are then effectively described by the corresponding minimally coupled relativistic wave equations, and a post-Newtonian expansion of these equations in the parameter c^{-1} yields a systematic post-Newtonian description of quantum systems in gravitational fields. In Sect. 16.2, we will explore this strategy in detail and discuss related issues.

2. *Inclusion of backreaction* in a 'classical-quantum scheme' must mean to consider the matter's energy-momentum tensor $T_{\mu\nu}$ on the right-hand side of the classical equation (16.1) as a classical field, like, e.g., the expectation value of the energy-momentum tensor in some state. In that case, the metric will acquire a dependence on that state, which also re-couples to the evolution of the state via the matter equations (which contain the metric). This will effectively cause a non-linearity in the dynamical evolution of the quantum state.

   Semiclassical models of this kind are widely believed to be inconsistent and/or unphysical. However, when making statements along this line, one needs to be particularly careful, and such claims of inconsistency often turn out to be founded in implicit assumptions. An important aspect is the precise theoretical description of the reduction of quantum states upon measurement (or, rather, the lack thereof). In Sect. 16.3, we will discuss these and related issues concerning the gravitational backreaction of quantum matter on a classical spacetime.

16.1.2 A Note on the Equivalence Principle

The equivalence principle is commonly viewed as the core assumption of GR. In the proper sense of the word, it is a *heuristic* principle, characterising the coupling of gravity to any sort of (classical) matter, the dynamical laws of which have already been formulated in a way compatible with Special Relativity, i.e. in a Poincaré invariant form. In connection with matter described by ordinary Quantum Mechanics (henceforth abbreviated QM) this principle loses its heuristic power, one obvious reason being the lack of Poincaré invariance of the Schrödinger equation. Speculations of whether QM violates the equivalence principle *per se* have a long

[3] The furthest developed mathematically rigorous approach to QFTCS uses methods from algebraic quantum field theory, and has led, for example, to the notion of *locally covariant quantum field theories* on curved spacetimes. Among other results, this allows to transfer (properly formulated versions of) the spin-statistics and CPT theorems from RQFT to QFTCS (see, in particular, chapter 4 of [3]). As is well-known, QFTCS is also of fundamental importance for the physics of black holes, leading to the emission of Hawking radiation [4], and is believed to play an important role in relativistic cosmology [5].

history and are still ongoing. In that situation we find it useful to reflect on the meaning of this principle.

The perhaps most concise way to state the implication of the equivalence principle, as it is realised in GR, is to say that the coupling to gravity is exclusively furnished via *one and the same geometry* that is common to all matter components. It does *not* say, e.g., that all 'pieces of matter' fall alike (in a way only depending on the initial conditions but not on the inner composition and nature of the 'piece'). This latter statement is only true for the highly idealised concept of a 'test particle', the degree of approximate realisation of which strongly depends on the physical context.

Indeed, a 'test particle' must have an extent much less than the curvature radius of the ambient spacetime, for otherwise it cannot probe the *local* gravitational field. On the other hand, its extent must be much larger than its own gravitational radius GM/c^2, for otherwise its own gravity dominates the ambient one and also its binding energy becomes of the order of its rest energy. Furthermore it may not have any appreciable spin angular momentum, higher mass-multipole moment (quadrupole and higher), charge, etc. We emphasise that such qualities need not disappear with vanishing size [6]. We leave it to the reader to come up with those physical properties that may still be varied within the set of 'test particles'.

Hence, according to GR, the centre of mass of a banana will not describe the same worldline as that of an apple for the same initial conditions, due to the higher quadrupole moment of the former, which couples to the spacetime curvature. Adding spin angular momentum with respect to the centre of mass will again influence its trajectory, again due to curvature couplings.[4] Nothing of that sort is, of course, in violation of the equivalence principle, since all these deviations can be fully accounted for by a single spacetime geometry.

Likewise, in QM, the 'centre' of a wave packet need not fall on the same trajectory as a point particle for equivalent initial conditions. Also, the wave behaviour of the former does depend on the inertial mass of the particle it represents (as the de Broglie wavelength does). Again, this does not *per se* contradict the equivalence principle, as sometimes suggested following a famous argument by Salecker (and contradicted by Feynman) made during the famous 1957 Chapel Hill conference [8, chapter 23]. In fact, in can be proven that *any* solution to the Schrödinger equation in a homogeneous (though arbitrarily time dependent) force field is obtained from a solution of the force-free equation by translating it along the integral flow of a classical solution curve and multiplying it with an appropriate space- and time-dependent phase factor [9]. This implies that the spatial probability distribution falls exactly like a continuous dust cloud of classical particles with identical initial velocities. If the particles' spatial paths only depend on the ratio of the gravitational to the inertial mass, then so does the 'path' of the probability

[4] The spin-curvature coupling is already present in the standard (lowest order) pole-dipole approximation of the Mathisson–Papapetrou–Dixon equation, the quadrupole coupling appears in the next order; see, e.g., the beautiful review [7].

distribution. Note that this statement is exact and holds for *all* solutions to the Schrödinger equation, not just those approximating classical behaviour.

? Exercise

16.1 More precisely, the above statement is as follows: if $\mathbf{F}(t)$ is a homogeneous force field and $V(t, \mathbf{x}) = -\mathbf{x} \cdot \mathbf{F}(t)$ the corresponding potential that appears in the Schrödinger equation, then any solution ψ of the latter is of the form $\left(\exp(\mathrm{i}\alpha)\,\psi'\right) \circ \Phi^{-1}$, where ψ' is a solution of the free Schrödinger equation, $\Phi : \mathbb{R}^4 \to \mathbb{R}^4$ is the flow map $(t, \mathbf{x}) \mapsto \Phi(t, \mathbf{x}) := (t, \mathbf{x} + \boldsymbol{\xi}(t))$, corresponding to a classical solution of $m\ddot{\boldsymbol{\xi}}(t) = \mathbf{F}(t)$ with initial condition $\boldsymbol{\xi}(t = 0) = \mathbf{0}$ and arbitrary initial velocity, and $\alpha(t, \mathbf{x})$ is a phase factor for which an explicit integral formula can be written down depending on $\boldsymbol{\xi}(t)$.

Prove this theorem and derive the integral formula for $\alpha(t, \mathbf{x})$. (If you get stuck, see [9, section 3].)

Finally we also emphasise that the often used 'equality of inertial and gravitational mass' expresses the equivalence principle only in restricted physical situations, and then only in the context of Newtonian gravity. Only if the Newtonian laws of free fall apply does that equality ensure the universality of free fall (for unstructured bodies). On the other hand, in GR, we do not have an unambiguous way to even define the notion of (passive) gravitational mass of a body interacting with others. Here, the equivalence principle should really be formulated in an invariant way that is independent of representation dependent definitions of 'masses'. We will have much more to say about this in Sect. 16.2.2.

16.1.3 Forces versus Inertial Structure

At this point we also wish to recall another aspect concerning the dichotomy between our theory of gravity on one hand, and our theories of fundamental matter on the other. The latter comprise the standard model, which is a field-theoretic description of the weak, strong, and electromagnetic forces. On the other hand, according to GR, gravity is not a *force* but rather unified with the inertial structure. For that reason one sometimes speaks of the 'gravito-inertial field' (rather than just 'gravitational field'). The geometric structure representing this field is the connection. Only *after* the connection is known, and the inertial motion thereby specified, does it make sense to speak of forces: a force is, by definition, the cause for deviations from inertial motion.

A mathematically distinguishing feature of the gravito-inertial field in comparison with other fundamental fields, that is often not sufficiently appreciated, is that the set of connections is an affine space, not a vector space (like the set of sections in a vector bundle). This means that there is no such thing as a 'zero connection' and hence no spacetime 'free' of gravitation/inertia. True, a flat connection may—in affine charts—be represented by identically vanishing coefficients, like in Minkowski space endowed with affine (inertial) coordinates, but that clearly does not mean that inertia—and hence gravity—is somehow 'absent' in any reasonably invariant way. Hence, whereas it makes sense to say that a certain region of spacetime is 'free' of electromagnetic fields, it makes no sense to make such a statement for gravity.

In passing we note that the notion of a 'geometric object' is not at all restricted to tensor fields; see, e.g., [10]. Tensor fields transform linearly under changes of coordinates (or diffeomorphisms) whereas connections have an affine (i.e. inhomogeneous) transformation property. Yet it clearly remains true in both cases that the coefficients are known in *all* coordinate systems if they are known in a single one. That, in fact, is often taken as a working definition of a 'geometric object'; e.g., [11, § 4.13, pp. 84–87]. In particular, this applies to connections on the tangent bundle (like the Levi–Civita connection) that can be used together with the metric to algebraically form other geometric objects, which need not be tensors.

In order for the above statement to make sense, that a certain region of spacetime is 'free of gravity', 'gravity' would have to be identified with a tensor field with all tensor-space values, in particular zero, as admissible states. Note that the metric itself does not fall into that class, since, e.g., the zero section is not admissible so that the set of gravitational states is not a vector space. Such a vector space structure is achieved if one chooses a fixed background connection as reference and represents the 'gravitational field' as the *difference* of the physical connection to that reference. But then the latter defines a background structure that explicitly breaks diffeomorphism invariance down to the subgroup of those diffeomorphisms preserving that background. Alternatively, 'zero gravity' is also sometimes taken to mean 'zero curvature' [12]. But that re-introduces the old dichotomy between gravity and inertia (inertial forces clearly exist in flat Minkowski space) that GR had so successfully overcome, their unification being the heart of the equivalence principle, just like background independence is.[5] For further discussions of these general aspects we refer to [14–16].

The inertial structure of flat Minkowski space is also deeply rooted in its symmetry properties and hence in all theories of interactions except gravity. In fact, the inertial structure endows Minkowski space with the structure of a four-dimensional real affine space in which an open subset in the set of all 'straight lines', namely the timelike ones, represent inertial motion of 'test particles'. It can be shown [17] that the subgroup within the group of bijections of Minkowski space that consists of those bijections that preserve this inertial structure (i.e. map timelike straight lines to timelike straight lines) is the group generated by Poincaré

[5] Einstein considered the unification of inertia and gravity to be *the* distinctive physical achievement of GR, not the fact that it can be formulated in purely geometric terms, which he regarded more a matter of semantics rather than physics; see [13].

transformations and homotheties.[6] The latter are eliminated if, e.g., a measure of length is provided at least along any of the inertial straight lines (a 'clock'). In that sense the inertial structure alone almost determines the Poincaré group without further assumptions. Note in particular that no continuity or other regularity assumption enters the proof of this result. Now, according to Felix Klein's 'Erlanger Programm' [18] geometries and their automorphism groups are two sides of the same coin. Hence, the Poincaré group, which lies at the heart of fundamental theories of matter, is the algebraic expression of the inertial structure, which is of geometric nature.

In special-relativistic theories the inertial structure is fixed once and for all in a way that is entirely independent of the matter content. In contrast, the dynamics of the matter clearly *does* depend on the inertial structure. Hence this dependence is unidirectional. In that sense Minkowski space and the Poincaré group are absolute structures, something to be abandoned according to the principles that led to GR. An alternative approach how to give up the absolute structure of Minkowski space is to 'gauge' Poincaré symmetry, as discussed in Chaps. 3 and 4. That abandoning global Poincaré symmetry will likely imply a major revision in the concepts of quantum field theory should be obvious by once more recalling the central role the Poincaré group and its representation theory plays there, e.g., regarding the concepts of 'particle' and 'elementary'.

For the highly idealised concept of test particles, an inertial structure reduces to the concept of a *path structure*. The latter assigns a unique path in spacetime to any pair consisting of a point and one-dimensional subspace in the tangent space at that point. In other words, the path is universal for all 'test particles', only depending on the initial conditions, but not on other contingent properties the particle may still have. The path structure in GR is special in a twofold way: first, the paths are geodesics for *some* linear connection; second, the connection is the Levi–Civita connection for *some* metric. It has been worked out in precise mathematical terms how to characterise such special path structures [19, 20].

The important point we wish to stress and keep in mind at this point is the distinction between 'forces' on one side, and 'inertial structure' on the other. The fields in the standard model account for forces, whereas the field in GR accounts for the inertial structure. The former have a natural 'zero' value, representing physical absence, but that does not apply for the latter; spacetime cannot be without inertial structure. Moreover, the former are quantised, the latter is—so far—only understood classically.

Different attitudes exist as to whether and how the conceptual difference just discussed should be overcome. Whereas some feel it would be desirable to reformulate gravity in a way more like the other 'forces', others believe just the opposite and stress the physical significance of that difference. For example, in the

[6] Homotheties are scalings about any centre point, i.e. maps of the form $x \mapsto x' := a(x - x_0) + x_0$, where x_0 is any point in Minkowski space, the 'centre' of the homothety, and a is a non-zero real number, its 'scaling' parameter.

first case, the 'graviton'—a word often epitomising the believed quantum nature of gravity—is taken to mediate gravity in the same sense as the other gauge bosons mediate the other forces. In the second case, this would not make sense and the programme to 'quantise gravity' is presumably no more plausible than that of 'general-relativising' RQFT or 'gravitizing Quantum Mechanics' [21].

Another attitude towards the relation between *gravity and matter* or *inertial structures and forces* is that gravity is subordinate to matter, an accompanying phenomenon based on certain collective matter degrees of freedom. This is also sometimes expressed by saying that gravity is an *emergent* phenomenon. In this instance, GR would have a status comparable to, say, the Navier–Stokes equation of hydrodynamics, which also describes collective degrees of freedom that eventually can be reduced to those of the fluid's constituents. In such a picture, the microscopic state of matter fully determines the state of the gravitational field, which has no own propagating degrees of freedom. Gravitational waves are then, like sound waves, collective excitations of an underlying and more fundamental field of matter. The idea that the state of matter and their interactions should completely determine the inertial structure predates GR and goes back to the critique that Ernst Mach voiced in his book [22] on mechanics on Newton's interpretation of his [Newton's] famous bucket experiment. Einstein made this idea into what he called 'Mach's Principle' on which he based many heuristic considerations during the formative years of GR. Remarkably, even in 1918, well more than two years after the final formulation of GR, Einstein explicitly named Mach's Principle as an essential and indispensable part of GR, next to the principle of relativity and the principle of equivalence [23].[7]

Clearly, the mathematical structure of GR does not support Einstein's version of Mach's principle. According to GR, the gravitational field has its own degrees of freedom that propagate causally, albeit the causal structure with respect to which this statement is true is not a fixed background structure but rather determined by the evolving field itself. That remark holds irrespective of gauge dependent appearances of Einstein's equations, in which certain components of the field may seem to propagate instantaneously due to the fact that they obey elliptic rather than hyperbolic equations. But that is just as deceptive as in ordinary electrodynamics, where, e.g., in the Coulomb gauge the scalar potential obeys an elliptic Poisson equation whereas the transversal part of the vector potential obeys a hyperbolic d'Alembert equation with respect to the transverse current density[8] as source; see, e.g., [25] and [26, exercise 6.20, pp. 291–292]. This remark becomes important in the recent debates on alleged inferences of quantisation of gravity from gravitationally induced entanglement; see [27] and references therein.

[7] For a comprehensive account on the meanings and significance of Mach's Principle see [24].

[8] Note that the transverse current density is a non-local and non-causal function of the physical current density.

16.1.4 Foundations of General Relativity and the Role of Quantum Mechanics

One way to 'understand' the foundations of a physical theory is to axiomatise it, that is, to rigorously deduce it from a minimal set of assumptions. The latter should be operationally meaningful though they will in most cases be highly idealised. Attempts to axiomatise GR face the problem to somehow represent 'clocks' and 'rods', which from a physical point of view are highly complex systems. The strongest simplification is to reduce them to point particles and light rays, by means of which one may define an inertial structure and a conformal structure as in the EPS scheme of Ehlers et al. [28, 29]. In that scheme the 'particles' represent the inertial structure and are given by unparameterised timelike worldlines, which the axioms force to be unparameterised geodesics (also known as autoparallels) of a torsion-free connection. The 'particles' are therefore just an equivalence class of torsion-free connections which share the same autoparallels.[9] The 'light rays' in the EPS scheme determine a conformal equivalence class of metrics. A compatibility axiom requires that the 'light rays' are suitable limits of the 'particles', i.e. that the generators of the metric light cones (which are conformal invariants) are suitable limits of the unparameterised timelike geodesics. All this does not yet lead to a semi-Riemannian structure but rather to a Weyl structure consisting of a triple $(\mathcal{M}, [g], \nabla)$, where $\mathcal{M}$ is a smooth 4-d differentiable manifold, $[g]$ is a conformal equivalence class of metrics, and ∇ is a torsion-free connection that satisfies $\nabla_\lambda g_{\mu\nu} = \varphi_\lambda g_{\mu\nu}$ for some 1-form φ depending on the representative g of $[g]$. If $g'_{\mu\nu} = \exp(\Omega) g_{\mu\nu}$ then $\varphi'_\lambda = \varphi_\lambda + \partial_\lambda \Omega$. Hence, we may also identify a Weyl structure with equivalence classes $[(g, \varphi)]$ of pairs (g, φ), where $(g, \varphi) \sim (g', \varphi')$ if and only if there exists a smooth function $\Omega : \mathcal{M} \to \mathbb{R}$ such that $g' = \exp(\Omega) g$ and $\varphi' - \varphi = d\Omega$. Such an equivalence class $[(g, \varphi)]$ determines a unique torsion-free connection ∇ such that for any representative (g, φ) one has $\nabla g = \varphi \otimes g$. This would be equivalent to a semi-Riemannian structure if and only if φ is exact, $\varphi = df$, in which case there is a representative (g, φ) so that $\varphi = 0$ and $\nabla g = 0$, i.e., ∇ is the Levi–Civita connection for g. EPS close this gap in physical terms by simply stating as an additional axiom that there are no so-called 'second clock effects', which amounts to just $d\varphi = 0$ and hence (at least local) exactness.

Now, a somewhat surprising result is that instead of this postulate, the gap can also be closed by the requirement that the WKB limit of the massive Dirac and/or Klein–Gordon field in a Weylian spacetime is such that the rays (integral lines of the gradient of the eikonal) are just the 'particles' [30]. This is an entirely new aspect concerning the relation between QM and GR. It suggests that QM can, in fact, also play a positive role in laying the foundations to GR, rather than just cause trouble. An unexpected twist to the story indeed!

[9] Two connections $\Gamma^\lambda_{\mu\nu}$ and $\hat{\Gamma}^\lambda_{\mu\nu}$ are equivalent in that sense if and only if there exists a covector field V_μ such that $\hat{\Gamma}^\lambda_{\mu\nu} - \Gamma^\lambda_{\mu\nu} = \delta^\lambda_\mu V_\nu - \delta^\lambda_\nu V_\mu$.

16.2 Quantum Matter on Classical Gravitational Backgrounds

In this section, we will discuss the description of quantum matter systems on a fixed classical background spacetime—i.e. we consider the matter fields as 'test systems' probing the gravitational field and neglect their backreaction onto it. As explained at the end of Sect. 16.1.1, the generally accepted theoretical framework for this setting is QFTCS. However, notwithstanding its formal and conceptual successes, QFTCS is mathematically and conceptually very 'heavy', and its relation to (approximately) Galilei-symmetric physics, which for example is relevant for the description of quantum-optical experiments (as discussed in Chap. 14), is not as well understood as one might suspect.

As long as one is only interested in the effect of background *Newtonian* gravity on quantum systems (the experimental study of which is discussed in Chap. 14 as well), this does not pose a problem: one may simply include the background Newtonian gravitational potential Φ into the Hamiltonian describing the system in the usual way, leading to a Schrödinger equation that for a single bosonic particle of mass m, only subject to gravity, is of the form

$$i\hbar\partial_t\Psi = \left(-\frac{\hbar^2}{2m}\nabla^2 + m\Phi\right)\Psi. \tag{16.2}$$

This simple coupling of Galilei-invariant QM to Newtonian gravity has been extensively tested in the gravitational field of the earth, beginning with neutron interferometry in the Colella–Overhauser–Werner (COW) experiment in 1975 [31] and leading to modern light-pulse atom interferometers, which for example provide the most sensitive gravimeters to date [32]. However, as soon as one is interested in *post*-Newtonian effects of gravity on quantum systems, one needs a method either to describe the respective situation in terms of QFTCS proper or to include 'post-Newtonian corrections' into the Schrödinger equation above. Note that such post-Newtonian effects can be of two different kinds: they either correspond to post-Newtonian couplings of the 'Newtonian' potential Φ, or are couplings to those parts of the multi-component Einsteinian gravitational field that—as a matter of principle—simply do not exist in the scalar Newtonian theory, i.e. couplings to the vectorial (gravitomagnetic) and tensorial (gravitational waves) components.

Apart from the obvious fundamental theoretical interest in post-Newtonian gravitational effects in quantum systems, recently there has also been increased interest from an experimentally-oriented point of view, since the ever-increasing precision of quantum experiments is expected to allow for the detection of novel 'relativistic effects' that were not considered before. In particular, for composite systems one expects post-Newtonian couplings between internal and external (i.e. centre-of-mass) degrees of freedom of both non-gravitational and gravitational origin, which might lead, e.g., to quantum dephasing [33–35].

The occurrence of such post-Newtonian couplings may on a heuristic level be understood based on the notion of 'relativistic effects' known from classical physics:

formally replacing, in the spirit of the 'mass defect', the mass parameter in the single-particle Hamiltonian for a quantum system under Newtonian gravity by the rest mass M of a composite system plus its internal energy H_{int} divided by c^2, we obtain a coupling between the internal and external degrees of freedom according to

$$H_{\text{single particle}}(\mathbf{p}, \mathbf{r}, m) = \frac{\mathbf{p}^2}{2m} + m\Phi(\mathbf{r}) \tag{16.3a}$$

$$\begin{aligned} \rightarrow H_{\text{composite}} &= H_{\text{single particle}}(\mathbf{P}, \mathbf{R}, M + H_{\text{int}}/c^2) \\ &= \frac{\mathbf{P}^2}{2(M + H_{\text{int}}/c^2)} + (M + H_{\text{int}}/c^2)\Phi(\mathbf{R}) \\ &= \frac{\mathbf{P}^2}{2M}\left(1 - \frac{H_{\text{int}}}{Mc^2}\right) + (M + H_{\text{int}}/c^2)\Phi(\mathbf{R}) + O\left(c^{-4}\right), \end{aligned} \tag{16.3b}$$

where $\mathbf{P}$ and $\mathbf{R}$ denote the 'central coordinates' of the composite system, i.e. total momentum and some appropriate centre of mass coordinate. Another way to heuristically motivate such a post-Newtonian internal–external coupling is the replacement, in the Schrödinger equation describing the internal dynamics, '$t \rightarrow \tau$' of coordinate time by proper time of the central worldline—since the internal degrees of freedom experience special-relativistic and gravitational time dilation, so the argument goes, they will become coupled to the central dynamics [33,34,36].

Of course, such heuristic motivations of post-Newtonian couplings have great suggestive value. They are, however, conceptually dangerous for several reasons: Firstly, such treatments guarantee neither completeness nor independence of the suggested 'novel relativistic effects'—one might, on the one hand, overlook some effects, while on the other hand double-counting others (as would have been the case in our example above, had we included *both* the 'mass defect' and 'relativistic time dilation'). Secondly, those descriptions rely fundamentally on semiclassical notions such as central worldlines, and thus presuppose (a) separability (at least approximately) of the total state of the system into an external and an internal part—even though interactions, leading to entanglement, are precisely the point of interest!—, and (b) a semiclassical nature of the external state. Of course, for some applications—e.g. in atom interferometry—these assumptions may be well-suited; nevertheless a fundamental understanding of the post-Newtonian coupling of quantum systems to gravitational fields should be *independent* of such assumptions.

Therefore, to correctly and completely describe quantum systems in gravitational fields, a systematic treatment is needed, starting from well-established first principles and properly *deriving* a full theoretical description. Only such a systematic approach is *complete* and *free of redundancies*, and therefore can allow for reliable predictions when applied, for example, to experimental situations. As said before, in the end such a properly relativistic description should emerge from QFTCS. However, as long as we are only interested in low-order 'relativistic corrections' to the Newtonian coupling to gravity, consider approximately stationary spacetimes (such that there is a consistent notion of particles), and stay below the energy threshold of pair production, we may avoid the use of the complex framework of QFTCS. Instead, we may restrict ourselves to a fixed-particle sector, in which the theory is effectively described by classical field equations, and in this sector employ a post-Newtonian expansion framework as an easier method for systematic description of quantum systems under gravity. This essentially seems to be the only straightforward way to apply the equivalence principle to usual Galilei-invariant QM: to first deform Galilei to Poincaré invariance (the deformation parameter being c^{-1}), then apply the minimal coupling scheme, dictated by the equivalence principle and providing the couplings to the gravitational field at least up to additional curvature terms, and then contract again by a post-Newtonian expansion in the deformation parameter c^{-1}.

In Sect. 16.2.1, we are going to describe such a systematic post-Newtonian expansion framework applicable to the description of model quantum systems. This framework enables us in particular to derive a full description of a simple toy model of a two-particle atom to order c^{-2}, which will be explained in Sect. 16.2.1.2. In particular, this provides a proper derivation of the above-mentioned internal–external couplings. However, we will encounter 'anomalous' couplings of the internal kinetic and potential energies to the Newtonian gravitational potential: depending on the choice of coordinates in which the internal degrees of freedom are described, one does not get the heuristically expected coupling from (16.3), but the different energy forms couple differently. This phenomenon of 'anomalous' couplings appears quite generally in the context of the post-Newtonian description of composite systems under gravity. In Sect. 16.2.2, we will discuss these issues from a more general point of view, in the context of diffeomorphism-invariant field theory: we will make precise and critically evaluate arguments by Carlip [37] about the emergence of such 'anomalous' couplings and their possible elimination by application of diffeomorphisms. The content of Sect. 16.2.1 already appeared in the cited literature, whereas that of Sect. 16.2.2 is new.

16.2.1 Systematic Description of Model Systems in Post-Newtonian Gravity

In order to be able to perform a post-Newtonian expansion of a (locally) Poincaré-invariant theory, which perturbatively includes post-Newtonian effects on top of a Newtonian description, we need some *background structures* that enable us to speak of 'weak gravitational fields', 'small velocities', and space and time as separate notions. For these background structures, we take a *background metric* in combination with a *background time evolution vector field*, which is hypersurface orthogonal, timelike, and of constant length with respect to the background metric: The time evolution field gives us a $3 + 1$ decomposition of spacetime into time (its integral curves) and space (the leaves of its orthogonal distribution), while also being a reference for the definition of 'slow movements'. Similarly, the background metric serves as reference point for the 'absence' of gravitational fields.[10]

Concretely, we take as background spacetime *Minkowski spacetime* $(\mathcal{M}, \eta)$ and as background time evolution vector field a timelike geodesic vector field u on $(\mathcal{M}, \eta)$, with Minkowski square $\eta(u, u) = -c^2$ (where c is the velocity of light). u can be interpreted as the four-velocity field of a family of inertial observers in background Minkowski spacetime to which we will refer our post-Newtonian expansions. The *physical* spacetime metric g will be a perturbation on top of the background Minkowski metric η. On the three-dimensional leaves of 'space', which are η-orthogonal to u, we have two Riemannian metrics: a flat metric δ induced by the background metric η, as well as the physical spatial metric ${}^{(3)}g$ induced by the physical metric g.

Introducing coordinates $(x^\mu) = (x^0 = ct, x^a)$ adapted to the background structures, i.e. such that in these coordinates the Minkowski metric takes its usual component form $(\eta_{\mu\nu}) = \mathrm{diag}(-1, 1, 1, 1)$ and the time evolution field is $u = \partial/\partial t$, we may express all the following post-Newtonian computations in those coordinates, which will make the results look like usual Newtonian results plus post-Newtonian corrections. Note, however, that the results themselves are of course independent of the coordinates chosen to express them.

As the expansion parameter in powers of which we organise our post-Newtonian expansion, we take the inverse of the velocity of light, c^{-1}. That is, we will expand all relevant quantities as formal power series in the parameter c^{-1}. The term of order c^0 in such a series corresponds to the Newtonian limit, while the higher-order terms are the post-Newtonian corrections. Formally, the Newtonian limit is obtained as the $c \to \infty$ limit of the series. Of course, analytically speaking, a 'Taylor expansion' in a dimensionful parameter like c does not make sense (and even less so any limit in which c is varied, since it is a constant of nature); only for dimensionless parameters can a meaningful 'small-parameter approximation' be made. Nevertheless, as a convenient device to keep track of post-Newtonian effects, such a formal expansion is perfectly fine, enabling us to view the post-Newtonian theory as a formal deformation of its Newtonian counterpart.[11] Note also that for the expansion of some objects, terms of negative order in c^{-1} will appear, such that strictly speaking we are dealing with formal Laurent series. For a series with non-vanishing terms of negative order, the formal Newtonian $c \to \infty$ limit does not exist.

[10] The size of deviations from these reference points is measured by means of the Euclidean metric defined by the (Lorentzian) background metric $g^{(0)}$ and the time evolution vector field u as $g_{\mathrm{Euc}} := g^{(0)} - 2\frac{u^\flat \otimes u^\flat}{g^{(0)}(u,u)}$, where $u^\flat = g^{(0)}(u, \cdot\,)$ is the one-form associated to u via $g^{(0)}$.

[11] In physical realisations of a Newtonian limit, the corresponding expansion parameter has to be chosen as the dimensionless ratio of some typical velocity of the system under consideration to the speed of light. For some discussion of the relationship of formal '$c \to \infty$' limits to actual physical approximations, see, e.g., section II B of reference [38].

In order to obtain a consistent Newtonian limit, we need to identify the time coordinate t (defined as the flow parameter along the background time evolution field u) with Newtonian absolute time. Therefore, we have to treat t as being of order c^0 in our formal post-Newtonian expansion. Note that this implies that the timelike coordinate $x^0 = ct$ with physical dimension of length is to be treated as of order c^1, differently to the spacelike coordinate functions x^a to which we assign order c^0. This necessity of treating the time direction differently from spacelike directions in the consideration of the formal relationship between Poincaré- and Galilei-symmetric theories is well-known: it arises, for example, in the context of Newton–Cartan gravity (i.e. geometrised Newtonian gravity) [39, 40], or in the Lie algebra contraction from the Poincaré to the Galilei algebra [41, 42]. Due to this, for example the Minkowski metric

$$\eta = \eta_{\mu\nu}\,\mathrm{d}x^\mu \otimes \mathrm{d}x^\nu = -c^2\mathrm{d}t^2 + \mathrm{d}\mathbf{x}^2 \tag{16.4}$$

consists of a temporal part of order c^2 and a spatial part of order c^0.

We expand the components of the inverse of the physical spacetime metric as formal power series

$$g^{\mu\nu} = \eta^{\mu\nu} + \sum_{k=1}^{\infty} c^{-k} g_{(k)}^{\mu\nu}\,, \tag{16.5}$$

with the lowest-order terms assumed to be given by the components of the inverse Minkowski metric, and the higher-order coefficients $g_{(k)}^{\mu\nu}$ being arbitrary. Likewise, this could have been given by the corresponding power series expansion of the metric; here we choose to expand the inverse metric for later notational simplicity.

An important example for a post-Newtonian metric is the *Eddington–Robertson parametrised post-Newtonian metric*, given by

$$g^{\text{ER-PPN}} = \left(-1 - 2\frac{\Phi}{c^2} - 2\beta\frac{\Phi^2}{c^4}\right)c^2\mathrm{d}t^2 + \left(1 - 2\gamma\frac{\Phi}{c^2}\right)\mathrm{d}\mathbf{x}^2 + O\!\left(c^{-4}\right). \tag{16.6}$$

Note that this is the power series expansion of the metric itself and not its inverse. The dimensionless *Eddington–Robertson parameters* β and γ account for possible deviations from GR. For the case of GR, which corresponds to the values $\beta = \gamma = 1$, the metric (16.6) solves the Einstein field equations in a c^{-1}-expansion for a static source, with Φ being the Newtonian gravitational potential of the source. The family of metrics for different values of β, γ then form a 'test theory', enabling tests of GR against possible different metric theories of gravity. The Eddington–Robertson metric is the most simple example of a metric in the general parameterised post-Newtonian (henceforth abbreviated by PPN) formalism, which provides a general framework of metric test theories of gravity in the weak-field regime. For a detailed discussion, see [43, 44].

16.2.1.1 Single Particles

Based on the post-Newtonian expansion framework introduced above, we now discuss the systematic description of single, free quantum particles in post-Newtonian gravitational fields. Performing a WKB-inspired formal expansion of classical relativistic field equations, we will arrive at a Schrödinger equation with post-Newtonian corrections [45–49]. We will motivate the consideration of the classical field equations from QFTCS, and briefly explain the expansion of the minimally coupled Klein–Gordon equation as an example. Details of the discussion may be found in [48, 49]. Related discussions for a massive Dirac field are contained in [50, 51], and in terms of a systematic post-Newtonian approximation in Fermi normal coordinates with respect to a rotating frame along an accelerated worldline in [52].

The consideration of the *classical* relativistic field theories for the description of the one-particle sector of the *quantised* theory may be motivated as follows. On a globally hyperbolic, stationary spacetime, for a given relativistic free field theory, there is a preferred Fock space representation of the field observables in QFTCS (and thus a preferred notion of particles). Loosely speaking,[12] the construction of this representation works as follows [1]: We consider classical solutions of the given relativistic field equation, and among those we pick out the subspace of 'positive-frequency solutions' with respect to the stationarity Killing field. Completing the positive-frequency subspace with respect to the 'field inner product' (i.e. the Klein–Gordon inner product for Klein–Gordon fields, or the Dirac inner product for Dirac fields, etc.), we obtain a 'one-particle Hilbert space'. The Hilbert space for the quantum field theory, on which the field operators can be represented, is then the bosonic or fermionic Fock space over this 'one-particle space', according to the spin of the field.

This means that, in the framework of QFTCS, the one-particle sector of a given free relativistic quantum field theory on a globally hyperbolic stationary spacetime is described by the positive-frequency solutions (in an appropriate sense) of the classical field equation, with the corresponding inner product. This is the underlying reason for the effective description of the one-particle sector of the quantum theory by the *classical* field theory—which in the literature is often called consideration of the 'first-quantised theory'—working so well, as long as one is concerned only with processes far enough below the threshold of pair production.

For a non-stationary spacetime, the above motivation of course breaks down: there is no time translation symmetry and, therefore, not even a natural notion of particles in QFTCS. However, as long as we assume the background time evolution vector field u to be *approximately* Killing for the physical spacetime metric g, on a heuristic level we can still expect perturbative 'positive-frequency' solutions of classical field equations to lead to approximately correct predictions regarding the

[12] For details and caveats of the construction, we refer to the extensive discussion in the monograph by Wald [1]; see Sect. 4.3 and references therein for the case of Klein–Gordon fields, as well as section 4.7 for fermionic and other higher-spin fields.

observations of observers moving along the orbits of u. Thus, from QFTCS we are led to the consideration of formal perturbative expansions of relativistic field equations on curved spacetimes as the natural means for the description of single quantum particles in post-Newtonian gravitational fields.

Assuming a formal power series expansion of the inverse physical spacetime metric as in (16.5), we are able to (in principle) obtain the post-Newtonian expansion of any relativistic field equation in complete generality, by making a WKB-inspired power series ansatz for positive-frequency solutions. Using this procedure, we may obtain the Schrödinger equation that describes the corresponding one-particle states to arbitrary post-Newtonian order, in a completely systematic fashion. Specifically, for a minimally coupled scalar field, a brief outline of this procedure is as follows (see [48, 49] for details):

The minimally coupled Klein–Gordon equation is

$$\left(\Box - \frac{m^2c^2}{\hbar^2}\right)\Psi_{\text{KG}} = 0, \tag{16.7}$$

where $\Box = g^{\mu\nu}\nabla_\mu\nabla_\nu$ is the d'Alembert operator of the spacetime metric g. Expressing the covariant derivatives in terms of the components of the metric and inserting the formal power series expansion (16.5), we obtain an expansion of the d'Alembert operator in powers of c^{-1}. In this process, we have to express the expansion coefficients of the *metric* components $g_{\mu\nu}$ in terms of those of the *inverse* metric components $g^{\mu\nu}$, which is possible by use of a formal Neumann series and the Cauchy product formula for products of infinite series. We then make the WKB-inspired ansatz

$$\Psi_{\text{KG}} = \exp\left(\frac{\mathrm{i}c^2}{\hbar}S\right)\psi\,, \qquad \psi = \sum_{k=0}^{\infty} c^{-k}a_k \tag{16.8}$$

for the Klein–Gordon field [47], where we assume S, a_k to be independent of the expansion parameter c^{-1}. Inserting this ansatz into the Klein–Gordon equation (16.7) with expanded d'Alembert operator, we may then compare coefficients of powers of c^{-1} in order to obtain equations for S and the a_k. At the lowest ocurring order c^4, we obtain that S depends solely on time; the equation at order c^3 is then identically satisfied. At order c^2, the equation gives us $(\partial_t S)^2 = m^2$; since we are interested in positive-frequency solutions, we choose

$$S = -mt \tag{16.9}$$

(discarding the constant of integration, which would lead to an irrelevant global phase). This means that for such solutions, the function ψ from (16.8) is the Klein–Gordon field with the 'rest-energy phase factor' $\exp(-\mathrm{i}mc^2t/\hbar)$ separated off. At order c^1, the Klein–Gordon equation then leads to the requirement

$$g^{00}_{(1)} = 0 \tag{16.10}$$

for the metric, which is satisfied in any metric theory of gravity that reproduces the correct equations of motion for test particles in the Newtonian limit.

? Exercise

16.2

(i) Compute the derivatives $\partial_\mu \Psi_{\text{KG}}$ and $\partial_\mu \partial_\nu \Psi_{\text{KG}}$ of the ansatz (16.8) for the Klein–Gordon field.

(ii) Using the derivatives of Ψ_{KG} and the d'Alembert operator $\Box = -c^{-2}\partial_t^2 + \Delta$ in Minkowski spacetime, determine the lowest-order terms in the c^{-1} expansion of the Klein–Gordon equation (16.7). Show that the leading-order equation, at order c^4, leads to $\partial_t S = 0$. Show that the equation at order c^3 is identically satisfied, and use the order c^2 equation to determine S. Finally, show that the c^0 term leads to the free Schrödinger equation for a_0.

(iii) Instead of Minkowski spacetime, we now consider a 'Newtonian' metric given by $g = -(1+2\frac{\phi}{c^2})c^2\mathrm{d}t^2 + \mathrm{d}\mathbf{x}^2 + O(c^{-2})$. For a general metric g, the d'Alembert operator acting on functions is given by

$$\Box f = \nabla^\mu \nabla_\mu f = \frac{1}{\sqrt{-g}} \partial_\mu (\sqrt{-g} g^{\mu\nu} \partial_\nu f),$$

where $\sqrt{-g}$ denotes the square root of minus the determinant of the matrix of metric components. Compute $\Box f$ for the Newtonian metric to order c^{-2} in terms of partial space and time derivatives ∂_a, ∂_t. (Be careful: $\partial_0 = \frac{\partial}{\partial(ct)} = c^{-1}\partial_t$!)

Analogously to the Minkowski calculation, show that the c^{-1} expansion of the Klein–Gordon equation in the Newtonian metric to order c^0 leads to a Schrödinger equation for a_0 including the Newtonian potential ϕ.

Hint: the inverse metric has components $g^{00} = -1 + 2\frac{\phi}{c^2} + (c^{-4})$, $g^{0a} = (c^{-3})$, $g^{ab} = \delta^{ab} + (c^{-2})$. *For the determinant term, use* $\frac{1}{\sqrt{-g}}\partial_\mu\sqrt{-g} = -\frac{1}{2} g_{\rho\sigma}\partial_\mu g^{\rho\sigma}$.

Using these results, we can then obtain a general, expanded version of the Klein–Gordon equation. This fully expanded equation is rather horrendously complicated (see the appendix of [48]), but it can nevertheless be written down in its entirety. From it, we may read off, order by order, equations for the coefficient functions a_k. After obtaining the equations for all the a_k up to a fixed order n,[13] we may recombine them into an equation for ψ to order c^{-n}, which takes the form of a Schrödinger equation plus higher-order post-Newtonian corrections. The inner product on the Hilbert space in which the 'wave functions' ψ described by this

[13] In this process, when considering higher orders, the equations for a_k begin to involve time derivatives of the lower-order functions a_l. To eliminate those, we have to re-use the already derived equations for the lower-order a_l, in order to in the end obtain an equation for ψ which takes the form of a Schrödinger equation.

Schrödinger equation live can be obtained by inserting the ansatz (16.8) into the Klein–Gordon inner product and expanding it to the desired order in c^{-1}. Since the Klein–Gordon inner product involves time derivatives of the fields, we have to use the derived Schrödinger equation in this process. After this has been done, we have a concrete representation of the (approximate) one-particle sector of the Klein–Gordon quantum field theory in our post-Newtonian spacetime in terms of a Schrödinger equation and an inner product for 'wave functions' ψ living on three-dimensional space as defined by our background structures. Regarding the interpretation of the natural position operator that we obtain for the post-Newtonian theory, which in this representation of the Hilbert space is given by multiplication of wave functions by coordinate position x^a, we note that it arises—up to higher-order corrections—from the operator in the one-particle sector of Klein–Gordon theory which multiplies the Klein–Gordon fields by coordinate position x^a. In the case of Minkowski spacetime, this operator is the well-known Newton–Wigner position operator [53, 54].

As a last step, one may perform a unitary transformation such that, in the new representation, the position operator is still given by multiplication with x^a, but the inner product of the theory takes the form of a 'flat' L^2 inner product

$$\langle \psi_\mathrm{f}, \varphi_\mathrm{f} \rangle_\mathrm{f} := \int \mathrm{d}^3x \, \overline{\psi_\mathrm{f}} \, \varphi_\mathrm{f} \tag{16.11}$$

over coordinate space $\{x^a\}$ (instead of the expanded Klein–Gordon inner product) [46, 48]. This means that the 'flat wave functions' ψ_f in this representation are in fact scalar *densities* on three-space. The necessary transformation may be read off order by order from the expanded Klein–Gordon inner product, and may then be used to compute the Schrödinger equation in this 'flat' representation.

Concretely, to first order in c^{-1} the general Schrödinger equation thus obtained, in the 'flat' representation, takes the form

$$\begin{aligned} \mathrm{i}\hbar\partial_t\psi_\mathrm{f} = \Bigg[&-\frac{\hbar^2}{2m}\nabla^2 - \frac{1}{2}\left\{g^{0a}_{(1)}, -\mathrm{i}\hbar\partial_a\right\} + \frac{m}{2}g^{00}_{(2)} \\ &+ c^{-1}\bigg(\frac{1}{2m}(-\mathrm{i}\hbar)\partial_a\Big(g^{ab}_{(1)}(-\mathrm{i}\hbar)\partial_b\Big) \\ &+ \frac{m}{2}g^{00}_{(3)} - \frac{1}{2}\left\{g^{0a}_{(2)}, -\mathrm{i}\hbar\partial_a\right\} - \frac{\hbar^2}{8m}[\nabla^2 \mathrm{tr}(\eta g^{-1}_{(1)})]\bigg) + O\Big(c^{-2}\Big)\Bigg]\psi_\mathrm{f}, \end{aligned} \tag{16.12}$$

where $\nabla^2 = \delta^{ab}\partial_a\partial_b$ is the 'flat' background Laplacian operator, and $\{A, B\} = AB + BA$ denotes the anticommutator. Comparing this result to the c^{-1}-expanded classical Hamiltonian of a point particle in a curved metric, it turns out that (adopting a specific ordering scheme) all terms apart from the last one may be obtained by naive canonical quantisation of that classical Hamiltonian.

For concrete post-Newtonian metrics, some of the generic expansion coefficients from (16.5) vanish, which simplifies the necessary computations. For the case of the Eddington–Robertson PPN metric (16.6), the derivation is not difficult (but somewhat tedious) to carry out completely, resulting in the Hamiltonian

$$\begin{aligned} H_{\mathrm{f}}^{\text{ER–PPN}} &= -\frac{\hbar^2}{2m}\boldsymbol{\nabla}^2 + m\Phi + \frac{1}{c^2}\bigg(-\frac{\hbar^4}{8m^3}\nabla^4 - \frac{3\hbar^2}{4m}\gamma(\boldsymbol{\nabla}^2\Phi) - \frac{\hbar^2}{2m}(2\gamma+1)\Phi\boldsymbol{\nabla}^2 \\ &\quad - \frac{\hbar^2}{2m}(2\gamma+1)(\partial_a\Phi)\delta^{ab}\partial_b + (2\beta-1)\frac{m\Phi^2}{2}\bigg) + O\big(c^{-4}\big) \end{aligned} \tag{16.13a}$$

in the 'flat' representation.[14] Comparing this to the classical Hamiltonian for a point particle in the Eddington–Robertson metric

$$H_{\text{class}}^{\text{ER–PPN}} = \frac{\mathbf{p}^2}{2m} + m\Phi + \frac{1}{c^2}\bigg(-\frac{\mathbf{p}^4}{8m^3} + \frac{2\gamma+1}{2m}\mathbf{p}^2\Phi + (2\beta-1)\frac{m\Phi^2}{2}\bigg) + O\big(c^{-4}\big), \tag{16.13b}$$

we see that by canonical quantisation we would not in general be able to reproduce the quantum Hamiltonian by using a specific ordering scheme, but would have to choose the ordering scheme depending on the value of γ. However, this ambiguity is due to the term in (16.13a) proportional to $\boldsymbol{\nabla}^2\Phi$, which by the Newtonian gravitational field equation is proportional to the mass density of the matter generating the gravitational field. Therefore, when the quantum system is localised outside of the generating matter, we may describe it by simple canonical quantisation of the classical point-particle theory, quantising the $\mathbf{p}^2\Phi$ term in the 'obvious' symmetric ordering $\mathbf{p}\cdot\Phi\mathbf{p}$.

Conclusion 16.1

Formal c^{-1}-expansions of classical relativistic wave equations in post-Newtonian spacetimes provide a systematic method for the description of single free quantum particles in post-Newtonian gravitational background fields, motivated from the framework of QFTCS.

[14] We note that this Hamiltonian may also be obtained by a very similar, but more explicitly WKB-like method in which the logarithm of the Klein–Gordon field is expanded in c^{-1}; see [46].

> **Conclusion 16.2**

For the description of single free spin-0 quantum particles in a background post-Newtonian gravitational field as described by the Eddington–Robertson PPN metric, naive canonical quantisation of free point particle dynamics gives the same result as QFTCS-inspired post-Newtonian expansion of the Klein–Gordon equation, up to terms that vanish outside the matter distribution generating the gravitational field.

16.2.1.2 Composite Systems

In this section, we will describe a systematic method for the derivation of a complete Hamiltonian description of an electromagnetically bound two-particle quantum system in a post-Newtonian gravitational background field described by the Eddington–Robertson PPN metric to order c^{-2}. Extending the systematic calculation by Sonnleitner and Barnett for the non-gravitational case [55] to the gravitational situation, this provides a properly relativistic derivation of the coupling of composite quantum systems to post-Newtonian gravity, without the need to 'guess' this coupling based on heuristic 'relativistic effects'. We aim to keep the present discussion rather short; full details may be found in chapter 4 of [49] and, in a somewhat less general form, in [56].

The idea of the non-gravitational derivation by Sonnleitner and Barnett, taking place in Minkowski spacetime, is as follows [55] (compare Fig. 16.1). Sonnleitner and Barnett start from the classical special-relativistic Lagrangian describing two point particles of arbitrary masses and opposite and equal electric charges, interacting with electromagnetic potentials. They then split the electromagnetic potentials into 'internal' (i.e. generated by the particles) and 'external' parts, employ the Coulomb gauge, and solve the Maxwell equations for the internal part to lowest order in c^{-1}, expressing the solutions in terms of the particles' positions

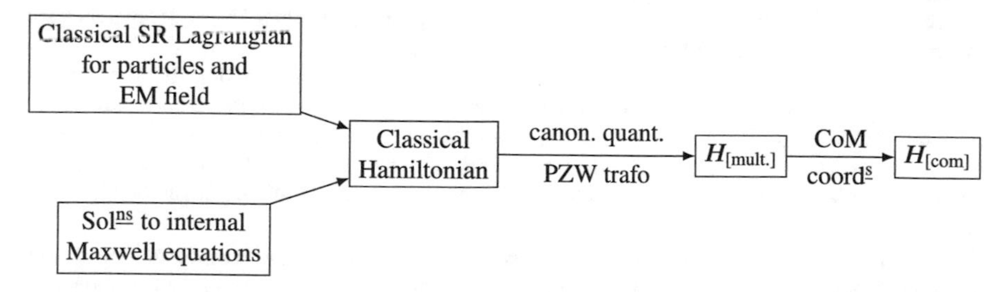

Fig. 16.1 Strategy of the derivation of an 'approximately relativistic' Hamiltonian for a toy hydrogenic system without gravity by Sonnleitner and Barnett in [55]

and velocities. Re-inserting the internal solutions into the classical Lagrangian,[15] expanding it to order c^{-2}, and performing a Legendre transformation, they arrive at a classical Hamiltonian describing the system in the post-Newtonian regime. By canonically quantising this classical Hamiltonian and performing a Power–Zienau–Woolley (PZW) unitary transformation together with a multipolar expansion of the external electromagnetic field, they then obtain a quantum 'multipolar Hamiltonian' $H_{[\text{mult.}]}$. This describes the two particles as quantum particles, interacting with each other through the (eliminated) internal and with the external electromagnetic fields. The PZW transformation is a standard method in quantum optics, used to transform a Hamiltonian describing interactions of particles with the electromagnetic field from a minimally coupled form in terms of the potentials to a multipolar form in terms of the field strengths. Finally, Sonnleitner and Barnett introduce Newtonian centre of mass and relative coordinates and the corresponding canonical momenta, arriving at a 'centre of mass Hamiltonian' $H_{[\text{com}]}$. This has an interpretation in terms of central and internal dynamics of a 'composite particle', coupled to each other via the formal replacement $m \to M + H_{\text{int}}/c^2$ in the central Hamiltonian as in the heuristic motivation (16.3), as well as to the external electromagnetic field.[16]

[15] In the method as presented by Sonnleitner and Barnett in [55], a small inconsistency arises at this point. If we want to keep the external vector potential as a dynamical variable in the action, then on the one hand, its equations of motion have to be the vacuum Maxwell equations, while on the other hand, it has to enter the equations of motion of the particles. For the Lagrangian which arises from directly inserting the internal potentials, this is indeed the case: variation of the corresponding action leads to the desired equations of motion. However, this Lagrangian contains second-order time derivatives of the particle positions, such that one cannot employ conventional Hamiltonian formalism.

Sonnleitner and Barnett disregard the second-order time derivative terms, arguing that they are related to formally diverging backreaction terms. However, *this* is again problematic: these terms would have been the ones ensuring the vacuum Maxwell equations as equations of motion for the external potential; without them, the Lagrangian gives, again, the *sourced* Maxwell equations for the external potential, and the formalism becomes inconsistent.

We may avoid this inconsistency by simply treating the external electromagnetic potentials as given *background* fields instead of dynamical variables. This ensures the consistency of the equations of motion while still allowing us to perform the internal–external field split. An extensive discussion of this point may be found in chapter 4 of [49]. Note that this inconsistency was addressed neither by Sonnleitner and Barnett in [55] nor in our (two of the present authors') article [56].

[16] In addition to the 'internal' part of the Hamiltonian describing the internal motion of the atom, the 'central' part describing the motion of the centre-of-mass degrees of freedom together with the expected internal–external coupling, and the part describing the interaction with the external electromagnetic field, there arises a 'cross term' Hamiltonian H_{X}. This contains additional couplings between the internal degrees of freedom and the central momentum. Sonnleitner and Barnett continue with the construction of a canonical transformation that eliminates these cross terms. Of course, for a thorough description of physical situations, e.g. in experiments, one cannot simply *assume* that the resulting new coordinates correspond to the physically realised observables, but has to be careful to express all results in terms of operationally clearly defined quantities. A similar issue will become relevant in the gravitational case, as discussed below.

We are now going to explain how to extend this method in order to include a weak gravitational background field, described by the Eddington–Robertson PPN metric (16.6). Our geometric post-Newtonian expansion scheme described in the beginning of Sect. 16.2.1, based on the background Minkowski metric η and the time evolution field u, allows us to include gravity very easily, at least conceptually speaking: in our adapted coordinates (ct, x^a) we may repeat all the steps of the derivation by Sonnleitner and Barnett, the only difference being that we now start from the general-relativistic action describing our particles and fields *in the curved spacetime geometry* as given by $g^{\text{ER-PPN}}$. This leads to 'gravitational corrections' being included in virtually all steps of the derivation, i.e. additional terms including the gravitational potential Φ. Finally, geometric operations appearing in the resulting Hamiltonian may be re-written in terms of the physical spacetime metric, whereby some terms obtain a somewhat more intuitive interpretation in terms of metric quantities.

The action we start from, in which our particles and fields are minimally coupled to the spacetime metric $g = g^{\text{ER-PPN}}$, reads as follows:

$$\begin{aligned} S_{\text{total}} &= -c^2 \sum_{i=1}^{2} m_i \int \mathrm{d}t \sqrt{-g_{\mu\nu} \dot{r}_i^{\mu} \dot{r}_i^{\nu} / c^2} \\ &\quad + \int \mathrm{d}t \, \mathrm{d}^3 x \, \sqrt{-g} \left(-\frac{\varepsilon_0 c^2}{4} F_{\mu\nu} F^{\mu\nu} + J^{\mu} A_{\mu} \right) . \end{aligned} \tag{16.14}$$

Here, m_i are the masses of the particles, $r_i^{\mu}(t)$ the coordinates of their worldlines, $\sqrt{-g}$ denotes the square root of minus the determinant of the matrix of metric components, A_{μ} are the components of the total electromagnetic four-potential, $F_{\mu\nu} = \partial_{\mu} A_{\nu} - \partial_{\nu} A_{\mu}$ are the components of the field strength, and $J^{\mu} = j^{\mu} / \sqrt{-g}$ are the components of the current four-vector field of the particles. The current density j, which is a vector field density, is given by

$$j^{\mu}(t, \mathbf{x}) = \sum_{i=1}^{2} e_i \delta^{(3)}(\mathbf{x} - \mathbf{r}_i(t)) \dot{r}_i^{\mu}(t), \tag{16.15}$$

where $e_1 = -e_2 =: e$ are the electric charges of the particles and $\delta^{(3)}$ is the three-dimensional Dirac delta distribution.

By inserting the Eddington–Robertson metric into the kinetic terms of the particles and expanding in c^{-1}, we obtain 'gravitational corrections' to the kinetic terms. These may then very easily be directly included into the derivation by Sonnleitner and Barnett. For the electromagnetic fields as well, the computation does not pose any intrinsic difficulties: the Maxwell equations in the gravitational field may be written in terms of the gravity-free Maxwell equations, which allows to perturbatively include the gravitational effects on top of the 'internal' potential solutions from the non-gravitational case. The derivation of this result however turns out to be quite lengthy, in particular if we do not neglect derivatives of the gravitational potential Φ. Full details of the calculations may be found in chapter 4 of [49].[17]

The resulting Hamiltonian takes the symbolic form[18]

$$H_{[\text{com}],\text{total}} = H_{\text{C}} + H_{\text{A}} + H_{\text{AL}} + O\left(c^{-4}\right), \tag{16.16a}$$

where H_{AL} describes the atom-light interaction, i.e. the interaction of our system with the external electromagnetic fields, H_{C} can be interpreted as describing the dynamics of the central degrees of freedom, and H_{A} the internal atomic motion. The latter two are given by

$$\begin{aligned} H_{\text{C}} = &\frac{\mathbf{P}^2}{2M}\left[1 - \frac{1}{Mc^2}\left(\frac{\mathbf{p}^2}{2\mu} - \frac{e^2}{4\pi\varepsilon_0 r}\right)\right] + \left[M + \frac{1}{c^2}\left(\frac{\mathbf{p}^2}{2\mu} - \frac{e^2}{4\pi\varepsilon_0 r}\right)\right]\Phi(\mathbf{R}) \\ &- \frac{\mathbf{P}^4}{8M^3c^2} + \frac{2\gamma+1}{2Mc^2}\mathbf{P}\cdot\Phi(\mathbf{R})\mathbf{P} + (2\beta - 1)\frac{M\Phi(\mathbf{R})^2}{2c^2}, \end{aligned} \tag{16.16b}$$

$$\begin{aligned} H_{\text{A}} = &\left(1 + 2\gamma\frac{\Phi(\mathbf{R})}{c^2}\right)\frac{\mathbf{p}^2}{2\mu} - \left(1 + \gamma\frac{\Phi(\mathbf{R})}{c^2}\right)\frac{e^2}{4\pi\varepsilon_0 r} \\ &- \frac{M - 3\mu}{M}\frac{\mathbf{p}^4}{8\mu^3c^2} - \frac{e^2}{4\pi\varepsilon_0}\frac{1}{2\mu Mc^2}\left(\mathbf{p}\cdot\frac{1}{r}\mathbf{p} + \mathbf{p}\cdot\mathbf{r}\frac{1}{r^3}\mathbf{r}\cdot\mathbf{p}\right) \\ &- \frac{2\gamma+1}{2Mc^2}\frac{\Delta m}{\mu}\mathbf{p}\cdot(\mathbf{r}\cdot\nabla\Phi(\mathbf{R}))\mathbf{p} + (\gamma+1)\frac{e^2}{4\pi\varepsilon_0 r}\frac{\Delta m}{2Mc^2}\mathbf{r}\cdot\nabla\Phi(\mathbf{R}), \end{aligned} \tag{16.16c}$$

where $\mathbf{R}, \mathbf{P}$ are the central position and momentum, $\mathbf{r}, \mathbf{p}$ are the relative position and momentum, M, μ are the total and reduced mass of the system, respectively, and $\Delta m = m_1 - m_2$ the mass difference [49]. In these expressions, 'dot products' of three-vectors are taken with respect to the *flat* metric δ induced on three-space by the background Minkowski metric η, e.g. $r = \sqrt{\delta_{ab}r^a r^b}$ and $\mathbf{p}^2 = \delta^{ab}p_a p_b$.

The 'physical spatial metric' ${}^{(3)}g$, i.e. the metric on three-space induced by the *physical* spacetime metric $g = g^{\text{ER–PPN}}$, is given as

$${}^{(3)}g = \left(1 - 2\gamma\frac{\Phi}{c^2}\right)\delta + O\left(c^{-4}\right) \tag{16.17a}$$

(compare the Eddington–Robertson metric (16.6)), from which we see that its inverse is

$${}^{(3)}g^{-1} = \left(1 + 2\gamma\frac{\Phi}{c^2}\right)\delta^{-1} + O\left(c^{-4}\right). \tag{16.17b}$$

[17] Note that for simplicity, in our article [56] we neglected derivatives of Φ in the treatment of the electromagnetic fields.

[18] To simplify the presentation, here we leave out the 'cross terms' H_{X} that arise in our gravitational case exactly as in the non-gravitational case; see footnote 16 on page 512. We also leave out an additional cross term $\frac{2\gamma+1}{2Mc^2}[\mathbf{P}\cdot(\mathbf{r}\cdot\nabla\Phi(\mathbf{R}))\mathbf{p_r} + \text{H.c.}]$ involving the derivative of the gravitational potential, since it is irrelevant for the following general discussion.

Thus we may rewrite the kinetic and Coulomb interaction terms from H_{A} in (16.16c) in terms of ${}^{(3)}g$ as

$$\left(1+2\gamma\frac{\Phi(\mathbf{R})}{c^2}\right)\frac{\mathbf{p}^2}{2\mu}=\frac{{}^{(3)}g^{-1}[\mathbf{R}](\mathbf{p},\mathbf{p})}{2\mu}+O\left(c^{-4}\right), \tag{16.18a}$$

$$-\left(1+\gamma\frac{\Phi(\mathbf{R})}{c^2}\right)\frac{e^2}{4\pi\varepsilon_0 r}=-\frac{e^2}{4\pi\varepsilon_0\sqrt{{}^{(3)}g[\mathbf{R}](\mathbf{r},\mathbf{r})}}+O\left(c^{-4}\right). \tag{16.18b}$$

In this form, these terms look like the usual kinetic and Coulomb interaction energies from atomic physics in Galilei-invariant QM, only with the Euclidean metric used to measure distance and momentum-squared replaced by the physical spatial metric. Thus the internal atomic Hamiltonian takes the form

$$H_{\mathrm{A}}=\frac{{}^{(3)}g^{-1}[\mathbf{R}](\mathbf{p},\mathbf{p})}{2\mu}-\frac{e^2}{4\pi\varepsilon_0\sqrt{{}^{(3)}g[\mathbf{R}](\mathbf{r},\mathbf{r})}}$$
$$+\left(c^{-2}\text{ SR \& `Darwin' corrections}+c^{-2}\nabla\Phi\text{ term}\right)+O\left(c^{-4}\right). \tag{16.19a}$$

The internal–external coupling terms that we have included into the central Hamiltonian H_{C} in (16.16b) may then be expressed in terms of H_{A}, leading to

$$H_{\mathrm{C}}=\frac{\mathbf{P}^2}{2M}\left(1-\frac{H_{\mathrm{A}}}{Mc^2}\right)+\left(M+\frac{H_{\mathrm{A}}}{c^2}\right)\Phi(\mathbf{R})$$
$$-\frac{\mathbf{P}^4}{8M^3c^2}+\frac{2\gamma+1}{2Mc^2}\mathbf{P}\cdot\Phi(\mathbf{R})\mathbf{P}+(2\beta-1)\frac{M\Phi(\mathbf{R})^2}{2c^2}+O\left(c^{-4}\right). \tag{16.19b}$$

Comparing this to the Hamiltonian for a single particle in the Eddington–Robertson metric (16.13), we see that this has the form

$$H_{\mathrm{C}}=H^{\text{ER–PPN}}_{\text{single particle}}(\mathbf{P},\mathbf{Q},M+H_{\mathrm{A}}/c^2)+O\left(c^{-4}\right). \tag{16.20}$$

Thus, starting from first principles our calculation supports the heuristic picture of a 'composite point particle' whose mass is given, according to the 'mass defect', by the total rest mass plus the internal energy divided by c^2.

Conclusion 16.3

Composite quantum systems in post-Newtonian gravity can be described in a systematic and properly relativistic way, starting from well-understood first principles. The results confirm, to some extent, the heuristically motivated internal–external couplings from the 'mass defect' picture as in (16.3).

Note that this interpretation in terms of a 'composite point particle' whose internal energy contributes to its mass—the 'inertial' one in the kinetic term as well as the 'gravitational' one coupling to Φ—depends crucially on the rewriting (16.18) of the internal atomic energies in terms of the momentum-squared and distance as measured with the 'physical spatial metric' ${}^{(3)}g$. Instead, we could have chosen to interpret as internal kinetic and interaction energies the terms

$$H_{\text{A,kin.}}^{\text{backgr.}} = \frac{\delta^{-1}(\mathbf{p},\mathbf{p})}{2\mu} = \frac{\mathbf{p}^2}{2\mu}, \tag{16.21a}$$

$$H_{\text{A,interact.}}^{\text{backgr.}} = -\frac{e^2}{4\pi\varepsilon_0\sqrt{\delta(\mathbf{r},\mathbf{r})}} = -\frac{e^2}{4\pi\varepsilon_0 r} \tag{16.21b}$$

expressed with the *background* spatial metric δ. This would have meant to take as internal energy the part

$$\begin{aligned} H_{\text{A}}^{\text{backgr.}} &= \frac{\mathbf{p}^2}{2\mu} - \frac{e^2}{4\pi\varepsilon_0 r} + (c^{-2}\text{ SR \& 'Darwin' corrections} \\ &\quad + c^{-2}\nabla\Phi\text{ term}) + O\left(c^{-4}\right) \end{aligned} \tag{16.22}$$

of the total Hamiltonian. Then we would have had to interpret the left-over γ terms from H_{A} as part of the central Hamiltonian, giving it the form

$$\begin{aligned} H_{\text{C}}^{\text{backgr.}} &= H_{\text{C}} + 2\gamma\frac{\Phi(\mathbf{R})}{c^2}\frac{\mathbf{p}^2}{2\mu} - \gamma\frac{\Phi(\mathbf{R})}{c^2}\frac{e^2}{4\pi\varepsilon_0 r} \\ &= \frac{\mathbf{P}^2}{2M}\left(1 - \frac{H_{\text{A}}^{\text{backgr.}}}{Mc^2}\right) + \left[M + \frac{H_{\text{A}}^{\text{backgr.}}}{c^2} + \frac{\gamma}{c^2}\left(2\frac{\mathbf{p}^2}{2\mu} - \frac{e^2}{4\pi\varepsilon_0 r}\right)\right]\Phi(\mathbf{R}) \\ &\quad - \frac{\mathbf{P}^4}{8M^3c^2} + \frac{2\gamma+1}{2Mc^2}\mathbf{P}\cdot\Phi(\mathbf{R})\mathbf{P} + (2\beta-1)\frac{M\Phi(\mathbf{R})^2}{2c^2} + O\left(c^{-4}\right). \end{aligned} \tag{16.23}$$

Naively looking at this equation, we could have concluded that not only does the internal energy (16.22) of the composite system contribute differently to the inertial and the (passive) gravitational masses of the composite particle, but also that internal kinetic and interaction energies (16.21) contribute differently to the (passive) gravitational mass. That is, we might have concluded an 'anomalous coupling' of internal energies to the gravitational potential Φ. Note that, due to the virial theorem, the time average of the additional coupling term $2H_{\text{A,kin.}}^{\text{backgr.}} + H_{\text{A,interact.}}^{\text{backgr.}}$ vanishes, which for stationary states solves the apparent interpretational tension regarding the 'composite point particle' picture [37, 57, 58].

One might be tempted to argue that 'real' physical distances and times as measured by 'rods and clocks' are the ones as defined by the physical spacetime metric g, and that therefore the 'correct' way to express internal energies is in terms of metric quantities with respect to g (as we did before in (16.18)), which makes the 'anomalous' couplings go away. However, this supposed argument is not sufficient to argue for or against the absence of such couplings in all physical situations: just because the *Hamiltonian* looks natural when expressed in terms of g, we do not know anything about the *state* of the system. The state might, a priori, have been prepared in a way that is sensitive not only to lengths as defined by g, but also to some other geometric structures.[19] For a proper analysis of physical situations, e.g. in experiments, one has (in principle) to describe the whole situation, including all preparation and measurement procedures, in terms of operationally defined quantities, and express all predicted results in terms of these operational quantities.

> **Conclusion 16.4**

Although metric lengths are usually interpreted as being measured by 'rods and clocks', they are not necessarily operationally realised in physical situations. Therefore, for a proper assessment if 'anomalous' couplings of internal energies to the gravitational potential are physically relevant, it is not sufficient to show that they are eliminated from the Hamiltonian by rewriting the coupling in terms of metric lengths—a proper analysis needs to describe the *whole* situation in terms of operationally defined quantities.

16.2.2 Seemingly Anomalous Couplings of Internal Energies—A Perspective from Diffeomorphism Invariance

Towards the end of the previous section, we have seen a specific example of the emergence of 'anomalous' couplings of internal energies to the Newtonian potential in the systematic (post-)Newtonian description of composite systems in gravitational fields: when the internal energies are written in terms of metric quantities with respect to the background metric, defining the 'absence' of gravitational fields, the 'gravitational mass' multiplying the Newtonian potential in lowest order contains, in addition to the rest mass, not only a contribution of total internal energy divided by c^2, but an additional contribution proportional to (2 kinetic energy + potential energy). When the internal energies are expressed in terms of metric quantities with respect to the *physical* metric, instead, these 'anomalous' couplings disappear from the Hamiltonian (since they are 'absorbed' into the definitions of the metric lengths). We also noticed that the time average of the additional coupling term vanishes due to the virial theorem.

[19] For example, spatial light propagation in a static spacetime is described by the so-called *optical metric*, which is determined by the spacetime metric g and the staticity Killing field.

Such apparently 'anomalous' coupling terms are not special to quantum-theoretic descriptions of composite systems. They typically appear in post-Newtonian descriptions of the dynamics of composite systems in GR. This was already noted early in the history of GR, like, e.g., in 1938 by Eddington and Clark in their approximate formulation of the dynamics of n gravitationally interacting point masses [59]. They found an expression for the total active gravitational mass of the system (by looking at the asymptotic form of the metric) that was given by the sum of the individual masses, *three* times the internal kinetic energy, and *twice* the internal potential energy. This differs from the simple sum of all three energies (rest, kinetic, and potential) by (2 kinetic energy + potential energy), so as if the active gravitational mass did not match the sum of all energies (divided by c^2). However, as they also immediately noted, and as was well known from the Newtonian theory of gravitating point masses [60], the apparent excess energy (2 kinetic energy + potential energy) equals the second time-derivative of the total moment of inertia, which (being a total time derivative) clearly vanishes for bounded motions upon forming the time average.

In this section we aim to give more general arguments on the appearance of such seemingly 'anomalous' couplings and the possibility to 'hide' them by coordinate redefinitions. Our exposition is based on a discussion by Carlip [37], in which he speaks of employing 'general covariance' to argue about the 'anomalous' couplings. We will make the underlying assumptions of this argumentation more explicit, in particular emphasising the importance of background structures for the arguments—namely, the background metric η and time evolution field u, as introduced in the beginning of Sect. 16.2.1. Apart from being essential for the very definition of the notion of 'weak fields', this background structure is also important when considering the matter energy-momentum tensor, which is only defined *with respect to a metric*.

In order to be more precise, we will avoid the somewhat diffuse terminology 'general covariance' and speak of *diffeomorphism invariance* instead. In fact, this is more than just linguistic pedantry: as we will see, it is an important conceptual point that we consider *active* diffeomorphisms rather than just passive changes of coordinates. The main difference is that the latter necessarily affect *all* fields, whereas the former may be selectively applied to *some* fields, while leaving invariant others. This we apply, e.g., to keep a meaningful distinction between dynamic fields, which get acted upon by the diffeomorphisms, and background fields, which are left invariant. Furthermore, we make precise the claim [37] that by arguments based on the vanishing of 'anomalous' coupling terms, one may prove the special-relativistic virial theorem.

Note that for the arguments in this section, quantisation of matter is not needed, and we will argue in the language of classical Lagrangian field theory.

16.2.2.1 The Mathematical Setting

Let $\mathcal{M}$ be the spacetime manifold. We will consider matter fields Ψ which are sections in some *natural* vector bundle E on $\mathcal{M}$. We emphasise the condition of naturality, which implies that each diffeomorphism $\varphi \in \mathrm{Diff}(\mathcal{M})$ has a naturally associated lift to the total space E [61]. More precisely, this means that associated to each $\varphi \in \mathrm{Diff}(\mathcal{M})$ we have a unique vector bundle isomorphism $(\hat{\varphi}, \varphi) : E \to E$ consisting of a family of linear isomorphisms

$$\hat{\varphi}_p : E_p \to E_{\varphi(p)} \tag{16.24}$$

depending smoothly on the point $p \in \mathcal{M}$, in such a way as to satisfy $\widehat{\mathrm{id}_{\mathcal{M}}} = \mathrm{id}_E$ and $\widehat{\varphi_1 \circ \varphi_2} = \hat{\varphi}_1 \circ \hat{\varphi}_2$. We also assume that $\hat{\varphi}$ depends 'smoothly' on φ in a weak sense that will be made precise below. Examples of natural bundles are all vector bundles associated to the (general linear) frame bundle of $\mathcal{M}$.[20] The space of sections $\Gamma(E)$ then carries a natural pushforward action of the diffeomorphism group,

$$\mathrm{Diff}(\mathcal{M}) \times \Gamma(E) \ni (\varphi, \Psi) \mapsto \varphi_* \Psi := \hat{\varphi} \circ \Psi \circ \varphi^{-1} \in \Gamma(E). \tag{16.25}$$

It is this pushforward that we need to be 'smooth' in φ in a weak sense: for φ_t the local flow of any vector field X, we assume $(\varphi_t)_* \Psi$ to depend smoothly on t, such that the Lie derivative $\mathcal{L}_X \Psi = \frac{\mathrm{d}}{\mathrm{d}t} (\varphi_{-t})_* \Psi \Big|_{t=0}$ exists and we have

$$(\varphi_t)_* \Psi = \Psi - t \mathcal{L}_X \Psi + O\big(t^2\big). \tag{16.26}$$

Let now $S : \mathrm{Lor}(\mathcal{M}) \times \Gamma(E) \times O(\mathcal{M}) \to \mathbb{R}$, $(g, \Psi, V) \mapsto S[g, \Psi; V]$ be the matter action, taking as inputs a Lorentzian metric g on $\mathcal{M}$, a matter configuration Ψ, and an open 'region of integration' V in $\mathcal{M}$. (We do not necessarily need the action be defined as an integral over some Lagrangian density, as long as the assumptions we make in the following make sense and are satisfied. However, action integrals in this usual sense provide the most important class of actions.) Usually we will take the integration region V infinite in space but finite in time, such as to render the action finite. We define the matter energy-momentum tensor by the functional derivative[21]

$$\frac{\delta S}{\delta g_{\mu\nu}}[g, \Psi; V] =: \frac{1}{2c} \sqrt{|\det g|}\, T^{\mu\nu}[g, \Psi], \tag{16.27}$$

[20] Note that a priori this excludes spinor fields, as spinor bundles do not admit natural lifts of general diffeomorphisms, but only of paths of isometries that start at the identity, being associated to the bundle of 'spin frames', i.e. a spin structure, which is a double cover of the bundle of *orthonormal* frames. Below, most of our arguments employ 'infinitesimal' diffeomorphisms, i.e. Lie derivatives of fields with respect to vector fields. Here the same comment applies. Lie derivatives of sections in E only exist if vector fields on $\mathcal{M}$ lift to vector fields on the total space of the frame bundle. For spinor fields, this is a priori not the case, unless the vector field generates an isometry (i.e., is a Killing field). Lie derivatives for spinor fields with respect to general vector fields can be defined relative to an extra prescription for lifting the vector field to the frame bundle [62, 63].

[21] This is simply the Hilbert energy-momentum tensor, i.e. the energy-momentum tensor which would appear on the right-hand side of Einstein's equations if we added the Einstein–Hilbert action as the dynamical action for the metric g.

i.e., by demanding that it satisfy

$$
\begin{aligned}
S[g+\tilde{g}, \Psi; V] &= S[g, \Psi; V] + \frac{1}{2c}\int_V \mathrm{d}^4x\sqrt{|\det g|}\, T^{\mu\nu}[g, \Psi]\tilde{g}_{\mu\nu} + O\left(\tilde{g}^2\right) \\
&= S[g, \Psi; V] + \frac{1}{2c}\int_V \mathrm{dvol}_g\, T^{\mu\nu}[g, \Psi]\tilde{g}_{\mu\nu} + O\left(\tilde{g}^2\right)
\end{aligned}
\tag{16.28}
$$

for any sufficiently small symmetric two-tensor $\tilde{g}$, where dvol_g denotes the Riemannian volume form of g. Note that the assumption of this functional derivative being well-defined and independent of the integration region V amounts to a restriction on the possible matter actions S—for example, double integrals over V are excluded. We also assume the energy-momentum tensor to depend *locally* on the matter fields, meaning that $\mathrm{supp}(T[g, \Psi]) \subseteq \mathrm{supp}(\Psi)$.[22] For example, this holds for any action that is an integral over a Lagrangian density containing at most finitely many derivatives of Ψ.

Our fundamental assumption is that the matter action be diffeomorphism invariant, i.e. that for any diffeomorphism $\varphi \in \mathrm{Diff}(\mathcal{M})$ we have

$$
S[\varphi_* g, \varphi_* \Psi; \varphi(V)] = S[g, \Psi; V] \tag{16.29}
$$

for all g, Ψ, V.[23] Note that for a theory formulated in differential-geometric terms, this will essentially always be the case, as long as no background structures enter the definition of the action S.

In order to analyse the coupling of the matter fields to weak post-Newtonian gravity, we again need background structures in order to define the notions of weak gravity and small velocities. As in Sect. 16.2.1, we take as background spacetime Minkowski spacetime $(\mathcal{M}, \eta)$, and a background time evolution vector field u on it that is geodesic with respect to η and has Minkowski square $\eta(u, u) = -c^2$. As before, for convenience we choose global Lorentzian coordinates $(x^0 = ct, x^a)$ on $(\mathcal{M}, \eta)$ adapted to u, i.e. coordinates such that $(\eta_{\mu\nu}) = \mathrm{diag}(-1, 1, 1, 1)$ and $u = \partial/\partial t$, in which we will work and to which we will refer all components

[22] Recall that $\mathrm{supp}(f)$ denotes the closure of the set of points at which f assumes a non-zero value. Hence, the complement of the support is open and consists of those points for which there exist open neighbourhoods restricted to which f is identically zero. Now, $\mathrm{supp}(T[g, \Psi]) \subseteq \mathrm{supp}(\Psi)$ is equivalent to the statement that the complement of $\mathrm{supp}(\Psi)$ is contained in the complement of $\mathrm{supp}(T[g, \psi])$, i.e. that whenever Ψ vanishes identically in some open neighbourhood, so does $\mathrm{supp}(T[g, \Psi])$. This is, e.g., the case if $T[g, \Psi](x)$ depends on $\Psi(x)$ and *finitely* many derivatives of Ψ at x.

[23] As is well-known, diffeomorphism invariance of the action implies diffeomorphism covariance of the energy-momentum tensor, i.e.

$$
T[\varphi_* g, \varphi_* \Psi] = \varphi_*(T[g, \Psi]).
$$

The proof of this is as follows. Let g be a Lorentzian metric on $\mathcal{M}$, $\tilde{g}$ a symmetric covariant tensor field, Ψ a matter field and φ a diffeomorphism. By the definition of the energy-momentum tensor,

of tensors in the following. However, the general discussion in the following will depend only on the geometric structures η and u, while being independent of the choice of coordinates.

16.2.2.2 The Reaction to Weak Gravity: Emergence of Seemingly Anomalous Couplings

We now consider a metric which is a (small) perturbation of the background Minkowski metric, $g = \eta + h$. The matter action then takes the form

$$S[\eta + h, \Psi; V] = S[\eta, \Psi; V] + \frac{1}{2c} \int_V \mathrm{dvol}_\eta \, T^{\mu\nu}[\eta, \Psi] h_{\mu\nu} + O\left(h^2\right) \tag{16.30}$$

of a 'non-gravitational' background term plus an interaction term. This is the interaction term considered in [37]. As integration region, we consider some 'sandwich' $V = I \times \mathbb{R}^3$ of spacetime, for a finite temporal interval I.

Let us now consider a perturbation $h_{\mu\nu} = -2\frac{\Phi}{c^2}\delta_{\mu\nu}$, which is just of the form determined by the linearised Einstein equations to leading order in a small-velocity approximation for the sourcing matter. The components $\delta_{\mu\nu}$ are those of the tensor field ${}^{(4)}\delta := \eta + 2u^\flat \otimes u^\flat / c^2$, where $u^\flat = \eta(u, \cdot)$ is the one-form associated (via η) to the vector field u that defines the reference with respect to which the motion of the source is considered slow. In background-adapted coordinates, $\delta_{\mu\nu}$ is just the Kronecker delta. The interaction term in the Lagrangian function then takes the form

$$L_{\mathrm{int}}(t) = -\int_{\{ct\}\times\mathbb{R}^3} \mathrm{d}^3x \, \frac{\Phi}{c^2} \delta_{\mu\nu} T^{\mu\nu}[\eta, \Psi]. \tag{16.31a}$$

---

for ε sufficiently small we have

$$\begin{aligned}
&S[\varphi_*(g + \varepsilon\tilde{g}), \varphi_*\Psi; \varphi(V)] - S[\varphi_* g, \varphi_*\Psi; \varphi(V)] \\
&= \varepsilon \frac{1}{2c} \int_{\varphi(V)} \mathrm{dvol}_{\varphi_* g} \, T^{\mu\nu}[\varphi_* g, \varphi_*\Psi](\varphi_*\tilde{g})_{\mu\nu} + O\left(\varepsilon^2\right) \\
&= \varepsilon \frac{1}{2c} \int_{\varphi(V)} \varphi_*\left(\mathrm{dvol}_g \left(\varphi^*(T[\varphi_* g, \varphi_*\Psi])\right)^{\mu\nu} \tilde{g}_{\mu\nu}\right) + O\left(\varepsilon^2\right) \\
&= \varepsilon \frac{1}{2c} \int_V \mathrm{dvol}_g \left(\varphi^*(T[\varphi_* g, \varphi_*\Psi])\right)^{\mu\nu} \tilde{g}_{\mu\nu} + O\left(\varepsilon^2\right)
\end{aligned}$$

where we used that the pushforward commutes with tensor products and contractions, and that the Riemannian volume form of the pushforward metric $\varphi_* g$ is the pushforward of the volume form of the original metric. If, now, the matter action is diffeomorphism invariant, the above expression is *also* equal to

$$S[g + \varepsilon\tilde{g}, \Psi; V] - S[g, \Psi; V] = \varepsilon \frac{1}{2c} \int_V \mathrm{dvol}_g \, T^{\mu\nu}[g, \Psi]\tilde{g}_{\mu\nu} + O\left(\varepsilon^2\right).$$

Since $\tilde{g}$ was arbitrary, this implies $\varphi^*(T[\varphi_* g, \varphi_*\Psi]) = T[g, \Psi]$, i.e., the covariance law $T[\varphi_* g, \varphi_*\Psi] = \varphi_*(T[g, \Psi])$.

Assuming that the Newtonian potential Φ is approximately constant over the extent of our system, this becomes

$$L_{\text{int}}(t) = -\Phi(t, \bar{\mathbf{x}})\frac{1}{c^2}\int_{\{ct\}\times\mathbb{R}^3} \mathrm{d}^3x\, \delta_{\mu\nu}T^{\mu\nu}[\eta, \Psi], \tag{16.31b}$$

where $\bar{\mathbf{x}}$ are the spatial coordinates of some point inside the system. Thus, the expression

$$M_g = \frac{1}{c^2}\int_{\{ct\}\times\mathbb{R}^3} \mathrm{d}^3x\, \delta_{\mu\nu}T^{\mu\nu}[\eta, \Psi] \tag{16.32}$$

features as the 'gravitational mass' of the system in linear order, i.e. the quantity coupling to the Newtonian potential.

At this point, we see how the 'anomalous' coupling terms arise: the energy of our system Ψ, with respect to the background structures η and u, is given as

$$E = \int_{\{ct\}\times\mathbb{R}^3} \mathrm{d}^3x\, T^{00}[\eta, \Psi], \tag{16.33}$$

while the 'gravitational mass' (16.32) receives the additional contribution

$$M_g c^2 - E = \sum_{a=1}^{3}\int_{\{ct\}\times\mathbb{R}^3} \mathrm{d}^3x\, T^{aa}[\eta, \Psi] \tag{16.34}$$

on top of that energy. For a massive point particle on a worldline $z(t) = (ct, \mathbf{z}(t))$ in a general metric, the energy-momentum tensor is given by

$$T^{\mu\nu}_{\text{part.}}[g, \mathbf{z}](x) = \frac{mc}{\sqrt{|\det g|(x)}}\delta^{(3)}(\mathbf{x} - \mathbf{z}(t))\frac{\dot{z}^\mu(t)\dot{z}^\nu(t)}{\sqrt{-g(\dot{z}(t), \dot{z}(t))}}, \tag{16.35}$$

so the energy (16.33) and the 'anomalous' term (16.34) evaluate to

$$\begin{aligned} E_{\text{part.}} &= mc^2 + \frac{m}{2}\dot{\mathbf{z}}(t)^2 + O\left(c^{-2}\right) \\ &=: mc^2 + E_{\text{kin.,Newt.}} + O\left(c^{-2}\right), \end{aligned} \tag{16.36a}$$

$$(M_g c^2 - E)_{\text{part.}} = 2E_{\text{kin.,Newt.}} + O\left(c^{-2}\right). \tag{16.36b}$$

? Exercise

16.3 Using the energy-momentum tensor for a point particle, derive (16.36).

In contrast, for 'ultra-relativistic' matter, e.g. the electromagnetic field, the energy-momentum tensor is traceless, $g_{\mu\nu}T^{\mu\nu}_{\text{ultra-rel.}}[g, \Psi] = 0$. For Minkowski spacetime, this directly implies $T^{00}_{\text{ultra-rel.}}[\eta, \Psi] = \sum_{a=1}^{3} T^{aa}_{\text{ultra-rel.}}[\eta, \Psi]$, resulting in

$$(M_g c^2 - E)_{\text{ultra-rel.}} = E_{\text{ultra-rel.}}. \tag{16.37}$$

Hence, for a system of slowly moving charged particles, interacting exclusively by their electromagnetic field, Eqs. (16.36b) and (16.37) combine to (2 kinetic energy+ potential energy), which is just the 'anomalous' coupling term that we encountered for the specific calculation in Sect. 16.2.1.2. Following [37] it is thus shown to be of general origin.

16.2.2.3 Anomalous Couplings and Diffeomorphism Invariance

We will now show that the apparently 'anomalous' gravitational coupling term (16.34) in the action depends on the representative of the diffeomorphism equivalence class of gravitational fields. In other words, it can be removed by actively transforming the field of metric perturbations by some appropriate diffeomorphism. To make this explicit, we consider a general diffeomorphism φ by which we transform the physical metric $g = \eta + h$ and matter field Ψ. The transformed metric $\varphi_* g$ is again to be considered as a perturbation of the *same* background metric η. Hence, $\varphi_* g =: \eta + h'$ with

$$h' = \varphi_* g - \eta = \varphi_* h + \varphi_* \eta - \eta. \tag{16.38}$$

Note that here, as already announced before, it is crucial that we think of the diffeomorphism φ as an active transformation of fields, and not just a passive coordinate change. Only an active transformation allows to transform one field while keeping others fixed.

Assuming diffeomorphism invariance of the matter action, we obtain

$$\begin{aligned} S[\eta + h, \Psi; V] &= S[\eta + h', \varphi_* \Psi; \varphi(V)] \\ &= S[\eta, \varphi_* \Psi; \varphi(V)] + \frac{1}{2c} \int_{\varphi(V)} \text{dvol}_\eta \, T^{\mu\nu}[\eta, \varphi_* \Psi] h'_{\mu\nu} + O\left(h'^2\right). \end{aligned} \tag{16.39}$$

Comparing to (16.30), we see that under the transformation the background term has changed, as well as the term coupling to the metric perturbation.

Now we specialise again to the situation considered in the previous section, where the initial metric perturbation was of the form $h_{\mu\nu} = -2\frac{\Phi}{c^2}\delta_{\mu\nu}$, with potential Φ approximately constant over the extent of the system. We define a (linearised) diffeomorphism φ that rescales space by a factor of $1 - \frac{\Phi}{c^2}$ in that part of spacetime containing the system, i.e. on $\mathrm{supp}(\Psi)$. This φ leaves invariant each 'spatial leaf' $\{ct\} \times \mathbb{R}^3 \subset V$ (as a set), and transforms the metric perturbation into the form[24]

$$h'_{\mu\nu} = -2\frac{\Phi}{c^2}\delta^0_\mu\delta^0_\nu \quad \text{on } \mathrm{supp}(\varphi_*\Psi). \tag{16.40}$$

Put differently, application of φ 'transforms away' all components of the metric perturbation apart from the 00 component over the extent of the system. Note that we have to assume that Φ be (approximately) constant over the system not only in space, but also in the temporal direction: if it changed over time, the diffeomorphism would have to scale space differently at different times, thereby introducing mixed spatio-temporal terms h'_{0a} in the metric perturbation.[25]

At this point, the assumption of locality of the energy-momentum tensor in its matter-field argument comes into play: due to this locality, the transformed perturbation has the 'purely temporal' form (16.40) also on $\mathrm{supp}(T[\eta, \varphi_*\Psi]) \subseteq \mathrm{supp}(\varphi_*\Psi)$. Therefore, transforming by φ, the action (16.39) may be written in the form

$$S[\eta+h, \Psi; V] = S[\eta, \varphi_*\Psi; V] - \frac{1}{c}\int_V \mathrm{dvol}_\eta \, \frac{\Phi}{c^2} T^{00}[\eta, \varphi_*\Psi] + O\left(h'^2\right), \tag{16.41}$$

---

[24] In coordinate-free language, we have $h' = -2\frac{\Phi}{c^2}\frac{u^\flat \otimes u^\flat}{c^2}$ on $\mathrm{supp}(\varphi_*\Psi)$.

[25] In detail, the (linearised) diffeomorphism φ accomplishing this transformation can in our setting be described as follows (we refrain from giving coordinate-independent constructions): Consider the vector field with components

$$X^0 = 0, \quad X^a = -\frac{\Phi(\bar{\mathbf{x}})}{c^2}\xi(\mathbf{x})(x^a - \bar{x}^a)\,.$$

Here $\bar{x}^a$ are the coordinates of a fixed spatial reference position within the body, i.e. within $\mathrm{supp}(\Psi)$, and ξ is a function that is constantly 1 on $\mathrm{supp}(\Psi)$ and falls off rapidly to zero outside. Defining φ to be the diffeomorphism generated by X (i.e. the flow for unit time) and writing $\varepsilon := \Phi(\bar{\mathbf{x}})/c^2$ for brevity, by the definition of the Lie derivative we have

$$\begin{aligned} h' + \eta &= \varphi_*(\eta + h) \\ &= \eta + h - \mathcal{L}_X(\eta + h) + O\left(\varepsilon^2\right) \\ &= \eta + h - \mathcal{L}_X\eta + O\left(\varepsilon^2\right), \end{aligned}$$

since h itself is of order ε. This means $h' = h - \mathcal{L}_X\eta$ (up to higher order terms), or in components

$$h'_{\mu\nu} = h_{\mu\nu} - \partial_\mu X_\nu - \partial_\nu X_\mu\,.$$

By our choice of X the new metric perturbation takes on the desired form (16.40) (up to terms of order ε^2) on $\mathrm{supp}(\Psi)$. To conclude that it has this form also on $\mathrm{supp}(\varphi_*\Psi) = \varphi(\mathrm{supp}(\Psi))$, note that for any function f we have $f \circ \varphi = f + \mathcal{L}_X f + O(\varepsilon^2)$, such that

$$f = O\left(\varepsilon^2\right) \text{ on } U \implies f = O\left(\varepsilon^2\right) \text{ on } \varphi(U).$$

Applying this to $U = \mathrm{supp}(\Psi)$ and $f = h_{\mu b}$, we obtain (16.40).

and the transformed interaction term is

$$L_{\text{int,new}}(t) = -\Phi(t, \bar{\mathbf{x}})\frac{1}{c^2}\int_{\{ct\}\times\mathbb{R}^3} \mathrm{d}^3x\, T^{00}[\eta, \varphi_*\Psi]. \tag{16.42}$$

The quantity that now appears as the 'gravitational mass' (after the transformation by φ) is the spatial integral of the energy density $T^{00}[\eta, \varphi_*\Psi]$ of the transformed system on the Minkowski background. We stress once more that the background fields η and u are always the same, so that in particular the notion of 'energy density' for the original and the transformed fields is the same, namely $T^{00} = T(u^\flat/c, u^\flat/c)$.

We see that in the new diffeomorphism-equivalent representation an 'anomalous' coupling as in (16.32) no longer exists. Moreover, we note that in the new representation the spatial part of the metric is just of Euclidean form. Thus, the new representation expresses the quantities which describe the state of the system in a form in which the spatial metric in the rest-frame of the observer u is Euclidean. This is precisely the representation which we observed in the concrete example of Sect. 16.2.1.2 to eliminate the disturbing coupling terms from the Hamiltonian. Hence we see that our previous example may be considered as a special case of a general law, according to which apparently 'anomalous' couplings disappear in particular, metrically preferred representations.[26]

16.2.2.4 Dynamical Consequences of Diffeomorphism Invariance: The Virial Theorem

Carlip [37] claimed that his discussion of the elimination of the seemingly anomalous coupling terms by 'general covariance' might even be seen as a *derivation* of the special-relativistic virial theorem: since by the application of spatial rescaling diffeomorphisms as in the previous section we may arbitrarily alter the term coupling $\int \mathrm{d}^4x \sum_{a=1}^{3} T^{aa}$ to the Newtonian potential, but the 'physical coupling' ought not depend on our 'choice of gauge', we are supposed to conclude that this integral must vanish. This is just the statement of the special-relativistic virial theorem, a discussion of which may be found in [64, § 34].

This argument seems too good to be true: it should be clear that a statement such as the virial theorem cannot be derived from kinematical assumptions (such as diffeomorphism invariance of the action) alone; some *dynamical* assumption is needed (usually, one takes local energy-momentum conservation, also called 'closedness' of the system under consideration, i.e. $\partial_\mu T^{\mu\nu} = 0$). Looking at the transformation behaviour (16.39) of the matter action in weak gravity under the action of a diffeomorphism, we see that the proposed argument breaks down due to the fact that not only the coupling term but also the 'background term' changes,

[26] 'Metrically preferred' means in our case that the pushforward φ_* transforms the physical spatial metric ${}^{(3)}g = g|_{\{t=\text{const.}\}}$ into the flat background spatial metric $\delta = \eta|_{\{t=\text{const.}\}}$, such that the η-length of a transformed spacelike vector $\varphi_* v$ is the same as the g-length of the original vector v.

from $S[\eta, \Psi; V]$ in (16.30) to $S[\eta, \varphi_*\Psi; \varphi(V)]$ in (16.39). Therefore, without any further assumptions, we cannot deduce equality of the two coupling terms.

However, the desired conclusion (virial theorem) can be drawn if

$$\frac{\delta S}{\delta \Psi}[\eta, \Psi; V] = 0 \text{ under variations vanishing on } \partial V, \tag{16.43}$$

i.e. if we assume that the matter field configuration Ψ *solve the equations of motion* (on the Minkowski background). The precise argument runs as follows: Fix an $\varepsilon > 0$, and let X be a vector field on V that vanishes on the boundary ∂V. Let φ be the flow-diffeomorphism associated to X for flow parameter ε. Writing $\varphi_*\Psi = \Psi - \varepsilon \mathcal{L}_X \Psi + \mathcal{O}(\varepsilon^2)$, we see that to leading order in ε the difference $\varphi_*\Psi - \Psi$ vanishes on ∂V, such that the field equation (16.43) implies

$$S[\eta, \varphi_*\Psi; V] = S[\eta, \Psi; V] + \mathcal{O}\left(\varepsilon^2\right). \tag{16.44}$$

Expanding in ε, we also directly obtain

$$T^{\mu\nu}[\eta, \varphi_*\Psi] = T^{\mu\nu}[\eta, \Psi] + \mathcal{O}(\varepsilon). \tag{16.45}$$

On the other hand, combining (16.30) and (16.39), we have

$$\begin{aligned} &S[\eta, \Psi; V] + \frac{1}{2c} \int_V \mathrm{dvol}_\eta \, T^{\mu\nu}[\eta, \Psi] h_{\mu\nu} + \mathcal{O}\left(h^2\right) \\ &= S[\eta, \varphi_*\Psi; V] + \frac{1}{2c} \int_V \mathrm{dvol}_\eta \, T^{\mu\nu}[\eta, \varphi_*\Psi] h'_{\mu\nu} + \mathcal{O}\left(h'^2\right), \end{aligned} \tag{16.46}$$

where h' is given by (16.38) in terms of η, h and φ. Assuming h, and therefore also h', to be of order ε, we may combine (16.46) with equality of the background terms up to quadratic terms in ε (16.44) and equality of the energy-momentum tensor up to linear terms (16.45), and obtain

$$0 = \int_V \mathrm{dvol}_\eta \, T^{\mu\nu}[\eta, \Psi](h_{\mu\nu} - h'_{\mu\nu}). \tag{16.47}$$

We now want to apply this equation to the situation considered in the previous section, taking $V = [0, ct] \times \mathbb{R}^3$, $h_{\mu\nu} = -2\frac{\Phi}{c^2}\delta_{\mu\nu}$ with a constant Φ, and φ a diffeomorphism that scales space on $\mathrm{supp}(\Psi)$, such as to obtain $h'_{\mu\nu} = -2\frac{\Phi}{c^2}\delta^0_\mu\delta^0_\nu$ there. If we were able to apply (16.47) to these ingredients, we could conclude a statement about $\sum_{a=1}^3 T^{aa}[\eta, \Psi]$. However, we are met with an obstacle: in the derivation of (16.47), we needed that the vector field generating the diffeomorphism vanish on the boundary ∂V. At the spatial boundary of V, i.e. at 'spatial infinity', this does not pose a problem if we assume that Ψ have spatially compact support—we then need φ to scale space only in a finite spatial region, and may take its generator X to fall rapidly to zero outside. At the temporal boundary of V however, we apply the same spatial rescaling as at all other times, and the generating vector field does not vanish. Therefore, in our situation, (16.47) only holds up to a boundary term on the temporal boundary of V, arising from the failure of (16.44). If, however, we assume this boundary term to be local in Ψ, it vanishes when we take the average over larger and larger time intervals (since $\mathrm{supp}(\Psi)$ is spatially compact). Thus, we arrive at the special-relativistic virial theorem:

$$\lim_{t\to\infty} \frac{1}{t} \int_{[0,ct]\times\mathbb{R}^3} \mathrm{d}^4x \sum_{a=1}^{3} T^{aa}[\eta, \Psi] = 0\,. \tag{16.48}$$

We thus have shown that by a suitable adaptation of Carlip's argument from [37], one can indeed prove the virial theorem.

Note that we may, in fact, easily deduce local energy-momentum conservation from (16.47): According to (16.38), for $h = O(\varepsilon)$ we have in general

$$h_{\mu\nu} - h'_{\mu\nu} = \varepsilon(\mathcal{L}_X\eta)_{\mu\nu} + O\left(\varepsilon^2\right) = 2\varepsilon\partial_{(\mu}X_{\nu)} + O\left(\varepsilon^2\right). \tag{16.49}$$

Thus, for vector fields X vanishing on the boundary ∂V, (16.47) implies

$$\begin{aligned} 0 &= \int_V \mathrm{dvol}_\eta\, T^{\mu\nu}[\eta, \Psi]\partial_{(\mu}X_{\nu)} \\ &= \int_V \mathrm{dvol}_\eta\, T^{\mu\nu}[\eta, \Psi]\partial_\mu X_\nu \\ &= -\int_V \mathrm{dvol}_\eta\, \partial_\mu T^{\mu\nu}[\eta, \Psi]X_\nu\,, \end{aligned} \tag{16.50}$$

which, due to X being arbitrary, implies $\partial_\mu T^{\mu\nu}[\eta, \Psi] = 0$.

> Conclusion 16.5

Seemingly anomalous coupling terms of internal energies of a composite system to the Newtonian gravitational potential depend on the chosen representative of the diffeomorphism equivalence class of fields. In other words, they are gauge dependent and may be eliminated from the action by choosing an appropriate representative (i.e. by 'choosing a gauge'). However, as already stressed in the discussion in Sect. 16.2.1.2, this is not sufficient to argue for or against the *physical* relevance of such coupling terms: in a sense, the coupling terms have simply been 'hidden' by the field redefinition $\Psi \to \varphi_*\Psi$. The question of *operational* significance in concrete physical situations is still the important one, which cannot be answered by simple 'kinematic' arguments as the one discussed in this section, without any assumptions about the complete physical situation.[27]

Notwithstanding the non-viability of such arguments for answering questions about the physical relevance of such couplings, by making the additional *dynamical* assumption of the field solving the equations of motion, arguments based on diffeomorphism invariance may be used to prove the special-relativistic virial theorem.

16.3 Quantum Matter with Gravitational Backreaction

The objective of most research in 'quantum gravity' focuses on the question *how* gravity can be quantised (and the consequences of this endeavour, as discussed in Chaps. 1 and 2). Taking a step back, before asking for the *how*, one may first ask *if* the gravitational field should be quantised at all. Of course, this presupposes a reasonable definition of what it means for gravity to be quantised—which is the defining feature that makes a theory a quantum theory?

Einstein's equations (16.1) *can* be understood as describing a *field* $g_{\mu\nu}(x)$ on spacetime. From this point of view, one may find it reasonable to apply the same quantisation rules to solutions $g_{\mu\nu}(x)$ of Einstein's equations that one applies, for instance, to the classical solutions ψ of the Dirac equation in order to arrive at fermionic quantum fields. However, $g_{\mu\nu}(x)$ is not simply a field *on* spacetime; it describes the metric and further differential-geometric properties of spacetime itself. Einstein's equations can only be understood as equations of a field *on* spacetime in a perturbative sense: by separating the metric $g_{\mu\nu} \to g_{\mu\nu}+h_{\mu\nu}$ into some background metric $g_{\mu\nu}$ and only treating the variation $h_{\mu\nu}$ with respect to that background as a field living on the *background* defined by $g_{\mu\nu}$.

From a more philosophical point of view, one may ask whether such an artificial splitting of the structure of spacetime into background and field is a more plausible approach (even if in the end physical predictions would turn out to be independent of the way in which the splitting is done) than the alternative that gravity is fundamentally different and spacetime cannot simply be quantised in the same way as matter fields. In any case, the perturbative quantisation in the *exact* same way as for matter fields cannot be ultimately correct, because it is known to result in non-renormalisable divergences [65, 66].

We can avoid the ambiguity about what it means to quantise gravity altogether by asking the opposite: can we construct a theory that consistently combines (classical) GR with quantum matter, specifically, that solves the problem of defining the right-

[27] In that respect we contradict the immediate conclusion drawn in [58, p. 6], that 'correctly defining internal energies yields the true and unique gravitational mass and exposes the validity of the equivalence principle'. We also contradict the implicit statement made in this quotation, namely that the validity of the equivalence principle hinges on the equality of various notions of 'mass', the definitions of which are—as we have just seen—gauge dependent.

hand side in Einstein's equations from quantum matter fields? Any theory that accomplishes this, we want to refer to as *semiclassical gravity*.

16.3.1 Semiclassical Gravity Sourced by Mean Energy-Momentum

Once a proper mathematical model for quantum fields in curved spacetime has been established, one can define an energy-momentum operator $\hat{T}_{\mu\nu}$ via canonical quantisation of the corresponding classical object. An obvious way to include the gravitational backreaction of these fields are the semiclassical Einstein equations [67, 68]

$$R_{\mu\nu} - \frac{1}{2} R \, g_{\mu\nu} = \frac{8\pi G}{c^4} \left\langle \hat{T}_{\mu\nu} \right\rangle , \tag{16.51}$$

where $\langle \hat{T}_{\mu\nu} \rangle = \langle \Psi | \, \hat{T}_{\mu\nu} \, | \Psi \rangle$ denotes the expectation value in the state $|\Psi\rangle \in \mathcal{F}_{\pm}(\mathcal{H})$ of the matter field. This choice of right-hand side ensures the correct classical limit to Einstein's equations (16.1).

The proper Hilbert space structure for modelling quantum states in curved spacetime provided, the expectation value is readily defined. The classical energy-momentum tensor, however, is generally quadratic in the fields. A straightforward substitution of classical fields by field operators $\phi \to \hat{\phi}$ results in an object $\hat{T}_{\mu\nu}(x)$ which includes two-point correlation functions such as $\langle \hat{\phi}(x)\hat{\phi}(x') \rangle$ at the *same* spacetime point $x = x'$, which diverge. Therefore, in addition to the classical definition of $T_{\mu\nu}$ an appropriate renormalisation procedure is required, which results in a certain ambiguity of $\hat{T}_{\mu\nu}(x)$. It has been shown by Wald [1, 69, 70] that the renormalised energy-momentum operator compatible with the semiclassical Einstein equations (16.51) can be defined in an axiomatic way, requiring:

1. For any two *orthogonal* states $\langle \Psi | \Phi \rangle = 0$ the matrix elements $\langle \Psi | \, \hat{T}_{\mu\nu} \, | \Phi \rangle$ agree with those obtained by the formal substitution $\phi \to \hat{\phi}$ of classical fields with field operators in the classical energy-momentum tensor.
2. In flat Minkowski spacetime the renormalised $\hat{T}_{\mu\nu}$ reduces to the normal ordered energy-momentum operator obtained by substituting $\phi \to \hat{\phi}$.
3. Expectation values of the renormalised energy-momentum operator are covariantly conserved: $\nabla^{\mu} \langle \hat{T}_{\mu\nu} \rangle = 0$.
4. Causality holds in the sense that only changes in the metric in the causal past of some spacetime point p can affect the value of $\langle \hat{T}_{\mu\nu} \rangle$ at p.

The renormalised $\hat{T}_{\mu\nu}$ satisfying these axioms is uniquely determined up to the addition of local curvature terms.

16.3.1.1 The Nonrelativistic Limit of Semiclassical Gravity

In order to arrive at a nonrelativistic[28] Schrödinger equation for semiclassical gravity, one can follow the usual procedure to derive the Newtonian limit of Einstein's equations via the linearised theory, combined with the assumption of slow velocities of the sourcing matter [71]. The metric is written as $g_{\mu\nu} = \eta_{\mu\nu} + h_{\mu\nu}$ with a perturbation $h_{\mu\nu}$ around flat Minkowski spacetime. Introducing the trace-reversed metric $\overline{h}_{\mu\nu} = h_{\mu\nu} - \frac{1}{2}\eta_{\mu\nu}\eta^{\rho\sigma}h_{\rho\sigma}$ and applying the de Donder gauge condition $\partial^\nu \overline{h}_{\mu\nu} = 0$, Einstein's equations to linear order in the metric perturbation yield the wave equations

$$\Box \overline{h}_{\mu\nu} = -\frac{16\pi G}{c^4}\left\langle \hat{T}_{\mu\nu}\right\rangle , \tag{16.52}$$

$\Box = \nabla^\mu \nabla_\mu$ being the d'Alembert operator. In the weak field, nonrelativistic limit, the behaviour is dominated by the 00-component, and the d'Alembert operator can be approximated by the flat space Laplace operator ∇^2, neglecting time derivative terms of order c^{-2}. Defining the Newtonian potential $\Phi = -\frac{c^2}{4}\overline{h}_{00}$ and the mass density operator $\hat{\rho} = \hat{T}_{00}/c^2$, one finds the Poisson equation

$$\nabla^2 \Phi = 4\pi G \left\langle \hat{\rho}\right\rangle . \tag{16.53}$$

For a single field of mass m particles, we can define the N-particle state

$$|\Psi_N\rangle = \frac{1}{\sqrt{N!}}\int \mathrm{d}^3x_1 \cdots \mathrm{d}^3x_N \Psi_N(t, \mathbf{x}_1, \ldots, \mathbf{x}_N)\hat{\psi}^\dagger(\mathbf{x}_1)\cdots\hat{\psi}^\dagger(\mathbf{x}_N)\,|0\rangle \tag{16.54}$$

with the nonrelativistic field operators $\hat{\psi}$ and N-particle wave function Ψ_N. The mass density operator[29] $\hat{\rho}(\mathbf{x}) = m\hat{\psi}^\dagger(\mathbf{x})\hat{\psi}(\mathbf{x})$ has the time dependent expectation value

$$\begin{aligned}\left\langle \hat{\rho}(\mathbf{x})\right\rangle_N &= \langle \Psi_N|\, m\hat{\psi}^\dagger(\mathbf{x})\hat{\psi}(\mathbf{x})\,|\Psi_N\rangle \\ &= m\sum_{i=1}^{N}\int\left(\prod_{j=1}^{N}\mathrm{d}^3x_j\right)\delta^{(3)}(\mathbf{x}-\mathbf{x}_i)\,|\Psi_N(t,\mathbf{x}_1,\ldots,\mathbf{x}_N)|^2 , \end{aligned}\tag{16.55}$$

[28] Note that here we employ the common, yet somewhat misleading, adjective 'nonrelativistic' to designate Galilei-invariant dynamical laws in distinction from 'relativistic' ones, which then are those obeying Poincaré invariance. Nevertheless, we want to emphasise that it is not the validity of the physical relativity principle that distinguishes both cases; rather, their difference lies in the way in which that principle is implemented.

[29] Note that, contrary to perturbatively quantised gravity, all expressions derived from $\hat{\rho}$ in semiclassical gravity are well-defined, and there is no need for renormalisation.

and specifically $\langle\hat{\rho}\rangle = m|\psi|^2$ for a single particle with wave function $\psi(t, \mathbf{x})$. Integrating the Poisson equation (16.53) results in the potential

$$\begin{aligned}\Phi(t, \mathbf{x}) &= -G \int \mathrm{d}^3x' \frac{\langle\hat{\rho}(\mathbf{x}')\rangle}{|\mathbf{x}-\mathbf{x}'|} \\ &= -G\,m \sum_{i=1}^{N} \int \left(\prod_{j=1}^{N} \mathrm{d}^3x_j\right) \frac{|\Psi_N(t, \mathbf{x}_1, \ldots, \mathbf{x}_N)|^2}{|\mathbf{x}-\mathbf{x}_i|}\,. \end{aligned} \tag{16.56}$$

Given the classical spacetime structure corresponding to the Newtonian gravitation potential (16.56), the problem is that of Sect. 16.2: what is the Schrödinger equation that follows for the dynamics of matter in said classical spacetime?

The plausible answer, confirmed for the external homogeneous potential in the earth's gravity [31], is that the potential should enter the Hamiltonian in the usual way, i.e. the Hamilton operator for the evolution of N particles in the position basis should be

$$\hat{H}_N = \sum_{i=1}^{N} \left(\frac{\hat{\mathbf{p}}_i^2}{2m} + m\Phi(t, \mathbf{x}_i) \right) + V_{\mathrm{matter}}\,, \tag{16.57}$$

where $\hat{\mathbf{p}} = -\mathrm{i}\hbar\nabla$ is the momentum operator and the potential V_{matter} contains all external and internal non-gravitational forces. The resulting Schrödinger equation $\mathrm{i}\hbar\partial_t\Psi_N = \hat{H}_N\Psi_N$ is called the (N-particle) *Schrödinger–Newton* equation. Due to the wave function dependence of the gravitational potential (16.56), it is a nonlinear Schrödinger equation which for the case $N = 1$ of a single particle reads

$$\mathrm{i}\hbar\partial_t\psi(t, \mathbf{x}) = \left[-\frac{\hbar^2}{2m}\nabla^2 - Gm^2 \int \mathrm{d}^3x' \frac{|\psi(t, \mathbf{x}')|^2}{|\mathbf{x}-\mathbf{x}'|} + V_{\mathrm{ext}} \right] \psi(t, \mathbf{x})\,. \tag{16.58}$$

The gravitational term describes a self-interaction: the particle is attracted by a distribution of its mass m with the probability density $|\psi|^2$. Furthermore, the balance between this gravitational self-attraction and the free spreading of the Schrödinger equation results in the existence of stationary solutions [72].

Despite its nonlinearity, the Schrödinger–Newton equation maintains many of the typical properties from linear QM. Specifically, the norm of the wave function is conserved, $\partial_t \int \mathrm{d}^3x\,|\psi(t, \mathbf{x})|^2 = 0$, allowing for a probabilistic interpretation as in standard QM. Note also that at any given time the uncertainty relation between non-commuting observables remains intact. Arguments that semiclassical gravity would violate position-momentum uncertainty [73], therefore, do not apply to semiclassical gravity based on the semiclassical Einstein equations. In any case, a violation of the uncertainty relation would only constitute a testable deviation from standard QM and not an inconsistency, as long as its magnitude is not in contradiction to experimentally established values.

? Exercise

16.4 The single particle Schrödinger–Newton equation can be derived from the Lagrangian density

$$\mathcal{L} = \frac{i\hbar}{2}\left(\psi^* \partial_t \psi - \psi \partial_t \psi^*\right) - \frac{\hbar^2}{2m}|\nabla\psi|^2 - V_{\text{ext}}|\psi|^2 - \frac{m}{2}\Phi|\psi|^2$$

by variation with respect to the independent variables ψ and ψ^*, taking into account that Φ is itself a convolution $\Phi = U * |\psi|^2$ of the potential $U(\mathbf{x}) = -Gm/|\mathbf{x}|$ with the magnitude squared of the wave function. Show this, and that the conserved Noether charge for the symmetry under phase transformations $\psi \to e^{i\alpha}\psi$ is the norm of the wave function.

It is sometimes claimed [74,75] that the Schrödinger–Newton equation would not follow as the weak field nonrelativistic limit of the semiclassical Einstein equations. The criticism is based on the observation that in analogy with quantum electrodynamics one would expect the nonrelativistic, second-quantised Hamiltonian acting on Fock space states to take the form

$$\hat{H}_{\text{qg}} = -\frac{\hbar^2}{2m}\int d^3x\, \hat{\psi}^\dagger(\mathbf{x})\nabla^2\hat{\psi}(\mathbf{x}) - G\iint d^3x\, d^3x'\, \frac{\hat{\rho}(\mathbf{x})\hat{\rho}(\mathbf{x}')}{|\mathbf{x}-\mathbf{x}'|}\,. \tag{16.59}$$

This Hamiltonian results in divergent matrix elements which can be cured either by the introduction of a regularised mass density operator [74] $\hat{\rho}_{\text{reg}}(\mathbf{x})$, smeared over some spatial region around $\mathbf{x}$, or by replacing the product of mass density operators by its normal ordered equivalent $:\hat{\rho}(\mathbf{x})\hat{\rho}(\mathbf{x}'): = m^2\hat{\psi}^\dagger(\mathbf{x})\hat{\psi}^\dagger(\mathbf{x}')\hat{\psi}(\mathbf{x})\hat{\psi}(\mathbf{x}')$. This procedure removes the self-interaction[30] at the level of single particles, and yields a linear Schrödinger equation. This is, in fact, the Hamiltonian one would expect from quantised gravity. With respect to semiclassical gravity, however, assuming an analogy to quantum electrodynamics amounts to circular reasoning. The two occurrences of $\hat{\rho}$ in Eq. (16.59) stem from the nonrelativistic limit of the coupling $\sim h_{\mu\nu}T^{\mu\nu}$ in the linearised combined action for gravity and matter. From a quantum gravity perspective, one expects both terms to be subject to canonical quantisation. In the semiclassical theory, on the other hand, one would only quantise the matter part $T^{\mu\nu}$, whereas the metric perturbation remains classical. Instead of the Hamiltonian (16.59) one has [71]

$$\hat{H}_{\text{sc}} = -\frac{\hbar^2}{2m}\int d^3x\, \hat{\psi}^\dagger(\mathbf{x})\nabla^2\hat{\psi}(\mathbf{x}) - G\iint d^3x\, d^3x'\, \frac{\hat{\rho}(\mathbf{x})\langle\hat{\rho}(\mathbf{x}')\rangle}{|\mathbf{x}-\mathbf{x}'|}\,, \tag{16.60}$$

with the first-quantised potential (16.56).

---

[30] Exactly as in quantum electrodynamics, the self-interaction does not appear at tree level but is reintroduced via higher loop orders.

Although the proper derivation of a nonrelativistic Schrödinger equation from quantum fields in curved spacetime is an unsolved question (cf. Sect. 16.2), the true conflict that makes many question the validity of the Schrödinger–Newton equation is not its derivability from the semiclassical Einstein equations. Reading between the lines, one finds that what the critique [74, 75] of the Schrödinger–Newton equation is actually based on are its 'problematic consequences' with regard to its connection to the observed reality.

16.3.2 Consistency of Semiclassical Gravity

Historically, the question whether semiclassical gravity is consistent already concerned physicists in the early days of quantum field theory. During the Chapel Hill conference [8], Feynman proposed the following thought experiment:

> Suppose we have an object with spin which goes through a Stern-Gerlach experiment. Say it has spin 1/2, so it comes to one of two counters. Connect the counters by means of rods, etc., to an indicator which is either up when the object arrives at counter 1, or down when the object arrives at counter 2. Suppose the indicator is a little ball, 1 cm in diameter.
>
> Now, how do we analyze this experiment according to quantum mechanics? We have an amplitude that the ball is up, and an amplitude that the ball is down. That is, we have an amplitude (from a wave function) that the spin of an electron in the first part of the equipment is either up or down. And if we imagine that the ball can be analyzed through the interconnections up to this dimension ($\approx$ 1 cm) by the quantum mechanics, then before we make an observation we still have to give an amplitude that the ball is up and an amplitude that the ball is down. Now, since the ball is big enough to produce a *real* gravitational field (we know there's a field there, since Coulomb measured it with a 1 cm ball) we could use that gravitational field to move another ball, and amplify that, and use the connections to the second ball as the measuring equipment.

Denoting with $|\uparrow\rangle$ and $|\downarrow\rangle$ the spin eigenstates and with $|U\rangle$ and $|D\rangle$ the corresponding final states of the ball, the experiment amounts to creating an entangled state

$$|\Psi\rangle = \frac{1}{\sqrt{2}} \left(|\uparrow\rangle \otimes |U\rangle + |\downarrow\rangle \otimes |D\rangle \right) . \tag{16.61}$$

For the second ball to move into a position consistent with measurement outcomes for the position of the first ball, according to Feynman, the gravitational field should possess an amplitude as well. And indeed, the gravitational potential according to semiclassical gravity would be that of half the mass at position U and the other half at position D, *regardless* of the measurement outcome for the position of the first ball.

This has been noticed and put to the test by Page and Geilker [76]. In their experiment, the link between the quantum states $|\uparrow\rangle$ or $|\downarrow\rangle$ and the position of the first ball was as classical as it can be: it consisted of measuring the emission of γ rays from a cobalt-60 source over 30 seconds, walking over to another table,[31] and

[31] Whether it was literally or only metaphorically another table is not clear from their paper.

setting the position of a torsion balance into one of two positions depending on the measured emission (above/below average). To nobody's surprise, the experiment confirms that whenever $|\uparrow\rangle$ is measured the gravitational field is that of a ball at position U and vice versa. This allows two possible conclusions:

1. the gravitational field must, in fact, possess an amplitude, or
2. the state of the system under consideration has *not* been the entangled state (16.61).

That (16.61) cannot be the full story should, however, be old news to anyone who has ever encountered the quantum measurement problem [77,78] before. Only in a many worlds interpretation [79] is the system expected to be in this state still after the spin measurement—with all the corresponding difficulties [80] of such an interpretation. The consequence of the Page–Geilker experiment is that semiclassical gravity is incompatible with a no-collapse interpretation of QM.

Conclusion 16.6

Semiclassical gravity, i.e. Eq. (16.51) for a single quantum field, is *incomplete*. There must be a dynamical process, connected with the spin measurement, that leads to an objective reduction of the wave function:

$$|\Psi\rangle \rightarrow \begin{cases} |\uparrow\rangle \otimes |U\rangle & \text{with probability } 50\% \\ |\downarrow\rangle \otimes |D\rangle & \text{with probability } 50\%\,. \end{cases} \tag{16.62}$$

Page and Geilker are well aware of this possibility. They refute it, because a wave function collapse described by Eq. (16.62) seems to blatantly contradict the covariant conservation of energy-momentum, Wald's third axiom.

Interestingly, a similar argument could be made to refute QM, at least if defined in the traditional way [81], including the postulate that the wave function after a measurement does not evolve from the wave function before measurement according to the Schrödinger equation, but rather through projection on the corresponding eigenstate. In the same way, one could simply *postulate* that the combined state (g, Ψ) for metric and quantum field collapses upon 'measurement', in violation of the semiclassical Einstein equations, and continues to evolve according to semiclassical gravity thereafter. Of course, one then faces the same measurement problem as in standard quantum theory, the crucial difference being that a many worlds or 'operational' interpretation is not only difficult to reconcile with Born's rule but also in conflict with the observed reality.

If one does not take an agnostic point of view about measurement (as commonly accepted in non-gravitational quantum physics), semiclassical gravity requires the introduction of an objective collapse, with the instantaneous collapse (16.62) being only an effective, nonrelativistic description. We conclude:

Conclusion 16.7

The objective reduction dynamics (16.62), required to render semiclassical gravity a complete theory (Conclusion 16.6), is the nonrelativistic limit of a relativistic dynamical law for the fields compatible with the conservation law

$$\nabla^{\mu}\left\langle \hat{T}_{\mu\nu}\right\rangle = 0\,. \tag{16.63}$$

This leaves us with three options, *none* of which can be excluded, as of yet:

1. A consistent model for collapse, compatible with the conservation law (16.63), the Born rule probabilities (16.62) (or rather the generalisation to arbitrary states), as well as all other observations in quantum theory, especially the violation of Bell's inequalities, is fundamentally *impossible*.
2. There is a *new* process, to be modelled outside the formalism for quantum fields on curved spacetime with backreaction, that provides an explanation of the collapse (16.62) compatible with (16.63).
3. A consistent explanation for the *effective* collapse (16.62) can be given *within* the theory of semiclassical gravity (by taking into account *all* matter fields and their interactions).

Clearly, the first possibility is the perspective taken by Page and Geilker, among many others, and would necessitate *some* sort of quantisation of GR if it were true. The second possibility includes relativistic generalisations of collapse models [82]. The last certainly constitutes the most interesting alternative, that the explanation of wave function collapse could somehow lie within semiclassical gravity itself, although it is also the most speculative one.

16.3.2.1 The Role of the Density Operator and Its Dynamics

The quantum mechanical measurement postulate (16.62) predicts stochastic outcomes. Semiclassical gravity by itself, on the contrary, is a deterministic model—as are QM and quantum field theory without the collapse postulate. On the other hand, even in classical, deterministic theories stochastic phenomena are a regular occurrence in situations of many degrees of freedoms with incomplete information about the precise initial conditions.

The quantum mechanical formalism deals with these twofold statistics with the introduction of the density operator: given an ensemble $\{(p_j, |\psi_j\rangle)\}_j$ of pure Hilbert space states $|\psi_j\rangle$ that are expected with classical probabilities p_j, the density operator of the system is given by

$$\hat{\varrho} = \sum_j p_j \left|\psi_j\right\rangle\!\left\langle\psi_j\right|\,. \tag{16.64}$$

Two ensembles are called *equivalent* if they have the same density operator. The probability for any outcome o of a projective measurement of an operator $\hat{O}$ is given by the trace $\operatorname{tr} |o\rangle\langle o|\varrho\rangle$ of the projector on the corresponding eigenstate[32] $|o\rangle$ to o and the density operator. The information accessible via projective measurements is, therefore, entirely encoded in $\hat{\varrho}$.

The density operator serves a second role in conventional QM. Pure states in a composite Hilbert space, $|\Psi\rangle \in \mathcal{H}_1 \otimes \mathcal{H}_2$, are generally entangled (non-separable). There is no pure state $|\psi_1\rangle \in \mathcal{H}_1$ from which one could obtain probabilities for measurement outcomes of an operator $\hat{O}_1 \in \operatorname{End}(\mathcal{H}_1)$. Instead, if $\hat{\varrho} = |\Psi\rangle\langle\Psi|$ is the (pure state) density operator, the probabilities for $\hat{O}_1$ can be derived from the partial trace $\hat{\varrho}_1 = \operatorname{tr}_{\mathcal{H}_2}\hat{\varrho}$ over the second Hilbert space.

For a linear Hamiltonian, the time evolution of the Hilbert space states $|\psi_j\rangle$ in Eq. (16.64) induces the time evolution law $\mathrm{i}\hbar\partial_t\hat{\varrho} = [\hat{H}, \hat{\varrho}]$ for the density operator. It has a closed form, implying that equivalence of ensembles is a property preserved under time evolution. Even the partial trace $\hat{\varrho}_1$ for a subsystem obeys a closed time evolution law—in the case of a Markovian dynamics a master equation in Lindblad form $\hbar\partial_t\hat{\varrho} = -\mathrm{i}[\hat{H}, \hat{\varrho}] + \hbar\mathcal{L}(\hat{\varrho})$—which is linear in $\hat{\varrho}$ (although not generally unitary).

Considering, instead, the nonlinear evolution law (16.58), the spatial density matrix $\varrho_t(\mathbf{x}, \mathbf{x}') = \sum_j p_j\psi_j(t, \mathbf{x})\psi_j^*(t, \mathbf{x}')$ evolves according to

$$\partial_t\varrho_t(\mathbf{x}, \mathbf{x}') = -\frac{\mathrm{i}}{\hbar}[\hat{H}_0, \varrho_t(\mathbf{x}, \mathbf{x}')] + \sum_j \frac{\mathrm{i}\,G m^2}{\hbar} p_j\psi_j(t, \mathbf{x})\psi_j^*(t, \mathbf{x}') \times \int \mathrm{d}^3x'' |\psi_j(t, \mathbf{x}'')|^2 \left(\frac{1}{|\mathbf{x} - \mathbf{x}''|} - \frac{1}{|\mathbf{x}' - \mathbf{x}''|}\right). \tag{16.65}$$

It does not have a closed form and, therefore, does not preserve equivalence of ensembles. Although it is possible to calculate a density operator *at any given time* in order to obtain probabilistic predictions, the dynamics cannot be described in terms of the density operator for nonlinear systems.

Example

Let $|\psi_{1,2}\rangle$ be stationary solutions of the Schrödinger–Newton equation (16.58) centred around $\mathbf{x}_{1,2}$, respectively, and define $|\psi_\pm\rangle = \frac{1}{\sqrt{2}}(|\psi_1\rangle \pm |\psi_2\rangle)$; note that those superposition states are time dependent. The ensemble $A = \{(\frac{1}{2}, |\psi_1\rangle), (\frac{1}{2}, |\psi_2\rangle)\}$ then has the constant density matrix $\varrho_A(\mathbf{x}, \mathbf{x}')$. The ensemble $B = \{(\frac{1}{2}, |\psi_+\rangle), (\frac{1}{2}, |\psi_-\rangle)\}$ has the same initial density matrix. Nonetheless, its density matrix evolves in time.

[32] We assume, for simplicity, that operators have non-degenerate spectra, although the generalisation to degenerate eigenvalues is straightforward.

The stationary wave functions $\psi_j(\mathbf{x})$ are real valued and assumed to be sharply peaked around $\mathbf{x}_j$. The initial wave functions $\psi_\pm(0, \mathbf{x})$ then are real valued, as well, and in absence of an external potential, $V_{\text{ext}} = 0$, one finds

$$\begin{aligned}
\psi_\pm(t, \mathbf{x}) &= \exp\left[-\frac{it}{\hbar}\left(\frac{\hat{\mathbf{p}}^2}{2m} - Gm^2 \int d^3x' \frac{|\psi_\pm(0, \mathbf{x}')|^2}{|\mathbf{x} - \mathbf{x}'|}\right)\right] \psi_\pm(0, \mathbf{x}) \\
&\approx \exp\left[-it\left(\mathcal{L}_1 + \mathcal{L}_2\right)\right] \psi_\pm(0, \mathbf{x}) \\
&\approx \left[1 - it\left(\mathcal{L}_1 + \mathcal{L}_2\right) - \frac{t^2}{2}\left(\mathcal{L}_1^2 + \mathcal{L}_2^2 + \{\mathcal{L}_1, \mathcal{L}_2\}\right)\right] \psi_\pm(0, \mathbf{x}),
\end{aligned} \tag{16.66}$$

where we approximated the ψ_j as sharply peaked in the first step, and used the Zassenhaus formula expanding to quadratic order in time in the second, defining

$$\mathcal{L}_j = \frac{\hat{\mathbf{p}}^2}{4\hbar m} - \frac{Gm^2}{2\hbar|\mathbf{x} - \mathbf{x}_j|}. \tag{16.67}$$

This approximates the nonlinear evolution by the application of a linear operator. Similarly, the stationarity of the $\psi_j(\mathbf{x})$ implies they must be close to states in the kernel of the operators $1 - \exp(-it\mathcal{L}_j)$. Linearising in $\mathbf{x}$ and ignoring constant phase contributions, we have

$$\begin{aligned}
\psi_\pm(t, \mathbf{x}) \approx{}& \psi_\pm(0, \mathbf{x}) - \frac{it}{\sqrt{2}}\left(\mathcal{L}_2\psi_1(\mathbf{x}) \pm \mathcal{L}_1\psi_2(\mathbf{x})\right) \\
&- \frac{t^2}{2\sqrt{2}}\left(\mathcal{L}_2^2\psi_1(\mathbf{x}) \pm \mathcal{L}_1^2\psi_2(\mathbf{x}) + [\mathcal{L}_1, \mathcal{L}_2]\left(\psi_1(\mathbf{x}) \mp \psi_2(\mathbf{x})\right)\right),
\end{aligned} \tag{16.68}$$

and thus the difference in the diagonal elements of the density matrix is

$$\begin{aligned}
\Delta P(\mathbf{x}) &= \varrho_B(\mathbf{x}, \mathbf{x}) - \varrho_A(\mathbf{x}, \mathbf{x}) \\
&\approx \frac{t^2}{2}\Big((\mathcal{L}_2\psi_1(\mathbf{x}))^2 - \psi_1(\mathbf{x})\mathcal{L}_2^2\psi_1(\mathbf{x}) - \psi_1(\mathbf{x})[\mathcal{L}_1, \mathcal{L}_2]\psi_1(\mathbf{x}) \\
&\quad + (\mathcal{L}_1\psi_2(\mathbf{x}))^2 - \psi_2(\mathbf{x})\mathcal{L}_1^2\psi_2(\mathbf{x}) - \psi_2(\mathbf{x})[\mathcal{L}_2, \mathcal{L}_1]\psi_2(\mathbf{x})\Big).
\end{aligned} \tag{16.69}$$

It is tedious and not particularly insightful to attempt to evaluate these expressions explicitly. Nonetheless, we can get a good idea of the dynamics, considering that the operators $\mathcal{L}_1$ acting on ψ_2 and vice versa induce a motion of the peaks comparable to the gravitational attraction between two particles, with both the gravitational mass and the kinetic energy split equally between the two peaks of the superposition. ◀

16.3.3 Causality of Semiclassical Gravity

The property discussed in the previous subsection, that equivalent ensembles evolve into distinguishable ones, has consequences for the possibility to send signals faster than light. Although the possibility of faster-than-light signalling has been discussed before [73], the argument is often attributed to Gisin [83] who introduces the following lemma:[33]

Lemma 16.1 *Let* $\{(p_j, |\psi_j\rangle)\}_{j=1,\ldots,n}$ *and* $\{(q_k, |\chi_k\rangle)\}_{k=1,\ldots,m}$ *be two equivalent ensembles of states* $|\psi_j\rangle, |\chi_k\rangle \in \mathcal{H}$*. Let* $\mathcal{K}$ *be a complex Hilbert space of dimension* $l \geq \max(n, m)$*. Then there are two orthonormal bases* $\{|\alpha_i\rangle\}_{i=1,\ldots,l}$, $\{|\beta_i\rangle\}_{i=1,\ldots,l}$ *of* $\mathcal{K}$ *and a state vector* $|\Psi\rangle \in \mathcal{H} \otimes \mathcal{K}$ *such that*

$$|\Psi\rangle = \sum_{j=1}^{n} \sqrt{p_j}\, |\psi_j\rangle \otimes |\alpha_j\rangle = \sum_{k=1}^{m} \sqrt{q_k}\, |\chi_k\rangle \otimes |\beta_k\rangle\ . \tag{16.70}$$

The state $|\Psi\rangle$ has the reduced density matrix

$$\hat{\varrho}_{\mathcal{H}} = \sum_{j=1}^{n} p_j\, |\psi_j\rangle\langle\psi_j| = \sum_{k=1}^{m} q_k\, |\chi_k\rangle\ \langle\chi_k| -3pt\rangle \tag{16.71}$$

when traced over the Hilbert space $\mathcal{K}$. Enter Harvey and Krista, both adept experimenters stationed in remote locations. At an earlier time, the state $|\Psi\rangle$ has been prepared and distributed such that Harvey is able to perform local measurements with respect to operators in $\mathcal{H}$. Krista, on the other hand, can decide to perform a measurement with respect to one of the self-adjoint operators $\hat{A}, \hat{B} \in \mathrm{End}(\mathcal{K})$, whose matrices are diagonal relative to the bases $\{|\alpha_i\rangle\}_i$ and $\{|\beta_i\rangle\}_i$, respectively. According to the quantum mechanical measurement postulate, the global state after Krista's measurement is one of the projections

$$|\Psi\rangle \to \frac{|\alpha_i\rangle\ \langle\alpha_i|\Psi\rangle}{|\langle\alpha_i|\Psi\rangle|} = |\psi_i\rangle \otimes |\alpha_i\rangle \qquad \text{with probability}\, p_i \tag{16.72}$$

if she chooses to measure with respect to $\hat{A}$, and one of the projections

$$|\Psi\rangle \to \frac{|\beta_i\rangle\ \langle\beta_i|\Psi\rangle}{|\langle\beta_i|\Psi\rangle|} = |\chi_i\rangle \otimes |\beta_i\rangle \qquad \text{with probability}\, q_i \tag{16.73}$$

[33] See appendix B of the preprint version of reference [84] for a complete proof.

if she chooses to measure with respect to $\hat{B}$. Due to the separability of the collapsed states (16.72) and (16.73), and without knowledge about Krista's measurement outcome, Harvey's subsystem is then represented by the ensembles $\{(p_j, |\psi_j\rangle)\}_{j=1,\dots,n}$ and $\{(q_k, |\chi_k\rangle)\}_{k=1,\dots,m}$, respectively. Being equivalent ensembles, these are described by the same density operator (16.71).

The crucial ascertainment is now that, because with a linear dynamical law equivalent ensembles remain equivalent, Harvey has no means to detect which ensemble he is dealing with. The probabilities of all possible measurements he can perform on the Hilbert space $\mathcal{H}$ are fully determined by $\hat{\varrho}_{\mathcal{H}}(t)$ at any time t. With a nonlinear evolution law, on the other hand, the two ensembles evolve differently, as discussed in the previous subsection for the example of the Schrödinger–Newton equation. After some finite time t they become distinguishable. If Krista and Harvey previously agreed to encode a binary signal with Krista's choice of basis, then this signal reaches Harvey before any light signal could, provided the distance between them is larger than ct.

Before jumping to the conclusion that this sort of faster-than-light signalling is detrimental to semiclassical gravity, one may at least consider a series of potential loopholes in the argument:

1. The fact that a state $|\Psi\rangle \in \mathcal{H} \otimes \mathcal{K}$ with the property (16.70) exists does not necessarily imply that this state can ever be created as the consequence of some dynamical law, given the initial conditions of the universe. There is at least a theoretical possibility that the dynamics are such that they prevent *physical* states from ever approaching such a state.
2. Standard quantum (field) theory permits only local interactions. Hence, the entangled state $|\Psi\rangle$ must ultimately be created locally (either directly or via some third system) and brought to Krista's and Harvey's locations at separation $d \leq ct_0$ in final time t_0. The nonlinear evolution that results in the distinguishability of the initially equivalent ensembles, therefore, already acts during the time interval $[-t_0, 0]$ of separation, and not only during the time interval $[0, t]$ after Krista's measurement. The initial state $|\Psi(t = 0)\rangle$ is then *itself* the result of the nonlinear evolution and cannot simply be assumed to be of the form (16.70).
3. Projective measurement with an instantaneous collapse as in Eqs. (16.72) and (16.73) is a frame dependent assumption, already in obvious contradiction with relativity. Without a Lorentz invariant formulation of the collapse, one could justifiably ask why one should even bother about the possibility of faster-than-light signalling in a nonrelativistic model.
4. Even if one accepts the possibility of faster-than-light signalling, notwithstanding the three previous points, it is not evident that this specific form of signalling would result in a conflict with causality. Such a conflict would occur if Harvey used a second entangled system to signal back to Krista *and* that second signal would arrive at Krista's location *before* she makes her choice of basis. This procedure requires *two* measurement processes which must be instantaneous in

different frames of reference. Hence, a complete thought experiment for causality violation goes beyond the scenario considered above.

Regarding the third point, specifically, and taking into account the requirement for an objective collapse of the wave function established above, one would like to consider a scenario for faster-than-light signalling including the collapse dynamics. Relativistic models for the wave function collapse are still only sparsely developed [85].

Penrose [86, 87] suggests that collapse should occur in such a way that a spatial superposition of two classical states decays into a classical state with a rate proportional to the gravitational self-energy between the mass distributions belonging to the two states in superposition. Diósi's nonrelativistic collapse model [88] implements this idea. It also requires the introduction of a length scale r_c, acting as a cutoff in order to prevent divergences. Conforming with other nonrelativistic collapse models, it is based on a stochastic evolution law for Hilbert space states and a *linear* evolution of the density operator.

If we maintain the idea of a cutoff length scale and only focus on superposition states of two wave functions $\psi_{1,2}$ narrowly peaked around $\mathbf{x}_{1,2}$, respectively, we can give a more ad hoc description of collapse:

Whenever a superposition $\psi(\mathbf{x}) = \alpha\psi_1(\mathbf{x}) + \beta\psi_2(\mathbf{x})$ of two narrowly peaked wave functions $\psi_1(\mathbf{x}) \sim \delta^{(3)}(\mathbf{x} - \mathbf{x}_1)$ and $\psi_2(\mathbf{x}) \sim \delta^{(3)}(\mathbf{x} - \mathbf{x}_2)$ for a particle of mass m exceeds the cutoff scale, $|\mathbf{x}_1 - \mathbf{x}_2| \geq r_c$, the state undergoes a *collapse*

$$\psi(\mathbf{x}) \to \begin{cases} \psi_1(\mathbf{x}) & \text{with probability } |\alpha|^2 \\ \psi_2(\mathbf{x}) & \text{with probability } |\beta|^2 \end{cases} \tag{16.74}$$

within the time scale

$$\tau_c = \frac{\hbar\, r_c}{G m^2}\,. \tag{16.75}$$

This prototype of a collapse dynamics remains agnostic about the dynamics of more complex states that are not simple superpositions of two localised peaks of a single particle wave function.

For a spatial separation below r_c, as well as in addition to the collapse dynamics for larger separations, an evolution according to the Schrödinger–Newton equation (16.58) is assumed. It predicts that the two peaks, initially at $\mathbf{x}_{1,2}$, shift towards

$$\widetilde{\mathbf{x}}_1 = \mathbf{x}_1 - \frac{\mathbf{x}_1 - \mathbf{x}_2}{|\mathbf{x}_1 - \mathbf{x}_2|} |\alpha|^2 \delta x \,, \qquad \widetilde{\mathbf{x}}_2 = \mathbf{x}_2 + \frac{\mathbf{x}_1 - \mathbf{x}_2}{|\mathbf{x}_1 - \mathbf{x}_2|} |\beta|^2 \delta x \,, \tag{16.76}$$

decreasing their initial distance $\Delta x = |\mathbf{x}_1 - \mathbf{x}_2|$ by

$$\delta x \approx \frac{Gmt^2}{2\Delta x^2} \,, \tag{16.77}$$

where we assume $\delta x \ll \Delta x$. In order to resolve this decrease against the expected value $\delta x = 0$ for a classical mixture of the states $\psi_{1,2}$ with probabilities $|\alpha|^2$ and $|\beta|^2$, it must be larger than the free spreading of the wave function. Due to the position-momentum uncertainty relation, this spreading for a state with initial position uncertainty $\delta\xi$ and momentum uncertainty δp is limited by

$$\delta\xi + \frac{t}{m}\delta p \geq \delta\xi + \frac{\hbar t}{2m\,\delta\xi} = \sqrt{\frac{\hbar t}{2m}} \left(\zeta + \frac{1}{\zeta} \right) \geq \sqrt{\frac{2\hbar t}{m}} \,, \tag{16.78}$$

and hence the minimal resolution after time t is

$$\delta x^4 \geq \frac{4\hbar^2 t^2}{m^2} \overset{(16.77)}{=} \frac{8\hbar^2\,\delta x\,\Delta x^2}{Gm^3} \qquad \Rightarrow \qquad m^3 \geq \frac{8\hbar^2\,\Delta x^2}{G\,\delta x^3} \gg \frac{8\hbar^2}{G\,\delta x} \,. \tag{16.79}$$

Therefore, a minimum mass is required in order to achieve the necessary resolution. However, if we account for a dynamical collapse of the wave function according to the above ad hoc description, a larger mass implies a faster collapse and we must take care that the superposition is maintained throughout the entire time t of the experiment by requiring $t < \tau_c$. One then has

$$r_c = \frac{G\,m^2\,\tau_c}{\hbar} > \frac{G\,m^2\,t}{\hbar} \overset{(16.77)}{=} \sqrt{\frac{2G\,m^3\,\Delta x^2\,\delta x}{\hbar^2}} \overset{(16.79)}{\geq} \frac{4\,\Delta x^2}{\delta x} \gg \Delta x \gg \delta x \,. \tag{16.80}$$

Any shift δx above the length scale r_c would always remain unobservable, because the collapse happens too fast. Combining Eqs. (16.80) and (16.79) yields

$$r_c \geq \frac{32\hbar^2}{Gm^3} \left(\frac{\Delta x}{\delta x} \right)^4 \gg \frac{32\hbar^2}{Gm^3} \,. \tag{16.81}$$

This limit only holds for separations above the size of the particle, $\Delta x > 2R$. For a spherical particle with homogeneous mass density one then finds

$$m = \frac{4\pi}{3}\rho\, R^3 < \frac{\pi}{6}\rho\, \Delta x^3 \qquad \Rightarrow \qquad r_c \gg \left(\frac{6912\, \hbar^2}{\pi^3\, G\, \rho^3}\right)^{1/10}, \tag{16.82}$$

which implies that any value for r_c below 140 nm makes it impossible to resolve the required separation even for the densest elements.

For separations $\Delta x < 2R$ one finds instead of Eq. (16.77) that

$$\delta x \approx \frac{2\pi}{3} G\, \rho\, \Delta x\, t^2 \left(1 + O\left(\frac{\Delta x}{R}\right)\right). \tag{16.83}$$

Inserting this into Eq. (16.79) yields

$$\delta x^4 \geq \frac{4\hbar^2 t^2}{m^2} = \frac{6\, \hbar^2\, \delta x}{\pi\, G\, \rho\, \Delta x\, m^2} \qquad \Rightarrow \qquad \delta x \geq \left(\frac{6\, \hbar^2}{\pi\, G\, \rho\, \Delta x\, m^2}\right)^{1/3} \tag{16.84}$$

and instead of Eq. (16.80) one finds, using $2R > \Delta x > \delta x$ and (16.84),

$$r_c > \sqrt{\frac{3\, G\, m^4\, \delta x}{2\pi\, \rho\, \hbar^2\, \Delta x}} > R\,. \tag{16.85}$$

The radius is limited by the condition (16.84), which with $\Delta x < 2R$ yields

$$R^3 > \frac{\delta x^3}{8} \geq \frac{3\, \hbar^2}{4\pi\, G\, \rho\, \Delta x\, m^2} = \frac{27\, \hbar^2}{64\pi^3\, G\, \rho^3\, \Delta x\, R^6} > \frac{27\, \hbar^2}{128\pi^3\, G\, \rho^3\, R^7} \tag{16.86}$$

implying

$$r_c > \left(\frac{27\, \hbar^2}{128\pi^3\, G\, \rho^3}\right)^{1/10} \gtrsim 50\,\text{nm}\,. \tag{16.87}$$

These minimum values required for r_c are orders of magnitude above the parameter range usually assumed in collapse models, as well as above parameters excluded by observation. For instance, levitated nanoparticles [89] only pose a limit at about $r_c \gtrsim 1$ fm. Even the recent underground tests [90], which consider radiation emission rather than spatial superposition states and are of limited use for constraining the prototype collapse dynamics used here, only restrict the cutoff parameter of the Diósi model to $r_c \gtrsim 500$ pm. If there is a consistent relativistic description of collapse that approximates the prototypical model here for nonrelativistic superpositions of two sharp peaks, there is a large parameter range for which it would effectively prevent at least the most simple ideas to use the nonlinearity of the Schrödinger–Newton equation for faster-than-light signalling, while being perfectly consistent with observation.

16.3.4 Other Schemes to Include Backreaction for Quantum Matter on a Classical Spacetime

So far, we have only discussed the specific model for semiclassical gravity described by the semiclassical Einstein equations (16.51). The definition of semiclassical gravity given at the beginning of Sect. 16.3 also allows for other ways to introduce quantum matter as the source of curvature in Einstein's equations.

One alternative presents itself in the context of collapse models [82]. In these models, the Fock space state obeys a stochastic differential equation, e.g.,

$$|\mathrm{d}\psi\rangle_t = \Big[-\frac{\mathrm{i}}{\hbar}\hat{H}\,\mathrm{d}t + \sqrt{\gamma}\int \mathrm{d}^3x\,\left(\hat{\rho}_{\mathrm{reg}}(\mathbf{x}) - \langle\hat{\rho}_{\mathrm{reg}}(\mathbf{x})\rangle_t\right)\mathrm{d}W(t,\mathbf{x}) \\ -\frac{\gamma}{2}\int \mathrm{d}^3x\,\left(\hat{\rho}_{\mathrm{reg}}(\mathbf{x}) - \langle\hat{\rho}_{\mathrm{reg}}(\mathbf{x})\rangle_t\right)^2 \mathrm{d}t\Big]|\psi\rangle_t \tag{16.88}$$

in the continuous spontaneous localisation (CSL) model, where $\hat{H}$ is the usual Hamiltonian, $\hat{\rho}_{\mathrm{reg}}$ the regularised mass density operator, $W(t,\mathbf{x})$ describes an ensemble of independent stochastic Wiener processes (one for every point $\mathbf{x}$), and γ is a free coupling parameter. A second free parameter comes from the regularisation length scale r_c for the mass density $\hat{\rho}_{\mathrm{reg}}$. The density operator satisfies the stochastic master equation

$$\mathrm{d}\hat{\varrho}(t) = -\frac{\mathrm{i}}{\hbar}\left[\hat{H},\hat{\varrho}(t)\right]\mathrm{d}t - \frac{\gamma}{8\hbar^2}\left[\hat{\rho}_{\mathrm{reg}}(\mathbf{x}),\left[\hat{\rho}_{\mathrm{reg}}(\mathbf{x}),\hat{\varrho}(t)\right]\right]\mathrm{d}t \\ +\frac{\gamma}{2\hbar^2}\left\{\hat{\rho}_{\mathrm{reg}}(\mathbf{x}) - \langle\hat{\rho}_{\mathrm{reg}}(\mathbf{x})\rangle,\hat{\varrho}(t)\right\}\mathrm{d}W(t,\mathbf{x})\,. \tag{16.89}$$

The characteristic feature of collapse models is that after averaging over the noise term in the second line of (16.89), the density operator satisfies the linear Gorini–Kossakowski–Sudarshan–Lindblad equation [91, 92] (i.e., a master equation in Lindblad form), which ensures that equivalent ensembles evolve into equivalent ensembles and no faster-than-light signalling can occur, as detailed in the previous subsection.

An intuitive way to couple the mass density to gravity, as pointed out by Tilloy and Diósi [93], is then to use the *signal*

$$\rho(t,\mathbf{x}) = \langle\hat{\rho}_{\mathrm{reg}}(\mathbf{x})\rangle_t + \delta\rho_t\,, \tag{16.90}$$

where $\delta\rho_t$ are the noise fluctuations resulting from the stochastic part in the second line of Eq. (16.89). One obtains a classical Newtonian potential

$$\Phi(t, \mathbf{x}) = -G \int \mathrm{d}^3x' \, \frac{\rho(t, \mathbf{x}')}{|\mathbf{x} - \mathbf{x}'|}\,, \tag{16.91}$$

entering into the Schrödinger equation instead of the potential (16.56). In the language of standard quantum physics, (16.90) can be interpreted as the information retrievable via weak measurement of the mass density, which is then fed back into the dynamics as the source of the gravitational field. The joint dynamics contain decoherence terms both from the stochastic noise and from the gravitational potential [93, 94]. In the language of collapse models, the gravitational field is sourced by the collapse events (flashes) in spacetime [95].

Related is the concept of hybrid classical-quantum dynamics [96]. Let $\mathcal{H}$ be a Hilbert space of some quantum system and $\mathcal{M} \cong \mathbb{R}^{2n}$ a classical phase space. Then a *hybrid state* is given by

$$\hat{\varrho}_{cq} : \mathcal{M} \to \mathrm{End}(\mathcal{H}) \qquad \text{with} \qquad \int_{\mathcal{M}} \mathrm{d}z \, \mathrm{tr}\hat{\varrho}_{cq}(z) = 1\,, \tag{16.92}$$

i.e., by assigning a subnormalised density matrix to every point in phase space. It can be understood as the product $\hat{\varrho}_{cq} = \int \mathrm{d}z \, \varrho_c(z) \, |z\rangle \, \langle z| \otimes \hat{\varrho}_q(z)$ of a density matrix $\hat{\varrho}_q$ for the quantum system and a probability distribution $\varrho_c(z)$ of the orthonormal classical phase space states $|z\rangle$. This formalism has been used by Albers et al. [97] in order to couple a scalar quantum field to classical scalar gravity. Oppenheim [98] recently proposed a model based on the ADM formalism [99] of GR, which describes a Hamiltonian evolution of 3-manifolds with a time parameter t. The classical phase space $\mathcal{M}$ is formed by the 3-metrics g and their canonical momenta π. The Hamiltonian and momentum constraints H and P^i in the ADM Lagrangian,

$$\mathcal{L} = -g_{ij}\partial_t\pi^{ij} - NH - N_i P^i - 2\partial_i\left(\pi^{ij}N_j - \frac{1}{2}\pi N^i + \nabla^i N\sqrt{g}\right), \tag{16.93}$$

where N, N_i are the lapse and shift function, respectively, $\pi = g_{ij}\pi^{ij}$ denotes the trace, g the metric determinant, and the covariant derivative ∇ is taken with respect to the 3-metric, are then replaced by corresponding operators,

$$H = -\sqrt{g}\left[R + g^{-1}\left(\frac{1}{2}\pi^2 - \pi^{ij}\pi_{ij}\right)\right]\hat{\mathrm{I}} + f^{\alpha\beta}\hat{L}_\alpha^\dagger \hat{L}_\beta \tag{16.94}$$

$$P^i = -2\nabla_j\pi^{ij}\,\hat{\mathrm{I}} + g_i^{\alpha\beta}\hat{L}_\alpha^\dagger \hat{L}_\beta\,, \tag{16.95}$$

acting on hybrid states, with $\hat{\mathrm{I}}$ the identity, $\hat{L}_\alpha$ a set of Lindblad operators, and $f^{\alpha\beta}$, $g_i^{\alpha\beta}$ coefficient functions on 3-space which depend on the classical state (g, π). The scalar curvature R is, again, to be taken with respect to the 3-metric g.

This formalism allows for a consistent semiclassical theory without the need to introduce any explicit assumptions about wave function collapse. It is, however, a probabilistic theory of *statistical ensembles* of 3-metrics rather than a deterministic theory for a single spacetime. The large freedom of choice for the proper Lindblad operators and coefficient functions makes it currently difficult to arrive at experimental predictions.

Finally, Kent [100, 101] has developed a framework of causal quantum theory. Consider a time orientable, globally hyperbolic Lorentzian 4-manifold $(\mathcal{M}, g)$ with a foliation $\mathcal{M} = \cup_{t\in\mathbb{R}} \mathcal{S}_t$ of spacelike hypersurfaces $\mathcal{S}_t$, that allows to define a unitary evolution law $\psi(t, \mathbf{x}) = U(t, t_0)\psi(t_0, \mathbf{x})$ with the initial wave function[34] defined on $\mathcal{S}_{t_0}$. Further assume a set of local measurement events $M_i = (t_i, \mathbf{x}_i; \hat{O}_i, |o_i\rangle)$ $(i = 1, \dots, n)$ of eigenstates $|o_i\rangle$ of operators $\hat{O}_i$ at spacetime points $(t_i, \mathbf{x}_i \in \mathcal{S}_{t_i}) \in \mathcal{M}$ which are time-ordered, $t_0 < t_1 < t_2 < \cdots < t_n < t$. In standard quantum theory, the final state is obtained as the unitary evolution with $U(t, t_0)$ conditioned over the measurement outcomes:

$$\psi(t, \mathbf{x}) = U(t, t_n) \left(\prod_{i=n,n-1,\dots,1} |o_i\rangle \langle o_i| \, U(t_i, t_{i-1}) \right) \psi(t_0, \mathbf{x}) \, . \tag{16.96}$$

In the causal theory proposed by Kent, for any point $\mathbf{x} \in \mathcal{S}_t$ the time evolution law (16.96) holds, however with the important distinction that only those measurement events M_i with $(t_i, \mathbf{x}_i)$ in the *past light cone* of $(t, \mathbf{x})$ are conditioned over. This results in a definition of *local states* and predictions that differ from standard quantum theory, although thus far not being in any obvious contradiction with observation.

A nonlinear modification of the dynamics based on these *local* states does not result in the problem of faster-than-light signalling discussed previously. One could, therefore, attempt to formulate a semiclassical theory of gravity in which spacetime curvature is sourced by the local states, in order to avoid issues with causality. To date, such a model has not been developed.

16.4 Concluding Remarks

In the winter semester 1923/24, Max Born delivered a lecture on the Bohr–Sommerfeld quantisation in atomic physics which he subsequently published as a book entitled 'Atommechanik' [102]. In that book he laid out with almost axiomatic precision the known formal principles from classical physics, like the Hamilton–Jacobi theory and its application to perturbation theory (which had proved very successful in astronomy), and their application to the Bohr–Sommerfeld

[34] More generally, one can consider a quantum field operator evolving between Cauchy surfaces according to the Tomonaga–Schwinger equation.

quantisation. At that time it was clear to everyone in the field that this loose collection of rather ad-hoc 'quantisation rules' would eventually be replaced by something with a proper logical foundation, a real *theory*. So why did Born put all his efforts into that book project, given the premature state of a real understanding? His own answer is this: '*... we are attempting a deductive presentation of atomic theory. The reservations, that the theory is not sufficiently developed, I wish to disperse with the remark that we are dealing with a test case, a logical experiment, the meaning of which just lies in the determination of the limits to which the principles of atomic and quantum physics succeed, and to pave the ways which shall lead us beyond those limits.*'

Our contribution to this collection should likewise be regarded as a 'logical experiment', the meaning of which lies in in the determination of the limits to which classical or semi-classical gravity can be combined with quantum mechanics. In the same vein, we believe that from these considerations we will receive useful hints as to where we really need to go beyond those limits.

On these grounds, we propose a pragmatic attitude which puts first things first: From an experimental point of view, the interface of classical gravity and quantum mechanics on the one hand as well as semiclassical gravity on the other hand present promising opportunities. Systematic post-Newtonian descriptions of the coupling between quantum matter and gravity make reliable testable predictions as to the influence of the gravitational field on the dynamics of quantum systems. Likewise, both the Schrödinger–Newton equation and models for objective wave function collapse offer concrete experimental possibilities. Experiments of this type require comparably small improvements over existing technology, and should be fully explored, with the hope that they yield effects that can guide us towards the correct fundamental theory. From the theoretical perspective, one should take conceptual challenges within the established fundamental theories—such as the ones presented by us—seriously in order to understand the precise breaking points of GR and RQFT. A more thorough understanding of where *exactly* they fail (and to which extent) when considered jointly is expected to reveal valuable information for the search of a fully consistent theory of gravitating quantum matter.

Acknowledgments A. G. acknowledges funding from the VolkswagenStiftung. D. G. and P. K. S. acknowledge support from the Deutsche Forschungsgemeinschaft via the Collaborative Research Centre 1227 'DQ-mat', projects A05 and B08, at Leibniz University Hannover. All three authors would like to thank the organisers of the 740th WE-Heraeus-Seminar 'Experimental Tests and Signatures of Modified and Quantum Gravity' for the opportunity to present and discuss the topics which laid the foundation for this work.

References

1. R.M. Wald, *Quantum Field Theory in Curved Spacetime and Black Hole Thermodynamics*. Chicago Lectures in Physics (The University of Chicago Press, Chicago, 1994)
2. C. Bär, K. Fredenhagen, *Quantum Field Theory on Curved Spacetimes* (Springer, Berlin, 2009). https://doi.org/10.1007/978-3-642-02780-2

3. R. Brunetti, C. Dappiaggi, K. Fredenhagen, J. Yngvason (Eds.), *Advances in Algebraic Quantum Field Theory* (Springer, Cham, 2015). https://doi.org/10.1007/978-3-319-21353-8
4. S.W. Hawking, Commun. Math. Phys. **43**(3), 199 (1975). https://doi.org/10.1007/BF02345020
5. V. Mukhanov, S. Winitzki, *Introduction to Quantum Effects in Gravity* (Cambridge University Press, Cambridge, 2007). https://doi.org/10.1017/CBO9780511809149
6. H. Ohanian, Am. J. Phys. **45**(10), 903 (1977). https://doi.org/10.1119/1.10744.
7. W.G. Dixon, in *Equations of Motion in Relativistic Gravity*, ed. by D. Puetzfeld, C. Lämmerzahl, B. Schutz (Springer International Publishing, Cham, 2015), pp. 1–66. https://doi.org/10.1007/978-3-319-18335-0_1
8. C. De Witt, D. Rickles (Eds.), *The Role of Gravitation in Physics: Report from the 1957 Chapel Hill Conference*. Wright Air Development Center, Technical Report 57-216 (The University of North Carolina, Chapel Hill, 1957). http://www.edition-open-access.de/sources/5/index.html
9. D. Giulini, in *Quantum Field Theory and Gravity. Conceptual and Mathematical Advances in the Search for a Unified Framework*, ed. by F. Finster, O. Müller, M. Nardmann, J. Tolksdorf, E. Zeidler (Birkhäuser Verlag, Basel, 2012), pp. 345–370. https://doi.org/10.1007/978-3-0348-0043-3_16. arXiv:1105.0749v2
10. S.E. Salvioli, J. Differ. Geom. **7**(1–2), 257 (1972). https://doi.org/10.4310/jdg/1214430830
11. A. Trautman, in *Lectures on General Relativity, Vol. 1*, ed. by S. Deser, K.W. Ford (Prentice-Hall, Englewood Cliffs, 1965), pp. 1–248. Lectures delivered 1964 at the Brandeis Summer Institute in Theoretical Physics
12. J.L. Synge, *Relativity: The General Theory* (North-Holland, Amsterdam, 1960)
13. D. Lehmkuhl, Stud. History Philos. Sci. B Stud. History Philos. Mod. Phys. **46**, 316 (2014). https://doi.org/10.1016/j.shpsb.2013.08.002
14. D. Giulini, Philos. Nat. **39**(2), 843 (2002)
15. D. Giulini, in *Approaches to Fundamental Physics*, ed. by E. Seiler, I.O. Stamatescu. Lecture Notes in Physics, vol. 721 (Springer, Berlin, 2007), pp. 105–120. https://doi.org/10.1007/978-3-540-71117-9_6
16. H. Pfister, M. King (Eds.), *Inertia and Gravitation. The Fundamental Nature and Structure of Space-Time*. Lecture Notes in Physics, vol. 897 (Springer, Cham, 2015). https://doi.org/10.1007/978-3-319-15036-9
17. A.D. Alexandrov, Annali di Matematica Pura ed Applicata (Bologna) **103**(8), 229 (1975). https://doi.org/10.1007/BF02414157
18. F. Klein, *Vergleichende Betrachtungen über Neuere Geometrische Forschungen*, 1st edn. (Verlag von Andreas Deichert, Erlangen, 1872). Reprinted in Mathematische Annalen (Leipzig) 43:43–100, 1892. English translation at arXiv:0807.3161
19. J. Ehlers, E. Köhler, J. Math. Phys. **18**(10), 2014 (1977). https://doi.org/10.1063/1.523175.
20. R.A. Coleman, H. Korte, J. Math. Phys. **21**(6), 1340 (1980). https://doi.org/10.1063/1.524598.
21. R. Penrose, Found. Phys. **44**, 557 (2014). https://doi.org/10.1007/s10701-013-9770-0.
22. E. Mach, *Die Mechanik in ihrer Entwickelung: Historisch-Kritisch Dargestellt* (F.A. Brockhaus, Leipzig, 1983)
23. A. Einstein, Ann. Phys. **360**(4), 241 (1918). https://doi.org/10.1002/andp.19183600402
24. J.B. Barbour, H. Pfister, *Mach's Principle: From Newton's Bucket to Quantum Gravity*. Einstein Studies, vol. 6 (Birkhäuser, Basel, 1995)
25. O. Brill, B. Goodman, Am. J. Phys. **35**(9), 832 (1967). https://doi.org/10.1119/1.1974261
26. J.D. Jackson, *Classical Electrodynamics*, 3rd edn. (Wiley, Hoboken, 1999)
27. V. Fragkos, M. Kopp, I. Pikovski, (2022). Preprint arXiv:2206.00558
28. J. Ehlers, F. Pirani, A. Schild, in *General Relativity. Papers in Honor of J.L. Synge*, ed. by L. O'Raifeartaigh (Clarendon Press, Oxford, 1972), pp. 63–84
29. J. Ehlers, F.A. Pirani, A. Schild, Gen. Relat. Gravit. **44**, 1587 (2012). https://doi.org/10.1007/s10714-012-1353-4. Reprint of the 1972 original as 'Golden Oldie'

30. J. Audretsch, Phys. Rev. D **27**(12), 2872 (1983). https://doi.org/10.1103/PhysRevD.27.2872
31. R. Colella, A.W. Overhauser, S.A. Werner, Phys. Rev. Lett. **34**(23), 1472 (1975). https://doi.org/10.1103/PhysRevLett.34.1472
32. R. Karcher, A. Imanaliev, S. Merlet, F. Pereira Dos Santos, New J. Phys. **20**(11), 113041 (2018). https://doi.org/10.1088/1367-2630/aaf07d
33. M. Zych, F. Costa, I. Pikovski, Č. Brukner, Nat. Commun. **2**, 505 (2011). https://doi.org/10.1038/ncomms1498
34. I. Pikovski, M. Zych, F. Costa, Č. Brukner, Nat. Phys. **11**, 668 (2015). https://doi.org/10.1038/nphys3366
35. Y. Bonder, E. Okon, D. Sudarsky, Phys. Rev. D **92**, 124050 (2015). https://doi.org/10.1103/PhysRevD.92.124050
36. S. Loriani, A. Friedrich, C. Ufrecht, F. Di Pumpo, S. Kleinert, S. Abend, N. Gaaloul, C. Meiners, C. Schubert, D. Tell, E. Wodey, M. Zych, W. Ertmer, A. Roura, D. Schlippert, W.P. Schleich, E.M. Rasel, E. Giese, Sci. Adv. **5**(10), eaax8966 (2019). https://doi.org/10.1126/sciadv.aax8966
37. S. Carlip, Am. J. Phys. **66**(5), 409 (1998). https://doi.org/10.1119/1.18885
38. W. Tichy, É.É. Flanagan, Phys. Rev. D **84**, 044038 (2011). https://doi.org/10.1103/PhysRevD.84.044038
39. J. Ehlers, in *Grundlagenprobleme der Modernen Physik: Festschrift für Peter Mittelstaedt zum 50. Geburtstag*, ed. by J. Nitsch, J. Pfarr, E.W. Stachow (Bibliographisches Institute, Mannheim, 1981), pp. 65–84. Republished as [40]
40. J. Ehlers, Gen. Relat. Gravit. **51**, 163 (2019). https://doi.org/10.1007/s10714-019-2624-0. Republication of original article [39] as 'Golden Oldie'
41. E. Inonu, E.P. Wigner, Proc. Natl. Acad. Sci. USA **39**(6), 510 (1953). https://doi.org/10.1073/pnas.39.6.510
42. H. Bacry, J.M. Lévy-Leblond, J. Math. Phys. **9**(10), 1605 (1968). https://doi.org/10.1063/1.1664490
43. C.M. Will, *Theory and Experiment in Gravitational Physics*, 2nd edn. (Cambridge University Press, Cambridge, 2018). https://doi.org/10.1017/9781316338612
44. E. Poisson, C.M. Will, *Gravity: Newtonian, Post-Newtonian, Relativistic* (Cambridge University Press, Cambridge, 2014). https://doi.org/10.1017/CBO9781139507486
45. C. Kiefer, T.P. Singh, Phys. Rev. D **44**, 1067 (1991). https://doi.org/10.1103/PhysRevD.44.1067
46. C. Lämmerzahl, Phys. Lett. A **203**(1), 12 (1995). https://doi.org/10.1016/0375-9601(95)00345-4
47. D. Giulini, A. Großardt, Class. Quant. Gravity **29**(21), 215010 (2012). https://doi.org/10.1088/0264-9381/29/21/215010
48. P.K. Schwartz, D. Giulini, Class. Quant. Gravity **36**(9), 095016 (2019). https://doi.org/10.1088/1361-6382/ab0fbd
49. P.K. Schwartz, Post-Newtonian Description of Quantum Systems in Gravitational Fields. Doctoral thesis, Gottfried Wilhelm Leibniz Universität Hannover (2020). https://doi.org/10.15488/10085
50. A. Ito, Class. Quant. Gravity **38**(19), 195015 (2021). https://doi.org/10.1088/1361-6382/ac1be9
51. R.T. Perche, J. Neuser, Class. Quant. Gravity **38**(17), 175002 (2021). https://doi.org/10.1088/1361-6382/ac103d
52. A. Alibabaei, Geometric post-Newtonian description of spin-half particles in curved spacetime (2022). Master's thesis, Gottfried Wilhelm Leibniz Universität Hannover
53. T.D. Newton, E.P. Wigner, Rev. Mod. Phys. **21**(3), 400 (1949). https://doi.org/10.1103/revmodphys.21.400
54. P.K. Schwartz, D. Giulini, Int. J. Geom. Methods Mod. Phys. **17**(12), 2050176 (2020). https://doi.org/10.1142/S0219887820501765
55. M. Sonnleitner, S.M. Barnett, Phys. Rev. A **98**, 042106 (2018). https://doi.org/10.1103/PhysRevA.98.042106

56. P.K. Schwartz, D. Giulini, Phys. Rev. A **100**, 052116 (2019). https://doi.org/10.1103/PhysRevA.100.052116
57. K. Nordvedt, Int. J. Theor. Phys. **3**(2), 133 (1970). https://doi.org/10.1007/BF02412754
58. M. Zych, L. Rudnicki, I. Pikovski, Phys. Rev. D **99**(10), 104029 (2019). https://doi.org/10.1103/PhysRevD.99.104029
59. S.A. Eddington, G.L. Clark, Proc. R. Soc. Lond. A Math. Phys. Sci. **166**, 465 (1938). https://doi.org/10.1098/rspa.1938.0104
60. S.A. Eddington, Month. Not. R. Astron. Soc. **76**(6), 525 (1916). https://doi.org/10.1093/mnras/76.6.525
61. I. Kolář, P.W. Michor, J. Slovák, *Natural Operations in Differential Geometry* (Springer, Berlin, Heidelberg, 1993). https://doi.org/10.1007/978-3-662-02950-3
62. Y. Kosmann, Annali di Matematica Pura ed Applicata **91**, 317 (1971). https://doi.org/10.1007/BF02428822
63. M. Godina, P. Matteucci, Int. J. Geom. Methods Mod. Phys. **2**(2), 159 (2005). https://doi.org/10.1142/S0219887805000624
64. L.D. Landau, E.M. Lifshitz, *The Classical Theory of Fields*, 4th revised english edn. (Butterworth–Heinemann, Massachusetts, 1980).
65. G. 't Hooft, M.J.G. Veltman, Annales de l'Institut Henri Poincaré, physique théorique **20**, 69 (1974)
66. M.H. Goroff, A. Sagnotti, A. Sagnotti, Phys. Lett. B **160**(1), 81 (1985). https://doi.org/10.1016/0370-2693(85)91470-4
67. C. Møller, in *Colloques Internationaux CNRS*, ed. by A. Lichnerowicz, M.A. Tonnelat, vol. 91 (CNRS, Paris, 1962)
68. L. Rosenfeld, Nucl. Phys. **40**, 353 (1963). https://doi.org/10.1016/0029-5582(63)90279-7
69. R.M. Wald, Commun. Math. Phys. **54**(1), 1 (1977). https://doi.org/10.1007/BF01609833
70. R.M. Wald, Phys. Rev. D **17**(6), 1477 (1978). https://doi.org/10.1103/PhysRevD.17.1477
71. M. Bahrami, A. Großardt, S. Donadi, A. Bassi, New J. Phys. **16**, 115007 (2014). https://doi.org/10.1088/1367-2630/16/11/115007
72. I.M. Moroz, R. Penrose, P. Tod, Class. Quant. Gravity **15**(9), 2733 (1998). https://doi.org/10.1088/0264-9381/15/9/019
73. K. Eppley, E. Hannah, Found. Phys. **7**(1–2), 51 (1977). https://doi.org/10.1007/BF00715241
74. C. Anastopoulos, B.L. Hu, New J. Phys. **16**, 085007 (2014). https://doi.org/10.1088/1367-2630/16/8/085007
75. B.L.B. Hu, E. Verdaguer, *Semiclassical and Stochastic Gravity: Quantum Field Effects on Curved Spacetime*. Cambridge Monographs on Mathematical Physics (Cambridge University Press, Cambridge, 2020). https://doi.org/10.1017/9780511667497
76. D.N. Page, C.D. Geilker, Phys. Rev. Lett. **47**, 979 (1981). https://doi.org/10.1103/PhysRevLett.47.979
77. J. Bell, Phys. World **3**(8), 33 (1990). https://doi.org/10.1088/2058-7058/3/8/26
78. T. Maudlin, Topoi **14**(1), 7 (1995). https://doi.org/10.1007/BF00763473
79. H. Everett, Rev. Mod. Phys. **29**(3), 454 (1957). https://doi.org/10.1103/RevModPhys.29.454
80. A. Kent, Int. J. Mod. Phys. A **05**(09), 1745 (1990). https://doi.org/10.1142/S0217751X90000805
81. E. Schrödinger, Naturwissenschaften **23**(49), 823 (1935). https://doi.org/10.1007/BF01491914
82. A. Bassi, K. Lochan, S. Satin, T.P. Singh, H. Ulbricht, Rev. Mod. Phys. **85**, 471 (2013). https://doi.org/10.1103/RevModPhys.85.471
83. N. Gisin, Helvetica Phys. Acta **62**(4), 363 (1989). https://doi.org/10.5169/seals-116034
84. A. Bassi, K. Hejazi, Eur. J. Phys. **36**, 055027 (2015). https://doi.org/10.1088/0143-0807/36/5/055027
85. D.J. Bedingham, in *Do Wave Functions Jump? Perspectives of the Work of GianCarlo Ghirardi*, ed. by V. Allori, A. Bassi, D. Dürr, N. Zanghi, Fundamental Theories of Physics (Springer, Cham, 2020), pp. 191–203. https://doi.org/10.1007/978-3-030-46777-7
86. R. Penrose, Gen. Relat. Gravit. **28**(5), 581 (1996). https://doi.org/10.1007/BF02105068

87. R. Penrose, Philos. Trans. R. Soc. A Math. Phys. Eng. Sci. **356**, 1927 (1998). https://doi.org/10.1098/rsta.1998.0256
88. L. Diósi, Phys. Rev. A **40**(3), 1165 (1989). https://doi.org/10.1103/PhysRevA.40.1165
89. U. Delić, M. Reisenbauer, K. Dare, D. Grass, V. Vuletić, N. Kiesel, M. Aspelmeyer, Science **367**(6480), 892 (2020). https://doi.org/10.1126/science.aba3993
90. S. Donadi, K. Piscicchia, C. Curceanu, L. Diósi, M. Laubenstein, A. Bassi, Nat. Phys. **17**(1), 74 (2021). https://doi.org/10.1038/s41567-020-1008-4
91. V. Gorini, A. Kossakowski, E.C.G. Sudarshan, J. Math. Phys. **17**(5), 821 (1976). https://doi.org/10.1063/1.522979
92. G. Lindblad, Commun. Math. Phys. **48**(2), 119 (1976). https://doi.org/10.1007/BF01608499
93. A. Tilloy, L. Diósi, Phys. Rev. D **93**, 024026 (2016). https://doi.org/10.1103/PhysRevD.93.024026
94. D. Kafri, J.M. Taylor, G.J. Milburn, New J. Phys. **16**, 065020 (2014). https://doi.org/10.1088/1367-2630/16/6/065020
95. A. Tilloy, Phys. Rev. D **97**(2), 021502 (2018). https://doi.org/10.1103/PhysRevD.97.021502
96. M.J.W. Hall, M. Reginatto, Phys. Rev. A **72**(6), 062109 (2005). https://doi.org/10.1103/PhysRevA.72.062109
97. M. Albers, C. Kiefer, M. Reginatto, Phys. Rev. D **78**, 064051 (2008). https://doi.org/10.1103/PhysRevD.78.064051
98. J. Oppenheim, (2021). Preprint arXiv:1811.03116
99. R. Arnowitt, S. Deser, C.W. Misner, Phys. Rev. **116**(5), 1322 (1959). https://doi.org/10.1103/PhysRev.116.1322
100. A. Kent, Phys. Rev. A **72**, 012108 (2005). https://doi.org/10.1103/PhysRevA.72.012108
101. A. Kent, Proc. R. Soc. A Math. Phys. Eng. Sci. **474**, 20180501 (2018). https://doi.org/10.1098/rspa.2018.0501
102. M. Born, Vorlesungen über Atommechanik. Erster Band, in *Struktur der Materie in Einzeldarstellungen*, vol. II (Springer, Berlin, 1925)

Printed in the United States
by Baker & Taylor Publisher Services